Geological Society of America
Memoir 184

Permian-Triassic Pangean Basins and Foldbelts Along the Panthalassan Margin of Gondwanaland

Edited by

J. J. Veevers
and
C. McA. Powell*
Australian Plate Research Group
School of Earth Sciences
Macquarie University
North Ryde, N.S.W. 2109, Australia

1994

*Present address: Department of Geology and Geophysics, University
of Western Australia, Nedlands, W.A. 6009, Australia.

Published by The Geological Society of America, Inc.
3300 Penrose Place, P.O. Box 9140, Boulder, Colorado 80301

Printed in U.S.A.

GSA Books Science Editor Richard A. Hoppin

Library of Congress Cataloging-in-Publication Data

Permian-Triassic Pangean basins and foldbelts along the Panthalassan
    margin of Gondwanaland / edited by J. J. Veevers and C. McA. Powell.
        p.   cm. — (Memoir / Geological Society of America ; 184)
    Includes bibliographical references and index.
    ISBN 0-8137-1184-3
    1. Geology, Stratigraphic—Permian.   2. Geology, Stratigraphic-
-Triassic.   3. Gondwana (Geology).   4. Sedimentary basins.
I. Veevers, J. J.   II. Powell, Chris McA., 1943–   .   III. Series:
Memoir (Geological Society of America) ; 184.
QE674.P465   1994
551.7'56—dc20   94-5322
                                                            CIP

10 9 8 7 6 5 4 3 2 1

Contents

Geological Society of America
Memoir 184
1994

Introduction

C. McA. Powell* and J. J. Veevers

Australian Plate Research Group, School of Earth Sciences, Macquarie University, North Ryde, N.S.W. 2109, Australia

The idea for this volume arose during the Third International Sedimentological Congress held in Canberra in August 1986. At that meeting, Powell and Collinson were discussing the tectonic framework for the Gondwana Series in Antarctica, when Powell remarked that the foreland basin, which Conaghan et al. (1982) had just recently documented for the Late Permian of the Sydney Basin in eastern Australia, appeared to extend through Antarctica to Argentina. Powell had, in 1985, spent two months visiting the Sierra de la Ventana and the Callingasta-Uspallata basins in Argentina at the invitation of Arturo Amos and colleagues from the University of Buenos Aires. Powell had thus seen, and remarked on, the great similarity between the gross framework of the sedimentary architecture at the two extremities of Gondwanaland. Collinson's reply was that although he agreed with Powell's conclusion, he felt they were probably the only people at the conference who shared that opinion. Accordingly, it was decided to analyze and synthesize the data to test the hypothesis.

The roots of the foreland-basin model for the Sydney Basin go back to the 1970s when our former colleague at Macquarie University, Gilbert Jones, advocated the application of precise comparisons of modern analogues with ancient basins. Papers on such comparisons were published as Jones (1976), Cas and Jones (1979), Jones and McDonnell (1981), Conaghan et al. (1982), and Veevers et al. (1982). The then-current state of understanding of the Sydney Basin was summarized in Jones et al. (1984) in our monograph, *Phanerozoic earth history of Australia* (Veevers, 1984).

We decided at the outset that the synthesis of the postulated Gondwanan Foreland Basin should be a summary of the current state of knowledge of each of the main sectors of its reconstructed 10,000 km extent from western Argentina to northern Queensland. Each sector was to be written or coauthored by people who had first-hand knowledge of the region, and the Macquarie group was to draw together all the threads to weave a picture of how the basin evolved. Four sectors, eastern Australia, Antarctica, southern Africa, and southern South America, were identified, and coauthors who had worked and published on each sector were sought and found.

The initial intention was to present the information in a standard format, so that for each sector there should be orogen-normal cross sections; orogen-normal time-space diagrams; selected columnar sections, coupled with paleocurrent and provenance data; stage maps showing local and sector paleogeography. The orogen-normal cross-sections and time-space diagrams were to extend from the Panthalassan orogenic margin across the foreland basin into the craton and were to include information about deformation, metamorphism, and igneous activity.

From the outset, we recognized that an understanding of a sedimentary basin requires an understanding of its surroundings and historical context. It is not sufficient to study just the component formations in the basin, no matter how thorough or elegant that study may be. In order to understand a sedimentary basin one needs to know the history of the basement on which it lies and the evolution of adjacent areas. No basin is an island unto itself, although very few texts or monographs give adequate coverage of the basin prehistory or the coeval evolution of the adjacent tectonic zones. This we have tried to do.

The initial outline was ambitious. Each sector was to be described thoroughly together with an introduction on the reconstruction of Gondwanaland and summary chapters on tectonostratigraphic cycles and the tectonic evolution of the basin. One of the main aims was to provide an integrated summary of the paleocurrent and provenance information in a common format.

Progress on the volume in 1987–1988 was good. Espejo and López-Gamundí spent six months at Macquarie University in 1987–1988 as Australian Research Council research associates, where they wrote the first draft of the southern South American chapter; at the same time Collinson spent a six months' sabbatical leave writing the first draft of the Antarctic

---

*Present address: Department of Geology and Geophysics, University of Western Australia, Nedlands, W.A. 6009, Australia.

Powell, C. McA., and Veevers, J. J., 1994, Introduction, *in* Veevers, J. J., and Powell, C. McA., eds., Permian-Triassic Pangean Basins and Foldbelts Along the Panthalassan Margin of Gondwanaland: Boulder, Colorado, Geological Society of America Memoir 184.

chapter. Powell, who designed the initial format of the volume, was unexpectedly seconded to the academic administrative role of head of the School of Earth Sciences in early 1987 and thus had much less time to devote to the monograph than had initially been planned. Nonetheless, by the end of 1989, drafts of all sectors except eastern Australia had been written, and a diagrammatic outline of the final chapter had been prepared.

Powell was appointed chair of Geology at the University of Western Australia in January 1990, and from then his participation in production of the volume was limited. Eastern Australia, the cradle of the Permian foreland basin idea, proved to be more challenging than first thought. Veevers and Conaghan set about examining the entire geological base, from primary well-log and outcrop sources, and a synthesis drawn from this primary data analysis has resulted. In many ways, it is difficult to describe a forest when you live in one of the trees; it is far easier to observe from a distant hill. In this case, however, the end justified the means, and if the eastern Australia chapter is longer than others it is because this sector had more new information generated during the compilation than the others. Major new reports were published, and many long-standing canons were challenged.

With Powell's departure to Western Australia, Veevers took over prime responsibility for producing and editing the monograph. To say that there were not frustrations in bringing the volume to press would be untrue. The final version was assembled in mid-1993, however, and we believe it represents a new level of understanding of the Gondwana basin that bordered the Panthalassan margin of Gondwanaland. As to whether it was a foreland basin, read on!

Du Toit (1937) had no such doubt, because he was using the asserted continuity of the basin along the Panthalassan margin as confirmation of the contentious issue of the day, the Gondwanaland reconstruction by continental drift. This volume, dedicated to Du Toit's genius, is a confirmation, in turn, of his Samfrau Orogenic Zone, Gondwanide "foredeep," and foreland (Fig. 1).

Benefits from the long period of gestation include catching up with (1) Briggs's (1991) brachiopod zonation of the eastern Australian Permian, (2) the map and bulletin on southeastern Queensland (Finlayson, 1990), (3) the dating of New Zealand granites (Kimbrough et al., 1992), (4) Grunow et al.'s (1991) reconstruction of West Antarctica, and (5) the clearer dating of the magmatic arc in southern South America (Gust et al., 1985; Kay et al., 1989; Pankhurst, 1990; and Storey and Alabaster, 1991).

Any originality the work may have stems from the integration in a uniform time scale of information from the Gondwanaland margin. In aiming at a precise timetable of events, we had to rely on the stratigraphic distribution of the almost ubiquitous palynomorphs in the dominantly nonmarine Gondwana strata, supplemented by the rare (except in eastern

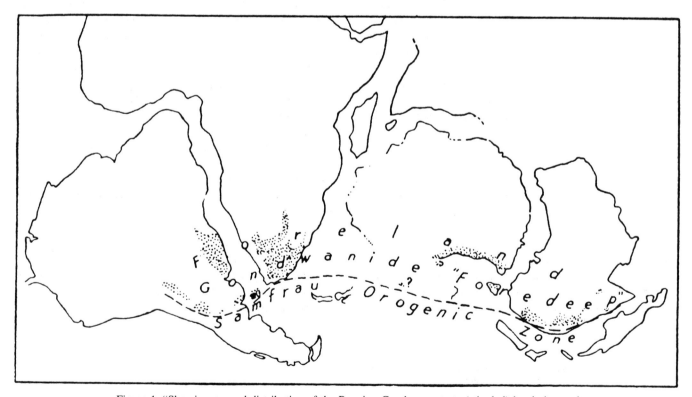

Figure 1. "Showing general distribution of the Permian Gondwana strata (stippled) bordering and connected with the Samfrau Orogenic Zone." Reproduced from figure 10 of *Our Wandering Continents* (Du Toit, 1937) with the permission of Oliver and Boyd.

Australia) marine invertebrates in the Permian (except Antarctica) and land vertebrates in the Triassic. Each of the regional chapters contains an appendix which aims to apply an integrated Gondwanaland-wide correlation. The correlation with the standard time scale and with the radiometric or numerical time scale given by Palmer (1983) allows us to cite ages in Ma, and the distinction between radiometric ages and biostratigraphic ages expressed in Ma is clear from the context. In the synthesis (Chapter 7), detached from the primary data, we use an asterisk to distinguish biostratigraphic ages (i.e., Ma*). We make the further distinction of the term "date" as the raw estimated age by the radiometric method, and "age" as the inferred best estimate of the geological event.

ACKNOWLEDGMENTS

Acknowledgments are made in each chapter, as appropriate. Here we wish to acknowledge the support of the Australian Research Grants Committee and its successor Australian Research Council, the School of Earth Sciences of Macquarie University, in particular the staff of the Drawing Office, Judy Davis, John Cleasby, and Ken Rousell, whose skills rendered our rough drafts into fair drawings. Powell acknowledges additionally the support of the Department of Geology and Geophysics, University of Western Australia.

We are grateful to our colleagues E. J. Cowan, K. L. McDonnell and S. E. Shaw, who contributed to appendices in Chapter 3, and to those who contributed personal communications.

Finally, our thanks to the GSA reviewers, individually acknowledged in each chapter, who unstintingly gave time and effort in helping improve this work, as did Richard A. Hoppin, GSA books science editor.

REFERENCES CITED

Briggs, D.J.C., 1991, Correlation charts for the Permian of the Sydney-Bowen Basin and New England Orogen: Newcastle Symposium on Advances in the Study of the Sydney Basin, Department of Geology, University of Newcastle, v. 25, p. 30–37.

Cas, R.A.F., and Jones, J. G., 1979, Paleozoic interarc basin in eastern Australia and a modern New Zealand analogue: New Zealand Journal of Geology and Geophysics, v. 22, p. 71–85.

Conaghan, P. J., Jones, J. G., McDonnell, K. L., and Royce, K., 1982, A dynamic fluvial model for the Sydney Basin: Journal of the Geological Society of Australia, v. 29, p. 55–70.

Du Toit, A. L., 1937: Our wandering continents: Edinburgh, Oliver and Boyd, 366 p.

Finlayson, D. M., ed., 1990, The Eromanga-Brisbane Geoscience Transect: A guide to basin development across Phanerozoic Australia in southern Queensland: Australian Bureau of Mineral Resources, v. 232, 261 p.

Grunow, A. M., Kent, D. V., and Dalziel, I.W.D., 1991, New paleomagnetic data from Thurston Island: Implications for the tectonics of West Antarctica and Weddell Sea opening: Journal of Geophysical Research, v. 96, p. 17935–17954.

Gust, D. A., Biddle, K. T., Phelps, D. W., and Uliana, M. A., 1985, Associated Middle to Late Jurassic volcanism and extension in southern South America: Tectonophysics, v. 116, p. 223–253.

Jones, J. G., 1976, Lachlan Fold Belt: An actualistic approach to environmental/tectonic interpretation: Australian Society of Exploration Geophysicists Bulletin, v. 7, p. 10–11.

Jones, J. G., and McDonnell, K. L., 1981, Papua New Guinea analogue for the Late Permian environment of northeastern New South Wales: Palaeogeography, Palaeoclimatology, Palaeoecology, v. 34, p. 191–205.

Jones, J. G., Conaghan, P. J., McDonnell, K. L., Flood, R. H., and Shaw, S. E., 1984, Papuan Basin analogue and a foreland basin model for the Bowen-Sydney Basin, *in* Veevers, J. J., ed., Phanerozoic Earth history of Australia: Oxford, England, Clarendon Press, p. 243–262.

Kay, S. M., Ramos, V. A., Mpodozis, C., and Sruoga, P., 1989, Late Paleozoic to Jurassic silicic magmatism at the Gondwana margin: Analogy to the middle Proterozoic in North America: Geology, v. 17, p. 324–328.

Kimbrough, D. L., Mattinson, J. M., Coombs, D. S., Landis, C. A., and Johnston, M. R., 1992, Uranium-lead ages from the Dun Mountain ophiolite belt and Brook Street terrane, South Island, New Zealand: Geological Society of America Bulletin, v. 104, p. 429–443.

Palmer, A. R., 1983, The decade of North American geology 1983 geologic time scale: Geology, v. 11, p. 503–504.

Pankhurst, R. J., 1990, The Paleozoic and Andean magmatic arcs of West Antarctica and southern South America: Geological Society of America Special Paper, v. 241, p. 1–7.

Storey, B. C., and Alabaster, T., 1991, Tectonomagmatic controls on Gondwana break-up models: Evidence from the Proto-Pacific margin of Antarctica: Tectonics, v. 10, p. 1274–1288.

Veevers, J. J., ed., 1984, Phanerozoic Earth history of Australia: Oxford, Clarendon Press, 418 p.

Veevers, J. J., Jones, J. G., and Powell, C. McA., 1982, Tectonic framework of Australia's sedimentary basins: Australian Petroleum Exploration Association Journal, v. 22, p. 283–300.

MANUSCRIPT ACCEPTED BY THE SOCIETY SEPTEMBER 9, 1993

Geological Society of America
Memoir 184
1994

Reconstruction of the Panthalassan margin of Gondwanaland

C. McA. Powell and Z. X. Li
Department of Geology and Geophysics, University of Western Australia, Nedlands, W. A. 6009, Australia

ABSTRACT

Gondwanaland was reconstructed by first forming East Gondwanaland (India, Antarctica and Australia) and then closing East Gondwanaland to Africa with Madagascar in a tight northern fit against Somalia. The poles of rotation to form Gondwanaland follow Powell et al. (1988), Lawver and Scotese (1987), Lawver et al. (1992), and Veevers et al. (1991). Paleolatitudes found from the Gondwanan apparent polar-wander path documented by Li et al. (1993a, 1993b) show that South America and southern Africa were in high latitudes in the Devonian and Early Carboniferous, while Australia was in low latitudes, and that during the Late Carboniferous to the end of the Permian, Australia and adjacent Antarctica were in high latitudes while southern South America and southern Africa were in middle-to-low latitudes. The movement of the paleopole along the Panthalassan margin toward the Antarctic Peninsula during the Triassic to Early Jurassic placed most of the Panthalassan margin in middle to high latitudes during the early Mesozoic.

BASIS OF RECONSTRUCTION

The reconstruction of Gondwanaland has been revised continuously since Wegener's (1915) first reconstruction, and there is now broad consensus about the fit of the major continental pieces. The reconstruction of South America against Africa (Carey, 1958; Bullard et al., 1965) has changed little, except in detail (Fairhead, 1988; Fairhead and Binks, 1991), and the fit of Australia and Antarctica (Veevers and Eittreim, 1988; Sandwell and McAdoo, 1988; Veevers, 1990; Lawver et al., 1991) is also broadly agreed to. The fit of India against Antarctica has in the past been contentious, but in the last two decades magnetic-anomaly (Sclater and Fisher, 1974; Norton and Sclater, 1979) and SEASAT (Haxby, 1985) data have supported the fit of India against Enderby Land in Antarctica. Thus, East (Australia, India and Antarctica) and West (South America and Africa) Gondwanaland can be pieced together with confidence. The major uncertainty has been how to fit East to West Gondwanaland, i.e., how to put Madagascar, India, and Antarctica against eastern Africa (see a summary of alternatives in Powell et al., 1980, fig. 2).

With new magnetic-anomaly, bathymetric, and satellite gravity data in the past few years a consensus has emerged.

Recent reconstructions (for example, Norton and Sclater, 1979; Lawver and Scotese, 1987; C. McA. Powell and C. R. Scotese, 1986 [unpublished] *in* Powell et al., 1988, Scotese and McKerrow, 1990) all place Madagascar against the Tanzanian-Kenyan margin of Africa, and put the southern half of the western Indian margin against the eastern margin of Madagascar. In many of the earlier fits the tip of India was shown as adjacent to, and slightly north of, the southern tip of Madagascar, following the fit of Smith and Hallam (1970), which was determined by matching continental-margin bathymetry, before seafloor magnetic anomalies and ocean-floor bathymetry were available. Powell et al. (1980) used bathymetry, transform faults, and magnetic anomalies data to place India about 500 km southward along the Madagascar margin from the Smith and Hallam bathymetric fit. The Powell and Scotese (1986 [unpublished], *in* Powell et al., 1988) reconstruction, which we have followed, places India further south along the eastern Madagascar margin than in the Smith and Hallam (1970) fit. It also provides for a tighter fit between East and West Gondwanaland in recognition of the continental stretching preceding breakup in many of the Gondwanaland fragments (cf. Lawver et al., 1992).

The reconstruction of the major Gondwanaland fragments

Powell, C. McA., and Li, Z. X., 1994, Reconstruction of the Panthalassan margin of Gondwanaland, *in* Veevers, J. J., and Powell, C. McA., eds., Permian-Triassic Pangean Basins and Foldbelts Along the Panthalassan Margin of Gondwanaland: Boulder, Colorado, Geological Society of America Memoir 184.

used in this study (Fig. 1) using the poles listed in Table 1 is a combination of the reconstructions of Powell et al. (1988) for East Gondwanaland, Powell and Scotese (1986 [unpublished], in Powell et al., 1988) for positioning East Gondwanaland against West Gondwanaland, and Lawver and Scotese (1987) for West Gondwanaland. Our reconstruction is almost identical to that of Scotese and McKerrow (1990). Lawver et al. (1992) presented a tight-fit reconstruction of Gondwanaland, taking into account the postulated continental extension between the Gondwanaland continents; a list of the poles for reconstructing their fit is presented for comparison in Table 2. The differences between Lawver et al.'s (1992) fit and ours are minor.

The reconstruction of the smaller continental pieces along the Paleo-Pacific margin of South America, Antarctica, and Australia is more subjective and has been based on geologic argument as well as paleomagnetic constraints. In South America, the positions of the Precordillera and the entire southern half of Chile and Argentina are suspect prior to the mid-Carboniferous (Ramos, 1988), but they appear to have been in place since then. The Falkland Islands off South America are reconstructed by a counterclockwise rotation of 100° to lie off southeast South Africa near East London, in conformity with paleomagnetic data (Mitchell et al., 1986; Taylor and Shaw, 1989), confirming Adie's (1952) geological reconstruction.

In Antarctica, the positions of the Antarctic Peninsula, Thurston Island-Eights Coast, Ellsworth-Whitmore Mountains, and Marie Byrd Land blocks can be constrained paleomagnetically (Grunow et al., 1991). The position of North Island New Zealand and the Lord Howe Rise are determined by closing the Tasman Sea (Veevers, 1984; Veevers et al., 1991). The position of South Island New Zealand with respect to North Island New Zealand is determined by Korsch and Wellman (1988, fig. 16). Because an unknown amount of crustal extension was involved in the later Gondwanaland breakup and was followed by substantial transpression, any more precise reconstruction of the New Zealand blocks is probably not justified by the available information.

In eastern Australia, the position of the only likely suspect terrane, the Kin Kin terrane of the Gympie composite terrane, is constrained by geological argument (Murray, 1990). Paleomagnetic work (Z. X. Li, P. W. Baillie, and C. McA. Powell, in preparation) shows that the Precambrian western two-thirds of Tasmania, which lies outboard of the Kanmantoo Fold Belt (Flöttmann et al., 1993), lay approximately in its present position adjacent to the Australian continent by the end of the Cambrian. Late Mesozoic extension of ~130 km in a NNE direction (Powell et al., 1988) has been restored in the Gondwanaland reconstruction. The northern margin of Australia stops at the Central Highlands of New Guinea because regions to the north are younger accreted terranes (Pigram and Davies, 1987). The resulting reconstruction (Fig. 1) represents Gondwanaland in the Late Triassic (230 Ma) prior to the rifting which led to breakup in the Middle Jurassic (160 Ma).

PALEOMAGNETIC POLES

The Paleozoic Gondwanaland apparent polar-wander path (APWP) has been a subject of considerable discussion in recent years (e.g., Bachtadse and Briden, 1991; Li et al., 1990, 1993a, b; Schmidt et al., 1990; Scotese and Barrett, 1990), with most attention focussing on how the paleopole, which was in northern Africa during the Cambro-Ordovician (Ripperdan and Kirschvink, 1992), traversed Gondwanaland to reach East Antarctica by the Permo-Carboniferous. Li et al. (1990, 1993a, 1993b) and Schmidt et al. (1987, 1990) consid-

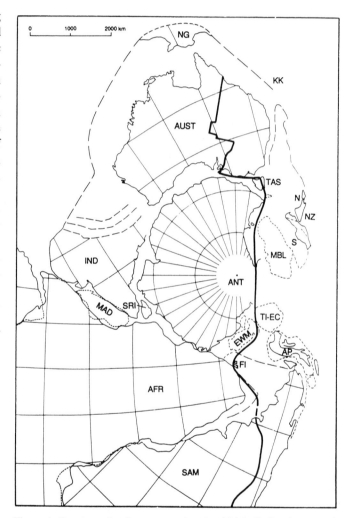

Figure 1. Gondwanaland reconstruction (see Table 1). Lambert equal-area projection centered on 30°S, 45°E, with Africa held fixed in modern coordinates. Modern longitude and latitude lines (10° × 10°) are shown on each continental block. The heavy line is the oceanward limit of Precambrian crust of the main blocks. NG—New Guinea, south of the Central Highlands, AUST—Australia, KK—Kin Kin, TAS—Tasmania, N NZ—North Island New Zealand, S NZ —South Island New Zealand, MBL—Marie Byrd Land, TI–EC—Thurston Island–Eights Coast, AP—Antarctic Peninsula, EWM—Ellsworth-Whitmore Mountains, FI—Falkland Islands, IND—India, MAD—Madagascar, SRI—Sri Lanka, AFR—Africa, SAM—South America.

TABLE 1. POLES USED TO RECONSTRUCT GONDWANALAND

Rotation Pair	Latitude (N°)	Longitude (E°)	Angle* (°)	Reference†
Antarctica to Australia	8.26	35.37	30.75	1
Tasmania to Australia	47.00	4.00	3.56	1
India to Australia	-14.79	15.35	-64.75	1
Sri Lanka to India	9.80	82.90	-24.29	2
Australia to Africa	29.94	119.84	-52.81	3
Madagascar to Africa	-3.33	-78.76	19.41	3
Arabia to Africa	26.50	21.50	-7.60	4
South America to Africa	45.50	-32.20	58.20	4
New Zealand northwest of the Alpine Fault to Australia	-14.00	142.00	-25.00	5

*Counterclockwise rotations positive.
†1 = Powell et al., 1988; 2 = Lawver et al., 1992, and personal communication, 1993; 3 = C. McA. Powell and C. R. Scotese, 1986, unpublished work; 4 = Lawver and Scotese, 1987; 5 = Johnson and Veevers, 1984, table 2, modified by Veevers et al., 1991, table 1.

TABLE 2. POLES FOR A TIGHTER GONDWANALAND FIT*

Rotation Pair	Latitude (N°)	Longitude (E°)	Angle† (°)	Reference§
Antarctica to Australia	-2.0	38.9	31.5	2
Tasmania to Australia	47.0	4.0	3.56	1
India to Australia	-13.1	9.7	-64.5	2
Sri Lanka to India	9.8	82.9	-24.29	2
Australia to Africa	21.3	112.9	-55.81	2
Madagascar to Africa	-5.5	-90.6	21.12	2
Arabia to Africa	30.9	17.5	-6.32	2
South America to Africa	50.0	-32.5	55.08	2
New Zealand northwest of the Alpine Fault to Australia	-24.3	160.8	-39.89	2

*After Lawver et al., 1992.
†Counterclockwise rotations positive.
§1 = Powell et al., 1988; 2 = Lawver et al., 1992, and personal communication, 1993.

ered that the paleopole was in southern South America by the Early Devonian, and that it then moved northward to central Africa by the Late Devonian before crossing Gondwanaland to reach East Antarctica by the Late Carboniferous. Bachtadse and Briden (1991) and Scotese and Barrett (1990) considered that the path was a more gradual movement toward southern South America during the Devonian and then across East Antarctica without the return to central Africa proposed by Li et al. (1990, 1993a, 1993b) and Schmidt et al. (1990).

Recent work by Chen et al. (1993a, 1993b), Hurley and VanderVoo (1987), Li et al. (1988, 1989, 1991a, 1991b), Schmidt et al. (1986), and Thrupp et al. (1991), summarized in

Li et al. (1993b), has documented a number of well-constrained paleomagnetic poles from eastern and central Australia, which show that the APWP moved from southern Argentina in the Middle Devonian to central Africa by the end of the Devonian and Early Carboniferous, before moving rapidly to East Antarctica in the Late Carboniferous (Fig. 2). There are two intervals of rapid apparent polar wander, one in the Middle and Late Devonian (375–355 Ma) and the other in the Visean to Westphalian interval (340–310 Ma) (Chen et al., 1993a, 1993b; Meert et al., 1993). The tight control on the age of the magnetization for most of these paleopoles, which were obtained from Givetian-Frasnian (Schmidt et al., 1986), Frasnian-Famennian (Hurley and VanderVoo, 1987), Famennian (Thrupp et al., 1991), Famennian-Tournaisian (Li et al., 1988, 1991a, 1991b; Chen et al., 1993a) and Visean (Chen et al., 1993b) rocks, all deformed in the mid-Carboniferous Kanimblan and Alice Springs orogenic movements, documents the central African loop in the Late Devonian and Early Carboniferous. It is this interpretation of the Gondwanan APWP we have used in our paleogeographic reconstructions (Fig. 3, and later chapters).

The Devonian to Permian part of the Gondwanaland APWP (Fig. 2) is adopted from Li et al. (1993a), which was based on combining high-quality paleomagnetic data from Australia with similar quality data from other Gondwanan continents. The Early Mesozoic part of the APWP is based on

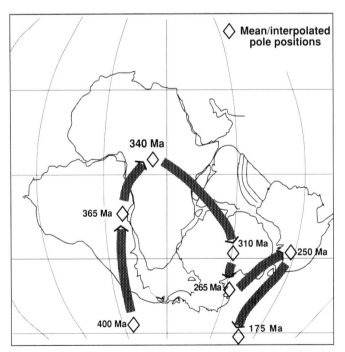

Figure 2. Gondwanaland APWP from ca. 400 Ma to ca. 175 Ma, after Embleton (1984) and Li et al. (1993a, 1993b). Map is centered at 0°S, 30°E; modern African lines of longitude and latitude (30° × 30°) are used.

Australian data compiled by Embleton (1984) and is supported by new results obtained by Lackie (1988) from the southern New England Fold Belt in northern N.S.W. The interpolated paleopoles used to define paleolatitudes in each tectonic stage are listed in Table 3.

ACKNOWLEDGMENTS

We thank the GSA reviewers, L. A. Lawver and M. R. McElhinny, for their thorough reviews, and Lawver additionally for making available his latest pole table (Table 2). This work was supported by the Australian Research Council.

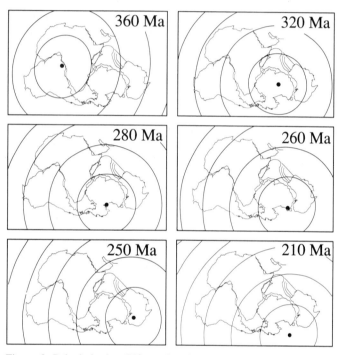

Figure 3. Paleolatitudes (30° spacing) in Gondwanaland at six intervals using the APWP in Fig. 2. Paleopoles used are listed in Table 3.

TABLE 3. PALEOPOLE POSITIONS USED FOR FIGURE 3*

Age (Ma)	Latitude (N°)	Longitude (E°)
360	-06	009
320	-25	060
280	-38	060
260	-35	075
250	-29	089
210	-47	080

*In modern African coordinates.

REFERENCES CITED

Adie, R. J., 1952, The position of the Falkland Islands in a reconstruction of Gondwanaland: Geological Magazine, v. 89, p. 401–410.

Bachtadse, V., and Briden, J. C., 1991, Palaeomagnetism of Devonian ring complexes from the Bayuda Desert, Sudan—New constraints on the apparent polar wander path for Gondwanaland: Geophysical Journal International, v. 104, p. 635–646.

Bullard, E., Everett, J. E., and Smith, A. G., 1965, The fit of the continents around the Atlantic: Philosophical Transactions, Royal Society of London, v. 258, p. 41–51.

Carey, S. W., 1958, A tectonic approach to continental drift, in Carey, S. W., ed., Continental drift, a symposium: Hobart, University of Tasmania, p. 177–355.

Chen, Z., Li, Z. X., Powell, C. McA., and Balme, B. E., 1993a, Palaeomagnetism of the Brewer Conglomerate in central Australia, and fast movement of Gondwanaland during the Late Devonian: Geophysical Journal International, v. 115, p. 564–574.

Chen, Z., Li, Z. X., Powell, C. McA., and Balme, B. E., 1993b, An Early Carboniferous paleomagnetic pole for Gondwanaland: New results from diamond-drill core materials of the Mount Eclipse Sandstone in the Ngalia Basin, central Australia: Journal of Geophysical Research (in press).

Embleton, B.J.J., 1984, Continental palaeomagnetism, in Veevers, J. J., ed., Phanerozoic earth history of Australia: Oxford, Clarendon, p. 11–16.

Fairhead, J. D., 1988, Mesozoic plate reconstructions of the central South Atlantic Ocean: The role of the West and Central African rift system: Tectonophysics, v. 155, p. 181–191.

Fairhead, J. D., and Binks, R. M., 1991, Differential opening of the Central and South Atlantic oceans and the opening of the west African rift system: Tectonophysics, v. 187, p. 191–203.

Flöttmann, T., Gibson, G. M., and Kleinschmidt, G., 1993, Structural continuity of the Ross and Delamerian orogens of Antarctica and Australia along the margin of the Paleo-Pacific: Geology, v. 21, p. 319–322.

Grunow, A. M., Kent, D. V., and Dalziel, I.W.D., 1991, New paleomagnetic data from Thurston Island: Implications for the tectonics of West Antarctica and Weddell Sea opening: Journal of Geophysical Research, v. 96, p. 17935–17954.

Haxby, W. F., 1985, Gravity field map of the World's Oceans: Palisades, New York, Lamont-Doherty Geological Observatory.

Hurley, N. F., and VanderVoo, R., 1987, Paleomagnetism of Upper Devonian reefal limestones, Canning Basin, Western Australia: Bulletin of the Geological Society of America, v. 98, p. 138–146.

Johnson, B. D., and Veevers, J. J., 1984, Oceanic palaeomagnetism, in Veevers, J. J., ed., Phanerozoic earth history of Australia: Oxford, Clarendon Press, p. 17–38.

Korsch, R. J., and Wellman, H. W., 1988, The geological evolution of New Zealand and the New Zealand region, in Nairn, A.E.M., Stehli, F. G., and Uyeda, S., eds., The ocean basins and margins, v. 7B: The Pacific Ocean: New York, Plenum Press, p. 411–482.

Lackie, M. A., 1988, The palaeomagnetism and magnetic fabric of the Late Permian Dundee Rhyodacite, New England, in Kleeman, J., ed., New England Orogen: Tectonics and metallogenesis: Armidale, New South Wales, Geology Department, University of New England, p. 157–169.

Lawver, L. A., and Scotese, C. R., 1987, A revised reconstruction of Gondwanaland, in McKenzie, G. D., ed., Gondwana Six: Structure, Tectonics and Geophysics: Washington, D.C., American Geophysical

Union, Geophysical Monograph, v. 40, p. 17–23.

Lawver, L. A., Royer, J. Y., Sandwell, D. T., and Scotese, C. R., 1991, Evolution of the Antarctic continental margins, *in* Thomas, M.R.A., Crame, J. A., and Thomson, J. W., eds., Geological evolution of Antarctica: Cambridge, U.K., Cambridge University Press, p. 533–539.

Lawver, L. A., Gahagan, L. M., and Coffin, M. F., 1992, The development of Paleoseaways around Antarctica, *in* The Antarctic paleoenvironment: A perspective on global change: Washington, D.C., American Geophysical Union, Antarctic Research Series, v. 56, p. 7–30.

Li, Z. X., Schmidt, P. W., and Embleton, B.J.J., 1988, Palaeomagnetism of the Hervey Group, central New South Wales and its tectonic implications: Tectonics, v. 7, p. 351–367.

Li, Z. X., Powell, C. McA., and Schmidt, P. W., 1989, Syn-deformational remanent magnetization of the Mount Eclipse Sandstone, central Australia: Geophysical Journal International, v. 99, p. 205–222.

Li, Z. X., Powell, C. McA., Thrupp, G. A., and Schmidt, P. W., 1990, Australian Palaeozoic palaeomagnetism and tectonics—II: A revised apparent polar wander path and palaeogeography: Journal of Structural Geology, v. 12, p. 567–575.

Li, Z. X., Powell, C. McA., Embleton, B.J.J., and Schmidt, P. W., 1991a, New palaeomagnetic results from the Amadeus Basin and their implications for stratigraphy and tectonics, *in* Korsch, R. J., and Kennard, J. M., eds., Geological and geophysical studies of the Amadeus Basin, central Australia: Bureau of Mineral Resources Bulletin 236, p. 349–360.

Li, Z. X., Powell, C. McA., and Morris, D. G., 1991b, Syn-deformational and drilling-induced remanent magnetizations from diamond drill cores of the Mt. Eclipse Sandstone, central Australia: Australian Journal of Earth Sciences, v. 38, p. 473–484.

Li, Z. X., Powell, C. McA., and Trench, A., 1993a, Palaeozoic global reconstructions, *in* Long, J. A., ed., Palaeozoic vertebrate biostratigraphy and biogeography: London, Belhaven Press, p. 25–53.

Li, Z. X., Chen, Z., and Powell, C. McA., 1993b, New Late Palaeozoic palaeomagnetic results from cratonic Australia, and revision of the Gondwanan apparent polar wander path: Australian Journal of Exploration Geophysics, v. 24 (in press).

Meert, J. G., VanderVoo, R., Powell, C. McA., Li, Z. X., McElhinny, M. W., and Symons, D.T.A., 1993, A plate-tectonic speed limit?: Nature, v. 363, p. 216–217.

Mitchell, C., Taylor, G. K., Cox, K. G., and Shaw, J., 1986, Are the Falkland Islands a rotated microplate?: Nature, v. 319, p. 131–134.

Murray, C. G., 1990, Comparison of suspect terranes of the Gympie province with other units of the New England Fold Belt, eastern Australia: Australasian Institute of Mining and Metallurgy, Pacific Rim 90 Congress, v. 2, p. 247–255.

Norton, I. O., and Sclater, J. G., 1979, A model for the evolution of the Indian Ocean and the breakup of Gondwanaland: Journal of Geophysical Research, v. 84, p. 6803–6830.

Pigram, C., and Davies, H., 1987, Terranes and the accretionary history of New Guinea orogen: Australian Bureau of Mineral Resources Journal, v. 10, p. 193–211.

Powell, C. McA., Johnson, B. D., and Veevers, J. J., 1980, A revised fit of East and West Gondwanaland: Tectonophysics, v. 63, p. 13–29.

Powell, C. McA., Roots, S. R., and Veevers, J. J., 1988, Pre-breakup continental extension in East Gondwanaland and the early opening of the eastern Indian Ocean: Tectonophysics, v. 155, p. 261–283.

Ramos, V., 1988, Late Proterozoic–Early Paleozoic of South America, a collisional history: Episodes, v. 11, p. 168-175.

Ripperdan, R. L., and Kirschvink, J. L., 1992, Paleomagnetic results from the Cambrian-Ordovician boundary section at Black Mountain, Georgina Basin, western Queensland, Australia, *in* Webby, B., and Laurie, J., eds., Global perspectives on Ordovician geology: Rotterdam, A. A. Balkema, p. 93–103.

Sandwell, D. T., and McAdoo, D. C., 1988, Marine gravity of the Southern Ocean and Antarctic Ocean from GEOSAT: Journal of Geophysical Research, v. 93, p. 10389–10396.

Schmidt, P. W., Embleton, B.J.J., Cudahy, T. J., and Powell, C. McA., 1986, Prefolding and pre-megakinking magnetization from the Devonian Comerong Volcanics, New South Wales, Australia, and their bearing on the Gondwana Pole Path: Tectonics, v. 5, p. 135–150.

Schmidt, P. W., Embleton, B.J.J., and Palmer, H. C., 1987, Pre- and post-folding magnetization from the Devonian Snowy River Volcanics and Buchan Caves Limestone, Victoria: Geophysical Journal of the Royal Astronomical Society, v. 91, p. 155–170.

Schmidt, P. W., Powell, C. McA., Li, Z. X., and Thrupp, G. A., 1990, Reliability of Palaeozoic palaeomagnetic poles and APWP of Gondwanaland: Tectonophysics, v. 184, p. 87–100.

Sclater, J. G., and Fisher, R. L., 1974, Evolution of the East Central Indian Ocean, with emphasis on the tectonic setting of the Ninety-east Ridge: Bulletin of the Geological Society of America, v. 85, p. 683–702.

Scotese, C. R., and Barrett, S. F., 1990, Gondwana's movement over the South Pole during the Palaeozoic: Evidence from lithological indicators of climate, *in* McKerrow, W. S., and Scotese, C. R., eds., Palaeozoic palaeogeography and biogeography: Geological Society of London Memoir 12, p. 75–85.

Scotese, C. R., and McKerrow, W. S., 1990, Revised world maps and introduction, *in* McKerrow, W. S., and Scotese, C. R., eds., Palaeozoic palaeogeography and biogeography: Geological Society of London Memoir 12, p. 1–21.

Smith, A. G., and Hallam, A., 1970, The fit of the southern continents: Nature, v. 225, p. 139–144.

Taylor, G. K., and Shaw, J., 1989, The Falkland Islands: New palaeomagnetic data and their origin as a displaced terrane from southern Africa, *in* Millhouse, J. W., ed., Deep structure and past kinematics of accreted terranes: Washington, D.C., American Geophysical Union, Geophysical Monograph, v. 50, p. 59–72.

Thrupp, G. A., Kent, D. V., Schmidt, P. W., and Powell, C. McA., 1991, Palaeomagnetism of red beds of the Late Devonian Worange Point Formation, SE Australia: Geophysical Journal International, v. 104, p. 179–201.

Veevers, J. J., ed., 1984, Phanerozoic earth history of Australia: Oxford, Clarendon Press, 418 p.

Veevers, J. J., 1990, Antarctica-Australia fit resolved by satellite mapping of oceanic fracture zones: Australian Journal of Earth Sciences, v. 37, p. 123–126.

Veevers, J. J., and Eittreim, S. L., 1988, Reconstruction of Antarctica and Australia at breakup (95 ± 5 Ma) and before rifting (160 Ma): Australian Journal of Earth Science, v. 35, p. 355–362.

Veevers, J. J., Powell, C. McA., and Roots, S. R., 1991, Review of seafloor spreading around Australia, 1: Synthesis of the patterns of spreading: Australian Journal of Earth Sciences, v. 38, p. 373–389.

Wegener, A., 1915, Die Entstehung der Kontinente und Ozeane: Braunschweig, Germany, Vieweg, 367 p.

Manuscript Accepted by the Society September 9, 1993

Geological Society of America
Memoir 184
1994

Eastern Australia

J. J. Veevers, P. J. Conaghan, and C. McA. Powell*
Australian Plate Research Group, School of Earth Sciences, Macquarie University, North Ryde, N.S.W. 2109, Australia

With contributions by
E. J. Cowan
Department of Geology, University of Toronto, Toronto, Ontario M5S 3B1, Canada
K. L. McDonnell and S. E. Shaw
School of Earth Sciences, Macquarie University, North Ryde, N.S.W. 2109, Australia

ABSTRACT

The Sydney-Gunnedah-Bowen Basin developed above the junction between (a) the western, early to mid-Paleozoic Lachlan and Thomson Fold Belts, terminally deformed and intruded in the mid-Carboniferous, and (b) the eastern, mid- to late Paleozoic New England Fold Belt (NEFB). Accordingly, the basement of the Sydney-Gunnedah-Bowen Basin varies along strike. In the south, the Sydney-Gunnedah Basin developed *above* the Late Devonian–Early Carboniferous Andean-type magmatic arc and fore arc, whereas in the north, the Bowen Basin developed *behind* the magmatic arc. The magmatic arc was displaced by crustal transtension during the latest Carboniferous–Early Permian. During transtension, the NEFB was intruded by S-type granitoids with co-magmatic ignimbrites and uplifted during right-lateral shearing to form the initial stage of an orocline.

Thereafter to the end of the Triassic, eastern Australia developed through seven stages: *Stage A (290–268 Ma), extension-volcanism* of the collapsed Kanimblan–NEFB upland with thick volcanics and sediment, was a local manifestation of the first release of Pangean-induced heat and is comparable with the vast magmatic province of the same age that developed after the Variscan Orogeny in Europe. A glacio-eustatic marine transgression at 277 Ma crossed the NEFB to reach the newly formed Bowen and Sydney Basins.

Stage B (268–258 Ma), a marine sag on the platform and embryonic magmatic arc/foreland basin, brought the sea to the western edge of the Bowen-Gunnedah-Sydney Basin and covered Tasmania. The first tuff attributable to the north-migrating convergent magmatic arc reached Tasmania 265 Ma, and the first convergent granitoid reached the NEFB also at 265 Ma and was followed by the deposition of coarse sediment in the embryonic foreland basin.

Stage C (258–250 Ma), orogenic piedmont coal/tuff, initiated the foreland basin between the uplift of the mature NEFB and a foreswell that bounded the Galilee, Coorabin, and Tasmania Basins of the western craton.

*Present address: Department of Geology and Geophysics, University of Western Australia, Nedlands, W. A. 6009, Australia.

Veevers, J. J., Conaghan, P. J., and Powell, C. McA., 1994, Eastern Australia, *in* Veevers, J. J., and Powell, C. McA., eds., Permian-Triassic Pangean Basins and Foldbelts Along the Panthalassan Margin of Gondwanaland: Boulder, Colorado, Geological Society of America Memoir 184.

In *Stage D (250–241 Ma), orogenic piedmont redbeds barren of coal and tuff,* the sea was driven from the foreland basin by climactic outpouring of volcanolithic sediment from granitoids in the orogenic upland.

Waning plutonic activity in the narrowing NEFB upland led to *Stage E (241–235 Ma), a cratonic quartz-sand sheet* that covered the entire foreland basin.

Stage F (235–230 Ma), orogenic paralic sediment, saw a final pulse of sediment shed from the overthrusted NEFB.

The terminal deformation and uplift of the NEFB represents the northern part of the Gondwanide event that involved the entire Panthalassan margin to South America in the final episode of the foreland basin/orogen. The final *Stage G (230–200 Ma) involved rifting of the orogen by right-lateral transtension* to form the Tarong and Ipswich coal and volcanic basins in a second release of Pangean heat. By the Early Jurassic, cratonic quartz sand breached the NEFB upland at the Queensland–New South Wales border. The modern New England upland dates from the mid-Cretaceous inception of the Eastern Highlands by the underplating of the lower-plate margin during the separation of the upper-plate Lord Howe Rise.

INTRODUCTION

This synopsis of the geologic history of eastern Australia focuses on the Permian-Triassic succession of sedimentary and igneous rocks of the NEFB and related basins: from north to south, the Bowen-Galilee Basins, Gunnedah-Gilgandra Basins, Sydney–sub-Murray Basins, and the Tasmania Basin, all constituents of the Gondwana facies of the Pangean Supersequence. Our overriding aim has been to establish a chronological framework of geological events along the Panthalassan margin of Gondwanaland (presented in Appendix 1) as a basis for presenting the geologic history in the form of time-space diagrams and stage maps.

The pre-Gondwanan history by Powell and the Gondwanan history by Veevers and Conaghan are updated expansions of material given in Phanerozoic Earth history of Australia (Veevers, 1984), in particular Powell's (1984a) account of the Uluru regime and the account of the Bowen-Sydney Basin by Jones et al. (1984) which has been expanded in space to include northern and southeastern Queensland and the sub-Murray and Tasmania Basins and in time to the latest Carboniferous start of the extensional stage. First reported here are many new field observations by P. J. Conaghan, who compiled the maps, and the Rb/Sr dates of plutons in the NEFB by S. E. Shaw (Appendix 3). David Briggs provided details of his Permian brachiopod zonation ahead of publication.

Our approach to the analysis of eastern Australian Permian-Triassic geology is through the stratigraphic/radiometric history of the flanking basin and the tracing of geological events from the basin to the orogen. Recently the NEFB has become more accessible to this kind of analysis from the higher resolution of geological events provided by Briggs's (1991) scheme of brachiopod zones of the Permian of eastern Australia, Price et al.'s (1985) and Helby et al.'s (1987) schemes of palynological zones, and by Shaw's (Appendix 3) 150 new isotopic dates of New England granitoids and ignimbrites. The current palynological schemes place the Permian/Triassic

boundary above the base of the Rewan or Narrabeen Groups (Balme, 1970). By means of ^{13}C chemostratigraphy, Morante (1993) and Morante et al. (1994) restore the boundary to its traditional place at the base of the Rewan or Narrabeen Group at their contact with the upper coal measures. With the help of this information, our account updates the analysis of the Bowen-Sydney Basin and the NEFB by Jones et al. (1984). Sources of basic information are Day et al. (1983) for Queensland, Balfe et al.'s (1988) map of the solid geology of the Bowen Basin, the set of papers and colored map in Finlayson (1990) for southeastern Queensland, Pogson (1972) for the geological map of New South Wales, and Banks and Clarke (1987) and Clarke and Forsyth (1989) for Tasmania. The timescale adopted is that of Palmer (1983). We make the following change: the Permian/Triassic boundary is renumbered from the rounded 245 Ma to the rounded 250 Ma, after Claoué-Long et al.'s (1991) direct determination of 251.2 ± 3.4 Ma by the sensitive high resolution ion microprobe (SHRIMP) U-Pb method on zircons from the Chinese Permian/Triassic stratotype.

PRE-GONDWANAN BASIN HISTORY

Introduction

The Sydney-Bowen Basin lies across the junction between two fold belts of contrasting structure and geological history, the Lachlan-Thomson Fold Belt (LTFB) to the west, and NEFB to the east (Fig. 1). As such, the Sydney-Bowen Basin is the overlap assemblage between the two fold belts (*see* Coney et al., 1990; Leitch and Scheibner, 1987; and Powell et al., 1990, for recent summaries of the terrane history). Accounts of the paleogeographical evolution of the Lachlan Fold Belt (LFB) (e.g., Cas, 1983; Leitch and Scheibner, 1987; Packham, 1987; Powell, 1983a, 1984a; Scheibner, 1986) vary in detail, but all have the LFB located craton-ward of a subduction zone that lay farther east for most

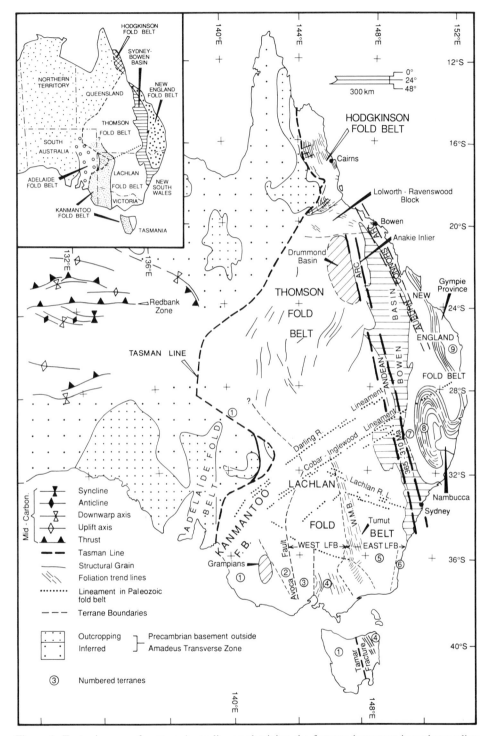

Figure 1. Tectonic map of eastern Australia, emphasizing the features important in understanding the Sydney-Gunnedah-Bowen Basin. Adapted from Powell et al. (1990, fig. 1). Inset shows the three main fold belts comprising the Tasman Fold Belt. The Tasman Line is the eastern limit of known Precambrian rocks.

of the Paleozoic. No unequivocal trench deposits have been identified in the LFB, but the arrangement of facies is consistent with its setting as a passive margin or back-arc region for most of the Paleozoic.

The position of the join between the Kanmantoo Fold Belt (KFB) and LFB has been taken as the Avoca fault (Fig. 2), in line with new interpretations of the age and extent of the KFB (*see* Wilson et al., 1992; Glen et al., 1992). The Stawell terrane (2) appears to have a more complex deformational history than the adjacent Bendigo terrane (3) to which it appears

provenance-linked in the Early Ordovician. Major deformation in the Stawell terrane is thus interpreted to be the same as in the Stavely-Glenelg terrane (Middle/Late Cambrian; Gibson, 1992). Provenance linking between the East and West LFB first occurred in the Ordovician, with a period of terrane dispersion and disruption in the Silurian. Provenance linking was re-established in the late Early Devonian (Powell et al., 1993) with the westerly shed of juvenile volcanogenic material from the uplifted Benambra terrane (5) into the Melbourne-Mathinna terrane (4).

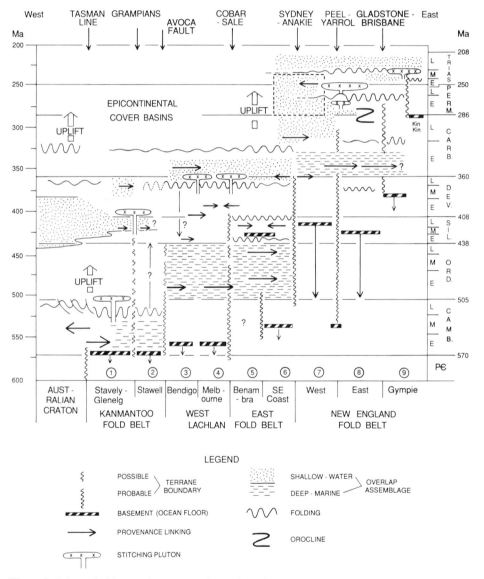

Figure 2. Schematic history of terrane amalgamation of the southern and eastern Tasman Fold Belt, modified from Powell et al. (1990, fig. 2) by addition of the Gympie Province in which the easternmost unit, the Kin Kin Terrane, is a possible exotic terrane (Murray, 1990a), and revision of information in the Kanmantoo, West Lachlan, and New England Fold Belts. Circled numbers refer to terranes, or groups of terranes shown on Figure 1 and described in Powell et al. (1990). Place names with arrows at the top of the diagram refer to location of major dislocations in the Tasman Fold Belt.

The Sydney-Anakie line is the presumed location of the join between the LFB and the NEFB, now concealed beneath the Sydney-Bowen Basin. The West NEFB (7) corresponds to the Tamworth-Yarrol terrane (Leitch and Scheibner, 1987), which is separated from the east NEFB (8) by the Great Serpentinite Belt along the Peel and Yarrol Faults. A 530 Ma ophiolite (Aitchison et al., 1992a), representing a fragment of Cambrian ocean floor, is incorporated in the Peel Fault. The east NEFB comprises a collage of terranes. Six are identified by Leitch and Scheibner (1987) in New South Wales, and at least a further three by Korsch and Harrington (1987) in Queensland. The Gympie Province contains six distinct units, with at least three suspect terranes, one of which is composite (Murray, 1990a). In the generalized representation of Figure 2 only the easternmost unit (Gympie Group, Brooweena and Keefton Formations, and Kin Kin Beds) is separated, as it is the most likely candidate for an exotic terrane (Murray, 1990a). The other terranes of the Gympie Province are linked to the main part of the NEFB in the Early Carboniferous (Murray, 1990a; Little et al., 1992, 1993).

The Sydney-Bowen Basin, dashed-line box in Figure 2, lies over the join between the LFB and the NEFB. Provenance linking (Powell, 1984a; Flood and Aitchison, 1992) indicates amalgamation as early as Devonian-Carboniferous, with the main consolidation during the mid-Carboniferous Kanimblan deformation.

By contrast, the NEFB (Figs. 3, 4) is characterized by sediments of fore-arc, trench-slope, and ocean-floor facies facing the Paleo-Pacific on the east (e.g., Day et al., 1978; Fergusson, 1984a, 1984b; Korsch and Harrington, 1981; Harrington and Korsch, 1985; Leitch and Scheibner, 1987; Murray, 1986; Murray et al., 1987; Roberts and Engel, 1980; Scheibner, 1985; Korsch et al., 1990a; Little et al., 1992, 1993). Furthermore, in contrast with the LFB, the NEFB has evidence of accretion of exotic terranes (e.g., Kin Kin terrane

in the Early Triassic and possibly the Calliope Island Arc in the Middle Devonian). There is also evidence of lateral displacements of several hundred kilometers (Harrington and Korsch, 1985; Murray et al., 1987).

Accordingly, the basement to the Sydney-Bowen Basin varies from place to place. At the time of initiation of the Sydney-Bowen Basin in the latest Carboniferous, the LTFB was a relatively stable continental margin along which, in northern New South Wales at least, an Andean-type magmatic arc had been erupting since the latest Devonian (Figs. 2 and 4).

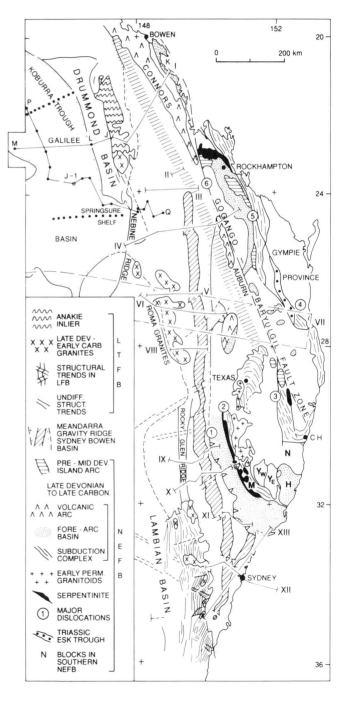

Figure 3. Simplified structural map of basement features important in the development of the Sydney-Bowen Basin. Numbered dislocations: 1—Hunter-Mooki Thrust System; 2—Peel-Manning Fault Zone; 3—Demon Fault; 4—Great Moreton Fault; 5—Yarrol Fault Zone; 6—Gogango-Baryulgil Fault Zone. Bold lettered blocks in the southern NEFB from Leitch (1988): H—Hastings, M—Macdonald and Manning, N—Nambucca, Y_E—Yarrowitch East, Y_W—Yarrowitch West. Map base 1:5,000,000 Tectonic Map of Australia (Geological Society of Australia, 1971) with units shown from Murray et al. (1987, 1989) and modifications to the Drummond Basin and Lachlan Fold Belt basement from Powell (1984b) and Powell et al. (1985). Boundaries of the Sydney-Gunnedah-Bowen-Galilee Basin: dotted—erosional edge; dashed—inferred subsurface; solid—inferred depositional or faulted edge. Open hachured line is a zone of inferred basement shearing associated with orocline development. Fine lines show the location of cross sections I to XIII (Figs. 20, 21, 15), cross sections PQ (Fig. 22) (J-1—Jericho-1 well), and time-space diagrams M-L-J (Fig. 27) and J-K (Fig. 28).

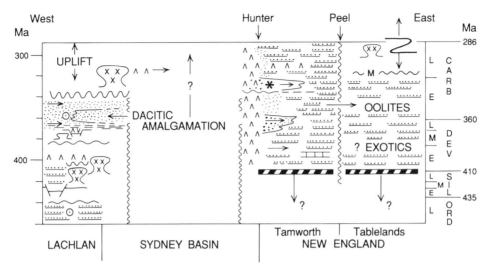

Figure 4. Time-space plot for the northeastern Lachlan Fold Belt and adjacent New England Fold Belt at the approximate latitudes of Sydney and Gunnedah Basins. Diagram shows that the Sydney-Gunnedah Basins developed over a zone inferred to have previously been a magmatic arc for 100 m.y., the last 50 m.y. being as an Andean-type arc along the eastern edge of the Lachlan Fold Belt. Symbols as in Figure 2.

Its northern extent into Queensland is more debatable, but its presence could be reflected by the subsurface Devonian-Carboniferous Roma granites in southern Queensland and the Late Devonian Retreat batholith in the southern Anakie Inlier (Figs. 1 and 3). The widespread Late Devonian Silver Hills Volcanics in the Anakie Inlier form the base to the Drummond Basin succession and could be either the extrusive equivalents of the Retreat Batholith or related to crustal extension preceding the onset of the Andean-type magmatism. Many of the formations in the overlying Late Devonian–Early Carboniferous Drummond Basin contain tuffaceous material (Olgers, 1972), indicating the proximity of a volcanic arc.

Paleocurrent and provenance patterns in the Lambie (Powell, 1983b, 1984a, and 1985, unpublished data) and Drummond (Olgers, 1972) Basins immediately to the west of the inferred magmatic arc suggest that they were foreland basins adjacent to active volcanic arcs. Moreover, coeval sediments in the Tamworth belt (West NEFB in Fig. 2) indicate a westerly juvenile volcanic source from the Late Devonian to the Late Carboniferous (Fig. 4) (Leitch, 1974; McPhie, 1987). The entire LTFB was deformed, uplifted and eroded in the mid-Carboniferous (350 to 325 Ma) with post-tectonic granites intruded between 325 and 310 Ma in the Late Carboniferous.

The NEFB was an active accretionary complex during the same interval, with imbricate stacking of trench-slope facies and emplacement of allochthonous slices in the fore-arc terranes (Fergusson, 1982, 1984a, 1984b; Cawood, 1982; Cawood and Leitch, 1985). Parts of the Devonian-Carboniferous magmatic arcs are preserved in the Queensland part of the fold belt, but in New South Wales there is no evidence of the volcanic chain other than provenance indications that it lay farther to the west.

We speculate that the Andean-type magmatic arc which bordered the LTFB was the same arc which belonged with the fore arc of the NEFB, at least for the 70 m.y. from the Late Devonian until the latest Carboniferous formation of the Sydney-Bowen Basin (Powell et al., 1990; Flood and Aitchison, 1992). Unequivocal evidence, however, that the two arcs were one and the same has yet to be found, and it is not until the latest Carboniferous (ca. 300 Ma) that the Sydney-Bowen Basin overlaps the join between the two fold belts.

Structure of the Lachlan-Thomson Fold Belt

The structure of the LFB beneath, and marginal to, the New South Wales sector of the Sydney-Bowen Basin is dominated by meridional folds formed during the mid-Carboniferous Kanimblan deformation. The structure in the Thomson Fold Belt is poorly known, as all of it, except the Anakie Inlier, is subsurface. Bouguer gravity trends in the Thomson Fold Belt (TFB) are dominantly northeasterly (Murray et al., 1989), and fold trends in the Drummond Basin are mainly meridional to northeasterly. The magmatic arc along the eastern edge of the LTFB in the Late Devonian and Early Carboniferous (Fig. 1), inferred from paleocurrent and provenance studies in the Late Devonian-Early Carboniferous Lambie (Powell, 1983b, 1984a) and Drummond (Olgers, 1972) Basins, trended 345°. Its roots are now concealed beneath the Sydney Basin. To the north, along strike in southern Queensland, the Roma Granite (Murray, 1986, fig. 5) is a two-mica S-type granite, which Cec Murray (personal communication, 1993) considers unlikely to represent part of such an arc. Powell (1984a, fig. 224) interpreted its extension further north to lie along the Anakie Inlier,

noting the 379–352 Ma Retreat Batholith (I-type, Cec Murray, 1993, personal communication) and the presence of proximal contemporaneous volcanic rocks in the latest Devonian Drummond Basin (Olgers, 1972) as evidence.

The mid-Carboniferous folding, which is most pronounced and tight in the eastern LFB (Powell et al., 1977), trends northerly in the LFB slightly oblique to the inferred NNW paleogeographic trend (Powell, 1984a). Postdating the meridional fold trends in the LFB is a series of megakinks with dextral offsets on ESE–trending kink planes, and sinistral offsets on ENE–trending kinks (Fig. 3; Powell, 1984b; Powell et al., 1985). These megakinks are parallel to prominent LANDSAT lineaments such as the ENE–trending Darling River and Cobar-Inglewood Lineaments, and the ESE–trending Lachlan River Lineament (Fig. 1; Scheibner and Stevens, 1974).

North-northwest trends are prominent in the structural grain of much of the eastern LFB and show up in the filtered Bouguer gravity map of Murray et al. (1989) where NNW–trending gravity highs and lows correspond to anticlinorial and synclinorial zones, respectively. In the TFB, the Bouguer gravity-residual trends are north-northeasterly, except along the eastern margin where they trend north-northwesterly (Murray et al., 1989).

A prominent gravity high near the inferred eastern margin of the LTFB is the Meandarra Gravity Ridge (MGR) (Qureshi, 1984; Murray et al., 1989), which runs almost due north from Sydney for almost 1,200 km to 24°S (Fig. 3). The ridge, almost 50 km wide, lies about the middle of the present outcrop belt of the Sydney-Bowen Basin. It appears to be offset by steps of 10 km or so, sympathetic in displacement sense and orientation with the sinistral megakinks in the LFB (Powell, 1984b, Powell et al., 1985). The likely southern extension of the MGR in the Blue Mountains west of Sydney has been modeled as a mafic body of density 2.9 tm^{-3} and a maximum thickness of 12 km (Qureshi, 1984). It has been interpreted as either the buried Late Devonian–Early Carboniferous magmatic arc (Day et al., 1978; Powell *in* Qureshi, 1984, p. 300), or a mafic to ultramafic keel to a volcanic graben which preceded the main fill of the Sydney Basin (Scheibner, 1973; Murray et al., 1989; Murray, 1990b).

Structure of the New England Fold Belt

The structure of the NEFB is complex, with steep fault zones along which hundreds of kilometers of strike-slip displacement could have occurred. There is also a large Z-fold (the Texas–Coffs Harbour Megafold) in the southern NEFB which is interpreted (Flood and Ferguson, 1982; Murray et al., 1987) to have formed by 500 km of dextral movement in the latest Carboniferous to mid-Permian interval. An alternative interpretation (Harrington and Korsch, 1985; Korsch and Harrington, 1987) is that there have been several hundreds of kilometers of dextral displacement of the entire NEFB with respect to the LTFB along the approximate line of the Late Devonian–Early Carboniferous Andean magmatic arc proposed here (Fig. 1). A smaller oppositely curved bend in the structure of the southern NEFB near Newcastle is interpreted (Leitch, 1988) to have been (a) the line of displacement of the Hastings Block from a position near Newcastle (Fig. 5a), and (b) the line along which the southern New England fold trends were bent in the latest Permian or Early Triassic (Fig. 3).

The principal grain of the southern NEFB follows the trace of the Peel Fault (Dislocation 2 in Fig. 3), which is concave to the east. The Tamworth fore-arc basin lies west of the Peel Fault, and is separated from the younger sediments of the Sydney Basin by the Hunter-Mooki Thrust system (Dislocation 1, Fig. 3). The Peel Fault is a fundamental structure of the NEFB and has a history which dates back to at least Early Devonian, and possibly mid-Cambrian (Aitchison et al., 1992a). Throughout the Devonian and Carboniferous, the Peel Fault separated relatively shallow-water fore-arc facies to the west from deep-marine trench-floor and trench-slope deposits to the east. Serpentinite crops out along its length, and several phases of movement are postulated (Offler and Hand, 1988), with both sinistral (Corbett, 1976) and dextral (Scheibner and Glen, 1972) displacement.

The Yarrol and Great Moreton Faults appear to be the equivalents of the Peel Fault in the Queensland sector of the NEFB, with a similar separation of Devonian and Carboniferous fore-arc facies to the west from trench-slope and basinal facies to the east (Figs. 3, 5). The Yarrol–Great Moreton Fault system is offset nearly 200 km to the east of the Peel Fault, an offset approximately equal to the amplitude of the Texas–Coffs Harbour Megafold, as originally noted by Bryan (1925). Inboard of most of the Yarrol fore-arc terrane, and separating it from the Late Devonian–Early Carboniferous Connors-Auburn volcanic arc, the Gogango-Baryulgil Fault Zone is postulated (Murray et al., 1987) to have borne the 500-km displacement implied by the Texas–Coffs Harbour Megafold. The Gogango-Baryulgil Fault Zone is expected to have had a long and complex history, acting as a dextral transcurrent fault from latest Carboniferous until mid-Permian, and then as a root zone for westward-directed thrusts during the Late Permian to Middle Triassic Hunter-Bowen Orogeny. It is likely that small deep-marine sedimentary basins (such as the Grantleigh Trough) formed on transtensional sections of the fault zone, and these were later everted during the Late Permian/Triassic.

In the interpretation presented here (Figs. 5 and 6), a deep crustal shear (a transform fault) cuts obliquely across the Andean magmatic arc so that dextral movement along the Panthalassan margin displaced the Auburn-Connors volcanic arc southward by almost 700 km, thereby duplicating the volcanic arc in central and southern Queensland (Fig. 6, *cf.* Powell, 1984a, fig. 224). Part of the postulated transform fault is shown as lying along the Gogango-Baryulgil Fault of Murray et al. (1987), but to the north it cuts westward of the volcanic arc. The model of Murray et al. (1987) is different in that their transcurrent movement does not cut the arc, which is thus not displaced with respect to the LTFB.

There are many other faults in the NEFB with significant

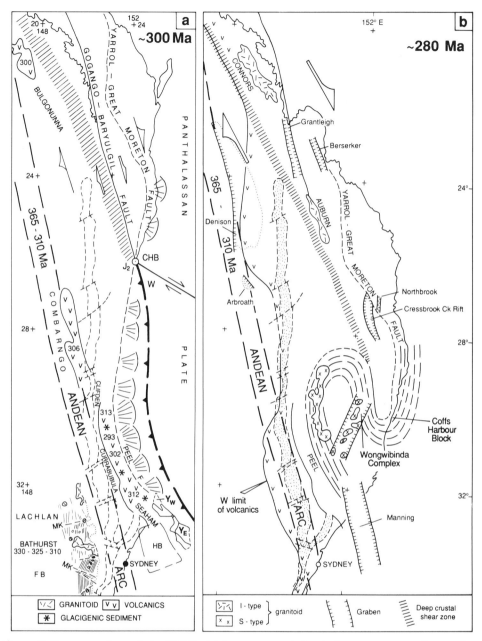

Figure 5. Reconstructions of the NEFB. (a) Last spasm of the Andean-type magmatic arc: Late Carboniferous (~310 Ma to 300 Ma) prior to the formation of the New England Orocline. The dashed lines show the Meandarra Gravity Ridge (MGR), which terminates against the basement shear zone associated with the future orocline. An Andean-type margin persisted south of 27°S (McPhie, 1987), related to subduction of the Panthalassan margin. The reconstruction is based on Murray et al. (1987, fig. 8b), but differs in the interpretation of the position of the deep crustal shear zone. East Yarrowitch (Y_E), West Yarrowitch (Y_W), and Hastings Block (HB) positioned after Leitch (1988). Note that coordinates shown east of the Yarrol–Great Moreton Fault relate to the eastern fault block. (b) First sign of extension: latest Carboniferous–Early Permian (300 Ma to 280 Ma) after formation of the New England Orocline. The NNW-trending deep crustal shear zone is shown connecting to the Texas–Coffs Harbour Megafold, with transtensional grabens opening in a more northerly orientation. S-type granitoids of the Bundarra (295 Ma) and Hillgrove (290 Ma, Watanabe et al., 1988; 300 Ma, R. H. Flood, personal communication, 1993; 304–298 Ma, Landenberger et al., 1992) suites are intruded across the megafold trend. Modified from Murray et al. (1987, fig. 9). Abbreviations: CHB—Coffs Harbour Block; J_2—triple junction; MK—megakink; W—Wongwibinda.

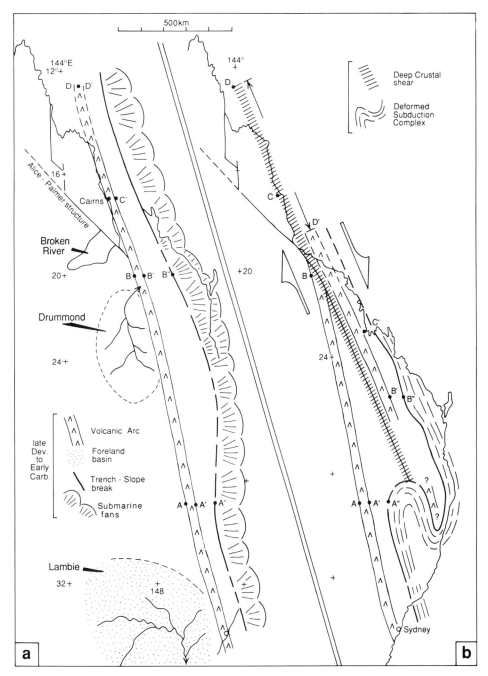

Figure 6. Tectonic sketch outlining essential elements (a) before and (b) after the postulated translation of the Late Devonian to Early Carboniferous volcanic arc approximately 650 km to the south along a deep-crustal dextral shear. The displacement led to the formation of the New England Orocline in the Late Carboniferous to Early Permian interval (300 Ma to 280 Ma). A, B, C and D are shown as matching points on the volcanic arc before and after formation of the orocline. The postulated deep-crustal shear corresponds with strong NNW-trending gravity ridges parallel to, and along, the north Queensland coast (Province XXV, Murray et al., 1989). This model for orocline formation differs from that of Murray et al. (1987) in that their dextral shear does not displace the magmatic arc.

displacement; e.g., the Demon Fault in the southern NEFB (Fig. 3) has 29 km of dextral offset. An important difference between the basements of the NEFB and LTFB is that, in the former, large lateral displacements, followed by thrusting, occurred during the development of the Sydney-Bowen Basin, whereas in the latter the basement was stable. These movements affected the style, thickness, and distribution of younger sediments.

Another feature of the NEFB is the extensive igneous activity which occurred before, and during, sedimentation in the Sydney-Bowen Basin. The calc-alkaline Connors-Auburn volcanic arc was active until around 315 Ma (Westphalian). Thereafter, there appears to have been a lull in magmatic activity in the northern NEFB, although calc-alkaline volcanism appeared to have continued in the New South Wales part at least until 305 Ma (McPhie, 1987; Murray et al., 1987) (Fig. 5a). Plutonic activity in the NEFB resumed around 300 Ma in the latest Carboniferous when the S-type Bundarra and Hillgrove suites were emplaced east of, and adjacent to, the Peel Fault; at the same time the Bulgonunna Volcanics were erupted in the northern part of the Bowen Basin as crustal extension commenced inboard of the east Australian margin (Fig. 5).

Late Devonian to Late Carboniferous tectonic evolution

The most important feature about the prehistory of the Sydney-Gunnedah Basin is that it developed above the site of a Late Devonian and Early Carboniferous Andean magmatic arc along the eastern edge of the LFB: an arc which could have been the same arc as that inferred to have bordered the western edge of the Tamworth Zone of the NEFB (Figs. 1, 2, 4). The arc thus lies across the junction of two major fold belts beneath the Sydney-Gunnedah Basin, which developed in part, if not entirely, by the thermal subsidence which must have occurred as the magmatic arc cooled and subsided. The Andean-type arc had the largely nonmarine Lambie Basin as a foreland basin to the west in the Late Devonian and Early Carboniferous, and the Tamworth Zone as a dominantly marine fore-arc basin to the east. Farther east were zones of deep-marine sediment likely to have accumulated on the ocean floor. The prehistory of the Sydney-Bowen Basin, shown schematically in Figure 4, suggests that docking of the LTFB and NEFB might have occurred as early as mid-Devonian in Queensland (Day et al., 1978; Murray, 1986) and latest Devonian in New South Wales (Powell, 1984a; Flood and Aitchison, 1992) but that consolidation was probably not complete until the Late Carboniferous when the LFB was stabilized after the 330 ± 10 Ma Kanimblan folding.

There was little noticeable effect of the mid-Carboniferous deformation in the Tamworth Zone, but regional high P/T metamorphism and deformation in the Tia Complex (Fukui et al., 1993) and metamorphism and deformation in the Coffs Harbour Block (Graham and Korsch, 1985) (Fig. 5) farther east occurred about 320 Ma. This metamorphism and deformation, thought to be related to subduction (Fukui et al., 1993), could have been associated with accretion of these areas as exotic

blocks in the NEFB subduction complex. After the Kanimblan deformation, the effects of which extended through the Amadeus Transverse Zone to Western Australia (Veevers and Powell, 1984), post-tectonic Late Carboniferous I-type granites were intruded in the Bathurst region west of Sydney and in the Coonabarabran region beneath the Great Artesian Basin (E. Scheibner, personal communication, 1993).

In Queensland, the Bowen Basin developed between the inferred extension of the Andean-type arc along the Anakie Inlier and the Connors-Auburn Arch to the east. In the model presented here (Figs. 5 and 6), the Bowen Basin lies above a broad zone of crustal extension across which the arc and fore-arc terranes in central Queensland were displaced by dextral transtension relative to the LTFB continental crust to the west.

The onset of the sedimentation in the Sydney-Bowen Basin coincides with the development of a transtensional regime, arguably of dextral displacement. The reconstruction (Fig. 5a) suggests that the tectonic setting of eastern Australia was analogous to the Neogene setting of the west coast of North America where the early Cenozoic subduction of a mid-ocean spreading ridge was responsible for the change to a transtensional tectonic regime. The former Andean margin in California and northern Mexico has been transformed into a zone of dextral shear with the transtensional Basin-and-Range lying continent-ward. Ocean-floor subduction continues today to the north in Oregon and Washington, and to the south in southern Mexico, so that different tectonic regimes coexist, and the change from one tectonic setting to the other is diachronous.

In eastern Australia, the reconstruction suggests that ocean-floor subduction ceased earlier in the north than in the south (Fig. 5a). Thus, the Bulgonunna Volcanics, arguably a bimodal suite related to continental extension (McPhie et al., 1990), are coeval with subduction-related magmatism in the Currabubula region near Tamworth (McPhie, 1987). The triple junction, J2, is depicted in Fig. 5 at 26°S, which is the northernmost occurrence of Late Carboniferous volcanism of possible calc-alkaline affinity; it is not known whether the Combarngo Volcanics (Fig. 5a) are calc-alkaline or bimodal. The Bulgonunna Volcanics, at 21°S, are related to dextral transtension, and are analogous to shoshonitic volcanics formed on the northern margin of Papua New Guinea in the past 5 m.y., when Miocene subduction was transformed into sinistral transtension.

In the model (Figs. 5a and 6a), the master shear zone is interpreted as trending NNW broadly parallel to, but slightly counterclockwise from, the 345°-trending Devonian-Carboniferous magmatic arc. Thus, north of the Lolworth-Ravenswood Block (Fig. 1), the master shear, probably a zone 50- to 100-km-wide rather than a single fault (*cf.* San Andreas system of the western United States), could have run into the magmatic arc and crossed to the continental side. Subsequent dextral displacement on this shear could have caused the northern segment of the Andean magmatic arc to be displaced southward (Fig. 6).

Secondary faults oriented more northerly than 330° would have lain in a transtensional orientation to the master shear.

Thus, features such as the Grantleigh Trough and the Berserker, Northbrook, Cressbrook Creek, Barnard, and Manning Basins (Fig. 5b) are interpreted as transtensional grabens. Blueschists in the North D'Aguilar Block of southern Queensland are interpreted to have formed about 306 Ma by extension in the oceanic Rocksburg Greenstone during this basin-forming event (Little et al., 1992, 1993). Where the marginal faults cut through to the asthenosphere, the grabens were floored by ocean crust, as suggested for the Barnard and Manning Basins (Leitch, 1988). The Denison and Arbroath Troughs are related half-grabens. The Hillgrove and Bundarra batholiths are magmatic rocks related to this transtensional phase, and presumably intruded some of the graben-fills.

The Texas-Coffs Harbour Megafold (the New England Orocline) also formed during this interval. Constraints on its age are loose. Oolite-bearing sediments of presumed Early Carboniferous age are involved in the megafold, and late Early Permian granites cut across its trend. Its formation during the latest Carboniferous or Early Permian is thus permissible from geological constraints. Estimates of the amount of displacement vary. Murray et al. (1987) suggested 450 km of dextral offset, and the reconstruction (Fig. 6b) shows approximately 650 km displacement. These figures are probably minima, because only the displacement recorded in the megafold trace is being measured. There could be additional displacement parallel to, and oceanward of, the megafold for which no offset markers have been recognized.

The displacement on the crustal shear and megafold has important consequences for the geology of North Queensland, where a Late Devonian to Early Carboniferous magmatic arc is inferred to have lain to the east (Lang, 1988), although none has been found yet. Many coastal outcrops are highly sheared metamorphic rocks (e.g., Barnard Metamorphics), largely of unknown or equivocal age. They represent the kinds of rocks expected to be found in a through-going crustal shear zone. If correct, this inference provides a possible explanation for the termination of the Sydney-Bowen Basin at the latitude of the Lolworth-Ravenswood Block (Fig. 1), because the northern part of the magmatic arc had been removed by the mid-Permian so that there was no former hot zone over which substantial subsidence could occur.

The Meandarra Gravity Ridge (MGR) is somewhat of an enigma in this scheme. It extends from just south of Sydney almost due north to 24°S (Fig. 5b). In the southern part, its position is broadly coincident with the position of the inferred Andean magmatic arc, to which origin it has previously been attributed (Day et al., 1978; Powell *in* Qureshi, 1984). To the north it trends more northerly, however, than the inferred Andean arc passing east of the thick Permian-Carboniferous sedimentary successions of the Denison Trough (Fig. 5b) along the axis of the Taroom Trough (*see* Figs. 17 and 30 in later sections). The continuity of the MGR and of the subparallel Namoi Gravity Ridge to the east suggests that the deep crustal shear does not cross either, providing they predate the shearing.

It is possible that the MGR is a composite representing relatively dense, mafic intrusions partly along the line of the old Andean arc in the south and along a transtensional zone in the north. It is also possible that the Andean arc in Queensland has been displaced eastward from its former position adjacent to the eastern edge of the LTFB. One striking feature of the ridge is that it stops at 24°S at the position where it is crossed by the postulated NNW–trending crustal shear. This position also coincides with the latitude north of which there is major younger westward thrusting in midcrustal levels (Korsch et al., 1990b; Cec Murray, personal communication, 1993). Interpretation of the Namoi Gravity Ridge is uncertain. Strong magnetic trends are parallel to the Namoi Gravity Ridge and the Peel Fault in the southern NEFB, but in Queensland the gravity ridge continues north, whereas the magnetic trends follow the New England Orocline (Wellman, 1990).

Summary

There are some important differences in the nature of the basement on which the latest Carboniferous to mid-Triassic Sydney-Bowen Basin developed along the Pacific margin of Australia. The Sydney Basin developed above the former crest and fore arc of an Andean-type magmatic arc. The Bowen Basin formed over a complex zone of transtension, interpreted as largely back arc to a displaced segment of the former continental-margin arc. Both basins have feather-edge onlaps onto the older deformed basement of Early and mid-Paleozoic fold belts deformed and stabilized in the mid-Carboniferous.

In both basins there was active transtension and dextral displacement in the coeval fold belt oceanward of the site of the basins, but the amount of extension appears to have been greater in the Bowen Basin. Consequently, the potential for subsidence in the Bowen Basin was greater than for the Sydney Basin, realized in the 10-km-thick Bowen Basin compared with the 6-km-thick Sydney Basin.

In both basins, the early transtensional history is succeeded later in the Permian by a foreland basin at the onset of subduction-related calc-alkaline magmatism, which commenced earlier in the south than in the north.

GONDWANAN HISTORY

Having sketched the pre-Gondwanan history of eastern Australia, we now come to a synopsis of its Gondwanan (Permian-Triassic) geologic history. Our chief aim is to place the Australian depositional and magmatic events in a chronology that can be extended along the rest of the Panthalassan margin. The radiometric time scale suffices for the igneous rocks; the biostratigraphical zones compiled in Appendix 1, drawn mainly from the diverse and well-known Australian sections, are an attempt at a Gondwana-wide scheme of time-correlation for the sedimentary rocks. For convenience, stability, and wide currency, we adopted the DNAG time scale (Palmer, 1983), es-

sentially that of Harland et al. (1982), and have made only one change to it: the Permian/Triassic boundary is shifted from 245 Ma to 250 Ma (Claoué-Long et al., 1991). We locate it in eastern Australia at the boundary between the upper coal measures and the overlying Rewan-Narrabeen Group (Morante, 1993; Morante et al., 1994).

We recognize seven stages of Gondwanan history in eastern Australia (Table 1) and divide eastern Australia into six regions: (1) the Sydney-Gunnedah-Bowen Basin and New England Fold Belt (NEFB) in New South Wales; (2) the Bowen Basin and NEFB in Queensland; (3) southeastern Queensland; (4) the Tasmania Basin; (5) north Queensland, and (6) the basins beneath the Murray Basin. The first four regions are each illustrated by maps, time-space diagrams, and stratigraphic columns. Cross sections of all the regions except north Queensland are presented stacked from north to south. The time-space diagrams provide the basis for the paleogeographical maps, first a set for each region, then combined for eastern Australia, which in turn contribute to the paleogeographical reconstructions in the final chapter.

Sydney-Gunnedah Basin and NEFB in New South Wales

The formations and plutons shown on the base map (Fig. 7) are grouped according to their position before, during, or

TABLE 1. EAST AUSTRALIAN PERMIAN-TRIASSIC STAGES

Stage	Ma	Presumed Mechanism
	200	
G. Rifting of the orogen		Rifting, sagging
3. COLLAPSE OF OROGEN		
---------------------- 230 ------------------		
F. Orogenic paralic sediment		Right-lateral transcurrence/ thrusting
	235	
E. Cratonic sand sheet		Quiescence
	241	
D. Orogenic piedmont redbeds		Thrusting
	250	
C. Orogenic: orogenic piedmont coal/tuff regressive marine sediment		Thrusting
2. MATURE MAGMATIC OROGEN/FORELAND BASIN		
---------------------- 258 ------------------		
B. Marine sag of the platform and embryonic magmatic orogen/foreland basin		Flexure
	268	
A. Extension-volcanism		Detachment faulting
1. COLLAPSE OF KANIMBLAN-COFFS HARBOUR OROGEN		
---------------------- 290 ------------------		

after the 7 stages listed in Table 1. The sedimentary rocks are dated according to the biostratigraphical scheme described in Appendix 1 and the igneous and metamorphic rocks by radiometric methods, notably Shaw's survey of the New England Batholith (Appendix 3).

Time relations eastern Sydney Basin–NEFB (Fig. 8, Table 2).

Events before the Sydney Basin. These events are the intrusion of the presumed equivalents of the Bathurst Granite (BA), dated as in Figure 12 (EF) (*see* Fig. 12 in a later section), and coeval megakinking and regional (Kanimblan) deformation (encircled dot and cross indicate main compressive stress), as documented in Figure 12 (EF). The Late Devonian Merrimbula Group (correlated with the Lambie Group) lies immediately west of an inferred magmatic arc, in the region of the granite of unknown age found at the bottom of Australian Oil & Gas Woronora-1 (3 in Fig. 7). The Merrimbula Group is underlain by the Comerong-Yalwal Volcanics. The Mogo Hill diatreme, between Sydney and Gosford, contains xenoliths of quartzofeldspathic sandstone with clasts of ignimbrite. The xenoliths resemble Silurian-Devonian rocks of the eastern Lachlan Fold Belt and are interpreted as indicating the eastward extension of the foldbelt terrane beneath the Sydney Basin (Emerson and Wass, 1980; O'Reilly, 1990), as found by Leaman (1990) from gravity and magnetic data, in place of Qureshi's (1984) model of a 12-km-thick mafic body within the upper crust.

Marine-nonmarine. Arrows (Fig. 8) indicate shoreline movements in the NNE–SSW plane; encircled crosses indicate movements WNW. The Early Jurassic (Pliensbachian *C. torosa* zone) nonmarine sediments are known only as clasts in diatremes (Helby and Morgan, 1979). Initial Stage 3a transgression (T) terminated by main regression (R), followed by marine incursions (I), from Herbert (1980a).

Glacigenic sediment. In Figure 8, marine glacigenic sediment is manifested by dropstones deposited from sea ice in the interval from the base of the lowermost formation through the Kulnura Marine Tongue (33), and in the Dempsey Formation (35) (Herbert and Helby, 1980, p. 278, 288) and glendonites, pseudomorphs after ikaite, in the interval including the Dalwood Group to the Kulnura Marine Tongue (33) (Carr et al., 1989). Conaghan (1984) reported periglacial (permafrost) structures in the floor of the Wongawilli Coal (14) and Bulli Coal at the top of the Illawarra Coal Measures.

Coal. Coal (Fig. 8) is restricted to the Clyde (c in 4 and 6) (Evans, 1991), Greta, and upper (Tomago, Newcastle, Illawarra) coal measures, and to the Farley Formation (27) (Garretts Seam). The occurrence in the Wianamatta Group amounts to nothing more than lenticles of banded light and dull coal up to 15 cm thick (Herbert 1979, p. 45, 50). At Bundanoon, the Wongawilli Coal is overlain disconformably by the Hawkesbury Sandstone (Wilson et al., 1958) (*cf.* Hawkesbury/Burralow disconformity in the west, Fig. 12 in a later section).

Redbeds. In the north, redbeds (Fig. 8) occur from near

the base of the Dooralong Shale (36) (Uren, 1980, fig. 9.2) to unit E of McDonnell (1972, p. 61); redbeds are confirmed in outcrop by Cowan (1985, p. 93, 94) in the middle Terrigal Formation (42); see also Helby (1973, p. 150). There are redbeds in the south from the Wombarra Claystone to the Bald Hill Claystone (Hanlon et al., 1954).

Tuff. Wandrawandian: Runnegar (1980b); Mulbring: Booker (1960, p. 35), Byrnes (1982b), New South Wales Department of Mineral Resources (1982), J. Brunton (personal communication, 1991); Tomago Coal Measures: Diessel (1980a, p. 105), Diessel et al. (1985), Hamilton, (1966), Joint Coal Board of New South Wales (1978), Shaw et al. (1991); Newcastle Coal Measures: Diessel (1985), Jones et al. (1987), Packham and Emerson (1975); Illawarra Coal Measures: Bowman (1974), Byrnes (1982a), Bamberry and Doyle (1987), Australian Gas Light (1982, 1983); Narrabeen: Byrnes (1983); Wianamatta: Byrnes (1974, 1981, 1982c), Herbert (1980b, p. 272).

According to Shaw et al. (1991, p. 48, 49):

Patterns of tephra development in the Sydney and Gunnedah Basins are generally the same, both in terms of stratigraphic positioning and frequency of occurrence. They differ, however, in the nature of the beds (i.e., tephra-flow versus tephra-fall, and thickness), a consequence of distance from source. In the Sydney Basin, the first appearance of tephra is at the top of the Wandrawandian Siltstone (in the south) and the Mulbring Siltstone (in the north). These tephra units, many less than 10 mm thick, increase in thickness and frequency in an irregular pattern upwards throughout the Tomago Coal Measures and correlatives and culminate in the Newcastle Coal Measures and correlatives with very thick units (up to 25 m; Diessel 1985), some of which are surge and tephra-flow deposits. They contain bentonitic clays and have a mineralogy of quartz, biotite and feldspar (Diessel, 1985). The thickness and bedding characteristics of some of these tephra units (e.g., Nobbys Tuff and Reids Mistake Tuff) suggest that they are at relatively close to moderate distances from the eruptive sources (McDonnell, 1983; Diessel, 1985; Mushenko, 1985). . . . The disappearance of tephra above the Upper Permian coal measures of the Sydney Basin is not considered to be due to non-preservation, as suitable low-energy (including lacustrine) environments conducive to their preservation are a feature of parts of the overlying succession, i.e., Narrabeen Group. In the Gunnedah Basin, however, the top of the Upper Permian coal measures succession was eroded during latest Permian diastrophism and it is therefore difficult to assess whether tephra deposits were present in the section that has been removed.

In both basins, the major tephra level reaches to the top of the Stage 5—*P. crenulata* palynological zone, calibrated as 250 Ma.

Again, according to Shaw et al. (1991, p. 49):

Younger tephra deposits in the Sydney Basin are presently known only from the Middle Triassic Wianamatta Group. The first occurrence is in the base of the Ashfield Shale and consists of a thin white tuff a few mm thick (Byrnes, 1982c). A second tephra unit is found at the base of the overlying Bringelly Shale (Cobbitty Claystone Bed) and contains grains of quartz, apatite, zircon, garnet (?), muscovite (?) and alkali feldspar (Byrnes, 1974), suggesting a rhyolitic source.

An earlier tuff comes from the Terrigal/Newport Formation (42 in Fig. 8) (Byrnes, 1983).

Maar volcanism. It is dated by the *Classopollis classoides* Zone of the Surat Basin (Helby and Morgan, 1979; Crawford et al., 1980, p. 322), now called the *Corollina torosa* Oppel Zone, and dated (Helby et al., 1987, p. 21) as Hettangian to Pliensbachian (208–193 Ma).

Hunter Thrust System. Movement started during deposition of the Newcastle Coal Measures (Collins, 1991, table 1) and continued past the deposition of the Narrabeen Group (Glen and Beckett, 1989).

Lochinvar Anticline. The fault and fold movements registered in the Lochinvar Anticline are shown in Figure 9. According to Stuntz (1972, p. 7):

The Lochinvar Anticline was a growing structure which probably came into existence just before "Upper Coal Measure" times as evidenced by the uppermost of the preceding strata (the Mulbring Sub-Group) thinning over it. [Note similar thinning in the equivalent strata of the Muswellbrook Anticline—unit 4 of the Ponds Creek Formation and in the overlying Saltwater Creek Sandstone Member of the Tomago (Wittingham) Coal Measures (Veevers 1960, figs. 5A and 5B).] The Tomago Coal Measures also thin over it, but towards the close of Tomago times a sudden uplift and tilting from the northeast occurred. This movement caused strike faults to develop along each flank, the anticline becoming a horst block plunging towards the southwest. The horst block was subsequently eroded and the remainder of the Tomago plus Newcastle Coal Measures were laid down with local erosional unconformity on the older Tomago strata around the nose and flanks of the anticline. The core of the anticline remained exposed [down to the Muree] however and there was a break in the deposition of these beds across its crest. This subsequently led to a slight unconformity of non-deposition between the Permian and the Triassic [= Munmorah Conglomerate at the base of the Narrabeen Group (Herbert, 1980a, p. 42)].

Herbert (1993) gives a more complete account of the growth of the Lochinvar Anticline.

In the Buchanan Tunnel on the eastern flank of the Lochinvar Anticline (Diessel, 1980b, p. 466, 474–476), the Dempsey Formation (253 Ma, topmost Tomago) rests with an angular discordance of 15° on the Mt Vincent Formation (lower Tomago) across a short time gap.

Figure 7 (on following page spread). Solid (bedrock) geology of central-northeastern New South Wales and adjacent southeast Queensland. Main sources are Pogson (1972), Scheibner (1974), Powell et al. (1985, fig. 2), Barnes and Willis (1989), Barnes (1990), Stroud (1989), Brownlow (1986), Gilligan et al. (1987), Barnes et al. (1991, fig. 1), Whitaker and Green (1980), Hawke and Cramsie (1984), and Thomson (1976). Additional data from Etheridge (1987); Flood et al. (1988); Hill (1986); Shaw et al. (1988); Pogson and Hitchins (1973); Murray et al. (1981); Crane and Hunt (1979, fig. 2); Campbell and McKelvey (1972); Bourke (1980); Holmes (1982); Hensel et al. (1985). Blocks in NEFB from Murray (1990a) (Fig. 14). Offshore Clarence Basin from O'Brien et al. (1990, p. 119). Offshore Sydney Basin from Grybowski (1992, figs. 2, 7).

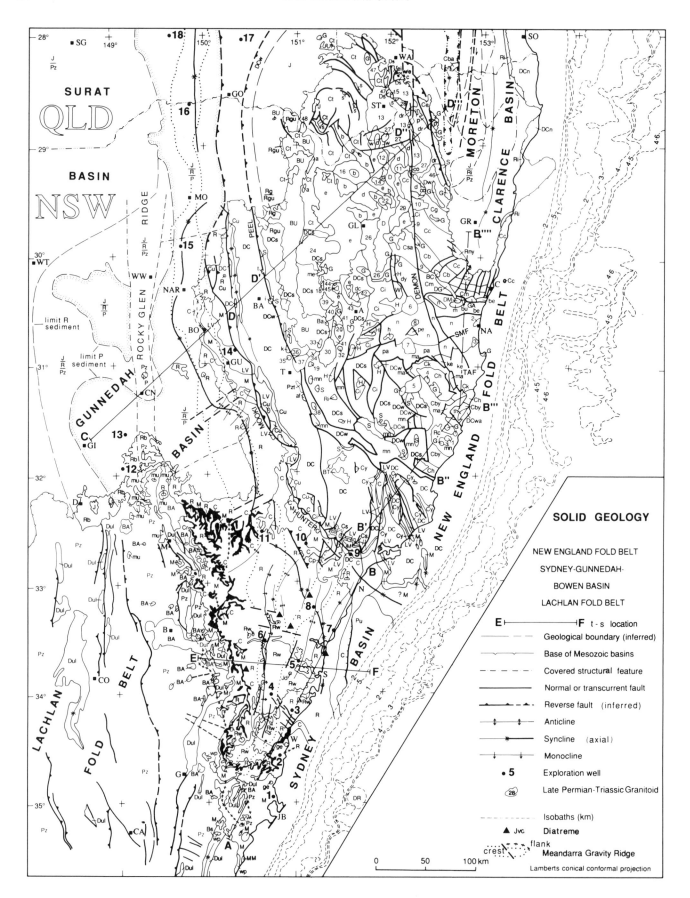

SOLID GEOLOGY

NEW ENGLAND FOLD BELT

SYDNEY-GUNNEDAH-

BOWEN BASIN

LACHLAN FOLD BELT

E ├─────────────┤ F   t - s location

Geological boundary (inferred)

Base of Mesozoic basins

Covered structural feature

Normal or transcurrent fault

Reverse fault (inferred)

Anticline

Syncline (axial)

Monocline

• 5   Exploration well

㉘   Late Permian-Triassic Granitoid

Isobaths (km)

▲ Jvc   Diatreme

crest   flank   Meandarra Gravity Ridge

0   50   100 km

Lamberts conical conformal projection

Key to Fig. 7, map of Sydney-Gunnedah Basin - New England Fold Belt, New South Wales.
Stages A - D: Units are Permian unless indicated otherwise.
T - Tertiary, K - Cretaceous, J - Jurassic, R - Triassic, Mz - Mesozoic,
P - Permian, C - Carboniferous, D - Devonian,S - Silurian, Ordovician, Pzl - Paleozoic. E - Early, M - Middle, L- Late.
LFB - Lachlan Fold Belt
Groups listed in CAPITALS. Numerals refer to New England Fold Belt granitoids, except 5 and 31, ignimbrites.

Stage unknown
G   granitoid, undiff.
Pu   undiff. Perm.(offshore Sydney)
S   serpentinite

Post-G (<200 Ma)
J   undiff. Jurassic
Jd   Prospect dolerite intrusion

Stage G (230 - 200 Ma)
J   undiff. Jurassic
Jvd   volcanic diatreme/intrusion
Ri   Ipswich Basin succession
Granitoids: 1 - 8
BC   Billip Creek

Stage F (235 - 230 Ma)
R   Napperby 'c' and Deriah
Rw   WIANAMATTA (also E)
Granitoids: 9 - 11

Stage E (241 - 235 Ma)
Rb   Ballimore (incl. Digby, Napperby) (D and E)
Rc   CAMDEN HAVEN (D and E)
Rg   Gragin
Rgu   Gunnee
Rny   NYMBOIDA
R   undiff. E-M Triassic Sydney/Gunnedah (D, E)
Granitoids: 12 - 14

Stage D (250 - 241 Ma)
R   undiff. E-M Triassic Sydney/Gunnedah (D, E)
Rb   Ballimore (incl. Digby and Napperby) (D and E)
Rc   CAMDEN HAVEN (D and E)
co   Coombadjha
d   Dundee (31), Tent Hill
e   Emmaville
gi   Gibraltar
Granitoids: 15 - 43, *D* =17,21-23,25-27

Stage C (258 - 250 Ma)
c   Condamine
C   Coal Measures: ILLAWARRA (black), NEWCASTLE, TOMAGO, BLACK JACK, GLOUCESTER

dr   Drake
ei   Eight Mile Creek
g   Gilgurry
ge   Gerringong (B and C)
me   Merrastone
r   Rhyolite Range
w   Wallangarra
Subsurface only: Cressbrook Creek
Granitoids: 44 - 48
DG   Dundurrabin
DR   Dampier Ridge
GA   Glenifer
MM   Milton, O'Hara
W   Wongwibinda

Stage B (268 - 258 Ma)
a   Ashford
b   Bondonga
bu   Buffers Creek (A and B)
dc   Dummy Creek
GB   Glenmore Beds
ge   Gerringong (B and C)
ke   KEMPSEY (A and B)
M   Marine (Sydney/Gunnedah)
ma   MACLEAY (A and B)
mn   MANNING (A and B)
n   Nambucca (A and B)
p   Plumbago Creek
pa   Parrabel (A and B)
w/e   Wallaby, Eurydesma
Granitoids:
BT   Barrington Tops
DM   Dorrigo Mountain

Stage A (290 - 268 Ma)
af   Andersons Flat
be   Bellingen
bu   Buffers Creek (A and B)
dy   Dyamberin
h   Halls Peak
i   Ironbark Creek
k   Kensington
ke   KEMPSEY (A and B)
LV   lower volcanics
m   Macgraths Hump
ma   MACLEAY (A and B)
mn   MANNING (A and B)
mu   Mudgee outliers
n   Nambucca (A and B)
pa   Parrabel (A and B)
pe   Petroi

s   Silver Spur, Alum Rock
t   Tarakan
th   Thrumster
wp   Wasp Head/Pebbley Beach,TALATERANG
Cr   Rylstone
Granitoids:
Ba   Balala
BU   Bundarra
H   Hillgrove
Ri   Rockisle

Pre-A (>290 Ma)
Cb   Brooklana
Cba   Mount Barney
Cby   Byabbara
Cc   Coramba
Ce   Emu Creek
Cg   Gundahl
Ch   undiff. E C sediments of the Hastings Block
Ci   Girrakool
Ck   Kullatine
Cm   Moombil
Cp   undiff. below Seaham at Pokolbin
Cs   Seaham
Csa   Sara
Ct   Texas
Cu   undiff. L C of Tamworth Belt
Cy   Yagon, Muirs Creek, and equivalents
DC   Undiff. D and E C in Tamworth Belt
DCn   Neranleigh - Fernvale
DCs   "Sandon Association",includes Sandon, Cara, Whitlow, in west; Tinoni and others in east
DCw   Woolomin
DCwa   Watonga
Ds   SILVERWOOD
Dul   LAMBIE and others (LFB)
Dw   Willowie Creek
Pz   O - D in LFB and Coffs Harbour Block; undiff.
Pzt   Cambro - Ordovician ?allochthonous of Tamworth Belt
Granitoids:
BA   Bathurst (LFB)

Wells, etc						
1	Genoa COONEMIA	5	Shell/AOG DURAL S	10	AOG LODER	16	UKA MACINTYRE
2	Farmout STOCKYARD MOUNTAIN	6	AOG / Exoil KURRAJONG HTS	11	AOG MARTINDALE	17	WPA BILLA BILLA
		7	Strevens TERRIGAL	12	DMR MIRRIE	18	UKA FLINTON
3	AOG WORONORA	8	AOG KULNURA	13	DMR PIBBON	**Faults**	
4	AOG KIRKHAM	9	Planet E. MAITLAND	14	Mid Eastern KELVIN	SMF	STOCKYARD MT
				15	DMR BELLATA	TAF	TAYLORS ARM

Towns							
A	ARMIDALE	CN	COONABARABRAN	JB	JERVIS BAY	SO	SOUTHPORT
B	BATHURST	D	DUBBO	M	MUDGEE	ST	STANTHORPE
BA	BARRABA	GI	GILGANDRA	MO	MOREE	T	TAMWORTH
BO	BOGGABRI	GL	GLEN INNES	N	NEWCASTLE	W	WOLLONGONG
C	COFFS HARBOUR	GO	GOONDIWINDI	NA	NAMBUCCA	WA	WARWICK
CA	CANBERRA	GR	GRAFTON	NAR	NARRABRI	WT	WALGETT
CO	COWRA	GU	GUNNEDAH	S	SYDNEY	WW	WEE WAA
		G	GOULBURN	SG	SAINT GEORGE		

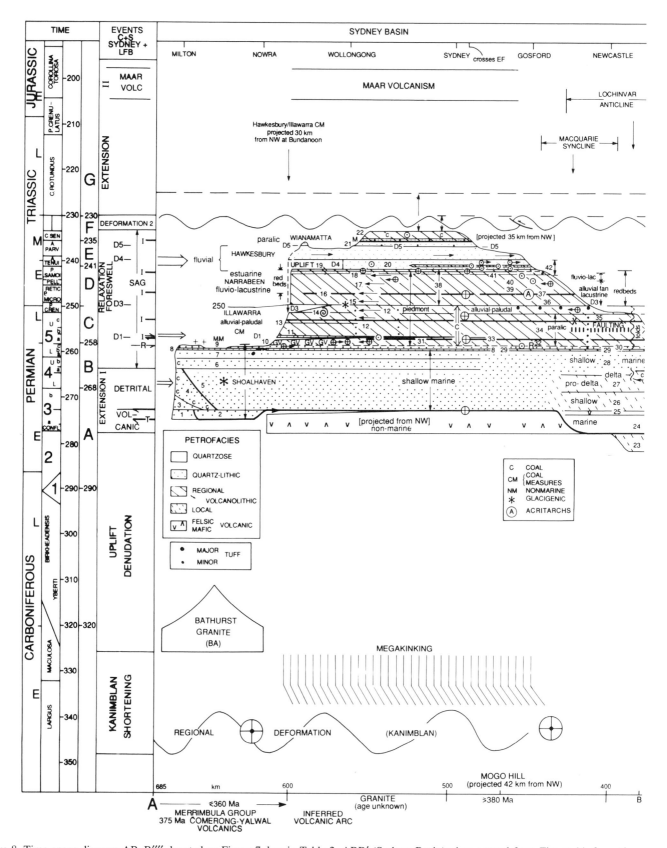

Figure 8. Time-space diagram AB–B'''', located on Figure 7, key in Table 2. ABB' (Sydney Basin): time control from Figure 44; formations and environments from Herbert and Helby (1980). The upper part of the chart (<265 Ma) is from Shaw et al. (1991, fig. 3).

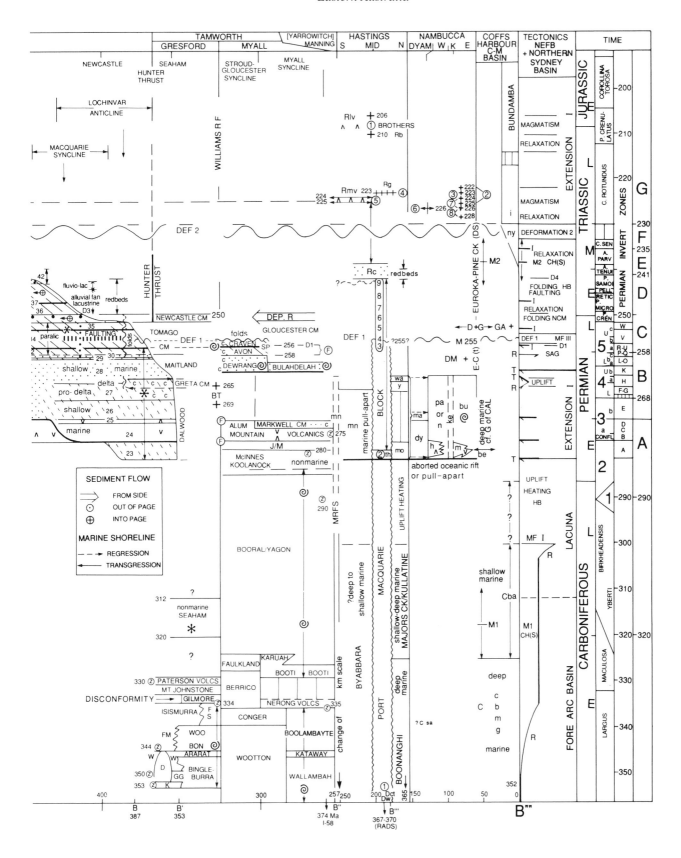

TABLE 2. KEY TO FIGURE 8

22 Bringelly Shale	42 Terrigal Formation
M Minchinbury Sandstone +	41 Patonga Claystone
lowermost 30 m of Bringelly Shale	40 Tuggerah Formation
21 Ashfield Shale	39 Munmorah Conglomerate
20 Gosford Subgroup	38 Clifton Subgroup
19 Bald Hill Claystone (Clifton Subgroup)	37 Marine ingression
+ Garie Formation (Gosford Subgroup)	36 Dooralong Shale
18 Bulgo Sandstone	35 Hexham Subgroup
17 Scarborough Sandstone	34 Four Mile Creek Subgroup
16 Wombarra Claystone	33 Kulnura Marine Tongue
15 Coalcliff Sandstone	32 Wallis Creek Subgroup
14 Wongawilli Coal	31 Cumberland Subgroup
13 Tongarra Coal	30 Mulbring Siltstone
12 Sydney Subgroup	29 Muree Sandstone
11 Marrangaroo Conglomerate	28 Branxton Formation
10 Broughton Formation	27 Farley Formation
GV Gerringong Volcanics (lavas)	26 Rutherford Formation
MM Milton Monzonite	25 Allandale Formation
9 Berry Formation	24 Lochinvar Formation
8 Nowra Sandstone	23 Glacial Beds
7 Wandrawandian Siltstone	DEF 2 = compression Sydney
6 Snapper Point Formation	and Lorne Basins
5 Pebbley Beach Formation	DEF 1 = east-west compres-
4 Jindelara Formation	sion in the northern Sydney
3 Yadboro Conglomerate	Basin and southern NEFB
2 Wasp Head Formation	
1 Pigeon House Siltstone	

Note: The 6-m-thick Mittagong Formation lies between the Hawkesbury Sandstone and the Wianamatta Group and is represented by the line that delimits the top of the Hawkesbury Sandstone. The Mittagong Formation is cut out locally at the disconformity (D5) below the Wianamatta Group (Jones and Clark, 1991, p. 10, 19).

Abbreviations in the Carboniferous of the Gresford Block are K = Kingsfield Formation; GG = Goonoo Goonoo Mudstone; D = Dangarfield Formation; W = Waverley Formation; WOO = Woolooma Formation; BON = Bonnington Siltstone. Other abbreviations are given in the text.

Deformation. In descending order, the regional deformations are the following:

DEF 2. Bounded by the youngest preserved Wianamatta (233 Ma) and Camden Haven (Rc in Fig. 8) (239 Ma) sediment and above by the Werrikimbe ignimbrite (Rmv in Fig. 8) (225 Ma). We place DEF 2 at 230 Ma between 225 and 233 Ma.

DEF 1. Manifested by faulting, is restricted to the northern Sydney Basin and southern NEFB, and is dated as 255 Ma: postdated by the flat-lying Hexham Subgroup (including the Dempsey Formation or Zone W—253 Ma, *see* Fig. 44 in a later section) of the Tomago Coal Measures and predated by the tilted Kulnura Marine Tongue (Zone U—256 Ma) and other strata of the lower Tomago Coal Measures and equivalent strata (Speldon Formation) of the Stroud-Gloucester Syncline. The Lochinvar Anticline and other folds in the northern Sydney Basin grew during deposition of the Greta Coal Measures and between the start of deposition of the Mulbring

Siltstone and the end of deposition of the Newcastle Coal Measures (Fig. 9), best seen in the Macquarie Syncline (vertical wiggly lines in Fig. 8).

Disconformities. In descending order, disconformities are the following:

D5. An uneven surface cut out of the Mittagong Formation and the top of the Hawkesbury Sandstone, capped by the Ashfield Shale (236 Ma) (Jones and Clark, 1991).

D4. South and west of Wollongong, capped by Hawkesbury Sandstone (238 Ma) resting on the Wongawilli Coal (14 in Fig. 8) at Bundanoon and elsewhere on the Bald Hill Claystone and remnants of the overlying 6-m-thick Garie Formation (19 in Fig. 8) (Sherwin and Holmes, 1986).

D3. Capped by Narrabeen Group (250 Ma) south of Wollongong and underlain by formations of the Sydney Subgroup down to the Wongawilli Coal (14) (Bowman, 1970, 1974). In the north, D3 is capped by the basal Narrabeen Group, namely, the Dooralong Shale (36)/Munmorah Conglomerate (39) (250 Ma), and in the Macquarie Syncline is underlain, at a low-angle unconformity, by the uppermost Vales Point Coal and Wallarah Coal of the Newcastle Coal Measures (Uren, 1980, fig. 9.2), and on the eastern flank of the Lochinvar Anticline by the Maitland Group (Herbert, 1980a, p. 42).

D2. Seen in Figure 12, EF (in a later section) is not present in AB (Fig. 8). D2 cuts out the topmost formations of the Charbon Subgroup and is capped by the quartzose Gap Sandstone of the Wallerawang Subgroup (Bembrick, 1983).

D1. Capped by Marrangaroo Conglomerate (11) in the south and the base of the Sydney Subgroup (12) to the north (Bowman, 1974, fig. 4) as far as Sydney (Havord et al., 1984). In the Stroud-Gloucester Syncline, D1 is cut into the Avon Coal Measures, which were syndepositionally faulted (F in Fig. 8) at 258 and 256 Ma, immediately before the deposition of the capping marine Speldon Formation (SP in Fig. 8) (Lennox and Wilcock, 1985).

In the Nowra area, D1 reflects relaxation of a foreswell synchronous with the folding of the Mulbring Formation in the Newcastle area and the syndepositional faulting of the Avon Coal Measures in the Stroud-Gloucester Syncline. D3 reflects relaxation of the foreswell after the terminal folding of the Newcastle Coal Measures, as seen in the Macquarie Syncline. D4 reflects the relaxation of the foreswell that followed deformation in the Hastings Block (9 in Fig. 8) and led to deposition of the Camden Haven Group, and in the Sydney Basin to the deposition of the cratonic Hawkesbury Sandstone at the expense of the regressive orogenic Narrabeen Group. D5 has no obvious correlation with events elsewhere, except possibly the static metamorphism of the southern Coffs Harbour Block (M2 in Fig. 8).

Petrofacies, sediment and lava transport directions, and inferred provenances. The vertical distribution of the petrofacies is shown in the columnar sections of Figure 10. For the fluvial sediments of the coal measures and the overlying Triassic

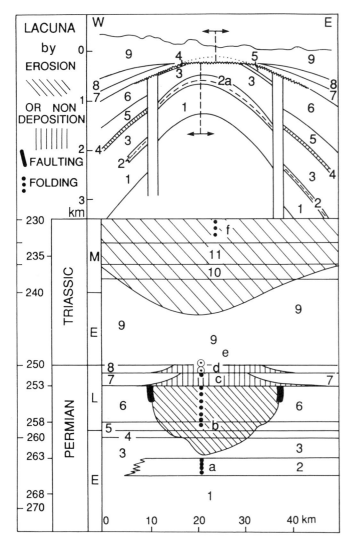

Figure 9. Lochinvar Anticline. *Above,* sketch cross section (from Stuntz, 1972, fig. 6) to which we have added an approximate vertical scale, and *below,* time-space diagram interpreted by us from the cross section. Formations are 1—Farley-Rutherford, 2—Greta Coal Measures, 3—Branxton, 4—Muree, 5—Mulbring, 6—"Lower" Tomago Coal Measures, 7—"Upper" Tomago Coal Measures (Dempsey and above), 8—Newcastle Coal Measures, 9—Narrabeen, 10—Hawkesbury, 11—Wianamatta.

a—Greta deposition (265–263 Ma): uplift in the north is indicated by the influx of coarse material in the Greta Coal Measures, coinciding with intrusion/cooling of the Barrington Tops Granodiorite (269–265 Ma), and initial growth (vertical dotted line) of the Lochinvar and Muswellbrook Anticlines, shown by coals splitting away from the crestal zones due to more rapid subsidence of trough zones (Britten et al., 1975, p. 233; Evans and Migliucci, 1991).

b—Mulbring thins on crest of Lochinvar Anticline (Stuntz, 1972, p. 7) and Muswellbrook Anticline (Veevers, 1960, fig. 5).

c—Growth of anticline to rise above depositional regression of the Newcastle Coal Measures.

d—Final growth of anticline, shown by encircled dots, as seen in east-flanking Macquarie Syncline (Uren, 1980, fig. 9.2) during deposition of Newcastle Coal Measures.

e—Base of Narrabeen (Dooralong Shale and Munmorah Conglomerate) laps over the structure (Herbert, 1980a, p. 42).

f—Final deformation of Lochinvar Anticline along an axis a few kilometers east of old axis after deposition of the Wianamatta Group, presumably before the deposition in the north of the 225 Ma Werrikimbe ignimbrite and the 230 Ma (Carnian) Ipswich sequence.

---

formations, the crossbed dip azimuth indicates the transport direction. Much of the Greta Coal Measures, the Dalwood Group, and the Shoalhaven Group were derived from a local source.

The Gerringong Volcanics (Table 3) were extruded at the Q/R zonal boundary (Appendix 1), equivalent to 258 Ma on the DNAG scale and matched by the oldest K/Ar date of 257.5 ± 5 Ma from the youngest (Berkley) member. All other dates are younger, including two (125 Ma and 193 Ma) rejected by Evernden and Richards (1962), and the others, 251 ± 5 and 249.5 ± 5 Ma, indicating mild argon leakage. The associated intrusives in the southern Sydney Basin are younger than the intruded zones L–Q, equivalent to 261 Ma to 259 Ma. The radiometric ages of 245 ± 6, 241, 239, and 238 Ma are consistent with the intrusion of shoshonitic magma at the same time as its extrusion 258 m.y. ago, at the start of Stage C. The Dampier Ridge microgranite and gabbro (DR) (Symonds et al., 1988) is restored to the Australian continental margin at 35°S off Jervis Bay from the seafloor spreading pattern in Veevers et al. (1991).

Its age is unknown but its fit against the southern Sydney Basin suggests that it is part of the southern magmatic province.

As pointed out by Shaw et al. (1991), the Currarong Orogen of Jones et al. (1984) (= offshore southern sector of the NEFB of Jones et al., 1987) must have been the center for latite extrusives in the southern Sydney Basin. Here we add to this center the presumed equivalent intrusives of the Milton Monzonite, Bawley or O'Hara Headland gabbro, and Dampier Ridge microgranite and gabbro.

Regional provenance. Linked studies of crossbed dip azimuth and thin-section petrography of sand and rudite clasts relate transport direction and provenance of fluvial and associated sediments (Conaghan et al., 1982). Six petrofacies (five sedimentary and one volcanic) are recognized (Fig. 10, key), as shown in Figures 8 and 12 (in a later section): (1) a quartzose facies (a in Fig. 10) derived from the craton to the southwest, represented by the Nowra Sandstone, the lower and upper Newport Formation, and the Hawkesbury Sandstone; (2) a quartz-lithic petrofacies (b), from mainly regional and local cratonic sources, with an overlay of ice-rafted debris, represented by the Wasp Head, Pebbley Beach, Snapper Point, Wandrawandian Formations of the Shoalhaven Group and equivalents in the northern Sydney Basin, as well as coarse sediment shed from the foreswell (Marrangaroo/Blackmans Flat Conglomerate) in the Sydney Basin and "Western bedload

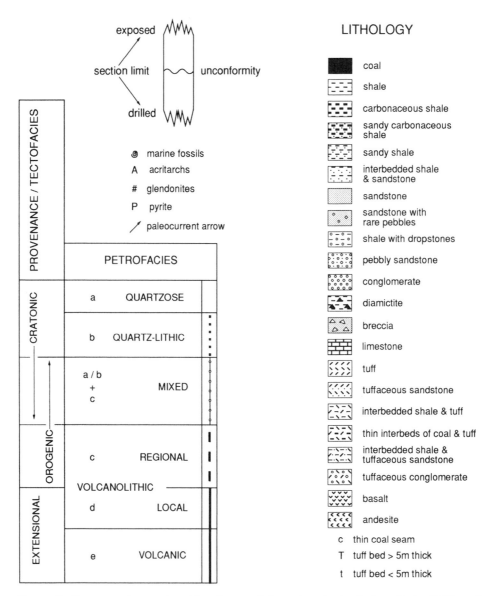

Figure 10. Representative stratigraphic columns of formations in the Sydney Basin (I–IX) and Gunnedah Basin (X–XII), showing lithology, provenance, and depositional environment; explanation precedes columns I and II. Thickness in meters. Petrofacies from Appendix 2.

I–IX Sydney Basin. I, II: Extensional-volcanic stage A. I: Lochinvar Anticline, from Osborne (1949, fig. 2) and Retallack (1980, fig. 21.3). II: Loder Dome, from Loder-1 (Nicholas, 1968, fig. 1; Mayne et al., 1974, p. 140–141). III, IV: Marine sag, stage B. III: North, composite section Lochinvar Anticline, Rutherford and Farley Formations from Osborne (1949, fig. 2) and Mayne et al. (1974, p. 146–147); Greta Formation from Booker (1960, fig. 6); Branxton, Muree, and Mulbring Formations from Mayne et al. (1974, p. 157, 163, 165, pl. 3). IV: South, from Woronora-1 (Mayne et al., 1974, p. 141, 149, 159, 163, 167, pl. 2). The stratigraphic position of the tuff in the Wandrawandian Formation from Runnegar (1980b, fig. 23.15). Paleocurrents in Snapper Point Formation from Gostin (1968, fig. 2.6), in Nowra Formation from Runnegar (1980c, fig. 4.1). V, VI: Orogenic coal measures, stage C. V: Newcastle area, from Diessel (1985), Brown and Preston (1985), and Brown (1978). Tuffs in the Tomago Coal Measures from Brown (1978) and White-house (1984); paleocurrent data from Diessel et al. (1985), Diessel (1985), and McDonnell (1983). VI: Singleton-Broke area, stratigraphy from Standing Committee on Coalfield Geology of New South Wales (1986a), Beckett and McDonald (1984), and Beckett (1988); paleocurrents from Cameron (1980). VII, VIII, IX: Orogenic redbeds, stage D. VII: North, Gosford, from Uren (1980, fig. 9.2), McDonnell (1980), and McDonnell (1983); ? paral environment in Terrigal Formation from Naing (1990); tuff from Byrnes (1983). VIII: Sydney, from Conaghan (1980, fig. 12.1) and Herbert (1980b, fig. 13.2); tuff in base of Ashfield Shale from Byrnes (1982c), and in base of

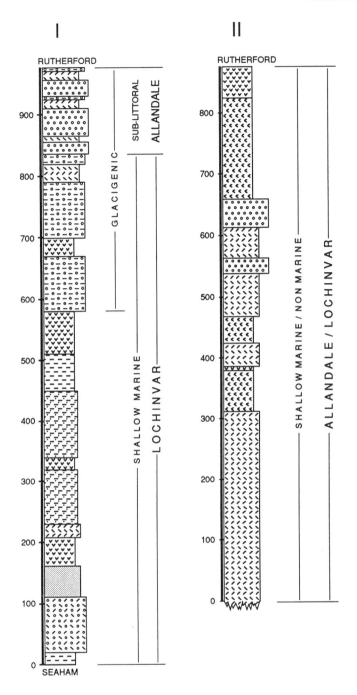

Bringelly Shale from Byrnes (1974); paleocurrent formation means of Hawkesbury from Conaghan and Jones (1975) and Standard (1969); Minchinbury-Bringelly from Herbert (1970). Petrofacies of Wianamatta Group from a synthesis of mineralogical data (Conaghan, unpublished). IX: South, Coalcliff-Garie, from McDonnell (1983); marine fossil (foraminiferid) from Scheibnerova (1980, table 22.1).

X, XI, XII: All stages in the Gunnedah Basin. X: from Bellata-1 (Etheridge, 1987, fig. 4). XI: from Springfield DDH-1 (Hamilton, 1985a, p. 64). XII: from Mount Pleasant-1 (Jones, 1985, p. 75) and Runnegar (1970, fig. 2). The Vickery Formation is now called the Maules Creek Formation (Oil Company of Australia, written communication, 1989).

sands" in the Gunnedah Basin; (3) a volcanolithic petrofacies of local derivation (d), represented by detrital sediment in the Dalwood Group, Greta Coal Measures, and the Muree Sandstone; (4) a volcanolithic petrofacies of regional derivation from the northeastern orogen (c), represented by the upper coal measures, and Narrabeen Group to below the base of the Newport Formation, and the upper part of the Wianamatta Group; (5) a mixed petrofacies of both cratonic and orogenic derivation, represented in the west by the Caley Formation and in the east by the Terrigal Formation, uppermost Bulgo Sandstone, middle Newport Formation, and the middle part of the Wianamatta Group; (6) a volcanic petrofacies, comprising the eruptive and pyroclastic rocks of extensional Stage A.

Other volcanic rocks, higher in the succession, are the Gerringong Volcanics and ubiquitous tuff in the upper coal measures of Stage C. The only indubitably coeval volcanogenic sediment is tuff, described by Diessel (1985) and Jones et al. (1987) in the Newcastle Coal Measures, by Beckett and McDonald (1984) in the Jerrys Plains Subgroup of the Wittingham Coal Measures, by Bowman (1970) in the Illawarra Coal Measures, by Packham and Emerson (1975) from the Wollombi Coal Measures of the central Sydney Basin, and by Runnegar (1980b) in the Wandrawandian Siltstone, all derived from the basin margin and beyond on the northeast and east. The first known tuff after the basal volcanics is in the Wandrawandian Siltstone. Next come the Gerringong Volcanics (mafic lavas and interlayered volcanogenic sediment), then minor tuffs in the Tomago Coal Measures, and finally major tuffs in the Newcastle Coal Measures (NCM) and the equivalent Black Jack Formation of the Gunnedah Basin. Jones et al. (1987, p. 211) describe the character and proximity of the volcanism that contributed to the NCM in these terms:

Not only are volcanic rock fragments the principal component of the conglomerates and sandstones, which form over 50% of the NCM, but about 20% of these measures is comprised of tuff.

David (1907) identified the coal measure "chert" as tuff by recognizing microscopic fragments of feldspar crystals interspersed with cusps and triangles and concave-convex pieces of glass. At Swansea Head, a 10-m-thick unit of tuff on a coal seam is interpreted as a pyroclastic surge and flow (Diessel, 1985) that came from the east as indicated by fallen trees with roots on the eastern side.

Clasts of ignimbrite dominate in the coarse fraction, but the relative contribution of juvenile and basement sources is unresolved. At present levels of exposure within the central sector of the New England orogen, Carboniferous ignimbrites are a conspicuous component along the southwestern perimeter, and Late Permian ones are widespread within the central orogen northeast of Tamworth. Both older (Carboniferous) and coeval (Permian) ignimbrites characteristically have phenocrysts of plagioclase, quartz, and biotite, as do NCM tuffs—

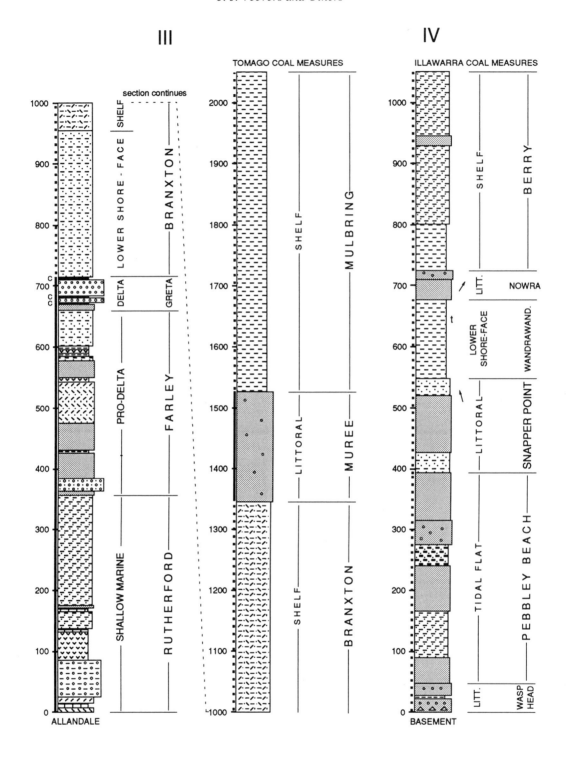

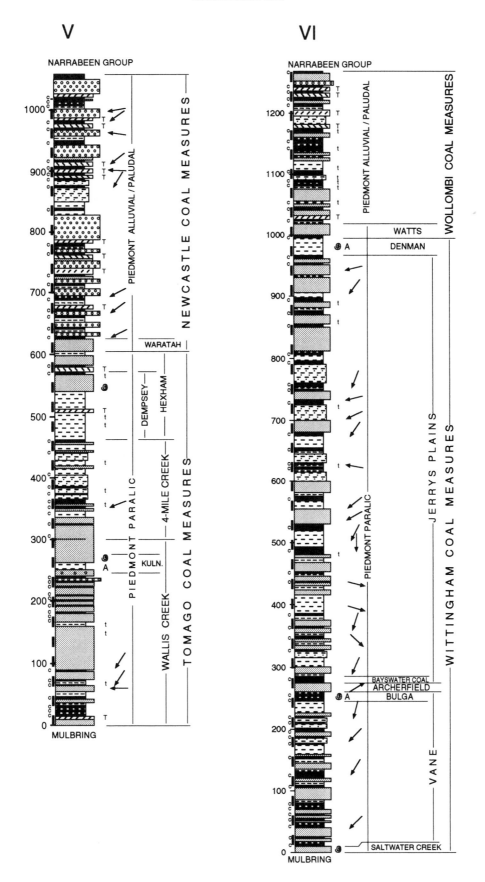

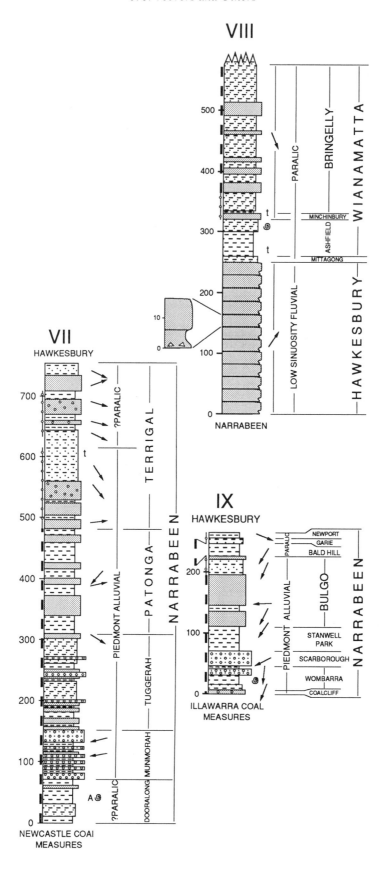

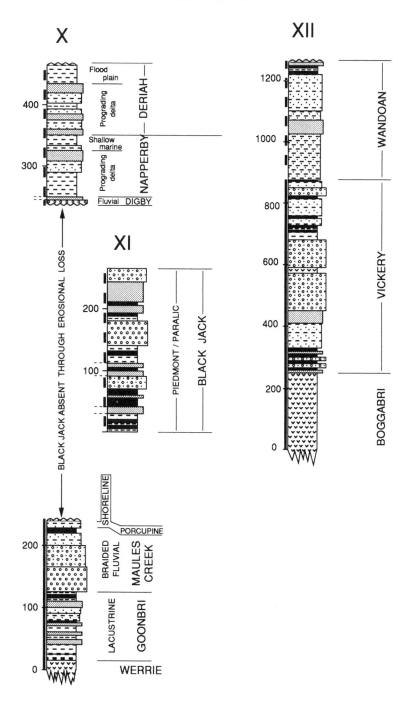

TABLE 3. AGE OF THE GERRINGONG VOLCANICS*

Locality		Member Body	Stratigraphic Age (Ma)		Radiometric Age		Reference[†]
					Method	Age (Ma)	
			EXTRUSIVES				
		(top)					
GA136	Mt Kiera	Berkley	Q/R[§]	258	K/Ar Feldspar	257.5 ± 5	1
GA139	Berry	Cambewarra	Q/R[§]	258	K/Ar Feldspar	125**	1
GA137	Pt. Kembla	Dapto	Q/R[§]	258	K/Ar Feldspar	193**	1
7843	Big Island	Dapto-Saddleback	Q/R[§]	258	K/Ar Whole Rock	251 ± 5	2
GA138	Kiama	Bumbo	Q/R[§]	258	K/Ar Feldspar	249.5 ± 5	1
		(bottom)					
			INTRUSIVES				
....	Milton	Milton Monzonite	<260 Ma, intrudes Wandrawandian Siltstone		K/Ar	245	3
7842	Milton	Milton Monzonite	<260 Ma, intrudes Wandrawandian Siltstone		K/Ar Whole Rock	245 ± 6	2
7841	Bawley	Gabbro	<261 Ma, intrudes Snapper Point Formation		K/Ar Whole Rock	241 ± 4	2
7840	Stockyard Mt.	Latite	<259 Ma, intrudes Berry Formation		K/Ar Whole Rock	238 ± 6	2
....	Currambene	Dolerite	<260 Ma, intrudes Nowra Sandstone		K/Ar Whole Rock	239	3

*Extrusive members and intrusive bodies. Stratigraphy from Carr, 1983; geochemistry of shoshonitic lavas from Carr, 1986.
[†]1 = Evernden and Richards, 1962; 2 = Facer and Carr, 1979; 3 = Mayne et al., 1974, table 6.
[§]Brachiopod zone (Briggs, 1991).
**Rejected by Evernden and Richards, 1962.

this phenocryst assemblage is common only in calc-alkaline dacite and rhyolite, and the absence of hornblende and pyroxene suggests rhyolite (Ewart, 1979). This mineralogy, however, does not discriminate basement from coeval volcanics.

According to Conaghan et al. (Appendix 2), a considerable proportion of the volcanic rock fragments of the Late Permian to Middle Triassic Sydney Basin fill were derived as epiclastics from the Carboniferous volcanic rocks of the NEFB. Evidence for contemporaneous volcanism during the Late Permian comes entirely from volcanic ash-flow/ash-fall deposits in the top of the Shoalhaven Group and in the Tomago and Newcastle Coal Measures and equivalents in the Gunnedah Basin, and they are not recorded evidently in the host composition of the lithic sandstones. Apart from local concentrations of fresh plagioclase phenoclasts that probably represent reworked tuff (McDonnell, 1983), the bulk of the volcanogenic material in the Sydney Basin is epiclastic from the erosion of ancient (Carboniferous) volcanics, and only a small amount is identifiable as juvenile pyroclastics from contemporaneous Late Permian volcanicity.

The clasts of the NCM conglomerates are uniformly well rounded and were probably derived not locally but from an area within the NEFB. Late Permian igneous rocks extend in a belt of granite and ignimbrite northeast from Tamworth. Volcanic activity climaxed in this region at about 250 Ma (Shaw et al., 1991; Appendix 3) with the eruption of the Dundee rhyodacitic ignimbrites. They lie 250 km distant from Newcastle, and Jones et al. (1987) argue that a proximal source to the east, as shown by the pyroclastic surge deposit at Swansea Head, is likely to have been more copious and that the inferred change in trend of the orogen, from southeast to south, means

that the area of thickest accumulation of the NCM lay in an orogenic recess at a site favorable for thick coal accumulation (Jones et al., 1987).

Except for a thin tuff in the Terrigal Formation in the Gosford area, the overlying Narrabeen Group up to the base of the Newport/Terrigal Formation has the same provenance but lacks tuff, presumably because tuff was not preserved in the oxidizing environment.

The rest of the volcanolithic sediment was contributed in unknown proportions from coeval and pre-existing (basement) volcanics. The basement volcanics are Devonian and older in the southwest and Late Carboniferous and older in the NEFB. The volcanics at the base of the Sydney Basin, in particular those in the exposed Lochinvar and Allandale Formations, are a local source of epiclastic sediment, and the Gerringong Volcanics are inferred to be the source of the Bald Hill Claystone (Ward, 1972; Herbert and Helby, 1980, p. 185).

Structural blocks of the NEFB in New South Wales. These blocks (Fig. 8) are from Murray (1990a, figs. 1 and 3) (*see* Fig. 14 in a later section) except where modified by us in detail, as in the Emu Creek Block (Murray et al., 1981) and from Flood and Aitchison (1988) and Aitchison (1990); the northern boundary of Hastings Block is modified to include the Kempsey Beds (Lennox and Roberts, 1988). Smaller blocks in southern NEFB are from Leitch (1988). Ages of igneous rocks younger than 255 Ma are from Shaw (Appendix 3); sedimentary ages are from Briggs (1991). The data on pre-Gondwanan autochthonous pelagic fossils in the NEFB (Table 4) indicate that (1) the youngest pelagic sediment is Namurian or younger: the youngest assemblage, in an olistolith, is Namurian so that the enclosing sediment is at most this young (<325–315 Ma); (2)

TABLE 4. AGE-DIAGNOSTIC ASSOCIATIONS OF AUTOCHTHONOUS PELAGIC FOSSILS IN EARLY CARBONIFEROUS AND OLDER FORMATIONS OF THE NEW ENGLAND FOLD BELT*

Locality	Formation	Block	Terrane	Fossils	Age	Ma	Reference†
WA-69	Wisemans Arm	Macdonald	Texas-Woolomin	Radiolarians	Namurian	333 to 315§	1
WA-50	Woolomin	Macdonald	Texas-Woolomin	Conodonts	Late Silurian	421 to 408	1
A-29	Woolomin	Macdonald	Texas-Woolomin	Radiolarians	Middle Tournaisian–Early Visean	356 to 350	1
I-58	DC Devonian and Early Carboniferous	Tamworth		Radiolarians	Late Devonian	374 to 360	1
I-41(A)	Sandon Association	Hastings		Radiolarians	Early Frasnian	374 to 370	1
39-a(B)	Sandon Association	Hastings		Radiolarians	Late Frasnian	370 to 367	1
114-g(C)	Sandon Association	Hastings		Radiolarians	Middle Famennian	365 to 362	1
119-c(D)	Sandon Association	Hastings		Radiolarians	Famennian-Tournaisian	362 to 357	1
I-54	Watonga	Port Macquarie		Conodonts	Silurian-Devonian	421 to 360	1
I-55	Watonga	Port Macquarie		Radiolarians	Late Frasnian	370 to 367	1
32	Sandon Association	Macdonald	Texas-Woolomin	Radiolarians	Early Visean	345	2
237	Woolomin	Macdonald	Texas-Woolomin	Radiolarians	Middle Tournaisian	356 to 350	2
238	Woolomin	Macdonald	Texas-Woolomin	Radiolarians	Middle Tournaisian	356 to 350	2
.....	Neranleigh-Fernvale	South D'Aguilar		Radiolarians	Middle Tournaisian	356	3
L1901	Yarramie	Gamilaroi	Tamworth	Radiolarians	Late Devonian	374 to 367	4
L1902	Yarramie	Gamilaroi	Tamworth	Radiolarians	Late Devonian	374 to 367	4
L1903	Woolomin	Djungati	Texas-Woolomin	Radiolarians	Middle Silurian**	420	4
L1904	Woolomin	Djungati	Texas-Woolomin	Radiolarians	End-Famennian	360	4
L1896	Sandon Association	Anaiwan	Texas-Woolomin	Radiolarians	Early Carboniferious	360 to 320	4
L1898	Sandon Association	Anaiwan	Texas-Woolomin	Radiolarians	Famennian	363 to 360	4
L1899	Gundahl Complex	Anaiwan	Coffs Harbour	Radiolarians	Tournaisian	360 to 352	4
L1900	Sandon Association	Anaiwan	Texas-Woolomin	Radiolarians	End-Tournaisian	352	4
L1911	Sandon Association	Anaiwan	Texas-Woolomin	Radiolarians	Tournaisian-Visean	350	5
L1912	Texas	Anaiwan	Texas-Woolomin	Radiolarians	Meramec	340	5
L1913	Texas	Anaiwan	Texas-Woolomin	Radiolarians	Visean	340	5
L1914	Texas	Anaiwan	Texas-Woolomin	Radiolarians	V1b	340	5
478	DCa(Pg)‡	Gympie	Amamoor	Radiolarians	Visean	345	6

*Shown on Figure 7, except Neranleigh Fernvale (DCn), which is shown on Figure 30. See discussion by Aitchison, 1988, and reply by Ishiga and Leitch, 1989.

†1 = Ishiga et al., 1988; 2 = Iwasaki, 1988; 3 = Aitchison, 1988; 4 = Aitchison, 1990; 5 = Aitchison and Flood, 1990; 6 = Ishiga, 1990.

§Olistolith, sediment younger.

**Succession continues to Late Devonian.

‡Locality within unit mapped as Permian Gympie Group, probably represents a tectonically embedded fragment. Other localities in units mapped as Amamoor Beds, which elsewhere contain Late Permian invertebrate fossils, yield Devonian-Carboniferous (360 Ma) radiolarians.

the oldest recorded planktonic assemblage is Middle Silurian (420 Ma); and (3) all but a few samples indicate Late Devonian and Early Carboniferous ages. Benthic fossils that range back to the Late Cambrian (Leitch and Cawood, 1987) occur in large bodies of sediment (Pzt in Fig. 7) adjacent to the Peel Fault, southeast of Tamworth in the Tamworth Block. According to Aitchison (1990), Aitchison and Flood (1990), and Aitchison et al. (1992b), Late Devonian or Early Carboniferous radiolarians in interbedded cherts indicate ages of deposition in a 34 m.y. range from Late Devonian (374 Ma) through mid-Visean (340 Ma), to suggest (but not conclusively) that the Cambrian and Ordovician bodies are allochthonous.

Nambucca Block (abbreviations given for Fig. 8). We accept the succession:

YOUNG

Buffers Creek Formation (bu)

McGraths Hump Metabasalt (m)

Bellingen Slate (be)

OLD

as in Leitch and Asthana (1985, p. 126), *not* the reverse, as in Leitch (1988, fig. 2). The succession is Permian because the Buffers Creek Formation (bu) contains fragments of the Permian *Atomodesma* (Leitch and Asthana, 1985, p. 130) before the regional deformation/metamorphism at 255 Ma. We place the base of the succession at Briggs's zone A (282 Ma) and the top at zone H in sympathy with adjacent columns. The succession was deposited in deep water, likened by Leitch and Asthana (1985) to incipient ocean-floor generation as in the Gulf of California. The next event is the intrusion of the Dorrigo Mountain Complex (DM), microdiorite and microadamellite, sheared and brecciated presumably during the 255-Ma regional deformation and metamorphism of the Nambucca Slate Belt (Leitch and McDougall, 1979), and intruded into the McGraths Hump Metabasalt (m) and Buffers Creek Formation (bu). We place its intrusion at 260 Ma. The Dundurrabin Granodiorite (DG) intruded the Dorrigo Mountain Complex (DM) and the Buffers Creek Formation (bu) in the Nambucca Block and, on the other side of the Euroka-Crossmaglen transcurrent fault [E-C(t)], the Moombil Siltstone (Cm) and the Brooklana Beds (Cb) in the Coffs Harbour Block. The Dundurrabin Granodiorite (DG) is only slighted strained and is therefore regarded as post-dating the 255-Ma deformation. It is an S-type granite of the same kind as those of the 300 Ma Hillgrove Suite but, contrary to Leitch and Asthana (1985, p. 133) and Graham and Korsch (1985, p. 52), it is apparently much younger because it is bracketed by the 255-Ma deformation below and by Tertiary basalt above. We regard the Dundurrabin Granodiorite as being emplaced soon after 255 Ma, as an example of melting during deformation. The Nambucca Block is intruded by 228–222 Ma granitoids (2, 3, 6–8, Appendix 3) during the Late Triassic relaxation of the orogen, at the same time as the initial deposition

in the Clarence-Moreton Basin (C-M B) and of the emplacement of ignimbrite (Rmv) and granitoids (Rg) in the Hastings Block (Fig. 11). Note that the Round Mountain (6) granitoid stitches the Dyamberin (DYAM) Block, containing the Dyamberin (dy) and Sara (Csa) Beds, to the rest of the Nambucca Block.

Coffs Harbour Block (abbreviations given for Fig. 8). The succession, Cm (Moombil)/b (Brooklana)/c (Coramba Beds), thought to be Carboniferous (Roberts, 1985, p. 41), youngs to the north. As seen in coastal outcrops, the Coramba Beds register a low-grade regional metamorphism (M$_1$) dated as 318 ± 8 Ma (Graham and Korsch, 1985), from a Rb/Sr date on schist [M1 CH(S)]. In the southern part of the Coffs Harbour Block, the Moombil Beds (Leitch, 1974; Leitch and Asthana, 1985) and the Coramba and Brooklana Beds register a static metamorphism (M$_2$), which produced thermal biotite on a regional scale, dated as 238 ± 5 Ma (Graham and Korsch, 1985), some 17 m.y. younger than the 255 Ma metamorphism in the adjacent Nambucca Block. The Gundahl (Cg) Complex contains Tournaisian (360–352 Ma) radiolarians (Aitchison, 1990) (Table 4). The Mt Barney Beds (CBa), brought to the surface within the Clarence-Moreton Basin by a Tertiary stock, are in the *L. levis* zone (324–286 Ma) (Murray et al., 1981), as are the Majors Creek and Kullatine Formations of the northern Hastings Block, and we regard them as the same age, with a top at 300 Ma. The Bellingen Slate (be) extends from the Nambucca Block across the Pine Creek dip-slip (DS) Fault to the Coffs Harbour Block (Brownlow, 1986). The Bellingen Slate is probably intruded by the Glenifer Adamellite (GA), which Leitch and Asthana (1985, p. 138) regard as similar in age to the Dundurrabin Granodiorite (DG), hence we place it at 252 Ma. The main movement along the Bellingen Fault system (Euroka [E], Pine Creek [PC], Crossmaglen [C] Faults) was transcurrent on the Euroka-Crossmaglen Faults [E-C(t)] before the intrusion of the Glenifer Adamellite (GA), and was followed by dip-slip (DS) movement, down to the south, along the Euroka-Pine Creek Fault after the emplacement of the Glenifer Adamellite. The Dundurrabin Granodiorite (DG) stitches the Coffs Harbour and Nambucca Blocks immediately after the transcurrent movement between the blocks. Final relaxation of the oroclinal upland led to the deposition of the Clarence-Moreton Basin with the Nymboida Coal Measures (ny) possibly in the Middle Triassic (= Esk Group) at the base, succeeded by the Carnian (230 Ma) Ipswich succession (i).

Kempsey Beds (ke) (abbreviations given for Fig. 8). According to Lennox and Roberts (1988, p. 71), "The Kempsey Beds may well have been deposited elsewhere and moved to their present position by later transcurrent faulting" within a Kempsey Block (K). Their provenance and paleobathymetric gradient is from the north, opposite the gradient from the shelf on the northern flank of the Hastings Block passing into deeper water within the Nambucca Block, shown in Figure 8 by the broken horizontal arrows, linking the Macleay Beds

(ma) with the Parrabel Beds (pa) or Nambucca Beds (n), and by vertical heavy broken lines separating the Kempsey Beds (ke) from the rest of the Nambucca Block.

Hastings Block (abbreviations given for Fig. 8). According to Lennox and Roberts (1988, p. 71, 75):

The Parrabel Beds [pa] were deposited on the continental slope beyond the marine platform which received the Early Permian Parrabel Anticline succession. The lithology of the Parrabel Beds is virtually the same as that of strongly deformed and metamorphosed sediments [n] in the Nambucca Block to the north (Leitch, 1978), inferring continuous deposition between the Hastings and Nambucca Blocks in the Early Permian. . . . Emplacement of the Hastings Block relative to the Nambucca Block therefore must have taken place between the termination of Carboniferous deposition in the Westphalian?, and the Early Permian (Fauna II) marine transgression; i.e., during the latest Carboniferous or earliest Permian . . . uplift and non deposition across the northern Hastings Block, possibly related with development of the Texas-Coffs Harbour Megafold, and movement of the block to its present position.

We mark this event as Megafold (MF) I, at 300 Ma.

Permian-Triassic events in the Hastings/Port Macquarie Blocks are shown in Figure 11. The Early and Middle Triassic Camden Haven Group of the Lorne Basin (Leitch and Bocking, 1980) and Late Triassic volcanics of the Forbes River or Werrikimbe area (Gilligan et al., 1987) provide a unique source of data "for assessing the nature and timing of movements late in the history of the southeastern part of the Fold Belt" (Leitch and Bocking, 1980, p. 89). Data from the Port Macquarie Block and the middle part of the Hastings Block, detailed in Figure 11, are shown in the middle of the column in Figure 8, between the successions in the north and south. Numerals 1–11 refer to events in the Port Macquarie Block (from Leitch and Bocking, 1980); in descending order, they are as follows:

11. Vertical movement (200 m) on the Sapling Creek Fault (SCF), the southwestern part of which marks the southern boundary between the Hastings and Port Macquarie Blocks.

10. Camden Haven Group (Rc) deposited in depressions between growing fault blocks as an alluvial fan-playa complex with thin (<0.3 m) coal beds (Herbert, 1975).

9. Local folding and generation of crenulation cleavage.

8. Burrawan (B) serpentinite emplaced and then displaced by 4 km of left-lateral movement on the Sapling Creek Fault (SCF) (Leitch and Bocking, 1980) from its southern extension west of Grants Head to displace the boundary between the Hastings and Port Macquarie Blocks.

7. Left-lateral movement on NNE-trending transcurrent faults.

6. Greenschist metamorphism of 5.

5. Continued intrusion of Karikeree Metadolerite.

4. Deformation in near-isoclinal folds and metamorphism to the greenschist facies of previously deposited rocks.

3. Karikeree Metadolerite cuts 1 and 2.

2. Thrumster Slate deposited (dated as 280 Ma from similarity with Nambucca Slate).

1. Deposition of the Watonga (late Frasnian) and probably the Touchwood Formations in the Late Devonian (Table 4) (Ishiga et al., 1988, localities I-54, I-55).

Events in the Hastings Block, in particular in the Grants Head district (Leitch and Bocking, 1980), in descending order, shown in Figure 11 are as follows:

IX. Open folding of the volcanics (Rlv) of VIII apparently involved with Camden Haven Group (Rc) in south-plunging syncline west of the Middle Brother.

VIII. Granitoids of the Laurieton area (Rb, The Brothers) emplaced 210–206 Ma (Shaw, Appendix 3, biotite Rb/Sr ages of 205.9 and 209.8 Ma from the Middle Brother; McDougall and Wellman, 1976, concordant hornblende potassium-argon age from the Middle Brother of 209.7 ± 6 Ma (converted in Dalrymple, 1979, from 205 ± 6 Ma in old constants). The rhyolitic volcanics (Rlv) (Gilligan et al., 1987) at the top of the Lorne Basin succession are regarded as co-genetic with the Brothers granitoids (Gilligan and Brownlow, 1987, p. 373).

VII. Emplacement and deposition of rhyolitic and rhyodacitic volcanics (Rmv) in the Forbes River (Werikimbe) area, dated by Shaw (Appendix 3) as 223.9 and 225.4 Ma, and emplacement of granitoids of the Gundle Belt (Rg), at Cascade Creek dated by Shaw (Appendix 3) as 222.9 Ma. The volcanics overlap the Yarras Suture (YS) and Kunderang Fault (KF) that separate the Hastings Block from the Yarrowitch Block so that the emplacement of the dolerite and serpentinite along the suture is dated as being before 225 Ma.

VI. Grants Head district.
   c. Final transcurrent movement on SCF.
   b. Left-lateral movement on NNE-trending faults.
   a. Folding (as in Bonny Hills Syncline [BHS]), including emplacement of serpentinite in the Waterloo Creek Anticline (WCA).

V. Camden Haven Group (Rc) deposited in a depression between fault blocks.

IVb. Mild deformation of I and III.

IVa. Moderate to intense deformation of I and III in the northwest, within the interval between the Macleay Group (Pma) in Fig. 11A (263 Ma) and the Werrikimbe area volcanics (Rmv in Fig. 11A) (225 Ma); we narrow the interval by limiting it to no older than the regional deformation (4) at 255 Ma and no younger than the Camden Haven Beds (V).

III. Early Permian deposition: marine Manning (Pmn) Group in the south and Macleay Group (Pma = Warbro Formation [wa]/Yessabah Limestone [y]/Mooraback Beds [mo]) in the north. Deep-water equivalents are the Parrabel Beds (pa in Fig. 7) in the northeastern part of the Hastings Block and the Kempsey Beds (Pke) in the southern part of the Nambucca Block. Lennox and Roberts (1988, p. 71) suggest that the

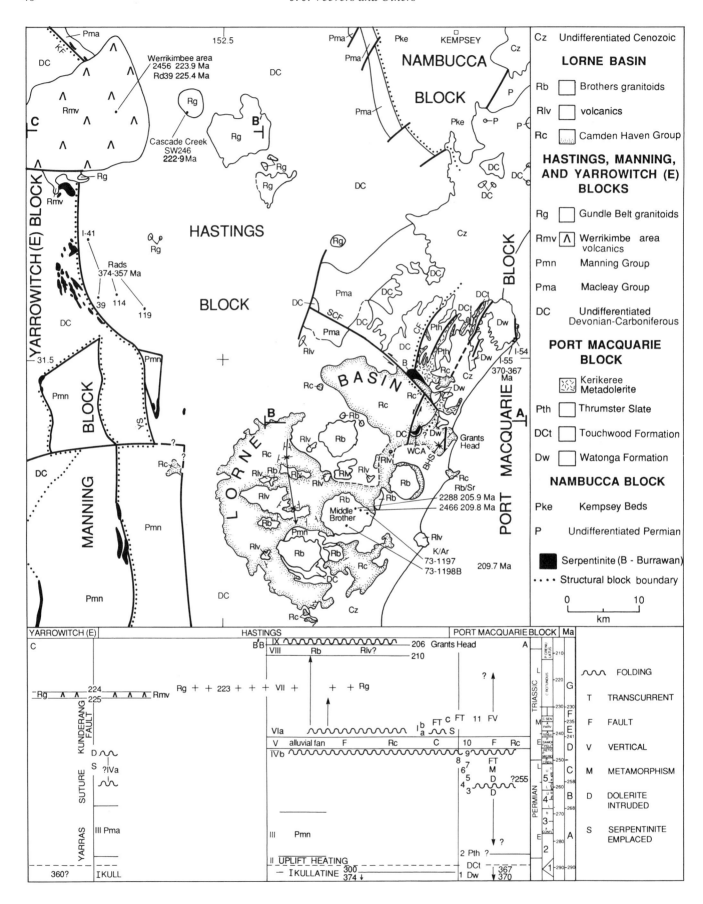

Kempsey Beds may have moved to their present position by later transcurrent faulting; we suggest that the Kempsey Beds may have been part of the Port Macquarie Block when it collided with the Hastings Block in the south and the Nambucca Block in the north at ?255 Ma.

II. Heating and mild uplift (Lennox and Roberts, 1988, p. 71).

I. Deposition of the marine Kullatine/Youdale/Majors Creek Formations in the *L. levis* zone, including the presumed equivalents in the Byabbara Beds (Cby in Fig. 7, part of DC in Fig. 11). Older deposits include chert (I-41, Table 4) near the Yarras Suture Zone.

In terms of the eastern Australian tectonic stages (abbreviations given for Fig. 11), (a) the granitoids (Rg) and rhyodacitic volcanics (Rmv) 225–223 Ma (VII) mark the onset of Stage G (Ipswich extension). Apparently unaffected by faulting, the volcanics postdate (b) the compressive events of Stage F, represented by VI. (c) The compressive event (4), whose age is constrained stratigraphically between 280 Ma and 243 Ma, is correlated with the 255 Ma deformation and metamorphism of the Nambucca Slate, interpreted as indicating the docking of the Port Macquarie Block. This 255 Ma compressive event coincides with the phase of faulting (DEF 1 in Fig. 8) (255–252 Ma) in the Tomago Coal Measures.

Data from the Port Macquarie Block and the middle part of the Hastings Block, detailed in Figure 11, are shown in the middle of the column in Figure 8 between the successions in the north and south. Numerals 1–11 (Fig. 11) refer to events in the Port Macquarie Block, including 1, deposition of the Watonga Formation (Dw) and Touchwood Formation (DCt) in the Late Devonian (370–367 Ma) (Table 4), and, in particular, 4, deformation in near-isoclinal folds and greenschist metamorphism, dated stratigraphically within the interval 280 Ma (2, Thrumster, th)–243 Ma (Camden Haven Group, Rc) and correlated by us with the 255 Ma deformation and metamorphism of the Nambucca Slate, and interpreted as reflecting the juxtaposition of the Port Macquarie Block against the rest of the NEFB. The main folding of the Camden Haven Group (Rc in Fig.

Figure 11. Map of the eastern part of the NEFB in New South Wales showing the chief Permian-Triassic events in the Hastings and Port Macquarie Blocks. Regional mapping and block boundaries from Gilligan et al. (1987), with details of the Port Macquarie Block from Leitch (1981) and of the Grants Head district from Leitch and Bocking (1980); radiometric ages of the volcanics in the Forbes River or Werrikimbe area and granite in the Cascade Creek area from Shaw (Appendix 3) and of the Brothers granitoids that intrude the Lorne Basin from Shaw (Appendix 3) and McDougall and Wellman (1976). Below, time-space diagram CB′BA across the region, compiled from information cited in the text. BHS—Bonny Hills Syncline; CF—Cowarra Fault; KF—Kunderang Fault; SCF—Sapling Creek Fault; WCA—Waterloo Creek Anticline; YS—Yarras Suture.

11A) probably took place before the emplacement of the Werrikimbe area ignimbrite (5 in Fig. 8), which furthermore overlaps the boundary of the Yarrowitch and Hastings Blocks.

In the north (Fig. 8), the Macleay Group (ma), comprising the Mooraback (mo), Yessabah (y), and Warbro (wa) Formations, disconformably overlies the Kullatine and older formations after an event of heating and uplift (Lennox and Roberts, 1988, p. 71), as noted previously, interpreted as reflecting Megafold I. In the south and west, the Byabbara Beds are faulted against the Manning Group (mn) (zones A–G), seen also in the core of the south Brother intrusion (Fig. 11).

Manning Block (Fig. 8) (Leitch, 1988, p. 63). This block, an Early Permian pull-apart basin within the Manning River Fault System (MRFS), comprises the Manning Group (zone A to H), probably with ocean floor at its base, with a fill that includes serpentinite breccia. As noted above, the Yarrowitch Block and the Hastings Block are overlapped by the Werrikimbe area ignimbrite.

Tamworth Block. Radiolarians at I-58 in the Myall area are Late Devonian (374–360 Ma) (Table 4). Correlation of the Carboniferous and basal Permian of the Myall Block is from Roberts et al. (1991, fig. 1), who report eight ion-probe (SHRIMP) U-Pb dates on single grains of zircon from the Myall and Gresford Blocks, shown in Figure 8 by encircled z's. Except for the nonmarine Nerong Volcanics, the section is mainly marine from the Conger/Boolambayte through the lower part of the Koolanock Sandstone, which marks the top of the *L. levis* zone (broken horizontal line), placed at the 286 Ma Carboniferous/Permian boundary between zircon ages of 280 and 290 Ma. Note another precise calibration of the radiometric and biostratigraphic time scales by the 275 Ma age of Stage 3a in the Alum Mountain Volcanics.

The Carboniferous succession of the Gresford and adjacent Rouchel Blocks (Fig. 8) is from Roberts et al. (1991, fig. 1). The entire succession from the basal Bingleburra Mudstone through the Flagstaff Formation (FS) is marine. The dominantly volcanogenic Waverley (W) Formation (western part only), Isismurra Formation, Gilmore Volcanics, Mt Johnstone Formation, and glacigenic Seaham Formation are nonmarine, and register a continuous retreat of the sea that started at 350 Ma and reached its maximum at 334 Ma. Roberts et al. (1991, p. 40) note that "the prominent disconformity beneath the Mt Johnstone Formation, previously regarded as having tectonic significance, is best regarded as a brief but widespread uplift within the volcanic arc." We show this event (Fig. 8) as taking place during the initial uplift during the Kanimblan shortening. In the Stroud-Gloucester and Myall Synclines and in the Myall Block, the Alum Mountain Volcanics and Markwell Coal Measures contain Stage 3a palynomorphs (Helby et al., 1986; McMinn, 1987) and, as has been noted, the Alum Mountain Volcanics contain 275 Ma zircons (Roberts et al., 1991), to make a precise calibration with the time scale of Palmer (1983). The Alum Mountain Volcanics are affected by syndepositional faults (F) (Lennox and Wilcock, 1985). The mar-

ginal marine Dewrang Group contains invertebrate zones K–Q, palynological stage lower 5a (Helby et al., 1986), and the regressive Thirty Foot Coal (c); the Bulahdelah Formation contains zones K–M; and together with the Dewrang Group marks the sag phase (Stage B) of the Sydney Basin. The regressive Gloucester (Avon, Speldon, Craven) Coal Measures are equivalent to the lower part of the Tomago Coal Measures of the Sydney Basin, with the marine ingressive (I) Speldon (SP) Formation (zone U) equivalent to the Kulnura Marine Tongue (33). We use "ingression" to signify a brief transgression. Faulting (F) accompanied deposition from the base of the Avon Coal Measures (258 Ma) to the top of the Wenham Formation (256 Ma), immediately above the Speldon Formation. The faulting amounted to a 400-m throw on a meridional fault in the Avon Subgroup as well as east-trending normal faults (Lennox and Wilcock, 1985, p. 39, 40) and gave rise to disconformity 1 (D1). The Craven Subgroup, the youngest preserved in the syncline, is upper Stage 5, and older than the Denman Formation (= Dempsey or Zone W) (McMinn, 1987, p. 156). We regard it as upper 5a. The folding of the Stroud-Gloucester and Myall Synclines is inferred to be part of deformation 1 (DEF 1).

Events in the central and southern Sydney Basin and Lachlan Fold Belt compared with those in the northern Sydney Basin and NEFB (Fig. 8, right-side column).

1. Mid-Carboniferous terminal (Kanimblan) compressive deformation, megakinking, and granitoid intrusion of the Late Devonian foreland basin and magmatic arc produced the end-state of the Lachlan Fold Belt. Shortly afterward, at about 300 Ma, transpressive deformation of the Devonian-Carboniferous fore-arc basin and subduction accretionary wedge (Fig. 5) formed the initial phase of the Texas Megafold (MF I in Fig. 8) of the NEFB. Gross marine regression across the fore-arc basin and accretionary wedge followed regional metamorphism (M1 in Fig. 8) in the southern Coffs Harbour Block and Texas-Woolomin Block and later oroclinal folding.

2. The subsequent lacuna (uplift and denudation), except in the Myall Block, which continued to subside, was followed by event 3.

3. Early Permian extension of the recently deformed region (Fig. 5b) occurred after the lacuna. The nonmarine to marine volcanic/detrital rift of the initial Sydney Basin formed within the denuded Lachlan Fold Belt and Tamworth Block of the NEFB; and a set of marine aborted oceanic rift (pull-apart) basins formed within the rest of the initial NEFB. The ensemble of volcanic environments superimposed on eastern Australia matches those of (a) the Neogene southwestern United States, from the Rio Grande Rift in the interior through the pull-apart basins of southern California and the young oceanic rift of the Gulf of California; and (b) of the coeval Early Permian rifts and basins superimposed on the newly formed Variscides of Europe. The uplifted Barrington Tops Granite (BT in Fig. 8) shed coarse detritus southward to form the Greta Coal Measures in a brief depositional regression before event 4.

4. Renewed transgression over a wider basin or sag marked by the westward onlap over basement of the Snapper Point Formation and the return of the sea to the Myall Block followed the Early Permian extension. North of the Myall Block, no further deposition is known until the Triassic. Metamorphism and intense deformation of the aborted oceanic rift and pull-apart basins at 255 Ma, interpreted as MF III (255 Ma), was accompanied in the northern Sydney Basin by faulting and folding of the Lochinvar Anticline and other folds. Immediately before the 255 Ma event, the first (D1; abbreviations given for Fig. 8) of four disconformities in the southern and western Sydney Basin indicates the initial activity of the foreswell with the shedding of the apron of the Marrangaroo Conglomerate (11); in the Myall Block, the coeval faulting that accompanied deposition of the Avon Coal Measures gave way to a marine ingression that swept southward through the Speldon (SP) and Kulnura Marine Tongues (33). Another ingression (Dempsey Formation, 35) followed deformation 1 (DEF 1). The ingressive and paralic deposition of the Tomago Coal Measures were replaced by the geologically instant depositional regression of the alluvial-paludal Newcastle Coal Measures that prograded 200 km southward by the great influx of sediment driven by the thrusting of the NEFB over the Sydney foreland basin. The end of deposition of the Newcastle Coal Measures at 250 Ma is marked by major tuffs (Fig. 10, V and VI), as in the equivalent coal measures in the Gunnedah and Bowen Basins, which reflect the magmatic climax of the Emmaville and Dundee ignimbrites in the NEFB (*see* Fig. 13 in a later section). Finally, a movement of the foreswell, indicated by disconformity D3, is correlated with the broad folding of the Newcastle Coal Measures before relaxation and the change of deposition from tuffaceous coal measures to redbeds barren of both coal and tuff. About 241 Ma, (a) deposition of the Camden Haven Group (Rc) accompanied faulting in the Hastings Block, (b) the southern Coffs Harbour Block was thermally metamorphosed to produce biotite and collapsed in places to accumulate the Nymboida Coal Measures (ny); and (c) the Hawkesbury Sandstone stepped back across the foreswell (D4) denuded down to the Wongawilli Coal (14).

5. Deformation 2 of the Camden Haven Group took place after 239 Ma and before the eruption of the Werrikimbe ignimbrite (Rmv, 5, in Fig. 8; 225 Ma) and associated granitoids from 228 Ma (8, Fig. 8), and probably before the wider relaxation of New England marked by the initial deposition of the Ipswich sequence at 230 Ma. This was the event responsible for the final deformation of the Sydney Basin immediately after the last deposition of the Wianamatta Group (*see* Fig. 16F in a later section).

6. Extension II (Fig. 8) was accompanied by (a) the initial deposition of the Ipswich sequence and eruption of granitoids (3, 4, 6–8) and ignimbrite (5) in the Hastings and Nambucca Blocks; (b) subsequent deposition of the Bundamba Group and eruption of the Brothers (1) granitoid and associated volcanics (Rlv); and (c) maar volcanism in the Sydney Basin.

Time relations central Sydney Basin–NEFB (Fig. 12, Table 5).

Events before the Sydney Basin. The first deposit of the Sydney Basin in the west—the dacitic and rhyolitic ignimbrites, coherent lavas, and reworked volcaniclastic rocks of the Rylstone Volcanics (Shaw et al., 1989), dated as 292 Ma (Rb/Sr on biotite)—nonconformably overlies the Bathurst Granite, which ranges in age from a mode of 325–320 Ma by Rb/Sr on biotite (S. E. Shaw, personal communication, 1990) to a tail of K/Ar dates of 312 Ma (Evernden and Richards, 1962, GA 197) and 310 Ma (Facer, 1978). The intrusion of the Bathurst Granite postdates the megakinking (Powell, 1984b) of the eastern Australian foldbelt, which underwent regional metamorphism and terminal east-west shortening (Kanimblan Orogeny, Powell et al., 1977) that formed the Hill End synclinorium immediately west of Hampton, with K/Ar ages from metamorphic biotite of 338 ± 10 Ma and 349 ± 10 Ma (Cas et al., 1976), and this intrusion affected the 405 ± 11 Ma Wologorong Batholith 25 km south of Hampton generating Rb/Sr dates of biotite of 339 Ma and 348 Ma (Shaw et al., 1982). The Lambie Group was deposited in a foreland basin during the Late Devonian, with its top at least latest Devonian (Powell et al., 1977, p. 416). The bulk of the Lambie Group in the eastern LFB was derived from an inferred volcanic arc on the east (Powell, 1984a, p. 333).

Glacigenic sediment. The subglacial striated pavements, tillites, and proglacial varves in the Mudgee outliers (mu in Fig. 7, plotted in Fig. 13) are nonmarine (Dickins, 1984, and references therein). Dropstones and glendonites are found from the base of the marine section through the Kulnura Marine Tongue (7 in Fig. 12) (Carr et al., 1989), and in the marine ingressive Baal Bone/Dempsey Formation (8A in Fig. 12).

Coal. It is confined to the Illawarra Coal Measures in the west, and the Tomago-Newcastle Coal Measures in the east, and to lenticles in the Wianamatta Group.

Redbeds. In the west (Goldbery and Holland, 1973; Bembrick, 1980; and Loughnan et al., 1974), redbeds are Mt York Claystone–Docker Head Claystone; in east (Hanlon et al., 1954, p. 116), redbeds are Wombarra Claystone (12)–Bald Hill Claystone (17).

Tuff. See the description of Figure 8 (timespace diagram AB). Occurrences shown in this diagram are in (1) the Wandrawandian Siltstone, (2) the coal measures, (3) the Newport Formation (18), and (4) the Ashfield Shale (20)–Minchinbury Sandstone (21) interval.

Maar volcanism. See Figure 8 (time-space diagram AB).

Disconformities. The four disconformities of AB (Fig. 8) are found in this diagram, together with a fifth one, D2. In descending stratigraphic order, the disconformities are the following:

D5. As seen also in AB (Fig. 8), the Mittagong Formation is cut out locally below the Wianamatta Group (Jones and Clark, 1991, p. 10, 19).

D4. In the west at the base of Hawkesbury Sandstone, dated as 239 Ma, cut into the Burralow Formation (Bembrick,

1980, fig. 8.11), echoes that beneath the Hawkesbury near Wollongong (Fig. 8, AB).

D3. In the west, capped by Narrabeen Group (250 Ma) and underlain by the Wallerawang Subgroup (Bembrick, 1983).

D2. In the west, capped by the quartzose wedge of the Gap Sandstone (9A in Fig. 12), dated as 251 Ma, and underlain by the Charbon Subgroup.

D1. Stretching eastward almost to Sydney (located at 8), is capped by Marrangaroo Conglomerate (9), dated as 256 Ma, and cuts down from the Kulnura Marine Tongue (7) in the east to the Berry Formation in the west (Havord et al., 1984; Hunt et al., 1986).

The disconformities in the west (D1–D4) indicate a very low, intermittent foreswell, with the capping sediment marking relaxation: D1 at 256 Ma (correlated with the faulting and folding in the northeast at 258–256 Ma), D2 at 251.5 Ma (correlated with the folding in the Lochinvar Anticline during the deposition of the Hexham Subgroup (35 in Fig. 8, AB), D3 at 250 Ma, and D4 at 239 Ma. The regional D5 at 236 Ma indicates wider movement.

Marine transgressive-regressive cycles and ingressions. The initial transgression reached to the line of the Lapstone Monocline (Fig. 7), which we interpret as the western wall of an extensional or rift depression. The overstep of the marine Snapper Point Formation west of the depositional hinge and across the line of the Lapstone Monocline at 263 Ma marked the change from extension to sagging, as seen also in the northeast in the transgression of the Maitland Group over the Greta Coal Measures and the Dewrang/Bulahdelah transgression over the Stage A volcanics of the Stroud-Gloucester and Myall Synclines. A eustatic rise in sea level cannot explain the coeval emergence of part of the NEFB, inferred from the age of the youngest preserved Permian sediment in the Hastings and Nambucca Blocks (Fig. 8), and probably marks the onset of uplift associated with the 255 Ma deformation (DEF 1 in Fig. 12). Later overstep at Marulan (1A in Fig. 12) and at Hampton and Kanangra Walls were at 260–259 Ma, and followed the Nowra Sandstone (2) regression. The terminal regression of the coal measures above the Berry Formation was followed by four subsequent ingressions.

Petro- and tecto-facies. The Kanimblan events led through a transition of reduced uplift and denudation to the next cycle of deposition, with preliminary extension marked by the felsic Rylstone Volcanics followed by widespread extension with concomitant deposition of the Lochinvar-Allandale felsic and mafic volcanics and volcanolithic sediment. Volcanics then gave way to quartz-lithic sediment derived mainly from the Lachlan Fold Belt. This petrofacies persisted into the lower part of the sag phase until it was replaced by the regional volcanolithic sediment of the coal measures derived from the magmatic New England orogen in the northeast and the Currarong orogen in the east. The first (indirect) sign of the eastern magmatic source is the tuff in the Wandrawandian Siltstone, the second (direct) one is the Gerringong Volcanics (GV) (abbrevi-

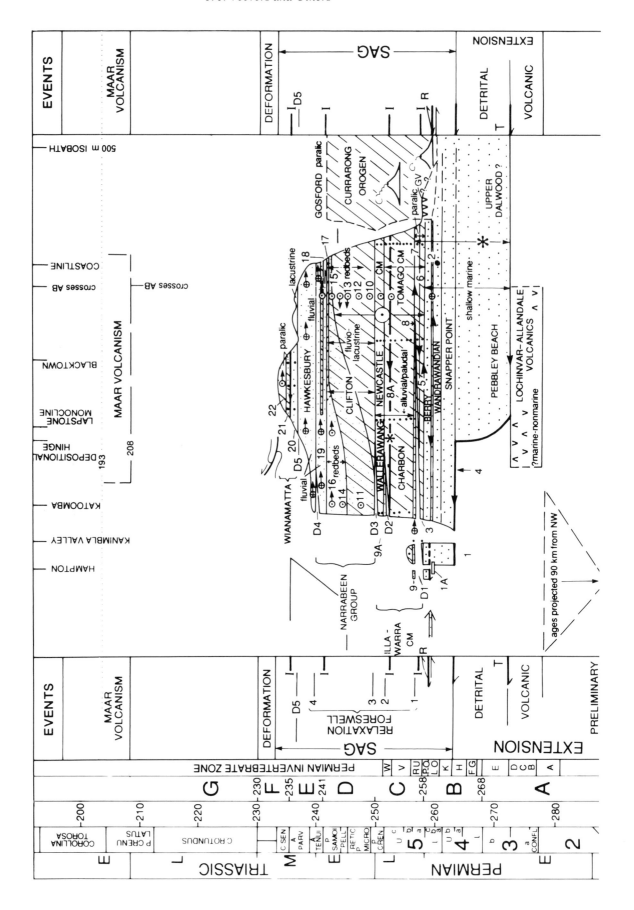

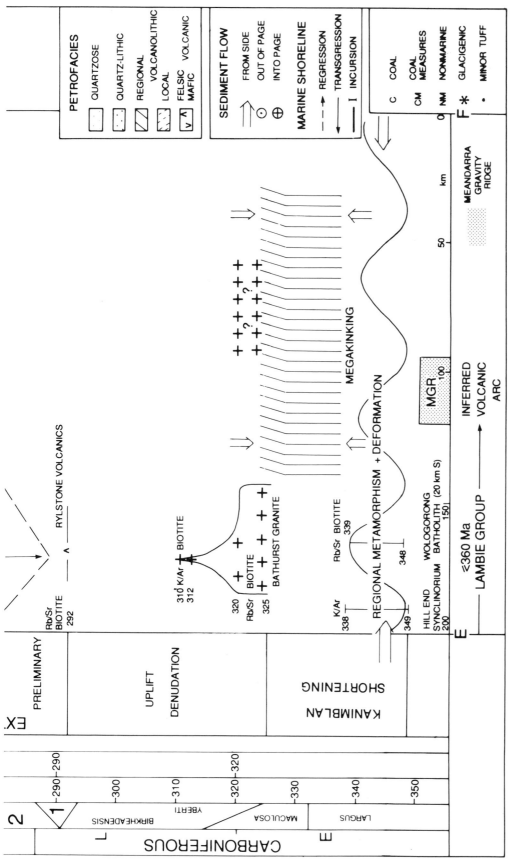

Figure 12. Time-space diagram EF, through the central Sydney Basin, from the Bathurst Granite eastward to the 500 m isobath (Fig. 7). Key in Table 5. Data from supporting references of Figure 15 (cross-section XII), together with Bembrick (1980, 1983), Bembrick and Holmes (1972), Brunker (1964), Conaghan et al. (1982), Cowan (1985), Crowley (1991), Goldbery (1972), Goldbery and Holland (1973), Havord et al. (1984), Herbert (1980c, 1983), Holland (1972), Hunt et al. (1986), Jones (1986), Jones and Clark (1991), Leaman (1990), Royce (1979), Shaw et al. (1989), Stuntz (1972), Ward (1971), and Wass and Gould (1969). Age of Bathurst Granite from S. E. Shaw (personal communication, 1990). For tuff, see sources in description of Figure 8 and additionally Australian Gas Light Company (1982, 1983), Herbert (1983), Packham and Emerson (1975), Bembrick (1983), Bradley (1980), Standing Committee on Coalfield Geology of New South Wales (1986b), Conaghan (personal observations, 1978), and E. K. Yoo (personal communication, 1991).

TABLE 5. KEY TO FIGURE 12

12 Wombarra Claystone	22 Bringelly Shale
11 Caley Formation	(minus the lowermost 30 m)
10 Coalcliff Sandstone	21 Minchinbury Sandstone
9 Marrangaroo/Blackmans	(+30 m of lowermost
Conglomerate	Bringelly Shale)
8 East limit Marrangaroo	20 Ashfield Shale
Conglomerate	19 Burralow Formation
8A Marine ingression—Dempsey	18 Newport Formation
in east, Baal Bone in west	17 Bald Hill Claystone
7 Kulnura Marine Tongue	(Clifton Subgroup) + Garie
6 Lower Wallis Creek Subgroup	Formation (Gosford
5 Pheasants Nest Formation	Subgroup)
4 Western pinchout of Kulnura	16 Banks Wall Sandstone
Marine Tongue	15 Bulgo Sandstone
3 Nile Subgroup	14 Burra Moko Sandstone
2 Nowra	13 Scarborough Sandstone
	GV Gerringong Volcanics
	(lavas)

Note: 1A, South Marulan outlier, projected 115 km from the south (Wass and Gould, 1969), dated by Briggs (1991) as zones N and O. 1, Mini Mini Range outlier, projected 7 km from south (Conaghan, personal observation, 1978), contains, above granite basement, 64 m of pebbly arkose (Snapper Point Formation) succeeded by 18 m of pebbly siltstone (Berry Formation), capped by 43 m of Illawarra Coal Measures, including at the base 4 m of pebbly siltstone with tuff comprising the Nile Subgroup (3); then, above a disconformity (D1), 14 m of Marrangaroo/Blackmans Flat Conglomerate (9) (Cullen Bullen Subgroup), which elsewhere in the region contains tuff, and finally 25 m of coal measures (Charbon Subgroup, also with tuff). The heavy broken line indicates the base of the section of Kanangra Walls, projected 19 km from the south, comprising 10 m of Megalong Conglomerate, identified as Snapper Point Formation by Herbert and Helby (1980, p. 289), overlain by the Berry Formation and Illawarra Coal Measures.

The Mittagong Formation (6 m thick) lies between the Hawkesbury Sandstone and the Wianamatta Group, and is represented by the line that delimits the top of the Hawkesbury Sandstone. The Mittagong Formation is cut out locally at the disconformity (D5) below the Wianamatta Group (Jones and Clark, 1991, p. 10, 19). In the west, the line separating the Burralow Formation (19) from the Banks Wall Sandstone (16) represents the thin stratigraphic interval of the 4-m-thick Wentworth Falls Claystone and 7 to 12 m of overlying pebbly quartzose sandstone, the western equivalent of the Bald Hill Claystone (17), in turn overlain by the 2-m-thick Docker Head Claystone, the western equivalent of the Garie Formation (Loughnan et al., 1974).

ation and numbers in Fig. 12), interpreted as part of the magmatic orogen itself, and the third the Bulgo Sandstone (15) and Bald Hill Claystone (17) that were shed westerly from the orogen. Most of the epiclastic volcanolithic sediment in the coal measures and Narrabeen Group near the coast came from the northeast. In the west, the Narrabeen Group is mixed quartzose and volcanolithic in the lower part (Caley Formation), and quartzose through the lower part of the Wianamatta (Ashfield Shale), before the mixed petrofacies reappears in the Minchinbury Sandstone (21) and finally the regional volcanolithic facies in the Bringelly Shale (22). Further details are given by Conaghan et al. in Appendix 2.

Time relations Gunnedah Basin–NEFB (Fig. 13).

Gunnedah Basin pre-Permian. Regional deformation, megakinking, and granitoid intrusion are all projected from the south (from Fig. 12, EF). The Lachlan Fold Belt basement of low-grade metasediment is known in the subsurface as far north as 30.9°S (and beyond the Gunnedah Basin to 29.3°S) and as far east as 149.5°E, near Narrabri. Inliers (Pz) along the Rocky Glen Ridge (Fig. 7) north of Coonabarabran and northeast of DM Mirrie DDH 1 are from Pogson (1972).

Permian and Triassic. The main sources of information for the Permian are McPhie (1984), Beckett et al. (1983), Etheridge (1987, p. 6, 13, 14), Tadros (1986, 1988), Yoo (1988), Hamilton et al. (1988), and Hamilton (1991); for the Triassic, additionally, Jian and Ward (1993); and for the Boggabri Volcanics and Werrie Basalt, Leitch and Skilbeck (1991) and Leitch (1993). According to Briggs (1991), the Porcupine Formation contains zones L–O (261–259 Ma) and the Watermark zones P and Q (259 Ma and 258 Ma). Note that two domains of time-space are projected into the line of section: (1) the Deriah Formation from the north, and (2) from 100 km in the south (outlined by dotted line), between 272 Ma and 235 Ma about Boggabri, where the Digby Formation rests on the Werrie/Boggabri Volcanics by erosion about 248 Ma. At Bellata-1, the Digby Formation rests on the Watermark Formation also as a result of 248 Ma erosion. In the Bellata area, the angular unconformity between the Permian-Triassic Gunnedah Basin and the Jurassic Surat Basin (Etheridge, 1987, fig. 7) developed during compression 3.

Marine transgressive-regressive (T,R) cycles and ingressions, including the Tamworth Block. The main transgression (1) Temi 277–275 Ma was followed by (2) Loders Mount (LM in Fig. 13), 275–273 Ma and (3) Borambil, 273 Ma, with the main regression at 268 Ma. Ingressions were (4) Porcupine-Watermark, 261–258 Ma; (5) Arkarula, 257 Ma, represented in the west by acritarchs (A in Fig. 13); and (6) in the Napperby, at 240 Ma, also represented by acritarchs (A).

Coal measures. They are found at four levels: (1) 275 Ma—Werris Creek; (2) 268–261 Ma—Maules Creek, the upper half equivalent to the Ashford Coal Measures (a CM in Fig. 13) of the Texas-Woolomin Block; (3) 258–250 Ma—Black Jack (BJ in Fig. 13), including the Melvilles, Hoskissons, and Wondoba Coals; and (4) 235 Ma—Deriah Formation (coal bands only).

Glacigenic sediment. (1) The glaciolacustrine Mudgee outliers, described in Figure 12 (EF) are projected 90 km from the south. (2) Porcupine and lower Watermark from Beckett et al. (1983, p. 6, 7) and Hamilton (1987, p. 150). (3) The Arkarula Sandstone Member near the base of the Black Jack (BJ) Formation contains pebbles in silty sandstone that are interpreted as dropstones (N. Z. Tadros, personal communication, 1993); it is the same age as the glacigenic Kulnura Marine Tongue of the Sydney Basin, and it is older than the youngest glacigenic Dempsey Formation.

Redbeds. The only obvious redbeds are the chocolate shales interbedded with quartzose sandstone in the Wollar Sandstone at 32°S near the boundary of the Gunnedah and Sydney Basins (Hawke and Cramsie, 1984, p. 45). Elsewhere, the only other possibility is the gray-brown mudstone in the Digby and Napperby Formations, interpreted as a paleosol (Jian and Ward, 1993).

Tuff. From Shaw et al. (1991), with additions from Jian and Ward (1993) in respect of the upper Digby and Deriah Formations. According to Shaw et al. (1991):

In the Gunnedah Basin, tephra first appears in the Watermark Formation as rare thin air-fall beds (max. 10–15 cm). Rare tephra beds of similar thickness are present in the lower part of the Black Jack Formation but from the Wondoba Coal Member upwards, tephra abundance increases dramatically, forming an interbedded tephra-coal sequence that culminates in the Tuffaceous Stony Coal Facies (Byrnes, 1982d; Tadros, 1986; N. Z. Tadros, personal communication, 1991). Individual tephra beds are up to 5-m thick but more commonly 30–40 cm thick. The distal nature of these tephra units is reflected in their fine grain size and the presence of biotite flakes as much as 1.2 mm across. The distances to probable Group 1 (Emmaville- or Dundee-related?) source vents are of the order of 100 km . . . the indicated interval from the Wondoba Coal Member to the eroded top of the Black Jack Formation is 251–248 Ma (and possibly younger because of the missing sequence), corresponding well with the radiometric age peak of Group I granitoids (249–246 Ma). As biotite radiometric data record pluton cooling ages, rather than ages of intrusion or extrusion, the age discrepancies are minor and are in the direction to be expected.

The coarse biotite flakes mentioned above come from 28.5 m from the preserved top of the Black Jack Formation in Department of Mines Clift Diamond Drill Hole 2 (which contains noteworthy zircon 76 m below the top of the Black Jack Formation) and in the neighboring DM Millie DDH 1 (Byrnes, 1982d; Hamilton, 1985a).

"Thin beds of pelletal flint-clayrock, interpreted by us to be tuffs, occur within the Napperby (previously Wallingarah) Formation (Higgins and Loughnan, 1973)" (Shaw et al., 1991, p. 49), shown as queried dots in Figure 13. Jian and Ward (1993) report flint-clay deposits from the upper part of the sandy interval of the Digby Formation, and we mark these also in Figure 13 as probably derived from tuff in situ. Loughnan (1991) interprets flint clays of the Sydney Basin as indicators of seasonal-humid warm to hot conditions but does not mention their derivation from a tuff parent, although he does consider this possibility elsewhere (Loughnan, 1978). Diessel (1985, p. 207–209) argues, however, that such flint-claystones in the Permian of the Sydney Basin had this origin.

The sandstone of the lower 100 m (interval D) of the Deriah Formation (R1 and R2 of Bourke and Hawke, 1977; T1 and T2 of Etheridge, 1987) contains green lithic fragments and perthitic feldspar grains that possibly reflect a thermal episode in the NEFB (Jian and Ward, 1993). In DM Moema DDH 1A, the upper part of the Deriah Formation is about 40-m thick, and comprises off-white lithic sandstone and mudstone with plant rootlets and coal bands, and a basaltic flow 24-m-thick (Bourke and Hawke, 1977).

Rocky Glen paleo-ridge (RGR). An influx of 55 m of conglomerate constituting the lower half of the Porcupine Formation in the west (McDonald and Skilbeck, 1991) and continued onlap through the Black Jack Formation (Yoo, 1988, fig. 4, AA'), which is seen also in the Bellata area 100 km to the north (Etheridge, 1987, fig. 7), indicates the Rocky Glen paleo-ridge.

Boggabri paleo-ridge (BR). Sediment from the Boggabri paleo-ridge was shed to either side during deposition of the Maules Creek Formation to the conglomeratic base of the Porcupine Formation (Thomson, 1986, fig. 2; Hill, 1986, p. 10, 13). In the Deriah Forest area, about 30°S, the paleo-ridge remained active until it was covered by the Napperby Formation (Hill, 1986, p. 11).

Provenance. According to Thomson and Flood (1984): (1) "The Maules Creek Formation sediments were sourced locally from the Boggabri Volcanics . . . with a minor component of quartzite cobbles and pebbles being derived from the Lachlan Fold Belt in the west."
(2) "The first indication of an influx of sediment derived from the late Permian New England Volcanic Arch is the Goran Conglomerate Member of the Black Jack Formation. This trend was continued during the deposition of the . . . Digby Formation," both influxes of regional volcanolithic sediment shown by arrows that lead back east to the Texas-Woolomin block. A third arrow indicates another influx to the Porcupine Formation. Beckett et al. (1983, p. 6) regard the Porcupine Formation as an earlier indication of this source: "The predominantly acid volcanic composition of the formation and its pronounced westward thinning suggest a source in the (then) New England Volcanic Arch to the east." As noted previously, at least part of

Figure 13 (on following page spread). Time-space CD D'D"D''', located on Figure 7. (1) CD (Gunnedah Basin). Data compiled from Hamilton (1985a, 1985b, 1987), Hamilton et al. (1988, fig. 7, BB'), Yoo (1988), Tadros (1986), Tadros and Hamilton (1991), McMinn (1983, 1993), Thomson (1986), Bourke and Hawke (1977), Hawke and Cramsie (1984), McDonald and Skilbeck (1991), Jian and Ward (1993), Higgins and Loughnan (1973), Loughnan and Evans (1978), Holmes (1982), and Pickett (1984). Top part (<265 Ma) modified from Shaw et al. (1991, fig. 3). Line of section from Hamilton et al. (1988, fig. 2, BB'), extended southwest across Rocky Glen Ridge and Gilgandra Trough, mainly from Yoo (1988). Age of Garrawilla Volcanics from Dulhunty (1972, 1976, 1986), Dulhunty et al. (1987), and Middlemost et al. (1992).

(2) DD''', NEFB in New South Wales. Blocks from Murray (1990a, figs. 1 and 3) (Fig. 14) except where modified by us in detail, as in the Emu Creek Block, and from Flood and Aitchison (1988) and Aitchison (1990); northern boundary of Hastings Block modified to include the Kempsey Beds (Lennox and Roberts, 1988). Smaller blocks in southern NEFB from Leitch (1988). Ages of igneous rocks younger than 255 Ma from Shaw (Appendix 3); Permian sedimentary ages from Briggs (1991); pre-mid-Carboniferous from Table 4.

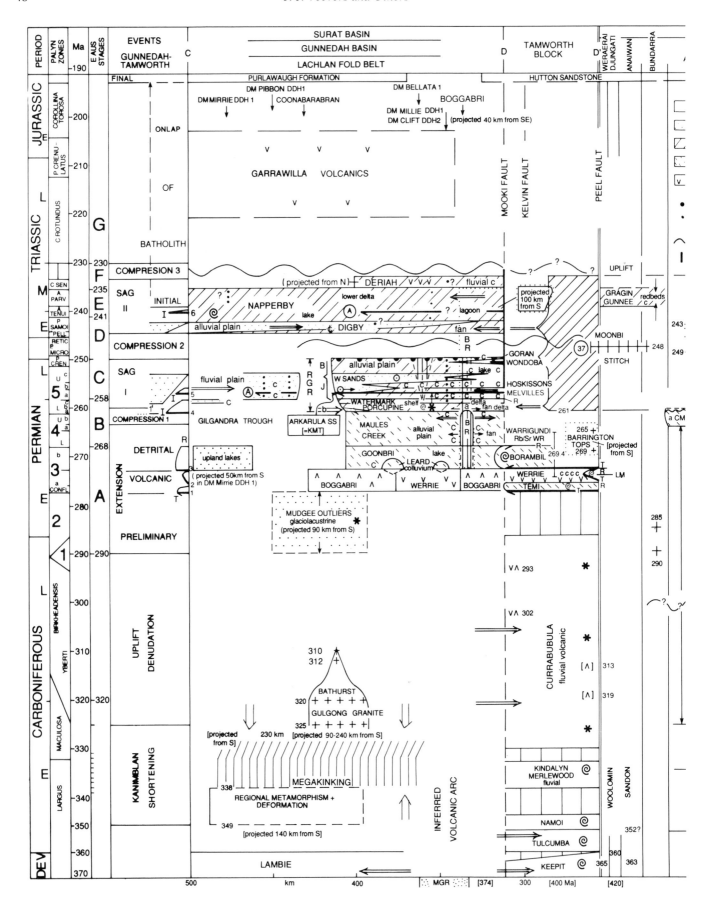

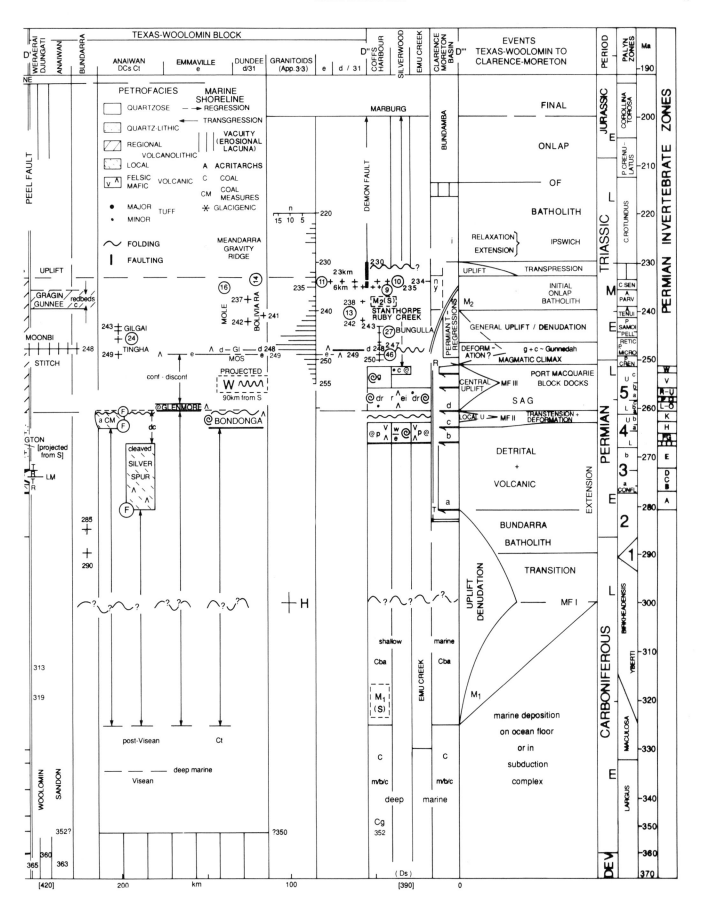

the Porcupine was derived from the Boggabri paleo-ridge, and at least the lower half in the west was derived from the Rocky Glen paleo-ridge.

The Goonbri and Maules Creek are volcanolithic sediments derived from the local sources of the underlying volcanics. Except the sheets of quartzose sediment from the west within the Gilgandra Trough and the Black Jack (BJ, W sands, both in Fig. 13), and in the Digby Formations, the section from the Porcupine Formation to the top is predominantly volcanolithic, including pyroclastic sediment (tuff), all derived from the NEFB.

The two conglomerate sequences that wedge out to the west on either side of the angular unconformity, the Goran Conglomerate Member and the eastern and lower part of the Digby Formation, reflect deposition at the foot of mountains probably generated by thrusting along the Mooki Fault, which we regard as active from the time of the first major pulse of sediment from the east at 258 Ma to final deformation at ca. 230 Ma.

Surat Basin. The overlying Late Triassic–mid Cretaceous Surat Basin sequence starts with the Late Triassic (221–208) and Jurassic (208–160 Ma) Garrawilla Volcanics of basaltic to trachytic flows and pyroclastics (Dulhunty, 1972, 1976, 1986; Dulhunty et al., 1987; Middlemost et al., 1992; Veevers, 1984, p. 263) overlain by the fluvial and lacustrine Purlawaugh Formation of Toarcian and younger age (Hamilton et al., 1988, p. 226).

Tamworth Block—Devonian and Carboniferous. The oldest known demonstrably autochthonous sediment is Late Devonian (Table 4). Benthic fossils that range back to the Late Cambrian (Leitch and Cawood, 1987) occur in large bodies of sediment southeast of Tamworth (Pzt in Fig. 7) in the Tamworth Block (Fig. 14). According to Aitchison (1990) and Aitchison and Flood (1990), Late Devonian or Early Carboniferous radiolarians in interbedded cherts indicate ages of deposition in a 34 m.y. range from Late Devonian (374 Ma) through mid-Visean (340 Ma), to suggest that the Cambrian and Ordovician bodies are allochthonous. The evidence is not conclusive, however, and it remains possible that the Cambrian and Ordovician bodies are autochthonous.

From 374 to 350 Ma, the marine Keepit Conglomerate and Tulcumba Sandstone in the Werrie Syncline were derived from an inferred volcanic arc immediately to the west. According to the paleogeography of Powell (1984a, fig. 222A), the inferred Late Devonian–earliest Carboniferous volcanic arc is located at the eastern proximal andesitic breccia exposed at Carroll Gap, immediately east of the line of the Mooki Fault. The rest of the Carboniferous is from McKelvey and McPhie (1985, p. 17). Ignimbrite from near the top of the glacigenic Currabubula Formation has K-Ar ages of 293 ± 4 Ma on hornblende and 302 ± 4 Ma on plagioclase (McPhie, 1984). Along strike, in the Rocky Creek Syncline, the Clifden Formation contains the Ermelo Dacite Tuff dated as 319 Ma and a higher tuff at 313 Ma (McKelvey and McPhie, 1985, p. 17, 19).

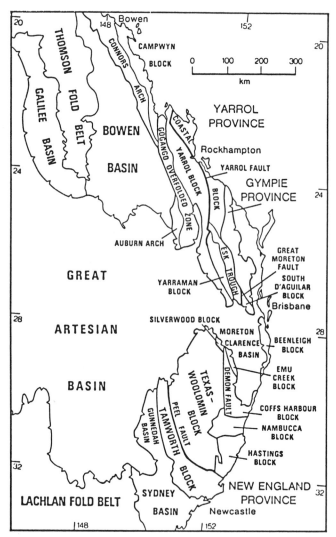

Figure 14. Structural blocks of the NEFB, from Murray (1990a).

The westerly provenance persisted during deposition of the Currabubula Formation with an older (perhaps Late Devonian and early Carboniferous) plutonic/metasedimentary/dormant volcanic terrain continuing to shed material eastward accompanied by volcanic debris eroded from the flanks of active large-scale silicic calderas (McKelvey and McPhie, 1985, p. 23).

Cherry (1989) identified the metasedimentary terrain by the discovery of a Late Devonian Lambie Group clast in the Currabubula Formation in the area southeast of Quirindi. We recognize this terrain as the doubly deformed Lachlan Fold Belt intruded by mid-Carboniferous granite that lies beneath the Gunnedah Basin.

Permian. Marine fossils (Runnegar, 1970, locality 35) in uppermost Temi Formation probably belong to the Allandale fauna, zone B, or Stage 3a. The overlying Werrie Formation (Briggs, 1991) comprises the Werrie Basalt and interlayered

Loders Mount (LM in Fig. 13) Beds (with marine fossils, Evan Leitch, personal communication, 1991) and the overlying Werris Creek–Willow Tree Coal Measures (Britten and Hanlon, 1975, p. 238). The Werrie Formation is overlain by the Borambil Creek Formation (= unnamed marine sediment in Zone E of Briggs, 1991). The Warrigundi Igneous Complex is dated by the Rb/Sr whole-rock method as 269.4 ± 4.6 Ma (Flood et al., 1988) that overlaps within analytical uncertainty the biostratigraphically determined age of the Werrie Basalt (*see* Fig. 44 in a later section).

Outliers of Early Permian (Allandale fauna, zone B of Briggs, 1991) marine sediments dot the eastern edge of the Tamworth Block along the Peel Fault. North of the line of Figure 13 is the Tarakan (t) Formation (Fig. 7) (Brown, 1987), and south of it are the Ironbark Creek Arenite (i) (Price, 1972), the Kensington Formation (k) with andesitic and basaltic lava and tuff (Price, 1972; Brown, 1987), and the Andersons Flat Beds (af) (Runnegar, 1970, locality 36).

The Moonbi Granite, dated as 248 Ma (37 in Appendix 3), stitched the Tamworth and Texas-Woolomin Blocks across the line of the Peel Fault during the deformation of the Black Jack and older formations in the Gunnedah Basin.

The Early Jurassic Hutton Sandstone of the Surat Basin followed terminal deformation (compression 3).

Events in the Gunnedah Basin and Tamworth Block (left-side column in Fig. 13). The Kanimblan orogenic and magmatic events in the Lachlan Fold Belt and related deposition in the Tamworth Block were followed at 290 Ma by the first of three phases of extension: (1) preliminary collapse of parts of the Lachlan Fold Belt to accumulate the Mudgee outliers in glacial lakes; (2) volcanic eruptions of mafic to felsic lava in the north beyond a marine shoreline that transgressed from the south, and (3) detrital deposition derived from the exposed volcanic terrain in lakes and in the regressing shallow sea in the southeast; peat-bearing fans were shed from the Boggabri Ridge, whose rise accelerated during Compression 1 before it was covered by the ingressive Porcupine Formation during Sag I. The abrupt uplift of the Rocky Glen paleo-ridge in the west, indicated by the influx of conglomerate into the lower half of the Porcupine Formation, is interpreted as due to compression 1′, which is the same age as the Porcupine ingression (I4 in Fig. 13). Volcanolithic coal measures shed from the recently deformed and uplifted NEFB now thrusting over the Gunnedah Basin along the Mooki Thrust System were interleaved with sheets of quartzose sand shed from the western craton. Airborne tuff from vents in the NEFB was deposited indiscriminately on volcanolithic and quartzose substrates and climaxed in the uppermost preserved layers of the Black Jack Formation during the eruption of the Emmaville/Dundee Volcanics and the emplacement of the Moonbi Granite to the east. Compression 2 then deformed and uplifted the Gunnedah Basin succession west of the Tamworth Block that was crumpled into the long meridional Werrie, Belvue, and Rocky Creek

folds behind the Mooki Thrust System. Uplift and stripping were followed in the earliest Triassic by initial deposition in Sag II of quartzose alluvial-plain facies in the west (upper Digby Formation) countering volcanolithic fan facies in the east (lower Digby Formation), and volcanolithic facies alone in lakes and deltas (Napperby Formation), and finally rivers, accompanied by mafic lava (Deriah Formation). Compression 3 terminally deformed the Gunnedah Basin during the lacuna between the Gunnedah and Surat Basins. From evidence elsewhere, we narrow the interval to 233–230 Ma.

Texas-Woolomin Block (Weraerai, Djungati, and Anaiwan terranes between the Peel Fault and the Bundarra pluton). The Woolomin Beds range from Late Silurian to Early Carboniferous; the Sandon Beds are latest Devonian and earliest Carboniferous (Table 4). In the earliest Triassic (248 Ma) (Appendix 3), the Moonbi, Attunga Creek, Inlet, and Bendemeer plutons stitched (1) the Peel Fault across the Tamworth, Weraerai, Djungati, and Anaiwan terranes, and (2) the Bundarra pluton and the Anaiwan terrane. Shortly after, in the Middle Triassic (*A. parvispinosus* zone), the volcanolithic Gunnee Formation, with redbeds in the form of dark rusty-brown shales, and the overlying Gragin Conglomerate of the Warialda Trough (Hawke and Cramsie, 1984, p. 4) overlap the stitched terranes but are unknown west of the Peel Fault (Bourke, 1980). Equivalent initial deposits on the flanks of the New England Batholith in the east are the basal deposits (Nymboida Coal Measures, ny in Fig. 13) of the Clarence-Moreton Basin. The Gunnee Beds and Gragin Conglomerate are disconformably onlapped by the base of the Surat Basin, represented by the Hutton Sandstone, equivalent to the Purlawaugh Formation (Bourke, 1980, p. 31). We regard the shallow stripping of the Gunnee/Gragin as due to mild uplift during compression 3. The Bundarra pluton on the west was therefore onlapped by sediment initially 239 Ma ago in the Middle Triassic and finally 193 Ma ago in the Early Jurassic.

Bundarra pluton. Shaw and Flood's (1982) 280 ± 5 Ma age is overlapped by the 287 ± 10 Ma age of Hensel et al. (1985); the best estimate (R. H. Flood, personal communication, 1991) is 290–285 Ma. The Bundarra pluton is overlain by the Gunnee/Gragin. Similar S-type granites are those of the 300 Ma Hillgrove (H in Fig. 13) Suite except the Bundarra is unsheared, probably because it escaped the later shearing along faults that affected the Hillgrove granites.

Rest of Texas-Woolomin Block. (All abbreviations refer to Fig. 13.) The Texas Beds (Ct) of the Anaiwan terrane contain Visean radiolarians (L1912–1914; Table 4). Allochthonous Visean limestone and ooliths in other localities of the Texas Beds indicate late Visean or younger deposition (Olgers, 1974, p. 11). Korsch (1977, p. 343) reported a late Visean age for the Ashford Limestone, which he regarded as autochthonous.

The Early Carboniferous Texas Beds are unconformably overlain by the Ashford Coal Measures, the Silver Spur Beds, and the Bondonga Beds. The Ashford Coal Measures (a CM),

upper Stage 4 (McKelvey and Gutsche, 1969; Briggs, 1991), unconformably overlie the Beacon Mudstone (Packham, 1969, p. 269; Britten, 1975), part of the Texas Beds, in a syndepositional transtensional half graben, the steep flanks of which are suggested by thick conglomerate. The half graben was deformed by being tilted 40° westward and overthrusted from the west, probably during Compression 1 at 261 Ma. The Ashford Coal Measures derived their detrital volcanogenic sediment from local basement (McKelvey and Gutsche, 1969, p. 18). The Tinghai/Gilgai plutons (24 in Appendix 3) intruded the Texas Beds (but not the Ashford Coal Measures) from 249 Ma to 243 Ma.

The Silver Spur Beds and equivalents in the Texas area of southeastern Queensland and southward (Olgers and Flood, 1970; Olgers et al., 1974; Day et al., 1983, p. 91, 106–107, 112) comprise conglomerate, pebbly arenite, mudstone, and limestone and, in the Alum Rock area, near Stanthorpe, rhyolitic volcanics. They unconformably overlie or are faulted (?syndepositionally) into the folded Texas Beds. The Silver Spur Beds were locally strongly folded and metamorphosed or cleaved but because they are nowhere covered by anything older than Cenozoic deposits, the age of deformation is uncertain. In Fig. 13, we project from outside the area of the preserved Silver Spur Beds and Ashford Coal Measures (and the nearby Bondonga Beds) the 260 Ma age of the subhorizontal conglomerate, sandstone, and acid volcanics of the Glenmore Beds (Olgers and Flood, 1970), with marine fossils of the *discinia* or M zone (D. Briggs, personal communication, 1991). If the Glenmore Beds were in fact originally deposited in the areas of preserved older Permian sediment that was already deformed, viz., the Ashford Coal Measures and Bondonga Beds, this deformation would be constrained to the interval about 261 Ma, equivalent to compression 1 in the Gunnedah Basin. The Glenmore Beds themselves are overlain conformably to disconformably by the Emmaville Volcanics (Stroud, 1989, p. 33) and therefore remained subhorizontal during the 255 Ma deformation in the southeastern part of New England, including the Wongwibinda (W) Complex and the Nambucca Slate Belt.

The Bondonga Beds of conglomerate, lithic sandstone, felsic volcanics, and minor basalt (Barnes and Willis, 1989, p. 25, 26) are overlain at a clear-cut unconformity by the "Mosman" Formation (MOS)/Gibraltar Ignimbrite (GI), part of the Emmaville Volcanics. According to Barnes and Willis (1989, p. 25, 33), the poorly bedded Bondonga Beds are subvertical to vertical to overturned and are unconformably overlain in the Mole River valley at Gibraltar Range by 10–25 m of fluviolacustrine interbedded paraconglomerate, lithic and feldspathic sandstone and mudstone with *Glossopteris* ("Mosman Formation") overlain by the lower part of a single ignimbritic flow of the Gibraltar Ignimbrite, 35 m to 80 m of which are preserved. The Bondonga Beds in the Mole River area contain marine fossils, including a long-ranging coral and

bryozoan, and the bivalve *Aviculopecten subquinquelineatus,* known also in the Irenian (260 Ma) Muree Formation of the Sydney Basin. Marine fossils in the Bondonga Beds are known also north of Emmaville (Wood, 1982, p. 335). Briggs (1991) plots the age of the Bondonga Beds as zones upper H and K (264–261 Ma), equivalent to that of the Dummy Creek Conglomerate, near Armidale. D.J.C. Briggs (personal communication, 1991) dated the Dummy Creek Conglomerate from its flora containing *Gangamopteris.* According to Barnes et al. (1991, p. 10), "Some marine faunas are present at Sunnyside," but this report may be incorrect (R. Barnes, personal communication, 1991). The age of deformation of the Bondonga Beds at Gibraltar Range is constrained therefore as no older than 264 Ma and no younger than the age of the Emmaville or 248 Ma. By extrapolating the 260 Ma age of the subhorizontal Glenmore Beds, we narrow the range to 261 Ma, as plotted on Figure 13.

This deformation is absent in the south near Armidale, where the subhorizontal Dummy Creek (dc) Conglomerate, ?coeval with the Bondonga Beds, is in places disconformably overlain by the Annalee Pyroclastics (McKelvey and Gutsche, 1969), which are regarded as equivalent to the Emmaville Volcanics (Barnes et al., 1991).

In the area north and northwest of Glen Innes (GL in Fig. 7), the Emmaville (e) Volcanics overlie, presumably unconformably (Barnes et al., 1991, p. 7), the Bondonga Beds. The age of the Emmaville Volcanics is inferred from their conformity (?continuity) with the Tent Hill Volcanics, in turn "Overlain by Dundee Rhyodacite, possibly with gradational contact" (Barnes et al., 1991, p. 7). Shaw (Appendix 3) dates the Dundee as 248 Ma (consistent with the intruding 247 Ma old Bungulla pluton, 27 in Figs. 7 and 51), and so the Emmaville Volcanics must be = or >248 Ma. What we call the Emmaville Volcanics south and west of Glen Innes are presumably intruded by the complex of the Tingha (249 Ma) and Gilgai (246.5–243 Ma) plutons. From their apparent continuity with the Tent Hill/Dundee, we plot the Emmaville Volcanics at 249 Ma. The Dundee Rhyodacite (d/31), as well as the Emmaville Volcanics, are intruded by the 242–237 Ma Mole Granite (16), which also metamorphosed the Gibraltar Ignimbrite (Barnes et al., 1991, p. 7, 9). The outcrop southwest of Armidale (A in Fig. 7), shown as Emmaville Volcanics, is intruded by and associated with the Uralla Granite (41, 249 Ma).

Granitoids. This column (Fig. 13) is a histogram of the age distribution of the Late Permian and Triassic granitoid plutons, including the Dundee (d) Rhyodacite, dated by Shaw (Appendix 3) for the entire southern NEFB (*see* Fig. 51 in Appendix 3). The peak values of 248 Ma and 247 Ma in the middle of the main cluster between 255 Ma and 240 Ma include the Dundee Rhyodacite at 248 Ma, and correlate within measuring uncertainty with the biostratigraphically dated tuff beds (250 Ma) at the preserved top of the Black Jack Formation (Shaw et al., 1991). The other major plutonic body in

addition to the Bundarra Suite in the west is the Hillgrove (H) Suite, 300 Ma (R. H. Flood, personal communication, September 1991).

Demon Fault (Shaw, 1969; McPhie and Fergusson, 1983). Demon Fault involves 6 km of dextral displacement of the 235 Ma Chaelundi pluton (9) and then 23 km of dextral displacement of the Coombadjha Volcanic Complex, effectively the Dandahra (10) (234 Ma, 235 Ma) and Billyrimba (11) (234 Ma) plutons (*see* Appendix 3, Fig. 51). Movement was younger than 235 Ma and older than the sedimentary cover of the Clarence-Moreton Basin, from the Marburg (200 Ma) or Bundamba (213 Ma) Groups, or, as we prefer, 230 Ma, immediately after compression 2 in the Bowen Basin (*see* Fig. 18 in a later section) between the deposition of the Moolayember Formation (233 Ma) and the rifting that accompanied the deposition of the Ipswich Group (230 Ma), marked by the 230 Ma Briggs Granodiorite and 228 Ma Hogback Granite that stitch the Venus Fault, possibly the northernmost part of the Demon Fault system.

Coffs Harbour Block. Pre-Permian rocks include the Willowie Creek Beds (Dw) and the succession (Moombil/ Brooklana/Coramba Beds (Cm/b/c in Fig. 13), thought to be Carboniferous (Roberts, 1985, p. 41), which youngs to the north. The Gundahl (Cg) Melange contains Tournaisian (360–352 Ma) radiolarians at locality L1899 (29.6°S, 152.55°E), included in the Anaiwan terrane (Aitchison, 1990) (Table 4). The Mt Barney Beds (Cba), brought to the surface within the Clarence-Moreton Basin by a Tertiary stock, are in the *L. levis* zone (324–286 Ma) (Murray et al., 1981). "The age of emplacement of the Gordonbrook [= Baryulgil] Serpentinite Belt [not shown in Fig. 13] remains unclear. The belt postdates the ?Siluro-Devonian Willowie Creek Beds and predates the ?Middle-Late Permian [260 Ma] Drake Volcanics [dr] and [250 Ma] Dumbudgery Creek Granodiorite [46]" (Barnes and Willis, 1989, p. 28) so that the age lies within the range 360–260 Ma. As seen in coastal outcrops, the Coramba Beds register a low-grade regional metamorphism $[M_1(S)]$ dated as 318 ± 8 Ma (Graham and Korsch, 1985).

Permian and younger rocks include marine sediments and dacitic to andesitic feldspathic crystal tuffs and andesitic to basaltic flows (Murray et al., 1981, p. 210). They are probably equivalent to the Plumbago (p) Creek Beds of the Emu Creek Block, are faulted against the Carboniferous Emu Creek (Ce) Beds on the east, are unconformably overlain by the Drake (dr) Volcanics (Murray et al., 1981, p. 207), and are intruded by the 242 Ma Stanthorpe Granite (13). The Drake Volcanics (zones L–U = Fauna IV, Runnegar, 1970; Telford, 1971), shallow marine volcanolithic sediments and rhyodacitic to andesitic flows and tuffs (Barnes et al., 1991, p. 10), and the overlying marine Gilgurry Mudstone (zones V and W) are intruded by the 247 Ma Bungulla pluton (27). The Drake and Gilgurry strata were gently folded (Telford, 1971), probably during the main movement along the Demon Fault at ca. 230

Ma. In the southern part of the Coffs Harbour Block, the Carboniferous Coramba and Brooklana Beds register a static metamorphism (M_2), which produced thermal biotite on a regional scale, dated as 238 ± 5 Ma (Graham and Korsch, 1985), some 17 m.y. younger than the 255 Ma metamorphism in the adjacent Nambucca Block.

Silverwood Block (Olgers, 1974; Olgers and Flood, 1970). In Figure 13, the Silverwood Block is projected between the Coffs Harbour Block and the Emu Creek Block. Pre-Permian is the Early Devonian (390 Ma) Silverwood Group (Ds) of sediment and volcanics, with blocks of Late Ordovician limestone (Day et al., 1983, p. 67, 77).

The Permian rocks are known from four isolated exposures (*see* Appendix 3, Fig. 51): (1) the marine Eurydesma Beds (e) of conglomerate, lithic sandstone, siltstone, and limestone, and the disconformably overlying Wallaby Beds (w) of conglomerate, lithic sandstone, and siltstone comprise a nonvolcanic equivalent to the Plumbago Creek Beds, and are down-faulted into the Silverwood Group; (2) the Eight Mile Creek Beds (ei) and equivalent (3) Rhyolite Range Beds (r) of marine sediments and overlying rhyolitic and dacitic volcanics (Olgers, 1974) are down-faulted into and unconformably overlie the Silverwood Group (Ds) by overlapping the Eurydesma and Wallaby Beds at the unconformity; (4) the marine Condamine Beds (c) of mudstone and sandstone, with minor crystal tuff, are down-faulted into the Silverwood Group. The Condamine Beds were probably initially deformed during the magmatic climax at 248 Ma and finally deformed and stripped during the 233–230 Ma transpression before being unconformably overlain by the Jurassic Marburg Formation.

Emu Creek Block (Murray et al., 1981). The Emu Creek Formation (Ce) occupies the *barringtonensis* and *levis* zones; it is unconformably overlain by or faulted against the Plumbago Creek Beds (p) with Fauna II (zones F–H), in turn, unconformably overlain by the Drake Volcanics (dr) of rhyodacitic to andesitic flows, tuffs, breccias, and volcanolithic sediment (Barnes et al., 1991, p. 10), all subsequently intruded by the Stanthorpe-Ruby Creek (13) granitoids (Murray, 1990a). The deformation that took place between deposition of the Emu Creek Formation and the Plumbago Creek Beds is probably marked by the 300 Ma intrusion of the Hillgrove Suite, during the first main movement of the Megafold (MF I).

The Coffs Harbour, Silverwood, and Emu Creek Blocks are onlapped by the Marburg Formation of the Clarence-Moreton Basin.

Clarence-Moreton Basin. The Red Cliff Coal Measures (i), with the Ipswich Flora of Late Triassic age, are presumably preceded by the Nymboida Coal Measures (ny), which, according to Retallack et al. (1977), are equivalent to the Middle Triassic Esk sequence. The Nymboida Coal Measures contain elements of the Ipswich Flora (*see* Willis, 1985, for correlation with Ipswich Coal Measures) but we show them as lying in the Middle Triassic. The overlying Bundamba Group extends into

the Jurassic. Before a lacuna that spans the Permian and parts of the adjacent periods are the Mt Barney (Cba) Beds, brought to the surface by a Tertiary intrusion. The Mt Barney Beds are in the *levis* zone and presumably overlie the succession of the Moombil (Cm), Brooklana (Cb) and Coramba (Cc) Beds that crop out in the underlying Coffs Harbour Block.

Summary of events in the Texas-Woolomin to Coffs Harbour Blocks. During the Devonian to mid-Carboniferous, deep marine deposits on the ocean floor were incorporated in a subduction complex which was initially metamorphosed at 318 Ma (M_1) during a change to shallow-dipping subduction or during collision of the Coffs Harbour Block and eastern Australia. The subduction complex was subsequently deformed and intruded by S-type granitoids at 300 Ma in the first phase of development of the Texas Megafold (MF I). Concomitant uplift and denudation continued into an episode of extension, marked initially by intrusion in the west of the S-type Bundarra Batholith and then by marine transgressions (T a and b) in the north that spread detrital sediment and interbedded felsic volcanics across the deformed basement. A deeper southward penetration of the sea (Tc) spread the Bondonga Beds of detrital sediment and interbedded felsic volcanics as far south as Glen Innes and the similar Plumbago Creek Beds to the southeast. In the west, along and near the eastern edge of the Bundarra Batholith, the Ashford Coal Measures (a CM) were deposited in a syndepositional transtensional half graben, which, with the rest of the region, was then subjected to ?transpressional stress, interpreted as marking the second phase of development of the Texas Megafold (MF II). In the Drake area, brief uplift and stripping during the deformation was followed by marine volcanic deposition (T d) in a sag. The rest of the region was uplifted during the third phase of growth of the Texas Megafold (MF III), marked by the low-pressure and -temperature metamorphism of the Wongwibinda (W) and Tia Complexes and Nambucca Block in the central-east, occasioned by the docking of the Port Macquarie Block. The final regression (R) of the sea from the NEFB at 251 Ma coincided with the onset of the magmatic climax of the New England Batholith, which peaked with the effusion of the Emmaville and Dundee ignimbrites at 249–248 Ma. Initial deformation in the Drake area may date from this time, equivalent to compression 2 in the Gunnedah Basin. Uplift during the intrusion of the New England Batholith decreased with waning magmatism so that the initial onlap of the batholith was Middle Triassic in both the east (Nymboida Coal Measures, ny) and northwest (Gunnee Beds). Just south of the Nymboida Coal Measures, the southern Coffs Harbour Block was heated (M_2) to produce biotite in the fine-grained sediments of the Carboniferous Brooklana and Coramba Beds. One of the youngest phases of the New England Batholith is the ca. 235 Ma Chaelundi (9) pluton, offset 6 km dextrally by the Demon Fault before intrusion of the ~234 Ma Dandahra (10) and Billyrimba (11) plutons, which were offset dextrally 23 km by movement along the Demon Fault before stitching by the 230

Ma Briggs Granodiorite in the north (Venus Fault). We suggest that the movement expressed by transpression across the Demon Fault took place at the same time as compression 3 elsewhere in eastern Australia and was followed by extension/relaxation expressed by the Ipswich succession and subsequent Bundamba and Marburg Groups.

Cross sections of the Sydney and Gunnedah Basins are shown in Figure 15 (IX–XIII) and are discussed in later sections together with cross sections across the Bowen Basin.

Stage A, 290–268 Ma: Extension/volcanism of the platform (Fig. 16A). (1) NEFB: This includes the marine Barnard Basin of Leitch (1988), the Texas area (151°E) about the Queensland/New South Wales border, and the Bundarra (BU) and Hillgrove (H) Plutonic Suites. Data comes from Figures 7, 8, 12, 13, and 44 (in Appendix 1). Age of Balala (Ba) pluton (275 Ma), southwest of Armidale (A), is from Shaw et al. (1991); the age of Rockisle (Ri) pluton (277 Ma) is from Hensel et al. (1985). (2) Gunnedah-Sydney Basin: (i) The Meandarra Gravity Ridge is from Murray et al. (1989); (ii) volcanics equivalent to Stage 3a or Tastubian (277–273 Ma) are from Hamilton (1985a), Yoo (1988), Hill (1986), Jones (1985), Hawke and Cramsie (1984, p. 33–36), Geary and Short (1989), Mayne et al. (1974), and Bradley et al. (1985). Selected thicknesses are in meters, from 17 m near Narrabri to an apparent maximum of 1,487+ m at 14. Silicic and mafic discriminations are by Leitch et al. (1988, fig. 2). In the Sydney Basin (Mayne et al., 1974; Bradley et al., 1985), the thickness ranges from 386+ m at 6 to 1,147+ m at 9. Rylstone Volcanics (Cr) on west are from Shaw et al. (1989). (iii) Nonmarine outliers (mu) north and west of Mudgee (M) include glacigenic rhythmites with Stage 2 palynomorphs (McMinn, 1983). Other nonmarine sediments are the Pigeon House Siltstone (1 in Fig. 8), Yadboro Conglomerate (3 in Fig. 8), and Jindelara Formation (4), all mapped in Figure 7 as wp, and subsequently covered by the marine transgression that starts at 277 Ma (Herbert, 1980a, p. 28, Wasp Head Formation). Shoreline on north skirts locality at 31.7°S of Temi Formation with Allandale fauna (Runnegar, 1970, loc. 35) and is then drawn halfway between exposed northern edge of Sydney Basin and nonmarine Gloucester and Myall Synclines on Tamworth Block.

Subsurface basement in southern Sydney Basin (Fig. 16A) is from Bradley et al. (1985): in 3, AOG Woronora-1 (p. 216), it is granite (pre-Permian); in 2, Farmout Stockyard Mountain-1 (p. 224), Late Devonian Lambie facies (Dul), as in the Budawang synclinorium (Bs); in Elecom Clyde River Diamond Drill Hole 1, 47 km east of Goulburn (G) (p. 221), quartzite (?Ordovician, not Devonian as in Bradley et al., 1985); in 1, Genoa Coonemia-1 (p. 226), metasediment (?Ordovician); in Elecom Clyde River Diamond Drill Hole 8, 25 km southwest of Jervis Bay (JB) (p. 227), schist (?Ordovician). Other basement features shown here are the Budawang synclinorium (Bs) and Late Devonian Lambie siliciclastics (Dul) and intruding granitoids (BA), dated as Carboniferous from their postdating the Late Devonian and predating the overlying Permian sediment.

According to Leitch et al. (1988), redeposited detrital rocks (diamictite, turbidite) up to at least 8-km-thick and very thick igneous rocks, including pillow basalt, silicic ash-fall tuff, and volcaniclastics, were deposited in the Barnard Basin over what became the NEFB. The western extent of the sea is indicated by the Allandale fauna in the Ironbark, Tarakan, and Kensington Beds immediately west of the Peel Fault.

Basaltic magma trapped at depth may have given rise to partial melting in the lower crust producing the Halls Peak silicic rocks (and perhaps the Early Permian "S-type" granites of southern New England [Bundarra and Hillgrove suites]). A similar assemblage of geological features—basalts of ocean floor type, silicic volcanics, and a thick sequence of clastic sediments—is found in the presently active Salton Trough at the head of the Gulf of California which is a possible modern analogue of the Barnard Basin (Leitch et al., 1988, p. 15).

In our interpretation, the marine Barnard Basin is separated by a narrow strip of land (the Gloucester and Myall Synclines) from the marine and volcanic Sydney Basin (to 31.7°S) (but *see* Leitch and Skilbeck, 1991, fig. 3, for another view). West of the Barnard Basin was the volcanic terrain of the Gunnedah Basin. Both Sydney-Gunnedah and Barnard Basins seem to have been the result of the same set of extensional stresses acting on different kinds of lithosphere.

Stage B, 268–258 Ma: Stage of marine sag on the platform and embryonic magmatic arc/foreland basin (Fig. 16B). The marine Barnard Basin continued into this stage, and the preserved marine deposits covered a wider area to the north. The formations, all marine except the Ashford Coal Measures, are the Plumbago Creek, the Wallaby and Eurydesma Beds, and the Bondonga. The deformed Bondonga Beds and Ashford Coal Measures are postdated by the marine Glenmore Beds (GB), and the Plumbago Creek Beds and Wallaby/Eurydesma Beds are unconformably overlain by the marine Drake Volcanics, Rhyolite Range Beds, and the Eight Mile Creek Beds, which show that the underlying beds were folded and faulted in the interval 263–261 Ma, in an area delineated by the double line. In the south, the Barrington Tops Granodiorite (BT), dated at 265 Ma, is the first granitoid in the southern NEFB that can be related to the northerly migrating convergent plate boundary.

In the Gunnedah Basin, non-marine alluvial fans in the Maules Creek Formation (a), (Thomson, 1986, p. 18) were shed (single arrows) off the contemporary Boggabri paleo-ridge, later overlapped by the Porcupine Formation and possibly the Napperby Formation (Hill, 1986); conglomerate at the base of the Porcupine Formation (b, c) was shed from the Rocky Glen paleo-ridge, inferred from McDonald and Skilbeck (1991, fig. 9). Coal was deposited on the western side of the Rocky Glen paleo-ridge as indicated by drilling at Mirrie-1 (12) (Yoo, 1988). The first tuff, 259 Ma (triangle with dot), related to the convergent magmatic arc occurs in the Watermark Formation (Shaw et al., 1991).

In the Sydney Basin, early in the stage, the Greta Coal Measures at d (Muswellbrook) and e (Lochinvar) were deposited from the north, as shown by southeast-decreasing clast size (Harrington, 1984, p. 3.7, 3.8) during the initial growth of the Muswellbrook and Lochinvar Anticlines (Fig. 9) and during the cooling (unroofing) of the Barrington Tops Granite, and the Greta Coal Measures manifest the first deposition in the foreland basin. The facies change from a paralic Dewrang Formation in the Gloucester Syncline to the marine Bulahdelah Formation indicates a southeastern slope. The reverse slope leads to the Barrington Tops Granite and defines an upland about the granite and the Gloucester Syncline. To the north, nonmarine sediment of the Ashford Coal Measures (a) and the poorly dated Dummy Creek (dc) Conglomerate near Armidale represent intermontane deposits within a central-western upland (McKelvey and Gutsche, 1969, p. 18). Garretts Seam (f) (David, 1907, p. 117) anticipates the Greta Coal Measures and contains the latest tuff of the extensional-volcanic stage. The arrow at g indicates the flow azimuth of basalt lava flows (Gerringong Volcanics) at Kiama (R. H. Flood, personal communication, 1988) and the transport direction of the Gerringong marine volcaniclastic sands (Broughton Formation) (Runnegar, 1980c, p. 76), which are interpreted as a sedimentary apron about a vent on the western magmatic edge of the arc of the Currarong Orogen. The Gerringong Volcanics are traced north of Kiama (i) to the northern limit (h) of the Triassic Bald Hill Claystone (derived from the exposed Gerringong Volcanics, Ward, 1972, and *in* Herbert and Helby, 1980, p. 157) and (ii) to Garie Beach (i) by the northern limit of the Triassic Bulgo Sandstone, likewise derived from the Gerringong Volcanics; the Wandrawandian Siltstone (j) (Runnegar, 1980b, p. 527) contains the first tuff (260 Ma) (Appendix 2) related to the convergent magmatic arc, seen (P. J. Conaghan, personal observation) also at k, Penguin Head, Culburra; outliers of the Clyde River Coal Measures (l) were transgressed by the marine Snapper Point Formation. The transport direction of the shallow-marine Nowra Sandstone (8 in Fig. 8) reflects uplift in the adjacent LFB at 260 Ma (Runnegar, 1980c, p. 76).

Figure 16B shows the maximum transgression at the end of the stage represented by the Berry Formation, Broughton Formation/Gerringong Volcanics, Mulbring and Dewrang Formations in the Sydney Basin and by the Watermark Formation in the Gunnedah Basin, followed in the next stage by the major regression represented by the coal measures.

Stage C, 258–250 Ma: Orogenic piedmont coal/tuff—mature foreland basin (Fig. 16C). In the NEFB, shallow marine deposits, the Drake (dr) Volcanics, Gilgurry (g) Mudstone, Rhyolite (r) Range Beds, Eight Mile (ei) Creek Beds, and Condamine (c) Beds, in the area south of Warwick (WA) unconformably overlie the rocks deformed 263–261 Ma. To the east, the subsurface Cressbrook Creek Beds extend the area of volcanism. The only other surface deposits are the nonmarine Merrestone (me) Beds and Wallangarra (w) Volcanics, indicating a regression of the shoreline some 250 km eastward over

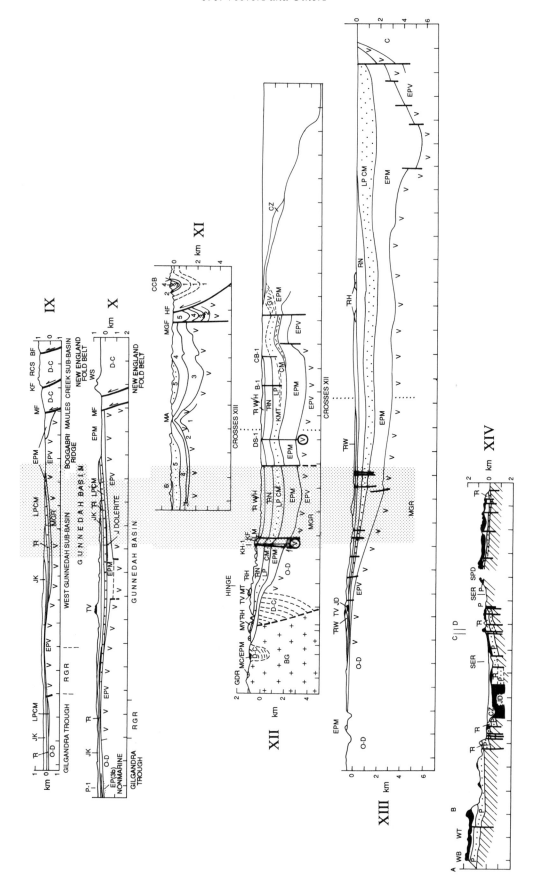

Figure 15. Cross-sections IX–XIII, the Gunnedah and Sydney Basins, located on Figure 3, and aligned on the Meandarra Gravity Ridge (MGR), with 10-km marks. Cross-section XIV, the Tasmania Basin. All sections with V:H = 5.

IX and X. Permo-Triassic Gunnedah Basin beneath the Jurassic-Cretaceous Surat Basin. Compiled by Conaghan from Chestnut et al. (1973), Hamilton (1985a), Offenberg et al. (1973), and Yoo (1988). Hamilton et al. (1988) is an additional source for IX, and Tadros (1988) for X. In X, Jurassic dolerite (solid black) and Early Permian (Stage 3b) nonmarine sediment were penetrated by Pibbon-1; Jurassic dolerite sills are shown in two places farther east; the solid black at the surface represents Tertiary volcanics.

The 45° dip of the Mooki and Kelvin Faults is from Liang (1991, fig. 6), who interprets the folding in the Tamworth Block as due to thin-skinned faults that step down 50 km eastward of the Mooki and Kelvin Faults to a décollement surface with a depth of 15 km.

In X, the Early Permian volcanics are shown with a thickness of at least 1.5 km. This thickness is projected southward from Mid-Eastern Kelvin-1, which, according to Runnegar (1970) was sited immediately west of the Mooki Fault. Plant microfossils from the upper part of the volcanics indicate an Early Permian (Stage 3) age. Accordingly we have re-located the Mooki Fault from its previously mapped position, and have also re-located part of the Kelvin Fault as mapped by Liang (1991, fig. 1). BF—Baldwin Fault, D-C—Devonian-Carboniferous rocks, EPM—Early Permian (mainly) marine sediment, EPV—Early Permian volcanics, JK—Jurassic and Early Cretaceous rocks of the Surat Basin, KF—Kelvin Fault, LPCM—Late Permian coal measures, MF—Mooki Fault; O-D—Ordovician-Devonian rocks of the Lachlan Fold Belt, P-1—Pibbon-1 well, R—Early and Middle Triassic sediment, RCS—Rocky Creek Syncline, RGR—Rocky Glen Ridge, TV—Tertiary volcanics, WS Werrie Syncline.

XI–XIII. Sydney Basin, aligned on the Meandarra Gravity Ridge (MGR). XI from Jones et al. (1987, fig. 4, CD). 6—Early and Middle Triassic sediment; 5—Late Permian coal measures (stippled); 4—Late Permian marine sediment; 3—Early Permian marine sediment, including nonmarine Greta Coal Measures; 2—Early Permian volcanics (v); 1—Carboniferous rocks. CCB—Cranky Corner Basin, HF—Hunter Fault, MA—Muswellbrook Anticline, MGF—Main Greta Fault.

XII from Bembrick and Lonergan (1976, fig. 3), extended by Conaghan westward from data in Powell and Fergusson (1979a, 1979b) and eastward: Cenozoic sediment from Phipps (1967) and Davies (1979), seismic reflectors on the base of the Early Permian marine sediment (EPM—Rutherford Formation) here and to the west from Mayne et al. (1974); the surface beneath the Cenozoic sediment is interpreted by Conaghan as Narrabeen Group, Late Permian coal measures, Gerringong Volcanics, and Early Permian marine sediment from information given in Hartman (1966), Kamerling (1966), Leah (1983), and Ringis et al. (1970). Cape Banks-1 (CB-1) from Bradley et al. (1985). Dip of Kurrajong Fault from Herbert (1989). CZ—Cenozoic sediment; B-1—Balmain-1; BG—Bathurst Granite; CB-1—Cape Banks-1; D-C—Late Devonian and (?)Early Carboniferous rocks of the Lachlan Fold Belt; DS-1—Dural South-1; EPM—Early Permian (mainly) marine sediment; EPV—Early Permian volcanics, encircled where known from drilling; GDR—Great Dividing Range; GV—Gerringong Volcanics; KF—Kurrajong Fault; KH-1—Kurrajong Heights-1; KMT—Kulnura Marine Tongue; LM—Lapstone Monocline; LPCM—Late Permian coal measures (stippled); MC/EPM—Marrangaroo Conglomerate on Berry Formation; MT—Mount Tomah; MV—Mount Victoria; O-D—Ordovician-Devonian rocks of the Lachlan Fold Belt; N—Triassic Narrabeen Group; RW—Triassic Wianamatta Group; TV—Tertiary volcanics.

XIII. NNE-SSW (oblique) cross section, from Jones et al. (1984, fig. 163, EE'). Added are the Early Permian volcanics (v) that are at least 1.5-km thick in the north. Abbreviations as in XII, and additionally JD—Jurassic dolerite.

XIV. Cross section (A–H) of the Tasmania Basin, located on Figure 34, compiled by David Hilyard from the following material provided by Clive Calver: Barton et al. (1969), Calver et al. (1988), Jennings and Burns (1958), Matthews (1979), Turner et al. (1984), and Gulline et al. (1991). JD—Jurassic Tasmanian Dolerite (solid black); P—Permian sediment (stippled); R—Triassic sediment; SER—South Esk River; SPD—St Pauls Dome; WB—Western Bluff; WT—Western Tiers. Diagonal ruling—pre-Permian.

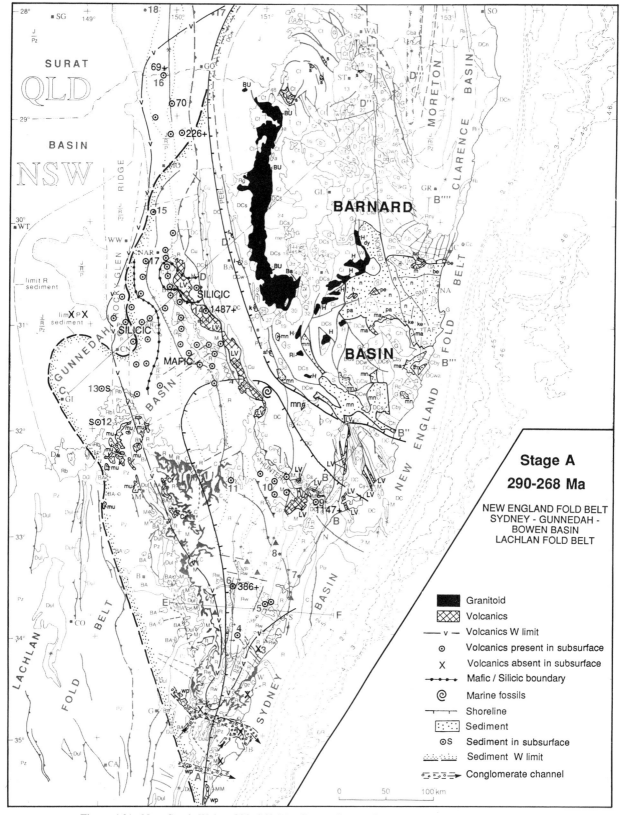

Figure 16A. New South Wales, 290–268 Ma: Stage of extension/volcanism of the platform. This and the following stage maps are drawn on the base map, Figure 7, and the subsurface information north of 29.5°S is from Exon (1976), Bourke (1980), Oil Company of Australia (written communication, 1989), and Geary and Short (1989).

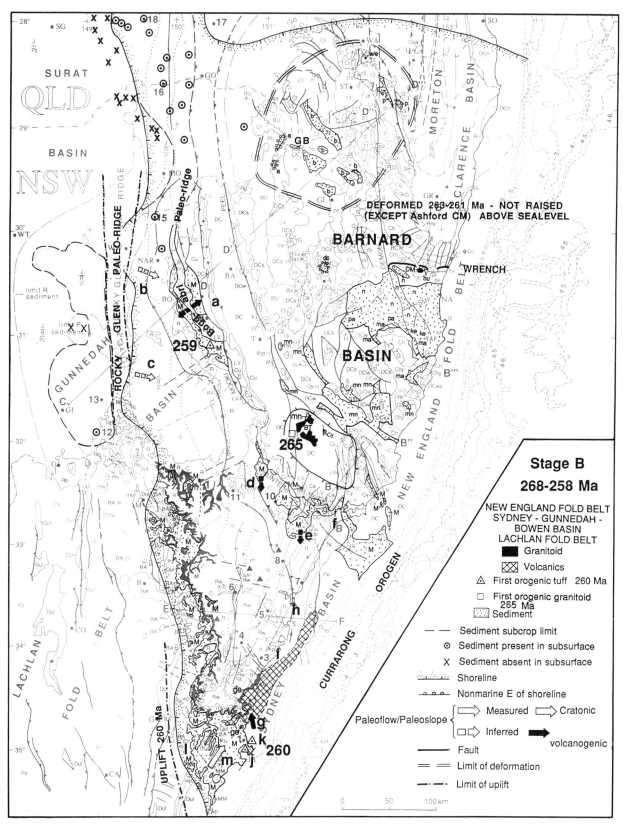

Figure 16B. New South Wales, 268–258 Ma: Stage of marine sag on the platform and embryonic magmatic arc/foreland basin.

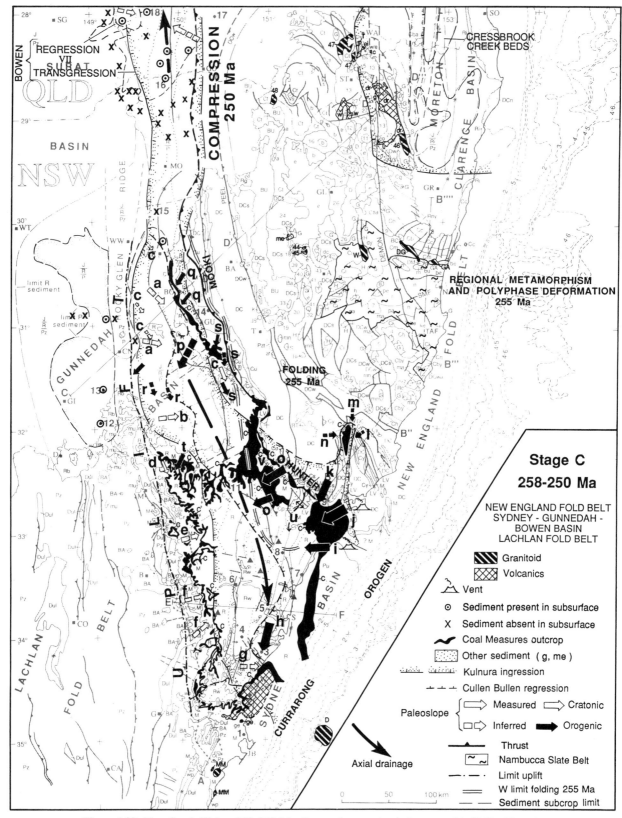

Figure 16C. New South Wales, 258–250 Ma: Stage of orogenic piedmont coal/tuff. D—Dampier Ridge.

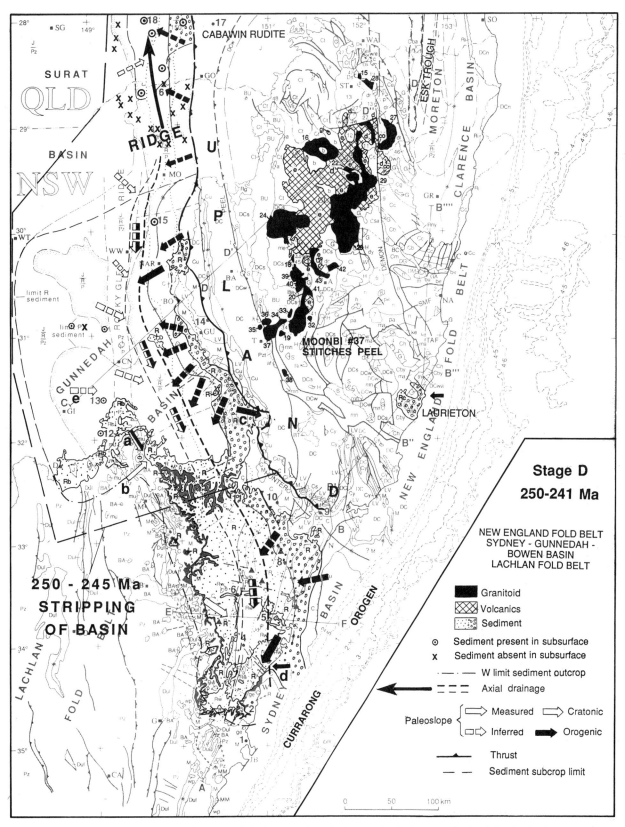

Figure 16D. New South Wales, 250–241 Ma: Stage of orogenic piedmont redbeds of the foreland basin.

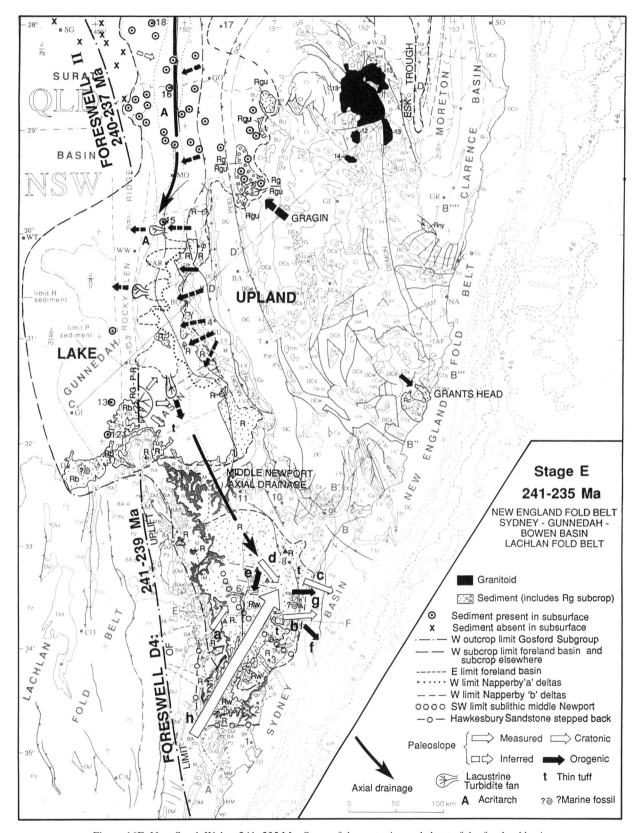

Figure 16E. New South Wales, 241–235 Ma: Stage of the cratonic sand sheet of the foreland basin.

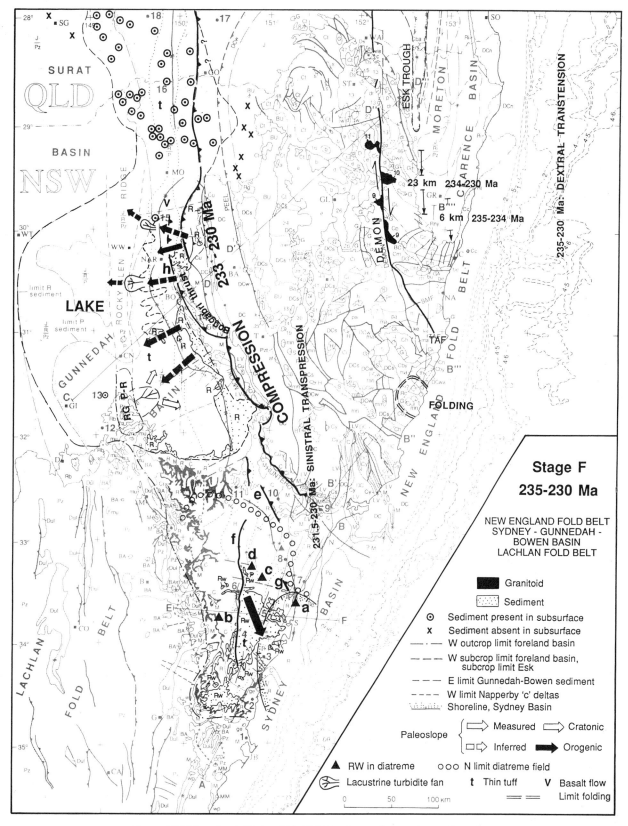

Figure 16F. New South Wales, 235–230 Ma: Stage of orogenic paralic sediment.

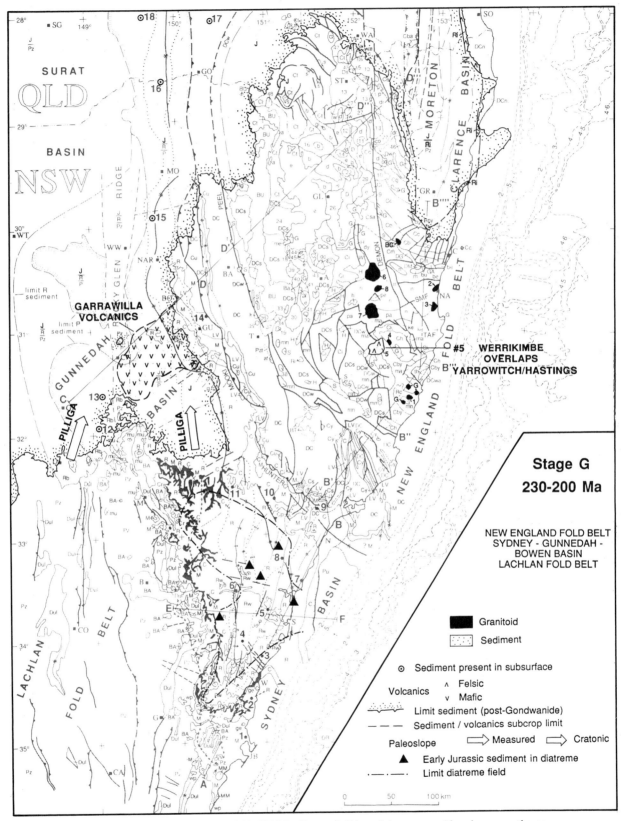

Figure 16G. New South Wales, 230–200 Ma: Stage of rifting of the orogen. Also shown are the paleoslope directions of the post-Stage G (Late Jurassic) Pilliga Sandstone (Arditto, 1982, fig. 3).

terrain intruded by granitoids (44/45, 47, 48). Most of the area southward to the northern Sydney Basin was folded during this stage (Lennox and Roberts, 1988; Collins, 1991, D1: east-west compression forming meridional folds). The age of deformation in the Lochinvar Anticline (u in Fig. 16C) in the Hunter valley is shown in Figure 9, with folding from 258 Ma to 253 Ma and faulting from 255 Ma to 253 Ma. The Muswellbrook Anticline (v in Fig. 16C) grew at the same time (Veevers, 1960). Besides polyphase deformation, the Nambucca Slate Belt was subjected to low-grade regional metamorphism, dated at 255 Ma (Leitch and McDougall, 1979). The metamorphic rocks at Wongwibinda (w) were reset by low P/T metamorphism at 260 Ma (Fukui et al., 1993).

In the Sydney-Gunnedah Basin, Stage C started with the major regression caused by the prograding paralic coal measures: the Cumberland Subgroup (31) and Wallis Creek Subgroup (32) (both in Fig. 8), the Nile Subgroup (3 in Fig. 12) in the Sydney Basin, and the Black Jack Formation in the Gunnedah Basin (Fig. 13). A brief marine transgression, called the Kulnura ingression, extended to the west (Hunt et al., 1986; Holmes, 1976; Yoo, 1991), and then north along the eastern edge of the Rocky Glen Ridge to the preserved limit west of Wee Waa (WW) (Hamilton, 1985b, 1991), thence northward into the Bowen Basin (*see* transgression V in Fig. 18). The eastern shoreline passes southward along the line of the Mooki Fault to the Gunnedah Basin (Hamilton, 1985, 1991; Uren, 1983; and Beckett, 1988) through an area of no data to envelop the growing Stroud-Gloucester Syncline (m), then southward to j. The ingression is represented in the Sydney Basin by the Kulnura Marine Tongue (KMT, 7 in Fig. 12; 33 in Fig. 8), by the Speldon Formation (SP in the Stroud-Gloucester Syncline, Fig. 8), and the Arkarula Sandstone in the Gunnedah Basin (Fig. 13). As seen in the southwestern Sydney Basin, the ingression was countered by a rising foreswell which was marked by an erosional lacuna and by pebbles in the KMT (Havord et al., 1984, fig. 1). On relaxation of the foreswell (D1 in Figs. 8 and 12), the ingression reversed to regression with deposition of a sheet of quartzose conglomerate and sand. In the Gunnedah Basin, the ingressive Arkarula Sandstone was succeeded by the regressive "western sands" without a break. The western sands were shed eastward from an uplifted craton, including the Rocky Glen paleo-ridge, a foreswell between the Gunnedah Basin and the Gilgandra Trough, which accumulated coal measures. In the Gunnedah Basin, the western sands (W SANDS in Fig. 13) are divided into lower and upper parts split by the Hoskissons Coal (Tadros and Hamilton, 1991). The lower part (a, b; letters and abbreviations throughout the rest of this section refer to Fig. 16C) wedges out at the line of the Cullen Bullen regression, and the upper part (c) extends to the eroded edge of the basin near Narrabri (NAR) and then south-southeastward along the outcrop to the Mooki Fault. Data for the lower part are a, Hamilton (1985b, fig. 26C); b, inferred from Hunt et al. (1986, fig. 2); c, Tadros and Hamilton (1991, fig. 7); d, Johnstone and

Bekker (1983, fig. 4). In the Sydney Basin, the sheet is called the Cullen Bullen Subgroup, comprising, in ascending order, the Marrangaroo Conglomerate, Lithgow Coal, Blackmans Flat Conglomerate, and Lidsdale Coal (Bembrick, 1983; Havord et al., 1984). The Blackmans Flat Conglomerate is restricted to the western periphery of the basin but the Marrangaroo Conglomerate, 20 m in the west, wedges out 100 km eastward, at the line of the Cullen Bullen regression (Havord et al., 1984, fig. 2). Paleoflow data for Marrangaroo Conglomerate are e, inferred from Havord et al. (1984, fig. 2); f, Jones (1986); g, P. J. Conaghan (1982, personal observation), and Hutton and Jones (1985).

Subsequently, the cratonic wedge was overrun by a westward-thinning orogenic wedge of labile coal measures that extended to the dot-and-dashed line that marks the limit of uplift. Sources of paleoflow arrows are h, McDonnell and Conaghan (1984, figs. 1b, 1c); i, Conaghan et al. (1982, fig. 5A), and McDonnell and Conaghan (1984, fig. 1B); j, Jones et al. (1987, fig. 2), paleoslope inferred from crossbedded sandstone/conglomerate and bedforms in ash-flow tuff, interpreted as within 30 km of the vent; k, Diessel et al. (1985); o, Cameron (1980); p, Hamilton (1985b, fig. 26A); q, r, and s, Tadros and Hamilton (1991, figs. 4, 10, 11, respectively); t, P. J. Conaghan (1990, personal observation); l, m, and n, fluvial sediments of the Gloucester-Stroud Syncline, Lennox and Wilcock (1985) (Fig. 8, faulting, F): l, Dog Trap Formation (258 Ma); m and n, Jilleon Formation and younger units (256 Ma). Growing folds of Lochinvar (Stuntz, 1972; summarized in Fig. 9) and Muswellbrook (Veevers, 1960) are represented by u and v. The vents at i and j are probably continuous along the Currarong Orogen through the older Gerringong Volcanics and Dampier Ridge (DR) granitoid to the Milton Monzonite and O'Hara gabbro (MM).

Stage D, 250–241 Ma: Orogenic piedmont redbeds of the foreland basin (Fig. 16D). The eastern limit of the Digby Formation subcrop is from Tadros (1988, fig. 8). In the Sydney Basin, most data pertain to 244 Ma near the base of the *samoilovichii* zone. At about 32.5°S, the southwestern limit of volcanogenic sediment is from Pogson and Rose (1969, fig. 2). In the Sydney Basin, the arrows (Fig. 16D) come from Conaghan et al. (1982, figs. 3A and 5C). The black-and-white arrows refer to the sublithic petrofacies along the axial drainage further constrained by the subsurface data in Ward (1971) and Brunker (1964). Letter d is a crossbed dip azimuth at Garie in the volcanic facies of the Bulgo Sandstone (Ward, 1972, p. 401), representing a pulse of westward-directed mafic detritus, uniquely attributable to the Gerringong Volcanics.

The paleoflow data (unbroken arrows are from field measurements; broken arrows are inferred from facies patterns) come from a, lower, and b, upper Wollar Sandstone at Ulan (Johnstone and Bekker, 1983, fig. 4), the correlative Digby Formation; c, quartz-lithic sandstone facies of the lower Digby Formation at Murrurundi (P. J. Conaghan, 1990, personal observation); e, quartzose upper Digby Formation (Hamilton et

al., 1988, p. 230); all other information in the Gunnedah Basin comes from Jian and Ward (1993, figs. 15.7, 15.12). Provenance in the southernmost Bowen Basin is inferred from data in Bastian (1965a, fig. 12), Exon (1976, p. 30 and fig. 24), and Butcher (1984, p. 341). The New England batholith culminated with the eruption of the ignimbrite sheets of the 250 Ma Emmaville Volcanics (e) and 248 Ma Dundee Rhyodacite (d) (Shaw, Appendix 3). The adjacent Gunnedah and Bowen Basins were folded, uplifted, and stripped from 250 Ma to 245 Ma. In the northeast, the southern part of the Esk Trough, mainly felsic volcanics, developed over the Cressbrook Creek rift. Elsewhere in the NEFB, the lower conglomeratic part of the Camden Haven Group (Rc) of the Lorne Basin was deposited in alluvial fans dominated by clasts of siliceous resistates (quartzite and jasper) in gravel that prograded over playa red mudstone (Pratt and Herbert, 1973, fig. 4) from the east (Evan Leitch, personal communication, 1993). The Stage C shoreline in the Drake area had retreated from eastern Australia, as had the marine incursion shoreline of the Gunnedah-Sydney Basin.

In the Sydney Basin, the coal measures, which had extended across the entire basin, were succeeded abruptly at 250 Ma by conglomerate, sandstone, and shale. This involved a change from coal measures with tuff to wholly epiclastic sediment that lacked coal and tuff but contained the first redbeds, not seen before.

The eastern part of the Gunnedah Basin was folded, uplifted, and stripped during the lacuna from 250 Ma (Blackjack) to 244 Ma (Digby), at the same time as the magmatic climax in the NEFB, during which the Moonbi pluton (37) stitched the Peel Fault. The Blackjack and older formations were tilted along the Mooki Fault and at Bellata were stripped down to the Porcupine Formation before being covered by the conglomeratic facies of the Digby Formation (Etheridge, 1987). The tilted area continues north of Narrabri to 26°S where the Early Triassic conglomerate of the Rewan Group and Cabawin Formation lapped against the tilted and uplifted coal measures to form a narrower basin during emplacement of Triassic granite and the first major overthrusting along the Moonie-Goondiwindi Fault (Exon, 1976, p. 30, 38). The folding in the Gunnedah Basin (compression 2 in Fig. 13) coincided with the magmatic climax in the NEFB. When deposition resumed at 245 Ma, the basal Digby Formation prograded transversely in a fan of orogenic sediment across the stripped surface, 100 km past the Rocky Glen Ridge to overstep the Permian sediment. Immediately afterward, the fan was overrun by a sheet of alluvial quartzose sand shed transversely from the craton. The quartzose sand blended with the orogenic sand in an axial or longitudinal valley that sloped southward along the Gunnedah-Sydney Basin. In the Sydney Basin, the cratonic sand overran the orogenic sand at 250 Ma (Fig. 12), some 5 m.y. before this event (b in Fig. 16D) in the Gunnedah Basin. The orogenic sediment (Jian and Ward, 1993) was deposited in alluvial fans that trended from west to south, and then, at c, recurved to the east. In Figure 16D, we show the axial drainage in its position

at 243 Ma, midway in its easterly migration toward the orogen. The initial position (at 244 Ma) of the axial drainage is marked by a. All the sediment, whether cratonic or orogenic, is inferred to have come from regional sources except the mafic "volcanic facies" of the Bulgo Sandstone in the southeastern Sydney Basin (d) (Fig. 10, column IX, 145 m above the base), which has a source that can be pinpointed in the Gerringong Volcanics exposed in the Currarong Orogen.

In the Sydney Basin, late Stage D was marked by the cessation of siliciclastic bed-load deposition by the widespread development of regolith. Regolith was transported westward from the weathered Currarong Orogen to form the Bald Hill Claystone (17 in Fig. 12, 19 in Fig. 8); an equivalent redbed unit in the Blue Mountains, west of Sydney, is the Wentworth Falls Claystone. Only in the west were quartz sands interbedded with the Bald Hill Claystone. The flood of orogenic sediment from the New England orogen terminated, as seen also in the Gunnedah Basin, where the influx of siliciclastic bed-load sediment almost ceased and a widespread flint-clay paleosol terminated the Digby depositional event (Jian and Ward, 1993). The northern Bowen Basin was separated from the Gunnedah Basin by a ridge which formed a saddle between north- and south-directed axial drainage. As in the south, so here the initial sediment from the upland was rudite.

Stage D concluded with a marine incursion at the top of the Bald Hill Claystone in the Sydney Basin (Naing, 1990) and with a swarm of acritarchs (A in Fig. 13) in the basal Napperby Formation.

Stage E, 241–235 Ma: Cratonic sand sheet of the foreland basin (Fig. 16E). In the NEFB, plutonism contracted to a focus around Stanthorpe (ST); volcanism in the Esk Trough continued from Stage D. Deposition continued in the Lorne Basin, where quartz-lithic sandstone of the Grants Head Formation entered the basin from the northwest (Evan Leitch, personal communication, 1993). At the latitude of Glen Innes (GL), deposition encroached eastward (Gunnee Formation and Gragin Conglomerate) beyond the Peel Fault to lap on to the Bundarra Batholith on the western flank of the NEFB and encroached westward (Nymboida Coal Measures) in an intramontane basin across the Coffs Harbour Block.

In the Bowen-Gunnedah-Sydney Basin, the paleogeography changed radically from Stage D to E. With the collapse of the ridge that separated the drainage in Stage D, drainage was continuous toward the south and southeast from the southern Bowen Basin through a lake that occupied the entire Gunnedah Basin, represented by the Napperby Formation, interval "a" (Jian and Ward, 1993), and overflowed a rim at 32.5°S to feed sublithic bed-load sediment of the middle Newport axial drainage (paleoflow arrows e, f, g). Lake overflow into the Sydney Basin also occurred during lower (b, c, d) and upper Newport Formation time but did not entrain appreciable volumes of lithic sediment from the Gunnedah Basin as in middle Newport time. The sea probably reached the Gunnedah Basin, as indicated by acritarchs (A) at 30°S and by the occurrence in

the Ballimore Beds, the correlative of the Napperby Formation in the southwestern Gunnedah Basin, of a merostome crustacean (Pickett, 1984), which we regard as probably marine.

An upland in the southern NEFB shed mainly fine-grained volcanolithic sediment into the Gunnedah Basin lake where two successive Napperby Formation delta systems and associated lacustrine turbidite fans (Napperby "a" and "b" of Jian and Ward, 1993) prograded westward toward the Rocky Glen paleo-ridge, manifested at least in the southern part of the lake in Napperby "b" time by the development of a cratonic-sourced debris fan that prograded eastward.

North of the map area, at 25°S, the Bowen Basin is delimited on the west by a foreswell (II) at 240–237 Ma, which we extend southward to 29.5°S. The axial drainage is represented by the quartzose Showgrounds/Clematis Sandstone.

In the Sydney Basin, the southward axial fluvial flow of Stage D was replaced, after a lull represented by the Bald Hill/Wentworth Falls Claystone and succeeding flint-claystones of the Garie Formation in the east and Docker Head Claystone in the west, by (a) entry of water from the southern Bowen Basin into the northern Gunnedah Basin and associated deposition south to about Moree (MO) of fluvio-lacustrine quartzose sand of the Showgrounds/Clematis Sandstone, and (b) a southeasterly axial flow that projected across the defunct Currarong Orogen (Cowan, 1993). Fine-grained, predominantly quartzose/sublithic sediment of the Newport/Terrigal Formation indicates that the regional topography was subdued and hence that the Hunter-Mooki fault did not mark a scarp at the mountain front. Thrust movement still continued (Glen and Beckett, 1989) but did not produce a mountain front that shed gravel as it had earlier.

Data in the western Sydney Basin are from Holland (1972) and Royce (1979); in the center from Gregory (1990) and Crowley (1991); in the east from Cowan (1985), Naing (1990), and Crowley (1991); and in the south from Bunny and Herbert (1971). Quartzose sediment from the southwest (a) is the Burralow Formation and the lower Newport Formation (b–d); at c the sediment is only 20 m thick near the northeastern wedge-out; a little later, sublithic sand of the middle Newport Formation (e–g), from the Gunnedah Basin, replaced the quartzose facies in the northeast. In reorganized drainage from Stage D, b, c, f, and g, all measured at the coast, indicate flow across the line of the defunct Currarong Orogen. Later still, the supply of orogenic sand from the southern NEFB failed, and the quartzose sand of the Hawkesbury Sandstone swept northward at least to its preserved edge. The long arrow, h, is the mean paleoflow vector (034°) of the Hawkesbury Sandstone, from Conaghan and Jones (1975), based on data in Standard (1969). In the south, the Hawkesbury Sandstone filled back beyond the Narrabeen Group over a surface that following tilting had been stripped down to the Wongawilli Coal of the Illawarra Coal Measures (Fig. 8).

Toward the end of Stage E, the lake in the Gunnedah Basin (Napperby Formation interval "c") expanded southward into the Sydney Basin, manifested by the Ashfield Shale (21 in Fig. 8), and northward into the southern and central Bowen Basin, manifested by the Snake Creek Mudstone Member of the Moolayember Formation.

Stage F, 235–230 Ma: Orogenic paralic sediment (Fig. 16F). As in Stage E, fine-grained sediment indicates that the regional topography was subdued and hence that the Hunter-Mooki Fault did not mark a scarp at the mountain front. Thrust movement still continued (Glen and Beckett, 1989) but did not produce a mountain front that shed gravel as it had earlier. Plutonic activity in New England (Appendix 3) is confined to the transcurrent Demon Fault, which we regard therefore as the conduit of the magma.

The subcrop limit of the Napperby Formation in the Gunnedah Basin is from Tadros (1988, fig. 10) and Yoo (1988, fig. 1). The top of the Napperby Formation, "interval c," equivalent to the lacustrine Ashfield Shale, or 235 Ma, represents a continuation of the previous lake environment but with an expanded delta plain behind a delta front that shed turbidite fans (Jian and Ward, 1993, fig. 15.22). The succeeding Deriah Formation, equivalent to the Bringelly Shale of the Sydney Basin (*see* Fig. 44 in Appendix 1), filled in the lake with the deposits of low-sinuosity rivers, with flow vectors from Jian and Ward (1993, fig. 15.24). Basalt (v), presumably a flow, was intersected in a drill hole in the Deriah Formation south of Moree (MO) (Bourke and Hawke, 1977).

The upper part of the Wianamatta Group accumulated in the Sydney Basin. Large inclusions of shale with *A. parvispinosus* in Jurassic diatremes (Helby and Morgan, 1979, table 1) include a, St Michaels Cave; b, Kedumba; c, Green Hill, and d, Angorawa. By interpreting the Wianamatta Group as a preferred host, the Wianamatta Group is extended further to the dense field of unsampled diatremes in the northwest (line of circles) (Crawford et al., 1980, p. 297). The regressive linear clastic shoreline is from Herbert (1980a, p. 51), and the mean fluvial facies paleocurrent is from Herbert (1970, p. 39). This is the terminal stage of the orogenic sediment in the foreland basin. The earlier flood of monotonously rhyodacitic material was replaced in the Minchinbury Sandstone and channel-fill sandstone of the overlying Bringelly Shale of the Wianamatta Group by intermediate-volcanic fragments (Herbert, 1980b, p. 264), reflecting a new source (Fig. 10, column VIII). Coaly intervals are the first since the Late Permian.

Mid-Triassic terminal deformation and uplift. After the last recorded deposit of the stage, at 233 Ma, the NEFB and adjacent parts of the foreland basin were locally deformed, uplifted, and eroded. In New England, the Demon Fault underwent first, from 235–234 Ma, 6 km of right-lateral displacement and then, from 234 to 230 Ma, another 23 km before being offset 8.5 km to the left by the Perry Fault by 230 Ma. The Lorne Basin was folded and the proximal Sydney (Glen and Beckett, 1989) and Gunnedah Basins thrusted and folded. Etheridge (1987, fig. 7) shows the termination of high-angle reverse faults at the unconformity between the Gunnedah and Surat Basins in the Bellata

area (15). According to Hamilton et al. (1988, p. 230), "Uplift in the Late Triassic was again most pronounced along the Boggabri Ridge and progressively older Gunnedah Basin sediments subcrop and outcrop to the east. Vitrinite reflectance data from boreholes in the southeast of the basin confirm the removal of up to 2,000 m of Triassic and Permian sediments during the Late Triassic period of erosion." Tadros (1988) suggested that the western side of the Boggabri Ridge may be marked by a thrust, the Boggabri Thrust, that upthrust the basin in the east and caused the uplift and subsequent erosion. We suggest that the thrust is a frontal splay similar to those described by Glen and Beckett (1989) from the Sydney Basin (e). That the Boggabri Thrust may have formed a scarp is suggested by the sand-percentage map of "interval c" (Jian and Ward, 1993, fig. 15.21), which shows an abrupt decrease in bedload in front of the fault line. According to Hamilton et al. (1988, p. 224), "Periodic compressive and left lateral strike slip movements along the main Hunter-Mooki Fault have resulted in the formation of a number of high relief (sometimes en-echelon) anticlines in front of the main thrust." The Wilga Park gas field, 40 km south of 15, is such an anticline developed in front of a frontal splay (h) of the main thrust.

In the Sydney Basin, many pieces of disparate evidence point to major thrusting at the end of the Middle Triassic. According to Herbert's (1989, p. 180) seismic study, the most probable period for the main movement on the Lapstone Monocline–Nepean Fault system (f), a set of en echelon high-angle reverse faults, appears to be during the Late Triassic after the last episode of deposition in the Sydney Basin. From Branagan and Pedram's (1990, p. 33) report of shale injections into the overlying sandstone in the Lapstone structure, we suggest that thrusting may have taken place sooner rather than later after deposition. Herbert (1989) remarks further that "the en echelon, discontinuous, left and right stepping characteristics of the major fault planes also indicates a degree of wrenching." From a study of northwest-trending thrust faults in the Hawkesbury Sandstone near Mooney Mooney Creek Bridge, 40 km north of Sydney (g), Mills et al. (1989, p. 223) determined the transpressive motion as sinistral.

Stage G, 230–200 Ma: Rifting of the orogen (Fig. 16G). Igneous activity was in four age groups, from young to old:

(4) Maar volcanic diatremes of the Sydney Basin, dated by palynomorphs as earliest Jurassic (208–193 Ma);

(3) The Brothers plutons (1) in the Lorne Basin, 210–206 Ma;

(2) The Garrawilla Volcanics (basalt) of the Surat Basin, 221–203 Ma; and

(1) West of Nambucca (NA), plutons 2–4, 6–8, the Werrikimbe ignimbrite (5), and the Billip Creek pluton (BC), 228–222 Ma.

The Werrikimbe ignimbrite is the first (preserved) surface igneous rock since the Permian-Triassic climax, and it straddles rocks last deformed in the mid-Triassic (Figs. 8, 11).

Deposition was confined to the intramontane Late

Triassic Ipswich/Clarence-Moreton Basin, which initially accumulated tuff and basalt then coal measures, in a return to abundant coal after the coal-free Early and Middle Triassic. Nearly 1,000 km to the west, above the Cooper Basin, the carbonaceous Peera Peera Formation initiated the Great Artesian Basin in the sump of the lowlands between the western-central craton and the New England mountain belt. Only by the end of this stage, at 200 Ma, did sediment (Precipice Sandstone and equivalents) encroach and, in places, cross the orogen during another jump eastward of the arc.

Bowen Basin and NEFB in Queensland

The formations and plutons shown on the base map (Fig. 17) are grouped according to their position before, during, or after the seven stages listed in Table 1. The sedimentary rocks have been dated according to the biostratigraphical scheme described in Appendix 1 (*see* Fig. 44) and the igneous and metamorphic rocks by radiometric methods.

In the following time-space diagrams, ages are based on Briggs (1989a, 1989b) and Price et al. (1985), as in Figure 44. General information is from Dickins and Malone (1973), Jensen (1975), Paten and McDonagh (1976), Day et al. (1983), and Draper (1985a, 1985b).

Time relations central Bowen Basin–NEFB (Fig. 18). EF corresponds with cross section IV (*see* Fig. 20 in a later section; the location of the cross section is shown in Fig. 3) and AB of Balfe et al. (1988) at lat. 25°S; FF′ extends across the Auburn Arch and Gogango Overfolded Zone to the Yarrol Block.

Auburn Arch. The Connors-Auburn Volcanic Arc (343 Ma) (Murray, 1986, p. 344) is intruded by I-type granitoids with a pooled Rb-Sr isochron of 316 ± 15 Ma (Webb and McDougall, 1968, p. 331; Murray, 1986, p. 353), and then succeeded by flows of the nonmarine Camboon Andesite, at least 3 km of andesitic and basaltic lava, agglomerate, lapilli tuff, and lithic and vitric tuff, and thick conglomerate in the lower part (Dear et al., 1971, p. 43), with K-Ar dates of 281 Ma and 294 Ma, succeeded at a transition by the Buffel Formation, 100 m of shallow marine bioclastic limestone (Draper, 1988), and equivalent Fairyland, Dresden, and Elvinia Formations. On the west, the Oxtrack/Brae/Pindari Formations at the base of the Blenheim Group disconformably overlie successively older members of the Buffel Formation to overlap the Camboon Volcanics (Dear et al., 1971, p. 51) at a vacuity (or erosional disconformity) eroded during uplift 263–258 m.y. ago. The succeeding Barfield and Flat Top Formations of the Blenheim Group are overlain by the Gyranda Formation, Baralaba Coal Measures, Rewan and Clematis Groups, and Moolayember Formation, succeeded after deformation in the mid-Triassic by the Early Jurassic Precipice Sandstone.

Gogango Overfolded Zone. According to Day et al. (1983, p. 105), the Late Permian and Triassic Gogango Overfolded Zone grew out of the Grantleigh Trough, the site of a marginal

sea or a pull-apart basin. The Camboon Volcanics interfinger with and are succeeded by a thin limestone with *Eurydesma* and then 5 km of marine flysch (Rannes Beds) of locally derived volcanolithic epiclastics and spilitic pillow basalts (RD = Rookwood Volcanics in Fig. 18) with the gross features of ocean floor basalt (Clare, 1993). All these rocks were initially deformed before the deposition of the flysch-like Boomer Formation and apparently at the same time as the cooling of the Ridgelands Granodiorite (+, 269–264 Ma) in the northern part of the Yarrol Block. "Deformation involved a substantial component of westward thrusting; tight to isoclinal folds are consistently overturned to the west and an east-dipping axial plane cleavage is developed in argillaceous rocks" (Day et al., 1983, p. 105). The Rawbelle Batholith, mainly granitoids, was then emplaced in the Gogango Overfolded Zone from 258 Ma to 218 Ma, with a peak from 247 Ma to 238 Ma (*see* Table 6 in a later section). A second Permian deformation (compression 1′) is dated at about 251 Ma by the occurrence of conformable sequences to the east (Yarrol Block) and west (Bowen Basin) in contrast to the complex tight folding in the Boomer Formation. At least some of this folding is due to a third (mid-Triassic) deformation (compression 2), involving folding in the west of the Permian to Middle Triassic sequence of the Bowen Basin and thrusting in the Baralaba Coal Measures (Hammond, 1988), and folding in the east in the Yarrol Block, followed by the deposition of the relatively undeformed Rhaetic (Price et al., 1985, fig. 8a) Callide Coal Measures and the ?Carnian-Norian Muncon Volcanics.

Yarrol Block. The Yarrol Block contains a rare record of marine shelf deposition through almost the entire Carboniferous (Roberts, 1985, p. 48) and extending, past a brief lacuna (286–280 Ma), into the Early Permian, in the form of the Burnett Formation and the equivalent volcanic fluvial to marginal marine 2-km-thick Youlambie Conglomerate, which contains diamictite and rhythmites with dropstones overlain by *Eurydesma* shale in the basal 150 m (Dear et al., 1971, p. 38, 39). The succeeding Yarrol Formation comprises 500 m of limestone, siltstone, and lithic sandstone, with locally interbedded andesitic lava. The abundant invertebrate fossils include an Aktaskian (268–265 Ma) goniatite (Dear et al., 1971). To the east, the equivalent "Berserker Beds . . . are intruded by serpentinite [SERP in Fig. 18],which in turn is cut by the Ridgelands Granodiorite" (Webb, 1969, p. 115), dated as 264 Ma and 269 Ma (Murray, 1986). We adopt the younger date. The upper limit of the deformation is marked by the base of the Rainbow Creek Formation (262 Ma) that rests unconformably on Devonian (DEV) volcanics in the south (Day et al., 1983, p. 106). In the Monto area, the Yarrol Formation grades up into the thick Owl Gully Volcanics, andesitic lavas and pyroclastics with interbedded sediment which contains rare marine invertebrates, all deposited apparently during the initial deformation of the region. The next formation in the succession is the Rainbow Creek Beds, 2 km of nonmarine lithic and feldspathic sandstone, siltstone, conglomerate, and acid volcanics with

Artinskian-Kungurian spores (Dear et al., 1971, p. 65). As noted previously, the Rainbow Creek Beds unconformably overlie Devonian volcanics and are unconformably overlain by the Muncon Volcanics in places and by the Callide Coal Measures in others. Next in the succession are the Moah Creek Beds (Kirkegaard et al., 1970, p. 66, 67), 2 km of conglomeratic mudstone containing invertebrates of Fauna IV (*see* Appendix 1, Fig. 44), and blocks of Carboniferous oolitic limestone, interpreted as nearshore mudflows at the western foot of Carboniferous terrain uplifted at this time. The mudflows grade westward to the turbidites of the Boomer Formation (Kirkegaard et al., 1970, p. 68). To the south, the first phases of the Rawbelle Batholith were emplaced. The 1-km-thick Dinner Creek Conglomerate disconformably overlies the Moah Beds, is nonmarine, and contains *Glossopteris*. The entire succession, from the Youlambie Conglomerate and Burnett Formation to the Dinner Creek Conglomerate contains volcanolithic and other locally sourced epiclastics intercalated with volcanics. The next events were the cooling of the Galloway Plains Tonalite, Wingfield Adamellite and younger parts of the Rawbelle Batholith at 247–233 Ma, followed by regional deformation. The Boolgal Granite, Glassford Complex, and Mt Saul Adamellite were then emplaced at 226–218 Ma (Day et al., 1983, p. 145; Table 6), probably at the same time as the eruption of the Muncon Volcanics, 200 m of tuff and andesite, with *Dicroidium,* regarded as Late Triassic (Day et al., 1983, p. 118); de Jersey (1974) correlated the Callide Coal Measures and the underlying tuffaceous shale, tuff, and volcanics, probably part of the Muncon Volcanics, with the topmost (Norian) part of the Ipswich Coal Measures; Price et al. (1985, fig. 8a) place the Callide Coal Measures (quartzose according to Svenson and Hayes, 1975) in the lower part of their palynological unit PT5 (Rhaetic), above the PT4 (Ipswich Coal Measures). Accordingly, we regard the Muncon Volcanics as ranging beneath the Rhaetic Callide Coal Measures down to the earliest (Carnian) part of the Late Triassic. The succession is completed by the Early Jurassic Precipice Sandstone.

Bowen Basin. According to Draper (1985b), the major events, from old to young, are the following:

I. The eruption of volcanics.

II. Deposition of the Reids Dome Beds, nonmarine graben-filling, coal-measure sequences.

III. First major transgression (MT in Fig. 18).

IV. First major regression (MR in Fig. 18), interrupted by deformation between the lower and upper parts of the Aldebaran Formation and onlap to the west.

V. Slow transgression.

V′. Regression at the top of the Freitag Formation.

VI. Rapid transgression, represents a major change in sedimentary style within the basin.

VII. Major regression.

VIII. Transgression.

IX. Regression, coincident with the onset of major tuff deposition.

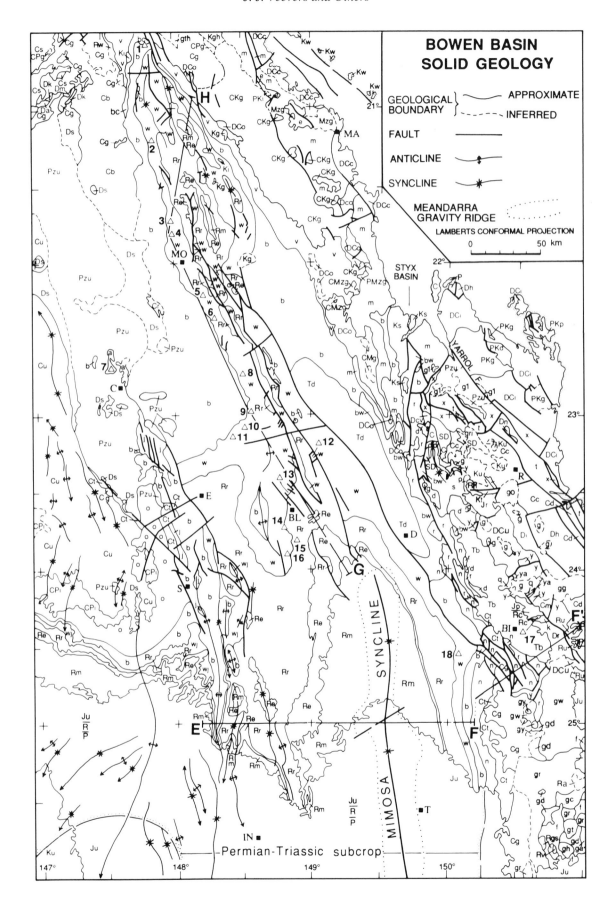

Key to Fig. 17, map of Bowen Basin.
Stages A - D: Units are Permian unless indicated otherwise.

T - Tertiary, K - Cretaceous, J - Jurassic, R - Triassic, Mz - Mesozoic, P - Permian,C - Carboniferous, D - Devonian, S - Silurian.   Groups listed in CAPITALS

<u>Stage unknown</u>

g	Granitoid, undiff., ?Permian
x	Serpentinite - "Permian"

<u>Post-G</u> <200 Ma

Tb	Biloela
Td	Duaringa
PKp	Peninsula Range
Ks	Styx
Kt	Stanwell
Ku	Undiff.
Kw	Whitsunday
Jp	Precipice
Jr	Razorback
Ju	Undiff.

Granitoids, etc

Ki	Undiff. igneous rocks
Kg	Granitoid, undiff.
Kgh	Hecate
Mzg	Granitoid, undiff.
PKd	Double Mountain
PKg	Granitoid, undiff.
PKi	Undiff. igneous rocks
PMzg	Granitoid, undiff.

<u>Stage G</u> 230-200 Ma

Jp	Precipice
Jr	Razorback
Ju	Undiff.
Ra	Mt Eagle
Rc	Callide
Rn	Native Cat
Ru	Muncon
Rv	volcanics

Granitoids

Rgs	Mount Saul
Rgb	Boolgal

<u>Stage F</u> 235-230 Ma

Rm	MOOLAYEMBER

<u>Stage E</u> 241-235 Ma

Re	CLEMATIS
Rw	Mount Wickham

<u>Stage D</u> 250-241 Ma

Rr	REWAN

Granitoids

gc	Coonambula
gd	Delubra (C and D)
gg	Galloway Plains
go	younger Bouldercombe
gr	Rawbelle batholith, undiff.
gt	Cheltenham Creek
gw	Wingfield
g1	intrusions Marlborough

<u>Stage C</u> 258-250 Ma

b	BACK CREEK (B and C)
bw	Boomer
o	Colinlea
p	Dinner Creek
s	Moah Creek
w	BLACKWATER
	wj Rangal, Bandanna, Baralaba

Granitoids

ga	Cadarga Creek
gd	Delubra (C and D)
gh	Hawkwood
gy	Crystal Vale

<u>Stage B</u> 268-258 Ma

a	Blair Athol
b	BACK CREEK (B and C)
bc	Collinsville
e	Calen
k	Rainbow Creek
q	Owl Gully
r	Rannes (A and B)
ya	Yarrol

Granitoids

gri	Ridgelands
go	older Bouldercombe

<u>Stage A</u> 290-268 Ma

d	Rookwood
f	Narayen, Nogo
j	Reids Dome *(mainly subsurface)*
m	Carmila
n	Camboon
P	Undiff. sediments of the Stanage Block
r	Rannes (A and B)
t	Berserker
v	Lizzie Creek
y	Youlambie
CPi	JOE JOE

Granitoids

gth	Thunderbolt

<u>Pre-A</u>

C	Undiff.
Cb	Bulgonunna
Cc	CASWELL CREEK
Cd	Crana
Cm	Three Moon
Cr	Mt Rankin
Cs	Star of Hope
Ct	Torsdale
Cu	Ducabrook
DCc	Campwyn
DCi	CURTIS ISLAND
DCo	Connors
DCu	Undiff.
DCv	Volcanics, undiff.
Da	St Anns
Dh	Mt Holly
Di	Capella Creek
Dk	Ukalunda
Dm	Mount Wyatt
Dn	Etna
Dr	Kroombit
Ds	Silver Hills
SD	Undiff.
Pzu	Undiff.

Granitoids

CKg	Urannah
CMzg	Granitoid, undiff.
CPg	Granitoid, undiff.
Cg	Granitoid, undiff.

Towns

BI	BILOELA	D	DUARINGA	MO	MORANBAH
BL	BLACKWATER	E	EMERALD	R	ROCKHAMPTON
C	CLERMONT	IN	INJUNE	S	SPRINGSURE
		MA	MACKAY	T	TAROOM

Coal Mines - Δ

1	Collinsville	7	Blair Athol	14	Blackwater
2	Newlands	8	Norwich Park	15	Cook
3	Riverside	9	German Creek	16	South Blackwater
4	Goonyella	10	Oaky Creek	17	Callide
5	Peak Downs	11	Gregory	18	Moura/Kianga
6	Saraji	12	Yarrabee		
		13	Curragh		

Figure 17. Bowen Basin base map. Main source Balfe et al. (1988), Bowen Basin Solid Geology, Queensland, 1:500,000, published by the Queensland Department of Mines. The Moorlands coal basin northwest of Clermont (C) is from Sorby and Scott (1988). The Meandarra Gravity Ridge, south of 24°S about longitude 149°30′E, is from Murray et al. (1989).

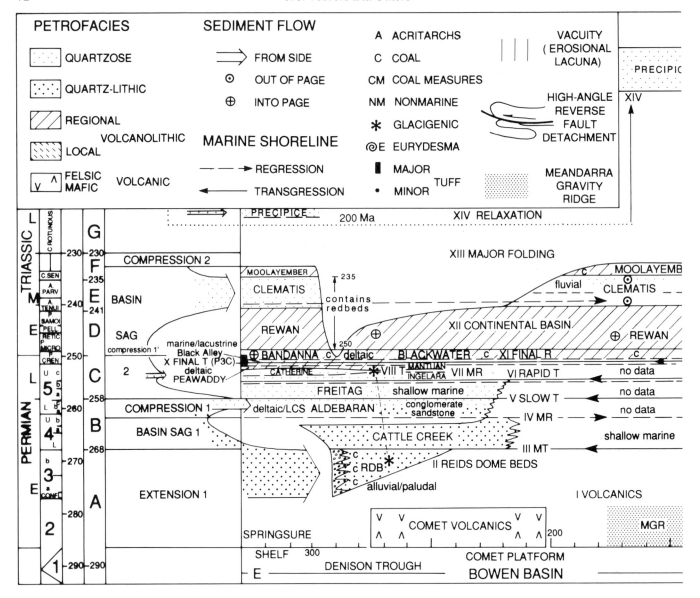

Figure 18. Time-space diagram of the southern Bowen Basin at 25°S, location shown on Figure 17.

X. Final transgression.

XI. Final regression, the culmination of coal-measure sedimentation.

XII. Filling of a continental basin, without coals except very thin ones in the Moolayember Formation.

XIII. Major folding: "Although tectonism continued throughout deposition of the Bowen Basin sequence, the present configuration of the basin resulted from a Late Triassic folding event."

We add a final event:

XIV. Relaxation that led to the deposition of the Early Jurassic Precipice Sandstone.

Denison Trough. General information comes from Paten et al. (1979) and Elliott (1985); tectonic regimes (Fig. 19) are from Ziolkowski and Taylor (1985). Pebbles are scattered through mudstone due to the incorporation, transportation, and deposition of fluvial sediment in seasonal river ice (Draper, 1983). The thick interval of mudstone with anhydrite (squares, column XIII, Fig. 24) (Mollan et al., 1969) reflects (cold) aridity, seen also in the Merrimelia Formation of the Cooper Basin (Williams et al., 1985). The Black Alley Shale and the lower part of the Bandanna Formation contain tuff.

Comet Platform. Murray (1986, p. 356) grouped the Comet Volcanics in the 306 Ma (Late Carboniferous) Combarngo Volcanics, which "range in composition from rhyolitic ignimbrite to altered andesite or basalt . . . and appear to be calc-alkaline." According to Kemp et al. (1977, p. 192), however, a palynoflora in the volcanics in AFO Comet-1 is probably Stage

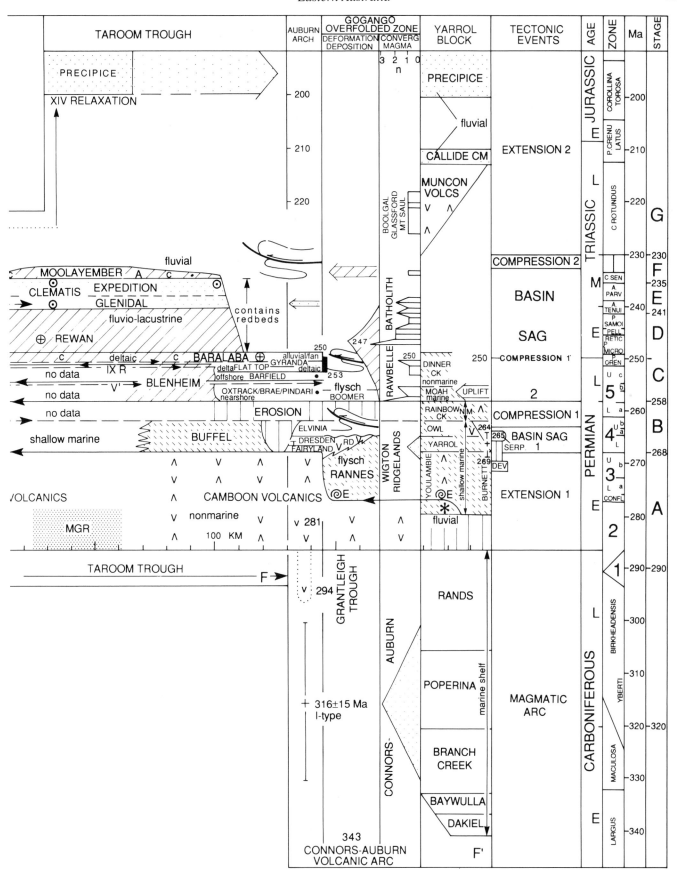

TABLE 6. RADIOMETRIC DATES (Ma) OF THE RAWBELLE BATHOLITH AND OTHER INTRUSIONS IN THE GOGANGO OVERFOLDED ZONE AND ADJACENT YARROL BLOCK*

Sample	Biotite	Hornblende	Sheet	Reference[†]
Wingfield Adamellite				
GA1249	241	**246**	56-1	1
GA1250	**239**	236	56-1	1
GA1369	**242**	236	56-1	1
GA5348	240	**247**	56-1	1
GA5353	233	**238**	56-5	1
GA5364	**239**	237	56-5	1
Crystal Vale Adamellite				
GA1370	**255**		56-1	2
Hawkwood Gabbro				
VIII (Ar/Ar)	**258**		56-5	3
Cadarga Creek Granodiorite				
QA46, 57	242	**257**	56-5	4
Delubra Quartz Gabbro				
QA48		**251**	56-5	4
QA55		**247**	56-5	4
V (Ar/Ar)		**258**	56-5	4
VI		**250**	56-5	4
VII (Ar/Ar)		**253**	56-5	4
Coonambula Granodiorite				
QA49		**247**	56-5	4
Cheltenham Creek Adamellite				
QA80		**245**	56-5	4
QA82		**244**	56-5	4
IV (Ar/Ar)	240	**241**	56-5	3
Rawbelle Batholith (undifferentiated)				
GA1168	**240**		56-5	1
GA1169	**245**		56-5	1
GA5342		243	56-5	1
GA5344/5	**234**		56-5	1
GA5350	233	**244**	56-5	1
GA5357	**233**		56-5	1
GA5362	**241**		56-5	1
QA61		226	56-5	4
QA81		**245**	56-5	4
QA83		**246**	56-5	4
Eidsvold Complex				
GA5337	**241**		56-5	1
GA5351		**250**	56-5	1
Boolgal Granite				
QA68 (arfvedsonite?)		**218**	56-5	4
Wigton Adamellite				
QA47		**264**	56-5	4
Mount Saul Adamellite				
QA54/62	219	**220**	56-5	4
Galloway Plains Tonalite				
GA887		**255**		1
Mulgildie-1: 24°58'S 151°08'E				
1166	244	**247**	56-1	1
1167	**244**	240	56-1	1

*Component plutons from Day et al. (1983, p. 144, 145). All by K/Ar method except a few, as noted, by Ar/Ar. Ages converted to new constants (Dalrymple, 1979), older (accepted) date **bold**. 1:250,000 sheet area: SG 56-1 = Monto (Dear et al., 1971, 1981); SG56-5 = Mundubbera (Whitaker et al., 1974).

[†]1 = Webb and McDougall, 1968, table XIV; 2 = *fide* Day et al., 1983, p. 145, for identification of pluton; 3 = Green and Webb, 1974, Table 1; 4 = Whitaker et al., 1974, table 2.

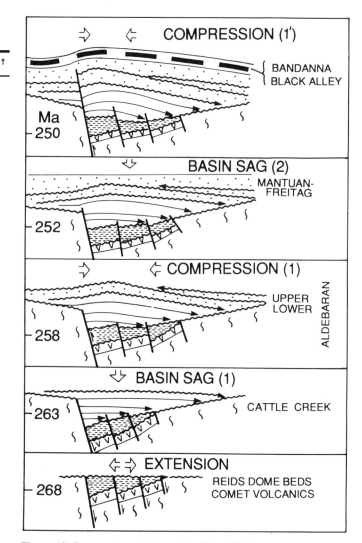

Figure 19. Structural evolution of the Warrinilla inverted half graben, located about Warrinilla North-3, 3 on Figure 29A, from figure 3 of Ziolkowski and Taylor (1985) with the addition of the Comet Volcanics.

2, and Day et al. (1983, p. 94) cite further palynological work that suggests similar ages elsewhere. The Comet Volcanics are therefore recognized as belonging to the Early Permian stage of extension, equivalent to the Camboon and Lizzie Creek volcanics, and postdate the Late Carboniferous calc-alkaline convergent Combarngo and extensional Bulgonunna Volcanics.

Southeast Bowen Basin and adjacent NEFB. According to Murray (1986, p. 370), rocks of the Camboon Volcanics (together with the Lizzie Creek Volcanics in the north) range from "basalt to rhyolite. Overall, andesitic volcanics are dominant, and pyroclastics are slightly more abundant than flows. Chemical data are almost completely lacking." The rocks were deposited in "subaerial and lacustrine environments close to active volcanoes." The eastward overlap of the Blenheim Group during event V (transgression) followed removal of the entire Buffel Formation during local deformation marked by

the IV regression. Another sign of proximal uplift is the alluvial-fan facies of the Baralaba Coal Measures. Abundant tuff extends through the Flat Top Formation, Gyranda Formation, and the basal Kaloola Member of the Baralaba Coal Measures, and is overlain by the upper part of the Baralaba Coal Measures, which has no tuff (Quinn, 1985).

Correlation of tectonic events. Events seen in the well-exposed east (Auburn Arch, Gogango Overfolded Zone, Yarrol Block) and the intensely drilled west (Denison Trough) correlate across the intervening Taroom Trough:

Extension 1, 286–268 Ma, accompanied by events I (volcanics) and II (Reids Dome Beds).

Basin sag 1, 268–263 Ma and III marine transgression.

Compression 1, 263–258 Ma and IV marine regression.

Basin sag 2, 258–233 Ma and V marine transgression, punctuated by compression 1' at or soon after 250 Ma.

Compression 2, 233–230 Ma.

Extension 2 starts at 230 Ma with the deposition of the Muncon Volcanics and Callide Coal Measures in the east, and is renewed here and to the west during the Early Jurassic (200 Ma) deposition of the Precipice Sandstone.

Provenances/tectofacies. Volcanics occur at the base (Lower Bowen Volcanics) and range from mafic (Owl Gully, Rookwood) to mixed felsic-mafic (Comet, Camboon). They are also intercalated in the Youlambie Conglomerate and Yarrol Formation with volcanolithic epiclastics derived from volcanics within the formation, as in the Early Permian part of the Yarrol Block succession, or from recently erupted proximal volcanics, as the Buffel Formation accumulated epiclastics from the Camboon Volcanics (Dickins and Malone 1973, p. 37). Quartz-lithic epiclastics are found in the Reids Dome Beds, Cattle Creek Formation, lower Aldebaran Formation, and Glenidal Formation in the west, and are derived from the adjacent craton. Quartzose epiclastics occur in the region of the western craton (upper Aldebaran to top of Catherine Formation) and upper Clematis Group (= Expedition Sandstone), the eastern part of which was deposited by a southward-flowing axial stream. The main volcanolithic epiclastics were derived from the region of the eastern magmatic orogen, and comprise the Oxtrack Formation in the east younging to the west to the Peawaddy Formation (Paten et al., 1979, p. 45), and its top (and bulk) the Rewan Group and (past the quartzose upper Clematis Group in the west) the Moolayember Formation. The Bandanna, Rewan, and Moolayember extend beyond the Bowen Basin to the western edge of the Galilee Basin (Veevers, 1984, p. 242). The part of the Rewan Group on the Comet Platform was deposited by axial streams that flowed north-northwest. A major tuff interval occupies the Flat Top–lower Baralaba (253–251 Ma) succession on the east, and the Mantuan–Black Alley (252–251 Ma) succession on the west (Paten and McDonagh, 1976, p. 410) (Quinn, 1985); in the east they are preceded by minor tuffs in the Pindari (Flood et al., 1981, p. 183) and Barfield (Dickins and Malone, 1973, p. 124) Formations. The contributary streams that flowed west from the orogen to join the southward-

flowing axial stream during deposition of the Clematis Group delivered semi-labile silicic volcanogenic sediment (including conglomerate) that reflects derivation from the adjacent craton (Bastian, 1965b, p. 4–5; Jensen, 1975, p. 60–62), as well as from the Baralaba Coal Measures and Gyranda Formation along the eastern margin of the basin.

Depositional environments. Data are from Fielding et al. (1990). Coal is from Dickins and Malone (1973), redbeds from Dickins and Malone (1973, p. 72, 73, 77) and Jensen (1975, p. 29, 34, 38, 39).

Ages of the Rawbelle Granite. Table 6 provides radiometric dates for the Rawbelle Granite.

Cross sections. Locations of Sydney-Bowen Basin cross sections are shown in Figure 3; the cross sections are given in Figures 15, 20, and 21. Regional cross sections are given in Figures 22 and 23.

Cross sections of the Bowen Basin (Fig. 20) show that the axis of the Meandarra Gravity Ridge (MGR, stippled) (Murray et al., 1989) lies 15 km west of the axis of the Mimosa Syncline (IV in Fig. 20) and is interpreted as an ultramafic feeder of the Early Permian volcanics of the region. The ultramafic feeder when later cold and unsupported by upward hydrostatic pressure of intrusion became a sinker about which the crust flexed downward so that thicker Permian strata were deposited over it. The Taroom Trough is thick therefore because of the volcanic feeder. Taking the data on thickness at face value, the strata appear to thin away from the axis of the trough toward the craton on the west and toward the later orogen on the east, indicating that the orogen did not bend down the craton at the foot of the thrust front as much as the sinker continued to draw down the crust. In the north (I), the incomplete section appears to thin (but only mildly) from the orogen to the craton. Regardless of the inconclusive data, the Late Permian and Early Triassic part of the Taroom Trough here and farther south in the Sydney Basin does not constitute a thick foreland wedge, as seen classically in western Canada or the Alps.

Segment BB' of cross-section IV (Fig. 20) (Hammond, 1988, fig. 2.3B) shows the eastern thrust-faulted flank of the Mimosa Syncline and Auburn Arch cored by the 316 Ma Glandore Granodiorite and the Prospect Creek Anticline. The section beneath the Buffel and Oxtrack Formations (Fig. 24, Column XXI) consists of the Camboon Andesite, Youlambie Conglomerate, and other Early Permian volcanic and volcanogenic rock. The Gogango Overfolded Zone (GOZ) lies between the Drumberle Fault (DF) and the Devonian Kroombit (K) Beds of the Yarrol Block, which we interpret, after Hammond's model, as separated from the GOZ by a high-angle reverse fault that comes off the basal thrust (? original detachment) fault. The eastern side of the Kroombit Beds is separated by a fault from the Youlambie (Y) Conglomerate and unconformably overlying Late Triassic Muncon Volcanics (MU) and Early Jurassic (J) sandstone. Cross-section IV″ in Figure 20, also from Hammond (1988, fig. 2.3A), is immediately north of the Auburn Arch and shows the interpretation of reverse-fault

splays coming off the thrust, notably concentrated at ramps and extending upward to the top of the Triassic section, hence being younger than Middle Triassic. The cleavage front extends a short distance west of the Banana Fault (BF).

The cross section of the Galilee Basin and Springsure Shelf (Fig. 22, PQ), location shown in Figure 3, shows the overstep at a low-angular unconformity of the Late Permian Colinlea Sandstone/Bandanna Coal Measures on deformed Early Permian Aramac Coal Measures, Jochmus Formation, and Jericho Formation and the Early Carboniferous Ducabrook Formation of the Drummond Basin. Stratigraphic columns of the Bowen Basin are grouped in Figure 24 and stratigraphic columns of the Galilee, sub-Murray, and Tasmania (epicratonic) Basins in Figure 25.

Time relations axial Bowen Basin (Fig. 26). The section line (Fig. 17) crosses the Nebo Synclinorium from Exmoor on the northeastern margin to the Goonyella mine and thence runs south-southeastward along the strike of the coal-producing Collinsville Shelf and Comet Platform along the line of the open-cut coal mines of Peak Downs, Saraji, Norwich Park, and German Creek, and east of AFO Cooroorah-1 Well (Malone et al., 1969, fig. 4) to the axis of the Mimosa Syncline. Sources are Staines and Koppe (1980) and Jones et al. (1984, p. 255), supported by the general sources cited previously, including Draper (1985b). With the direction of main sediment transport to the south-southeast along the basin axis, the sea was driven progressively southward by the deposition of the Moranbah and Fort Cooper Coal Measures to the final regression of the Rangal Coal Measures.

The main tuff sequence is confined to the Fair Hill Formation and Fort Cooper Coal Measures (Staines and Koppe, 1980, p. 285), in particular the Yarrabee Tuff Member immediately beneath the Rangal Coal Measures. Minor tuffs are (1) in the base of the Tiverton Formation (Dickins and Malone, 1973, p. 41), probably the last sign of activity from the Lizzie Creek Volcanics only 30 m below (Jensen et al., 1966, p. 20), (2) across the facies boundary between the German Creek/Macmillan Formations on the south and the Moranbah Coal Measures on the north (Staines and Koppe, 1980, p. 284), and (3) in the Blenheim and Moranbah at the northern outcrop.

The older limit of age of the Lizzie Creek Volcanics is given by the 299 Ma age of a nonconformably underlying granite (Day et al., 1983, p. 104), part of the complex of the Bulgonunna Volcanics shown in the time-space diagram JK (*see* Fig. 28 in a later section).

Provenances. (1) Volcanics (Lizzie Creek and Comet Volcanics) erupted from below; (2) volcanolithics (Tiverton Formation) were derived from exposed Lizzie Creek Volcanics; (3) quartzose and quartz-lithic petrofacies came from the western craton, in two sequences: (i) Gebbie in the north, Back Creek, German Creek, and Macmillan in the south, and (ii) Clematis, which is, however, depicted as having a mixed provenance in the axial stream in the Nebo Synclinorium because of the blending of quartz-lithic sand derived from the craton and volcano-lithic sand derived from the western edge of the orogen (*cf.* Fig. 18); (4) regional volcanolithic petrofacies came from the eastern orogen, also in two sequences: (i) Blenheim and Fair Hill through Rewan, containing the major tuff, and (ii) Moolayember, containing thin tuffs in the Mimosa Syncline in the south.

Paleocurrents. Our information comes from Jensen (1975), augmented by other vectors shown in Figure 29 C–F. The diagram runs along the axis of drainage between west-flowing streams from the orogen (encircled +) and east-flowing streams from the craton (encircled ●). The interfingering boundary of the German Creek Formation and the Moranbah Coal Measures at Saraji marks the axis; in the Rangal Coal Measures the axis lies west of the line of the diagram so that the drainage is wholly from the east. The Triassic succession in the Nebo Synclinorium occupies the axis of southern drainage with side streams in the northeast supplying the volcanolithic sediment of the Rewan through the Moolayember. At the southern end of the diagram, in the Mimosa Syncline, the Rangal and Rewan register a reversal to a northern axial drainage before the Glenidal and Expedition Sandstones register a resumption of southern drainage, and finally the upper Moolayember Formation registers a return to northwesterly drainage.

Time relations Galilee-Drummond Basin (Fig. 27). The diagram, located in Figure 3, shows the Galilee Basin with Permian data sketched and projected from the Springsure Shelf in the south and Triassic data from the Koburra Trough in the north. The exposed Drummond Basin lies between L and J.

Environments and provenance. Data are from Vine (1976), Hawkins (1978), Evans (1980), and Veevers (1984, p. 242, 243). Spinose acritarchs (A) in Jericho-1 suggest an ephemeral transgression-regression (Evans, 1980). Redbeds are from Vine (1976, p. 319). The Dunda Sandstone is interpreted as having a mixed quartzose-regional volcanolithic provenance, the Colinlea as mixed quartzose-local volcanolithic. Moolayember Formation is thickest in northeast, is indubitably volcanolithic, and indicates a change in the northern source area from quartzose to volcanolithic.

Paleocurrents. Dunda information is from Vine (1976); Glenidal and Expedition data are from Jensen (1975); Rewan information comes from Figure 41D (in a later section); and Moolayember information is from Figure 41E.

Structure. Data are from Gray (1976) and Gray and Swarbrick (1975) (*see* cross-section PQ, Fig. 22).

Tectonic events. These are the same as in the southern Bowen Basin (Fig. 18) except (a) there is evidence in this area of the mid-Carboniferous deformation in the lacuna between the Visean deposition of the Ducabrook Formation of the Drummond Basin and the onlap of the Lake Galilee Sandstone over the angular unconformity; (b) there is no direct evidence in this profile of Sag 1—any succession deposited during Sag 1 was presumably removed during compression 1 during the uplift of the wide Foreswell I; and (c) Extension 2 is indicated by the deposition of the Peera Peera Formation at the base of

the Eromanga Basin 250 km to the west (*see* Veevers, 1990b). The eastward source of the Glenidal Formation indicates the uplift of the narrow Foreswell II.

The lost section at the erosional vacuity of the Aramac Coal Measures/Jochmus Formation/Jericho Formation, shown also in cross-section PQ (Fig. 22), is shown in Figure 27 by vertical lines.

Time relations northern Bowen Basin (Fig. 28). This diagram crosses from the western side of the Anakie Inlier, and follows the line of cross-section I of Fig. 20 (location is shown in Fig. 3) to pass across the northern Bowen Basin (crosses Fig. 26, GH near Goonyella) to Connors Arch, Strathmuir Synclinorium and, at the coast, the Campwyn Block.

Calen Coal Measures. Spores are carbonized, the best estimate of age is equivalent to Gebbie Subgroup (Day et al., 1983, plate 6).

Igneous rocks. The Urannah Complex (with the Auburn Complex) is best estimated as 316 ± 15 Ma (Webb and McDougall, 1968, p. 331), intruded in the north by the Thunderbolt Granite, 281 Ma, possibly ranging to 265 Ma (Webb and McDougall, 1968, p. 325). Projected 200 km into the diagram from the northwestern tip of the basin are the Bulgonunna Volcanics, dated by Rb-Sr isochron (Webb and McDougall, 1968, p. 319) as 297 ± 12 Ma, the intruding granites as 308 ± 25 Ma, and from the pooled data as 299 ± 9 Ma, all confirmed by K-Ar average age of 289 Ma. McPhie et al. (1990) cite a U-Pb ion-probe analysis of single zircon grains from ignimbrites in the Bulgonunna Volcanics as giving a date close to 300 Ma. Accordingly we plot the Bulgonunna Volcanics as 300 Ma old, and show the intruding granites as ranging from 300–290 Ma. Subsequent intrusions are CP4 (Day et al., 1983, p. 144), adamellite and microgranite dated by K-Ar as 286 Ma; and P3 (p. 143), another adamellite near Townsville, 273 Ma and 275 Ma. The Mt Wickham Rhyolite (Day et al., 1983, p. 119) is dated by Rb-Sr as 238 ± 15 Ma. Not shown are the Cretaceous intrusions (e.g., Hecate Granite) that intrude the Urannah Complex and the northern Bowen Basin.

Small basins and outliers in the west. The section at

Figure 20 (on following page). Cross-sections I–IV of the Bowen Basin, aligned on the Meandarra Gravity Ridge (MGR), from Murray et al. (1989), all with V:H = 5, located on Fig. 3. Surface data from Balfe et al. (1988); main (upper, Blackwater) coal measures stippled.

I. Northern Bowen Basin (JK), adapted from Jones et al. (1984, fig. 163, CC′) by adding the Early Permian Lizzie Creek Volcanics within the Nebo Synclinorium and extending the section northeastward to the coast across the NEFB comprising the Connors Arch, Strathmuir Synclinorium, and Campwyn Block, and westward to the Anakie Inlier. Pre-Permian rocks identified by letter symbols of the solid geology (bedrock) map by Balfe et al. (1988). Basins on the west are Moorlands (Sorby and Scott, 1988), Blair Athol (Preston, 1985), and Wolfang (Cook and Taylor, 1979; Osman and Wilson, 1975). On the east, the Calen Coal Measures are from Koppe (1975). The Eungella Gravity Ridge (EGR) on the east and its interpolated western branch (Murray et al., 1989) straddle the axis of the Nebo Synclinorium. The reverse faults seen at the surface in the middle part of the section are interpreted by us as leading back to a listric fault that crops out along the eastern edge of the basin.

II. Line drawing of part of the unmigrated seismic profile BMR89.BO1 converted to depth by 1s TWT = 2 km (Korsch et al., 1990b, fig. 3) and scaled to V:H = 5. The Jellinbah (JF) and Yarrabee (YF) Faults that delimit the structural zones of Blackwater (BZ), Yarrabee (YZ), and the Dawson Fold Zone (DFZ) are interpreted as reverse listric faults that are rooted in a major detachment that flattens east of this part of the section in a ductile zone 20 km deep in the middle crust. R̅ = Triassic, LP = Late Permian coal measures (stippled).

III. Line drawing of a seismic section (from Elliott, 1989, fig. 4a) converted to depth by 1 s TWT = 2 km. V's are our interpretation of the Comet Volcanics beneath the main layers of Permian (P) and Triassic (R̅) sediment.

IV. Cross section DD′/BB′/D″D‴ across the southern part of the exposed Bowen Basin, through the Nebine Ridge/Springsure Shelf, Denison (Westgrove) Trough, Comet Platform, Mimosa Syncline (Taroom Trough), and NEFB (Auburn Arch, Gogango Overfolded Zone, and Yarrol Block). Adapted from Jones et al. (1984, fig. 163, DD′) and extended eastward by the addition of the interpretative BB′ (from Hammond, 1988, fig. 2.3B) and D″D‴ by interpretation of the surface geology (Balfe et al., 1988). BB′ is supplemented by the interpretative AA′ (IV″) (from Hammond 1988, fig. 2.3A), located some 30 km to the north. The part of the section across the Denison Trough–Comet Platform is supplemented by IV′ (CC′), which crosses the main section on the western side of the Mimosa Syncline. Abbreviations are A-C—Aldebaran/Cattle Creek Formations; BF (IV″)—Banana Fault; C—Camboon Volcanics; CF (IV″) Cooper Range Fault; DF—Drumberle Fault; F-B—Freitag–Black Alley Formations; J—Jurassic; K—Kroombit Beds (Devonian); M—Moolayember Formation; MU—Muncon Volcanics (Late Triassic); R—Rewan Group; RDB—Reids Dome Beds; V—Comet Volcanics—those known by drilling are circled; Y—Youlambie Conglomerate.

IV′ (CC′) is from Paten et al. (1979, fig. 4), and we have added the interpretation of Hammond (1987, fig. 4B; Hammond, 1990) that the main faults are listric (heavy lines) and are countered by antithetic faults (broken lines). Note here and in the section below that the early normal displacement along the Merivale Fault System was later reversed before being covered in CC′ by Jurassic strata and in the main section by Early Triassic strata (Elliott, 1985, fig. 5), presumably by movement during compression on pre-existing extensional structures. Ages are (from top) J—Jurassic, R̅—Early and Middle Triassic, LP—Late Permian, EP—Early Permian. Formations are RDB—Reids Dome Beds, V—Comet Volcanics, and M—metamorphics (slates), encircled where known from drilling. DC stands for Devonian-Carboniferous. Fault systems (FS) are A—Arcadia, M—Merivale, P—Purbrook, W—Warrinilla. The Denison Trough and Comet Platform section in IV is modified from the seismic data of Bauer and Dixon (1981), which indicate that the Denison Trough is wholly represented by the Westgrove Trough. The slate (S) was penetrated in Purbrook-1.

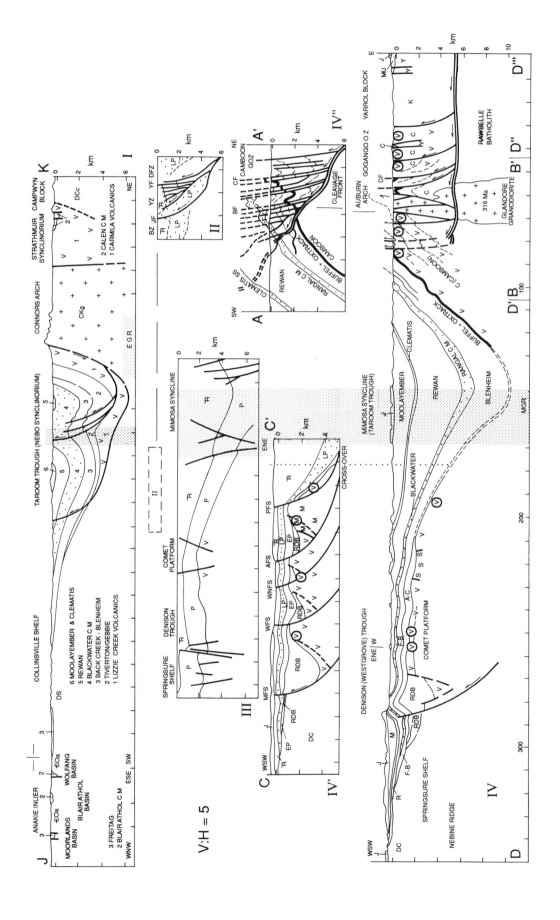

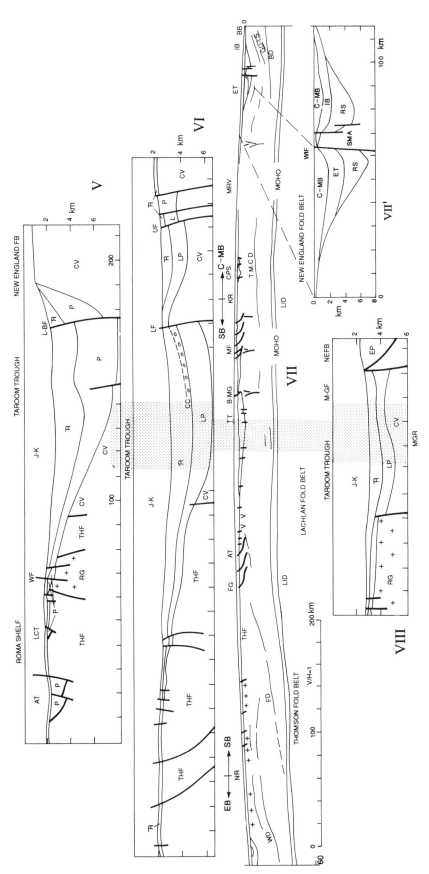

Figure 21. Cross sections of the Permian-Triassic Bowen Basin beneath the Jurassic-Cretaceous Surat Basin, aligned on the Meandarra Gravity Ridge (MGR). V:H = 5 except VII (V:H = 1). Located on Figure 3. V, VI, and VIII are line drawings of seismic profiles (Elliott, 1989, figs. 4b, 5a, 5b, with general location only) converted to depth by 1 s TWT = 2 km as in Korsch et al., 1990b, fig. 3) and scaled with V:H = 5.

V. Elliott's (1989, fig. 4b) section extended 30 km eastward from Exon (1976, plate A, section CD). Abbreviations are AT—Arbroath Trough, L-BF—Leichhardt-Burunga Fault, LCT—Lyndon Caves Trough, RG—Roma Granite (Devonian-Carboniferous), THF—Timbury Hills Formation (pre-Carboniferous), WF—Wallumbilla Fault; also in VI and VIII, CV—Combarngo Volcanics (Late Carboniferous), J-K—Jurassic-Cretaceous, R—Triassic, P—Permian.

VI. From Elliott (1989, fig. 5a). The 1-km-thick Cabawin Conglomerate extends westward from the foot of the Leichhardt Fault. Abbreviations are CC—Cabawin Conglomerate, CV—Combarngo Volcanics, LF—Leichhardt Fault, LP—Late Permian, THF—Timbury Hills Formation, UF—Undulla Fault.

VIII. From Elliott (1989, fig. 5b), with basement from Exon (1976). Abbreviations are CV—Combarngo Volcanics or Kuttung Series (Late Carboniferous), M-GF—Moonie-Goondiwindi Fault, RG—Roma Granite.

Section curved proportional to earth radius. Detail of eastern part of Taroom Trough (TT) from Finlayson (1990, p. 171), Cecil Plains Syncline (CPS) (p. 172), and Ipswich Basin (IB) (p. 173). Additional abbreviations are AT—Arbroath Trough, BB—Beenleigh Block, BD—Brisbane detachment, B-MG—Burunga-Mooki Geosuture, C-MB—Clarence-Moreton Basin, CPS—Cecil Plains Syncline, EB—Eromanga Basin, ET—Esk Trough, FD—Foyleview detachment, FG—Foyleview Geosuture, GD-LS—Greenbank deep-layered sequence, IB—Ipswich Basin, KR—Kumbarilla Ridge, MF—Moonie Fault, MRV—Main Range volcanics, NR—Nebine Ridge, SB—Surat Basin, THF—Timbury Hills Formation, TM-CD—Texas midcrustal detachment, TT—Taroom Trough, WD—Westgate detachment, WIF—West Ipswich Fault.

VII'. Enlarged from VII, V:H = 4. From Korsch et al. (1989, fig. 8). Abbreviations are C-MB—Clarence-Moreton Basin, RS—rift sediment, WIF—West Ipswich Fault, SMA—South Moreton Anticline.

VII. Regional lithospheric section of eastern Australia, reproduced with permission from Finlayson (1990, fig. 5. Note reduced horizontal scale and V:H = 1.

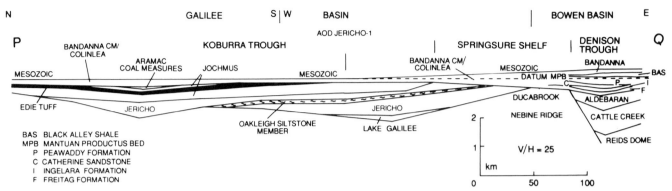

Figure 22. Cross-section PQ of the Galilee Basin and Springsure Shelf, located in Figure 3, showing the overstep at a low-angular unconformity of the Late Permian Colinlea Sandstone/Bandanna Formation on deformed Early Permian Aramac Coal Measures, Jochmus Formation, and Jericho Formation and the Early Carboniferous Ducabrook Formation of the Drummond Basin. Section north of AOD Jericho-1 from Gray and Swarbrick (1975, fig. 2), east of it from Gray (1976, fig. 5), as reproduced also by Evans (1980, figs. 4 and 5) but with a reduced vertical exaggeration of 25:1. The section is hung on a datum at the base of the Mesozoic (Triassic Mimosa Group or Jurassic Hutton Sandstone) north and west of Jericho-1, and on the Mantuan *Productus* bed east of Jericho-1.

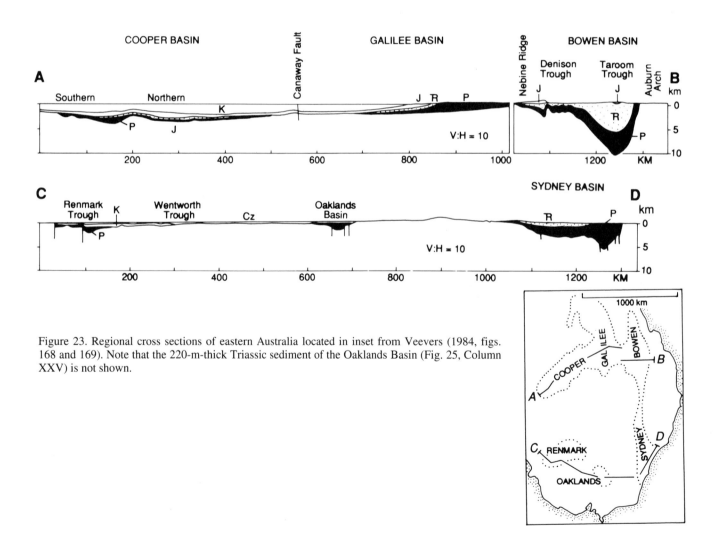

Figure 23. Regional cross sections of eastern Australia located in inset from Veevers (1984, figs. 168 and 169). Note that the 220-m-thick Triassic sediment of the Oaklands Basin (Fig. 25, Column XXV) is not shown.

Capella contains sporomorphs very tentatively identified as Stages 3 (Wilson, 1975; Edenborough, 1985), 4, and Lower 5 (Wilson, 1975). Coal occurs in stages 3 and Upper 4/Lower 5. This interval, outlined by dots in Figure 44 in Appendix 1 (Bowen, W central column), contains strata in the Karin and Wolfang (KW) outliers (Preston, 1985) and the Blair Athol (BA) outlier, dated as Upper 4a by Foster (1979). The Clermont outlier contains the *clarkei*-bed (Veevers et al., 1964). The Moorlands outlier (Sorby and Scott, 1988) contains palynomorphs that straddle Upper 5a/b (Price, 1983, p. 162).

Provenance. Moorlands data come from Sorby and Scott (1988), Blair Athol from Preston (1985). The Tiverton Formation is quartz-lithic, derived from the underlying Lizzie Creek Volcanics. The Calen Coal Measures are quartzose, according to Koppe (1975, p. 250).

Tectonics. The Carboniferous magmatic arc persists through the Urannah Complex (316 ± 15 Ma) to the 300 Ma Bulgonunna Volcanics and related granitoids that stretch to the end of the Carboniferous at 286 Ma. The remaining tectonic events are the same as those elsewhere in the Bowen Basin: extension 1 from 290 Ma to 268 Ma, basin sag 1 during transgression III, compression 1 marked by the prograding coastal plain of the Collinsville Coal Measures (regression IV), basin sag 2 starting with transgression V/regression V′ and continuing to the youngest preserved sediment of the Moolayember Formation until compression 2 (233–230 Ma) and succeeding extension 2.

Stage A: Volcanic-extension, 290–268 Ma. This stage followed the Late Carboniferous magmatic arc (Fig. 29A).

Reids Dome Beds (RDB, Fig. 29A; column XIII, Fig. 24). The isopachs (km) in the Denison Trough (Paten et al., 1979, fig. 5) are extended into the Capella area (Edenborough, 1985, p. 53) and through the Karin (not shown), Wolfang (W), and Blair Athol (BA) basins. The thickest penetrated RDB are 2.76+ km in Westgrove-3 (1), in the axis of the Westgrove Trough. The RDB overlie Devonian or older basement in the northwest, in Purbrook-1 (2) slate, presumably of the Timbury Hills Formation (Exon, 1976, plate 9), and elsewhere the Comet Volcanics, above which the Denison Trough succession is entirely sedimentary. The Denison Trough and the Arbroath Trough to the south constitute the only nonvolcanic rifts of this stage. Equivalents in the Galilee Basin west of the Denison Trough are the Joe Joe Group (CPi), 300–1,700 m of siliciclastics, with a 100-m-thick tuff, the Edie Tuff Member of the Jochmus Formation (Fig. 27). The eastern subcrop of the Joe Joe Group (CPi) is from Gray (1976, fig. 5) and Wells and Newhouse (1988, fig. 22) who supply subcrop data in subsequent figures.

Comet Volcanics (Fig. 24, column XIV). Three other wells (Warrinilla North-3 [3], Morella-1 [4], and AAO-7 Arcadia [5]) penetrated the RDB to recover mafic volcanics (of unknown thickness) similar to those found at the bottom of Comet-1 (6) (with a Stage 2 palynoflora, Kemp et al., 1977), Cooroorah-1 (7), and Glenhaughton-1 (8). We draw the west-

ern limit of these volcanics at the subsurface Merivale Fault south of 25°S; northward we draw the line 20 to 30 km west of Comet-1 (6) and Cooroorah-1 (7) to pass along the western outcrop of the Lizzie Creek Volcanics in the north.

Lizzie Creek Volcanics (v) and others (Fig. 24, column XV). These volcanics extend around the northern tip of the basin to occupy the eastern limb of the Nebo Synclinorium above the Urannah Complex mainly of Late Carboniferous (316 Ma) granitoid, itself intruded during this stage (281–265 Ma) by the Thunderbolt Granite (gth in Fig. 29A). On the east, the Urannah Complex and the Devonian-Carboniferous Connors Volcanics are overlain by the Carmila Beds (m), which continue south to 23°S. The Carmila Beds comprise 7,500 m of acid (rhyolitic, dacitic, andesitic) volcanics and volcaniclastic sediments that range from conglomerate to shale; they are entirely nonmarine except where they contain brachiopods, bryozoans, and crinoid stems (Day et al., 1983, p. 105) in the area around the Styx Basin. The nonmarine equivalent in the southern Bowen Basin and Gogango Overfolded Zone (GOZ) is the Camboon Andesite (n), 3 km or more of andesite and basalt flows, andesitic-dacitic welded tuff, sandstone and mudstone. In the Yarrol Block is the Youlambie Conglomerate (y), 2,000 m of conglomerate, sandstone, and siltstone, with bivalves, gastropods, and plants, and rhyolite and dacite. In the Gogango Overfolded Zone are the marine Narayen Beds (f), andesite and basalt, with conglomerate and farther north the Rannes Beds (r), thick marine flysch overlain by the Rookwood Volcanics (d), very thick pillow basalt, breccia, and siltstone. To the southeast, the Berserker Beds (t) comprise 3 km of acid to intermediate volcanics in the lower part overlain by volcaniclastic sediments with rich invertebrate faunas.

We draw the shoreline along the eastern edge of the Bowen Basin, with an embayment in the north that contains the marine parts of the Lizzie Creek Volcanics.

Stage B: Stage of marine sag on the platform and embryonic magmatic arc/foreland basin, 268–258 Ma (Fig. 29B). The initial (268 Ma) shoreline is drawn landward past the eroded edge of the marine formations at the base of the Back Creek Group (b), such as the Cattle Creek Formation on the west and in front of the coal measures at Blair Athol (a) and Calen (e) near Mackay (MA). The Aldebaran delta and the Collinsville coastal plain indicate depositional regression of the shoreline (broken barbed line) (Fielding et al., 1990, fig. 2; Gray, 1980, fig. 7) that accompanied folding in the west and thrusting of the thick sediments and volcanics of the Grantleigh Trough (Gogango Overfolded Zone) during the last 3 m.y. of this stage (compression 1, Figs. 18, 28). To the east, in the Yarrol Block, the marine Yarrol (ya) Formation and Owl Gully (q) Volcanics were replaced by the nonmarine Rainbow Creek Beds (k), which were deposited within the orogenic upland that emerged after the 261–258 Ma thrusting. The Ridgelands pluton (gri), the first orogenic pluton, was emplaced at 264 Ma.

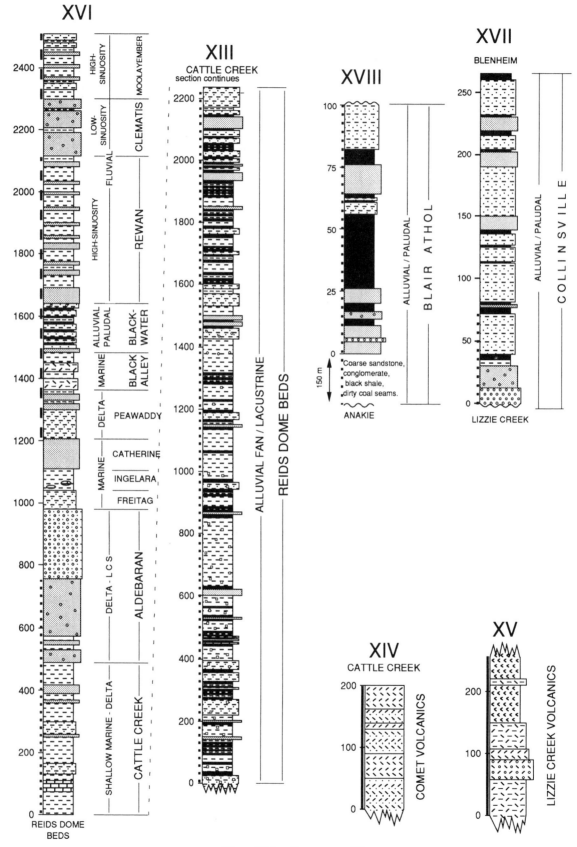

Figure 24 (caption on page 84).

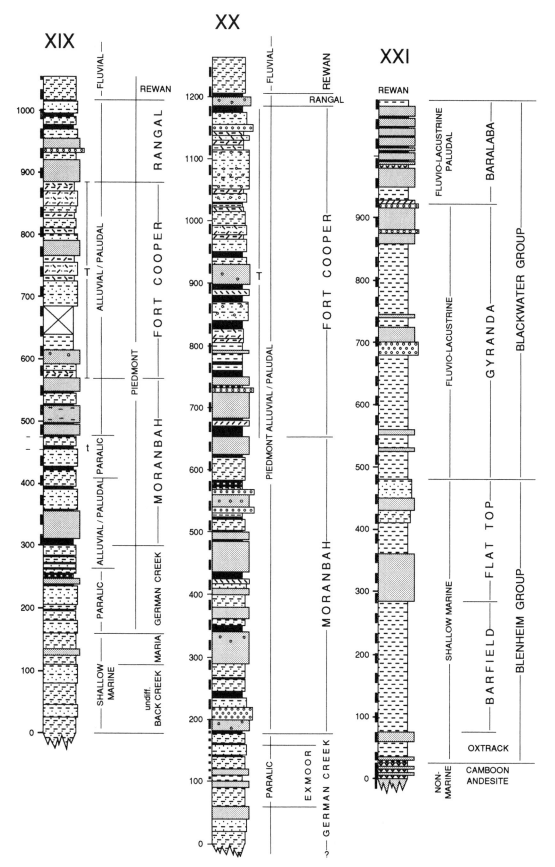

Figure 24 (caption on page 84).

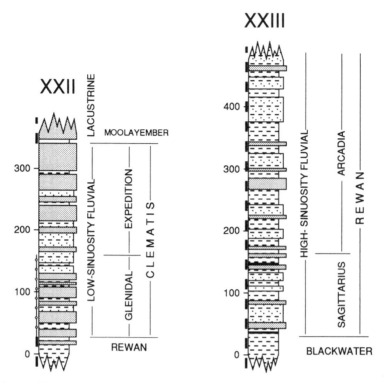

Figure 24 (on this page and preceding page spread). Representative stratigraphic columns of formations in the Bowen Basin (XIII–XXIII), showing lithology, provenance, and depositional environment. Thickness in meters; explanation is shown in Figure 10 preceding columns I and II.

XIII, XIV, XV. Extensional-volcanic stage A. XIII: from Reids Dome-1 (Mollan et al., 1969, pl. 8; Fielding et al., 1990). XIV: from Malone et al. (1969, pl. 4). XV: from Malone et al. (1969, pl. 2).

XVI: Marine (B), coal measure (C), and piedmont redbeds (D) stages: from Reids Dome-1 (Mollan et al., 1969, pl. 8A) and Fielding et al. (1990).

XVII, XVIII: stage B. XVII: Collinsville Coal Measures (within the Gebbie Subgroup) from Mengel (1975, fig. 2). XVIII: Blair Athol Coal Measures from Preston (1985, fig. 1).

XIX, XX: Coal measure stage C. XIX: Central: from Saraji-1 (Koppe, 1978, fig. 3). XX: northeast: from Exmoor-1 (Koppe, 1978, fig. 3).

XXI: Marine (B) and coal measure (C) stages from Banana-1 (Dear et al., 1971, fig. 9).

XXII, XXIII: Piedmont redbeds stage D. XXII: from Glenhaughton-1 (Jensen, 1975, fig. 11), XXIII: from Planet Warrinilla-1 (Jensen, 1975, fig. 10).

Stage C: Stage of orogenic piedmont coal/tuff: mature orogen/foreland basin, 258–250 Ma (Fig. 29C). The sea encroached eastward across the western flank of the Yarrol upland, and the Moah Creek Beds (s), 2 km of conglomeratic mudstone with blocks of Carboniferous oolitic limestone, were deposited at the foot of scarps. To the south, the older parts of the Rawbelle Batholith (ga, gd, gh, gy) were emplaced, and to the west near Taroom (T), the first orogenic tuff, in the Oxtrack Formation (256 Ma), was deposited. On the western side of the basin, the shoreline (Freitag Formation and basal Colinlea Sandstone) advanced (a) (Fielding et al., 1990, fig. 3) over the Aldebaran delta west of its 268 Ma position, and thereafter retreated (initial regression V' + VII) from an upland during deposition of the deltaic-marine or paralic quartzose upper Freitag Formation (regression V'), Catherine Sandstone and German Creek Coal Measures (regression VII).

Data from Staines and Koppe (1980, fig. 4c), Fielding et al. (1990, fig. 3), Paten et al. (1979, figs. 8, 9), Prouza and Park (1973), and Balfe et al. (1988). Fielding and McLoughlin (1992, figs. 5, 7) obtained a vector mean of 097° in distributary-channel crossbedded sandstone within coastal-plain craton-sourced sediment of the Freitag Formation southwest of Emerald (E), shown as b in Fig. 29C. The direction of regression (c) is from Paten et al. (1979, figs. 8, 9). Jensen (1971, p. 12, 15) obtained a grand vector mean of 075° of crossbed dip azimuths in the upper part of the German Creek Coal Measures (e) in the Cherwell Range area southwest of Moranbah. P. J. Conaghan (personal observation, 1980) measured a grand vector mean crossbed dip azimuth of 100° (f) in the German Creek Coal Measures in road and railway cuttings in the Peak Downs/Saraji mine areas, and Falkner and Fielding (1990, fig. 3a) measured 090° at the Gregory Mine

(g). In the equivalent Crocker Formation near Comet (d), Jensen (1971, p. 12) measured a vector mean of 054°. In the east, thick shelf mud and coarse turbidites of the Barfield Formation were succeeded by the first pulse of coarse tuffaceous volcanolithic sandstone (Flat Top Formation) in fandeltaic complexes derived from the emergent volcanic arc (Fielding et al., 1990, p. 22, 23). In the Saraji-Peak Downs area, the interfingering facies boundary between the German Creek Coal Measures and the Macmillan Formation on the south and the Moranbah Coal Measures on the north (Fig. 26), tied by tuff bands, defines the confluence of quartzose sediment from the west (e, f) and orogenic volcanolithic sediment from the east, indicated by Conaghan's (personal observation, 1980) measured grand mean crossbed dip azimuth of 252° at the Peak Downs mine (h). A final transgression (X) carried the shoreline into the southern Galilee Basin at the base of the Peawaddy Formation to its farthest west point (Jensen, 1971, p. 15) at 146°10′E (Gray, 1976, figs. 3, 5; Wells and Newhouse, 1988, figs. 27, 28; Fig. 41C) and at the base of the overlying Black Alley Shale with the P3c acritarch swarm (Evans, 1980, p. 302). The rest of the Black Alley Shale was deposited in a lake during the final marine regression XI, which culminated in the tuffaceous coal-measure deposition of the Blackwater Group (Upper Stage 5—*P. crenulata* palynological zone or 252–250 Ma), equivalent to the Newcastle Coal Measures of the Sydney Basin and tuffaceous stony coal facies of the Black Jack Formation of the Gunnedah Basin. At 250 Ma, thrusting and uplift in the east (compression 1′ in Fig. 18) and folding in the west (compression 1′ in Fig. 19) augmented the depositional regression. Alluvial-fan rudites were deposited northeast of Taroom (T) (Jensen 1971, p. 9; Fielding et al., 1990, p. 23) and west of Mackay (MA) in the Fort Cooper Coal Measures (Jensen 1975, p. 11) "where the Bowen Basin was probably cut off from the sea by uplift on the Eungella-Cracow Mobile Belt" (Jensen, 1971, p. 9; Dickins and Malone, 1973, p. 66, 67). In the north, crossbed dip azimuths in the Fair Hill and Burngrove Formations and Rangal Coal Measures are southwestward (i) and south-southeastward (j) (Jensen, 1975, fig. 7). In the Blackwater (BL) area, Jensen (1975, fig. 7) and Burgis (1975, fig. 26) measured a crossbed dip azimuth of west-northwest, and P. J. Conaghan (personal observation, 1980) estimated a westerly delta-lobe progradation. From the Blackwater Group, Jensen (1975, fig. 7) measured a northerly azimuth (m) near Springsure (S) and a northwesterly azimuth (n) in the south. The broken arrow from the Baralaba Coal Measures at the Moura-Kianga mine (o) is the azimuth of the estimated westerly regional dispersal of sediment in a lacustrine delta complex (P. J. Conaghan, personal observation, 1980, 1993) shown also by Fielding et al., 1990, fig. 5. In this area, Jensen (1975, fig. 7) measured a north-northeast dip azimuth, and Flood and Brady (1985, fig. 7) found a range of intra-set crossbed dip azimuths within sigmoidal crossbeds from north-northeast (as in Jensen, 1975) to southward (thin arrows within a semicircle). We interpret the

pattern of azimuths as manifesting recurved easterly progradation of delta lobes back into the foredeep in front of the thrusted upland of the Eungella-Cracow Mobile Belt. Flow vector p is from Fielding et al. (1990, fig. 5).

The residual physiographic basin was finally infilled by prograding deltaic and alluvial systems directed southwards (Bandanna Formation, Rangal Coal Measures) and westwards (Baralaba Coal Measures), and in the central-western part of the basin the Bandanna Formation prograded westward across the Springsure Shelf towards the Galilee Basin (Fielding et al., 1990, p. 23).

We add another direction. In the Taroom Trough region south of 24°S, northwesterly regional sediment dispersal is defined by the paleoflow information at m, n, and o. And this paleoflow united with the southerly flow into the westward progradation north of the folded area of the Denison Trough across the Springsure Shelf into the Galilee Basin (Conaghan et al., 1982, fig. 7A; Wells and Newhouse, 1988, figs. 7, 28) over the Peawaddy marine gulf (*see* Fig. 41C in a later section).

Stage D: Orogenic piedmont redbeds of the foreland basin 250–241 Ma (Fig. 29D). According to Fielding et al. (1990, p. 23),

In latest Permian times this uplift [Cracow-Eungella Mobile Belt], coupled with an inferred climatic change, gave rise to the accumulation of reddened alluvial sediments throughout the basin, devoid of coal (Rewan Group). Unconformity separates the Rewan Group from the underlying Blackwater Group on the basin margins, while in the basin centre the relationship is conformable.

We add that the Rewan Group (Rr) and succeeding Clematis Group (Re) are devoid of tuff also. The southern part of the Eungella-Cracow Mobile Belt was intruded by the main phase of the Rawbelle Batholith (gc, gd, gg, gr, gt, gw) and plutons to the north (go, gl) but we infer that tuff was absent in the basin because the plutons did not reach the surface. The Eungella-Cracow Mobile Belt shed even larger volumes of sediment, including the rudite of the Cabawin Formation in the Taroom Trough between 27°S and 28°S (Fig. 33D) (Exon, 1976, fig. 15).

All the mean measured crossbed dip azimuths are from the upper part of the Rewan Group, called the Arcadia Formation (Jensen, 1975, fig. 77) except Kassan and Fielding's (1991, fig. 5, p. 155) northeastward azimuth from the Arcadia Formation at Lonesome National Park (a in Fig. 29D), Bastian's (1965a, fig. 12) from Taroom (T), and ours from 23°S. The regional drainage pattern maintains that of Stage C (Conaghan et al., 1982, fig. 7B).

Stage E: Cratonic sand sheet of the foreland basin, 241–235 Ma (Fig. 29E). As seen in the regional map (Fig. 41E in a later section), a drainage divide (Foreswell II) in the interval 240–237 Ma arose between the Bowen foreland basin and the Galilee epicratonic basin, as indicated by the divergent sediment paleoflow pattern in the lower part of the Clematis Group, called the Glenidal Formation (240–237 Ma) (Jensen,

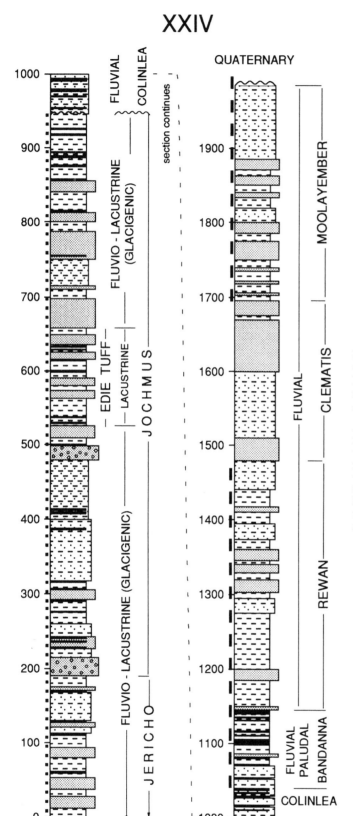

Figure 25. Representative stratigraphic columns of formations in the Galilee Basin (XXIV), Sub-Murray Basin (XXV), and Tasmania Basin (XXVI), showing lithology, provenance, and depositional environment. Thickness in meters; explanation in Figure 10 preceding columns I and II. Galilee Basin: XXIV: all stages, from Lake Galilee-1 (Pemberton, 1965) and Gray and Swarbrick (1975). Sub-Murray Basin, Ovens Graben: XXV: Jerilderie Formation and Coorabin Coal Measures in Waloona-1, from Yoo (1982, fig. 3); Urana Formation in Jerilderie-1 from O'Brien (1986, fig. 14). Tasmania Basin: XXVI: Base to top of upper marine from Banks (1962, fig. 26d [Maydena area], and 26c [Hobart]); Cygnet Coal Measures and Midlands Formation from Forsyth (1984, fig. 15). Base of Ross Formation to top of Tiers Formation from Collinson et al. (1990, fig. 3, Poatina) and Mount Nicholas Coal Measures (fig. 3, St Marys).

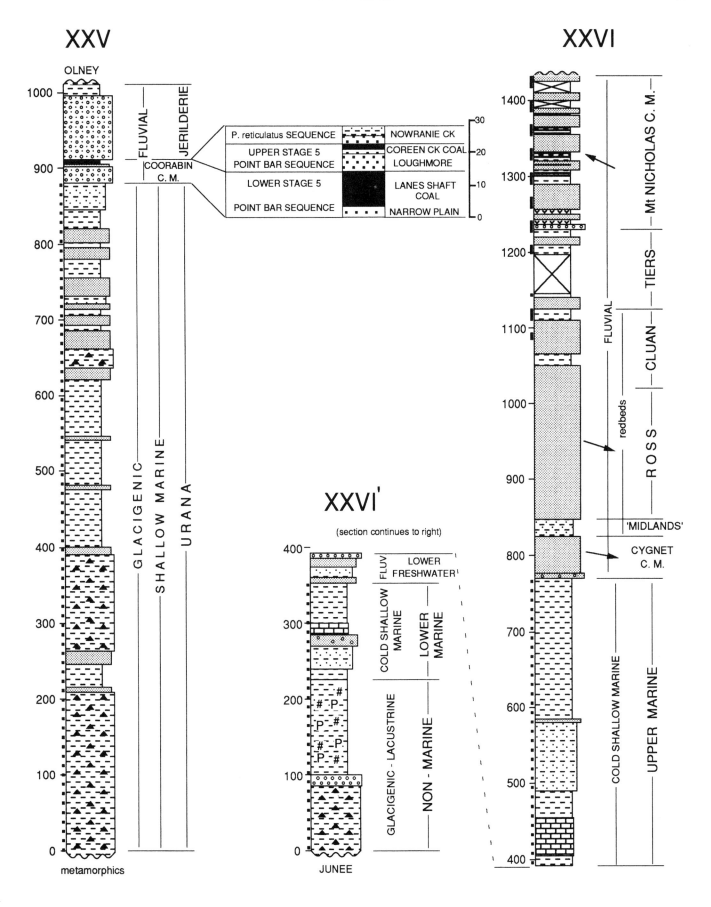

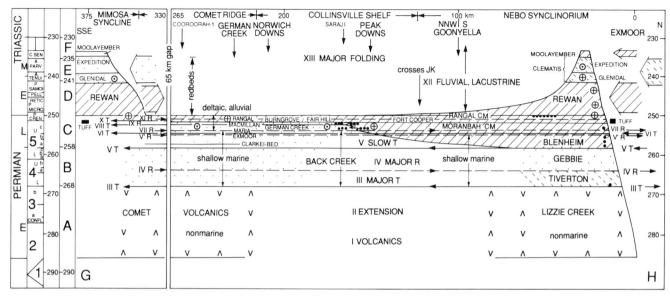

Figure 26. Time-space diagram GH, location is shown on Figure 17, symbols as in Figure 18.

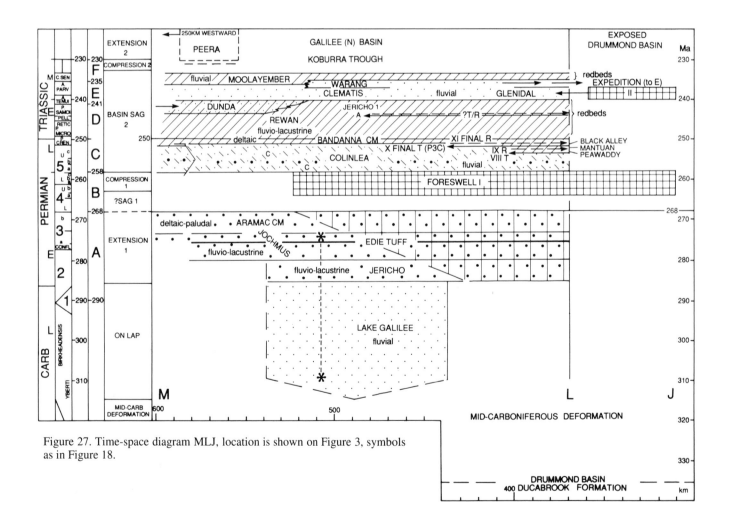

Figure 27. Time-space diagram MLJ, location is shown on Figure 3, symbols as in Figure 18.

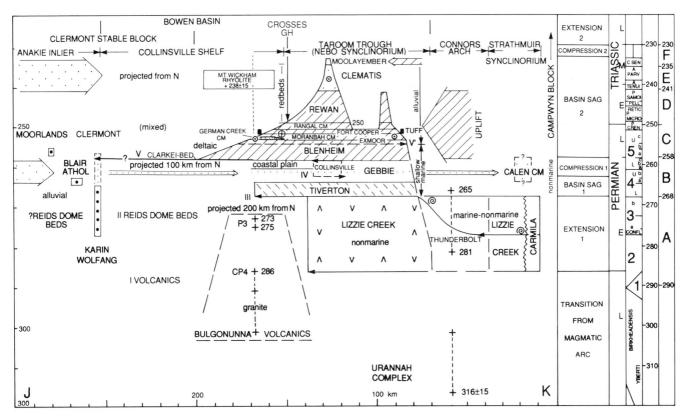

Figure 28. Time-space diagram JK, location is shown on Figure 3, symbols as in Figure 18. The quartz-lithic petrofacies of the Moorlands, Blair Athol, and Karin-Wolfang basins is denoted by big dots.

1975, fig. 78, vector mean trend) (Fig. 26). Another change from the pattern of Stages C and D is that the axial flow of the Bowen Basin is wholly southerly, presumably by blockage of the westerly drainage by Foreswell II along the line of the Nebine Ridge (thin arrows). The upper part of the Clematis Group, called the Expedition Sandstone, reversed paleoflow in the Galilee Basin (Fig. 41E) to cross the former Foreswell II (wide arrows) (Jensen, 1975, fig. 80, moving average trend, upper Expedition Sandstone) and join the southerly axial drainage of the Bowen Basin at about 25°S. To the north, the drainage divide is indicated by influx of quartzose sand (wide broken arrows) (Bastian, 1965b). The lithic (L), eastern, provenance of the Expedition Sandstone is from Bastian (1965a, 1965b), Jensen (1975), and Olgers et al. (1964).

Stage F: Orogenic paralic sediment, 235–230 Ma (Fig. 29F). Measured crossbed dip azimuths in the upper Moolayember Formation are from Alcock (1970). Paleoflow vectors from the major orogenic source on the east and from the minor cratonic source in the west are from Bastian (1965b). The axial drainage (Fielding et al., 1990, fig. 6) is constrained in the west by sand-isolith data (Wells and Newhouse, 1988, fig. 19). In addition to the westerly thinning and fining volcanolithic conglomerate, tuff was deposited in the Mimosa Syncline (Mollan et al., 1972, p. 35).

The Cracow-Eungella Mobile Belt was terminally thrusted

over the eastern half of the Bowen Basin, including the Nebo Synclinorium and the Dawson Folded Zone (Dickins and Malone, 1973) (Fig. 20) at the same time as the Denison Trough was folded. The deformation affected all up to the youngest in the basin, i.e., up to 233 Ma, and was followed by deposition of Carnian (230 Ma) sediment and volcanics in southeastern Queensland.

Stage G: Rifting of the orogen, 230–200 Ma (Fig. 29G). The deformed and uplifted area relaxed and accumulated first the ?230 Ma Muncon Volcanics (Ru) and overlying Callide Coal Measures (Rc) and then the quartzose sheet of the 200 Ma Precipice Sandstone (Jp) (Martin, 1981, fig. 4) and Razorback Sandstone (Jr) that unconformably overlie the folded and eroded terrain.

Southeastern Queensland

A map of southeastern Queensland (Fig. 30, p. 98), from 25°S to 28°S, bridges the gap between the Bowen Basin map (Fig. 17) and the New South Wales map (Fig. 7). The map and letter symbols were drawn from the colored map 1 in Finlayson (1990). Table 7 lists the structural elements, their ages, constituent formations, plutons, and metamorphics, and source of information, mainly Day et al. (1983), supported by Aitchison (1988), Briggs (1991), Elliott (1989), Exon (1976), Murray (1986, 1990a), Roberts (1985), and the information in

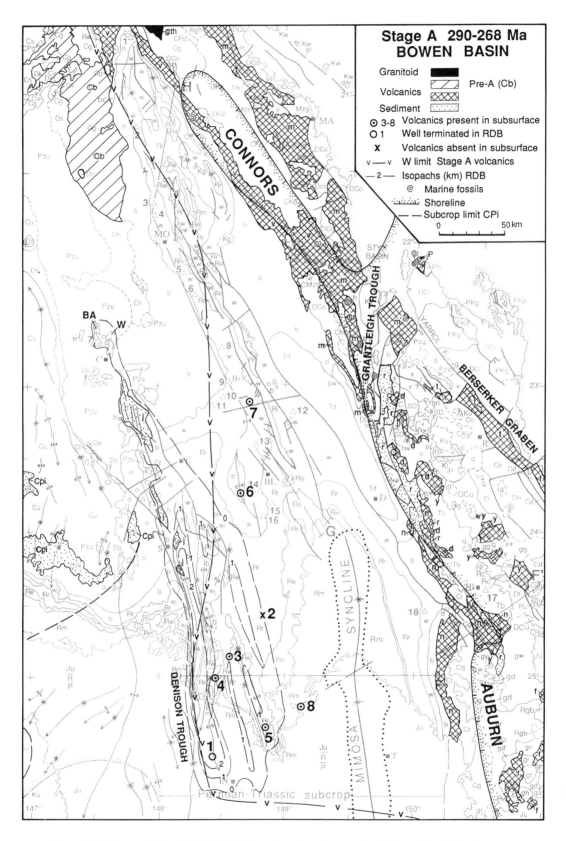

Figure 29A. Bowen Basin, 290–268 Ma: Stage A, volcanic-extension. This and the following stage maps are drawn on the base map shown in Figure 17.

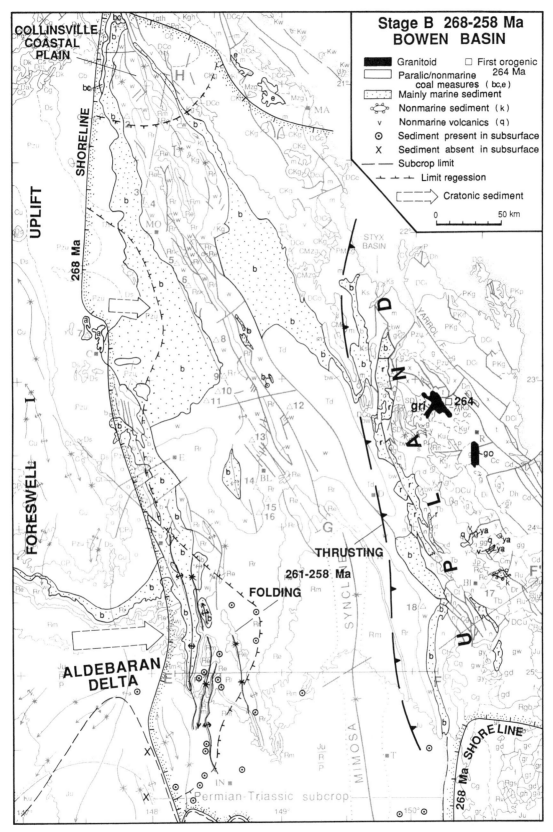

Figure 29B. Bowen Basin, 268–258 Ma: Stage B, marine sag on the platform and embryonic magmatic arc/foreland basin. Surat Basin subcrop from Exon (1976). Limit of Aldebaran regression from Gray (1976, fig. 7).

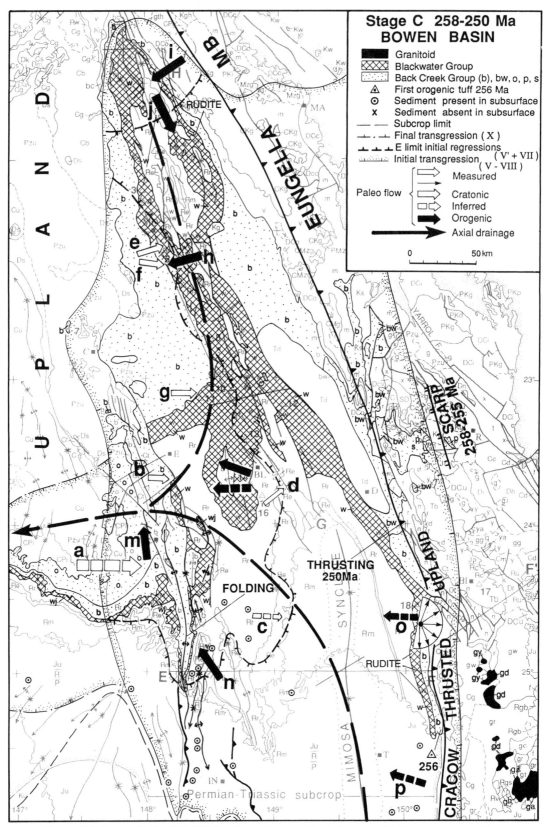

Figure 29C. Bowen Basin, 258–250 Ma: Stage C, orogenic piedmont coal/tuff: mature orogen/fore-land basin.

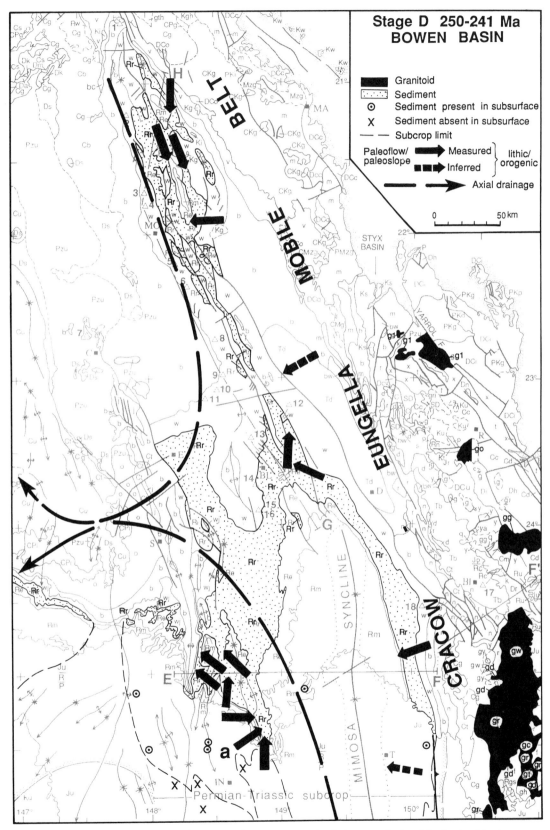

Figure 29D. Bowen Basin, 250–241 Ma: Stage D, orogenic piedmont redbeds of the foreland basin.

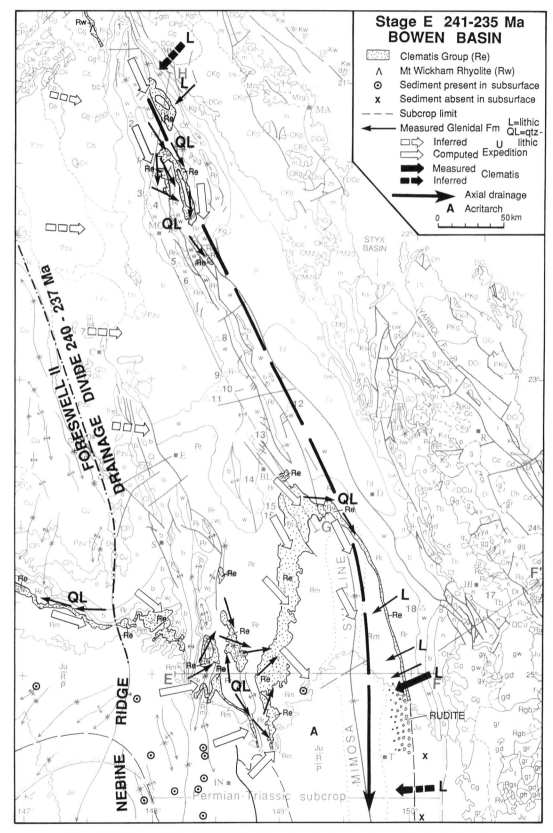

Figure 29E. Bowen Basin, 241–235 Ma: Stage E, cratonic sand sheet of the foreland basin.

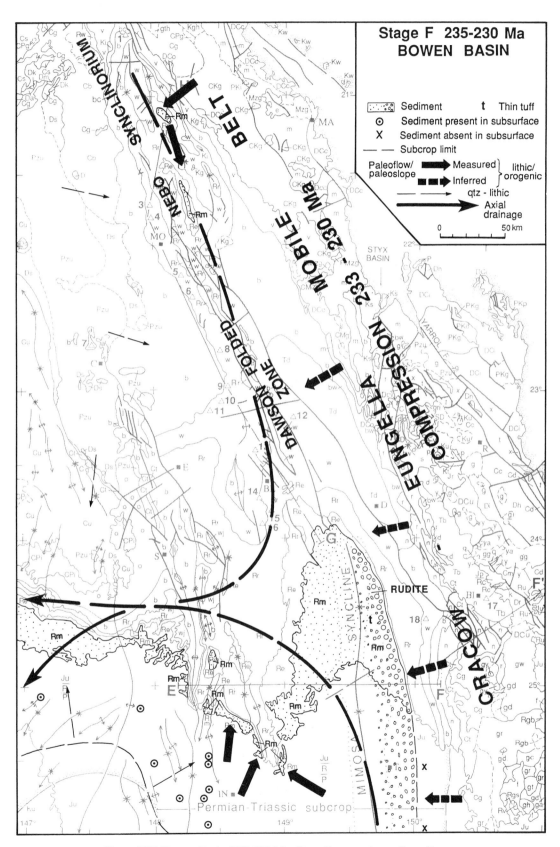

Figure 29F. Bowen Basin, 235–230 Ma: Stage F, orogenic paralic sediment.

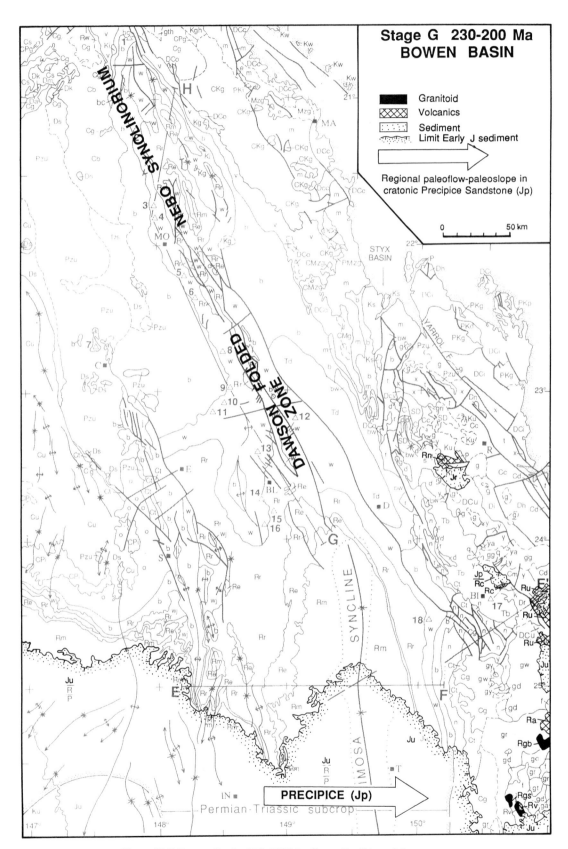

Figure 29G. Bowen Basin, 230–200 Ma: Stage G, rifting of the orogen.

Table 4 and Fig. 18. Table 8 lists granitoids and other igneous rocks and their ages, again with the principal source being Day et al. (1983), augmented by Murray (1986) and Webb (1981), and supported by Cranfield and Murray (1989), Cranfield et al. (1976), Ellis (1968), Evernden and Richards (1962), Green and Webb (1974), Horton (1972), Mathison (1967), Murphy et al. (1976), Webb and McDougall (1967, 1968), Webb and McNaughton (1978), and Whitaker et al. (1974).

Time relations: southeastern Queensland (Figs. 31, 32). The location of the diagram shown in Figure 31 is shown in Figure 30 by XX′–ZZ′. The locations of the diagrams in Figure 32 are shown by VV′–V″–V‴–WW′ in Figure 30, drawn from all the sources given above. Finally, the interpretation is given in the stage maps, Figures 33A–G. Information for the area north of 26°S comes from the Bowen Basin maps (Figs. 29A–G). Cross-sections V and VI of the Surat/Bowen Basin are given in Figure 21. The Demon Fault is traced tentatively into Southeastern Queensland from New South Wales.

Gympie Province or Composite Terrane. The Gympie Province (Murray, 1990a, 1990b) or Composite Terrane (Cranfield, 1990) lies east of the Yarrol Province along a boundary marked by the North Pine Fault, the Western Border Fault of the Esk Trough, and northward of 26°S the meridian of 151°45′E. The easternmost exposed strip of the Gympie Composite Terrane, between Gympie and Gigoomgan is a structural domain in which the Gympie Group (Pg in Fig. 31) units are "generally broadly folded, not regionally metamorphosed, and lack a slaty cleavage." In contrast, the undivided Gympie Group west of Gigoomgan is "tightly folded, regionally metamorphosed [lower greenschist facies], and has a well defined slaty cleavage" (Cranfield, 1990, p. 99). According to Murray (1990a, p. 247),

The [conformable] Permian to Early Triassic Gympie Group [Pg; abbreviations are for Fig. 31], Keefton [Rk] and Brooweena Formations, and Kin Kin [Rk] beds form a probable exotic terrane because they differ markedly in lithological sequence and deformational history from the remainder of the NEFB, and because basal volcanics represent an immature tholeiitic stage of island arc development. Docking of this [Kin Kin] terrane must have occurred in mid-Triassic time.

We regard the zones of shearing and intense deformation in the Gympie Composite Terrane as lines of weakness along which the stress produced in docking was relieved.

Folding of the Kin Kin beds predates the covering ~213 Ma North Arm Volcanics (Rn; abbreviations are for Fig. 31) (Murray, 1990a, p. 254) and accompanied the intrusion of the mid-Triassic (240 Ma) Goomboorian Diorite (Rg9),as shown in Murray (1990a, fig. 3; 1990b, fig. 2). The docking of the Kin Kin terrane probably induced the first phase of transpressive movement that generated the Middle Triassic (240–233 Ma) Esk Trough with its fill of the Toogoolawah Group (Rt). A second phase of transpressive movement at the end of the Middle Triassic (233–230 Ma) deformed the Toogoolawah

Group before extension led to the deposition of the adjacent Carnian Ipswich Coal Measures. This mid-Triassic change from compressive to extensional deformation is seen in the northern Canning Basin of Western Australia, the Cape Fold Belt and Karoo Basin of South Africa, and the Gondwana basins of peninsular India. The change to Carnian (230 Ma) coal measures, as from the barren Toogoolawah Group to the Ipswich Coal Measures, is seen also at Leigh Creek in South Australia, in Tasmania, in South Africa (Veevers, 1990b, 1990c), and in peninsular India. Veevers (1990a) characterized this event as the Carnian singularity in Pangean history: the change from its greatest aggregation after the collision of Cimmeria with Asia (Indosinian orogeny) to incipient dispersal with the onset of rifting that prefigured the Atlantic and Indian Oceans.

Demon Fault. Wellman (1990, p. 31) traced the Demon Fault from is exposed southern part in New South Wales northward along 152°07′E to the latitude of Kingaroy (K). "The gravity and magnetic anomalies show that this fault, and its northern continuation, has a nearly northern strike, and 22 km of dextral displacement" between offsets of the inferred Woolomin association at 28°S and between strong magnetic highs at 27°45′S and 26°45′S, depicted as parallel lines in Figure 30. "The movement must have occurred when the igneous rocks in the south were displaced [dextrally] by about 23 km" (Fig. 13). The 23 km of dextral offset was measured from the boundary between the Coombadjha Volcanic Complex (McPhie and Fergusson, 1983) and the Dandahra pluton (*see* 10 in Fig. 51, Appendix 3) on the east with dates of 233.7 Ma and 235.1 Ma and the Billyrimba pluton (11) on the west with dates of 233.0 Ma, 233.8 Ma, and 234.7 Ma (Appendix 3, Fig. 51), with a pooled mean of 234.0 Ma. Movement along this southern part of the fault was therefore younger than 234 Ma and older than the sedimentary cover of the Clarence-Moreton Basin, from the Marburg Formation (200 Ma) or Bundamba Group (213 Ma). "The gravity and magnetic data show that the troughs below the Clarence-Moreton Basin sequences were displaced by the fault, so the movement was likely to be after the Esk Trough activity [about 230 Ma] and before the Clarence-Moreton Basin sedimentation" [about 213 Ma].

A closer upper limit is found along the northern continuation of the fault. From the northernmost point of the meridional Demon Fault at the offset of a geophysical anomaly southeast of Kingaroy (Wellman, 1990), we trace the Demon Fault north northwestward (dotted line, Fig. 30) into the Runnymede Fault (RF in Fig. 30), northward past its junction with the Stephenton Fault (SF) to another north-northwestward part that passes through 26°S, 152°E. The fault resumes its meridional trend at latitude 25°43′S, is itself offset obliquely 7 km to the left or west by the Perry Fault (PF), and then continues northward along 151°48′E as the Venus Fault (VF), before disappearing under the stitching Hogback pluton (Rh) (*see* fig. 1 of Cranfield and Murray, 1989). The VF is not seen north of the Hogback pluton in the Bundaberg 1:250,000

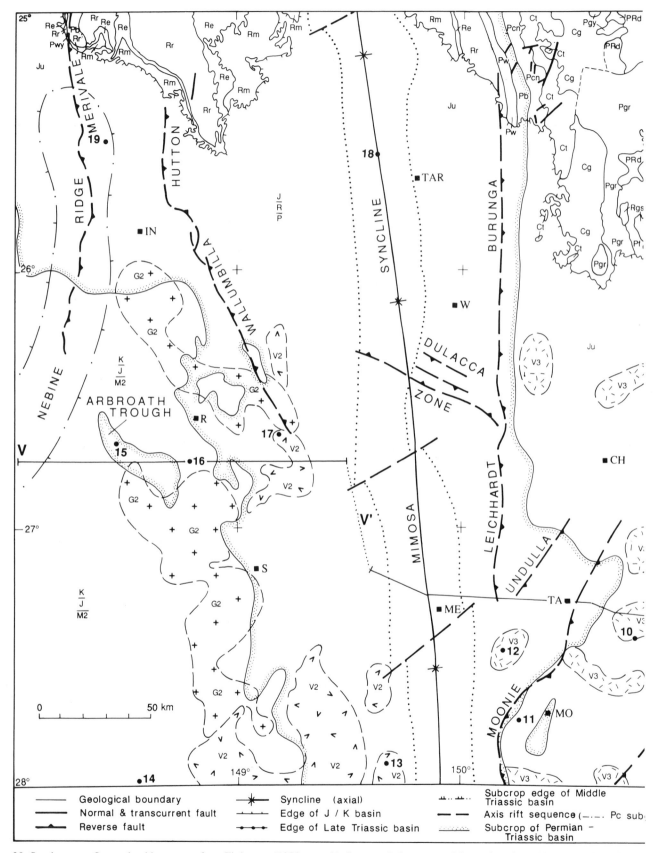

Figure 30. Southeastern Queensland base map, from Finlayson (1990, map 1). Structural elements and formations in Table 7, age of granitoids in Table 8. Key on page following this page spread.

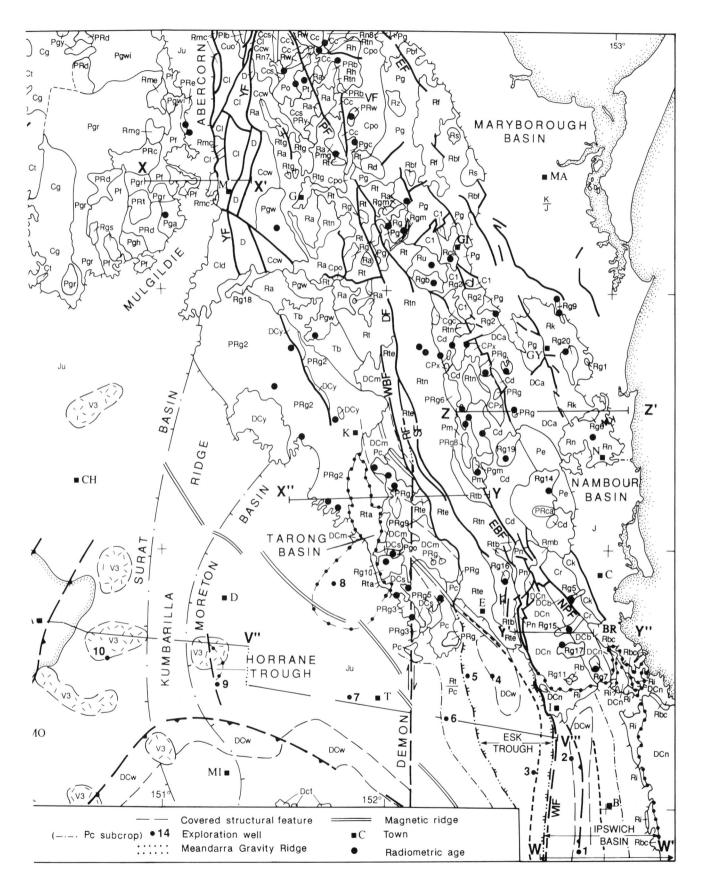

Key to Fig. 30, map of southeastern Queensland. Groups listed in CAPITALS. g = undated or unnamed granitoid.

Stage unknown
CPx	serpentinite (? A and/or Pre-A)

Volcanics
PRa	undiff. volcs

Granitoids
Rg	granitoid, undiff.
Rg6	Eerwahdale
Rg11	Karana
PRg	granitoid, undiff.
PRg8	Kimbala

Post-G (<200 Ma)
Tb	Tertiary basalt
K	Cretaceous undiff.
Ju	Jurassic undiff.

Stage G (230-200 Ma)
Ri	IPSWICH CM
Rta	Tarong

Volcanics
Ra	Aranbanga
Rbc	Brisbane Tuff, Chillingham
Ri	IPSWICH CM includes Sugars Basalt and other volcanics
Rmb	Mount Byron
Rme	Mt Eagle
Rn	North Arm (?)

Granitoids
Rd	Degilbo
Rf	Broomfield
Rg5	Dayboro
Rg7	Enoggera
Rg14	Neurum
Rg15	Mt Samson
Rg16	Somerset Dam

Rg17	Samford
Rg18	Toondahra
Rg19	Tungi Ck
Rg20	Woondum
Rgm	Mungore
Rgs	Mt Saul
Rh	Hogback
Rmg	Boolgal
Rs	Musket Flat
Ru	Boogooramunya
Rw	Wonbah
Rz	Tawah
PRb	Briggs
PRy	Yenda

Stage F (235-230 Ma)
Rm	MOOLAYEMBER
Rt	TOOGOOLAWAH (E,F)
Rte	Esk (E,F)

Volcanics
Rtn	Neara (D-F)

Granitoids
Rg2	Station Ck (E, F)
Rg10	Djuan
Rn7	Greenbank/Nour Nour (E,F)
Rn8	New Moonta
Rgb	Boonara
Rca	Calgoa

Stage E (241-235 Ma)
Re	CLEMATIS
Rmc	Cynthia
Rt	TOOGOOLAWAH (E,F)
Rtb	Bryden
Rte	Esk (E,F)
Rtg	Gayndah

Volcanics

Rtn	Neara (D-F)

Granitoids
Rg1	undiff.
Rg2	Station Ck (E,F)
Rg9	Goomboorian
Rn7	Greenbank/Nour Nour (E,F)
PRg6	Kingaham

Stage D (250-241 Ma)
Rbf	Brooweena
Rk	Kin Kin/ Keefton/Traveston
Rr	REWAN

Subsurface only: Cabawin
Pg	GYMPIE (A-D)

Volcanics
Rb	Brookfield (C,D)
Rtn	Neara (D-F)
PRca	Bellthorpe

Granitoids
PRd	Delubra (C,D)
PRe	Eidsvold
PRg	undiff.
PRg2	younger Boondooma
PRg3	Crows Nest
PRg5	Eskdale
PRg6	Kingaham Ck
PRg9	Taromeo (B-D)
PRw	Wateranga
PRc	Coonambula
Pgr	Rawbelle batholith, undiff.
PRt	Cheltenham Creek
Pgwi	Wingfield

Stage C (258-250 Ma)
Pg	GYMPIE (A-D)
Pb	BACK CREEK (B,C)
Pm	Marumba

Pw	BLACKWATER
Pwy	Rangal/Bandanna/ Baralaba

Volcanics
Pc	CRESSBROOK CREEK
Pn	Northbrook
Rb	Brookfield (C,D)

Granitoids
PRe	Eidsvold
PRg1	undiff.
PRg3	Crows Nest
PRg9	Taromeo (B-D)
Pga	Cadarga Creek
PRd	Delubra (C,D)
Pgh	Hawkwood
Pgm	Monsildale
Pgy	Crystal Vale
Pt	Tenningering

Stage B (268-258 Ma)
Pb	BACK CREEK (B,C)
Pg	GYMPIE (A-D)

Granitoids
PRg2	older Boondooma
PRg9	Taromeo (B-D)
Pgc	Chowey (A,B)
Pgo	Woolshed Mt
Pgw	Wigton
Pmg	Mingo
Po	Wolca

Stage A (290-268 Ma)
Pg	GYMPIE (A-D)
Plb	Burnett

Subsurface only: Arbroath, Reids Dome

Volcanics
Pe	Cedarton
Pcn	Camboon

Pf	Narayen/Nogo
Pg	GYMPIE

Granitoids
Pgc	Chowey (A,B)

Pre-A (>290 Ma)
C1	undiff.
Cc	CURTIS ISLAND
Ccs	Shoalwater
Ccw	Wandilla
Cd	Booloumba/Durundar
Ck	Kurwongbah
Cl	CASWELL CK/ CRANA
Cld	Derrarabungy
Cpo	Good Night
Cr	Rocksberg
Cuo	BOILING CREEK
DCa	Amamoor
DCb	Bunya
DCm	Maronghi Ck
DCn	Neranleigh-Fernvale
DCs	Sugarloaf
DCt	Texas
DCw	Woolomin
DCy	schist, gneiss
D	undiff.
M2	Timbury Hills

Volcanics
V2	Combarngo
Ct	Torsdale
V3	Kuttung

Granitoids
Cg	granitoid, undiff.
Cgc	Claddagh, Gallangowan
?CPx	Serpentinite
G2	Roma

Wells
1	Tamrookum Creek	11	Moonie -1
2	The Overflow	12	Cabawin-1
3	Boonah	13	Flinton
4	Lockrose	14	St George
5	Baylam	15	Arbroath
6	Ropely	16	Maffra
7	Ipswich 19	17	Wallumbilla South
8	Ipswich 14	18	Taroom - 3
9	Cecil Plains	19	Westgrove
10	Kumbarilla		

Towns
B	Beaudesert	MA	Maryborough
BR	Brisbane	ME	Meandarra
C	Caboolture	MI	Milmerran
CH	Chinchilla	MO	Moonie
D	Dalby	N	Nambour
E	Esk	R	Roma
G	Gayndah	S	Surat
GI	Gilgoomgan	T	Toowoomba
GY	Gympie	TA	Tara
I	Ipswich	TAR	Taroom
K	Kingaroy	W	Wandoan
M	Mundubbera		

Faults
DF	Demon	PF	Perry
EF	Electra	RF	Runnymede
EBF	Eastern Border	SF	Stephenton
WBF	Western "	VF	Venus
NPF	North Pine	WIF	West Ipswich
		YF	Yarrol

Sheet area (Ellis and Whitaker, 1976). Either the VF dies out in this area or it was not found in the original mapping, as its continuation north of the Perry Fault was not mapped (Ellis, 1968) until later (Cranfield and Murray, 1989).

Tasmania Basin

Pre-Gondwanan history. According to Williams (1989, p. 468), two contrasting regions, the Western and Eastern Tasmania Terranes of pre-Carboniferous (Late Cambrian through Devonian) folded rocks and middle Paleozoic granitoids are believed to have been juxtaposed during the Carboniferous at a north-northwest trending dislocation called the Tamar Fracture System (Fig. 34, p. 117). Concealed today by post-Devonian cover rocks, the Tamar Fracture System is inferred to be approximated (a) by an underground zone of maximum electrical conductivity between Launceston and Maria Island (stippled outlines) or (b) by the axis of thickest Early Permian Lower Parmeener Supergroup (solid line, from Fig. 37). Regardless of its precise location beneath the Tasmania Basin, the Tamar Fracture System marks a profound boundary between a western column of rocks that extends through Precambrian metamorphics and Ordovician to Early Devonian marine shelf deposits intruded by Late Devonian granitoids and an eastern column, above an unknown base, of Early Ordovician and Early Devonian continental slope, rise, and bathyal basin deposits (Powell et al., 1993) covered by Middle Devonian rhyolitic flows, all intruded by Late Devonian grani-

toids (Banks and Baillie, 1989, p. 182, 183). In dividing an apparently exclusively Phanerozoic terrane from a Precambrian and Phanerozoic terrane, the Tamar Fracture System constitutes part of the Tasman Line (Veevers, 1984, p. 94–96).

Gondwanan (Parmeener) time relations (Fig. 35, Table 9). The time-space diagram (Fig. 35) was compiled from information given in Clarke and Forsyth (1989, figs. 8.1, 8.10, 8.15) set against the palynological and faunal zones shown in Figure 44 (Appendix 1). A stratigraphic column (XXVI) is given in Figure 25 and a cross section (XIV) in Figure 15.

Environment and provenance. The Parmeener Supergroup is divided into two parts: (1) a lower succession of lacustrine shale and associated glacigenic mixtite followed by marine sediment (with ubiquitous dropstones and, at the base, the alga *Tasmanites*) split by a "Lower Freshwater" interval of coal measures; and (2) an upper succession of four units of exclusively nonmarine coal measures split by barren measures containing rare redbeds.

Three provenance groups are recognized: (1) quartzolithic, derived from local cratonic and immediately older glacigenic sources of the immature feldspathic to lithic wackes of the marine parts of the Lower Parmeener Supergroup and of the glacigenic shale and mixtite of the basal Parmeener, additionally supplied from distant (Antarctic) basement floor and nunataks; (2) quartzose, derived from western (including Antarctic) cratonic sources; and (3) volcano-lithic, derived from eastern orogenic sources of volcanogenic sediment, including tuff and basalt.

This information is put together in the following series of paleogeographic maps that correspond with the stages found from the rest of eastern Australia.

Eastern Australia Stage A (platform extension), 290–268 Ma (Fig. 36).

Rocks of the Parmeener Supergroup rest unconformably on Late Devonian granites and older folded rocks on a surface with a relief of about 1000 m. Highly metamorphosed Precambrian rocks of the Tyennan region and the folded Devonian rocks and granites of northeastern Tasmania stood up as positive features while the metamorphosed rocks of the Arthur Lineament, the faulted Linda Disturbance and the Tamar Fracture System [Fig. 34] were preferentially eroded by glaciers and formed low lying areas. Early deposition of glacigenic rocks [up to 510-m-thick] followed by carbonaceous muds then fossiliferous beds essentially filled the earliest hollows but the Tamar Fracture System and its prolongation seemed to sink more rapidly than other areas as shown by the isopach maps (Banks, 1989, p. 293).

The initial marine transgression lapped around lowlands in the east and west and an island south of Burnie; in the north, swarms of the alga *Tasmanites* (TA in Fig. 36) near the shore produced a distinctive 2-m-thick oil shale near the base of the marine section. During the 3 m.y. span of Faunizone 1 (277–274 Ma), 250 m of mainly pyritic and carbonaceous siltstone accumulated in the axis 40 km north of Hobart. With the retreat of the sea to the area south of Hobart except during a brief incursion (dotted line), 40 m of fluvial coal measures of

the lower freshwater interval accumulated in the north, including parts of the previous lowlands.

Eastern Australian Stage B (sag of marine platform and embryonic magmatic arc/foreland basin), 268–258 Ma (Fig. 37). Shallow seas transgressed in two northwesterly lobes; thin metabentonite layers suggest distant volcanism. Further transgressions from Faunizones 6 to 10 covered all Tasmania except possibly the northwestern and northeastern tips. The shoreline finally retreated to the southeast in a barrier-bar complex including silicic volcanic ash near Cygnet from nearby vents and the final deposition of ice-rafted dropstones up to 2 m across.

Eastern Australian Stage C (orogenic piedmont coal/tuff), 258–250 Ma (Fig. 38). This comprises Unit 1 less the "Midlands" interval of the Upper Parmeener Supergroup. In the Oatlands/Midlands area (Forsyth, 1984), Unit 1 of the Upper Parmeener Supergroup rests with light incision on Faunizone 10 siltstone. The lower member of the fluvial Cygnet Coal Measures correlative is pebble conglomerate and sandstone with rare very thin coals; the middle member is fine- to medium-grained sandstone with *Glossopteris,* which elsewhere in Australia is no younger than 250 Ma; and the upper member is mainly lutite, light bluish-gray with spotted purple-red mudstone overlain by layers with red-purple mud pellets; palynomorphs including a form not recorded below the Rewan Group indicate the upper *P. microcorpus* zone. The lower and middle members are equivalent to the Cygnet Coal Measures *sensu stricto*; the upper member that lacks coal or *Glossopteris* is distinguished here as "Midlands." Fluvial sand with coal around Cygnet in the south and in an arc from the coast north of Launceston to east of Zeehan and then to the south pinched out against presumed uplands in the northeast, northwest, and southwest, with a short-lived drainage divide east of Zeehan. The environment was a broad sandy plain with an easterly paleoslope; the only sign of the orogen, so prominent at this time in the Sydney and Bowen Basins, is the tuff near Southeast Cape, but the northeast upland probably acted as the initial foreswell (I) (Fig. 35).

Eastern Australian Stage D (orogenic redbeds), 250–241 Ma (Fig. 39). Rare redbeds in "Midlands" continue upward to the base of the quartz sandstone near the top of Unit 3 (Fig. 35). The interval with redbeds (250–235 Ma) is the same here as in the rest of eastern Australia. A well-drained plain of quartz sand was crossed by rivers of low sinuosity that flowed southeasterly from a presumed upland in the Tyennan region. "Northeastern Tasmania was either an area of non-deposition or of deposition and subsequent erosion. Distant vulcanism is suggested by the occurrence of smectite in some sandstones of this cycle" (Banks, 1989, p. 294). In either case, northeastern Tasmania remained elevated in a presumed foreswell.

Eastern Australian Stage G (Pangean rift phase), 230–200 Ma (Fig. 40). Unit 3 (Stages E and F) (not shown) registers a change in slope and provenance from the eastern orogen (Fig. 35). In the northeast, quartz and lithic sand was

TABLE 7. SOUTHEAST QUEENSLAND STRUCTURAL ELEMENTS AND FORMATIONS

Element	Age (Ma)	Formations (GROUP)	Letter Symbol*	Reference†
Surat Basin	Jurassic-Early Cretaceous	Griman Formation (Albian)	Klg	
		Precipice Sandstone	Jp	126–127
Mulgildie Basin (offshoot of Surat)	Early and Middle Jurassic	Mulgildie Coal Measures	Jmc	127–128; Exon, 1976
		Hutton Sandstone	Jlh	
		Evergreen Formation	Jle	
Maryborough Basin	Cretaceous	Burrum Coal Measures to	Kb	132
	Triassic, Jurassic-Cretaceous	Myrtle Creek	RJm	
Nambour Basin	Early Jurassic–Rhaetic (213)	Landsborough	RJl	132
		Precipice Sandstone	Jp	
Ipswich Basin	Late Triassic	IPSWICH COAL MEASURES BRASSAL SUBGROUP KHOLO SUBGROUP	Ri	118
		Brisbane Tuff	Rbc	
		Chillingham Volcanics	Rbc	
Tarong Basin	Late Triassic	Tarong Beds	Rta	118
Abercorn Trough	Middle Triassic	Cynthia Beds	Rmc	117–118
Esk Trough	(?Early) Middle Triassic	TOOGOOLAWAH GROUP	Rt	117–118
		Esk Formation	Rte	
		Neara Volcanics	Rtn	
		Bryden Formation	Rtb	
Clarence-Moreton Basin	Jurassic/Cretaceous	Woodenbong Beds	JKw	125–126
	Early Jurassic to Rhaetic (213)	BUNDAMBA GROUP	RJb	
Kumbarilla Ridge (Surat/Moreton)	213-200	Covered by Marburg Formation	Jlm	125, 127
Taroom Trough	Early Permian–Middle Triassic 268–233	BACK CREEK to MOOLAYEMBER	Pb to Rm	Elliott, 1989
	Early Permian 286–268	?Camboon	Pcn	
	Devonian 408–374	Timbury Hills	M2	
Roma Shelf	Late Carboniferous 306	Combarngo Volcanics	V2	Murray, 1986, p. 356
	Early Carboniferous 355	Roma granites	G2	Murray, 1986, p. 351
	Devonian 408-374	Timbury Hills	M2	
Arbroath Trough	Early Permian 281–273	Arbroath Beds		Elliott, 1989
Arbroath Trough	Devonian 408–374	Timbury Hills	M2	
Bowen Basin (exposed)	Middle Triassic 233	MOOLAYEMBER	Rm	Figure 17
		CLEMATIS	Re	
		REWAN	Rr	
		BLACKWATER	Pw	
	Early Permian 268	BACK CREEK	Pb	
Gogango Overfolded Zone		Narayen/Nogo	Pf	
		Camboon	Pcn	
		Torsdale	Ct	
Cressbrook-Buaraba Block	Late Permian (zones T + U)	CRESSBROOK CREEK	Pc	
		?Seq. 2 rift seds		Briggs, 1991 Korsch et al., 1989
Northbrook Block	Late Permian (zones T + U)	Northbrook Beds	Pn	112
South D'Aguilar Block	Late Permian (zones T + U)	Brookfield Volcanics	PRv	121
	Middle Tournaisian 356	Neranleigh-Fernvale	DCn	Aitchison, 1988, Table 4
	Early Carboniferous	Oolitic arenite		
	Devonian/Carboniferous	Bunya Phyllite	DCb	
Yarrol Block	Late Devonian to Early Cretaceous	Burnett Formation	Plb	Murray, 1986
		BOILING CREEK	Cuo	
		CASWELL CREEK	Cl	
		Derrarabungy Beds	Cld	
		Schist, gneiss	DCy	
Coastal Block	Devonian/Carboniferous	CURTIS ISLAND	DCc	
		Shoalwater Formation	DCs	87
		Wandilla	DCw	

TABLE 7. SOUTHEAST QUEENSLAND STRUCTURAL ELEMENTS AND FORMATIONS (continued)

Element	Age (Ma)	Formations	Letter Symbol*	Reference†
Yarraman Block	Late Triassic	Aranbanga Beds	Ra	120
	Middle Triassic	Gayndah Beds	Rtg	
	Devonian/Carboniferous	Maronghi Creek	DCm	87
		Sugarloaf Metamorphics	DCsm	
Gympie Block	Early Triassic	Brooweena	Rbf	Murray, 1990a
		Kin Kin/Keefton/Traveston	Rk	
	Permian	GYMPIE	Pg	
		Cedarton Volcanics	Pe	
		Amamoor Beds	DC(P)a	
		Marumba Beds	Pm	
	Early Namurian	Good Night Beds	CPo	Roberts, 1985, p. 54
	Devonian/Carboniferous	Bouloumba Beds	Cd	
		Durundur Shale	Cd	
		Kurwongbah Beds	Ck	
		Rocksberg Greenstone	Cr	
		(rocks mapped as GYMPIE and Amamoor)		Murray, 1990a
Beenleigh Block	Devonian/Carboniferous	Neranleigh-Fernvale Beds	DCn	

*Letter symbols (from Finlayson, 1990, map 1) refer to the pre-Jurassic units on Figure 30.
†Page numbers in reference column from Day et al., 1983.

deposited directly on marine Permian sediment and in turn was covered by basalt, shown a little before its time in Figure 40. Coal-measure deposition returned after the "dry" gap of the redbed interval.

After a brief lacuna, Unit 4 (Fig. 35) of volcanic lithic sandstone and coal measures was deposited from highly sinuous streams with a median paleocurrent azimuth to the west of northwest. The volcanogenic sand is mainly intermediate but also to a less extent silicic and mafic. Rhyolitic ash-flow tuff is known near the exposed top. An increase in the volume and caliber of conglomerate in the northeast suggests another uplift, interpreted as a recrudescent foreswell (II). The eastern half of Tasmania south of 41.5°S was covered by a coal-forming environment.

North Queensland

Stage A of extension-volcanism, (290–268 Ma) (see Fig. 41A in a later section). Radiometric dating of the volcanics near Nychum Homestead (Black et al., 1972) suggested Late Carboniferous, and Balme (1973) regarded the macroflora as Early Permian. The following information is from Day et al. (1983, p. 93). The Agate Creek Volcanics are taken as Early Permian. Granitoids of this age are, from north to south, the Mareeba Granite, Herbert River Granite (Day et al., 1983, p. 141), Tully Granite, Boori and Tuckers Igneous Complex (Day et al., 1983, p. 142), others cited by Day et al. (1983, p. 143), and the I-type Thunderbolt Granite and other intrusions nearby (CP6). Between 23° and 27°S, Murray (1986, fig. 8) distinguished one I-type and five S-types.

Stage C of orogenic piedmont/tuff (258–250 Ma) (see Fig. 41C in a later section). The main source of information is De Keyser and Lucas (1968, p. 59–65). Coal measures are found in Fig. 41C at Mt Mulligan (unconformably overlain by the Pepper Pot Sandstone of unknown age), Little River, and in the Normanby Formation, all with rhyodacitic tuff, in particular the Nychum Volcanics at Mitchell River, which contain a Late Permian flora (Black et al., 1972). Also of this age are the carbonaceous shale and crystal tuff with *Glossopteris* and slightly glauconitic sands in Marina-1 (De Keyser and Lucas, 1968, p. 64). There is carbonaceous shale also in Breeza Plains-1 (Carr, 1975, p. 251–254, 309). The Normanby Formation (De Keyser and Lucas, 1968, p. 63, 64) contains "very small moulds of *Thamnopora* sp. in the dark clay matrix of the basal(?) calcareous breccia or conglomerate." Whether the fossils are in place or were derived from an adjacent Devonian limestone is unclear. Together with the report of glauconite in Marina-1, the fossils point to the possibility that the Late Permian sea may have encroached on the area. The age of the coal measures seems to be that of the Blackwater Group.

Of similar age are the Trevethan Granite (263–251 Ma, K/Ar) and Finlayson Granite (251 Ma, K/Ar) (Day et al., 1983, p. 141).

Sub-Murray Basins

Data on the Sub-Murray Basins are given in column XXV (Fig. 25) and in the regional cross section (Fig. 23), and are discussed in the following synthesis.

TABLE 8. RADIOMETRIC AGES OF SOUTHEASTERN QUEENSLAND GRANITOIDS*

Symbol	Unit	Latitude (S)	Longitude (E)	Age (Ma)	Reference
Rg 16	Somerset Dam Igneous Complex	27°06.7'	152°32.2'	220	Mathison, 1967; Webb and McDougall, 1967; Cranfield et al., 1976
	Somerset Dam Igneous Complex	27°07.5'	152°32.8'	212[†]	
Rg 17	Samford Granodiorite	27°23.4'	152°47.7'	226/227	Cranfield et al., 1976
Rg 20	Woondum Granite	26°12.9'	152°48.5'	223/226	Webb and McDougall, 1967; Murphy et al., 1976
PRw	Wateranga Gabbro	25°20.0'	151°50.0'	244	Ellis, 1968
PRg2	Boondooma Igneous Complex	26°35.2'	151°47.0'	222[†]	Webb and McDougall, 1968
		26°21.8'	151°30.3'	239[†]	
		25°56.7'	151°22.1'	245/246	
		26°12.3'	151°33.0'	244/245	
		26°50.0'	155°45.0'	249	
		26°28.8'	151°46.2'	259	
		26°21.8'	151°30.3'	237[†]	Murphy et al., 1976
		26°51.5'	151°47.8'	261	
PRg3	Crows Nest Granite	27°10.4'	152°02.9'	239[†]	Webb and McDougall, 1968
		27°16.1'	152°06.7'	242[†]	
				251	Cranfield et al., 1976, Rb/Sr
PRg5	Eskdale Tonalite	27°11.0'	152°15.5'	238[†]	Webb and McDougall, 1968; Cranfield et al., 1976
		27°09.2'	152°05.8'	243[†]	
				248	
PRg6	Kingaham Creek Granodiorite	26°31.5'	152°21.5'	220[†]	Webb and McDougall, 1967; Murphy et al., 1976
		26°30.9'	152°21.0'	221[†]	
		26°33.5'	152°26.0'	239	
		26°29.0'	152°20.9'	242	
				248	Webb and McNaughton, 1978
PRg9	Taromeo Tonalite	26°45.7'	152°01.8'	243[†]	Webb and McDougall, 1968
		26°43.6'	152°00.7'	241[†]	
				250	
		26°41.2'	151°56.7'	264	Murphy et al., 1976, Ar/Ar
				248[†]	
Pgw	Wigton Granite	26°10.3'	151°41.0'	263	Ellis, 1968; Webb and McDougall, 1968
				250[†]	
		25°46.2'	151°29.8'	264	Whitaker et al., 1974; Murphy et al., 1976
Pgm	Monsildale Granodiorite	26°41.7'	152°24.0'	253	Murphy et al., 1976
Pgo	Woolshed Mountain Granodiorite	27°00.7'	152°01.6'	259	Cranfield et al., 1976
Cgc	Claddagh Granodiorite	26°15.5'	152°14.5'	298	Murphy et al., 1976
Cgc	Callengowan Granodiorite	26°27.0'	152°19.5'	320	Murphy et al., 1976
PRg1	Undifferentiated granitoid	26°29.0'	152°39.2'	251	Murphy et al., 1976
				242[†]	
PRg1	Undifferentiated granitoid	26°20.0'	152°27.5	217[†]/227[†]	Horton, 1972; Green and Webb, 1974, Ar/Ar
				247	
		26°19.2'	152°26.7'	231	Horton, 1972; Green and Webb, 1974, Ar/Ar
		26°17.9'	152°32.4'	222	Horton, 1972; Murphy et al., 1976, Ar/Ar
Rg1	Undifferentiated granitoid	26°11.9'	152°16.4'	238	Murphy et al., 1976
Rg10	Djuan Tonalite	27°03.4'	152°00.1'	235	Cranfield et al., 1976
Rg18	Toondahra Granite	27°03.4'	152°00.1'	215	Whitaker et al., 1974
Rg19	Tungi Creek Granodiorite	26°38.5'	152°32.2'	226	Murphy et al., 1976
Rd	Degilbo Granodiorite			220	Webb and McDougall, 1967; Ellis, 1968
				226 ± 16	Rb/Sr
Rf	Broomfield Granite			226 ± 16	Webb and McDougall, 1967; Ellis, 1968, Rb/Sr
	See Cranfield and Murray, 1989, for good dates.				
Rgm	Mungore Granite			212/219	Webb and McDougall, 1967
				226 ± 16	Ellis, 1968, Rb/Sr
Rmg	Mungore Granite (Mt Walsh)	25°39.4'	152°04.8'	221.8 ± 1	Stephens, 1992, Rb/Sr WR
Ps	Musket Flat Granodiorite			222	Webb and McDougall, 1967
				226 ± 16	Ellis, 1968, Rb/Sr
R2	Tawah Granite			220	Webb and McDougall, 1967
Rg2	Station Creek Granodiorite	26°05.4'	152°28.7'	231	Webb and McDougall, 1967; Murphy et al., 1976
		26°12.6'	152°20.7'	236	

TABLE 8. RADIOMETRIC AGES OF SOUTHEASTERN QUEENSLAND GRANITOIDS* (continued)

Symbol	Unit	Latitude (S)	Longitude (E)	Age (Ma)	Reference
Rg5	Dayboro Tonalite	27°10.4'	152°50.6'	238	Cranfield et al., 1976
Rg7	Enoggera Granite	27°27.3'	152°57.0'	224	Evernden and Richards, 1962; Cranfield et al., 1976
Rg9	Goomboorian Diorite	26°04.3'	152°45.6'	234† 235†	
		26°02.0'	152°45.2'	240	Green and Webb, 1974; Murphy et al., 1976
Rg14	Neurum Tonalite	26°45.2'	152°43.8'	228	Webb and McDougall, 1967; Murphy et al., 1976
Rtn	Neara Volcanics, North d'Aguilar Block			232	Murphy et al., 1976
	Neara Volcanics, Esk Trough	26°14.2'	152°10.9'	242 ± 8	Webb, 1981
		26°11.8'	152°08.1'	239	Cranfield et al., 1976, Ar/Ar
				236	Webb, 1981
Rgi	Dyke intruding Neara Volcanics	26°11.9'	152°07.6'	224	Irwin, 1976, Ar/Ar
Rgi	Dyke intruding Neara Volcanics	26°12.8'	152°07.6'	223	Irwin, 1973
Rn	North Arm Volcanics	26°54.5'	152°55.5'	213	Green and Webb, 1974; Webb, 1981, p. 114
Rn	North Arm Volcanics	26°30.05'	152°59.5'	217 ± 2	Ashley and Andrew, 1992
PRe	Eidsvold Complex	25°22.4'	151°07.2'	241†	Webb and McDougall, 1968
		25°20.0'	151°07.1'	250	
PRag	Cardaga Creek Granodiorite	25°42.7'	151°01.1'	258 242†	Whitaker et al., 1974
PRgi	"Bouraba quartz diorite"	27°22.4'	152°15.5'	191	Cranfield et al., 1976
Rgi	"Brisbane Valley porphyrites"	27°09.8'	152°23.7'	223/224	Webb and McDougall, 1967
?Rgi		27°00.8'	152°23.4'	138/141	Green and Webb, 1974
Ra	Aranbanga Volcanics			221	Ellis, 1968
Rmb	Mount Byron Volcanics			228	Cranfield et al., 1976; Murphy et al., 1976
——	Sugars Basalt, in Ipswich C.M			229/232	
Rg15	Mt Samson Granodiorite	27°18.5'	152°50.3'	222	Webb and McDougall, 1967; Cranfield et al., 1976
Rn7	Nour Nour Granodiorite	25°09.7'	151°31.9'	235 ± 3	Cranfield and Murray, 1989, table 1, hornblende, DA16
				238 ± 3	biotite DA17
Pt	Tenningering Granodiorite	25°11.5'	151°37.0'	256 ± 4	muscovite DA23
Po	Wolca Granite	25°12.7'	151°34.7'	261 ± 4	biotite DA21
Pmg	Mingo Granite	25°28.0'	151°45.3'	262 ± 2	biotite G1
Pgc	Chowey Granite	25°26.0'	151°52.0'	274 ± 3	biotite DA14
				266 ± 3	hornblende DA15
Rh	Hogback Granite	25°06.6'	151°46.5'	228 ± 4	biotite DA20, table 2
Rw	Wonbah >228, <232	Neighboring pluton		Inferred 230	
Rn8	New Moonta Diorite	25°03.8'	151°43.0'	232 ± 4	DA19
Ru	Boogooramunya Granite	25°54.0'	152°09.5'	226 ± 2	hornblende G5
Rgb	Boonara Granodiorite	25°57.5'	152°11.5'	233 ± 2	hornblende G6
Rca	Calgoa Diorite	25°52.0'	152°15.5'	234 ± 2	biotite G3

*All K/Ar except where noted otherwise. General reference: Day et al., 1983. Other key references are Webb, 1981, and Murray, 1986.
†Age inferred to be reset.

SYNTHESIS OF EASTERN AUSTRALIA

Some aspects of the pre-Permian history of the region are obscure and controversial. We take the view that the Meandarra Gravity Ridge, which initiates the meridional trend of the Sydney-Gunnedah-Bowen Basin, subtends a low angle with the NNW-trending Late Devonian–Early Carboniferous Andean-type arc (Fig. 5) so as to divide the region south of 30°S into an arc to fore-arc terrain on the west and an accretionary wedge complex of Devonian and Early Carboniferous shale and radiolarian chert on the east. Mid-Carboniferous (~320 Ma) deformation was effected by shallowing subduction or collision of the Coffs Harbour Block and Tia Complex with Australia (Graham and Korsch, 1985), or both, at the same time as or soon after the Kanimblan folding to the west. A right-lateral megafold had developed in the already deformed accretionary complex by the end of the Carboniferous, and minor movement continued during the Permian and Early and Middle Triassic until it received its final shape about 230 Ma after right-lateral movement along the Demon Fault.

In what follows, we outline, from north to south, the paleogeography and paleotectonics of eastern Australia during the

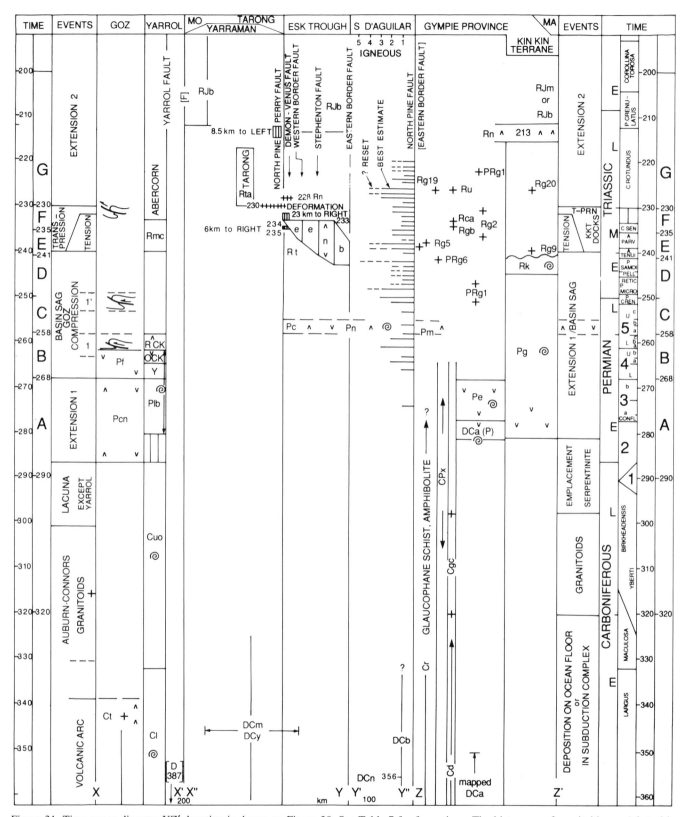

Figure 31. Time-space diagram XZ′, location is shown on Figure 30. See Table 7 for formations. The histogram of granitoid ages (plotted in the D'Aguilar column) was compiled from Table 8. Reset ages shown by broken line. In Yarrol Block, abbreviations are Y—Yarrol Formation, OCK—Owl Gully Volcanics, RCK—Rainbow Creek Formation. In Gympie Province, CPx stands for serpentinite. Other abbreviations are found in the key to Figure 30.

seven stages of Permian and Triassic development listed in Table 1 and shown in the accompanying set of figures (p. 122), which show Tasmania in its restored position (Veevers et al., 1991).

Stage A: Extension/volcanism of the platform, 290–268 Ma (Fig. 41A, Table 10)

The "North Queensland volcanic and plutonic province" extends south from the Cape York–Oriomo Ridge beneath Torres Strait at 9°S past outcrops in the Coen province about 13°S to the area between 16°S and 21°S, and comprises I- and S-type granitic batholiths and comagmatic volcanics, mainly rhyodacitic ash-flow tuffs (Richards, 1980). These include the Nychum and Agate Creek Volcanics north of 19°S and, from Townsville (T) southward, the Featherbed and Newcastle Range Volcanics. The western limit of volcanics runs past the 4.5-km-thick Bulgonunna Volcanics (BV) in the northern part of the Bowen Basin and continues south to 26°S to the subsurface Merivale Fault (MF), which marks the western subsurface limit of the mafic Comet Volcanics and the margin of the half graben of the Denison Trough (DT), filled with 3 km of the Reids Dome Beds. All these rocks are nonmarine, but in the east the basaltic to rhyolitic Lizzie Creek Volcanics, Carmila Beds, Camboon Andesite, and Youlambie Conglomerate contain shallow marine shelly fossils. The Carmila Beds are 7.5 km of rhyolitic, dacitic, and andesitic volcanics and volcaniclastic sediment ranging from conglomerate to shale. They are entirely nonmarine except at 22°S where they contain brachiopods, bryozoans, and crinoid stems (Day et al., 1983, p. 105). The nonmarine equivalent in the southern Bowen Basin and Gogango Overfolded Zone (GOZ) is the Camboon Andesite, 3 km or more of andesitic and basaltic flows, andesitic-dacitic welded tuff, sandstone and mudstone. East of the GOZ, the Yarrol Block contains the Youlambie Conglomerate, 2 km of conglomerate, sandstone, and siltstone, with bivalves, gastropods, and plants, and rhyolite and dacite. In the GOZ are the marine Narayen Beds, andesite and basalt with conglomerate, and farther north, in the Grantleigh Trough (GT) (Clare, 1993) are the Rannes Beds, thick marine flysch overlain by the Rookwood Volcanics, 4.5 km of pillow basalt, breccia, and siltstone. To the southeast, the Berserker Beds comprise 3 km of acid to intermediate volcanics in the lower part overlain by volcaniclastic sediments with rich invertebrate faunas.

The Denison Trough (DT) and the Arbroath Trough (AT) to the south are the only nonvolcanic rifts of this stage. Equivalent sediments in the Galilee Basin (GB) west of the Denison Trough are the Joe Joe Group, 300–1,700 m of siliciclastics, with a 100-m-thick tuff, the Edie Tuff Member of the Jochmus Formation. We draw the shoreline along the eastern edge of the Bowen Basin, with an embayment in the north that contains the marine parts of the Lizzie Creek Volcanics.

S-type granites (Murray, 1986) and the limit of volcanics approach the shoreline at about 26°S, which then extends west of Brisbane (B) to cross the state border at 150.4°E. Leitch et al. (1988) describe redeposited detrital rocks (diamictite, tur-

bidite) at least 8 km thick and very thick pillow basalt, silicic ash-fall tuff, and volcaniclastics of the Barnard Basin (BB). The widespread Allandale marine fauna occurs in the Ironbark and Kensington Beds immediately west of Peel Fault and at Halls Peak at 152°E, and indicates shallow water. Partial melting in the lower crust from ocean floor basalt in pull-apart basins may have produced the Halls Peak silicic rocks and the "S-type" Bundarra (BU) and Hillgrove (H) granites of southern New England. "A similar assemblage of geological features—basalts of ocean floor type, silicic volcanics, and a thick sequence of clastic sediments—is found in the presently active Salton Trough at head of the Gulf of California which is a possible modern analogue of the Barnard Basin" (Leitch et al., 1988). The marine Barnard Basin is separated by a narrow strip of land (Gloucester and Myall Synclines) from the marine and volcanic Sydney Basin (to 31.7°S), and bordered in the west by the basal 1.5-km-thick silicic and mafic Boggabri Volcanics and Werrie Basalt of the Gunnedah Basin.

Both Sydney-Gunnedah and Barnard Basins seem to have been the result of the same set of stresses acting on different kinds of basement—an accretionary complex on the east and a fore-arc region on the west. The Gunnedah-Sydney Basin lies above the Meandarra Gravity Ridge (Murray et al., 1989). In the Sydney Basin (Mayne et al., 1974; Bradley et al., 1985), the basal volcanics range in thickness to 1.5 km and are locally marine. In the west, they are exposed as the Rylstone Volcanics (Shaw et al., 1989). Nonmarine outliers (stipple) north and west of Mudgee include glacigenic rhythmites with Stage 2 palynomorphs. Other nonmarine sediments stretch south-southeast to the (present and ancient) coast and were subsequently covered by the marine transgression that started with deposition of the Wasp Head Formation at 277 Ma (Herbert, 1980a).

In Tasmania, "Rocks of the Parmeener Supergroup rest unconformably on Late Devonian granites and older folded rocks on a surface with a relief of about 1000 m. . . . Early deposition of glacigenic rocks followed by carbonaceous muds then fossiliferous beds essentially filled the earliest hollows" (Banks, 1989). In the north, swarms of the alga *Tasmanites* near the shore produced a distinctive 2-m-thick oil shale near the base of the marine section. With the retreat of the sea to the area south of Hobart (H in Fig. 41) except during a brief incursion, 40 m of fluvial coal measures of the lower freshwater interval accumulated in the north, including parts of the previous lowlands. In the Melbourne area of the mainland, north-moving ice was succeeded by a broad post-glacial marine transgression.

Stage A represents the base of the Pangean supersequence (Veevers, 1990a), and its 0.5×10^6 km^2 of magmatic rocks is comparable in extent and content to the coeval deposits of Europe (Veevers et al., 1994b). These are the only regions in the early Pangea that are dominated by magmatic rocks. Their common tectonic setting above the mid-Carboniferous Variscan and Kanimblan/Coffs Harbour/Tia orogens suggests

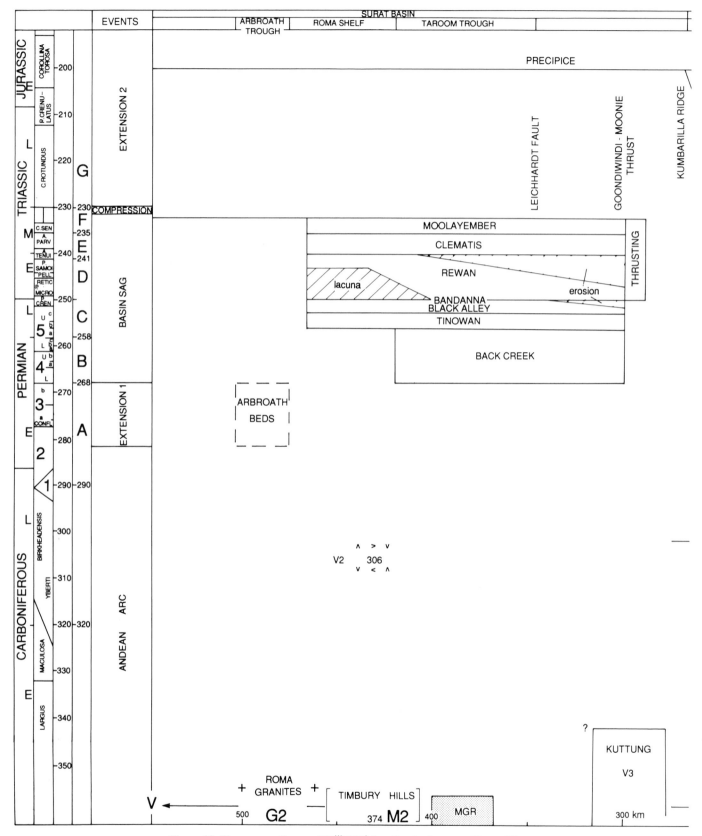

Figure 32. Time-space diagram VV″″WW′, location is shown on Figure 30.

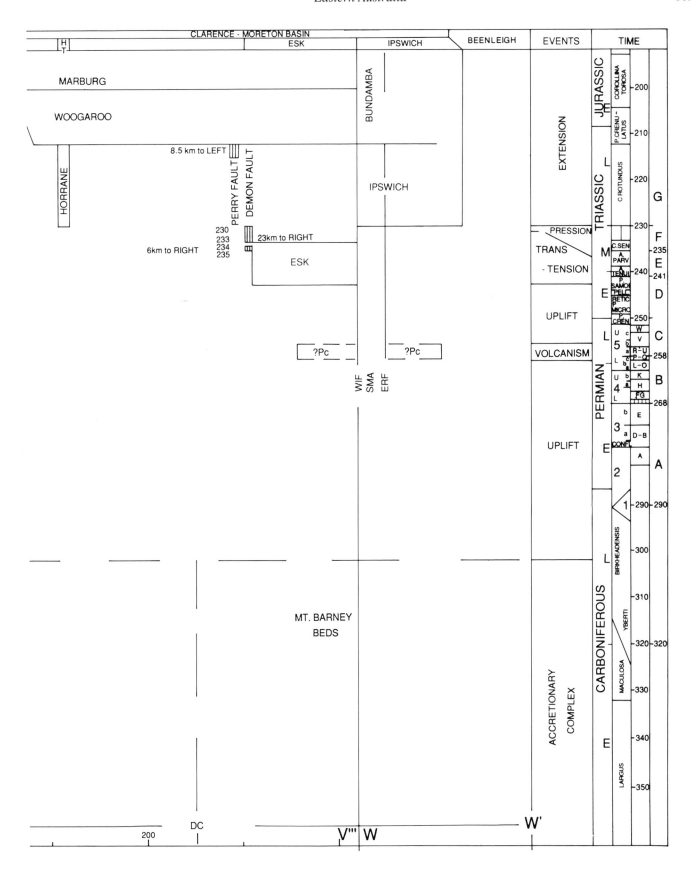

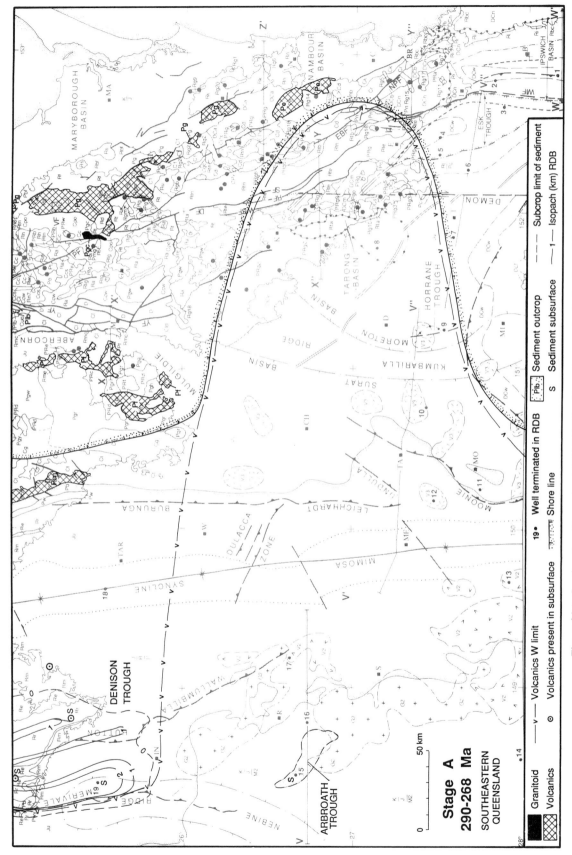

Figure 33A. Southeastern Queensland, 290–268 Ma: Stage A, extension/volcanism of the platform.

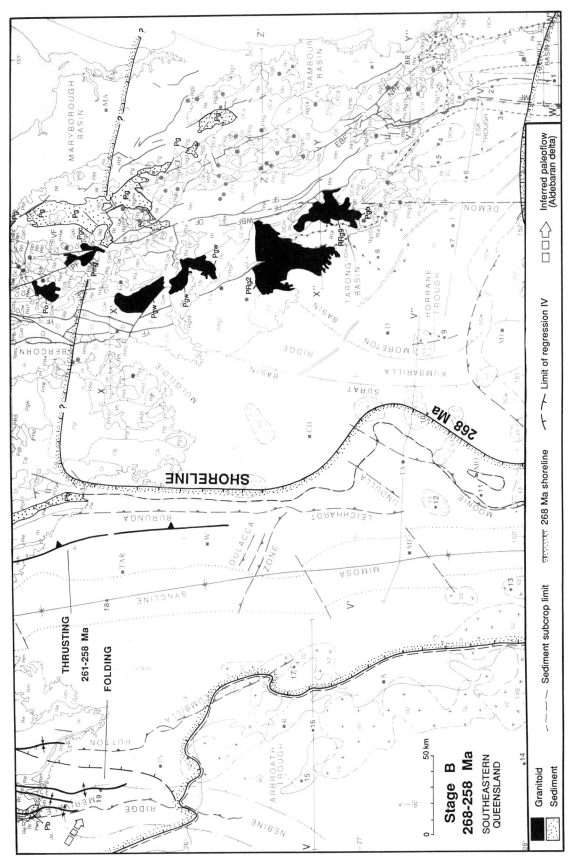

Figure 33B. Southeastern Queensland, 268–258 Ma: Stage B, marine sag on the platform and embryonic magmatic arc/foreland basin. Subsurface data from Exon (1976, fig. 13).

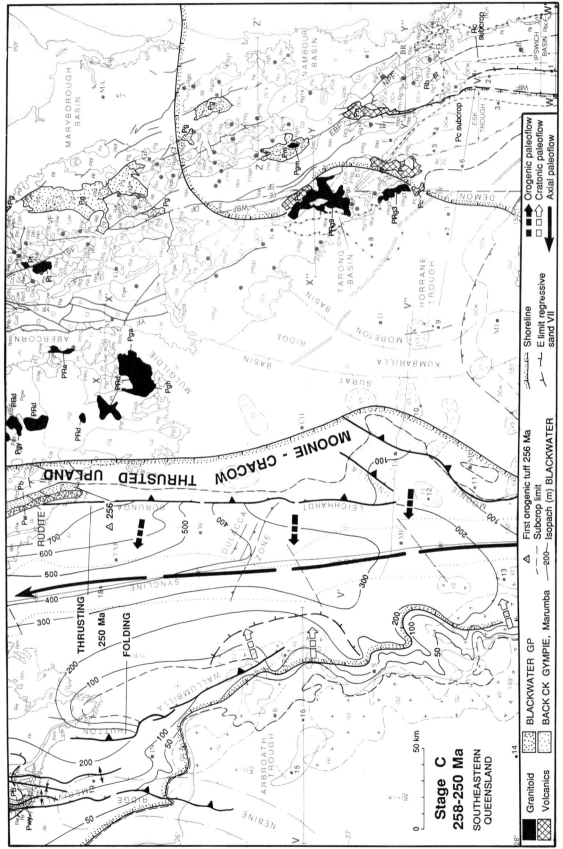

Figure 33C. Southeastern Queensland, 258–250 Ma: Stage C, orogenic piedmont coal/tuff. In this figure and in Figures 33D–F, subsurface data beneath the Surat Basin are from Exon (1976) and Butcher (1984). Eastern limit of cratonic quartzose sand (regression VII, Fig. 18) from Paten et al. (1974) and Harrington (1984, fig. 8.5). Provenance pattern for eastern sources is from Bastian (1965a, fig. 11) and from western sources from Paten and Groves (1974). The western shoreline is the limit of transgression X.

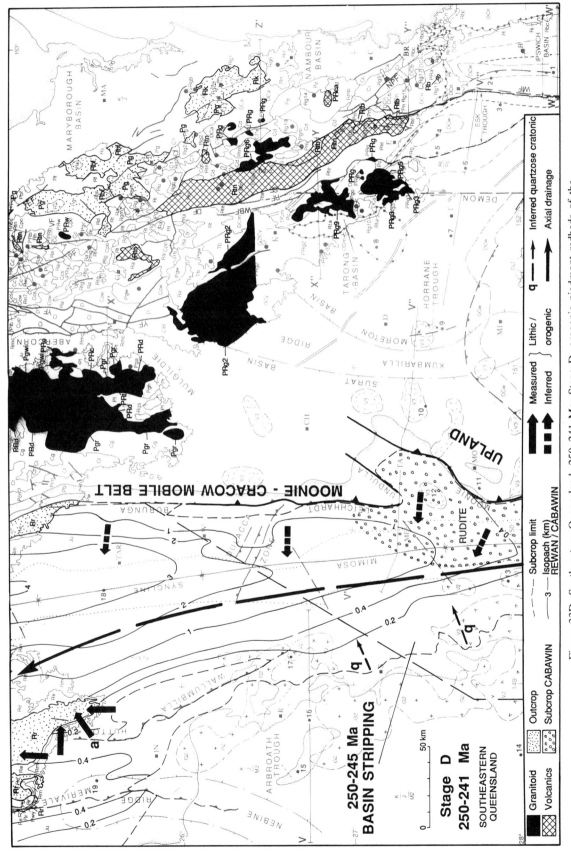

Figure 33D. Southeastern Queensland, 250–241 Ma: Stage D, orogenic piedmont redbeds of the foreland basin. The major (volcanolithic) sediment influx was from the eastern upland (Bastian, 1965a; Bastian and Arman, 1965); the minor (quartzose) sand from the craton in the Roma Shelf area is from Bastian (1965a, fig. 12). The northerly-directed axial drainage (Conaghan et al., 1982, fig. 7B) is cratonward of the depocentral axis approximated by the Mimosa Syncline, in keeping with the "progradational-fan/restricted axial-floodplain" model of Burbank and Beck (1991, fig. 4B).

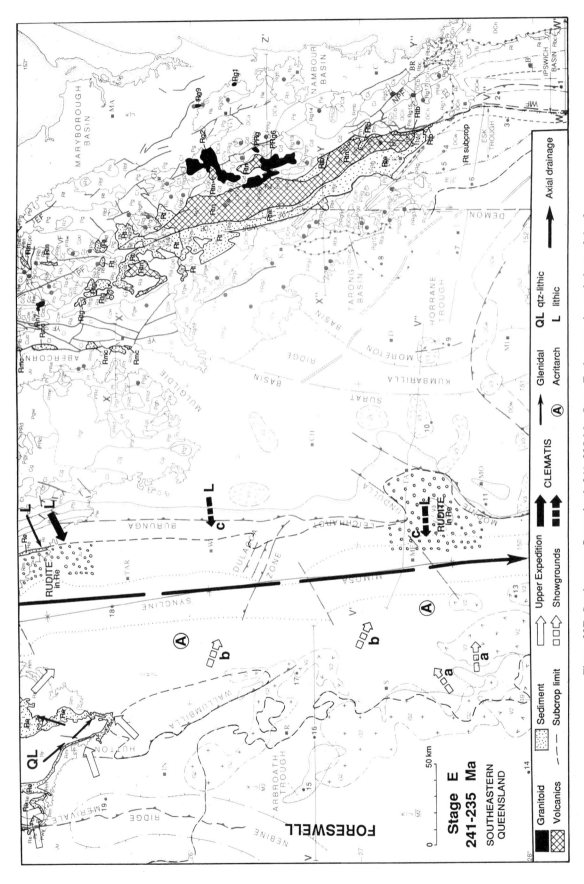

Figure 33E. Southeastern Queensland, 241–235 Ma: Stage E, the cratonic sand sheet of the foreland basin. Rudite in the Expedition/Clematis Sandstone on the eastern margin of the Taroom Trough at 25°S is from Bastian (1965b), and between 27°S and 28°S from Bastian (1965a, p. 37). Paleoflow beneath the Surat Basin at a is from Butcher (1984, fig. 22), at b and c from Bastian (1965a, fig. 13). Acritarchs (A) in the Showgrounds Formation are from Butcher (1984, p. 351).

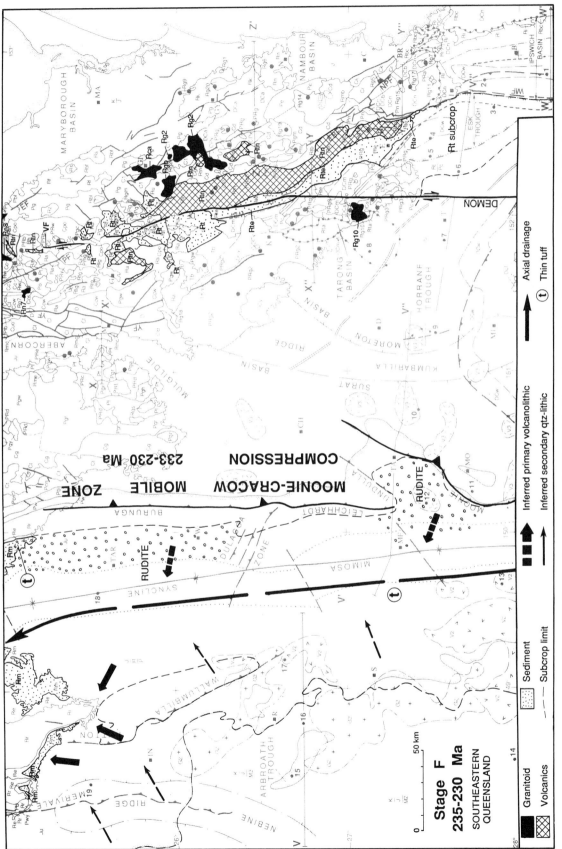

Figure 33F. Southeastern Queensland, 235–230 Ma: Stage F, orogenic paralic sediment. Rudite in the Wandoan (W) area from Bastian and Arman (1965, p. 4), and in the Meandarra (ME) area from Bastian (1965a, p. 41) and Fehr (1965, p. 6, 7). Paleoflow vectors are inferred from petrography (Bastian, 1965a, 1965b), Bastian and Arman (1965), and Allen and Houston (1964). The axial-drainage pattern is from Fielding et al. (1990, fig. 6). The major influx of volcanolithic sediment was from the orogenic upland, and the subordinate influx of quartz-lithic (plutono-metamorphic) sediment was from the Roma Shelf and Nebine Ridge.

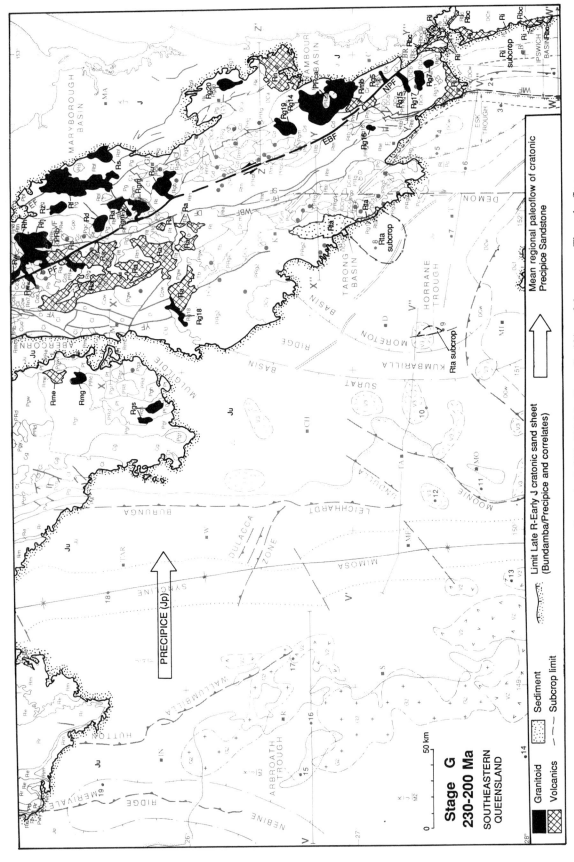

Figure 33G. Southeastern Queensland, 230–200 Ma: Stage G, rifting of the orogen. The paleoflow for the Precipice Sandstone is summarized from Martin (1981, fig. 4).

a common origin by eruption into a hot region undergoing dextral transtension.

Stage B: Marine sag on the platform and embryonic magmatic arc/foreland basin, 268–258 Ma (Fig. 41B)

The initial shoreline (dotted line) is drawn landward past the eroded edge of the marine formations at the base of the Back Creek Group, such as the Cattle Creek Formation on the west and in front of the coal measures at Blair Athol and Calen (CA). The Ridgelands granitoid (264 Ma) was the first pluton that can be related to the convergent arc. The Aldebaran delta (A) and the Collinsville coastal plain (CCP) indi-

cate depositional regression of the shoreline (Fielding et al., 1990) that accompanied folding in the west and thrusting in the Gogango Overfolded Zone (GOZ) of the thick sediments and volcanics of the Grantleigh Trough during the last 3 m.y. of this stage. To the east, in the Yarrol Block, the marine Yarrol Formation and Owl Gully Volcanics were replaced by the nonmarine Rainbow Creek Beds (R), which were deposited within the orogenic upland that emerged during the 261–258 Ma thrusting.

Southeastern Queensland was a region of plutonic intrusion (Boondooma batholith). Lacking any record of marine deposition, it is shown as lying above sea level.

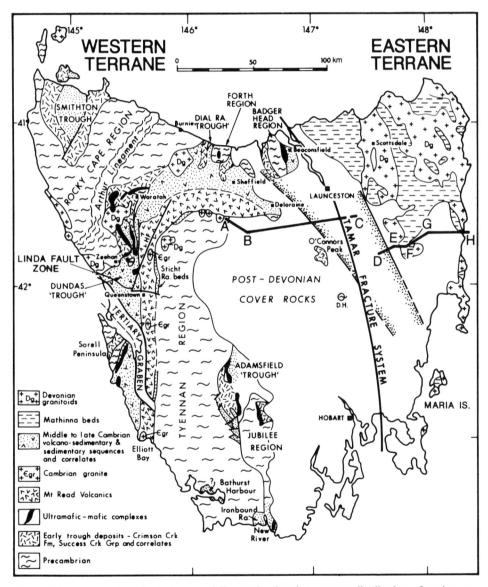

Figure 34. Simplified geological map of Tasmania showing present distribution of major pre-Carboniferous tectonic elements within the Eastern and Western Tasmania Terranes. Ordovician to Early Devonian rocks and much of younger cover not shown. Reproduced from Corbett and Turner (1989, fig. 5.1) with the permission of the authors and the Geological Society of Australia. The stippled lines about Launceston enclose a zone of maximum conductivity (Parkinson and Richardson, 1989, fig. 13.4). A–H is the line of cross-section XIV in Figure 15.

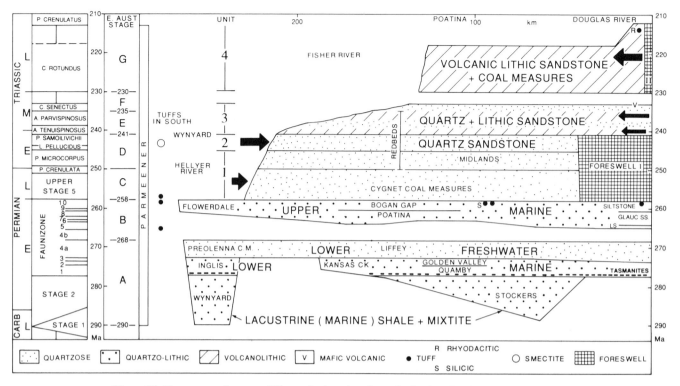

Figure 35. Time-space diagram of Tasmania, location shown by broken line in Figure 40. The interval with redbeds (250–235 Ma) is from information in Forsyth (1984). Information on provenance is listed in Table 9.

TABLE 9. SOURCE OF INFORMATION GIVEN IN FIGURE 35

Material—Location	Age (Ma)	Reference*—Page
發способWysavнародወőiьországօדania	---	---
Unit 4: Tuff between *rotundus* + *crenulata* zones from rhyolitic/rhyodacitic volcanoes	214 ± 1	1—334
Sandstone from calcalkaline volcanic and plutonic rock (magmatic arc)		1—332
Unit 3: Basalt (St Marys)	233 ± 5	1—333–334
Unit 2: Sandstone from cratonic block (granite pegmatite and some sedimentary and volcanic rocks) in Western Tasmania; clay minerals from basic volcanic tuff		1—319–320
Unit 1: Sandstone with garnet of eclogitic origin; quartz in north, arkose in south and center		2—27 1—311
Crystal tuff (Mt La Perouse)	257	3—13
Crystal tuff (Woodbridge)	Faunizone 10 = 258	3—13
Montmorillonite (St Marys)	Faunizone 10 = 258	3—13
Silicic volcanic ash	Faunizone 10 = 258	1—307
Metabentonite in CO_3	Faunizone 5a = 265	3—13

*1 = Clarke and Forsyth, 1989; 2 = Forsyth, 1984; 3 = Banks and Clarke, 1987.

The marine Barnard Basin continued into this stage, and the sea extended west into the Gunnedah Basin. The formations, all marine except the Ashford Coal Measures (ACM) and Dummy Creek Conglomerate (DC), are the Plumbago Creek Beds, the Wallaby and Eurydesma Beds, and the Bondonga Beds. The deformed Bondonga Beds and Ashford Coal Measures are postdated by the marine Glenmore Beds, and the Plumbago Creek Beds and Wallaby and Eurydesma Beds are unconformably overlain by the marine Drake Volcanics, Rhyolite Range Beds, and the Eight Mile Creek Beds, which show that the underlying beds were folded and faulted in the interval 263–261 Ma in an area delineated by the broken line. In the south, the unroofing of the 265 Ma Barrington Tops Granodiorite (BT) together with the paralic Dewrang Formation indicate an upland west of the open marine Bulahdelah Formation of the Myall Syncline.

In the Gunnedah Basin, nonmarine alluvial fans in the Maules Creek Formation (Thomson, 1986, p. 18) were shed off the contemporary Boggabri paleo-ridge (BP-R). Conglomerate at the base of the Porcupine Formation was shed from

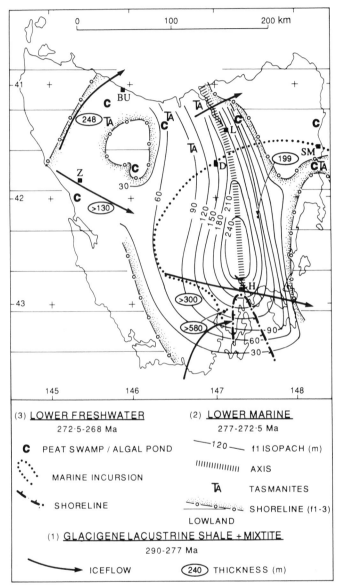

Figure 36. Tasmania, Eastern Australian Stage A (platform extension), 290–268 Ma, from Clarke and Forsyth (1989, figs. 8.2–8.5) and additionally (peat swamp and algal pond) Banks and Clarke (1987, fig. 6b). Thickness (m) from Hand (1990, fig. 1). Three episodes are superimposed: (3) the Lower Freshwater interval on (2) the Lower Marine interval on (1) the glacigenic interval of shale and mixtite. Ice flowed along the Arthur Lineament and Linda Fault Zone, shown on Figure 34. The axis of thickest Faunizone 1 is inferred to mark the Tamar Fracture system. Lowlands on east and west probably contracted during the Lower Freshwater interval. Abbreviations are BU—Burnie, D—Deloraine, H—Hobart, SM—Saint Marys, Z—Zeehan.

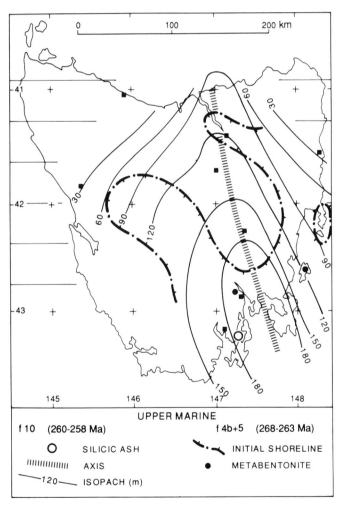

Figure 37. Tasmania, Eastern Australian Stage B (sag of marine platform and embryonic magmatic arc/foreland basin), 268–258 Ma. Upper marine interval of Lower Parmeener Supergroup: initial transgression for Faunizones 4b and 5 (268–263 Ma) is over lower freshwater sandy coastal plain, thin metabentonite layers in the limestone at Hobart and on Maria Island. The isopachs and axis are for Faunizone 10 (260–258 Ma), including silicic volcanic ash near Cygnet. From Clarke and Forsyth (1989, figs. 8.6, 8.8, 8.9).

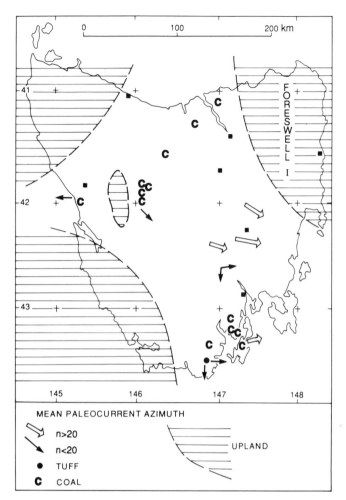

MEAN PALEOCURRENT AZIMUTH

⇨ n>20

➤ n<20

● TUFF

C COAL

Figure 38. Tasmania, Eastern Australian Stage C (orogenic piedmont coal/tuff), 258–250 Ma. Start of nonmarine Upper Parmeener Supergroup (Cygnet Coal Measures and correlatives) is shallowly incised into the upper marine interval. From Clarke and Forsyth (1989, fig. 8.12a) and Banks and Clarke (1987, fig. 10b).

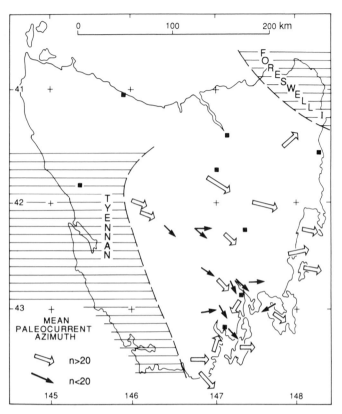

MEAN PALEOCURRENT AZIMUTH

⇨ n>20

➤ n<20

Figure 39. Tasmania, Eastern Australian Stage D (orogenic redbeds), 250–241 Ma. "Midlands" and Unit 2 of quartz sandstone. Azimuths from Clarke and Forsyth (1989, fig. 18b) and Collinson et al. (1990, fig. 4).

the Rocky Glen paleo-ridge (RG P-R), as inferred from McDonald and Skilbeck (1991). To the west, coal measures were deposited in the Gilgandra Trough (GT) toward the end of the stage. The first tuff related to the convergent magmatic arc appeared at 259 Ma in the Watermark Formation.

Early in the stage, the Greta Coal Measures (G) were deposited from the north, as shown by decreasing clast size toward the south (Harrington, 1984), during the initial growth of the Muswellbrook and Lochinvar Anticlines. The arrow south of Sydney represents the flow azimuth of basalt lava flows in the Gerringong Volcanics at Kiama (R. H. Flood, personal communication, 1988) and the transport direction of Gerringong marine volcaniclastic sands (Broughton Formation) (Runnegar, 1980c), interpreted as a sedimentary apron about a vent on the magmatic arc of the Currarong Orogen. The Wandrawandian Siltstone (Runnegar, 1980b) contains the first tuff (260 Ma) related to the convergent magmatic arc.

The maximum transgression at the end of the stage is represented by the Berry Formation, Broughton Formation/Gerringong Volcanics, Mulbring, and Dewrang Formations in the Sydney Basin, and by the Watermark Formation in the Gunnedah Basin and is followed in the next stage by the major regression represented by the coal measures.

Shallow seas covered all but the northeast and northwest corners of Tasmania. Thin metabentonite layers (265 Ma) near Hobart are the first sign of the convergent magmatic arc. The shoreline finally retreated to the southeast in a barrier-bar complex including silicic volcanic ash near Cygnet from nearby vents and the final deposition of ice-rafted dropstones up to 2 m across.

Stage C: Orogenic piedmont coal/tuff:mature orogen/ foreland basin, 258–250 Ma (Fig. 41C)

In north Queensland, between 14°S and 18°S, poorly dated coal measures with rhyodacitic tuff (De Keyser and Lucas, 1968) and plutons on the coast (Day et al., 1983) probably belong in this stage. Isolated occurrences of glauconitic sand and marine fossils point to the possibility that the Late Permian sea may have encroached on the area.

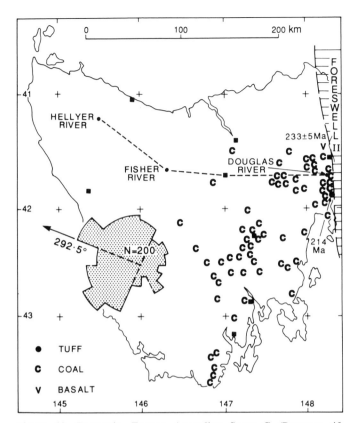

Figure 40. Tasmania, Eastern Australian Stage G (Pangean rift phase), 230–200 Ma. Interval of lithic sandstone and coal measures in eastern Tasmania. Highly sinuous streams have a median paleocurrent azimuth within the west-northwest semicircle of 292.5°, calculated from the compass rose in figure 8.12c of Clarke and Forsyth (1989). Broken line is location of Figure 35.

Farther south, the sea encroached eastward across the western flank of the Yarrol upland, and the Moah Creek (MC) Beds, 2 km of conglomeratic mudstone, with blocks of Carboniferous oolitic limestone, were deposited at the foot of scarps. To the south, the earliest plutons of the Rawbelle Batholith were emplaced, and to the west the first tuff related to the convergent magmatic arc, in the Oxtrack Formation (256 Ma), was deposited. On the western side of the basin, the shoreline (Freitag Formation) advanced over the Aldebaran delta past its 268 Ma position and thereafter retreated from an upland during deposition of the deltaic-marine or paralic quartzose Catherine Sandstone and German Creek Coal Measures. In the east, thick shelf mud and coarse turbidites of the Barfield Formation were succeeded by the first pulse of coarse tuffaceous volcanolithic sandstone (Flat Top Formation) in fan-deltaic complexes derived from the emergent volcanic arc (Fielding et al., 1990). In the Saraji-Peak Downs area, the interfingering facies boundary between the German Creek Coal Measures and the MacMillan Formation on the south and west and the Moranbah Coal Measures on the northeast, linked by tuff bands, defines the boundary between quartzose sediment from

the west and orogenic volcanolithic sediment from the east. A final transgression (X) carried the shoreline from the base of the Peawaddy Formation through the base of the overlying Black Alley Shale with the P3c acritarch swarm (Evans, 1980) to its farthest west almost to 146°E (Gray, 1976; Wells and Newhouse, 1988). The rest of the Black Alley Shale was deposited in a lake behind the shoreline. With the deposition of the tuffaceous coal measures of the Blackwater Group, the shoreline retreated to the east. Thrusting and uplift in the east and folding in the west at 250 Ma augmented the depositional regression. The uplift of the "Eungella-Cracow Mobile Belt" (Jensen, 1971; Dickins and Malone, 1973) led to the definitive retreat of the sea from the Bowen Basin.

The residual physiographic basin was finally infilled by prograding deltaic and alluvial systems directed southwards (Bandanna Formation, Rangal Coal Measures) and westwards (Baralaba Coal measures), and in the central-western part of the basin the Bandanna Formation prograded westward across the Springsure Shelf towards the Galilee Basin" (Fielding et al., 1990).

We add another direction. In the Taroom Trough between 24°S and at least 29°S, northerly regional sediment dispersal united with the southerly flow into the westward progradation north of the folded area of the Denison Trough across the Springsure Shelf into the Galilee Basin (Conaghan et al., 1982; Wells and Newhouse, 1988) over the former (Peawaddy) marine gulf.

Plutons were emplaced between 25°S and 30.5°S. The comagmatic Cressbrook Creek (CC) Volcanics west of Brisbane contain marine interbeds. Other shallow marine deposits (Drake Volcanics, Gilgurry Mudstone, Rhyolite Range Beds, Eight-mile Creek Beds, and Condamine Beds) in the area across the border unconformably overlie the rocks deformed in the period 263–261 Ma. The only other surface deposits, the nonmarine Merrestone Beds (M) and Wallangarra Volcanics (WV), indicate a regression of the shoreline some 250 km eastward over terrain intruded by granitoids along 30°S. Most of the area southward to and including part of the northern Sydney Basin was folded during this stage (Lennox and Roberts, 1988; Collins, 1991, D1: east-west compression forming meridional folds). Deformation in the Lochinvar Anticline in the Hunter valley—the Hunter Structural Movement—comprises folding from 258 to 253 Ma and faulting from 255 to 253 Ma (Stuntz, 1972), including transtensional faulting after 253 Ma (Herbert, 1993). The Muswellbrook Anticline started to grow at 258 Ma (Veevers, 1960). In addition to polyphase deformation, the Nambucca Slate Belt (NSB) was subjected to low-grade regional metamorphism and, at Wongwibinda (W), low P/T metamorphism, all dated as 255 Ma (Leitch and McDougall, 1979).

Detrital sediment and ash-fall tuff at Newcastle point eastward to vents in the Currarong Orogen (Jones et al., 1984) that trends south-southwest through the older Gerringong Volcanics and Dampier Ridge (DR) granitoid to the Milton Mon-

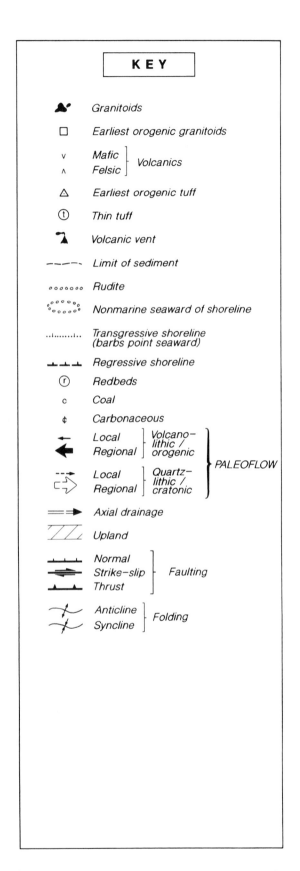

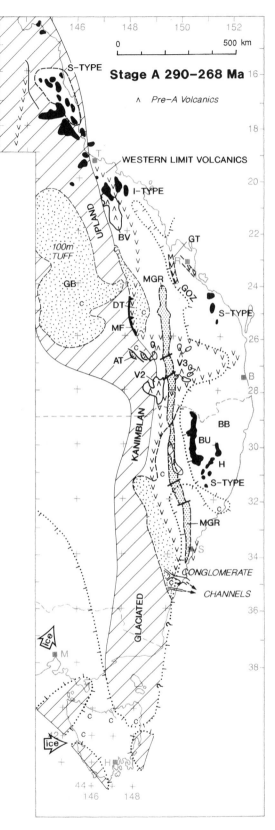

Figure 41A. Paleogeographical/paleotectonic map of eastern Australia: Stage A (290–268 Ma). Abbreviations given in Table 10. Tasmania (with present-day coordinates) is shown in its restored position.

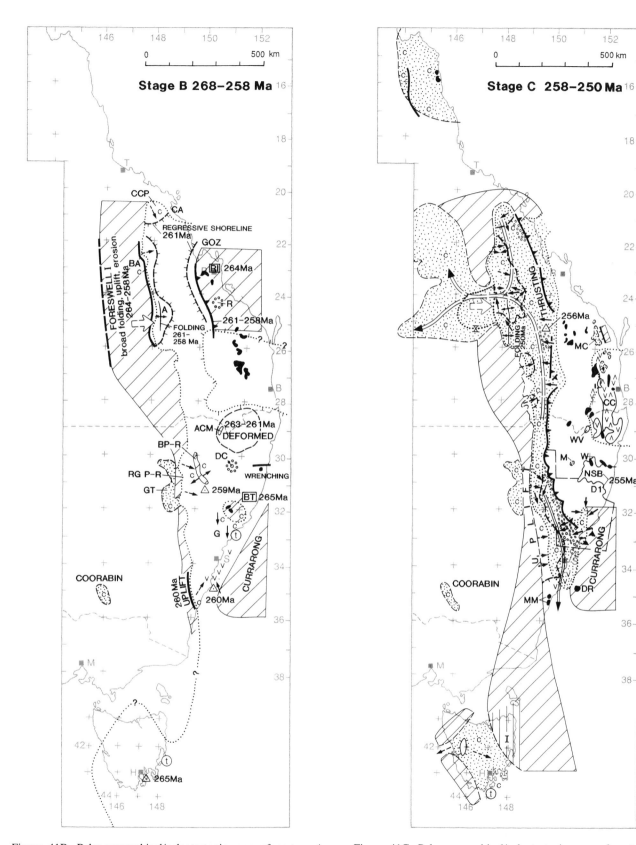

Figure 41B. Paleogeographical/paleotectonic map of eastern Australia: Stage B (268–258 Ma).

Figure 41C. Paleogeographical/paleotectonic map of eastern Australia: Stage C (258–250 Ma).

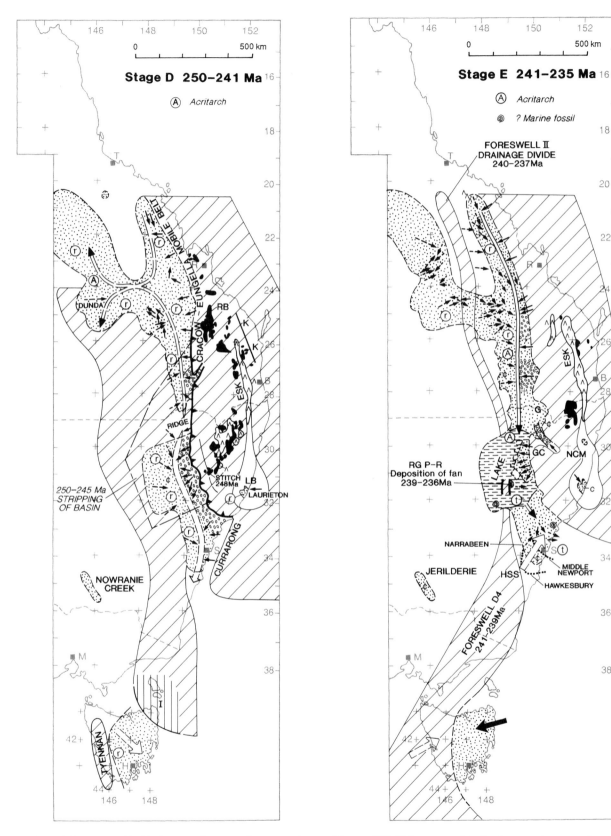

Figure 41D. Paleogeographical/paleotectonic map of eastern Australia: Stage D (250–241 Ma).

Figure 41E. Paleogeographical/paleotectonic map of eastern Australia: Stage E (241–235 Ma).

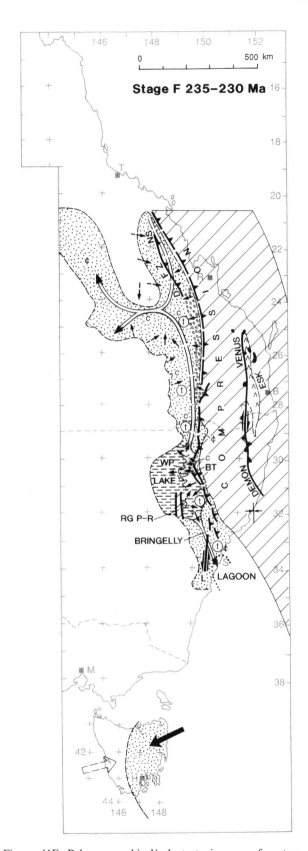

Figure 41F. Paleogeographical/paleotectonic map of eastern Australia: Stage F (235–230 Ma).

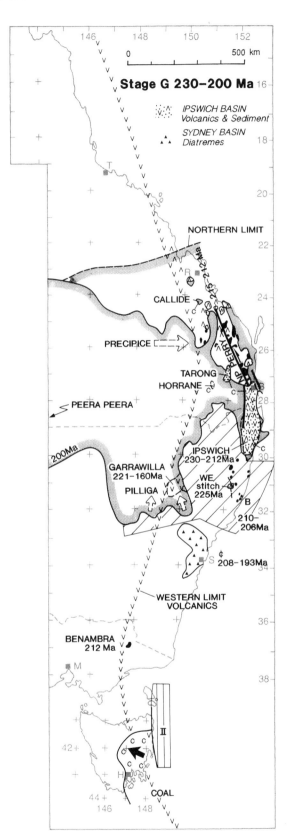

Figure 41G. Paleogeographical/paleotectonic map of eastern Australia: Stage G (230–200 Ma).

TABLE 10. ABBREVIATIONS USED IN FIGURE 41

Stage A			GOZ	Gogango Overfolded Zone		Stage E	
AT	Arbroath Trough		GT	Gilgandra Trough		G	Gunnee Formation
BB	Barnard Basin		R	Rainbow Creek Beds		GC	Gragin Conglomerate
BU	Bundarra Batholith		RG P-R	Rocky Glen paleo-ridge		HSS	Hawkesbury Sandstone
BV	Bulgonunna Volcanics		RI	Ridgelands Granodiorite		NCM	Nymboida Coal Measures
DT	Denison Trough					RG P-R	Rocky Glen paleo-ridge
GB	Galilee Basin		**Stage C**				
GOZ	Gogango Overfolded Zone		CC	Cressbrook Creek Group		**Stage F**	
GT	Grantleigh Trough		D1	Deformation 1		BT	Boggabri Thrust
H	Hillgrove Plutonic Suite		DR	Dampier Ridge granitoid		DFZ	Dawson Folded Zone
MF	Merivale Fault		I	Foreswell I		L	Lapstone Monocline
MGR	Meandarra Gravity Ridge		M	Merrastone Formation		NS	Nebo Synclinorium
V2	Combarngo Volcanics		MC	Moah Creek Beds		RG P-R	Rocky Glen paleo-ridge
V3	Kuttung Volcanics		MM	Milton Monzonite and O'Hara		WP	Wilga Park Anticline and gas field
				Head Gabbro			
			NSB	Nambucca Slate Belt		**Stage G**	
Stage B			W	Wongwibinda Complex		II	Foreswell II (Tasmania)
A	Aldebaran delta		WV	Wallangarra Volcanics		B	Brothers granitoid
ACM	Ashford Coal Measures		X	Transgression X		NP	North Pine Fault
BA	Blair Athol Coal Measures					WE	Werrikimbe Volcanic Complex
BP-R	Boggabri paleo-ridge		**Stage D**				
BT	Barrington Tops Granodiorite		I	Foreswell I			
CA	Calen Coal Measures		KK	Kin Kin Terrane			
CCP	Collinsville coastal plain		LB	Lorne Basin			
DC	Dummy Creek Conglomerate		RB	Rawbelle Batholith			
G	Greta Coal Measures						

zonite (MM) and O'Hara gabbro, and in interval B to the basalt at Kiama.

In the Gunnedah and northeast Sydney Basins and Gloucester Syncline, Stage C started with the major regression caused by paralic coal measures that prograded southwest from the inferred upland of the NEFB, followed by a brief marine ingression (dotted line in Fig. 41C) indicated by the Kulnura Marine Tongue in the Sydney Basin, the Speldon Formation in the Stroud-Gloucester Syncline, and the Arkarula Sandstone in the Gunnedah Basin. As seen in the southwestern Sydney Basin, the ingression was countered by a rising foreswell, marked by an erosional lacuna and by pebbles in the Kulnura Marine Tongue (Havord et al., 1984). On relaxation of the foreswell, the ingression reversed to regression with deposition of a sheet of quartzose conglomerate and sand. In the Gunnedah Basin, the ingressive Arkarula Sandstone was succeeded by the regressive "western sands," shed eastward from an uplifted craton, including the Rocky Glen paleo-ridge. In the Sydney Basin, the sheet is called the Cullen Bullen Subgroup, comprising, in ascending order, the Marrangaroo Conglomerate, Lithgow Coal, and the Blackmans Flat Conglomerate. Subsequently, the cratonic wedge was overrun by the orogenic facies of labile coal measures that extended to the western limit of the basin (broken line). Coal measures continued to accumulate in the Gilgandra Trough. North of Melbourne, the Coorabin Coal Measures were deposited on the craton.

In the Midlands area of Tasmania (Forsyth, 1984), the Upper Parmeener Supergroup rests with shallow incision on marine siltstone. Fluvial sand with coal around Cygnet in the south and in an arc from the coast north of Launceston to east

of Zeehan and then to the south pinched out against presumed uplands in the northeast, northwest, and southwest, with a short-lived drainage divide east of Zeehan. The environment was a broad sandy plain with an easterly paleoslope; the only sign of the orogen, so prominent at this time in the Sydney and Bowen Basins, is the tuff near Southeast Cape, but the northeast upland probably acted as the initial Foreswell I.

Stage D: Orogenic piedmont redbeds of the foreland basin, 250–241 Ma (Fig. 41D)

According to Fielding et al. (1990),

In latest Permian times this uplift [Cracow-Eungella Mobile Belt], coupled with an inferred climatic change, gave rise to the accumulation of reddened alluvial sediments throughout the basin, devoid of coal (Rewan Group). Unconformity separates the Rewan Group from the underlying Blackwater Group on the basin margins, while in the basin centre the relationship is conformable.

We add that the Rewan Group and succeeding Clematis Group are devoid of tuff also. The southern part of the Eungella-Cracow Mobile Belt was intruded by the main phase of the Rawbelle Batholith (RB) and the area west of Brisbane was split by the volcanic Esk Trough. We infer that tuff was absent in the basin because the adjacent plutons did not reach the surface. The Eungella-Cracow Mobile Belt shed even larger volumes of sediment, including the rudite of the Cabawin Formation in the Taroom Trough between 27°S and 28°S (Exon, 1976).

The regional drainage pattern maintains that of Stage C (Conaghan et al., 1982), with an east-trending ridge at 29.3°S separating the northward and southward drainage. From

32.5°S in the Gunnedah Basin northward to 26.5°S in the Bowen Basin, the Mooki-Goondiwindi-Moonie thrust fault became active, the Black Jack and Blackwater coal measures were tilted and stripped from 250–245 Ma, and they were overlapped by the conglomeratic base of the Early Triassic Digby Formation, Cabawin Conglomerate, and Rewan Group. In the Gunnedah Basin, quartzose sand from the west and a reduced supply of volcanogenic sand from the northeast and east reached equilibrium in a drainage axis that trended southeasterly and then southerly.

The New England batholith approached culmination with the eruption of the ignimbrite sheets of the 249 Ma Emmaville Volcanics and 248 Ma Dundee Rhyodacite. Elsewhere in New England, the Laurieton Conglomerate of the Camden Haven Group of the Lorne Basin (LB) was deposited in alluvial fans dominated by clasts of quartzite and chert, including jasper, that prograded westward over red playa mudstone from the uplifted sides of the basin. The Stage C shoreline in the Drake area had retreated from eastern Australia, as had the marine incursion shoreline of the Gunnedah-Sydney Basin.

In the Sydney Basin, the coal measures, which had extended across the entire basin, were succeeded abruptly at 250 Ma by conglomerate, sandstone, and shale. This involved a change from coal measures with ubiquitous tuff to wholly epiclastic sediment that lacked coal and tuff but contains the first redbeds.

The eastern part of the Gunnedah Basin was gently folded, uplifted, and stripped during the lacuna from 250 Ma (Blackjack) to 245 Ma (Digby) at the same time as the magmatic climax in the NEFB. The Blackjack and older formations were tilted along the Mooki Fault and at Bellata were stripped down to the Porcupine Formation before being covered by the conglomeratic facies of the Digby Formation (Etheridge, 1987). The tilted area continues north of Narrabri to 26°S where the Early Triassic conglomerate of the Cabawin Formation and the basal Rewan Group lapped against the tilted and uplifted coal measures to form a narrower basin during emplacement of Triassic granite and the first major overthrusting along the Moonie-Goondiwindi Fault (Exon, 1976). The folding in the Gunnedah Basin coincided with the magmatic climax in the NEFB. When deposition resumed at 245 Ma, the basal Digby Formation prograded transversely in a fan of orogenic sediment (arrows) across the stripped surface, 100 km past the Rocky Glen Ridge to overstep the Permian sediment to the broken line. Immediately afterward, the fan was overrun by a sheet of alluvial quartzose sand shed transversely from the craton (eastward- and southeastward-directed arrows). The quartzose sand blended with the orogenic sand in the trunk stream that flowed southward along the axis of the Gunnedah-Sydney Basin. In the Sydney Basin, the cratonic sand overran the orogenic sand at 250 Ma, some 5 m.y. earlier than in the Gunnedah Basin. The stage terminated a pause in the deposition of bed-load sediment derived from craton and orogen and the development first of redbed, and then of flint-clay paleosols.

In Tasmania, the interval with redbeds (250–235 Ma) is the same age as it is in the rest of eastern Australia. A well-drained plain of quartz sand was crossed by rivers of low sinuosity that flowed southeasterly from a presumed upland in the Tyennan region. "Northeastern Tasmania was either an area of non-deposition or of deposition and subsequent erosion. Distant vulcanism is suggested by the occurrence of smectite in some sandstones of this cycle" (Banks, 1989). In either case, northeastern Tasmania remained elevated in a presumed foreswell.

Stage E: Cratonic sand sheet of the foreland basin, 241–235 Ma (Fig. 41E)

A drainage divide (Foreswell II) in the interval 240–237 Ma arose between the Bowen foreland basin and the Galilee epicratonic basin, as indicated by the divergent pattern of sediment flow in the lower (Glenidal) part of the Clematis Group (240–237 Ma) (Jensen, 1975). Another change from the pattern of Stages C and D is that the axial flow of the Bowen Basin is wholly southerly, presumably caused by blockage of the westerly drainage by Forewell II along the line of the northern Nebine Ridge. The upper (Expedition) part of the Clematis Group reversed paleoflow in the Galilee Basin to cross the former Foreswell II (Jensen, 1975) and join the southerly axial drainage of the Bowen Basin at about 25°S. The southward-flowing stream system deposited the fluvio-lacustrine Showgrounds/Clematis Sandstone and debouched at about 29.5°S into a large lake that occupied the Gunnedah Basin.

In the NEFB, the volcanic Abercorn and Esk Troughs and associated plutons remained active, but to the south plutonism contracted to a focus about 29°S, 152°E. Deposition continued in the Lorne Basin, and at 30°S, the nonmarine Gunnee Formation encroached over the Peel Fault to the Bundarra Batholith on the western flank of the NEFB and the Nymboida Coal Measures (NCM) encroached westward in an intramontane basin across the Coffs Harbour Block. Part of the onlapping Gunnee Formation was itself buried by a large northwest-trending paleochannel of the Gragin Conglomerate (GC).

In the Bowen-Gunnedah-Sydney Basin, the paleogeography changed radically from Stage D to E. With the collapse of the ridge that separated the drainage in Stage D, drainage was continuous toward the south and southeast from the southern Bowen Basin through a lake that occupied the entire Gunnedah Basin and overflowed a rim at 32.5°S to feed sublithic bed-load sediment of the middle Newport axial drainage. The sea probably reached the Gunnedah Basin, as indicated by acritarchs (A) at 30°S and by the occurrence in the southwestern Gunnedah Basin of a merostome crustacean.

An upland in the southern NEFB shed mainly fine-grained volcanolithic sediment into the Gunnedah Basin lake where two successive delta systems and associated lacustrine turbidite fans prograded westward toward the Rocky Glen paleo-ridge, manifested by the development of a cratonic-sourced debris-fan that prograded eastward.

The Bowen Basin was delimited on the west by a fore-swell (II) at 240–237 Ma, which we extend southward to 29.5°S. The axial drainage is represented by the quartzose Showgrounds/Clematis Sandstone.

In the Sydney Basin, the southward axial fluvial flow of Stage D was replaced, after a lull represented by the Bald Hill/Wentworth Falls Claystone and succeeding flint-clay-stones of the Garie Formation in the east and Docker Head Claystone in the west, by (a) entry of water from the southern Bowen Basin into the northern Gunnedah Basin and associated deposition southward of fluvio-lacustrine quartzose sand and (b) a southeasterly axial flow that projected across the defunct Currarong Orogen. Fine-grained, predominantly quartzose/sublithic sediment indicates that the regional topography was subdued and hence that the Hunter-Mooki fault did not mark a scarp at the mountain front. Thrust movement still continued but did not produce a mountain front that shed gravel as it had earlier. Later still, the supply of orogenic sand from the southern NEFB failed, and the quartzose sand of the Hawkes-bury Sandstone swept northward at least to its preserved edge. In the south, the Hawkesbury Sandstone filled back beyond the Narrabeen Group over a surface that following tilting had been stripped down to the Wongawilli Coal of the Illawarra Coal Measures.

Toward the end of Stage E, the lake in the Gunnedah Basin expanded southward into the Sydney Basin and north-ward into the southern and central Bowen Basin.

Foreswell II between the Bowen and Galilee Basins (240–237 Ma), the Rocky Glen paleo-ridge (239–236 Ma), and D4 on the southwest edge of the Sydney Basin (241–239 Ma) are regarded as individual parts of a structure 1,500 km long that probably continued southward past Tasmania, as indicated there by an influx of quartzose sand from the southwest, countered by lithic sand from the northeast.

Stage F: Orogenic paralic sediment, 235–230 Ma (Fig. 41F)

Terminal thrusting affected the eastern half of the Bowen Basin, from the eastern edge through the Dawson Folded Zone (DFZ) (Dickins and Malone, 1973) and the Yerribee Zone. As in Stage E, dominantly fine-grained sediment suggests that the regional topography was subdued and hence that the Hunter-Mooki Fault did not mark a scarp at the mountain front. Thrust movement still continued (Glen and Beckett, 1989) but did not produce a mountain front that shed widespread gravel as it had earlier. Plutonic activity in New England is confined to the Demon Fault, which we regard therefore as the transpressional conduit of the magma.

In the Bowen Basin, the south-flowing drainage of Stage E changed by a reversal of slope in the southern Bowen Basin such that the lake contracted to the Gunnedah Basin and over-flowed northward to meet a southerly flowing axial stream at 24.5°S to flow westward to 147.5°E and thence divided into

northeast and southwest distributaries. Rudite was shed from the rising orogen.

In the Gunnedah Basin, the top of the Napperby Formation represents a continuation of the previous lake environment but with an expanded delta plain behind a delta front indicated by turbidite fans. The succeeding Deriah Formation filled the lake with the deposits of a low-sinuosity river, with flow vectors, indicated by the arrows, countered by fans shed from the Rocky Glen paleo-ridge (RG P-R).

In the Sydney Basin, the Minchinbury Sandstone and the Bringelly Shale of the Wianamatta Group were deposited from water that overflowed from the Gunnedah Basin lake. Large inclusions of shale with *A. parvispinosus* occur in Jurassic dia-tremes (Helby and Morgan, 1979; Crawford et al., 1980). We interpret the Wianamatta Group as the preferred host of the di-atremes, and extend its original occurrence to the northwest. The mean paleocurrent (arrow near Sydney) indicates axial drainage across a coastal plain up-slope from the regressive linear clastic shoreline (dotted line). This is the terminal stage of the orogenic sediment in the foreland basin. The previous flood of monotonously rhyodacitic material was replaced in the Minchinbury Sandstone by intermediate-volcanic fragments, reflecting a new intermediate source. Coaly intervals and tuff in all basins are the first since the Late Permian.

In Tasmania, lithic sand continued to be shed from the northeast.

Mid-Triassic terminal deformation and uplift. After the last recorded deposit of the stage, at 233 Ma, the NEFB and adjacent parts of the foreland basin were deformed, uplifted, and eroded. In New England, the Demon Fault underwent first, from 235–234 Ma, 6 km of right-lateral displacement and then, from 234 to 230 Ma, another 23 km. The Lorne Basin was folded and the proximal Sydney (Glen and Beckett, 1989) and Gunnedah Basins thrusted and folded. Etheridge (1987) shows the termination of high-angle reverse faults at the unconformity between the Gunnedah and Surat Basins in the Bellata area. According to Hamilton et al. (1988, p. 230),

Uplift in the Late Triassic was again most pronounced along the Boggabri Ridge [inferred Boggabri Thrust (BT) of Tadros, 1988, fig. 3] and progressively older Gunnedah Basin sediments subcrop and outcrop to the east. Vitrinite reflectance data from boreholes in the southeast of the basin confirm the removal of up to 2,000 m of Triassic and Permian sediments during the Late Triassic period of erosion. . . . . Periodic compressive and left-lateral strike-slip movements along the main Hunter-Mooki Fault have resulted in the formation of a number of high relief (sometimes en-echelon) anticlines in front of the main thrust [as for example, the Wilga Park (WP) gas field] (Hamilton et al., 1988, p. 218).

In the Sydney Basin, many pieces of disparate evidence point to major thrusting at the end of the Middle Triassic. According to Herbert's (1989) seismic study, the most probable period for the main movement on the Lapstone (L) Monocline-Nepean Fault system, a set of en echelon high-angle reverse

faults, appears to be during the Late Triassic after the last episode of deposition in the Sydney Basin. From Branagan and Pedram's (1990) report of shale injections into the overlying sandstone in the Lapstone structure, we suggest that thrusting may have taken place soon after deposition. Herbert (1989) remarks further that "the en echelon, discontinuous, left and right stepping characteristics of the major fault planes also indicate a degree of wrenching." From a study of northwest-trending thrust faults in the Hawkesbury Sandstone near Mooney Mooney Creek Bridge, 40 km north of Sydney, Mills et al. (1989) determined the transpressive motion as sinistral.

Stage G: Right-lateral transtensional rifting of the orogen, 230–200 Ma (Fig. 41G)

The only other known mid-Triassic sinistral transpression in eastern Australia is that along the Perry Fault, in southeastern Queensland, dated 215–212 Ma. The 222–215 Ma Aranbanga Volcanic Group (Cranfield and Murray, 1989) stitched the Perry Fault, whose final movement displaced the north-trending Venus Fault from the northern end of the Demon Fault. The Venus Fault itself was stitched by the 228 ± 4 Ma Hogback Granite. The transition from the previous stage was therefore marked by a change in sense, from dextral to sinistral, of transcurrent motion on faults.

Deposition was confined to the intramontane half grabens of the Late Triassic Ipswich/Clarence-Moreton Basin and Tarong Basin, which accumulated initially tuff and basalt then coal measures, in a return to abundant coal after the coal gap of the Early and Middle Triassic. Nearly 1,000 km to the west, above the Cooper Basin, the carbonaceous Peera Peera Formation initiated the Great Artesian Basin in the sump of the lowlands between the western-central craton and the NEFB. Only by the end of this stage, at 200 Ma, did the sheet of quartzose sediment of the Precipice Sandstone and equivalents (Martin, 1981) encroach and, in places, cross the eroded foreland basin and orogen.

In the south, igneous activity was in four age groups, from young to old:

(4) Maar volcanic diatremes of the Sydney Basin, dated by palynomorphs as earliest Jurassic, 208–193 Ma;

(3) The Brothers (B) plutons in the Lorne Basin, 210–206 Ma;

(2) The Garrawilla Volcanics (basalt) of the Surat Basin, 221–203–?160 Ma; and

(1) Plutons and the Werrikimbe ignimbrite, 228–222 Ma. The Werrikimbe ignimbrite is the first (preserved) surface igneous rock since the Permian-Triassic climax.

In Tasmania, after a brief lacuna, volcanolithic sands and coal measures were deposited from highly sinuous streams with a median paleocurrent azimuth west of northwest. The volcanogenic sand is mainly intermediate and to a less extent silicic and mafic. Rhyolitic ash-flow tuff is known near the exposed top. An increase in the volume and calibre of conglomerate in the northeast suggest another uplift, interpreted as a recrudescent Foreswell II. The eastern half of Tasmania south of 41.5°S was covered by coal measures.

Principal events—time-space composite

Changing sea/land levels (right-hand columns in Fig. 42). After the Late Carboniferous lacuna in all areas except the northern NEFB, which had continuous marine deposition, the marine record starts with the main transgression (MT in Fig. 42) at 280–277 Ma and the start of the main regression (MR) at 251 Ma. In between were the following:

(a) A regression in Tasmania and in the southern NEFB at 273 Ma,

(b) A transgression in the Bowen and Tasmania Basins at 268 Ma countered by a regression in the southern NEFB,

(c) A transgression in the Gunnedah-Sydney Basin at 261 Ma preceded by a regression at 262 Ma in the Bowen Basin,

(d) A transgression in the Bowen Basin at 258 Ma countered by a regression in the other three basins.

The regressive Late Permian was punctuated by marine ingressions (I in Fig. 42), shown by broad arrows, at 257–255 Ma in all basins except Tasmania and at 252 Ma in the Sydney and Bowen Basins. The Early and Middle Triassic record is apparently nonmarine except for incursions at 241–240 Ma in the Gunnedah-Sydney Basin and at 235 Ma in the Bowen and Sydney Basins. An incursion in the Galilee Basin at 244 Ma is suspected from the occurrence of acritarchs in Jericho-1 (Evans, 1980, p. 302).

The 280–277 Ma main transgression (MT) in Eastern Australian palynological zone 3a corresponds with the Gondwanaland transgression of the Tastubian *Eurydesma* zone (Dickins, 1984) and the Laurussian transgression of the Tastubian Neal-Lenox third-order cycles. The 251 Ma main regression (MR) at the boundary between the palynological zone 5 and the *P. crenulata* zone, Chhidruan and earliest Djulfian (Foster, 1982) but according to Morante (1993) and Morante et al. (1994) at or close to the P/Tr boundary, corresponds with the initial part of the minimum stand of sea level at the Permian/Triassic boundary of 250 Ma on our time scale. The correspondence of these large and well-resolved features in both Laurussian-Tethys and eastern Australia suggests that they are global or eustatic. The Tastubian transgression immediately follows the main melting of the Gondwanaland ice sheets and is interpreted as glacioeustatic (e.g., Veevers and Powell, 1987). Regression a corresponds with the sharp drop of sea level between the Lenox and Sterlitamakian. The smaller features b, c, and d in the eastern Australian 270–255 Ma transgressive-regressive record are poorly resolved and cannot be securely correlated with either the transgressive or regressive part of the 2-m.y.-long cycles that characterize this interval. Possible associations are the following: transgression b (countered by a regression in the southern NEFB) with the

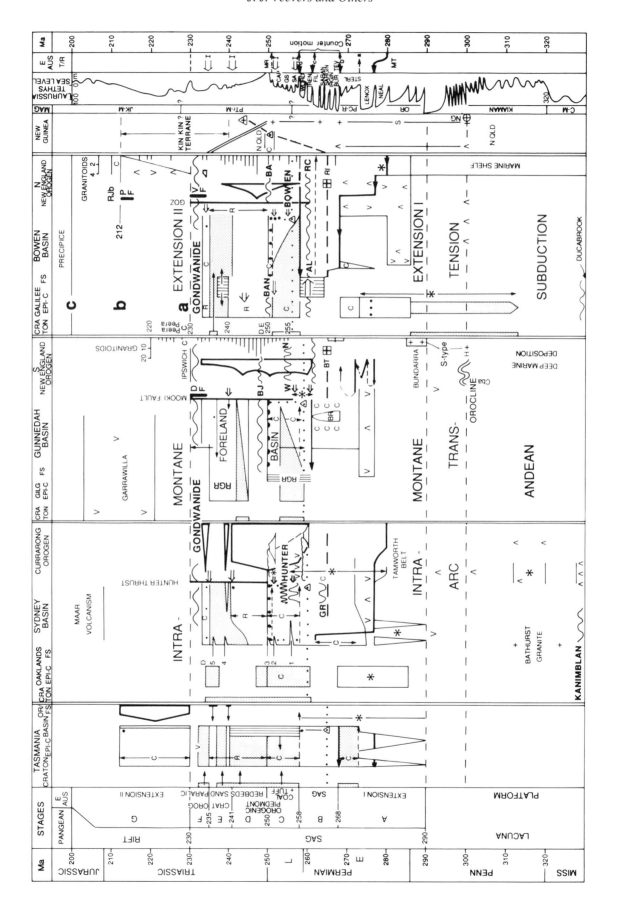

sharp rise of sea level at the base of the Burstev; regression c with the end of Filippov and transgression c with the Irenian transgression; and regression d with that of the Word and the apparently coeval transgression d in the Bowen Basin and slightly younger incursion in the Gunnedah Basin with the transgression of the San Andreas (SA).

Magnetic-polarity bias superchrons (fourth column from the right in Fig. 42). Menning (1986) and Steiner et al. (1989) place the boundary (the Illawarra Reversal) between the Permo-Carboniferous (PC-R) or Kiaman reversed superchron and the Permo-Triassic mixed superchron (PTr-M) at about the middle of the Capitanian equivalent to the base of the Tatarian, dated on the DNAG scale (Palmer, 1983) as 253 Ma, which supersedes Harland et al.'s (1990, p. 165–167) placement "at the mid-point of the Tatarian Stage." Larson and Olson (1991) link the change in magnetic reversal frequency at the long Cretaceous magnetic normal superchron to a change in core convective activity that accompanied the initiation of mantle plumes some few million years before their appearance at the surface as a hotspot. The estimated age of the

Illawarra Reversal—253 Ma—is consistent with the initiation of the plume at the core-mantle boundary and the appearance at the surface of the Siberian Traps 248.4 ± 2.4 Ma ago (Renne and Basu, 1991), some few to many million years later, with attendant immediate SO_2 cooling and subsequent climatic (greenhouse) warming about the Permian-Triassic boundary (Conaghan et al., 1994).

Permian-Triassic deformational events. From young to old order of age, the deformational events (Fig. 42)—extensional, compressive, and transcurrent—are as follows:

230 Ma. Intra-montane extension II, with individual phases a, b, and c.

233–230 Ma. Gondwanide, terminal overthrusting of the Hunter-Mooki Fault system; right-lateral movement on the Demon Fault (DF)–Venus Fault (VF).

250 Ma. Newcastle (NE)-Blackjack (BJ)-Bandanna (BAN)-Baralaba (BA), broad folding at end of coal-measure deposition, in Sydney Basin associated with sinistral transcurrent faults (Herbert, 1993).

255 Ma. Hunter-Wongwibinda (W)-Nambucca (N)-Bowen, intense folding/faulting and metamorphism in south NEFB, moderate folding in adjacent part of basins.

260 Ma. Aldebaran (AL)–Rainbow Creek (RC), intense folding/faulting in north NEFB and adjacent Bowen Basin, moderate folding in southwest Bowen Basin.

265 Ma. Greta (GR), coarse deposits of the embryonic foreland basin derived from inferred deformational uplift of the embryonic orogenic arc.

290 Ma. Intra-montane extension I.

Conclusions

1. During the Early Carboniferous, the Lachlan Fold Belt was terminally folded (Kanimblan event) during shallowing subduction and emplacement of a terrane within the subduction complex of the embryo NEFB, followed in the LFB by mega-kinking and intrusion of I-type granitoids. During the Late Carboniferous, the NEFB was uplifted, denuded, and intruded by S-type granitoids with co-magmatic ignimbrites during right-lateral transtension to form the initial stage of an orocline.

2. Thereafter to the end of the Triassic, eastern Australia developed through seven stages:

Stage A (290–268 Ma), extension-volcanism of the collapsed Kanimblan-NEFB upland, accumulated a thick succession of volcanics and sediment.

Stage B (268–258 Ma), a marine sag on the platform and embryonic magmatic arc/foreland basin, involved wider subsidence so that the sea reached the western edge of the Bowen-Gunnedah-Sydney Basin and most of Tasmania. The first granitoid of the magmatic arc, the 265 Ma Barrington Tops Granodiorite, was followed by deposition of the coarse Greta Coal Measures in the early foreland basin. The first arc-granitoid in Queensland was the Ridgelands Granite. The first tuff attributable to the north-migrating convergent magmatic arc

Figure 42. Time-space diagrams across parts of eastern Australia. Column head abbreviations are EPI-C—epi-cratonic basin, OR—orogen, FS—foreswell, GILG—Gilgandra, MAG—magnetic-polarity bias superchrons, E Aus T/R—Eastern Australian Permian and Triassic transgressive/regressive record. Deformation events are shown in bold: AL—Aldebaran, BA—Baralaba, BAN—Bandanna, BJ—Black Jack, DF—Demon Fault, GR—Greta, N—Nambucca, PF—Perry Fault, RC—Rainbow Creek, VF—Venus Fault, W—Wongwibinda. Other abbreviations are BR—Boggabri Ridge, BT—Barrington Tops Granodiorite, C—coal, Cba—Mount Barney Beds, D—Dundee Rhyodacite, E—Emmaville Volcanics, H—Hillgrove plutonic suite, N QLD—north Queensland, R—redbeds, RGR—Rocky Glen Ridge. D1–5 on the western side of the Sydney Basin signify disconformities, and the triangles on the east represent proximal volcanic vents. The asterisk indicates glacigenic sediment. The close vertical lines indicate the foreswell; the stippling on the left indicates the cratonic source of quartzose sediment; the heavy lines on the right show the orogenic source. The coarse dot within the triangle indicates the first tuff from the magmatic arc; later tuffs are shown by fine dot (minor) and coarse (major). Pangean stages (on far left) are from Veevers (1990a) and are arranged alongside eastern Australian stages. On far right, magnetic-polarity bias superchrons are from Harland et al. (1990, p. 166), with the base of the Permo-Triassic mixed superchron changed to the base of the Tatarian, in our scale 253 Ma, after Menning (1986). The Laurussian/Tethyan sea-level curve was compiled for 330–280 Ma from Ross and Ross (1987, p. 143), for 280–250 Ma from Ross and Ross (1987, p. 145), both from the North American and eastern Russian platform, and for 250–200 Ma, mainly from Tethys and North America, from Haq et al. (1988, p. 100, fig. 17). Abbreviations of third-order cycles, from the top down, are CAP—Capitan, GS—Goat Seep, SA—San Andreas, FIL—Filippov, STERL—Sterlitamak. The last column on the right is the Permian and Triassic marine transgressive/regressive record of eastern Australia. Heavy arrow to the left indicates transgressive; broken or short arrow to the right indicates regressive.

reached Tasmania at 265 Ma, the Sydney Basin at 260 Ma, and the Bowen Basin at 256 Ma.

Stage C (258–250 Ma), orogenic piedmont coal/tuff and regressive marine sediment, initiated the foreland basin between the initial uplift of the mature NEFB and a foreswell that bounded the Galilee, Gilgandra, Coorabin (Oaklands), and Tasmania Basins of the western craton.

In Stage D (250–241 Ma), orogenic piedmont redbeds, the sea retreated from the foreland basin during a climactic outpouring of volcanolithic sediment associated with granitoid emplacement in the orogenic upland.

In Stage E (241–235 Ma), a sheet of cratonic quartz sand swept across the Sydney and Bowen Basins, and a lake covered the Gunnedah Basin, which continued to receive volcanogenic sediment.

Stage F (235–230 Ma), orogenic paralic sediment, saw a final pulse of sediment shed into the foreland basin from the terminally overthrusted NEFB, which itself was wrenched to the right.

Stage G (230–200 Ma) involved rifting of the orogen by right-lateral transtension to form the Tarong and Ipswich coal and volcanic basins. By the Early Jurassic, cratonic quartz sand breached the NEFB upland at the Queensland–New South Wales border.

3. The extensional Stages A and G are local manifestations of Pangean stages of heat release.

4. Stage F is a local manifestation of the Gondwanide shortening event registered along the Panthalassan margin of the Gondwanaland province of Pangea.

5. The inception of the foreland basin along the Panthalassan margin, indicated by the onset of arc-related magmatism in the orogen, ranges from 286 Ma in Argentina, 277 Ma in southern Africa (Powell et al., 1991), 265 Ma in New Zealand, off Tasmania, and in the NEFB, 260 Ma tuff in the Sydney Basin, 259 Ma tuff in the Gunnedah Basin, 256 Ma tuff in the Bowen Basin, 250 Ma tuff in North Queensland, and 244 Ma granitoid in New Guinea.

6. The modern New England upland dates from the mid-Cretaceous inception of the Eastern Highlands (Jones and Veevers, 1983), probably by the underplating of the upper-plate margin during the separation of the lower-plate Lord Howe Rise (Lister et al., 1986). The New England upland was lifted higher than most other parts of the Eastern Highlands possibly because of its orogenic crustal structure.

ACKNOWLEDGMENTS

We acknowledge the help of David Briggs in supplying information ahead of publication on his Permian zonation, and we thank Feng Xu Jian and Colin Ward for Gunnedah Basin data, also ahead of publication. Cec Murray, John Roberts, Erwin Scheibner, and Colin Ward provided valuable discussion at a workshop on this chapter at Macquarie University in February 1992. Personal communications were received from Robert Barnes, John Brunton, Fernando Della-Pasqua, Dick Flood, David Grybowski, Chris Herbert, Brian Jones, Evan Leitch, Andrew McMinn, Cec Murray, Ric Morante, Ian Percival, Stirling Shaw, N. Z. Tadros, Colin Ward, and E. K. Yoo. Clive Calver supplied the materials for David Hilyard to compile the cross section of Tasmania (XIV in Fig. 15); David Durney helped with re-scaling a time-space diagram; and Adam Bryant, David Hilyard, Stuart McCracken, Ian Percival, and Andrew Roach helped with computing. Most of the figures were drafted by Judy Davis; the rest were done by Ken Roussell and John Cleasby. We are grateful to the GSA reviewers, Evan Leitch, Cec Murray, and Erwin Scheibner, for the effort they put into their constructive criticism of the manuscript. This work was supported by grants from the Australian Research Grants Scheme and the Australian Research Council.

APPENDIX 1. LATE CARBONIFEROUS-PERMIAN-TRIASSIC CORRELATION CHARTS FOR AUSTRALIA

J. J. Veevers

Biostratigraphical schemes
Primary anchor points.

(1) Invertebrate Stage A of Western Australia, including the marine *Eurydesma* fauna, distributed in all the Gondwanaland continents except Antarctica (Dickins, 1984), and correlated to the Tastubian Sub-stage of the international standard of the Urals by Tastubian ammonoids in the Western Australian basins (Archbold, 1982).

(2) The boundary between palynological zones VIII/*P. reticulatus* in the Canning Basin, traced by $\delta^{13}C$ (Morante, 1993; Morante et al., 1994) to the P/Tr boundary in the Chinese stratotype (Chen et al., 1991), dated as 251.2 ± 3.4 Ma (Claoué-Long et al., 1991). The radiometric estimate of the P/Tr boundary is confirmed by data from eastern Australia. Using the U/Pb method on bulk zircons and K/Ar on hornblende, Gulson et al. (1990) dated the Awaba Tuff, stratigraphically 50 m below the P/Tr boundary at the Newcastle Coal Measures/Narrabeen Group boundary, as 256 ± 4 Ma. This approaches the value of 250 Ma, which I take as the age of the P/Tr boundary in the Chinese stratotype, direct dated as 251.2 ± 3.4 Ma (Claoué-Long et al., 1991), and confirmed by the Rb/Sr biotite dates of 249.6 ± 1 Ma and 250.1 ± 1 Ma (duplicates) and 252.4 ± 1 Ma from the topmost Blackjack Formation of the Gunnedah Basin (Veevers et al., 1994a). The 250 Ma estimate supersedes the 245 Ma P/Tr date in Palmer (1983). Other dates from eastern Australia are cited below.

Secondary Anchor points. The palynological zones of Kemp et al. (1977) and Balme (1980), with Stage 3a correlated with Tastubian invertebrates of Stage A (Archbold, 1982); the *G. confluens* zone of Foster and Waterhouse (1988), correlated with the lower part of Stage 3a is hence early Tastubian. A sharp increase in palynological diversity is found in Euramerica at the base of the Permian and in Australia between Stages 2 and 3 (Balme, 1980) but correlating this event is uncertain because it creates a gap throughout the Asselian (Archbold, 1982, 1984); moreover, a change of this kind is likely to be diachronous due to the different latitudes of Euramerica and Gondwanaland (Foster, 1983). Accordingly, I adopt Archbold's scheme whereby the boundary between the Carboniferous and Permian is approximated by the boundary between palynological stages 1 and 2. Stage 1 is dominated by monosaccate pollen and is probably controlled by environmental factors or facies rather than age (Foster and Waterhouse, 1988, p. 143). In places where Stage 1 is missing,

Stage 2 locally overlies the *D. birkheadensis* zone of Powis (1984), equivalent to Biozone C of Jones and Truswell (1992), suggesting the wedge shown in Figure 44 in the section on eastern Australia (R. Helby, personal communication, 1989).

Western Australia (Fig. 43)

A comprehensive listing of formations is given in the Phanerozoic correlation chart in Memoir 3 of the Geological Survey of Western Australia (1990).

Palynostratigraphic units, Late Carboniferous. As arranged in Veevers and Powell (1987) except Carboniferous/Permian boundary near Stage 2/Stage 1, and zones before I, from Jones and Truswell (1992).

Palynostratigraphic units, latest Carboniferous and Permian. I to VIII: Kemp et al. (1977, figs. 5, 6), modified by Backhouse (1991, fig. 10). *Granulatisporites confluens* zone (Foster and Waterhouse, 1988) at top of II, equivalent to the lower part of Stage 3a (R. J. Helby, personal communication, 1989), as further indicated by the association in the Grant Formation with *Strophalosia* cf. *subcircularis,* also in the Wasp Head Formation (= Briggs's zone D), and of other invertebrates indicating "correlation with Faunizones 1–2 of Tasmania and the Allandale (*s.l.*) faunas of New South Wales" (Foster and Waterhouse, 1988, p. 135), all of which overlap the lower part of Stage 3a.

Permian/Triassic (P/Tr) boundary. Morante (1993) and Morante et al. (1994) found a downward offset in the $\delta^{13}C_{org}$ curve in core holes in the Paradise area of the Canning Basin. The offset occurs between the Upper Stage 5 and *Protohaploxypinus reticulatus* palynomorph zones, within the lower third of what the well-site geologist regarded as the Hardman Formation but above the shelly horizons, with the base of the overlying Blina Shale 150 m above the P/Tr boundary. I identify the entire column of mudstone and shale from 0–250 m as the Blina Shale, and from 250–330 m, the base of the shelly horizons, I identify as the Hardman Formation, so that the P/Tr boundary lies 5 m below the Blina/Hardman boundary.

Palynostratigraphic units, Triassic. K. saeptatus to M. crenulatus (Dolby and Balme, 1976, Helby et al., 1987). Age of zones above early Anisian *T. playfordii* (Dolby and Balme, 1976):

S. quadrifidus. Via Malagasy and Europe, "From lower Anisian to some horizon within the Carnian"; mid-Ladinian to end Carnian (Helby et al., 1987).

S. speciosus. Via Europe, upper Carnian; via Malagasy, late Middle Triassic to Rhaetian; Norian (Helby et al., 1987).

A. reducta. Rhaetian to Hettangian (Helby et al., 1987).

M. crenulatus. Late Triassic, possibly older than Rhaetian.

Faunal stages A–F (Dickins, 1976). Preceded by the Visean/Namurian brachiopod zones of Roberts (1985, p. 107). A problem of correlation arises from the paucity, if not absence, of stratigraphically significant fossils about the Permian/Carboniferous boundary—"Marine Asselian faunas are poorly known in the Gondwana realm" (Archbold, 1982, p. 272). Palynomorphs are less restricted and on arguments adduced from their relation with invertebrate faunas (Archbold, 1982, 1984), as given above, can be traced back into the Stephanian.

Anchor points (shown by vertical bars). (A) Ammonoids (Archbold, 1982, table 2): Urals ammonoid zones 4 and 5 equivalent to Tastubian (Holmwood), 6 equivalent to Sterlitamakian (Callytharra, Nura Nura). (b) Stage A (*Eurydesma*) fauna (Dickins, 1984, fig. 2), Tastubian. (c) Stage F (Hardman fauna) (= VIII, Kemp et al., 1977) correlated with the Djulfian/Chhidruan Kalabagh Member and Chhidru Formation of the Salt Range (Dickins and Shah, 1979), the lower part of the Hardman with the Chhidruan (Archbold, 1988a), and the ammonoid *Cyclolobus* in the upper part of the Hardman Formation indicates Djulfian/Chhidruan (Glenister et al., 1990). Morante (1993) and Morante et al. (1994) show by isotope correlation that the Hardman extends to the P/Tr boundary. (d) Kockatea fauna (= *K. saepta-*

tus Zone, Dolby and Balme, 1976): on the basis of ammonoids regarded as Griesbachian to Smithian (McTavish and Dickins, 1974). (e) Locker Shale conodonts (McTavish, 1973, McTavish and Dickins, 1974): Late Dienerian, Smithian, and Spathian. *T. playfordii* Zone associated with late Smithian, Spathian, and possibly early Anisian conodonts (Dolby and Balme, 1976, p. 120). (f) Ashmore Block conodonts (Jones and Nicoll, 1984): earliest Norian.

Collie Basin. Kemp et al. (1977), Backhouse (1991).

Perth Basin. Dolby and Balme (1976), in particular (p. 136) Lesueur equivalent to Ipswich Microflora; Kemp et al. (1977), Playford et al. (1975, 1976); in addition, for south, Foster (1979, p. 128), and north, Archbold (1982).

Carnarvon Basin onshore. Archbold (1982), Kemp et al. (1977), Dickins (1976). In addition, offshore (Exmouth Plateau): Late Triassic/Early Jurassic (Barber, 1982, p. 135). Also Norian lacuna in Hampton-1 (20°S) and widespread nonmarine deposition in the Dampier and Barrow Basins during the Carnian-Norian (Crostella and Barter, 1980, fig. 17). Brigadier Beds: early-middle Rhaetian. Middle Brigadier Beds to basal Dingo Siltstone: mid-Rhaetian to Pliensbachian (?early Toarcian) (Helby et al., 1987).

Canning Basin onshore. Kemp et al. (1977), Dickins (1976), Purcell (1984); Nura Nura/Cuncudgerie extended upward (*M. australe*) to Aktaskian (Archbold et al., 1993). The ammonoid *Cyclobus* in the upper part of the Hardman Formation indicates Djulfian/Chhidruan (Glenister et al., 1990), and brachiopods in the lower Hardman (Archbold, 1988a) indicate Chhidruan. Offshore Triassic from Powell (1976, fig. 10), including a lacuna at Lynher-1 from the Tatarian to the Norian, and, on the eastern flank of the Browse Basin, from the Artinskian to the Carnian (Powell, 1976, fig. 11).

Bonaparte Gulf Basin. Laws and Kraus (1974), Kemp et al. (1977), Brown (1980), Jones and Nicoll (1984), Roberts (1985). Forrest and Horstmann (1986) for Carnian lacuna in Petrel-2. Brachiopods and bivalves in the Hyland Bay Formation at bottom of Sahul Shoals-1 (SS-1) indicate Chhidruan (Archbold, 1988b). Helby et al. (1987) for Ashmore Block and Sahul Shoals-1 (SS-1): continuous succession to top of Norian.

Eastern Australia (Fig. 44)

Palynological zones, Late Carboniferous. As arranged by Veevers and Powell (1987), except Carboniferous/Permian boundary equivalent to Stage 1/Stage 2 (R. Helby, personal communication; Foster and Waterhouse, 1988, p. 143), and before Stage 1/2 from Jones and Truswell (1992).

Permo-Carboniferous palynological (mainly range) Zones 1–5. Kemp et al. (1977), Price (1983), Price et al. (1985); P3C acritarch swarm: Price (1983); Tatarian: Foster (1982, 1983).

Permo-Triassic. The *Playfordiaspora crenulata* zone straddles the Baralaba Coal Measures/Rewan boundary (Foster, 1982, 1983) and lies between the Upper 5 and *P. microcorpus* equivalent to *reticulatus* zones, all now shown as straddling the P/Tr boundary by palynological correlation with the P/Tr boundary in the Canning Basin determined by C isotopes (Morante, 1993; Morante et al., 1994), superseding the correlation by palynomorphs to the Salt Range of Pakistan (Balme, 1969, Foster, 1982, 1983), whereby the P/Tr boundary was placed some 200 m above the base of the Narrabeen or Rewan Group.

Triassic palynoflora. Helby (1973), de Jersey (1975), de Jersey and McKellar (1981), with *D. problematicus* zone equivalent to the combined *A. parvispinosus* zone and overlying *C. senectus* zone (A. McMinn, personal communication), summarized in Helby et al. (1987, p. 5) and Powis (1989, p. 268). *C. rotundus* zone, dated as Carnian (de Jersey, 1975), extends into the Norian (Price et al., 1985); (d) base of *Polycingulatisporites crenulatus* zone dated as Rhaetian from correlation with marine sequence of New Zealand (de

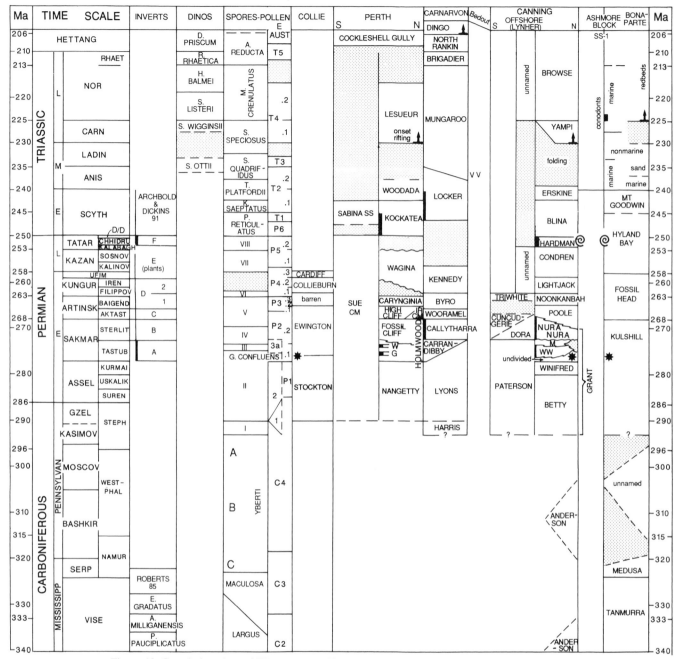

Figure 43. Correlation chart of Western Australia. Anchor points indicated by black bar, *G. conflu-ens* zone by asterisk, onset of rifting by arrow, lacuna by stipple, and marine fossils by coil.

Jersey, 1975), later regarded as (1) Norian- "Rhaetian" (McKellar, 1982), (2) Norian, probably extending to Hettangian (Helby et al., 1987). I accept Rhaetian to Hettangian (Price et al., 1985).
Carboniferous brachiopod zones. Roberts (1985, p. 17, 25, 30, 48, not 106), calibrated by ion-probe (SHRIMP) U-Pb ages on single zircon grains (Roberts et al., 1991). Also calibrated are Stage 3a (275 Ma in the Alum Mountain Volcanics) in agreement with the range of 277–273 Ma in the DNAG scale (Palmer, 1983).
Bowen Basin Permian faunal succession. Dickins (1976).
Bowen-Sydney Basin productid zones (Briggs, 1989a, 1989b, 1991, 1993). Here labeled A–W, correlated with Sydney Basin Permian

brachiopod zones (Runnegar and McClung, 1975, McClung, 1980b) and Tasmanian Faunizones 1–10 (Clarke and Banks, 1975, Clarke and Forsyth, 1989) by Briggs (1989a, 1989b, and personal communication, 1989). Briggs's (1989a, 1989b, and personal communication, 1989; V and W added from Briggs, 1991, and personal communication, 1991) zones are the following:

W, *Echinolosia ovalis*
V, *E.* n. sp. F
U, *"Wyndhamia" ingelarensis*
T, *E. minima*
S, *"W." clarkei*

R, *"W." blakei*
Q, *E.* n. sp. 6
P, *E.* n. sp. 5
O, *E.* n. sp. 4
N, *E.* n. sp. 3
M, *E. discinia*
L, *E.* n. sp. 1
K, *E. maxwelli*
J, *W. typica*
I, *E. preovalis-Ingelarella plana*
H, *E. preovalis-I. ovata*
G, *E. warwicki*
F, *E. curtosa*
E, *Bandoproductus* n. sp.
D, *Strophalosia subcircularis*
C, *I. elongata*
B, *"Trigonotreta campbelli"*
A, *Lyonia* n. sp.

Anchor points.

(a) Allandale fauna (in Lochinvar and Allandale Formations) equivalent to zones B–D and Tasmanian Faunizones 1, 2, and 3, Tastubian by correlation with Stage A of Western Australia (Archbold, 1982, Dickins, 1984, fig. 2).

(b) *Playfordiaspora crenulata* palynofloral zone, according to Foster (1982), is equivalent to the uppermost Chhidru Formation (latest Chhidruan or possibly early Djulfian), now regarded as straddling the P/Tr boundary from palynological correlation with the isotopically determined P/Tr boundary in the Canning Basin (Morante, 1993; Morante et al., 1994).

(c) *C. rotundus* zone, dated as Carnian (de Jersey, 1975), extends into the Norian (Price et al., 1985).

(d) Base of *Polycingulatisporites crenulatus* zone, dated as Rhaetian from correlation with marine sequence of New Zealand (de Jersey, 1975), later regarded as (1) Norian-"Rhaetian" (McKellar, 1982), (2) Norian, probably extends to Hettangian (Helby et al., 1987). I accept Rhaetian to Hettangian (Price et al., 1985).

The invertebrate zones of Waterhouse et al. (1983) are superseded by those of Briggs (1989a, 1989b; 1991). Briggs (personal communication, 1989) supplied the correlation with the zones of Runnegar and McClung (1975) and of the Tasmanian faunizones. Rare ammonoids in eastern Australia (Armstrong et al., 1967; Waterhouse et al., 1983, p. 124) do not provide anchor points as they do in Western Australia (Archbold, 1982), except the Aktaskian goniatite in the Yarrol Formation.

As noted previously, I follow Archbold's (1982) correlation of palynological Stages 1, 2, and 3a via invertebrates with the international standard in place of the schemes of Kemp et al. (1977) and Balme (1980).

Tasmania Basin.

Lower Parmeener Supergroup (Playford, 1965, Clarke and Banks, 1975, Clarke and Farmer, 1976, Kemp et al., 1977, Banks, 1978, Truswell, 1978, Clarke, 1985) and specific succession in southern Tasmania (Clarke and Forsythe, 1989) correlated by Briggs (1989, personal communication) such that the Lower Marine interval with Faunizone 1 starts in the Tastubian.

Upper Parmeener Supergroup (upper Stage 5 and younger) from Clarke and Forsyth (1989). In the Oatlands/Midlands area (Forsyth, 1984), Unit 1 of the Upper Parmeener Supergroup rests with light incision on Faunizone 10 siltstone. The lower member of the unit, correlated with the fluvial Cygnet Coal Measures, is pebble conglomerate and sandstone with rare very thin coals; the middle member is fine- to medium-grained sandstone with *Glossopteris,* which elsewhere in Australia is no younger than Permian or 250 Ma; and the upper member is mainly lutite, light bluish-gray with spotted pur-

ple-red mudstone overlain by layers with red-purple mud pellets; palynomorphs including a form not recorded below the Rewan Group indicate the upper *P. microcorpus* zone. In Figure 44, I have shown the lower and middle members as equivalent to the Cygnet Coal Measures *sensu stricto*; the upper member that lacks coal or *Glossopteris* is distinguished as "Midlands." Rare redbeds in "Midlands" continue upward to the base of the quartz sandstone and coal measures near the top of Unit 3. The interval with redbeds (250–235 Ma) is the same age as in the rest of eastern Australia.

Volcanism (Banks and Clarke, 1987, p. 13). (i) Metabentonite in limestone near Hobart, on Maria Island, and at Saltwater Lagoon, in Faunizone 5a (= Artinskian or 265 Ma). (ii) Cuspate shards, plagioclase crystals, feldspar-phyric clasts in Ferntree Group, Woodbridge, and montmorillonite near St Marys: Faunizone 10 (= Ufimian or 258 Ma). (iii) Cuspate fragments and volcaniclastic feldspar crystals in sandstone close to base of Upper Freshwater Group at Mt La Perouse—base Cygnet, Kalinovian, 257 Ma. (iv) Basalt flow in the coal measures at St Marys (K-Ar 233 ± 5 Ma), from Calver and Castleden (1981). (v) Tuff (Bicheno) between *rotundus* and *crenulatus* in coal measures (Bacon et al., 1989, p. 333–334), 214 ± 1 Ma.

Sydney Basin.

General (Retallack, 1980). Productid correlation by Briggs (1989a, 1989b), such that productid zones parallel the first appearances of the palynomorph *Dulhuntispora granulata* (Dg) (Price), *D. parvithola* (Dp) (Balme and Hennelly), and *Microreticulatisporites bitriangularis* (M) Balme and Hennelly. McMinn (1985) indicates Upper Stage 5a (first appearance of *D. parvithola*) in the Kulnura marine tongue (kmt), and Upper Stage 5c (first appearance of *M. bitriangularis*) in the Dempsey marine (De m) tongue.

South. Marine Shoalhaven Group (Runnegar, 1980a). Nonmarine Pigeon House (PH) Creek Siltstone to Clyde River Coal Measures range from Stage 3a to end of 4 (Evans et al., 1983; Evans, 1991). Illawarra Coal Measures (ICM), including (at top of Cumberland [CU] Subgroup) Erins Vale Formation with *D. parvithola* (McMinn, 1985), underlain by Gerringong Volcanics, dated as 258 Ma (*see* Table 3 and related text) and Broughton (BR) Formation. Weathered ash (Runnegar, 1980b, fig. 23.15) at Montague Point confirmed by Scott (1984) (Appendix 2) at top of Wandrawandian Siltstone (WA) equivalent to Filipp/Iren, 261 Ma. Other abbreviations are BE—Berry Formation, N—Nowra Sandstone, PE BCH—Pebbley Beach Formation, SP—Snapper Point Formation, W—Wasp Head Formation, Y—Yadboro Conglomerate.

In the west, an outlier of glacial beds in MacDonald Creek near Mudgee contains Stage 2 palynomorphs (McMinn, 1983).

North, Hunter region. Carboniferous from Roberts and Engel (1980) and Roberts (1985, p. 25). Seaham Formation formerly confined to Stage 1 (Helby, 1969, p. 264, 265) now moved down by the 320 Ma and 312 Ma zircon (z) ages given by Roberts et al. (1991); likewise Paterson Volcanics moved to 330 Ma. Permian from McClung (1980a, 1980b), with Lochinvar Formation, mainly mafic volcanics overlying glacial beds, and facies equivalents of the Cranky Corner sequence, and Allandale (AL in Fig. 44) Formation range from Stage 2 to 3a (Kemp et al., 1977; Archbold, 1982). Garretts Seam, near Raymond Terrace, is dated as *branxtonensis* zone (Retallack, 1980, p. 388), equivalent to the Farley Formation. Garretts Seam (Fig. 8) contains a tuff (David, 1907, p. 117) in continuation of the volcanics lower in the Dalwood Group. The oldest (259 Ma) tuff of extrabasinal origin is in the Mulbring Formation. Coal measures include the marine (m) Dempsey (DE) Formation and Kulnura marine tongue (kmt) (McMinn, 1985, 1986). Early Triassic from Helby and Morgan (1979), Helby et al. (1987). Igneous: Bathurst Granite 325 Ma (main) tailing off to 310 Ma (S. E. Shaw, personal communication, 1988); Rylstone Volcanics 292 Ma (Shaw et al., 1989). Diatremes have a black tuffaceous matrix that contains a diverse microflora of the

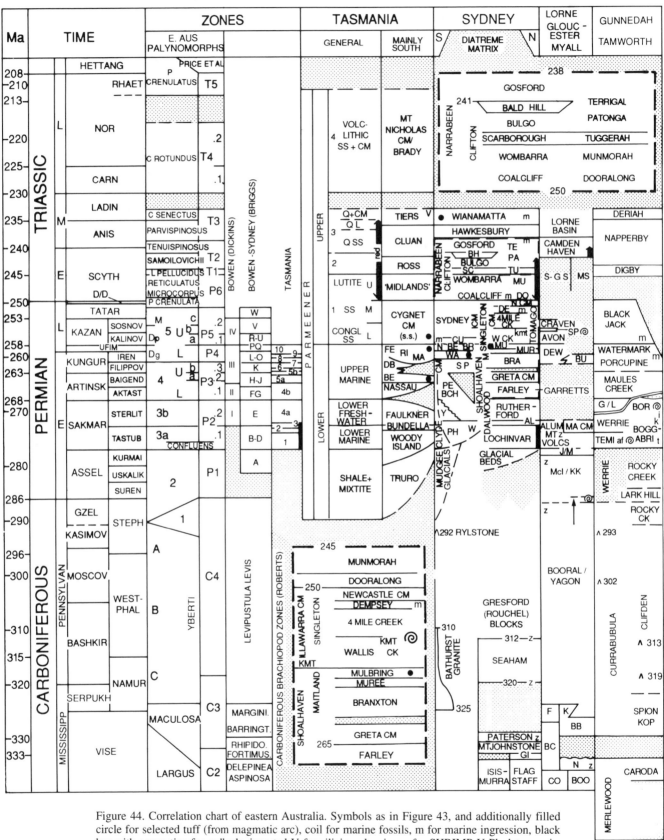

Figure 44. Correlation chart of eastern Australia. Symbols as in Figure 43, and additionally filled circle for selected tuff (from magmatic arc), coil for marine fossils, m for marine ingression, black bar with arrow tips for redbeds, inverted V for silicic volcanics, z for SHRIMP U-Pb date on zircon, plus sign for granitoid. Other abbreviations are given in Figures 7, 17, and 30.

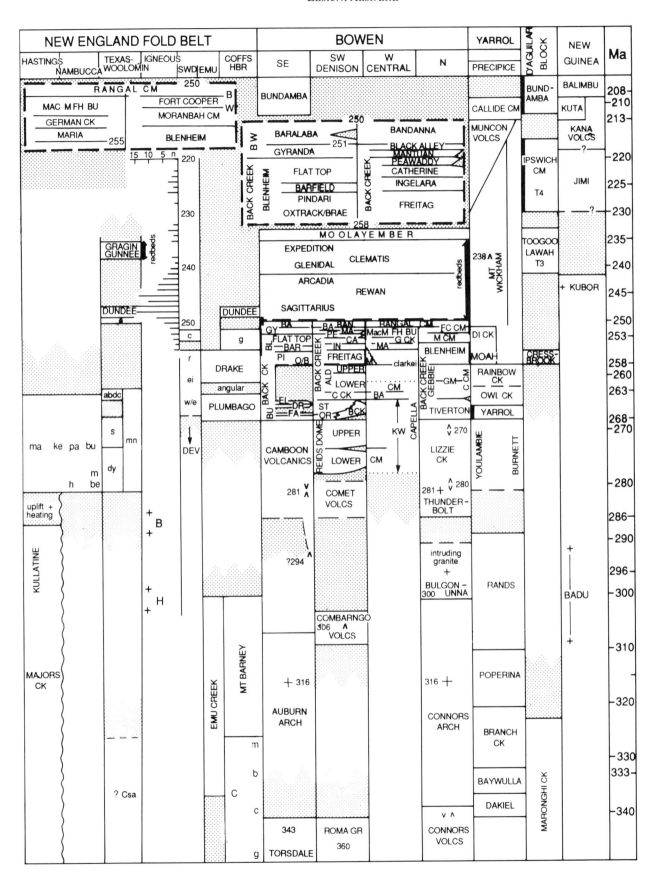

Classopollis classoides zone (= J1 interval zone of Price et al., 1985 = Sinemurian), derived from a surface layer of sediment present at the time of intrusion (Crawford et al., 1980). Abbreviated names are shown in full in the inset boxes at the bottom of the Tasmania column and at the top of the Sydney-Gunnedah columns.

Lorne Basin (Helby, 1973).

Stroud-Gloucester Syncline (S-GS) (Helby et al., 1986; McMinn, 1987).

Myall Syncline (MS) (McMinn, 1987; Roberts and Engel, 1987). The Alum Mountain Volcanics and Markwell Coal Measures contain Stage 3a palynomorphs (Helby et al., 1986; McMinn, 1987) and the Alum Mountain Volcanics 275 Ma zircons (z) (Roberts et al., 1991), to make a precise calibration with the time scale of Palmer (1983). The marginal marine Dewrang (DEW) Group contains invertebrate zones K–Q, palynological Stage Lower 5a (Helby et al., 1986), and the regressive Thirty Foot Coal; the Bulahdelah (BU) Formation contains K–M. In the Gloucester Coal Measures, the Craven and Avon Sub-Groups sandwich the marine Speldon Formation (Zone U), equivalent to the Kulnura Marine Tongue. Other abbreviations are J/M—Johnsons Creek/Muirs Creek, McI/KK—McInnes/Koolanock, F—Faulkland, K—Karuah, BB—Booti Booti, BC—Berrico Creek, N— Nerong, CO—Conger, BOO—Boolambayte.

Gunnedah Basin above and west of Tamworth Block.

Carboniferous. Ignimbrite from near top of Currabubula Formation has K/Ar ages of 293 ± 4 Ma on hornblende and 302 ± 4 Ma on plagioclase (McPhie, 1984). The Clifden Formation contains the Ermelo Dacite Tuff dated as 319 Ma and a higher tuff at 313 Ma (McKelvey and McPhie, 1985, p. 17, 19). See also Roberts and Engel (1980). Plutonic: Bundarra, 280 ± 5 Ma (Shaw and Flood, 1982) overlapped by 287 ± 10 Ma of Hensel et al. (1985); best estimate (R. H. Flood, personal communication, 1991) is 290–285 Ma. Hillgrove about 300 Ma (Hensel et al., 1985), 304–298 Ma by SHRIMP U/Pb on zircon (Landenberger et al., 1992), the same age as Bulgonunna Volcanics in northern Bowen Basin (McPhie et al., 1990). Barrington Tops Granodiorite: 265 Ma, 269 Ma. New England Batholith: 142 Rb/Sr biotite ages (Shaw, Appendix 3 in Chapter 3; Shaw et al., 1991) in 3 clusters: (1) from 255–mode 247–238 Ma, (2) 235–222 Ma, and (3) 210 Ma and 206 Ma.

Permian and Triassic. McPhie (1984), Beckett et al. (1983), Etheridge (1987, p. 6, 13, 14), and Hamilton et al. (1988). *Tamworth Block:* Runnegar (1970)—marine fossils in uppermost Temi Formation suggest Allandale fauna or Stage 3a. The overlying Werrie Formation comprises the Werrie Basalt and interlayered Loders Mount Beds (with marine fossils, Evan Leitch, personal communication, 1991) and the overlying Werris Creek/Willow Tree Coal measures (Britten and Hanlon, 1975, p. 238). The Werrie Formation is overlain by the Borambil Creek Formation (= unnamed marine sediment in Zone E of Briggs, 1991). Other abbreviations are BOR—Borambil Formation, G/L—Goonbri/Leard Formations.

Bowen Basin.

Palynology: Price et al. (1985), Price (1983). According to Briggs (1989a, p. 31), in the southwest Bowen Basin, the first appearance of *D. granulata (dulhuntyi)* is in the middle part of the Aldebaran Sandstone, of *D. parvithola* in uppermost Aldebaran, and *M. bitriangularis* in the middle Ingelara and, as implied in his table 1, in the Barfield Formation. The *M. evansii* acme-zone (P3c) lies near the base of the Black Alley Shale. Marine (productid) invertebrate zones are from Briggs (1989a, 1989b; 1991).

Southeast. Permian: Foster (1979, p. 126). Abbreviations in Figure 44, southeast Bowen Basin, are BU—Buffel Formation, comprising FA—Fairyland, DR—Dresden, and EL—Elvinia Members; others spelled out in inset box at top of column, except BW—Blackwater Group. Triassic: Foster (1982), de Jersey and McKellar (1981). $^{40}Ar/^{39}Ar$ ages of Camboon Andesite are 281 Ma and 294 Ma

(Runnegar, 1979, p. 271). The Oxtrack/Brae/Pindari Formations are plotted in the P5 zone (Price et al., 1985, fig. 9), equivalent to the Freitag Formation, against Briggs's (1989a) correlation with the lower Aldebaran Formation.

Southwest Bowen Basin or Denison Trough. Price et al. (1985). Part of column enlarged in inset at top of column, right side. Other abbreviations are BCK—Basin Creek, OR—Orion, ST—Stanleigh, C CK—Cattle Creek, ALD—Aldebaran.

West-Central. Foster (1979, 1983). The section at Capella contains sporomorphs very tentatively identified as Stages 3 (Wilson, 1975; Edenborough, 1985), 4, and Lower 5 (Wilson, 1975). Coal (CM in Fig. 44) occurs in Stages 3 and Upper 4/Lower 5. This interval, outlined by dots in Fig. 44, includes strata in the Karin (K) and Wolfang (W) outliers (Preston, 1985) and the Blair Athol (BA) outlier, dated as Upper 4a by Foster (1979). The Clermont inlier contains the *clarkei* bed (Veevers et al., 1964). The Moorlands (M) outlier contains palynomorphs that straddle Upper 5a/b (Price, 1983, p. 162).

North. Briggs (1989a, 1989b) supersedes Waterhouse and Jell (1983) and Waterhouse et al. (1983). Palynology by Foster (1982) and de Jersey and McKellar (1981). Radiometric ages of Lizzie Creek Volcanics (K/Ar, 270 Ma and 280 Ma) and Bulgonunna Volcanics (Rb/Sr, 297 ± 12 Ma) from Day et al. (1983), augmented by the 300 Ma age (U/Pb on single zircons) of the Bulgonunna Volcanics by McPhie et al. (1990). Stage 2 palynomorphs in Comet Volcanics from Kemp et al. (1977). C CM Collinsville Coal Measures, including GM marine Glendoo Member.

Yarrol Block.

Information up to and including Youlambie Conglomerate/Burnett Formation comes from Roberts (1985), above this from Dear et al. (1971) and Kirkegaard et al. (1970). Price et al. (1985, fig. 8a) place the Callide Coal Measures in the lower part of their palynological unit PT5 (Rhaetic), above PT4 (Ipswich Coal Measures). The Muncon Volcanics are probably the same age as the Ipswich Coal Measures (Carnian-Norian) and bracket the age of folding of the Bowen Basin to the interval 230 Ma above to 233 Ma below, the age of the folded Moolayember Group. The succession is completed by the Early Jurassic Precipice Sandstone. In Figure 44, Dinner Creek is abbreviated DI CK.

D'Aguilar Block.

Ipswich Coal Measures (= *C. rotundus*) are correlated with Western Australian Carnian *S. speciosus* and Norian *M. crenulatus* zones (Helby et al., 1987). Most importantly, the Ipswich Coal Measures mark resumption of deposition after folding of the early Ladinian Moolayember and older strata to the west.

New Guinea.

Skwarko et al. (1976). Badu Granite of Cape York-Oriomo Inlier from Day et al. (1983, p. 141).

Southern and central–eastern Australia (Fig. 45)

The *G. confluens* zone is recorded in the Stuart Range Formation and Cape Jervis Beds (Foster and Waterhouse, 1988).

Denman Basin.

Harris (1981), Wopfner (1981).

Arckaringa Basin.

Cooper (1981, 1983), Harris (1981), Price et al. (1985).

Pedirka Basin.

Harris (1981), Price et al. (1985), with Walkandi and basal Peera Peera T2 and T3, and then, after a lacuna, rest of Peera Peera T4 and T5 (Lower).

Troubridge Basin.

Cooper (1981), Wopfner (1981). Leigh Creek Coal Measures: lower part *C. rotundus* (de Jersey, 1975, p. 168), upper part *P. crenulatus* (p. 170).

Victoria.

Mainly Bacchus Marsh area. Archbold (1982) and Clark and

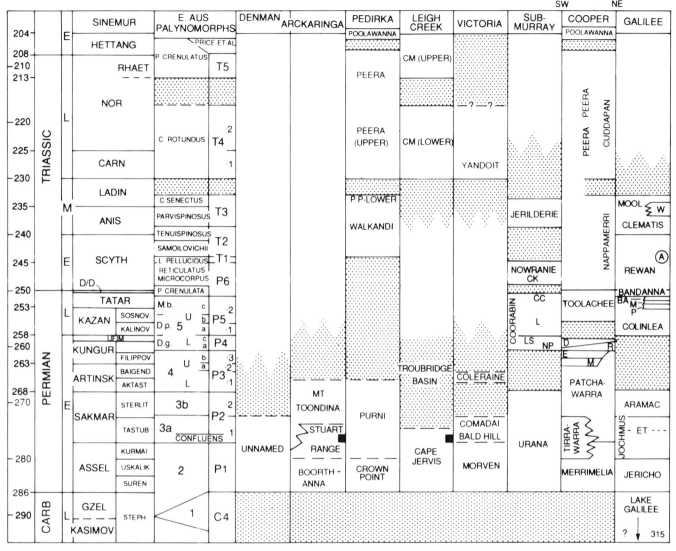

Figure 45. Correlation chart of southern and central-eastern Australia. Solid square indicates *G. confluens* zone. Abbreviations are A—acritarchs, BA—Black Alley, CC—Coreen Coal, CM—coal measures, D—Daralingie Beds, E—Epsilon Formation, ET—Edie Tuff, L—Loughmore Formation, LS—Lanes Shaft Coal, M—Murteree Shale, Mool—Moolayember, NP—Narrow Plain Formation, P—Peawaddy Formation, PP—Peera Peera Formation, R—Roseneath Shale, W—Warang Sandstone.

Cook (1988, p. 37, 38): Tastubian marine fossils (*Brachythyrinella* and *Notoconularia* at Comadai Creek. Bowen and Thomas (1976), Thomas (1969): general succession and *Notoconularia inornata* at Bald Hill regarded as late Artinskian-Kungurian. N. Archbold (personal communication, 1989): regarded as Tastubian by association with *Brachythyrinella* at Comadai. Kemp et al. (1977, p. 199): Stage 2 at Morven Bridge, recycled Stage 3 at Old Nuggetty Gully, Stage 3 in Duck Bay-1, and Stage 4 at Coleraine. Douglas and Ferguson (1988, p. 213, 214): Ipswich equivalent near Yandoit.

Sub-Murray Basin.

Urana Formation, Stages 2 and 3 (O'Brien, 1986); Coorabin Coal Measures comprise, from the bottom, the Narrow Plain Formation culminating in the Lanes Shaft Coal Member (Lower Stage 5) and the Loughmore Formation and Coreen Coal Member (Upper

Stage 5) (Morgan, 1977; Yoo, 1982). I exclude from the Coorabin Coal Measures the overlying Nowranie Creek Formation (*P. reticulatus* zone), which "culminates only in a carbonaceous shale"(Morgan, 1977, p. 4). The Jerilderie Formation is in the *A. parvispinosus* zone.

Cooper Basin.

Price et al. (1985), Powis (1989). Abbreviations in Figure 45 are M—Murteree, E—Epsilon, R—Roseneath, D—Daralingie.

Galilee Basin.

Price et al. (1985); Warang Sandstone from de Jersey and McKellar (1981); Jericho and Lake Galilee Sandstone (base is off scale in the Namurian, 315 Ma) from Jones and Truswell (1992). Abbreviations in Figure 45 are ET—Edie Tuff, P—Peawaddy, M—Mantuan, BA—Black Alley.

Acknowledgments

I thank David Briggs for letting me see his brachiopod zonation ahead of publication and for many valuable discussions, Robin Helby for discussion of palynological zones, and Neil Archbold and David Briggs for their advice on Permian correlation in Victoria.

APPENDIX 2. SANDSTONE PETROLOGY AND PROVENANCE

P. J. Conaghan, E. Jun Cowan, and K. L. McDonnell

Introduction

In this section we use the analytical approach of Dickinson and Suczek (1979) and Dickinson et al. (1983) to summarize the provenance evolution of the Sydney-Gunnedah-Bowen Basin (Figs. 46A–I) and to illustrate photographically representative volcanolithic grains types that characterize its piedmont basin fill (Figs. 47–50). Although a considerable body of petrographic information on this basin system has accumulated over recent decades, including sizable data bases of quantitative modal analyses (e.g., Jensen, 1975), few of these data are amenable to the Dickinson analytical approach because of insufficient detail regarding the quantitative proportions of the various lithic grain types and mono- versus poly-crystalline quartz varieties present. Consequently, the petrographic data base presently available for the purposes of this exercise is small, and in the case of rocks of Stages A–B of the Bowen Basin, altogether absent as far as we are aware.

We add (in proof) that Baker et al. (1993) and Ahmad et al. (1994) redress this deficiency and provide additional analyses of Bowen Basin sandstones of Stages B and C.

Comments on the discrimination of the sedimentary petrofacies

Five sedimentary petrofacies were defined in Figure 10, elaborated in the associated text, and depicted on the various time-space diagrams of Figures 8, 12, 13, and 18. The qualitative and quantitative basis for the discrimination of several of these petrofacies is documented by various authors, including Standard (1969), Ward (1972), Jensen (1975), Conaghan et al. (1982) and Cowan (1985, 1993). Most of this work focused on the petrofacies that characterize the stage of piedmont advance and retreat between 250 Ma and 241 Ma (Stages C + D) and the subsequent phase of inundation of the Sydney and Bowen Basins by the cratonic sand sheet of the Hawkesbury and Expedition Sandstones (Stage E). Unpublished petrographic data covering the equivalent time period in the Gunnedah Basin are documented by Hamilton (1987) for rocks of Stage C and by Jian (1991) for rocks of Stages D–F. Little analytical data for sedimentary rocks of Stages A and B (290–258 Ma) exist, however, so that the discrimination of petrofacies in these strata is somewhat subjective. And the problem is compounded by the facts that (1) most of these strata contain a conspicuous proportion of glacigenic ice-rafted debris of diverse lithological composition and of both local and distant derivation (Gostin, 1968), (2) many of the sandstones typically contain abundant amounts of siliceous resistates (Gostin, 1968, figs. 3.6, 3.7, 3.10) which makes qualitative resolution of their bulk composition in the field difficult, and (3) the bulk compositions of the main transgressive units (i.e., the Stage A Wasp Head Formation, and the Stage B Snapper Point Formation) show considerable geographic variation as a consequence of local provenance influences at and within close stratigraphic proximity to the basal unconformity.

Because of these complications, any attempt to discriminate petrofacies in these rocks on the basis of a rigid approach to the use of their megaquartz content as the main criterion of subdivision is fraught with difficulty. In these circumstances, the Dickinson approach of allocating microquartz as the main microcrystalline siliceous-lithic resistate to the Q pole in the QFL plot is therefore most appropriate and is followed here in Figure 46A, where Shoalhaven Group modal analysis data amenable to the Dickinson approach are plotted on the Folk et al. (1970) QFL diagram to illustrate the main variation between the constituent formations. Although the data base for several of the constituent formations of the Shoalhaven Group is small, general field observation of these stratigraphic units provides reasonable confidence in the likely validity of the main pattern of compositional separation present in this plot. In particular, field observation demonstrates that the quartzose Nowra Sandstone contains only a small amount of ice-rafted debris in contrast to all other formations of the Shoalhaven Group and the correlative Maitland Group in the northern Sydney Basin, so that its bulk composition remains relatively simple and presumably reasonably uniform. Its Dickinson mean Q% is 87% (Fig. 46A), which is almost 10% higher than the mean Q% of the Snapper Point Formation (mean = 79%), and 16% higher than the mean Q% of the Wasp Head Formation (mean = 71%). These general relationships are in agreement with modal analyses of these formations by Ozimic (1979), which provide minimum estimates of the proportion of siliceous resistates present through the nonspecification of the microquartz content. Notwithstanding the small sample size for the Nowra Sandstone, petrographic analyses by Ozimic (1979, table 2; not amenable to the Dickinson QFL plot) and general field observation of the formation indicate that it is the most quartzose formation in the Shoalhaven Group, lacking also the abundant glacigenic dropstones of diverse lithological character that characterize the other formations, including the quartzose Snapper Point Formation (cf. Gostin, 1968, fig. 3.6). Accordingly, the petrofacies affinities of the Nowra Sandstone are designated "quartzose." The volcaniclastic Broughton Sandstone is sourced locally from the coeval Gerringong Volcanics (basalt flows) regarded by Jones et al. (1984, 1987) as manifesting the western flank of the Currarong Orogen. Consequently, and notwithstanding the local volcanic source of the Broughton Sandstone, its volcanolithic petrofacies affinity is designated as regional rather than local (cf. Fig. 10, key). All other formations in the Shoalhaven Group, including the Snapper Point Formation, are designated "quartz-lithic" petrofacies.

In the present analysis, therefore, formations that have an average Dickinson Q% of about 90% are designated as quartzose and those with Q% between about 85% and about 50.0% are designated as quartz-lithic (or sublithic). On the Dickinson LsLvLm plot of the Shoalhaven Group data (excluding the Broughton Sandstone), the individual data points (not shown on Fig. 46H) show wide scatter but fall predominantly along the Lv-Ls baseline and the mean falls close to the Lv pole (Fig. 46H), the siliceous resistate grain types comprising predominantly sedimentary chert derived from the underlying Lower Paleozoic basement and siliceous and felsic volcanics derived from the Middle Devonian Edel-Yalwal-Comerong Volcanics in the hinterland to the west.

Grain types in the regional volcanolithic petrofacies

In stages C–D, the regional volcanolithic petrofacies is dominated by grains of volcanic origin (Fig. 46H), principally of felsic/siliceous character rather than of intermediate/andesitic character. This predominant volcanolithic character is demonstrated in terms of the rock-fragment content of the sandstones in Figure 46H where the relevant Sydney Basin formations/macrofacies, namely the Broughton Sandstone, and macrofacies Lsw/Lse and Sse all fall close to the Lv pole.

The same pattern holds in the Bowen Basin where the relevant formations for which data are available (namely the Baralaba and Moranbah Coal Measures and the Rewan, and Glenidal Formations) all plot close to this pole. Other grain types of volcanic origin includ-

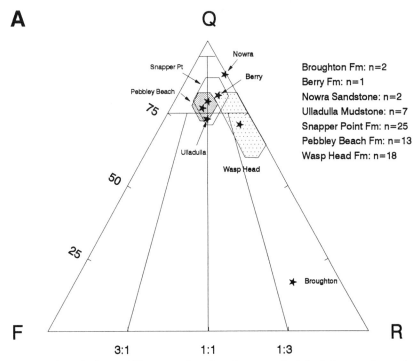

Figure 46A. Dickinson-QFL means (stars) and one-standard-deviation envelopes for sandstone modal analyses from each formation of the Shoalhaven Group in the southern Sydney Basin plotted on the Folk et al. (1970) QFR sandstone classification triangle, illustrating general relationship to the Folk compositional fields. Data sources are Gostin (1968) and P. J. Conaghan (1993, unpublished modal analyses).

ing feldspar (predominantly plagioclase) and beta-form quartz also occur but the volcanic rock fragments dominate.

Tectonite grains (phyllite and schist) and grains of clastic sedimentary and metasedimentary origin (argillite, siltstone, labile sandstone) are also present, and in the Sydney Basin, sedimentary chert, commonly containing radiolarians (Fig. 50H), is an important constituent in rocks of late Stage C age (Newcastle Coal Measures and correlatives) and in rocks of Stage D age (Narrabeen Group) and persists into the sublithic parts of the Gosford Subgroup (Middle Newport Formation and Terrigal Formation). These radiolarian chert clasts/grains are commonly red and crisscrossed by quartz microveins that heal fractures (Fig. 50H) and in these terms are identical to the radiolarian jaspers of the Woolomin Complex of the Texas-Woolomin Structural Block in the New South Wales segment of the adjacent New England Fold Belt.

Associated ash-fall and ash-flow/tephra interbeds are typically plagioclase-quartz-biotite tuffs (Diessel, 1985; Fig. 47, C and D), a phenocryst assemblage that Ewart (1979) suggests is common only in calc-alkaline dacite and rhyolite, and the absence of hornblende and pyroxene in these tuffs is, according to the data in Ewart, suggestive of rhyolite (Jones et al., 1987). The phenocryst mineralogy of Stage C tuffs in the Bowen Basin is similar (Jensen, 1975, p. 12–14).

Grain types characteristic of the regional volcanolithic petrofacies of Stage F are not documented here but these rocks are characteristically rich in feldspar and biotite and fine-grained volcanic rock types with lathwork plagioclase texture, and are commonly dark colored. According to various workers, e.g., Olgers et al. (1964) and Bastian (1965b) in the Bowen Basin, Jian (1991, p. 234–236) in the Gunnedah Basin, Herbert (1980b, p. 264) and our own thin-section observations in the Sydney Basin, the regional volcanolithic petrofa-

cies of Stage F rocks is of more intermediate/mafic origin than is the volcanolithic petrofacies of Stage C.

Provenance evolution of the Sydney-Gunnedah-Bowen Basin

A detailed account of this is beyond the scope of the present report but the first-order elements of this history are evident in the various triangular plots of Figure 46C–I. The Sydney Basin sandstones in Stages C–F are categorized genetically on a macrofacies basis of linked composition and paleoflow direction using the scheme of Conaghan et al. (1982), refined and extended here to accommodate new macrofacies subdivisions within the Gosford Subgroup in the central-eastern part of the basin (upper part of the Narrabeen Group) by Cowan (1985, 1993) and the inland extension of this subdivision to the northwest and southwest respectively by Gregory (1990) and Crowley (1991). The macrofacies subdivisions shown for Stages C–F rocks of the Sydney Basin are an extension of the petrofacies categories defined in Figure 10 (key) and depicted in the time-space sections A–B–B' (Fig. 8) and E–F (Fig. 12) for the Sydney Basin, C–D–D' (Fig. 13) for the Gunnedah Basin, and EF (Fig. 18) and others for the Bowen Basin. Bowen Basin data and Sydney Basin Shoalhaven Group data are categorized on a stratigraphic rather than macrofacies basis.

Macrofacies codes, in ascending stratigraphic order, are as follows.

Lsw/Lse: Upper coal measures (all areas) minus the Cullen Bullen Subgroup, plus the following parts of the Clifton Subgroup of the Narrabeen Group: (northeast) Clifton Subgroup (Munmorah Conglomerate, Tuggerah Formation and Patonga Claystone); (south) Clifton Subgroup minus the uppermost 40 m of the Bulgo Sandstone; (west) Beauchamp Falls Shale Member of the Caley Formation.

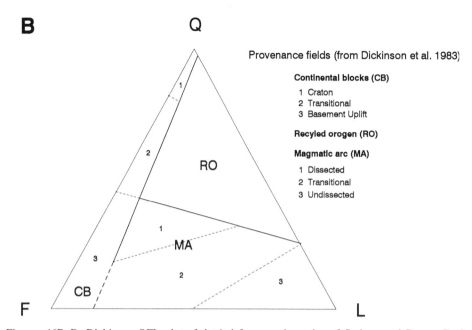

Figures 46B–D. Dickinson QFL plot of detrital framework modes of Sydney and Bowen Basin sandstones. (B) Definition diagram from Dickinson et al. (1983) showing genetic provenance fields. (C) Plot for Sydney and Bowen Basin sandstones of Stages A–D (290–241 Ma). (D) Plot for Sydney and Bowen Basin sandstones for stages E–F (241–230 Ma).

Petrographic data sources are as follows. *Sydney Basin:* Shoalhaven Group: Gostin (1968), Conaghan (1993, unpublished analyses); Lsw/Lse and Sne: McDonnell (1983), Conaghan (unpublished analyses, 1993); Sse(1) McDonnell (1983), Cowan (1985), Crowley (1991); Qse(1): Conaghan (unpublished analyses), Crowley (1991); Reg. In.: McDonnell (1983), Crowley (1991); Qse(2): McDonnell (1983), Cowan (1985), Gregory (1990); Sse(2): McDonnell (1983), Cowan (1985), Gregory (1990), Crowley (1991); Qse/ne: McDonnell (1983), Cowan (1985), Gregory (1990), Crowley (1991); Qne: McDonnell (1983), Cowan (1985), Gregory (1990), Crowley (1991); Lse(2): Conaghan (unpublished analyses). *Bowen Basin:* all formations (Conaghan, unpublished analyses).

Sne: Marrangaroo/Blackmans Flat Conglomerate (Cullen Bullen Subgroup of Illawarra Coal Measures).

Sse(1): The following upper parts of the Clifton Subgroup of the Narrabeen Group: (northeast) lower 128 m of the Terrigal Formation (Interval A of Cowan, 1993, fig. 5b); (south) uppermost 40 m of the Bulgo Sandstone (Interval A of Cowan, 1993, fig. 5a); (west) Caley Formation minus Beauchamp Falls Shale Member.

Qse(1): (west only; see Fig. 12) Burra-Moko Head Sandstone, Mount York Claystone, and Banks Wall Sandstone (to base of Wentworth Falls Claystone).

Reg. In.: stands for Regolith Interlude; analyses are from predominantly quartzose sandstone interbeds within the regolith units: (north) Bald Hill Claystone equivalent (Interval B of Cowan, 1993, fig. 5b); (south) Bald Hill Claystone (topmost formation of Clifton Subgroup) plus Garie Formation (basal formation of Gosford Subgroup, comprising flint-claystone) (Interval B of Cowan, 1993, fig. 5a); (west) Wentworth Falls Claystone (equivalent to the Bald Hill Claystone) plus Docker Head Claystone (the flint-claystone equivalent of Garie Formation).

Qse(2): (north) Basal 12 m of upper part of Terrigal Formation above Bald Hill Claystone equivalent (Interval C of Cowan, 1993, fig. 5b); (central and eastern) lower Newport Formation (Interval C of Cowan, 1993, fig. 5a); (south) not differentiated (mudrocks only); (west) probably not represented (cf. Crowley, 1991, enclosure 1).

Sse(2): (north) Interval D of Cowan (1993, fig. 5b) (section within the upper part of the Terrigal Formation); (central and eastern)

middle Newport Formation (Interval D of Cowan, 1993, fig. 5a); (south) not differentiated (mudrocks only); (west-central, eastern Blue Mountains) basal 30–35 m of Burralow Formation (cf. Crowley, 1991, enclosure 1); (far-west) possibly correlative with the basal part of "Western Burralow Facies" of Bembrick (1980, figs. 8.1, 8.11) and Crowley (1991, enclosure 1) but presently not readily differentiable; probably not represented.

Qse/ne: (north) Topmost 41.5 m of Terrigal Formation (Interval E of Cowan, 1993, fig. 5b); (north-central) Interval III of Gregory (1990, fig. 6.1); (central-eastern) upper Newport Formation (Interval E of Cowan, 1993, fig. 5a); (west-central) upper Newport Formation of Crowley (1991); (far-west) probably correlative with the full section of the "Western Burralow Facies" of Bembrick (1980), but presently not readily differentiable.

The lithofacies and macrofacies characteristics of the "Western Burralow Facies" in the western Blue Mountains, as a whole, match those of the upper Newport Formation in the north-central and coastal exposures of the basin (Conaghan, personal observation, 1985). For this reason and because the "Western Burralow Facies" are not readily differentiable, this entire stratigraphic unit in the Western Blue Mountains has been treated in the present petrological analysis as equivalent to the Upper Newport Formation.

Qne: Hawkesbury Sandstone (all areas).

Sse(3): Minchinbury Sandstone and lower 30 m of Bringelly Shale (*see* Fig. 10, column VIII); this macrofacies not represented in the Dickinson sandstone plots of Appendix 2 because usable data are

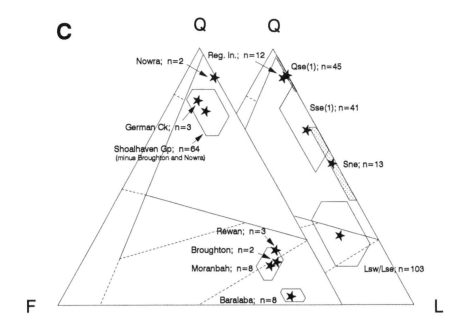

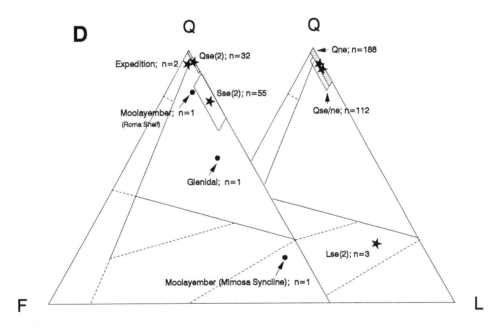

not available, but it is depicted in the time-space diagrams in Figures 8 and 12 and in column VIII of Figure 10.

Lse(2): Bringelly Shale minus lower 30 m.

In the Sydney Basin, for which data concerning Stages A and B strata are available, the sandstones plot in the Recycled Orogen field of the QFL triangle (Fig. 46C) and in the transitional subfield of the Recycled Orogen province in the QmFLt triangle (Fig. 46F, left triangle). This circumstance reflects derivation of the detritus from the adjacent recently cratonized Lachlan Fold Belt and the influx of mixed detritus of both supracrustal metasedimentary and volcanic origin as well as some detritus from unroofed plutons. The Nowra Sandstone, deposited toward the end of Stage B, straddles the Recycled Orogen and Continental Block fenceline (Fig. 46C, left triangle, and Fig. 46F, left triangle) and reflects increasingly deeper levels of erosional stripping of the Lachlan Fold Belt and unroofing of granites. The onset of the phase of regional volcanolithic sediment influx at the termination of Stage B at 258 Ma is registered in the southern Sydney Basin by the appearance of mafic lava flows of the Gerringong Volcanics and the reworking of these coherent volcanics by shallow marine agencies to form the Broughton Sandstone, whose plot in Figure 46C and F, like the ensuing piedmont macrofacies Lsw/Lse of the upper coal measures and lower Narrabeen Group, straddles the fenceline separating the Recycled Orogen and Transitional Magmatic Arc, the influx of recycled orogen detritus in this case originating from the now fully emergent New England Fold Belt (Figs. 16C, 16D).

In the Bowen Basin, the Stage C data from the Baralaba and

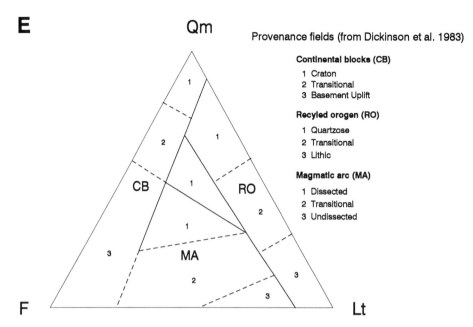

Figure 46E–G. Dickinson QmFLt plot of detrital framework modes of Sydney and Bowen Basin Sandstones. (E) Definition diagram from Dickinson (1985) showing genetic provenance fields. (F, G) Plots of Sydney Basin sandstones of Stages A and B (290–258 Ma) (F, left) and Stages C–F (258–230 Ma) (F, right and G, right) and Bowen Basin sandstones of Stages C–F (G, left). Sandstones of both basins are categorized in the same manner as in Figure 46C–D. Petrographic data sources for the QmFLt plots are the same as for the QFL plots (Fig. 46C, 46D) except for the following macrofacies categories, where the data sources are as follows: Qse(1): Crowley (1991); Qse(2): McDonnell (1983), Crowley (1991); Sse(2), Qse/ne and Qne: McDonnell (1983), Cowan (1985), Crowley (1991).

Moranbah Coal Measures plot in much the same field of Figures 46C and 46G as does the Broughton Sandstone of the Sydney Basin. The Baralaba Coal Measures data plot in the Undisected Magmatic Arc field and the Moranbah Coal Measures data plot in Figure 46C on the fenceline between the Undissected and Transitional Arc fields, and in Figure 46G on the fenceline between the Lithic Recycled Orogen and Transitional Magmatic Arc.

The progressive retreat of the piedmont is recorded in both the Sydney and Bowen Basins by the increasing shift toward the Q and Qm poles in Figures 46C, 46D, 46F, and 46G of the plots for the Triassic Rewan and Narrabeen Group sediments and the ensuing sublithic sandstones of the uppermost Clifton Subgroup in the Sydney Basin, macrofacies Sse(1), and Glenidal Formation in the Bowen Basin, followed in the Sydney Basin by the influx of the first phase of the cratonic sand sheet of macrofacies Qse(1) (Fig. 46F; *see also* time-space diagram E–F, Fig. 12). This phase of retreat of the piedmont culminated at 241 Ma with the virtual cessation of influx of siliciclastic bedload sediment into the Sydney Basin, and probably also the Bowen Basin, i.e., Arcadia Formation and lowermost Glenidal Formation (Jensen, 1975), and concomitant deep weathering to produce the regional regolith of the Bald Hill Claystone and overlying flint-claystone Garie Formation and their equivalents in the Gunnedah Basin (Jian and Ward, 1993). This lull in events is designated "Regolith Interlude" (Reg. In.) in the triangular plots of Figures 46C, 46F, and 46H; the minor sandstone interbeds that occur in this regolith plot in Figure 46F in the Quartzose Recycled Orogen field

but straddle the fenceline between this field and that of the cratonic subfield of the Continental Block field. Individual data points for the Regolith Interlude concentrate toward the Lv pole along the Lv-Lm baseline in the LsLvLm plot (Reg. In. mean shown in Fig. 46H), the more lithic- and volcanic rock-fragment-rich sandstones occurring in the northeastern rather than other parts of the basin and the lithic grain types in sandstones of the western part of the basin being predominantly of metamorphic origin. This pattern evidently reflects the survival and redistribution of more resistant volcanic rock fragments in the areas close to the New England Fold Belt and the minor influx or recycling of more resistant metamorphic lithic grains in the area adjacent to the Lachlan Fold Belt.

The ensuing early phase of Stage E in the Sydney Basin (i.e., Newport Formation time) witnessed a brief initial period of quartzose sediment influx [macrofacies Qse(2) = Lower Newport Formation] sourced from the craton but evidently tagged by a signature of volcanic-lithic grain types that cause it to plot close to the Lv pole in the LsLvLm triangle of Figure 46H. An influx of sublithic sediment comprising the Middle Newport Formation [macrofacies Sse(2), plotting in the Recycled Orogen field of Fig. 46D, left triangle, and in the Quartzose Recycled Orogen field of Fig. 46G, right-triangle] evidently records a minor resurgence of tectonic activity in the New England Fold Belt, with sediment delivered to the Sydney Basin via bedload overflow from the Gunnedah Basin (during Napperby Formation, Interval 'a' time; Jian and Ward, 1993; *see also* Fig. 16E). A return to cratonic quartzose sand influx in the Upper Newport (Qse/ne) and

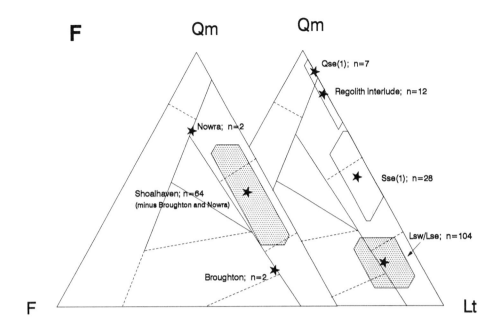

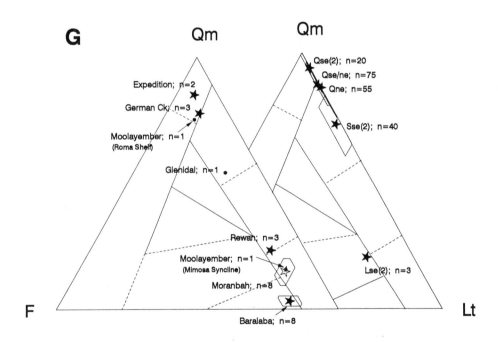

Hawkesbury Sandstone (Qne) in the Sydney Basin (Fig. 46D, right triangle) and the deposition of the analogous, cratonic, quartzose Expedition Sandstone in the Bowen Basin (Fig. 46D, left triangle) completed this trend toward increasing tectonic stability during Stage E, the Hawkesbury and Expedition Sandstones both plotting in Figure 46D on the fenceline dividing the field of the Recycled Orogen and the Craton subfield of the Continental Block province.

Stage F in the Sydney and Bowen Basins witnessed a return to regional volcanolithic sedimentation [Lse(2) in the Sydney Basin, Fig. 46D, right triangle; and Fig. 46G, right triangle] and in the Bowen Basin (= Moolayember Formation; Fig.46D, left triangle; and Fig. 46G, left triangle), and a continuation of volcanolithic sedimentation in the Gunnedah Basin (Deriah Formation; Jian, 1991; Jian and Ward, 1993). Macrofacies Lse(2) (= Bringelly Shale) in the Sydney Basin plots in the Transitional Magmatic Arc field in Figures 46D and 46G and the Moolayember Formation in the eastern, more proximal, part of the Bowen Basin (i.e., Mimosa Syncline area) plots in the Undissected Magmatic Arc field of Figure 46D and in Figure 46G (left triangle), in the lithic subfield of the recycled Orogen province, but close to the fenceline of this province with that of the Magmatic Arc. Although the

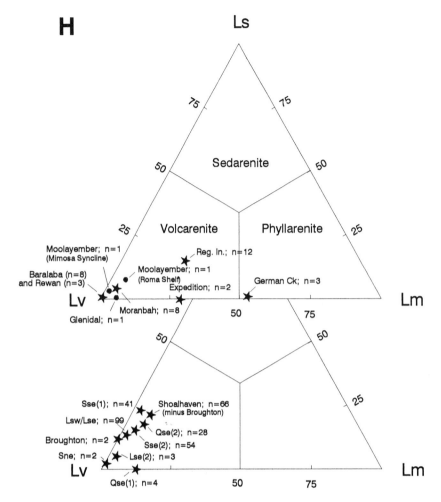

Figure 46H. Dickinson LsLvLm plot for Sydney Basin sandstones (lower triangle) excluding macrofacies Regolith Interlude (Reg. In.), plotted (for reasons of space) with Bowen Basin sandstones in the upper triangle. Petrographic data sources are the same as for Figure 46C, except for the Sydney Basin category Qse(1), from Crowley (1991).

present plot of the Moolayember Formation in the eastern part of the Bowen Basin involves only a single data point, the position of this point in the lithic subfield of the Recycled Orogen close to the fenceline of the Magmatic Arc province supports Bastian's (1965b, p. 10) conclusion that "the Moolayember Formation sediments were not derived predominantly from volcanic sources . . . , but instead they originated from many source areas." Bastian also remarked that "Volcanic detritus is more prominent in the eastern outcrops of the Moolayember Formation, indicating that the volcanic source was to the east." Another data point for the Moolayember Formation from the Roma Shelf area to the west (Fig. 46D, left triangle; and Fig. 46G, left triangle) reflects a more continental block influence.

Other petrofacies depicted in Figures 46C–46I but not discussed thus far include the coarse-grained Stage C quartz-lithic sandstones and conglomeratic sandstones of the Cullen Bullen Subgroup (macrofacies Sne) in the Sydney Basin (Fig. 46C, right triangle) and the cor-

relative Western Bedload Sands in the Gunnedah Basin (Fig. 46I). Both plot in approximately the same position on the triangle (Recycled Orogen) and reflect influx of Lachlan Fold Belt detritus from the foreswell along the western edge of the Basin. Similarly, the German Creek Coal Measures in the northwestern Bowen Basin plot in the Recycled Orogen field of Fig. 46C (left triangle) but close to the fenceline of the Continental Block province. In the QmFLt diagram of Fig. 46G (left triangle), the plot straddles this boundary and reflects uplift of the relatively stable cratonic block of the adjacent Anakie Inlier (Fig. 29C).

Acknowledgments

We thank Hanif Hawlader and Fuxiang He for help with point-count analyses of sandstones, and Hawlader for processing the data.

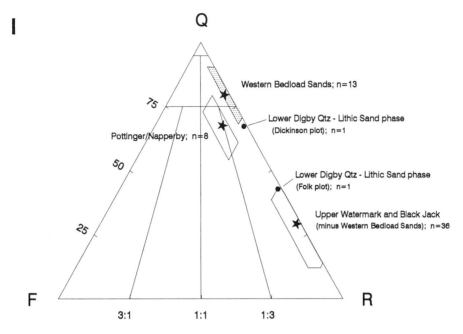

Figure 46I. Plots of modal analyses of Gunnedah Basin sandstones (Stages C–E, 258–235 Ma) on the Folk et al. (1970) QFR triangle. Except for one modal analysis of the Digby Formation, the data are not amenable to the Dickinson QFL plot because lithic grains and quartz categories are not specified quantitatively. Note that because microquartz/chert grains are plotted at the Q pole in the Dickinson plot, this results in a shift of the data points toward that pole, as illustrated by the lower Digby Formation sample which is plotted in both ways. Notwithstanding this overall shift in the data, note the general similarity of the positions in which the petrofacies of the Gunnedah Basin plot relative to the correlative petrofacies in the Sydney Basin, e.g., compare the plots of (1) the upper Watermark Formation and the Black Jack Formation minus the Western Bedload Sands with macrofacies Lsw/Lse (Fig. 46C) and (2) the Western Bedload Sands with macrofacies Sne (= Cullen Bullen Subgroup). Plots of the Digby and Napperby Formations shown here, although the sample size is small, are consistent with those of Jian (1991), which have a much larger modal analysis database. Data sources are: uppermost Watermark Formation and Black Jack Formation (which includes the Western Bedload Sands Facies): Hamilton (1987); Digby: Conaghan (1993, unpublished data); Pottinger Beds: Manser (1967, fig. 11-4); Napperby: Byrnes and Zlotkowski (1975).

APPENDIX 3. LATE PERMIAN-TRIASSIC RADIOMETRIC DATES OF GRANITOIDS AND ASSOCIATED VOLCANICS FROM THE SOUTHERN NEW ENGLAND FOLD BELT

S. E. Shaw

Table 11 (p. 156) is a compilation of radiometric dates by the biotite Rb/Sr method of Late Permian and Triassic granitoids and associated volcanics from the NEFB (Fig. 51). Late Carboniferous and Early Permian granitoids (Shaw and Flood, 1981; Roberts and Engel, 1987) are not treated. From the age, composition, and distribution of plutonic and volcanic rocks in the Fold Belt, Shaw et al. (1991) arranged the rocks in three groups according to age and location. A fourth group—the high-level intrusions in the Lorne Basin—is included here.

Isotope analyses were done on a multi-sample fully automated VG54E mass spectrometer fitted with a single collector at the CSIRO Division of Exploration Geoscience, North Ryde, as part of the Centre for Isotope Studies, supported by a consortium of universities with the financial assistance of the Australian Research Council; the analyses followed the procedures of Williams et al. (1975). Biotite dates were calculated using an initial $^{87}Sr/^{86}Sr$ ratio of 0.705, an average of the observed ratios of Late Permian and Triassic rocks (Shaw and Flood, 1981). Decay constant and standard values are lambda $^{87}Rb = 1.42 \times 10^{-11}a^{-1}$, $^{86}Sr/^{88}Sr = 0.1194$, and E and A $SrCO_3 = 0.708039 \pm 0.000065$ (2s external precision, population of 48). Analytical uncertainties at the 95% confidence level were calculated to be ±0.5%, but replicate analyses of coarse-grained (>0.1 mm) biotite have considerable isotopic differences, which could be eliminated by fine grinding. Geological variations exceed analytical precision, and I consider that realistic Rb/Sr age determinations are no better than ±1.5%. Dates of the same rock body by the K/Ar method (new constants) (Table 12) agree within analytical uncertainty except the older dates on hornblende at Carrai and Botumburra.

Figure 47. Photomicrographs of tuffs (A–D and representative grain types (E–H) of the regional volcanolithic petrofacies that characterizes the orogenic piedmont phase (Stages C and D, 258–241 Ma) of the eastern Australian foreland basin. These photomicrographs and those that follow in Figures 48–50 focus on the volcanic rock-fragment grain types.

(A) Plane-polarized light view of coarse-grained pumiceous tuff located 30 m below the top of the Wandrawandian Siltstone (correlative of the Ulladulla Mudstone) at Montague Point, Jervis Bay, New South Wales, documented by Runnegar (1980b, fig. 23.15), and Scott (1984, fig. 27). This is the 260 Ma (Stage B, Sydney Basin) "earliest orogenic tuff" referred to in this chapter. It consists of large (about 1 cm maximum) subangular to rounded vesicular pumice clasts, resorbed beta-form quartz and minor degraded mica. Many pumice clasts show collapsed/compacted relationships against neighboring, more rigid grains. Macquarie University sample and thin-section MU 39932. Scale bar: 5 mm.

(B) Plane-polarized light higher-magnification view of a vesicular pumice clast in the tuff illustrated in A. Scale bar: 1 mm.

(C, D) Plane-polarized (C) and partially cross-polarized (D) views of a 2.4-m-thick biotite tuff from the "tuffaceous stony coal facies" at the top of the late Permian (Stage C) Black Jack Formation, Gunnedah Basin. Scale bar: 1 mm (same for D). Tuff has the following composition: quartz, 1.3%; plagioclase, 12.4%; potash feldspar, 3.0%; volcanic rock fragments, 1.2%; biotite, 3.6%; ash matrix, 78.5%. All light-colored grains are plagioclase except grain at top left coded "q," which is quartz. Biotite crystals occur at center. Biotite from this tuff gives a Rb/Sr age of 252.4 Ma (Veevers et al., 1994a). Location: DM Clift DDH 2, located at 31°17′S, 150°17′E, 33 km south of Gunnedah, New South Wales; sample comes from 27 m below the eroded top of the Black Jack Formation (depth 73 m) beneath the Triassic Digby Formation.

(E, F) Plane-polarized (E) and partially cross-polarized (F) light views of an angular grain of flow-layered volcanic glass in volcanolithic sandstone of the Foybrook Formation, basal part of the Wittingham Coal Measures (Tomago Coal Measures correlative), Sydney Basin, upper Hunter Valley, New South Wales. Scale bar: 0.25 mm (same for F).

(G, H) Plane-polarized (G) and cross-polarized (H) view of spherulitic volcanic rock fragment. Scale bar: 0.5 mm (same for H). Sample location details as for E.

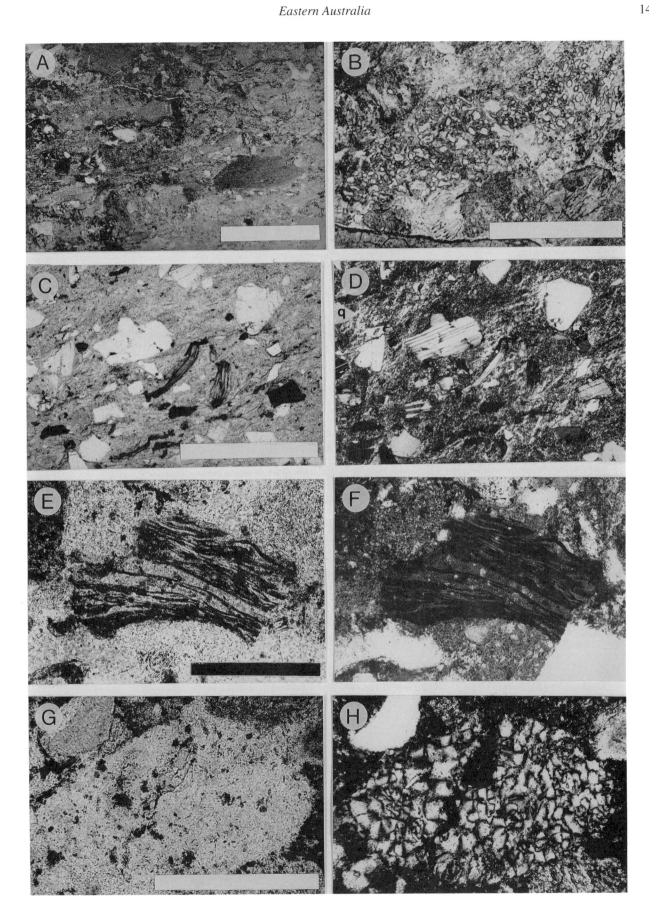

Figure 48. Photomicrographs of representative volcanic rock fragments of the regional volcano-lithic petrofacies that characterizes the orogenic piedmont phase (Stages C and D) of the eastern Australia foreland basin. Scale bar in all photographs equals 0.5 mm (same for right-hand paired photomicrographs).

(A) Plane-polarized view of a glassy volcanic rock fragment containing conspicuous shards with axiolitic structure. Late Permian Baralaba Coal Measures, Moura Open-cut Mine, southeast central Bowen Basin, Queensland.

(B) Plane-polarized view of a glassy volcanic rock fragment containing glass shards with axiolitic structure. Foybrook Formation, basal part of Wittingham Coal Measures, Sydney Basin, upper Hunter Valley, New South Wales.

(C, D) Plane-polarized (C) and cross-polarized (D) light views of a microcrystalline volcanic rock fragment (devitrified volcanic glass) showing flow layering (most evident in C). Sample locality details as for B.

(E, F). Plane-polarized (E) and cross-polarized (F) views of a fragment of a spherulite showing uniform radial fibrous structure and concentric pattern in peripheral zone. Sample locality details as for B.

(G) Plane-polarized view of part of a very coarse grain of volcanic glass containing numerous dark-colored microlites. Sample locality details as for B.

(H) Plane-polarized view of a dark-colored glassy volcanic rock fragment containing plagioclase phenocrysts. Sample locality details as for B.

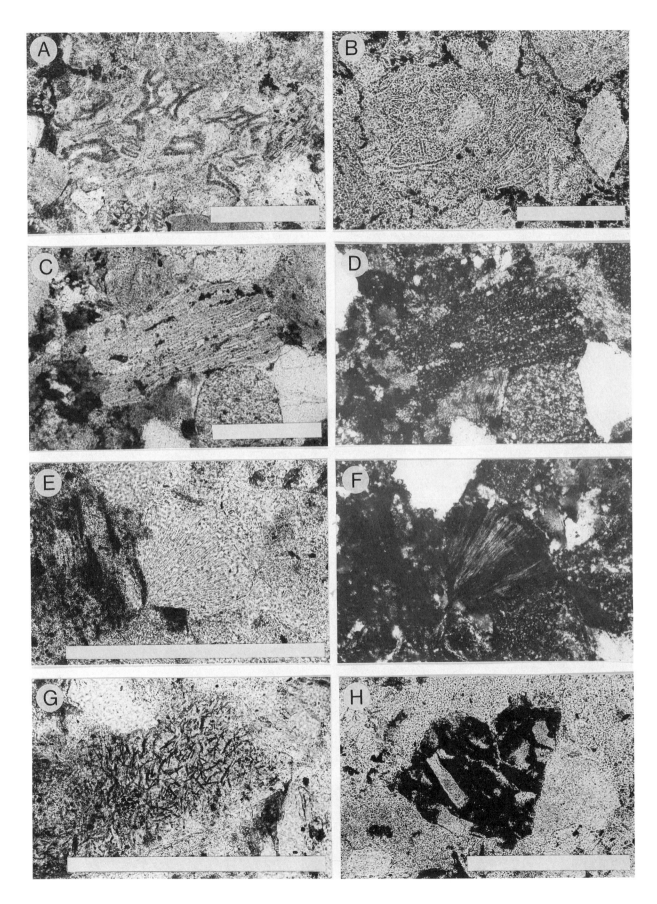

Figure 49. Photomicrographs of representative volcanic rock fragments of the regional volcanolithic petrofacies that characterizes the orogenic piedmont phase (Stage C and D) of the eastern Australia foreland basin. Scale bar in all photographs equals 0.5 mm (same for right-hand paired photomicrographs), and sample locality details of all grains illustrated are the same as for Figure 48B.

(A, B) Plane-polarized (A) and cross-polarized (B) views of a grain of ignimbritic vapor-phase quartz. Such grains are characterized by abundant vacuoles that manifest a fuzzy brown color in plane-polarized light. In cross-polarized light the grains are characterized by a (commonly moderately coarsely crystalline) mosaic of blocky quartz crystals with relatively sharp borders (evident in B).

(C, D) Plane-polarized (C) and partially cross-polarized (D) views of a coarse grain of felsitic volcanic rock. Such grains are characterized by a mosaic of cloudy quartz and feldspar crystals with fuzzy, irregular borders.

(E, F) Plane-polarized (E) and cross-polarized (F) views of a volcanic rock fragment characterized by a granular quartz-feldspar groundmass and sporadic larger feldspar phenocrysts.

(G, H) Plane-polarized (G) and partially cross-polarized (H) light views of a grain of radiaxial quartz-feldspar intergrowth.

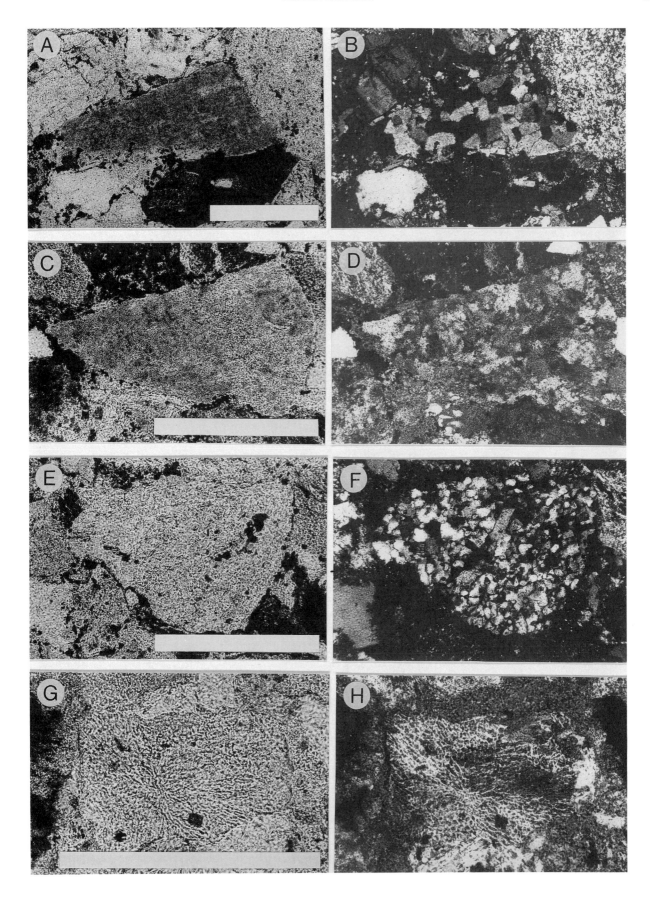

Figure 50. Photomicrographs of representative rock-fragment types of the regional volcanolithic petrofacies and (sublithic) cratonic-orogenic mixed petrofacies of the eastern Australia foreland basin. Sample locality details of grains in A–G are the same as those given for Figure 48B. Scale bar in all photographs equals 0.5 mm (same for right-hand paired photomicrographs).

(A, B) Plane-polarized (A) and cross-polarized (B) views of a large skeletized feldspar crystal (light area in B) that has been cannibalized/resorbed in a cuneiform-pattern/manner by the surrounding microgranular quartz-feldspar groundmass. Such textures occur in the Late Carboniferous ignimbrites of the New England Fold Belt in New South Wales.

(C, D) Plane-polarized (C) and cross-polarized (D) light views of a grain comprising an intergrown quartz crystal (clear areas in A) and cloudy, altered feldspar crystal (dark areas in A). Such alteration of feldspar is characteristic of the vapor-phase alteration zone in ignimbrites and is common in Late Carboniferous ignimbrites in the New England Fold Belt in New South Wales.

(E) Cross-polarized view of a medium sand grain that comprises a micrographic cuneiform (granophyric) quartz-feldspar intergrowth.

(F) Cross-polarized view of a very fine-textured volcanic rock fragment with abundant (lathwork) plagioclase crystallites and sporadic small phenocrysts.

(G) Plane-polarized view of a tectonic grain (quartz-mica schist).

(H) Cross-polarized view of a very coarse grain of radiolarian chert crisscrossed by microveins of quartz. The radiolarians are the circular light-colored, more coarsely crystalline areas. Such radiolarian chert clasts (commonly red colored) are abundant in the Sydney Basin in the conglomerates and sandstones of the Newcastle Coal Measures and correlatives and are identical to the radiolarian jasper of the Woolomin Complex of the southern New England Fold Belt. Such clasts are also common in the overlying Triassic Narrabeen Group. Specimen locality: sublithic middle Newport Formation (latest Early Triassic), Sydney Basin, Northshore Sydney area (Section 6A of Cowan, 1985).

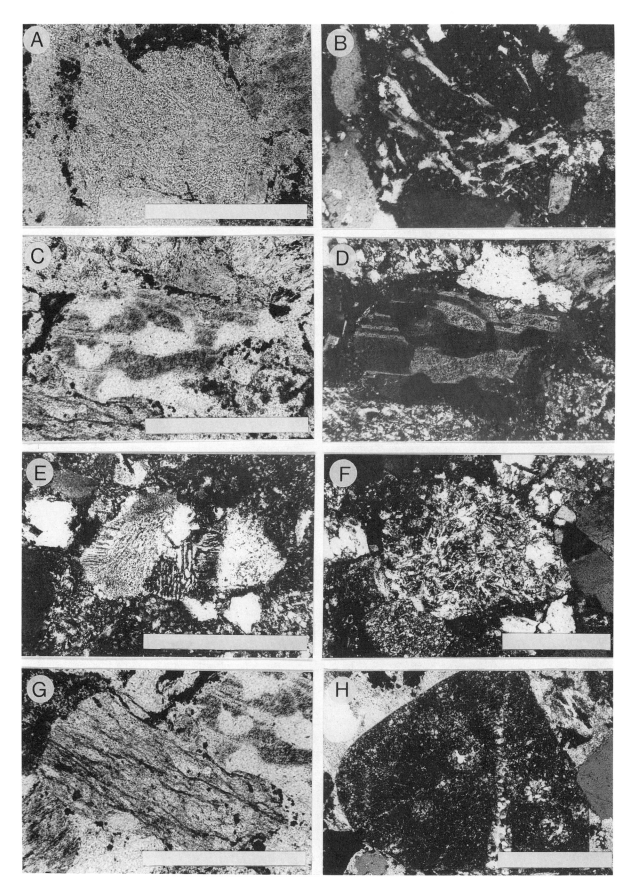

TABLE 11. AGES OF GRANITOIDS

Number	Unit / Specimen	Latitude (S)	Longitude (E)	Age (Ma)	Number	Unit / Specimen	Latitude (S)	Longitude (E)	Age (Ma)
1	**Middle Brother**				14	**Bolivia Range**			
	FS2288	31°41.2'	152°41.9'	205.9		FS2403	29°20.5'	151°53.6'	240.6
	FS2466	31°41.4'	152°42.2'	209.8	15	**Maryland**			
2	**Valla**					FS2520	28°32.4'	152°00.2'	241.7
	FS2304	30°35.0'	152°58.2'	222.1		FS2519	28°32.7'	152°00.0'	242.3
	FS2303	30°35.7'	152°57.2'	224.7	16	**Mole**			
3	**Yarrahapinni**					FS214	29°14.5'	151°42.0'	237.4
	FS2299	30°46.0'	152°57.4	222.6		FS2531	29°11.5'	151°41.3'	237.4
	FS2300	30°46.2'	152°56.5'	224.3		FS213	29°18.6'	151°42.2'	242.1
4	**Cascade Creek**					FS2529	29°20.4'	151°41.6'	242.3
	SW246	31°11.8'	152°26.7'	222.9	17	**Mt Jonblee**			
5	**Werrikimbe**					FS216	29°17.7'	152°00.4'	243.0
	FS2456	31°11.0'	152°20.5'	223.9		**Petries Sugarloaf**			
	Rd39	31°11.0'	152°20.5'	225.4		FS217	29°17.4'	152°04.6'	242.4
6	**Round Mountain**				18	**The Basin**			
	FS233	30°30.0'	152°15.0'	226.4		FS2480	30°17.5'	151°16.9'	245.2
	FS234	30°29.0'	152°17.5'	226.4	19	**Limbri**			
7	**Carrai**					FS1234	31°00.2'	151°11.2'	245.4
	FS2449	30°45.3'	152°12.5'	224.2	20	**Glenburnie**			
	Rd5	30°45.5'	152°14.0'	224.4		FS2580	30°44.5'	151°24.6'	245.5
	FS2496	30°53.8'	152°14.5'	226.4	21	**Red Range**			
8	**Botumburra**					FS1621	29°49.8'	151°53.7'	245.2
	FS2492	30°37.3'	152°16.2'	226.5	22	**Kingsgate**			
	SW174	30°37.2'	152°17.3'	227.8		FS2549	29°35.7'	152°00.8'	242.9
9	**Chaelundi**					FS1584	29°38.4'	152°02.2'	245.9
	CC57	30°03.5'	152°21.4'	232.8	23	**Sandy Flat**			
	CC100	30°05.2'	152°20.4	232.8		4810	29°12.8'	152°00.7'	244.5
	CC9	30°06.4'	152°26.0'	233.0		FS2408	29°13.9'	152°00.4'	246.8
	CC20	30°06.2'	152°20.4'	234.6	24	**Tingha**			
10	**Dandahra**					FS146	29°57.4'	151°07.6'	248.6
	FS2552	29°33.0'	152°15.5'	233.7		**Gilgai**			
	FS2554	29°31.7'	152°17.2'	235.1		FS145	29°50.9'	151°07.0'	242.9
11	**Billyrimba**					FS2482	30°04.9'	151°04.8'	243.9
	FS2429	29°13.7'	152°06.7'	233.0		FS2483	30°03.6'	151°06.7'	246.1
	FS2428	29°12.8'	152°07.0'	233.8		FS2486	29°51.0'	151°07.0'	246.5
	FS2424	29°09.9'	152°06.7'	234.7	25	**Oban River**			
12	**Mackenzie**					FS1580	29°55.4'	152°00.0'	247.5
	C291	29°04.5'	151°58.0'	238.2	26	**Wards Mistake**			
	FS2532	29°07.0'	151°53.4'	243.8		FS227	29°53.0'	151°56.7'	244.6
	Nonnington					FS1491A	29°38.7'	151°59.2'	245.0
	C2914	29°05.6'	151°56.0'	240.5		FS1616	29°47.5'	151°58.9'	245.6
13	**Stanthorpe**					FS1489	29°38.2'	151°58.6'	247.2
	FS2446	28°37.7'	151°56.8'	238.1		FS1628	29°56.4'	151°48.7'	247.9
	FS2445	28°40.5'	151°55.1'	238.4	27	**Bungulla**			
	FS2517	28°48.1'	151°46.1'	240.6		D439	28°57.7'	152°11.5'	242.9
	FS203	28°46.3'	152°05.3'	240.8		D435X	28°57.7'	152°11.2'	243.0
	D352	28°52.8'	152°06.2'	241.2		229	29°06.8'	152°03.3'	244.0
	FS2521	28°32.9'	152°02.2'	242.2		FS2540	29°04.8'	152°21.2'	244.0
	Ruby Creek					FS2536	29°11.5'	152°19.0'	244.1
	FS205	28°38.5'	152°01.3'	239.2		FS2534	29°09.4'	152°12.3'	245.5
	FS2518	28°36.5'	151°57.0'	239.7		FS2533	29°09.0'	152°11.9'	247.5
	FS2522	28°29.0'	151°52.4'	240.5	28	**Undercliffe Falls**			
						FS204	28°38.4'	152°07.9'	245.4

TABLE 11. AGES OF GRANITOIDS (continued)

Number	Unit / Specimen	Latitude (S)	Longitude (E)	Age (Ma)	Number	Unit / Specimen	Latitude (S)	Longitude (E)	Age (Ma)
29	**Mount Mitchell**				40	**Yarrowyck**			
	FS224	29°36.4'	152°05.8'	243.1		FS2475	30°28.3'	151°22.3'	245.8
	FS2551	29°34.7'	152°11.7'	248.6		FS2474	30°29.9'	151°22.5'	249.6
30	**Walcha Road**					**Dykes in Yarrowyck**			
	FS814	30°56.2'	151°19.5'	243.3		FS1564D	30°33.5'	151°22.5'	245.3
	FS827	30°59.0'	151°23.1'	244.3		FS1564M	30°33.7'	151°22.3'	248.2
	FS962	30°59.3'	151°23.4'	244.5	41	**Uralla**			
	WR1	30°57.6'	151°21.3'	245.4		FSNEH	30°38.0'	151°31.1'	246.8
	FS828	30°59.7'	151°23.1'	246.2		UQ	30°38.3'	151°29.5'	248.5
	Back Creek					WCR	30°38.3'	151°25.4'	250.7
	MG2	31°02.4'	151°20.3'	244.8		**Kentucky**			
	Spring Creek					DA1	30°47.2'	151°26.6'	245.3
	FS2597	30°46.3'	151°19.1'	247.5		FS2582	30°47.2'	151°26.6'	251.3
31	**Dundee and associated rocks**					**Terrible Vale**			
	FS2504	28°54.0'	151°53.7'	245.8		FS1906	30°43.6'	151°31.4'	247.7
	FS2505	29°00.9'	151°51.9'	246.0		FS2581	30°44.3'	151°25.9'	250.2
	FS2556	29°26.6'	152°22.5'	246.6	42	**Highlands**			
	FS2558	29°26.2'	152°23.1'	246.8		H2.0	30°23.0'	151°42.8'	245.2
	FS1299A	29°28.2'	151°38.2'	248.2		H2.15	30°18.9'	151°47.1'	246.1
	FS2172	29°26.0'	151°41.1'	248.7		H2.17	30°19.1'	151°46.9'	246.9
	FS1253	29°33.4'	151°46.3'	248.7		**Pine Tree**			
32	**Shalimar**					H2.7	30°20.8'	151°45.5'	247.7
	E20A	30°52.3'	151°27.9'	245.9	43	**Mount Duval**			
	E20B	30°52.3'	151°27.9'	246.8		FS241A	30°22.7'	151°33.4'	245.5
	E29A	30°52.5'	151°28.7'	246.9		FS241B	30°22.7'	151°33.4'	246.9
	E29B	30°52.5'	151°28.7'	247.0		FS2565	30°16.6'	151°33.6'	247.4
33	**Looanga**					FS2569	30°20.5'	151°32.2'	251.8
	LA1	30°48.5'	151°15.3'	246.7		**Glenore**			
34	**Bendemeer**					FS2562	30°12.4'	151°33.0'	247.6
	FS2469	30°52.0'	151°09.3'	247.1		FS2563	30°11.5'	151°33.3'	249.2
	FS2470	30°52.8'	151°13.2'	247.1		**Booralong**			
35	**Inlet**					FS2568	30°29.7'	151°31.9'	249.8
	FS2473	30°55.9'	150°54.5'	247.8	44	**Baldeslie stock**			
	FS2472	30°56.6'	150°54.7'	247.9		FS2592	30°16.2'	151°22.6'	250.2
36	**Attunga Creek**				45	**Parlour Mountain**			
	FS2571	30°52.6'	151°01.9'	248.9		FS2591	30°15.7'	151°26.4'	251.2
37	**Moonbi**				46	**Dumbudgery Creek**			
	FS2468	30°55.6'	151°07.6'	247.0		FS1595	29°12.9'	152°33.7'	249.3
	FS2467	30°59.5'	151°05.7'	247.1		FS1594	29°13.0'	152°34.9'	249.7
	FS2471	30°01.8'	151°00.1'	249.3	47	**Herries**			
38	**Duncans Creek**					FS2525	28°20.1'	151°50.8'	247.5
	FSDC	31°23.5'	151°11.5'	245.7		FS2526	28°18.8'	151°52.0'	252.6
	FS253	31°25.6'	151°13.0'	248.4		FS2447	28°23.2'	151°54.7'	252.8
39	**Gwydir River**					**Dykes in Herries**			
	FS2476	30°27.7'	151°21.2'	244.7		FS2524	28°20.4'	151°50.6'	246.1
	FS2477	30°26.3'	151°20.6'	247.0	48	**Mascotte**			
	FS2478	30°23.6'	151°17.9'	247.0		FS137	28°52.6'	151°05.3'	251.6
	FS240A	30°27.8'	151°21.2'	247.6		FS136	28°50.2'	151°04.3'	251.8
	FS240B	30°27.8'	151°21.2'	248.4		FS135	28°50.1'	151°05.8'	254.2
						FS135c	28°50.1'	151°05.8'	254.7

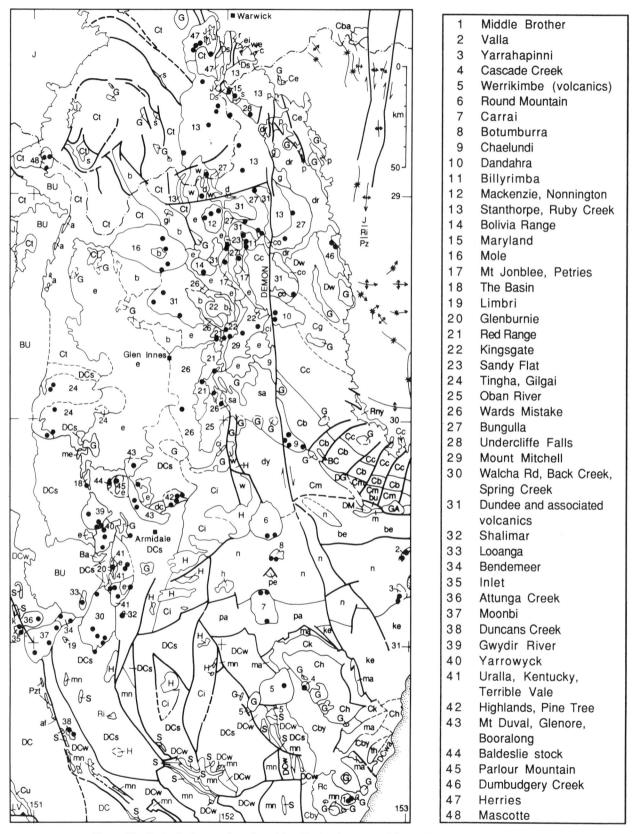

1	Middle Brother
2	Valla
3	Yarrahapinni
4	Cascade Creek
5	Werrikimbe (volcanics)
6	Round Mountain
7	Carrai
8	Botumburra
9	Chaelundi
10	Dandahra
11	Billyrimba
12	Mackenzie, Nonnington
13	Stanthorpe, Ruby Creek
14	Bolivia Range
15	Maryland
16	Mole
17	Mt Jonblee, Petries
18	The Basin
19	Limbri
20	Glenburnie
21	Red Range
22	Kingsgate
23	Sandy Flat
24	Tingha, Gilgai
25	Oban River
26	Wards Mistake
27	Bungulla
28	Undercliffe Falls
29	Mount Mitchell
30	Walcha Rd, Back Creek, Spring Creek
31	Dundee and associated volcanics
32	Shalimar
33	Looanga
34	Bendemeer
35	Inlet
36	Attunga Creek
37	Moonbi
38	Duncans Creek
39	Gwydir River
40	Yarrowyck
41	Uralla, Kentucky, Terrible Vale
42	Highlands, Pine Tree
43	Mt Duval, Glenore, Booralong
44	Baldeslie stock
45	Parlour Mountain
46	Dumbudgery Creek
47	Herries
48	Mascotte

Figure 51. Geological map of southern New England (extracted from Fig. 7), showing plutons and associated ignimbrites.

TABLE 12. COMPARISON OF DATES OBTAINED BY DIFFERENT METHODS FROM THE SAME PLUTONS OR IGNIMBRITE SHEETS

No.	Name **Age (Ma)**
1	Middle Brother **209.8** (Table 11) **209.7** (K/Ar pooled hornblende: McDougall and Wellman, 1976)
6	Round Mountain **226.4, 226.4** (Table 11) **228** (K/Ar biotite: Binns, 1966)
7	Carrai **224.2, 224.4, 226.4** (Table 11) **226, 227** (K/Ar biotite: Leitch and McDougall, 1979) **228 to 231** (K/Ar hornblende: Leitch and McDougall, 1979)
8	Botumburra **226.5, 227.8** (Table 11) **226, 227** (K/Ar biotite: Leitch and McDougall, 1979) **229 to 231** (K/Ar hornblende: Leitch and McDougall, 1979)
31	Dundee Rhyodacite Least affected by Stanthorpe Granite = **248.7** (Table 11, FS2172, FS1253)*

*Compare the K/Ar date of **253 ± 2.5** (F. Della-Pasqua, personal communication, 1991) on the same biotite (GA236) used by Evernden and Richards (1962, p. 38) to give K/Ar date (old constants) of **242** Ma, since recalculated with a corrected value for the original spike (J. Richards, personal communication, 1990, to C. Murray) to a value similar to that of FS2504 (245.8 Ma, Table 11), the youngest of seven Rb/Sr dates.

REFERENCES CITED

Ahmad, R., Tipper, J. C., and Eggleton, R. A., 1994, Compositional trends in the Permian sandstones from the Denison Trough, Bowen Basin, Queensland reflect changing provenance and tectonics: Sedimentary Geology, v. 89, p. 197–217.

Aitchison, J. C., 1988, Early Carboniferous radiolaria from the Neranleigh-Fernvale beds near Brisbane: Queensland Government Mining Journal, v. 89, p. 240–241.

Aitchison, J. C., 1989, Discussion: Radiolarian and conodont biostratigraphy of siliceous rocks from the New England Fold Belt: Australian Journal of Earth Sciences, v. 36, p. 141–142.

Aitchison, J. C., 1990, Significance of Devonian-Carboniferous radiolarians from accretionary terranes of the New England Orogen, eastern Australia: Marine Micropaleontology, v. 15, p. 365–378.

Aitchison, J. C., and Flood, P. G., 1990, Early Carboniferous radiolarian ages constrain the timing of sedimentation within the Anaiwan terrane, New England orogen, eastern Australia: Neues Jahrbuch für Geologie und Paläontologie Abhandlungen, v. 180, p. 1–19.

Aitchison, J. C., Ireland, T. R., Blake, Jr., M. C., and Flood, P. G., 1992a, 530 Ma zircon age for ophiolite from the New England Orogen: Oldest rocks known from eastern Australia: Geology, v. 20, p. 125–128.

Aitchison, J. C., Flood, P. G., and Spiller, F.C.P., 1992b, Tectonic setting and paleoenvironment of terranes in the southern New England orogen, east-ern Australia as constrained by radiolarian biostratigraphy: Palaeogeography, Palaeoclimatology, Palaeoecology, v. 94, p. 31–54.

Alcock, P. J., 1970, A report on the sedimentology of the Moolayember Formation, Bowen Basin, Queensland: Australian Bureau of Mineral Resources Record 1970/25, 43 p.

Allen, R. J., and Houston, B. R., 1964, Petrology of Mesozoic sandstones of Carnarvon Highway section, western Bowen and Surat Basins: Brisbane, Queensland Geological Survey Report, v. 6, 25 p.

Archbold, N. W., 1982, Correlation of the Early Permian faunas of Gondwana: Implications for the Gondwanan Carboniferous-Permian boundary: Journal of the Geological Society of Australia, v. 29, p. 267–276.

Archbold, N. W., 1984, Early Permian marine faunas from Australia, India and Tibet: An update on the Gondwanan Carboniferous-Permian boundary: Bulletin of the Indian Geological Association, v. 17, p. 133–138.

Archbold, N. W., 1988a, Studies on Western Australian Permian brachiopods 8: The Late Permian brachiopod fauna of the Kirkby Range Member, Canning Basin: Proceedings of the Royal Society of Victoria, v. 100, p. 21–32.

Archbold, N. W., 1988b, Permian Brachiopoda and Bivalvia from Sahul Shoals No. 1, Ashmore Block, northwestern Australia: Proceedings of the Royal Society of Victoria, v. 100, p. 33–38.

Archbold, N. W., Dickins, J. M., and Thomas, G. A., 1993, Correlation and age of Permian marine faunas in Western Australia, in Skwarko, S. K., ed., Palaeontology of the Permian of Western Australia: Perth, Geological Survey of Western Australia Bulletin, v. 136, p. 11–18.

Arditto, P. A., 1982, Deposition and diagenesis of the Jurassic Pilliga Sandstone, New South Wales: Journal of the Geological Society of Australia, v. 29, p. 191–203.

Armstrong, J. D., Dear, J. F., and Runnegar, B., 1967, Permian ammonoids from eastern Australia: Journal of the Geological Society of Australia, v. 14, p. 87–97.

Ashley, P. M., and Andrew, A. S., 1992, The Mount Ninderry acid sulphate alteration zone and its relation to epithermal mineralization in the North Arm Volcanics, southeast Queensland: Australian Journal of Earth Sciences, v. 39, p. 79–98.

Australian Gas Light Company, 1982, Well completion report Bootleg 2, 2A: New South Wales Department of Mineral Resources Open-File Report.

Australian Gas Light Company, 1983, Well completion report Bootleg 8: New South Wales Department of Mineral Resources Open-File Report.

Backhouse, J., 1991, Permian palynostratigraphy of the Collie Basin, Western Australia: Review of Palaeobotany and Palynology, v. 67, p. 237–314.

Bacon, C. A., Calver, C. R., and Everard, J. L., 1989, Volcanic rocks: Geological Society of Australia Special Publication, v. 15, p. 333–334.

Baker, J. C., Fielding, C. R., De Caritat, P., and Wilkinson, M. M., 1993, Permian evolution of sandstone composition in a complex back-arc extensional to foreland basin: The Bowen Basin, eastern Australia: Journal of Sedimentary Petrology, v. 63, p. 881–893.

Balfe, P. E., Draper, J. J., Scott, S. G., and Belcher, R. L., 1988, Bowen Basin solid geology: Brisbane, Queensland Department of Mines, scale 1:500,000.

Balme, B. E., 1969, The Permian-Triassic boundary in Australia: Geological Society of Australia Special Publication, v. 2, p. 99–112.

Balme, B. E., 1970, Palynology of Permian and Triassic strata in the Salt Range and Surghar Range, West Pakistan, in Kummel, B., and Teichert, C., eds., Stratigraphic boundary problems: Permian and Triassic of West Pakistan: University of Kansas Special Publication, v. 4, p. 305–453.

Balme, B. E., 1973, Correspondence: Age of a mixed Cardiopteris-Glossopteris flora: Journal of the Geological Society of Australia, v. 20, p. 103-104.

Balme, B. E., 1980, Palynology and the Carboniferous-Permian boundary in Australia and other Gondwana continents: Palynology, v. 4, p. 43–55.

Bamberry, W. J., and Doyle, R. P., 1987, A tuffaceous origin for the Burragorang Claystone Member and its lateral equivalents, in Proceedings, Newcastle Symposium on Advances in the Study of the Sydney Basin: University of Newcastle, Department of Geology, v. 21, p. 123–127.

Banks, M. R., 1962, Permian, in Spry, A., and Banks, M. R., eds., The Geology of Tasmania: Journal of the Geological Society of Australia,

v. 9, p. 189–215.

Banks, M. R., 1978, Correlation chart for the Triassic System of Australia: Australian Bureau of Mineral Resources Bulletin, v. 156, part C, 39 p.

Banks, M. R., 1989, Summary and structural development: Geological Society of Australia Special Publication, v. 15, p. 293–294.

Banks, M. R., and Baillie, P. W., 1989, Late Cambrian to Devonian: Geological Society of Australia Special Publication, v. 15, p. 182–237.

Banks, M. R., and Clarke, M. J., 1987, Changes in the geography of the Tasmania Basin in the Late Paleozoic, *in* McKenzie, G. D., ed., Gondwana six: Stratigraphy, sedimentology, and paleontology: Washington, D.C., American Geophysical Union Monograph 41, p. 1–14.

Barber, P. M., 1982, Palaeotectonic evolution and hydrocarbon genesis of the central Exmouth Plateau: Australian Petroleum Exploration Association Journal, v. 22, p. 131–144.

Barnes, R. G., 1990, Manilla 1:250,000 geological sheet SH56-9: New South Wales Geological Survey, Plan 18810 (unpublished).

Barnes, R. G., and Willis, I. L., 1989, The Geology of the Grafton and Maclean 1:250,000 sheet areas: Geological Survey of New South Wales Report GS 1989/117 (unpublished).

Barnes, R. G., Brown, R. E., Brownlow, J. W., and Stroud, W. J., 1991, Late Permian volcanics in New England—The Wandsworth Volcanic Group: New South Wales Geological Survey Quarterly Notes, v. 84, p. 1–36.

Barton, C. M., Bravo, A. P., Gulline, A. B., Longman, M. J., Marshall, B., Matthews, W. L., Moore, W. R., Naqvi, I. G., and Pike, G. P., 1969, Quamby, Tasmania: Tasmania Department of Mines Geological Atlas 1 Mile Series, Sheet 46 (8214N).

Bastian, L. V., 1965a, Petrological report on the basement to Lower Jurassic sections of some subsidized wells in the Surat Basin: Australian Bureau of Mineral Resources Record 1965/120, 69 p.

Bastian, L. V., 1965b, Petrographic notes on the Clematis Sandstone and Moolayember Formation, Bowen Basin, Queensland: Australian Bureau of Mineral Resources Record 1965/240, 14 p.

Bastian, L. V., and Arman, M., 1965, Petrographic notes on some Triassic sediments in U.K.A. Wondoan Number 1 well and in adjoining areas: Australian Bureau of Mineral Resources Record 1965/227.

Bauer, J. A., and Dixon, O., 1981, Results of a seismic survey in the southern Denison Trough, Queensland, 1978–79: Australian Bureau of Mineral Resources Journal, v. 6, p. 213–222.

Beckett, J., 1988, The Hunter Coalfield: Notes to accompany the 1:100,000 geological map: New South Wales Department of Mineral Resources Geological Survey Report GS 1988/051, unpublished.

Beckett, J., and McDonald, I., 1984, The depositional geology of the Jerrys Plains Subgroup, Upper Hunter region, New South Wales: New South Wales Geological Survey Quarterly Notes, v. 57, p. 1–22.

Beckett, J., Hamilton, D. S., and Weber, C. R., 1983, Permian and Triassic stratigraphy and sedimentation in the Gunnedah-Narrabri-Coonabarabran region: New South Wales Geological Survey Quarterly Notes, v. 51, p. 1–16.

Bembrick, C. S., 1980, Geology of the Blue Mountains, western Sydney Basin: New South Wales Geological Survey Bulletin, v. 26, p. 134–161.

Bembrick, C. S., 1983, Stratigraphy and sedimentation of the Late Permian Illawarra Coal Measures in the Western Coalfield, Sydney Basin, New South Wales: Journal of the Royal Society of New South Wales, v. 116, p. 105–117.

Bembrick, C. S., and Holmes, G. G., 1972, Further occurrences of the Nile Subgroup: New South Wales Geological Survey Quarterly Notes, v. 6, p. 5–10.

Bembrick, C. S., and Lonergan, A. D., 1976, Sydney Basin: Melbourne, Australasian Institute of Mining and Metallurgy Monograph Series, v. 7, p. 426–438.

Binns, R. A., 1966, Granitic intrusions and regional metamorphic rocks of Permian age from the Wongwibinda District, north-eastern New South Wales: Journal of the Royal Society of New South Wales, v. 99, p. 5–36.

Black, L. P., Morgan, W. R., and White, M. E., 1972, Age of a mixed

Cardiopteris-Glossopteris flora from Rb-Sr measurements on the Nychum Volcanics, north Queensland: Journal of the Geological Society of Australia, v. 19, p. 189–196.

Booker, F. W., 1960, Studies in Permian sedimentation in the Sydney Basin: Technical Report of the Department of Mines, New South Wales, v. 5, p. 10–62.

Bourke, D. J., 1980, Stratigraphy of the Mesozoic sequence in the Warialda-Goondiwindi area: Records of the Geological Survey of New South Wales, v. 19, p. 1–75.

Bourke, D. J., and Hawke, J. M., 1977, Correlation of sequences in the eastern side of the Coonamble Embayment and the Gunnedah Basin: New South Wales Geological Survey Quarterly Notes, v. 29, p. 7–18.

Bowen, R. L., and Thomas, G. A., 1976, Permian, *in* Geology of Victoria: Geological Society of Australia Special Publication, v. 5, p. 397–422.

Bowman, H. N., 1970, Palaeoenvironment and revised nomenclature of the upper Shoalhaven Group and Illawarra Coal Measures in the Wollongong-Kiama area, New South Wales: Records of the Geological Survey of New South Wales, v. 12, p. 163–182.

Bowman, H. N., 1974, Geology of the Wollongong, Kiama, and Robertson 1:50 000 sheets 9029-II, 9028-I and -IV: Sydney, Geological Survey of New South Wales, 179 p.

Bradley, G. M., 1980, Notes on the stratigraphy of the Illawarra Coal Measures in the Western Coalfields: New South Wales Geological Survey Open-File Report GS 1980/180, 31 p.

Bradley, G. M., Yoo, E. K., Moloney, J., Beckett, J., and Richardson, S. J., 1985, Petroleum data package, Sydney Basin, New South Wales: Geological Survey of New South Wales Report GS 1985/004, 229 p.

Branagan, D. F., and Pedram, H., 1990, The Lapstone Structural Complex, New South Wales: Australian Journal of Earth Sciences, v. 37, p. 23–36.

Briggs, D.J.C., 1989a, Can macropalaeontologists, palynologists and sequence stratigraphers ever agree on Eastern Australian Permian correlations? *in* Proceedings, Newcastle Symposium on Advances in the Study of the Sydney Basin, 23rd: Newcastle, University of Newcastle, Department of Geology, v. 23, p. 29–36.

Briggs, D.J.C., 1989b, Report to Australian Bureau of Mineral Resources, Canberra (unpublished).

Briggs, D.J.C., 1991, Correlation charts for the Permian of the Sydney-Bowen Basin and New England Orogen, *in* Proceedings, Newcastle Symposium on Advances in the Study of the Sydney Basin: University of Newcastle, Newcastle, Department of Geology, v. 25, p. 30–37.

Briggs, D.J.C., 1993, Time control in the Permian of the Sydney-Bowen Basin and the New England Orogen, *in* Findlay, R. H., Unrug, R., Banks, M. R., and Veevers, J. J., eds., Gondwana eight: Assembly, evolution and dispersal: Rotterdam, A. A. Balkema, p. 371–384.

Britten, R. A., 1975, Ashford Coal-field, New South Wales: Australasian Institute of Mining and Metallurgy Monograph Series, v. 6, p. 258–260, 381–382.

Britten, R. A., and Hanlon, F. N., 1975, North-western Coalfield: Australasian Institute of Mining and Metallurgy Monograph Series, v. 6, p. 236–244.

Britten, R. A., Smyth, M., Bennett, A.J.R., and Shibaoka, M., 1975, Environmental interpretations of Gondwana coal measure sequences in the Sydney Basin of New South Wales, *in* Campbell, K.S.W., ed., Gondwana geology: Canberra, Australian National University Press, p. 233–247.

Brown, C. M., 1980, Bonaparte Gulf Basin: Economic Social Committee of Asia Pacific (ESCAP) Atlas of Stratigraphy, 2, v. 7, p. 42–51.

Brown, K., and Preston, B., 1985, A revision of the stratigraphy of the Tomago Coal Measures, *in* Proceedings, Newcastle Symposium on Advances in the Study of the Sydney Basin: Newcastle, Department of Geology, University of Newcastle, v. 19, p. 50–55.

Brown, K. J., 1978, Bore log of Miller Ironbark DDH 54: Joint Coal Board of New South Wales, Sydney, reference number RN 4591, 39 p. (unpublished).

Brown, R. E., 1987, Newly defined stratigraphic units from the western New England Fold Belt, Manilla 1:250,000 sheet area: New South Wales Geological Survey Quarterly Notes, v. 69, p. 1–9.

Brownlow, J. W., 1986, Dorrigo-Coffs Harbour metallogenic series sheet SH/56 10-11,1:250 000, preliminary edition: Sydney, Geological Survey of New South Wales (unpublished).

Brunker, R. L., 1964, The Bedford Creek Bore: New South Wales Department of Mines Geological Report, v. 43.

Bryan, W. H., 1925, Earth movements in Queensland: Proceedings of the Royal Society of Queensland, v. 37, p. 3–82.

Bunny, M. R., and Herbert, C., 1971, The Lower Triassic Newport Formation, Narrabeen Group, southern Sydney Basin: Geological Survey of New South Wales Records, v. 13, p. 61–81.

Burbank, D. W., and Beck, R. A., 1991, Models of aggradation versus progradation in the Himalayan foreland: Geologische Rundschau, v. 80, p. 623–638.

Burgis, W. A., 1975, Environmental significance of folds in the Rangal Coal Measures at Blackwater, Queensland: Australia Bureau of Mineral Resources, Geology and Geophysics Report, v. 171, 64 p.

Butcher, P. M., 1984, The Showgrounds Formation, its setting and seal, Queensland: Australian Petroleum Exploration Association Journal, v. 24, p. 336–357.

Byrnes, J. G., 1974, Thin marker tuff band near base of the Bringelly Shale, Sydney Basin: New South Wales Geological Survey Report, v. GS 1974/473 (unpublished).

Byrnes, J. G., 1981, Further evidence for volcanic ash bands in Triassic strata of the Sydney Basin: New South Wales Geological Survey Report, v. GS 1981/175 (unpublished).

Byrnes, J. G., 1982a, Tuff bands in the Wongawilli Coal: New South Wales Geological Survey Report, v. GS 1982/228 (unpublished).

Byrnes, J. G., 1982b, Argillized vitric tuff in DM Dora Creek DDH 4: New South Wales Geological Survey Report, v. GS 1982/259 (unpublished).

Byrnes, J. G., 1982c, Kaolinized lithic tuffs in Triassic and Permian formations of the Sydney Basin: New South Wales Geological Survey Report, v. GS 1982/333 (unpublished).

Byrnes, J. G., 1982d, Tuffaceous claystones, volcanic-influenced sandstone and quartz-lithic sandstones from the Black Jack Formation in DM Clift DDH 2: New South Wales Geological Survey Report, GS 1982/361 (PR 82/46).

Byrnes, J. G., 1983, Thin kaolin band from the Terrigal Formation in North-West Oil and Minerals Longley DDH 1 near Somersby: New South Wales Geological Survey Report, v. GS 1983/037 (unpublished).

Byrnes, J. G., and Zlotkowski, T., 1975, Sandstones from the Talbragar Formation in DM Yaminba DDH1 near Coonabarabran: New South Wales Geological Survey Petrological Report 75/29 (unpublished).

Calver, C. R., and Castleden, R. H., 1981, Triassic basalts from Tasmania: Search, v. 12, p. 40–41.

Calver, C. R., Everard, J. L., Findlay, R. H., and Lennox, P. G., 1988, Ben Lomond, Tasmania: Hobart, Tasmania, Department of Mines Geological Atlas 1:50,000 Series, Sheet 8414N.

Cameron, R. G., 1980, Sedimentology of the Late Permian Wittingham Coal Measures in the Liddell-Warkworth-Broke region, northern Sydney Basin [B.Sc. thesis]: Sydney, Macquarie University, 129 p.

Campbell, K.S.W., and McKelvey, B. C., 1972, The geology of the Barrington District, New South Wales: Pacific Geology, v. 5, p. 7–48.

Carr, A. F., 1975, Little River–Oaky Creek District, Queensland: Melbourne, Australasian Institute of Mining and Metallurgy Monograph Series, v. 6, p. 251–254, 309.

Carr, P. F., 1983, A reappraisal of the stratigraphy of the upper Shoalhaven Group and the lower Illawarra Coal Measures, southern Sydney Basin, New South Wales: Linnean Society of New South Wales Proceedings, v. 106, p. 287–297.

Carr, P. F., 1986, Geochemistry of Late Permian shoshonitic lavas from the southern Sydney Basin: Sydney, Geological Society of Australia New South Wales Division Publication, v. 1, p. 165–183.

Carr, P. F., Jones, B. G., and Middleton, R. G., 1989, Precursor and formation of glendonites in the Sydney Basin: Australian Mineralogist, v. 4, p. 3–12.

Cas, R.A.F., 1983, A review of the facies patterns, palaeogeographic development and tectonic context of the Palaeozoic Lachlan Fold Belt of southeastern Australia: Geological Society of Australia Special Publication, v. 10, 104 p.

Cas, R.A.F., Flood, R. H., and Shaw, S. E., 1976, Hill End Trough: New radiometric ages: Search, v. 7, p. 205–207.

Cawood, P. A., 1982, Structural relations in the subduction complex of the Paleozoic New England Fold Belt, eastern Australia: Journal of Geology, v. 90, p. 381–392.

Cawood, P. A., and Leitch, E. C., 1985, Accretion and dispersal tectonics of the southern New England Fold Belt, Eastern Australia, *in* Howell, D. G., ed., Tectonostratigraphic terranes in the circum-Pacific region: Houston, Circum-Pacific Council of Energy and Mineral Resources Earth Science Series, v. 1, p. 481–492.

Chen J.-S., Chu X.-L., Shao M.-R., and Zhong H., 1991, Carbon isotope study of the Permian-Triassic boundary sequences in China: Chemical Geology (Isotope Geoscience Section), v. 89, p. 239–251.

Cherry, D., 1989, The palaeogeographic significance of a Late Devonian Lambie Group clast in the Late Carboniferous Currabubula Formation, New South Wales: Australian Journal of Earth Sciences, v. 36, p. 139–140.

Chestnut, W. S., Flood, R. H., McKelvey, B. C., and Cameron, G., 1973, Manilla 1:250 000 geological series sheet SH 56-9: Sydney, Geological Survey of New South Wales.

Claoué-Long, J. C., Zichao, Z., Gougan, M., and Shaohua, D., 1991, The age of the Permian-Triassic boundary: Earth and Planetary Science Letters, v. 105, p. 182–190.

Clare, A. P., 1993, Subaqueous basaltic volcanism in the Early Permian Grantleigh Trough, central eastern Australia, *in* Flood, P. G., and Aitchison, J. C., eds., New England Orogen, eastern Australia: NEO '93 Conference Proceedings: Armidale, University of New England, Department of Geology and Geophysics, p. 599–608.

Clark, I., and Cook, B., 1988, Victorian geology excursion guide: Canberra, Australian Academy of Science, 489 p.

Clarke, M. J., 1985, Tasmania, *in* Martinez Dias, C., ed., The Carboniferous of the World, II: Australia, Indian Subcontinent, South Africa, South America and North Africa: Madrid, Instituto Geológico y Minero de España, p. 85–88.

Clarke, M. J., and Banks, M. R., 1975, The stratigraphy of the lower (Permo-Carboniferous) parts of the Parmeener Super-Group, Tasmania, *in* Campbell, K.S.W., ed., Gondwana geology: Canberra, Australian National University Press, p. 453–467.

Clarke, M. J., and Farmer, N., 1976, Biostratigraphic nomenclature for Late Palaeozoic rocks in Tasmania: Royal Society of Tasmania Paper and Proceedings, v. 110, p. 91–109.

Clarke, M. J., and Forsyth, S. M., 1989, Late Carboniferous-Triassic: Geological Society of Australia Special Publication, v. 15, p. 293–338.

Collins, W. J., 1991, A reassessment of the 'Hunter-Bowen Orogeny': Tectonic implications for the southern New England Fold Belt: Australian Journal of Earth Sciences, v. 38, p. 409–423.

Collinson, J. W., Eggert, J. T., and Kemp, N. R., 1990, Triassic sandstone petrology of Tasmania: Evidence for a Tasmania-Transantarctic Basin: Royal Society of Tasmania Papers and Proceedings, v. 124, p. 61–75.

Conaghan, P. J., 1980, The Hawkesbury Sandstone: Gross characteristics and depositional environment: New South Wales Geological Survey Bulletin, v. 26, p. 188–253.

Conaghan, P. J., 1984, Aapamire (string-bog) origin for stone-roll swarms and associated 'fluvio-deltaic' coals in the Late Permian Illawarra Coal Measures of the southern Sydney Basin: Climatic, geomorphic, and tectonic implications: Geological Society of Australia Abstracts, v. 12, p. 106–109.

Conaghan, P. J., and Jones, J. G., 1975, The Hawkesbury Sandstone and the Brahmaputra: A depositional model for continental sheet sandstones: Journal of the Geological Society of Australia, v. 22, p. 275–283.

Conaghan, P. J., Jones, J. G., McDonnell, K. L., and Royce, K., 1982, A dynamic fluvial model for the Sydney Basin: Journal of the Geological

Society of Australia, v. 29, p. 55–70.

Conaghan, P. J., Shaw, S. E., and Veevers, J. J., 1994, Sedimentary evidence of the Permian/Triassic global crisis induced by the Siberian hotspot: Canadian Society of Petroleum Geologists, Memoir 17 (in press).

Coney, P. J., Edwards, A., Hine, R., Morrison, F., and Windrim, D., 1990, The regional tectonics of the Tasman orogenic system, eastern Australia: Journal of Structural Geology, v. 12, p. 519–543.

Cook, F. W., and Taylor, C. P., 1979, Permian strata of the Wolfang Basin: Queensland Government Mining Journal, v. 80, p. 342–349.

Cooper, B. J., 1981, Carboniferous and Permian sediments in South Australia and their correlation: Geological Survey of South Australia Quarterly Notes, v. 79, p. 2–6.

Cooper, B. J., 1983, Late Carboniferous and Early Permian stratigraphy and microfloras in Eastern Australia—Some implications from a study of the Arckaringa Basin, *in* Permian Geology of Queensland: Brisbane, Geological Society of Australia, Queensland Division, p. 215–220.

Corbett, G. J., 1976, A new fold structure in the Woolomin Beds suggesting a sinistral movement on the Peel Fault: Journal of the Geological Society of Australia, v. 23, p. 401–406.

Corbett, K. D., and Turner, N. J., 1989, Early Palaeozoic deformation and tectonics: Geological Society of Australia Special Publication, v. 15, p. 154–181.

Cowan, E. J., 1985, A basin analysis of the Triassic System, central coastal Sydney Basin [B.Sc. thesis]: Sydney, Macquarie University, 255 p.

Cowan, E. J., 1993, Longitudinal fluvial drainage patterns within a foreland basin-fill: Permo-Triassic Sydney Basin, Australia: Sedimentary Geology, v. 85, p. 557–577.

Crane, D. T., and Hunt, J. W., 1979, The Carboniferous sequence in the Gloucester-Myall Lake area, New South Wales: Journal of the Geological Society of Australia, v. 26, p. 341–352.

Cranfield, L. C., 1986, The geology of the South Burnett District, *in* Willmott, W. F., ed., Field conference in the South Burnett District: Brisbane, Geological Society of Australia Queensland Division, p. 1–12.

Cranfield, L. C., 1990, The Gympie Group and other Permian strata of the Gympie Composite Terrane, *in* Proceedings, Pacific Rim Congress, v. III: Australasian Institute of Mining and Metallurgy, p. 99–108.

Cranfield, L. C., and Murray, C. G., 1989, New and revised intrusive units in the Maryborough 1:250,000 sheet area southeast Queensland: Queensland Government Mining Journal, v. 90, p. 369–378.

Cranfield, L. C., Schwarzbock, H., and Day, R. W., 1976, Geology of the Ipswich and Brisbane 1:250 000 Sheet areas: Geological Survey of Queensland Report, v. 95.

Crawford, E., Herbert, C., Taylor, G., Helby, R., Morgan, R., and Ferguson, J., 1980, Diatremes of the Sydney Basin: New South Wales Geological Survey Bulletin, v. 26, p. 294–323.

Crostella, A., and Barter, T. P., 1980, Triassic-Jurassic depositional history of the Dampier and Beagle Sub-basins, Northwest Shelf of Australia: Australian Petroleum Exploration Association Journal, v. 20, p. 25–33.

Crowley, J., 1991, Geology of the upper Narrabeen Group between coastal outcrop and the Blue Mountains [B.Sc. thesis]: Sydney, Macquarie University.

Dalrymple, G. B., 1979, Critical tables for conversion of K-Ar ages from old to new constants: Geology, v. 7, p. 558–560.

David, T.W.E., 1907, The geology of the Hunter River Coal Measures, New South Wales: Sydney, Geological Survey of New South Wales (Geology) Memoir, v. 4, 372 p.

Davies, P. J., 1979, Marine geology of the continental shelf off southeast Australia: Australian Bureau of Mineral Resources, v. 195.

Day, R. W., Murray, C. G., and Whitaker, W. G., 1978, The eastern part of the Tasman orogenic zone: Tectonophysics, v. 48, p. 327–364.

Day, R. W., Whitaker, W. G., Murray, C. G., Wilson, I. H., and Grimes, K. G., 1983, Queensland Geology: Queensland Geological Survey Publication, v. 383, 194 p.

Dear, J. F., McKellar, R. G., and Tucker, R. M., 1971, Geology of the Monto

1:250 000 Sheet area: Queensland Geological Survey Report, v. 46, 124 p.

Dear, J. F., McKellar, R. G., Tucker, R. M., and Murphy, P. R., 1981, Monto, Queensland: Brisbane, Queensland Geological Survey, Australia 1:250,000 Geological Series, sheet SG 56-1.

de Jersey, N. J., 1974, Palynology and age of the Callide Coal Measures: Queensland Government Mining Journal, v. 75, p. 249–255.

de Jersey, N. J., 1975, Miospore zones in the Lower Mesozoic of Southeastern Queensland, *in* Campbell, K.S.W., ed., Gondwana geology: Canberra, Australian National University Press, p. 159–172.

de Jersey, N. J., and McKellar, J. L., 1981, Triassic palynology of the Warang Sandstone (Northern Galilee Basin) and its phytogeographic distribution, *in* Cresswell, M. M., and Vella, P., eds., Gondwana five: Rotterdam, A. A. Balkema, p. 31–37.

De Keyser, F., and Lucas, K. G., 1968, Geology of the Hodgkinson and Laura Basins, north Queensland: Australian Bureau of Mineral Resources Bulletin, v. 84, 254 p.

Dickins, J. M., 1976, Correlation chart for the Permian System of Australia: Australian Bureau of Mineral Resources Bulletin, v. 156B.

Dickins, J. M., 1984, Late Palaeozoic glaciation: Australian Bureau of Mineral Resources Journal, v. 9, p. 163–169.

Dickins, J. M., and Malone, E. J., 1973, Geology of the Bowen Basin: Australian Bureau of Mineral Resources Bulletin, v. 130.

Dickins, J. M., and Shah, S. C., 1979, Correlation of the Permian marine sequence of India and Western Australia, *in* Proceedings, Fourth International Gondwana Symposium, Calcutta: Delhi, Hindustan Publishing Corporation, p. 1–44.

Dickinson, W. R., and Suczek, C. A., 1979, Plate tectonics and sandstone compositions: American Association of Petroleum Geologists Bulletin, v. 63, p. 2164–2182.

Dickinson, W. R., and eight others, 1983, Provenance of North American Phanerozoic sandstones in relation to tectonic setting: Geological Society of America Bulletin, v. 94, p. 222–235.

Diessel, C.F.K., 1980a, Newcastle and Tomago Coal Measures: New South Wales Geological Survey Bulletin, v. 26, p. 100–114.

Diessel, C.F.K., 1980b, Excursion guide, day 2, stop 1 (Buchanan Tunnel): New South Wales Geological Survey Bulletin, v. 26, p. 473–476.

Diessel, C.F.K., 1985, Tuffs and tonsteins in the coal measures of New South Wales, Australia, *in* Proceedings, 10th International Congress on Carboniferous stratigraphy and geology, Volume 4: Madrid, Instituto Geológico y Minero de España, p. 197–210.

Diessel, C.F.K., Brown, K. J., and Preston, B., 1985, Excursion notes for the 19th Newcastle Symposium: University of Newcastle, Department of Geology (unpublished).

Dolby, J., and Balme, B. E., 1976, Triassic palynology of the Carnarvon Basin, Western Australia: Reviews of Palaeobotany and Palynology, v. 22, p. 105–168.

Douglas, J. G., and Ferguson, J. A., eds., 1988, Geology of Victoria: Melbourne, Geological Society of Australia Victorian Division, p. 1–664.

Draper, J. J., 1983, Origin of pebbles in mudstones in the Denison Trough, *in* Permian geology of Queensland: Brisbane, Geological Society of Australia Queensland Division, p. 305–316.

Draper, J. J., 1985a, Stratigraphy of the south-eastern Bowen Basin, *in* Bowen Basin Coal Symposium: Geological Society of Australia Abstracts, v. 17, p. 27–31.

Draper, J. J., 1985b, Summary of the Permian stratigraphy of the Bowen Basin, *in* Bowen Basin Coal Symposium: Geological Society of Australia Abstracts, v. 17, p. 45–49.

Draper, J. J., 1988, Permian limestone in the southeastern Bowen Basin, Queensland: An example of temperate carbonate deposition: Sedimentary Geology, v. 60, p. 155–162.

Dulhunty, J. A., 1972, Potassium-Argon dating and occurrence of Tertiary and Mesozoic basalts in the Binnaway District: Journal and Proceedings of the Royal Society of New South Wales, v. 105, p. 71–76.

Dulhunty, J. A., 1976, Potassium-Argon ages of igneous rocks in the Wollar-Rylstone district, New South Wales: Journal and Proceedings of the Royal Society of New South Wales, v. 109, p. 35–39.

Dulhunty, J. A., 1986, Mesozoic Garrawilla lavas beneath Tertiary Volcanics of the Nandewar Range: Journal and Proceedings of the Royal Society of New South Wales, v. 119, p. 29–32.

Dulhunty, J. A., Middlemost, E.A.K., and Beck, R. W., 1987, Potassium-Argon ages, petrology and geochemistry of some Mesozoic igneous rocks in Northeastern New South Wales: Journal and Proceedings of the Royal Society of New South Wales, v. 120, p. 71–90.

Edenborough, S. J., 1985, The Reids Dome Beds—Capella, central Queensland, *in* Bowen Basin Coal Symposium: Geological Society of Australia Abstracts, v. 17, p. 53–57.

Elliott, L., 1985, The stratigraphy of the Denison Trough, *in* Bowen Basin Coal Symposium: Geological Society of Australia Abstracts, v. 17, p. 33–38.

Elliott, L., 1989, The Surat and Bowen Basins: Australian Petroleum Exploration Association Journal, v. 29, p. 398–416.

Ellis, P. L., 1968, Geology of the Maryborough 1:250,000 sheet area: Queensland Geological Survey Report, v. 26.

Ellis, P. L., and Whitaker, W. G., 1976, Geology of the Bundaberg 1:250,000 sheet area: Queensland Geological Survey Report, v. 90.

Emerson, D. W., and Wass, S. Y., 1980, Diatreme characteristics—Evidence from the Mogo Hill intrusion, Sydney Basin: Australian Society of Exploration Geophysicists Bulletin, v. 11, p. 121–133.

Etheridge, L. T., 1987, New stratigraphic and structural data from the Surat and Gunnedah Basins and implications for petroleum exploration: New South Wales Geological Survey, Quarterly Notes, v. 66, p. 1–21.

Evans, P. R., 1980, Geology of the Galilee Basin, *in* Henderson, R. A., and Stephenson, P. J., eds., The Geology and Geophysics of Northeastern Australia: Geological Society of Australia, Queensland Division, p. 299–305.

Evans, P. R., 1991, Early Permian palaeogeography of the southern Sydney Basin: Newcastle Symposium on Advances in the Study of the Sydney Basin, University of Newcastle, Department of Geology, v. 25, p. 202–206.

Evans, P. R., and Migliucci, A., 1991, Evolution of the Sydney Basin during the Permian as a foreland basin to the Currarong and New England Orogens: Newcastle Symposium on Advances in the Study of the Sydney Basin, University of Newcastle, Department of Geology, v. 25, p. 22–29.

Evans, P. R., Seggie, R. J., and Walker, M. J., 1983, Early Permian stratigraphy and palaeogeography, southern Sydney Basin: Newcastle Symposium on Advances in the Study of the Sydney Basin, University of Newcastle, Department Geology, v. 17, 3 p.

Evernden, J. F., and Richards, J. R., 1962, Potassium-argon ages in eastern Australia: Journal of the Geological Society of Australia, v. 9, p. 1–49.

Ewart, A., 1979, A review of the mineralogy and chemistry of Tertiary-Recent dacitic, latitic, rhyolitic, and related salic volcanic rocks, *in* Barker, F., ed., Trondhjemites, dacites, and related rocks: Amsterdam, Elsevier, Developments in Petrology, v. 6, p. 13–21.

Exon, N. F., 1976, Geology of the Surat Basin in Queensland: Australian Bureau of Mineral Resources Bulletin, v. 166.

Facer, R. A., 1978, New and recalculated radiometric data supporting a Carboniferous age for the emplacement of the Bathurst Batholith, New South Wales: Journal of the Geological Society of Australia, v. 25, p. 429–432.

Facer, R. A., and Carr, P. F., 1979, K-Ar dating of Permian and Tertiary igneous activity in the southeastern Sydney Basin, New South Wales: Journal of the Geological Society of Australia, v. 26, p. 73–79.

Falkner, A., and Fielding, C. R., 1990, German Creek Formation and Moranbah Coal Measures: A transition from marine shelf to upper delta plain, *in* Proceedings, Newcastle Symposium on Advances in the Study of the Sydney Basin: Newcastle, University of Newcastle, Department of Geology, p. 41–48.

Fergusson, C. L., 1982, Structure of the late Palaeozoic Coffs Harbour beds, northeastern New South Wales: Journal of the Geological Society of Australia, v. 29, p. 25–40.

Fergusson, C. L., 1984a, Tectono-stratigraphy of a Palaeozoic subduction complex in the central Coffs Harbour Block of northeastern New South Wales: Australian Journal of Earth Sciences, v. 31, p. 217–236.

Fergusson, C. L., 1984b, The Gundahl Complex of the New England Fold Belt, eastern Australia: A tectonic melange formed in a Paleozoic subduction complex: Journal of Structural Geology, v. 6, p. 257–271.

Fielding, C. R., and McLoughlin, S., 1992, Sedimentology and palynostratigraphy of Permian rocks exposed at Fairbairn Dam, central Queensland: Australian Journal of Earth Sciences, v. 39, p. 631–649.

Fielding, C. R., Gray, A.R.G., Harris, G. I., and Salomon, J., 1990, The Bowen Basin and overlying Surat Basin: Australian Bureau of Mineral Resources Bulletin, v. 232, p. 105–116.

Finlayson, D. M., ed., 1990, The Eromanga-Brisbane Geoscience Transect: A guide to basin development across Phanerozoic Australia in southern Queensland: Australian Bureau of Mineral Resources, v. 232, 261 p.

Flood, P. G., and Aitchison, J. C., 1988, Tectonostratigraphic terranes of the southern part of the New England orogen, *in* Kleeman, J. D., ed., New England Orogen tectonics and metallogenesis: Armidale, University of New England, Department of Geology and Geophysics, p. 7–10.

Flood, P. G., and Aitchison, J. C., 1992, Late Devonian accretion of Gamilaroi Terrane to eastern Gondwana: Provenance linkage suggested by the first appearance of Lachlan Fold Belt–derived quartzarenite: Australian Journal of Earth Sciences, v. 39, p. 539–544.

Flood, P. G., and Brady, S. A., 1985, Origin of large-scale crossbeds in the Late Permian coal measures of the Sydney and Bowen Basins, eastern Australia: International Journal of Coal Geology, v. 5, p. 231–245.

Flood, P. G., and Fergusson, C. L., 1982, Tectonostratigraphic units and structure of the Texas Coffs Harbour region, *in* Flood, P. G., and Runnegar, B., eds., New England Geology: Armidale, University of New England, Department of Geology, v. p. 71–78.

Flood, P. G., Jell, J. S., and Waterhouse, J. B., 1981, Two Early Permian stratigraphic units in the southeastern Bowen Basin, central Queensland: Queensland Government Mining Journal, v. 82, p. 179–184.

Flood, R. H., Craven, S. J., Elmes, D. C., Preston, R. J., and Shaw, S. E., 1988, The Warrigundi Igneous Complex: Volcanic centres for the Werrie Basalt New South Wales, *in* Kleeman, J. D., ed., New England Orogen tectonics and metallogenesis: Armidale, University of New England, Department of Geology and Geophysics, p. 166–171.

Folk, R. L., Andrews, P. B., and Lewis, D. W., 1970, Detrital sedimentary rock classification and nomenclature for use in New Zealand: New Zealand Journal of Geology and Geophysics, v. 13, p. 937–968.

Forrest, J. T., and Horstmann, E. L., 1986, The Northwest Shelf of Australia—Geologic review of a potential major petroleum province of the future: American Association of Petroleum Geologists Memoir, v. 40, p. 457–485.

Forsyth, S. M., 1984, Oatlands, Tasmania: Geological Survey of Tasmania, Explanatory Report Geological Atlas 1:50,000 Sheet, v. 68.

Foster, C. B., 1979, Permian plant microfossils of the Blair Athol Coal Measures and basal Rewan Formation of Queensland: Queensland Geological Survey, Palaeontological Papers, v. 45.

Foster, C. B., 1982, Spore-pollen assemblages of the Bowen Basin, Queensland (Australia): Their relationship to the Permian/Triassic boundary: Review of Palaeobotany and Palynology, v. 36, p. 165–183.

Foster, C. B., 1983, Review of the time frame for the Permian of Queensland, *in* Permian Geology of Queensland: Brisbane, Geological Society of Australia Queensland Division, p. 107–120.

Foster, C. B., and Waterhouse, J. B., 1988, The Granulatosporites confluens Oppel-zone and Early Permian marine faunas from the Grant Formation on the Barbwire Terrace, Canning Basin, Western Australia: Australian Journal of Earth Sciences, v. 35, p. 135–157.

Fukui, S., Watanabe, T., and Itaya, T., 1993, High P/T and low P/T metamorphism in the Tia Complex, southern New England Fold Belt, *in* Proceedings, New England Orogen 93 Conference: Armidale, University of New England, Department of Geology and Geophysics, p. 215–221.

Geary, G. C., and Short, D. A., 1989, Oil Company of Australia N.L., Chester-1, Well Completion Report PEL 182, New South Wales: New South Wales Department of Mineral Resources, Open File Report WCR233, 101 p.

Geological Society of Australia, 1971, Tectonic Map of Australia and New Guinea: Sydney, Geological Society of Australia, scale 1:5,000,000.

Geological Survey of Western Australia, 1990, Geology and mineral resources of Western Australia: Western Australia Geological Survey Memoir, v. 3, 827 p.

Gibson, G. M., 1992, Yarramyljup Fault Zone: Eastern boundary of Glenelg River Complex and possible crustal suture in western Victoria: Geological Society of Australia Abstracts, v. 32, p. 232–233.

Gilligan, L. B., and Brownlow, J. W., 1987, Tamworth-Hastings 1:250,000 metallogenic map, mineral deposit data sheets and metallogenic study: Sydney, New South Wales Geological Survey, 438 p.

Gilligan, L. B., Brownlow, J. W., and Cameron, R. G., 1987, Tamworth-Hastings 1:250,000 metallogenic map: Sydney, New South Wales Geological Survey.

Glen, R. A., and Beckett, J., 1989, Thin-skinned tectonics in the Hunter Coalfield of New South Wales: Australian Journal of Earth Sciences, v. 36, p. 589–593.

Glen, R. A., Scheibner, E., and VandenBerg, A.H.M., 1992, Paleozoic intraplate escape tectonics in Gondwanaland and major strike-slip duplication in the Lachlan orogen of southeastern Australia: Geology, v. 20, p. 795–798.

Glenister, B. F., Baker, C., Furnish, W., and Dickins, J. M., 1990, Late Permian ammonoid cephalopod *Cyclobus* from Western Australia: Journal of Paleontology, v. 64, p. 399–402.

Goldbery, R., 1972, Geology of the western Blue Mountains: New South Wales Geological Survey Bulletin, v. 20, p. 1–172.

Goldbery, R., and Holland, W. N., 1973, Stratigraphy and sedimentation of redbed facies in Narrabeen Group of the Sydney Basin, Australia: American Association of Petroleum Geologists Bulletin, v. 57, p. 1314–1334.

Gostin, V. A., 1968, Stratigraphy and sedimentology of the Lower Permian sequence in the Durras-Ulladulla area, Sydney Basin, New South Wales [Ph.D. thesis]: Canberra, Australian National University, 160 p.

Graham, I. J., and Korsch, R. J., 1985, Rb-Sr geochronology of coarse-grained greywackes and argillites from the Coffs Harbour Block, eastern Australia: Chemical Geology (Isotope Geoscience Section), v. 58, p. 45–54.

Gray, A.R.G., 1976, Stratigraphic relationships of Late Palaeozoic sediments between Springsure and Jericho: Queensland Government Mining Journal, v. 77, p. 147–163.

Gray, A.R.G., 1980, Stratigraphic relationships of Permian strata in the southern Denison Trough: Queensland Government Mining Journal, v. 81, p. 110–130.

Gray, A.R.G., and Swarbrick, C.F.J., 1975, Nomenclature of Late Palaeozoic strata in the northeastern Galilee Basin: Queensland Government Mining Journal, v. 76, p. 344–352.

Green, D. C., and Webb, A. W., 1974, Geochronology of the northern part of the Tasman Geosyncline, *in* Denmead, A. K., Tweedale, G. W., and Wilson, A. F., eds., The Tasman Geosyncline: Brisbane, Geological Society of Australia, Queensland Division, p. 275–293.

Gregory, W. J., 1990, A basin analysis of the Triassic succession of the north-central Sydney Basin [M.Sc. thesis]: Sydney, Macquarie University, 434 p.

Grybowski, D. A., 1992, Exploration in Permit NSW/P10 in the offshore Sydney Basin: Australian Petroleum Exploration Association, v. 32, p. 251–263.

Gulline, A. B., Forsyth, S. M., Everard, J. L., Calver, C. R., and Matthews, W. L., 1991, Snow Hill, Tasmania: Hobart, Department of Mines, 1:50,000 geological sheet 56.

Gulson, B. L., Diessel, C.F.K., Mason, D. R., and Krogh, T. E., 1990, High precision radiometric ages from the northern Sydney Basin and their implication for the Permian time interval and sedimentation rates: Australian Journal of Earth Sciences, v. 37, p. 459–469.

Hamilton, D. S., 1985a, Petroleum Data Package—Gunnedah Basin, New South Wales: New South Wales Geological Survey Report GS 1985/005, 84 p.

Hamilton, D. S., 1985b, Deltaic depositional systems, coal distribution and quality, and petroleum potential, Permian Gunnedah Basin, New South Wales: Sedimentary Geology, v. 45, p. 35–75.

Hamilton, D. S., 1987, Depositional systems of Upper Permian sediments and their application to fuel mineral exploration, Gunnedah Basin, New South Wales [Ph.D. thesis]: University of Sydney, 475 p.

Hamilton, D. S., 1991, Genetic stratigraphy of the Gunnedah Basin, New South Wales: Australian Journal of Earth Sciences, v. 38, p. 95–113.

Hamilton, D. S., Newton, C. B., Smyth, M., Gilbert, T. D., Russel, N., McMinn, A., and Etheridge, L. T., 1988, The petroleum potential of the Gunnedah Basin and the overlying Surat Basin sequence, New South Wales: Australian Petroleum Exploration Association Journal, v. 28, p. 218–241.

Hamilton, J. D., 1966, Petrography of some Permian sediments from the lower Hunter Valley of New South Wales: Journal and Proceedings of the Royal Society of New South Wales, v. 98, p. 221–237.

Hammond, R., 1990, Modelling the geological structure of the Bowen Basin, present and future, *in* Proceedings, Bowen Basin Symposium: Brisbane, Geological Society of Australia Queensland Division, p. 7–9.

Hammond, R. L., 1987, The Bowen Basin, Queensland, Australia: An upper crustal extension model for its early history: Australian Bureau of Mineral Resources Record 1987/51, p. 131–139 (unpublished).

Hammond, R. L., 1988, The geological structure of the Bowen Basin, *in* Mallett, C. W., Hammond, R. L., Leach, J.H.J., Enever, J. R., and Mengel, C., eds., Bowen Basin—Stress, structure and mining conditions assessment for mine planning: Project No. 901, Final Report: Brisbane, Commonwealth Scientific and Industrial Research Organisation Division of Geomechanics, National Energy Research Demonstration and Development Committee, p. 10–69.

Hand, S. J., 1990, Palaeogeography of Tasmania's Permo-Carboniferous glacigene deposits: A tentative reconstruction, *in* Baillie, P. W., Geology in Tasmania, a generalist's influence—The M. R. Banks Symposium: Geological Society of Australia Tasmanian Division, p. 47–51.

Hanlon, F. N., Osborne, G. D., and Raggatt, H. G., 1954, Narrabeen Group: Its subdivisions and correlations between the South Coast and Narrabeen-Wyong Districts: Journal and Proceedings of the Royal Society of New South Wales, v. 87, p. 106–120.

Haq, B. U., Hardenbol, J., and Vail, P. R., 1988, Mesozoic and Cenozoic chronostratigraphy and cycles of sea-level change: Society of Economic Paleontologists and Mineralogists Special Publication, v. 42, p. 71–108.

Harland, W. B., Cox, A. V., Llewellyn, P. G., Pickton, C.A.G., Smith, A. G., and Walters, R., 1982, A geologic time scale: Cambridge, Cambridge University Press, 131 p.

Harland, W. B., Armstrong, R. L., Cox, A. V., Craig, L. E., Smith, A. G., and Smith, D. G., 1990, A geological time scale 1989: Cambridge, Cambridge University Press, 263 p.

Harrington, H. J., ed., 1984, Permian coals of eastern Australia: Canberra, Bureau of Mineral Resources and Commonwealth Scientific and Industrial Research Organisation, Division of Fossil Fuels, 500 p.

Harrington, H. J., and Korsch, R. J., 1985, Tectonic model for the Devonian to middle Permian of the New England Orogen: Australian Journal of Earth Sciences, v. 32, p. 163–179.

Harris, W. K., 1981, Permian diamictites of South Australia, *in* Hambrey, M. J., and Harland, W. B., eds., Earth's pre-Pleistocene glacial record: Cambridge, Cambridge University Press, p. 469–473.

Hartman, R. R., 1966, Magnetic evidence for a volcanic zone near the edge of the New South Wales continental shelf off Sydney, *in* Proceedings, Eighth Commonwealth Mining and Metallurgical Congress: Melbourne, Australasian Institute of Mining and Metallurgy, v. 5, p. 95–97.

Havord, P., Herbert, C., Conaghan, P. J., Hunt, J. W., and Royce, K., 1984, The Marrangaroo Conglomerate—Its distribution and origin in the Syd-

ney Basin: Geological Society of Australia Abstracts, v. 12, p. 221–223.

Hawke, J. M. and Cramsie, J. N., eds., 1984, Contributions to the geology of the Great Australian Basin in New South Wales: New South Wales Geological Survey Bulletin, v. 31, 295 p.

Hawkins, P. J., 1978, Galilee Basin—Review of petroleum prospects: Queensland Government Mining Journal, v. 79, p. 96–112.

Helby, R. J., 1969, Stratigraphic palynology: Journal of the Geological Society of Australia, v. 16, p. 264–265.

Helby, R. J., 1973, Review of Late Permian and Triassic palynology of New South Wales: Geological Society of Australia Special Publication, v. 4, p. 141–155.

Helby, R. J., and Morgan, R., 1979, Palynomorphs in Mesozoic volcanics of the Sydney Basin: New South Wales Geological Survey Quarterly Notes, v. 35, p. 1–15.

Helby, R. J., Lennox, M. and Roberts, J., 1986, The age of the Permian Sequence in the Stroud-Gloucester Trough: Journal and Proceedings of the Royal Society of New South Wales, v. 119, p. 33–42.

Helby, R. J., Morgan, R. and Partridge, A. D., 1987, A palynological zonation of the Australian Mesozoic: Association of Australasian Palaeontologists Memoir, v. 4, p. 1–94.

Hensel, H.-D., McCulloch, M. T., and Chappell, B. W., 1985, The New England Batholith: Constraints on derivation from Nd and Sr isotopic studies of granitoids and country rocks: Geochimica and Cosmochimica Acta, v. 49, p. 369–384.

Herbert, C., 1970, The sedimentology and palaeoenvironment of the Triassic Wianamatta Group sandstones, Sydney Basin: New South Wales Geological Survey Records, v. 12, p. 29–44.

Herbert, C., 1975, Lorne Basin, *in* Markham, N. L., and Basden, H., eds., The mineral deposits of New South Wales: Sydney, New South Wales Geological Survey, p. 507–509.

Herbert, C., 1979, The geology and resource potential of the Wianamatta Group: New South Wales Geological Survey Bulletin, v. 25, 203 p.

Herbert, C., 1980a, Depositional development of the Sydney Basin: New South Wales Geological Survey Bulletin, v. 26, p. 10–52.

Herbert, C., 1980b, Wianamatta Group and Mittagong Formation: New South Wales Geological Survey Bulletin, v. 26, p. 254–272.

Herbert, C., 1980c, Southwestern Sydney Basin: New South Wales Geological Survey Bulletin, v. 26, p. 82–99.

Herbert, C., 1983, Sydney Basin Stratigraphy, *in* Herbert, C., ed., Geology of the Sydney 1:100,000 sheet, 9130: Sydney, New South Wales Geological Survey, p. 7–34.

Herbert, C., 1989, The Lapstone Monocline–Nepean Fault—A high angle reverse fault system: Advances in the Study of the Sydney Basin, 23rd Symposium, University of Newcastle, p. 179–186.

Herbert, C., 1993, Tectonics of the New England Orogen reflected in the sequence stratigraphy of the Narrabeen Group/coal measures of the Sydney Basin, *in* Flood, P. G., and Aitchison, J. C., eds., New England Orogen, New England Orogen 93 Conference Proceedings: Armidale, University of New England, Department of Geology and Geophysics, p. 127–136.

Herbert, C., and Helby, R. J., eds., 1980, A guide to the Sydney Basin: New South Wales Geological Survey Bulletin, v. 26, 603 p.

Higgins, M. L., and Loughnan, F. C., 1973, Flint clays of the Merrygoen-Digilah area: Australasian Institute of Mining and Metallurgy Proceedings, v. 246, p. 33–40.

Hill, M.B.L., 1986, Geology of the Deriah Forest area—Implications for the structure and stratigraphy of the Gunnedah Basin: New South Wales Geological Survey Quarterly Notes, v. 62, p. 1–16.

Holland, W. N., 1972, New findings on the Narrabeen Group in the Blue Mountains: Search, v. 3, p. 176–180.

Holmes, G. G., 1976, Goulburn Valley coal drilling: New South Wales Geological Survey Quarterly Notes, v. 23, p. 2–14.

Holmes, W.B.K., 1982, The Middle Triassic flora from Benalong, near Dubbo, central-western New South Wales: Alcheringa, v. 6, p. 1–33.

Horton, P., 1972, Geology of the Mary Creek area, Upper Glastonbury, south-

east Queensland [B.Sc. thesis]: Brisbane, University of Queensland.

Hunt, J. W., Brakel, A. T., and Smyth, M., 1986, Origin and distribution of the Bayswater Seam and correlatives in the Permian Sydney and Gunnedah Basins, Australia: Australian Coal Geology, v. 6, p. 59–75.

Hutton, A. C., and Jones, B. J., 1985, The Woonona Seam—A reappraisal: Newcastle Symposium on Advances in the Study of the Sydney Basin, University of Newcastle, Geology Department, v. 19, p. 91–94.

Irwin, M. J., 1976, Aspects of Early Triassic sedimentation in the Esk Trough, southeast Queensland: University of Queensland, Papers of the Department of Geology, v. 7, p. 46–62.

Ishiga, H., 1990, Radiolarians from the Gympie Province, Eastern Australia: Melbourne, Australasian Institute of Mining and Metallurgy, Pacific Rim 90 Congress, v. III, p. 187–189.

Ishiga, H., and Leitch, E. C., 1989, Reply to discussion: Radiolarian and conodont biostratigraphy of siliceous rocks from New England Fold Belt: Australian Journal of Earth Sciences, v. 36, p. 142–143.

Ishiga, H., Leitch, E. C., Watanabe, T., Naka, T., and Iwasaki, M., 1988, Radiolarian and conodont biostratigraphy of siliceous rocks from the New England Fold Belt: Australian Journal of Earth Sciences, v. 35, p. 73–80.

Jennings, I. B., and Burns, K. L., 1958, Middlesex, Tasmania: Tasmania Department of Mines Geological Atlas 1 Mile Series, Zone 7, Sheet 45, scale 1″ = 1 mi.

Jensen, A. R., 1971, Regional aspects of the Upper Permian regression in the northern part of the Bowen Basin, *in* Davis, A., ed., Proceedings, Second Bowen Basin Symposium, Brisbane, 7–9 October 1970: Geological Survey of Queensland Report, v. 62, p. 7–20.

Jensen, A. R., 1975, Permo-Triassic stratigraphy and sedimentation in the Bowen Basin, Queensland: Australian Bureau of Mineral Resources Bulletin, v. 154, 187 p.

Jensen, A. R., Gregory, C. M., and Forbes, V. R., 1966, Geology of the Mackay 1:250 000 sheet area: Australian Bureau of Mineral Resources Report, v. 104.

Jian, F. X., 1991, Genetic stratigraphy, depositional systems, sandstone petrography and diagenesis of the Triassic in the Gunnedah Basin, NSW, Australia [Ph.D. thesis]: Sydney, University of New South Wales, 377 p.

Jian, F. X., and Ward, C. R., 1993, Triassic depositional episode of the Gunnedah Basin: Sydney, New South Wales Geological Survey Memoir Geology, v. 12, p. 297–326.

Johnstone, M. A., and Bekker, C., 1983, Final geological report on Ulan Coal Mines Ltd Authorisations 58 and 202: New South Wales Department of Mines Report (unpublished).

Joint Coal Board of New South Wales, 1978, Lithological log of Miller Ironbark DDH 54: Sydney, Joint Coal Board of New South Wales (unpublished).

Jones, D. C., 1985, Petroleum data package—Surat Basin, New South Wales: Sydney, New South Wales Geological Survey Report GS 1985/006, 79 p.

Jones, D. C., and Clark, N., eds., 1991, Geology of the Penrith 1:100 000 sheet, 9030: Sydney, New South Wales Geological Survey, 201 p.

Jones, I.W.O., 1986, Basement-cover relationships of the western margin of the Sydney Basin with special reference to the Marrangaroo Conglomerate [B.Sc. thesis]: Sydney, Macquarie University, 72 p.

Jones, J. G., and Veevers, J. J., 1983, Mesozoic origins and antecedents of the Eastern Highlands of Australia: Journal of the Geological Society of Australia, v. 30, p. 305–322.

Jones, J. G., and Veevers, J. J., 1984, Eastern Highlands, *in* Veevers, J. J., ed., Phanerozoic Earth history of Australia: Oxford, England, Clarendon Press, p. 115–143.

Jones, J. G., Conaghan, P. J., McDonnell, K. L., Flood, R. H., and Shaw, S. E., 1984, Papuan Basin analogue and a foreland basin model for the Bowen-Sydney Basin, *in* Veevers, J. J., ed., Phanerozoic Earth history of Australia: Oxford, England, Clarendon Press, p. 243–262.

Jones, J. G., Conaghan, P. J., and McDonnell, K. L., 1987, Coal measures of an orogenic recess: Late Permian Sydney Basin, Australia: Palaeogeog-

raphy, Palaeoclimatology, Palaeoecology, v. 58, p. 203–219.

Jones, M. J., and Truswell, E. M., 1992, Late Carboniferous and Early Permian palynostratigraphy of the Joe Joe Group, southern Galilee Basin, Queensland, and implications for Gondwanan stratigraphy: Australian Bureau of Mineral Resources Journal, v. 13, p. 143–185.

Jones, P. J., and Nicoll, R. S., 1984, Late Triassic conodonts from Sahul Shoals No. 1, Ashmore Block, northwestern Australia: Australian Bureau of Mineral Resources Journal, v. 9, p. 361–364.

Kamerling, P., 1966, Sydney Basin—offshore: Australian Petroleum Exploration Society Journal, v. 6, p. 76–80.

Kassan, J., and Fielding, C. R., 1991, Triassic depositional environments in the southwest Bowen Basin, in Proceedings, Newcastle Symposium on Advances in the Study of the Sydney Basin: University of Newcastle, Department of Geology, v. 25, p. 154–161.

Kemp, E. M., Balme, B. E., Helby, R. J., Kyle, R. A., Playford, G., and Price, P. L., 1977, Carboniferous and Permian palynostratigraphy in Australia and Antarctica: A review: Australian Bureau of Mineral Resources Journal, v. 2, p. 177–208.

Kirkegaard, A. G., Shaw, R. D., and Murray, C. G., 1970, Geology of the Rockhampton and Port Clinton 1:250,000 sheet areas: Geological Survey of Queensland Report, v. 38, 155 p.

Koppe, W. H., 1975, Calen Coal Measures, Queensland: Australasian Institute of Mining and Metallurgy Monograph Series, v. 6, p. 250–251.

Koppe, W. H., 1978, Review of the stratigraphy of the upper part of the Permian succession in the northern Bowen Basin: Queensland Government Mining Journal, v. 79, p. 35–45.

Korsch, R. J., 1977, A framework for the Palaeozoic geology of the southern part of the New England Geosyncline: Journal of the Geological Society of Australia, v. 25, p. 339–355.

Korsch, R. J., and Harrington, H. J., 1981, Stratigraphic and structural synthesis of the New England Orogen: Journal of the Geological Society of Australia, v. 28, p. 205–226.

Korsch, R. J., and Harrington, H. J., 1987, Oroclinal bending, fragmentation and deformation of terranes in the New England orogen, eastern Australasia, in Leitch, E. C., and Scheibner, E., eds., Terrane accretion and orogenic belts: Washington, D.C., American Geophysical Union Geodynamic Series, v. 19, p. 129–139.

Korsch, R. J., O'Brien, P. E., Sexton, M. J., Wake-Dyster, K. D., and Wells, A. T., 1989, Development of Mesozoic transtensional basins in easternmost Australia: Australian Journal of Earth Sciences, v. 36, p. 13–28.

Korsch, R. J., Harrington, H. J., Murray, C. G., Fergusson, C. L., and Flood, P. G., 1990a, Tectonics of the New England Orogen: Australian Bureau of Mineral Resources Bulletin, v. 232, p. 35–52.

Korsch, R. J., Wake-Dyster, K. D., and Johnstone, D. W., 1990b, Deep seismic profiling across the Bowen Basin, in Proceedings, Bowen Basin Symposium: Brisbane, Geological Society of Australia Queensland Division, p. 10–14.

Landenberger, B., Farrell, T. R., Collins, W. J., and Offler, R., 1992, Tectonic significance of U-Pb and Rb-Sr ages from granitoids of the Hillgrove Suite, New England Batholith, New South Wales, in Collins, W. J., ed., Specialist group in geochronology, mineralogy, and petrology workshop: Alice Springs, Application of geochronology to field related problems: Newcastle, University of Newcastle, Department of Geology, 2 p.

Lang, S. C., 1988, Devonian to Early Carboniferous history of the Broken River and Bundock Creek Groups, in Withnall, I. W., and 11 others, Stratigraphy, sedimentology, biostratigraphy and tectonics of the Ordovician to Carboniferous, Broken River Province, North Queensland: Sydney, Geological Society of Australia, Australian Sedimentologists Group, Field Guide Series, v. 5, p. 97–104.

Larson, R. L., and Olson, P., 1991, Mantle plumes control magnetic reversal frequency: Earth and Planetary Science Letters, v. 107, p. 437–447.

Laws, R., and Kraus, G. P., 1974, The regional geology of the Bonaparte Gulf Timor Sea area: Australian Petroleum Exploration Association Journal, v. 14, p. 77–84.

Leah, J., 1983, Geophysics, in Herbert, C., ed., Geology of the Sydney 1:100,000 sheet, 9130: Sydney, New South Wales Geological Survey, p. 127–135.

Leaman, D. E., 1990, The Sydney Basin: Composition of basement: Australian Journal of Earth Sciences, v. 37, p. 107–108.

Leitch, E. C., 1974, The geological development of the southern part of the New England Fold Belt: Geological Society of Australia Journal, v. 21, p. 133–156.

Leitch, E. C., 1978, Structural succession in a Late Palaeozoic slate belt and its tectonic significance: Tectonophysics, v. 47, p. 311–323.

Leitch, E. C., 1981, Rock units, structure and metamorphism of the Port Macquarie Block, eastern New England Fold Belt: Linnean Society of New South Wales Proceedings, v. 104, p. 273–292.

Leitch, E. C., 1988, The Barnard Basin and the Early Permian development of the southern part of the New England Fold Belt, in Kleeman, J. D., ed., New England Orogen tectonics and metallogenesis: Armidale, University of New England, Department of Geology and Geophysics, p. 61–67.

Leitch, E. C., 1993, The floor of the Gunnedah Basin north of the Liverpool Range: Sydney, New South Wales Geological Survey Memoir Geology, v. 12, p. 335–348.

Leitch, E. C., and Asthana, D., 1985, The geological development of the Thora district, northern margin of the Nambucca Slate Belt, eastern New England Fold Belt: Linnean Society of New South Wales Proceedings, v. 108, p. 119–140.

Leitch, E. C., and Bocking, M. A., 1980, Triassic rocks of the Grants Head district and the post-Permian deformation of the southeastern New England Fold Belt: Journal and Proceedings of the Royal Society of New South Wales, v. 113, p. 89–93.

Leitch, E. C. and Cawood, P. A., 1987, Provenance determination of volcaniclastic rocks: The nature and tectonic significance of a Cambrian conglomerate from the New England Fold Belt, eastern Australia: Journal of Sedimentary Petrology, v. 57, p. 630–638.

Leitch, E. C., and McDougall, I., 1979, The age of orogenesis in the Nambucca Slate Belt: A K-Ar study of low-grade regional metamorphic rocks: Journal of the Geological Society of Australia, v. 26, p. 111–119.

Leitch, E. C., and Scheibner, E., 1987, Stratotectonic terranes of the Eastern Australian Tasmanides, in Leitch, E. C., and Scheibner, E., eds., Terrane accretion and orogenic belts: Washington, D.C., American Geophysical Union Geodynamic Series, v. 19, p. 1–19.

Leitch, E. C., and Skilbeck, C. G., 1991, Early Permian volcanism and Early Permian facies belts at the base of the Gunnedah Basin, in the northern Sydney Basin and in the southern part of the New England Fold Belt: Newcastle Symposium on Advances in the study of the Sydney Basin, University of Newcastle, Department of Geology, v. 25, p. 59–66.

Leitch, E. C., Morris, P. A., and Hamilton, D. S., 1988, The nature and tectonic significance of Early Permian volcanic rocks from the Gunnedah Basin of the southern New England Fold Belt, in Proceedings, Newcastle Symposium on Advances in the study of the Sydney Basin: University of Newcastle, Department of Geology, v. 22, p. 9–15.

Lennox, M., and Wilcock, S., 1985, The Stroud-Gloucester Trough and its relation to the Sydney Basin, in Proceedings, Newcastle Symposium on Advances in the study of the Sydney Basin: University of Newcastle, Department of Geology, v. 19, p. 37–41.

Lennox, P. G., and Roberts, J., 1988, The Hastings Block—A key to the tectonic development of the New England Orogen, in Kleeman, J. D., ed., New England Orogen tectonics and metallogenesis: Armidale, University of New England, Department of Geology and Geophysics, p. 68–77.

Liang, T.C.K., 1991, Fault-related folding:Tulcumba Ridge, western New England: Australian Journal of Earth Sciences, v. 38, p. 349–355.

Lister, G. S., Etheridge, M. A., and Symonds, P. A., 1986, Detachment faulting and the evolution of passive continental margins: Geology, v. 14, p. 246–250.

Little, T. A., Holcombe, R. J., Gibson, G. M., Offler, R., Gans, P. B., and McWilliams, M. O., 1992, Exhumation of late Paleozoic blueschists in

Queensland, Australia, by extensional faulting: Geology, v. 20, p. 231–234.

Little, T. A., Holcombe, R. J., and Sliwa, R., 1993, Extensional exhumation of blueschist-bearing serpentinite-matrix melange in the New England Orogen of southeastern Queensland, Australia: Tectonics, v. 12, p. 536–549.

Loughnan, F. C., 1978, Flint clays, tonsteins and the kaolinite claystone facies: Clay Minerals, v. 13, p. 387–400.

Loughnan, F. C., 1991, Permian climate of the Sydney Basin—Cold or hot?: Journal and Proceedings of the Royal Society of New South Wales, v. 124, p. 35–40.

Loughnan, F. C., and Evans, P. R., 1978, The Permian and Mesozoic of the Merriwa-Binnaway-Ballimore area, New South Wales: Journal and Proceedings of the Royal Society of New South Wales, v. 111, p. 107–119.

Loughnan, F. C., Goldbery, R., and Holland, W. N., 1974, Kaolinite clayrocks in the Triassic Banks Wall Sandstone of the western Blue Mountains, New South Wales: Journal of the Geological Society of Australia, v. 21, p. 393–402.

Malone, E. J., Olgers, F., and Kirkegaard, A. G., 1969, The geology of the Duaringa and Saint Lawrence 1:250,000 sheet area, Queensland: Australian Bureau of Mineral Resources Report, v. 121, 133 p.

Manser, W., 1967, Stratigraphic studies of the Upper Palaeozoic and post-Palaeozoic succession in the Hunter Valley [Ph.D. thesis]: Armidale, University of New England, 246 p.

Martin, K. R., 1981, Deposition of the Precipice Sandstone and the evolution of the Surat Basin in the Early Jurassic: Australian Petroleum Exploration Association Journal, v. 21, p. 16–23.

Mathison, C. I., 1967, The Somerset Dam layered basic intrusion, southeastern Queensland: Journal of the Geological Society of Australia, v. 14, p. 57–86.

Matthews, W. L., 1979, Longford Basin geology: Tasmania Department of Mines, scale 1:100,000, map sheet.

Mayne, S. J., Nicholas, E., Bigg-Wither, A. L., Rasidi, J. S., and Raine, M. J., 1974, Geology of the Sydney Basin—A review: Australian Bureau of Mineral Resources Bulletin, v. 149, 229 p.

McClung, G., 1980a, Permian marine sedimentation in the northern Sydney Basin: New South Wales Geological Survey Bulletin, v. 26, p. 54–72.

McClung, G., 1980b, Permian biostratigraphy of the northern Sydney Basin: New South Wales Geological Survey Bulletin, v. 26, p. 360–375.

McDonald, S. J., and Skilbeck, C. G., 1991, Marine facies and their distribution, Permian Porcupine and Watermark Formations, Gunnedah Basin: Newcastle Symposium on Advances in the study of the Sydney Basin, University of Newcastle, Department of Geology, v. 25, p. 138–145.

McDonnell, K. L., 1972, Depositional environments of the Triassic Terrigal Formation, Sydney Basin [M.Sc. thesis]: Sydney, Macquarie University, 130 p.

McDonnell, K. L., 1980, Notes on the depositional environment of the Terrigal Formation, *in* Herbert, C., and Helby, R., eds., A guide to the Sydney Basin: Sydney, New South Wales Geological Survey Bulletin, v. 26, p. 170–176.

McDonnell, K. L., 1983, The Sydney Basin from Late Permian to Middle Triassic, a study focused on the coastal transect [Ph.D. thesis]: Sydney, Macquarie University, 347 p.

McDonnell, K. L., and Conaghan, P. J., 1984, Late Permian to Middle Triassic alluvial depositional environments of the Sydney Basin: Coastal transect: Geological Society of Australia Abstracts, v. 12, p. 367, 369.

McDougall, I., and Wellman, P., 1976, Potassium-argon ages for some Australian Mesozoic igneous rocks: Journal of the Geological Society of Australia, v. 23, p. 1–9.

McKellar, J. L., 1982, Palynostratigraphy of samples from GSQ Ipswich 26: Queensland Government Mining Journal, v. 83, p. 509–513.

McKelvey, B. C., and Gutsche, H. W., 1969, The geology of some Permian sequences on the New England Tablelands, New South Wales: Geological Society of Australia Special Publication, v. 2, p. 13–20.

McKelvey, B. C., and McPhie, J., 1985, Tamworth Belt, *in* Martinez Dias, C., ed., The Carboniferous of the world, II: Australia, Indian subcontinent,

South Africa, South America and North Africa: Madrid, Instituto Geológico y Minero de España, p. 15–23.

McMinn, A., 1983, Permo-Carboniferous palynology of a sample from MacDonalds Creek, Mudgee district: New South Wales Geological Survey Palynology Report 83/14, 4 p.

McMinn, A., 1985, Palynostratigraphy of the Middle Permian coal sequences of the Sydney Basin: Australian Journal of Earth Sciences, v. 32, p. 301–309.

McMinn, A., 1986, Reply to discussion: Australian Journal of Earth Sciences, v. 33, p. 370–371.

McMinn, A., 1987, Palynostratigraphy of the Stroud-Gloucester Trough, New South Wales: Alcheringa, v. 11, p. 151–164.

McMinn, A., 1993, Permian to Jurassic palynostratigraphy of the Gunnedah-Surat Basins in northern New South Wales: Sydney, New South Wales Geological Survey Memoir Geology, v. 12, p. 135–143.

McPhie, J., 1984, Permo-Carboniferous silicic volcanism and palaeogeography on the western edge of the New England Orogen, north-eastern New South Wales: Australian Journal of Earth Sciences, v. 31, p. 133–146.

McPhie, J., 1987, Andean analogue for Late Carboniferous volcanic arc and arc flank environments of the western New England Orogen, New South Wales, Australia: Tectonophysics, v. 138, p. 269–288.

McPhie, J., and Fergusson, C. L., 1983, Dextral movement on the Demon Fault, northeastern New South Wales: A reassessment: Journal and Proceedings of the Royal Society of New South Wales, v. 116, p. 123–127.

McPhie, J., Black, L. P., Law, S. R., Mackenzie, D. E., Oversby, B. S., and Wyborn, D., 1990, Distribution, character and setting of mineralised Palaeozoic volcanic sequences, Burdekin Falls region, northeastern Queensland: *in* Proceedings, Pacific Rim 90 Congress, Volume 2: Melbourne, Australasian Institute of Mining and Metallurgy, p. 465–471.

McTavish, R. A., 1973, Triassic conodont faunas from Western Australia: Neues Jahrbuch Geologisches Paläontologisches Abhandlungen, v. 143, p. 275–303.

McTavish, R. A., and Dickins, J. M., 1974, The age of the Kockatea Shale, Lower Triassic, Perth Basin—A reassessment: Journal of the Geological Society of Australia, v. 21, p. 195–202.

Mengel, D. C., 1975, Collinsville Coal Measures, *in* Traves, D. M., and King, D., eds., Economic geology of Australia and Papua New Guinea, 2. Coal: Melbourne, Australasian Institute of Mining and Metallurgy Monograph Series, v. 6, p. 83–89.

Menning, M., 1986, Zur Dauer des Zechsteins aus magnetostratigraphischer Sicht: Berlin, Zeitschrift der geologischen Wissenschaften, v. 14, p. 395–404.

Middlemost, E.A.K., Dulhunty, J. A., and Beck, R. W., 1992, Some Mesozoic igneous rocks from northeastern New South Wales and their tectonic setting: Journal and Proceedings, Royal Society of New South Wales, v. 125, p. 1–11.

Mills, K., Moelle, K., and Branagan, D., 1989, Faulting near Mooney Mooney Bridge, New South Wales, *in* Proceedings, Newcastle Symposium on Advances in the Study of the Sydney Basin: University of Newcastle, Department of Geology, v. 23, p. 217–224.

Mollan, R. G., Forbes, V. R., Jensen, A. R., Exon, N. F., and Gregory, C. M., 1972, Geology of the Eddystone, Taroom, and western part of the Mundubbera sheet areas, Queensland: Australian Bureau of Mineral Resources Report, v. 142, 137 p.

Morante, R., 1993, Determining the Permian/Triassic boundary in Australia through C-isotope chemostratigraphy, *in* Flood, P. G., and Aitchison, J. C., eds., New England Orogen, *in* Proceedings, New England Orogen 93 Conference: Armidale, University of New England, Department of Geology and Geophysics, p. 293–298.

Morante, R., Veevers, J. J., Andrew, A. S., and Hamilton, P. J., 1994, Determination of the Permian-Triassic boundary in Australia from carbon isotope stratigraphy: Australian Petroleum Exploration Association Journal, v. 34, p. 330–336.

Morgan, R., 1977, Stratigraphy and palynology of the Oaklands Basin, New South Wales: New South Wales Geological Survey Quarterly Notes, v. 29, p. 1–6.

Murphy, P. R., Schwarzbock, H., Cranfield, L. C., Withnall, I. W., and Murray, C. G., 1976, Geology of the Gympie 1:250,000 sheet area: Queensland Geological Survey Report, v. 96.

Murray, C. G., 1986, Metallogeny and tectonic development of the Tasman Fold Belt System in Queensland: Ore Geology Reviews, v. 1, p. 315–400.

Murray, C. G., 1990a, Comparison of suspect terranes of the Gympie Province with other units of the New England Fold Belt, Eastern Australia, *in* Proceedings, Pacific Rim 90 Congress, Volume 2: Melbourne, Australasian Institute of Mining and Metallurgy, p. 247–255.

Murray, C. G., 1990b, Summary of geological developments along the Eromanga-Brisbane Geoscience Transect: Australian Bureau of Mineral Resources Bulletin, v. 232, p. 11–20.

Murray, C. G., McClung, G. R., Whitaker, W. G., and Degeling, P. R., 1981, Geology of Late Palaeozoic sequences at Mount Barney, Queensland and Paddys Flat, New South Wales: Queensland Government Mining Journal, v. 82, p. 203–213.

Murray, C. G., Fergusson, C. L., Flood, P. G., Whitaker, W. J., and Korsch, R. J., 1987, Plate tectonic model for the Carboniferous evolution of the New England Fold Belt: Australian Journal of Earth Sciences, v. 34, p. 213–236.

Murray, C. G., Scheibner, E., and Walker, R. N., 1989, Regional geological interpretation of a digital coloured residual Bouguer gravity image of eastern Australia with a wavelength cut-off of 250 km: Australian Journal of Earth Sciences, v. 36, p. 423–449.

Mushenko, C. M., 1985, Silicic tuffs in the Newcastle Coal Measures [B.Sc. thesis]: Sydney, Macquarie University, 79 p.

Naing, T., 1990, Palaeoenvironmental studies of the Middle Triassic uppermost Narrabeen Group, Sydney Basin: Palaeoecological constraints with particular emphasis on trace fossil assemblages [Ph.D. thesis]: Sydney, Macquarie University, 629 p.

Nicholas, E., 1968, Petrological study of Loder (A.O.G.) No. 1 well, Sydney Basin, New South Wales: Australian Bureau of Mineral Resources Record 1968/130.

New South Wales Department of Mineral Resources, 1982, Lithological log of DM Dora DDH4: Microfiche No. 29884 (unpublished).

O'Brien, P. E., 1986, Stratigraphy and sedimentology of Late Carboniferous glaciomarine sediments beneath the Murray Basin, and their palaeogeographic and palaeoclimatic significance: Australian Bureau of Mineral Resources Journal, v. 10, p. 53–63.

O'Brien, P. E., Korsch, R. J., Wells, A. T., Sexton, M. J., and Wake-Dyster, K. D., 1990: Mesozoic basins at the eastern end of the Eromanga-Brisbane Geoscience Transect: Strike-slip faulting and basin development: Australian Bureau of Mineral Resources Bulletin, v. 232, p. 117–132.

Offenberg, A. C., West, J. L., and Hitchins, B. L., 1973, Coonamble 1:500,000 geological series sheet: Sydney, New South Wales Geological Survey.

Offler, R., and Hand, M., 1988, Metamorphism in the forearc and subduction complex sequences of the southern New England fold belt, *in* Kleeman, J. D., ed., New England orogen tectonics and metallogenesis: Armidale, University of New England, Department of Geology and Geophysics, p. 78–86.

Olgers, F., 1972, Geology of the Drummond Basin, Queensland: Australian Bureau of Mineral Resources Bulletin, v. 132, 78 p.

Olgers, F., 1974, Warwick, Queensland and New South Wales: Australian Bureau of Mineral Resources and Geological Survey of Queensland 1:250,000 Geological Series—Explanatory notes SH/56-2.

Olgers, F., and Flood, P. G., 1970, An angular Permian/Carboniferous unconformity in southeastern Queensland and northeastern New South Wales: Journal of the Geological Society of Australia, v. 17, p. 81–86.

Olgers, F., Webb, A. W., Smit, J.A.J., and Coxhead, B. A., 1964, The geology of the Baralaba 1:250,000 sheet area, Queensland: Australian Bureau of Mineral Resources Record 1964/26.

Olgers, F., Flood, P. G., and Robertson, A. D., 1974, Palaeozoic geology of the Warwick and Goondiwindi Sheet areas, Queensland and New South Wales: Australian Bureau of Mineral Resources Report, v. 164, 109 p.

O'Reilly, S. Y., 1990, Discussion—The Sydney Basin: Composition of basement: Australian Journal of Earth Sciences, v. 37, p. 485–486.

Osborne, G. D., 1949, The stratigraphy of the Lower Marine Series of the Permian System in the Hunter River Valley, New South Wales: Linnean Society of New South Wales Proceedings, v. 74, p. 203–223.

Osman, A. H., and Wilson, R. G., 1975, Blair Athol coal-field, *in* Traves, D. M., and King, D., eds., Economic geology of Australia and Papua New Guinea, Volume 2, Coal: Melbourne, Australasian Institute of Mining and Metallurgy, p. 376–380.

Ozimic, S., 1979, Petrological and petrophysical study of Permian arenites for potential subsurface storage of natural gas, Sydney Basin, New South Wales, Australia: Australian Petroleum Explorations Association Journal, v. 19, p. 115–130.

Packham, G. H., ed., 1969, The geology of New South Wales: Journal of the Geological Society of Australia, v. 16, 654 p.

Packham, G. H., 1987, The eastern Lachlan Fold Belt of southeast Australia: A possible Late Ordovician to Early Devonian sinistral strike slip regime, *in* Leitch, E. C., and Scheibner, E., eds., Terrane accretion and orogenic belts: Washington, D.C., American Geophysical Union Geodynamic Series, v. 19, p. 67–82.

Packham, G. H., and Emerson, D. W., 1975, Upper Permian coal measures of the central Sydney Basin, New South Wales—Seismic data analysis and drilling results: Australian Society of Exploration Geophysicists Bulletin, v. 6, p. 4–13.

Palmer, A. R., 1983, The decade of North American geology 1983 geologic time scale: Geology, v. 11, p. 503–504.

Parkinson, W. D., and Richardson, R. G., 1989, The magnetic field: Geological Society of Australia Special Publication, v. 15, p. 455–458.

Paten, R. J., and Groves, R. D., 1974, Permian stratigraphic nomenclature and stratigraphy Roma area, Queensland: Queensland Government Mining Journal, v. 75, p. 344–354.

Paten, R. J., and McDonagh, G. P., 1976, Bowen Basin: Australasian Institute of Mining and Metallurgy Monograph, v. 7, p. 403–420.

Paten, R. J., Brown, L. N., and Groves, R. D., 1979, Stratigraphic concepts and petroleum potential of the Denison Trough, Queensland: Australian Petroleum Exploration Association Journal, v. 19, p. 43–52.

Pemberton, R. L., 1965, Well completion report, ENL Lake Galilee No. 1: Queensland Geological Survey, unpublished report 1537/1965.

Phipps, C.V.G., 1967, The character and evolution of the Australian continental shelf: Australian Petroleum Exploration Association Journal, v. 7, p. 44–49.

Pickett, J. W., 1984, A new freshwater Limuloid from the Middle Triassic of New South Wales: Palaeontology, v. 27, p. 609–621.

Playford, G., 1965, Plant microfossils from Triassic sediments near Poatina, Tasmania: Journal of the Geological Society of Australia, v. 12, p. 173–210.

Playford, P. E., Cope, R. N., Cockbain, A. E., Low, G. H., and Lowry, D. C., 1975, Phanerozoic: Western Australian Geological Survey Memoir, v. 2, p. 223–432.

Playford, P. E., Cockbain, A. E., and Low, G. H., 1976, Geology of the Perth Basin, Western Australia: Western Australian Geological Survey Bulletin, v. 124, 311 p.

Pogson, D. J., 1972, Geological map of New South Wales: Sydney, New South Wales Geological Survey, scale 1:1,000,000.

Pogson, D. J., and Hitchins, B. L., 1973, New England geological sheet: Sydney, New South Wales Geological Survey, scale 1:500,000.

Pogson, D. J., and Rose, D. M., 1969, Preliminary investigations of the stratigraphy of the Triassic Narrabeen Group in the northwestern section of the Sydney Basin, New South Wales: New South Wales Geological Survey Records, v. 11, p. 61–78.

Powell, C. McA., 1983a, Tectonic relationship between the Late Ordovician and Late Silurian palaeogeographies of southeastern Australia: Journal of the Geological Society of Australia, v. 30, p. 353–373.

Powell, C. McA., 1983b, Geology of the New South Wales South Coast and adjacent Victoria with emphasis on the pre-Permian structural history:

Sydney, Geological Society of Australia, Specialist Group on Tectonics and Structural Geology, Field Guide 1, 118 p.

Powell, C. McA., 1984a, Ordovician to Carboniferous, *in* Veevers, J. J., ed., Phanerozoic Earth history of Australia: Oxford, Clarendon Press, p. 290–340.

Powell, C. McA., 1984b, Terminal fold-belt deformation: relationship of mid-Carboniferous megakinks in the Tasman fold belt to coeval thrusts in cratonic Australia: Geology, v. 12, p. 546–549.

Powell, C. McA., and Fergusson, C. L., 1979a, The relationship of structures across the Lambian unconformity near Taralga, New South Wales: Journal of the Geological Society of Australia, v. 26, p. 209–219.

Powell, C. McA., and Fergusson, C. L., 1979b, Analysis of the angular discordance across the Lambian unconformity in the Kowmung River–Murruin Creek area, eastern New South Wales: Journal and Proceedings of the Royal Society of New South Wales, v. 112, p. 37–42.

Powell, C. McA., Edgecombe, D. R., Henry, N. M., and Jones, J. G., 1977, Timing of regional deformation of the Hill End Trough: A reassessment: Journal of the Geological Society of Australia, v. 23, p. 407–421.

Powell, C. McA., Cole, J. P., and Cudahy, T. J., 1985, Megakinking in the Lachlan Fold Belt, Australia: Journal of Structural Geology, v. 7, p. 281–300.

Powell, C. McA., Li, Z. X., Thrupp, G. A., and Schmidt, P. W., 1990, Australian Palaeozoic palaeomagnetism and tectonics—1: Tectonostratigraphic terrane constraints from the Tasman Fold Belt: Journal of Structural Geology, v. 12, p. 553–565.

Powell, C. McA., Veevers, J. J., Collinson, J. W., Conaghan, P. J., and López-Gamundí, O. R., 1991, Late Carboniferous to Early Triassic foreland basin behind the Panthalassa margin of Gondwanaland, *in* Findlay, R. H., ed., Eighth International Symposium on Gondwana, Abstracts volume: Hobart, University of Tasmania, p. 67.

Powell, C. McA., Baillie, P. W., Conaghan, P. J., and Turner, N. J., 1993, Turbiditic Mathinna Group, Northeast Tasmania: Australian Journal of Earth Sciences, v. 40, p. 169–196.

Powell, D. E., 1976, The geological evolution of the continental margin of northwest Australia: Australian Petroleum Exploration Association Journal, v. 16, p. 13–23.

Powis, G. D., 1984, Palynostratigraphy of the Late Carboniferous sequence, Canning Basin, Western Australia, *in* Purcell, P. G., ed., The Canning Basin, Western Australia: Perth, Petroleum Exploration Society of Australia, p. 429–438.

Powis, G. D., 1989, Revision of Triassic stratigraphy at the Cooper Basin to Eromanga Basin transition, *in* O'Neil, B. J., ed., The Cooper and Eromanga Basins, Australia: Proceedings of Petroleum Exploration Society of Australia: Adelaide, Society of Petroleum Engineers, Australian Society of Exploration Geophysicists, p. 265–277.

Pratt, G. W., and Herbert, C., 1973, A reappraisal of the Lorne Basin: New South Wales Geological Survey Records, v. 15, p. 205–212.

Preston, K. B., 1985, The Blair Athol Coal Measures: Geological Society of Australia Abstracts, v. 17, p. 59–64.

Price, I., 1972, A new Permian and Upper Carboniferous(?) succession near Woodsreef, New South Wales, and its bearing on the palaeogeography of western New England: Linnean Society of New South Wales Proceedings, v. 97, p. 202–210.

Price, P. L., 1983, A Permian palynostratigraphy for Queensland, *in* Permian Geology of Queensland: Brisbane, Geological Society of Australia Queensland Division, p. 155–211.

Price, P. L., Filatoff, J., Williams, A. J., Pickering, S. A., and Wood, G. R., 1985, Late Palaeozoic and Mesozoic palynostratigraphical units: Colonial Sugar Refiners Oil and Gas Division Palynology Facility Report 274/25 (unpublished).

Prouza, V., and Park, W. J., 1973, Revision of Permian stratigraphy of the Emerald-German Creek-Comet area: Queensland Government Mining Journal, v. 74, p. 432–438.

Purcell, P. G., ed., 1984, The Canning Basin: Perth, Geological Society of Australia, Western Australian Branch, 582 p.

Quinn, G. W., 1985, Geology of the Rangal Coal Measures and equivalents: Geological Society of Australia Abstracts, v. 17, p. 92–99.

Qureshi, I. R., 1984, Wollondilly-Blue Mountains gravity gradient and its bearing on the origin of the Sydney Basin: Australian Journal of Earth Sciences, v. 31, p. 293–302.

Renne, P. R., and Basu, A. R., 1991, Rapid eruption of the Siberian Traps flood basalts at the Permo-Triassic boundary: Science, v. 253, p. 176–179.

Retallack, G., 1980, Late Carboniferous to Middle Triassic megafossil floras from the Sydney Basin: New South Wales Geological Survey Bulletin, v. 26, p. 384–430.

Retallack, G., Gould, R. E., and Runnegar, B., 1977, Isotopic dating of a Middle Triassic megafossil flora from near Nymboida, northeastern New South Wales: Linnean Society of New South Wales Proceedings, v. 101, p. 77–113.

Richards, D.N.G., 1980, Palaeozoic granitoids of northeastern Australia, *in* Henderson, R. A., and Stephenson, P. J., eds., The geology and geophysics of northeastern Australia: Geological Society of Australia, Queensland Division, p. 229–246.

Ringis, J., Hawkins, L. V., and Seedsman, K., 1970, Offshore seismic and magnetic surveys of the southern coalfields off Stanwell Park: Australasian Institute of Mining and Metallurgy Proceedings, v. 234, p. 7–16.

Roberts, J., 1985, Australia, *in* Martinez Dias, C., ed., The Carboniferous of the world, II: Australia, Indian subcontinent, South Africa, South America, and North Africa: Madrid, Instituto Geológico y Minero de España, p. 9–145.

Roberts, J., and Engel, B. A., 1980, Carboniferous palaeogeography of the Yarrol and New England Orogens, eastern Australia: Journal of the Geological Society of Australia, v. 27, p. 167–186.

Roberts, J., and Engel, B. A., 1987, Depositional and tectonic history of the southern New England Orogen: Australian Journal of Earth Sciences, v. 34, p. 1–20.

Roberts, J., Claoué-Long, J. C., and Jones, P. J., 1991, Calibration of the Carboniferous and Early Permian of the southern New England Orogen by SHRIMP ion microprobe zircon analyses, *in* Proceedings, Newcastle Symposium on Advances in the Study of the Sydney Basin: University of Newcastle Department of Geology, v. 25, p. 38–43.

Ross, C. A., and Ross, J.R.P., 1987, Late Paleozoic sea levels and depositional sequences: Cushman Foundation for Foraminiferal Research Special Publication, v. 24, p. 137–168.

Royce, K., 1979, Stratigraphy and sedimentology of the Blue Mountains [B.Sc. thesis]: Sydney, Macquarie University, 264 p.

Runnegar, B. N., 1970, The Permian faunas of northern New South Wales and the correlation between the Sydney and Bowen Basins: Journal of the Geological Society of Australia, v. 16, p. 697–710.

Runnegar, B. N., 1979, Ecology of Eurydesma and the Eurydesma fauna, Permian of eastern Australia: Alcheringa, v. 3, p. 261–285.

Runnegar, B. N., 1980a, Biostratigraphy of the Shoalhaven Group: Sydney, New South Wales Geological Survey Bulletin, v. 26, p. 376-382.

Runnegar, B. N., 1980b, Excursion guide, Day 7, Stop 3, Montague Roadstead, Jervis Bay (Early Permian, Shoalhaven Group): Sydney, New South Wales Geological Survey Bulletin, v. 26, p. 523–530.

Runnegar, B. N., 1980c, Marine Shoalhaven Group, southern Sydney Basin: Sydney, New South Wales Geological Survey Bulletin, v. 26, 74–81.

Runnegar, B. N., and McClung, G., 1975, A Permian time scale for Gondwanaland, *in* Campbell, K.S.W., ed., Gondwana Geology: Canberra, Australian National University Press, p. 425–441.

Scheibner, E., 1973, A plate tectonic model of the tectonic history of New South Wales: Journal of the Geological Society of Australia, v. 20, p. 405–426.

Scheibner, E., 1974, Tectonic map of New South Wales: New South Wales Geological Survey, scale 1:1,000,000.

Scheibner, E., 1978a, Tasman Fold Belt System or Orogenic System—Introduction: Tectonophysics, v. 48, p. 153–157.

Scheibner, E., ed., 1978b, The Phanerozoic structure of Australia and variations in tectonic style: Tectonophysics, v. 48, 153–427.

Scheibner, E., 1985, Suspect terranes in the Tasman Fold Belt system, Eastern Australia, *in* Howell, D. G., ed., Tectonostratigraphic terranes in the circum-Pacific region: Circum-Pacific Council of Energy and Mineral Resources Earth Science Series, v. 1, p. 493–514.

Scheibner, E., ed., 1986, Metallogeny and tectonic development of eastern Australia: Ore Geology Reviews, v. 1, p. 147–412.

Scheibner, E., and Glen, R. A., 1972, The Peel Thrust and its tectonic history: Sydney, New South Wales Geological Survey Quarterly Notes, v. 8, p. 2–14.

Scheibner, E., and Stevens, B.P.J., 1974, The Lachlan River lineament and its relationship to metallic deposits: New South Wales Geological Survey Quarterly Notes, v. 14, p. 8–18.

Scheibnerova, V., 1980, A review of Permian foraminifera in the Sydney Basin: New South Wales Geological Survey Bulletin, v. 26, p. 432–445.

Scott, A., 1984, The geology of the Warden Head area, New South Wales [B.Sc. thesis]: Sydney, Macquarie University, 137 p.

Shaw, S. E., 1969, Granitic rocks from the northern portion of the New England Batholith: Journal of the Geological Society of Australia, v. 16, p. 285–290.

Shaw, S. E., and Flood, R. H., 1981, The New England Batholith, eastern Australia: Geochemical variations in time and space: Journal of Geophysical Research, v. 86, p. 10,530–10,544.

Shaw, S. E., and Flood, R. H., 1982, The Bundarra Plutonic Suite: A summary of new data, *in* Flood, P. G., and Runnegar, B. N., eds., New England geology: Armidale, University of New England, Department of Geology, p. 183–192.

Shaw, S. E., Flood, R. H., and Riley, G. H., 1982, The Wologorong Batholith, New South Wales, and the extension of the I–S line of the Siluro-Devonian granitoids: Journal of the Geological Society of Australia, v. 29, p. 41–48.

Shaw, S. E., Flood, R. H., and Vernon, R. H., 1988, Structure of the Dundee Ignimbrite—Dundee, N.S.W., *in* Kleeman, J. D., ed., New England Orogen tectonics and metallogenesis: Armidale, University of New England, Department of Geology and Geophysics, p. 150–156.

Shaw, S. E., Flood, R. H., and Langworthy, P. J., 1989, Age and association of the Rylstone Volcanics: New isotopic evidence, *in* Proceedings, Newcastle Symposium on Advances in the Study of the Sydney Basin: University of Newcastle, Department of Geology, v. 23, p. 45–51.

Shaw, S. E., Conaghan, P. J., and Flood, R. H., 1991, Late Permian and Triassic igneous activity in the New England Batholith and contemporaneous tephra in the Sydney and Gunnedah Basins, *in* Proceedings, Newcastle Symposium on Advances in the Study of the Sydney Basin: University of Newcastle, Department of Geology, v. 25, p. 44–51.

Sherwin, L., and Holmes, G. G., eds., 1986, Geology of the Wollongong and Port Hacking 1:100,000 sheets 9029, 9129: Sydney, New South Wales Geological Survey, 179 p.

Skwarko, S. K., Nicoll, S., and Campbell, K.S.W., 1976, The Late Triassic molluscs, conodonts, and brachiopods of the Kuta Formation, Papua New Guinea: Australian Bureau of Mineral Resources Journal, v. 1, p. 219–230.

Sorby, L. A., and Scott, S. G., 1988, Geology and coal resources: The Moorlands Basin central Queensland: Queensland Government Mining Journal, v. 89, p. 347–353.

Staines, H.R.E., and Koppe, W. H., 1980, The geology of the north Bowen Basin, *in* Henderson, R. A., and Stephenson, P. J., eds., The geology and geophysics of Northeastern Australia: Brisbane, Geological Society of Australia, Queensland Division, p. 279–298.

Standard, J. C., 1969, Hawkesbury Sandstone, *in* Packham, G. H., ed., The geology of New South Wales: Journal of the Geological Society of Australia, v. 16, p. 407–415.

Standing Committee on Coalfield Geology of New South Wales, 1986a, Stratigraphy of the Jerrys Plains Subgroup of the Wittingham Coal Measures in the Singleton-Muswellbrook Coal District of the Hunter Valley: Sydney, New South Wales Geological Survey Records, v. 22, p. 129–143.

Standing Committee on Coalfield Geology of New South Wales, 1986b, Stratigraphic subdivision of the Illawarra Coal Measures in the Western Coalfield: Sydney, New South Wales Geological Survey Records, v. 22, p. 145–158.

Steiner, M., Ogg, J., Zhang, Z., and Sun, S., 1989, The Late Permian/Early Triassic magnetic polarity time scale and plate motions of South China: Journal of Geophysical Research, v. 94, p. 7,343–7,363.

Stephens, C. J., 1992, Geochemical constraints on fractionation and source models for an A-type granite: The Mt Walsh granite in the New England Fold Belt, eastern Australia: Transactions of the Royal Society of Edinburgh, Earth Sciences, v. 83, p. 500–501.

Stroud, W. J., 1989, Notes on revised geological compilations, Inverell and Goondiwindi 1:250,000 geological sheets, northern New England, New South Wales: Sydney, New South Wales Geological Survey Report, GS 1989/351, 56 p.

Stuntz, J., 1972, The subsurface distribution of the "Upper Coal Measures," Sydney Basin, New South Wales: Newcastle, Australasian Institute of Mining and Metallurgy Conference, p. 1–9.

Svenson, D., and Hayes, S., 1975, Callide Coal Measures, Queensland: Australasian Institute of Mining and Metallurgy Monograph Series, v. 6, p. 283–287.

Symonds, P. A., Willcox, J. B., and Kudrass, H. R., 1988, Dampier Ridge in the Tasman Sea—A continental fragment: Geological Society of Australia Abstracts, v. 21, p. 393–394.

Tadros, N. Z., 1986: Sedimentary and tectonic evolution, upper Black Jack Formation, Gunnedah Basin: Sydney, New South Wales Geological Survey Quarterly Notes, v. 65, p. 20–34.

Tadros, N. Z., 1988, Structural subdivision of the Gunnedah Basin: New South Wales Geological Survey Quarterly Notes, v. 73, p. 1–20.

Tadros, N. Z., and Hamilton, D. S., 1991, Utility of coal seams as sequence boundaries in the non-marine upper Black Jack Formation, Gunnedah Basin, *in* Proceedings, Newcastle Symposium on Advances in the Study of the Sydney Basin: University of Newcastle, Department of Geology, v. 25, p. 128–137.

Telford, P. G., 1971, Stratigraphy and palaeontology of the Drake area, New South Wales: Linnean Society of New South Wales Proceedings, v. 95, p. 232–245.

Thomas, G. A., 1969, Notoconularia, a new conulariid genus from the Permian of Eastern Australia: Journal of Paleontology, v. 43, p. 1,283–1,290.

Thomson, J., 1976, Geology of the Drake 1:100,000 sheet 9340: Sydney, New South Wales Geological Survey, 185 p.

Thomson, S., 1986, Early Permian sedimentation and provenance studies in the Gunnedah Basin, New South Wales, *in* Proceedings, Newcastle Symposium on Advances in the Study of the Sydney Basin: University of Newcastle, Department of Geology, v. 20, p. 15–18.

Thomson, S., and Flood, P. G., 1984, Reassessment of the tectonic setting and depositional environment of the Early Permian Maules Creek coal measures, Gunnedah Basin, New South Wales, *in* Proceedings, Newcastle Symposium on Advances in the Study of the Sydney Basin: University of Newcastle, Department of Geology, v. 18, p. 77–80.

Truswell, E. M., 1978, Palynology of the Permo-Carboniferous in Tasmania: An interim report: Tasmanian Geological Survey Bulletin, v. 56, 37 p.

Turner, N. J., Calver, C. R., Castleden, R. H., and Baillie, P. W., 1984, St Marys, Tasmania: Tasmania Department of Mines Geological Atlas, 1:50 000 Series, Sheet 44 (8514N).

Uren, R. E., 1980, Notes on the Clifton Sub-Group, northeastern Sydney Basin: New South Wales Geological Survey Bulletin, v. 26, p. 162–168.

Uren, R. E., 1983, Facies and sedimentary environments of the Wittingham Coal Measures, near Muswellbrook, *in* Proceedings, Newcastle Symposium on Advances in the study of the Sydney Basin: University of Newcastle, Department of Geology, v. 17, p. 30–33.

Veevers, J. J., 1960, Geology of the Howick area, Singleton-Muswellbrook district, New South Wales: Australian Bureau of Mineral Resources Report, v. 53, 23 p.

Veevers, J. J., ed., 1984, Phanerozoic Earth history of Australia: Oxford, Clarendon Press, 418 p.

Veevers, J. J., 1990a, Tectonic-climatic supercycle in the billion-year plate-tectonic eon: Permian Pangean icehouse alternates with Cretaceous dispersed-continents greenhouse: Sedimentary Geology, v. 68, p. 1–16.

Veevers, J. J., 1990b, Development of Australia's post-Carboniferous sedimentary basins: Petroleum Exploration Society of Australia Journal, v. 16, p. 25–32.

Veevers, J. J., and Powell, C. McA., 1984, Comparative tectonics of the transverse structural zones of Australia and North America, *in* Veevers, J. J., ed., Phanerozoic Earth history of Australia: Oxford, Clarendon Press, p. 340–348.

Veevers, J. J., and Powell, C. McA., 1987, Late Paleozoic glacial episodes in Gondwanaland reflected in transgressive-regressive depositional sequences in Euramerica: Geological Society of America Bulletin, v. 98, p. 475–487.

Veevers, J. J., Mollan, R. G., Olgers, F., and Kirkegaard, A. G., 1964, The geology of the Emerald 1:250,000 sheet area, Queensland: Australian Bureau of Mineral Resources Report, v. 68, 71 p.

Veevers, J. J., Powell, C. McA., and Roots, S. R., 1991, Review of seafloor spreading around Australia I: Synthesis of the patterns of spreading: Australian Journal of Earth Sciences, v. 38, p. 373–389.

Veevers, J. J., Conaghan, P. J., and Shaw, S. E., 1994a, Turning point in Pangean environmental history at the Permian/Triassic (P/Tr) boundary, *in* Klein, G. D., ed., Pangea: Paleoclimate, tectonics, and sedimentation during accretion, zenith, and breakup of a supercontinent: Geological Society of America Special Paper 288, p. 187–196.

Veevers, J. J., Clare, A., and Wopfner, H., 1994b, Neocratonic magmatic-sedimentary basins of post-Variscan Europe and post-Kanimblan Eastern Australia generated by right-lateral transtension of Permo-Carboniferous Pangea: Basin Research (in press).

Vine, R. R., 1976, Galilee Basin: Australasian Institute of Mining and Metallurgy Monograph, v. 7, p. 316–321.

Ward, C. R., 1971, Mineralogical changes as marker horizons for stratigraphic correlation in the Narrabeen Group of the Sydney Basin, New South Wales: Journal and Proceedings of the Royal Society of New South Wales, v. 104, p. 77–88.

Ward, C. R., 1972, Sedimentation in the Narrabeen Group, southern Sydney Basin, New South Wales: Journal of the Geological Society of Australia, v. 19, p. 393–409.

Wass, R. E., and Gould, I. G., 1969, Permian faunas and sediments from the South Marulan district, New South Wales: Linnean Society of New South Wales Proceedings, v. 93, p. 212–226.

Watanabe, T., Iwasaki, M., Ishiga, H., Ilizumi, S., Kawachi, Y., Morris, P., Naka, T., and Itaya, T., 1988, Late Carboniferous orogeny in the southern New England Fold Belt, New South Wales, Australia, *in* Kleeman, J. D., ed., The New England Orogen: Tectonics and metallogenesis: Armidale, University of New England, p. 93–98.

Waterhouse, J. B., and Jell, J. S., 1983, The sequence of Permian rocks and faunas near Exmoor Homestead, south of Collinsville, north Bowen Basin, *in* Permian Geology of Queensland: Brisbane, Geological Society of Australia, Queensland Division, p. 231–267.

Waterhouse, J. B., Briggs, D.J.C., and Parfrey, S. M., 1983, Major faunal assemblages in the Early Permian Tiverton Formation near Homevale Homestead, northern Bowen Basin, Queensland, *in* Permian Geology of Queensland: Brisbane, Geological Society of Australia, Queensland Division, p. 122–138.

Webb, A. W., 1969, Isotopic age determinations in Queensland and their relation to the geochronological time scale for the Permian: Geological Society of Australia Special Publication, v. 2, p. 113–116.

Webb, A. W., and McDougall, I., 1967, Isotopic dating evidence on the age of the Upper Permian and Middle Triassic: Earth and Planetary Science Letters, v. 2, p. 483–488.

Webb, A. W., and McDougall, I., 1968, The geochronology of the igneous rocks of eastern Queensland: Journal of the Geological Society of Australia, v. 15, p. 313–346.

Webb, J. A., 1981, A radiometric time scale of the Triassic: Journal of the Geological Society of Australia, v. 28, p. 107–121.

Webb, J. A., and McNaughton, N. J., 1978, Isotopic age of the Sugars Basalt: Queensland Government Mining Journal, v. 79, p. 591–595.

Wellman, P., 1990, A tectonic interpretation of the gravity and magnetic anomalies in southern Queensland: Australian Bureau of Mineral Resources, v. 232, p. 21–34.

Wells, A. T., and Newhouse, S., 1988, Late Palaeozoic and early Mesozoic rocks in the northeastern part of the central Eromanga Basin: Australian Bureau of Mineral Resources Record 1988/33, 21 p.

Whitaker, W. G., and Green, P. M., 1980, Moreton Geology 1:500,000: Brisbane, Queensland Geological Survey.

Whitaker, W. G., Murphy, P. R., and Rollason, R. G., 1974, Geology of the Mundubbera 1:250,000 sheet area: Brisbane, Queensland Geological Survey Report, v. 84, 113 p.

Whitehouse, J., 1984, Geology and coal resources of the Tomago area, north of Newcastle: Sydney, New South Wales Geological Survey Quarterly Notes, v. 54, p. 1–11.

Williams, B.P.J., Wild, E. K., and Suttill, R. J., 1985, Paraglacial aeolianites: Potential new hydrocarbon reservoirs, Gidgealpa Group, southern Cooper Basin: Australian Petroleum Exploration Association Journal, v. 25, p. 291–310.

Williams, E., 1989, Summary and synthesis: Geological Society of Australia Special Publication, v. 15, p. 468–499.

Williams, I. S., Compston, W., Chappell, B. W., and Shirahase, T., 1975, Rubidium-strontium age determinations on micas from a geologically controlled, composite batholith: Journal of the Geological Society of Australia, v. 22, p. 497–505.

Willis, I. L., 1985, Petroleum data package, Clarence-Moreton Basin, New South Wales: Sydney, New South Wales Geological Survey Report, GS 1985/010, 125 p.

Wilson, C.J.L., Will, T. M., Cayley, R. A., and Chen, S., 1992, Geologic framework and tectonic evolution in western Victoria, Australia: Tectonophysics, v. 214, p. 93–127.

Wilson, R. G., 1975, Capella District: Australasian Institute of Mining and Metallurgy Monograph Series, v. 6, p. 78–82.

Wilson, R. G., Wright, E. A., Taylor, B. L., and Probert, D. H., 1958, Review of the geology of the Southern Coalfield, New South Wales: Australasian Institute of Mining and Metallurgy Proceedings, v. 187, p. 81–104.

Wood, B. L., 1982, The geology and mineralization of the Emmaville tin field, New South Wales, *in* Flood, P. G., and Runnegar, B. N., eds., New England Geology: Armidale, University of Armidale, Department of Geology, p. 335–344.

Wopfner, H., 1981, Development of Permian intracratonic basins in Australia, *in* Cresswell, M. M., and Vella, P., eds., Gondwana five: Rotterdam, A. A. Balkema, p.185–190.

Yoo, E. K., 1982, Geology and coal resources of the northern part of the Oaklands Basin: Sydney, New South Wales Geological Survey Quarterly Notes, v. 49, p. 15–27.

Yoo, E. K., 1988, The Rocky Glen Ridge and Gilgandra Trough, beneath the Surat Basin: Sydney, New South Wales Geological Survey Quarterly Notes, v. 72, p. 17–27.

Yoo, E. K., 1991, Geology and coal resources of the northern part of the Western Coalfield: New South Wales Department of Mineral Resources, Coal Geology Report 1991-003, Open-File Report GS 1991/201 (unpublished).

Ziolkowski, V., and Taylor, R., 1985, Regional structure of the North Denison Trough, *in* Bowen Basin Coal Symposium: Geological Society of Australia Abstracts, v. 17, p. 129–135.

MANUSCRIPT ACCEPTED BY THE SOCIETY SEPTEMBER 9, 1993

Geological Society of America
Memoir 184
1994

Permian-Triassic Transantarctic basin

James W. Collinson, John L. Isbell,* and David H. Elliot
Department of Geological Sciences and Byrd Polar Research Center, Ohio State University, Columbus, Ohio 43210
Molly F. Miller and Julia M.G. Miller
Department of Geology, Vanderbilt University, Nashville, Tennessee 37235

With a contribution by
J. J. Veevers
Australian Plate Research Group, School of Earth Sciences, Macquarie University, North Ryde, N.S.W. 2109, Australia

ABSTRACT

The Permian-Triassic Transantarctic basin, which occupied the Panthalassan margin of the East Antarctic craton, including the present Transantarctic and Ellsworth Mountains, evolved above a mid-Paleozoic passive continental margin basement through the following stages: (1) Carboniferous/Permian extension, (2) late Early Permian back-arc basin, (3) Late Permian and Triassic foreland basin, and (4) Jurassic extension and tholeiitic volcanism. A mid-Paleozoic (Devonian) wedge of coastal-to-shallow marine quartzose sandstone developed on the eroded roots of the Late Cambrian-Early Ordovician Ross orogen. A lacuna in East Antarctica during the Carboniferous was followed by the inception of Gondwanan deposition in a wide Carboniferous/Permian extensional basin. Volcanic detritus at the base of the late Early Permian post-glacial marine(?) shale and sandstone sequence in the Ellsworth Mountains is the first sign of a volcanic arc and subduction along the Panthalassan margin. A similar but much thinner non-volcaniclastic sequence accumulated in the Transantarctic Mountains. The introduction of abundant volcanic detritus to the cratonic side of the basin and a 180° paleocurrent reversal in the Late Permian in the Beardmore Glacier area are the earliest indicators of tectonism along the outer margin of the basin and the inception of a foreland basin that accumulated thick Late Permian and Triassic braided stream deposits of mixed volcanic and cratonic provenance. The Permian sequences in the Ellsworth and Pensacola Mountains were folded in the Triassic. The foreland basin was succeeded in the Early Jurassic by extension and initial silicic and then tholeiitic volcanism that led to the breakup of Gondwanaland.

INTRODUCTION

The geology of Antarctica is poorly known compared to the other Gondwanan continents. Until the means are found to obtain more geologic information from beneath the ice, the Earth history of Antarctica can be known only in general terms from the limited outcrop (<2%). The Transantarctic and Ellsworth Mountains share a similar geologic history with their southern American, southern African, and eastern Australian neighbors. Herein we interpret Antarctic geology by comparison with the better known geology of other Gondwanan continents.

This chapter presents a synthesis of Gondwanan geology based on the hypothesis that the Panthalassan margin of the

---
*Present address: Department of Geology, University of Wisconsin, Milwaukee, Wisconsin 53201.

Collinson, J. W., Isbell, J. L., Elliot, D. H., Miller, M. F., and Miller, J.M.G., 1994, Permian-Triassic Transantarctic basin, *in* Veevers, J. J., and Powell, C. McA., eds., Permian-Triassic Pangean Basins and Foldbelts Along the Panthalassan Margin of Gondwanaland: Boulder, Colorado, Geological Society of America Memoir 184.

East Antarctic craton was a single major (Transantarctic) basin during the late Paleozoic and early Mesozoic. This concept was first outlined in a summary paper by Elliot (1975). Collinson et al. (1981) and Vavra et al. (1981) recognized this basin as a foreland basin and used paleocurrent data and compositional data to identify a calc-alkaline source in West Antarctica. By the term "foreland basin" we mean a sedimentary basin lying between the front of a mountain range and the adjacent craton (Allen et al., 1986, p. 3). Dalziel and Elliot (1982) discussed the basin as a "retroarc" and "foreland" basin in their paper on the problems of terranes in West Antarctica.

The authors of this chapter were members of a major expedition to the Beardmore Glacier region (Fig. 1) from 1985 to 1986. Many of the ideas presented in this chapter developed through the cooperative efforts of the members of this expedition. The editors of this volume invited Collinson to visit

Macquarie University for several months in 1988 and again in July 1991 to prepare this synthesis. The individual authors' specific contributions are as follows: Julia M. G. Miller, Permo-Carboniferous glacial deposits; Molly F. Miller, Early Permian shale deposits in the Beardmore Glacier region; John Isbell, Permian fluvial deposits in the Beardmore Glacier region and southern Victoria Land; and David H. Elliot, Jurassic stratigraphy and general Antarctic geology.

Geologic setting

More than 98% of Antarctica is covered by ice. Rocks are exposed in isolated areas along the present coastline and in mountain belts. Although the ice cover gives the continent the appearance of a continuous single land mass, it obscures the fact that much of West Antarctica is below sea level. Melting

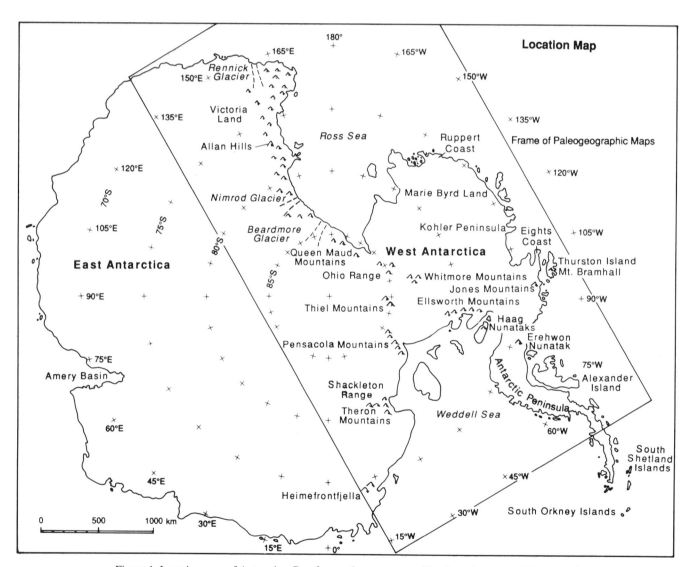

Figure 1. Location map of Antarctica. Box frames the area covered by the paleogeographic maps of Figure 6 and Figures 14–22.

of the ice would leave an archipelago of four large islands along the Pacific margin of East Antarctica. These islands comprise four major crustal blocks: Marie Byrd Land, Thurston Island–Eights Coast, Ellsworth-Whitmore Mountains, and the Antarctic Peninsula (Dalziel and Elliot, 1982). Major structural depressions separate these blocks from each other and East Antarctica (Fig. 2).

The margin of the East Antarctic craton is well exposed in the Transantarctic Mountains, which form a belt 50 to 100 km wide and 3,000 km long (Fig. 2). The chain of peaks, many of which are more than 4,000 m high, is almost continuous along the western margin of the Ross Sea, broken only by major outlet glaciers. The range continues toward and along the eastern margin of the Weddell Sea as a series of isolated ranges including the Thiel, Pensacola, and Theron Mountains (Fig. 1). In general, the Transantarctic Mountains represent the

upwarped fault margin of an extensive plateau that is tilted toward East Antarctica. The Transantarctic Mountains front represents a major crustal boundary where crustal thickness abruptly thins from 40 km under East Antarctica (Bentley, 1983) to 17 to 21 km under the Ross Sea and 20 km under parts of West Antarctica (Behrendt et al., 1991). Uplift of the Transantarctic Mountains began in the Cretaceous–early Cenozoic based on fission-track age versus elevation profiles (Gleadow and Fitzgerald, 1987; Stump and Fitzgerald, 1992). The high elevation of the Transantarctic Mountains is probably related in part to thermal uplift associated with lateral heat conduction from the thinned and extended lithosphere under West Antarctica into the thicker lithosphere of East Antarctica (Stern and Ten Brink, 1989).

Late Proterozoic to Ordovician folded and metamorphosed sedimentary rocks and granitic batholiths form the

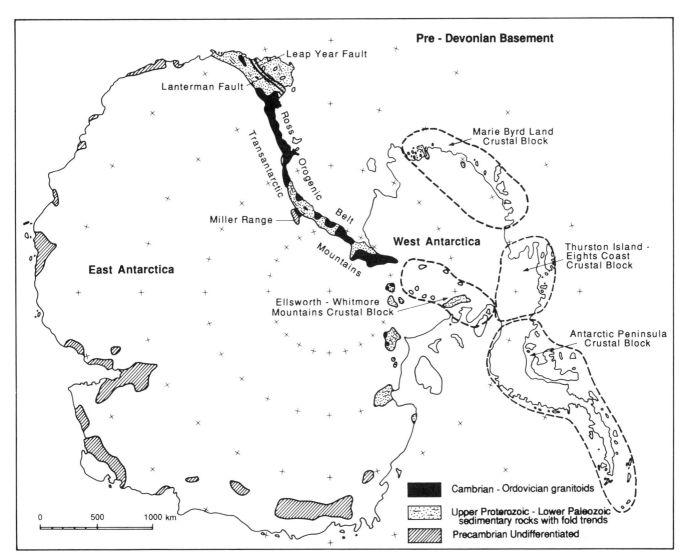

Figure 2. Present distribution of pre-Devonian basement rocks and West Antarctic crustal blocks.

basement along the mountain front (Fig. 2). Relatively unde-formed mid-Paleozoic to Jurassic cover rocks occur at higher elevations. Resistant Jurassic dolerite sills cap many of the high peaks. From radio-echo soundings, Drewry (1972, 1983) has shown that the bedrock surface exposed in the Trans-antarctic Mountains dips under the ice of the polar plateau at a very low angle, eventually merging into a gently undulating interior lowland. This surface is dissected by deep glacial val-leys that join the present outlet glaciers flowing across the drainage divide to West Antarctica or the Ross Sea.

The geologic history of the Panthalassan margin of East Antarctica was characterized by episodes of compressive tec-tonism associated with the convergent margin. These include the Late Proterozoic Beardmore orogeny, the Cambro-Ordo-vician Ross orogeny, and the Late Permian-Triassic Gond-wanide orogeny. Accretion of terranes may have occurred during the first two events (Rowell and Rees, 1989; Borg et al., 1990) and possibly in a mid-Paleozoic event (Borch-grevink Orogeny) that may have affected northern Victoria Land and Marie Byrd Land (Findlay, 1989).

Subsidence along the Transantarctic Mountains margin of East Antarctica resumed in the mid-Paleozoic on a passive continental margin on the roots of the Ross orogenic belt (Fig. 2). This unconformity forms a prominent surface called the Kukri peneplain (Gunn and Warren, 1962). The earliest post-Ross evidence of a back-arc basin and associated volcanic arc in West Antarctica is provided by the Early Permian calc-alka-line volcaniclastic sandstone and tuffs in the Ellsworth Mountains (Collinson et al., 1992). We suggest that a foreland basin formed during the Late Permian and Triassic with the onset of folding and thrusting. The thrust-fold belt is known only from the Ellsworth and Pensacola Mountains.

Since the Jurassic, the Transantarctic Mountains margin of East Antarctica and adjoining West Antarctica, particularly the Ross Sea embayment, have undergone major episodes of extension and uplift during the Cretaceous (Stump and Fitzgerald, 1992), Eocene (Gleadow and Fitzgerald, 1987), and Neogene (Behrendt and Cooper, 1991), culminating in the present Transantarctic Mountains. Convergence along the Pa-cific margin stopped along the Marie Byrd Land crustal block in the mid-Cretaceous, but continued into the Cenozoic out-board of the Antarctic Peninsula. Extension and vertical tec-tonism in the Early to Middle Jurassic preceded the intrusion and extrusion of large volumes of tholeiitic basaltic rocks along the Transantarctic Mountains belt (Elliot, 1991). Jurassic faulting is associated with the intrusion of diabase (Collinson and Elliot, 1984a, 1984b; Wilson, 1993). Evidence for relative uplift includes the development of considerable re-lief before the extrusion of lavas (Collinson et al., 1983). Seis-mic profiles along the western part of the Ross Sea have been interpreted as indicating extensive Mesozoic block faulting (Cooper et al., 1987, 1991).

Relating the geology of the various crustal blocks in West Antarctica to each other and fitting them into a Gondwanaland

reconstruction is more complex than restoring the effects of simple extension. Removing extension brings the probable po-sitions of the Permo-Triassic fold belt closer to the Trans-antarctic Mountains (Fig. 3). Left-lateral restoration of all the crustal blocks is required to reduce the overlap of the Ant-arctic Peninsula with Patagonia and the Malvinas (Falkland) Plateau in Gondwanaland reconstructions. Structural analysis of the Antarctic Peninsula suggests that dextral strike-slip de-formation, generated by oblique subduction, has characterized the Panthalassan margin since the Jurassic (Storey and Nell, 1988) and may have controlled the extension in the backarc region including the Weddell Sea embayment and the Antarctic Peninsula. Wilson (1990, 1993), however, in a de-tailed study of structures associated with emplacement of

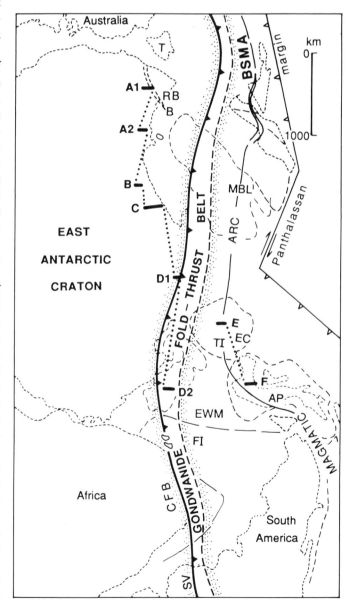

Jurassic dikes, found no evidence for translational movement in southern Victoria Land.

The Ellsworth-Whitmore Mountains crustal block, which contains an almost complete sequence of Paleozoic rocks and is probably underlain by Precambrian basement, requires more than simple closure and translation to account for its geology. A 90° rotation of the Ellsworth Mountains, first suggested by Schopf (1969) and later supported by paleomagnetic data (Watts and Bramall, 1981; Grunow et al., 1991), is geologically sensible, and we use the 230 Ma reconstruction of Grunow et al. (1991, fig. 5a). This aligns the Whitmore Mountains crustal block and the geologically similar Pensacola Mountains (Figs. 1 and 3). Paleomagnetic work on the Ellsworth-Whitmore Mountains crustal block suggests that it was in its present position relative to other crustal blocks by the mid-Jurassic (Grunow et al., 1991). To the reconstruction of Figure 3 (the complete Gondwanaland reconstruction is given in Powell and Li, this volume, Fig. 1) are added (1) mainland Australia and Tasmania (Powell et al., 1988, fig. 6; Veevers and Eittreim, 1988, fig. 3); (2) New Zealand, including the Brook Street Magmatic Arc (Korsch and Wellman, 1988, figs. 13 and 16) in the position indicated by Grunow et al. (1991, fig. 5a) but restored (by us) by removing the subsequent ocean floor and twofold continental extension; and (3) Africa and South America, from Grunow et al. (1991, fig. 5a). Finally, the Gondwanide Fold-Thrust Belt is drawn from New Zealand across Antarctica to southern Africa (Cape Fold Belt) and into South America (Sierra de la Ventana). This reconstruction is used in all the paleogeographic maps up to the Late Triassic. The Jurassic reconstruction of Grunow et al.

Figure 3. Late Triassic (230 Ma) palinspastic reconstruction, from Grunow et al. (1991, fig. 5a). Crustal blocks are restored to their position at 230 Ma during the terminal Gondwanide deformation and before the Jurassic breakup of Gondwanaland. The Ellsworth-Whitmore Mountains crustal block (EWM) is rotated 90° to lie against the East Antarctic craton. The post-Jurassic extension in West Antarctica has been removed by closing the gaps between each crustal block and East Antarctica. This reconstruction leads to a "gap" between the Thurston Island (TI) and Marie Byrd Land (MBL) blocks, which implies a break in the inferred fold-thrust belt. This "gap" could be eliminated by right-lateral translation of the MBL–New Zealand panthalassan margin (*see* Elliot, 1991); alternatively the "gap" may have existed and involved a plate margin such as is illustrated by Grunow et al. (1991), the fold-thrust belt being discontinuous. Lines A–F show segments used to construct the orogen-normal cross-section in Figure 4. The geographic grid of East Antarctica and coastlines (fine dotted lines) are for reference only. Abbreviations of crustal blocks (broken lines) and other features are AP = Antarctic Peninsula; B = Bowers; **BSMA** = Brook Street magmatic arc, from Korsch and Wellman (1988, p. 450), an Early Permian intra-oceanic tholeiitic arc intruded by 265–210 Ma calc-alkaline granite. The BSMA is located palinspastically (as in Figs. 17–22) by removing an assumed twofold subsequent extension; CFB = Cape Fold Belt; EC = Eights Coast; FI = Falkland Islands; RB = Robertson Bay; SV = Sierra de la Ventana.

(1991, fig. 5b) is used for our final map (see Fig. 22 later in the chapter).

The Transantarctic basin and cross section

The Transantarctic basin occupied the Panthalassan margin of the East Antarctic craton from the mid-Paleozoic until the breakup of Gondwanaland in the mid-Jurassic. Evidence of this major depositional basin is found throughout the Transantarctic Mountains that border East Antarctica and in the Ellsworth Mountains in West Antarctica (Collinson, 1991). Exposures along the Transantarctic Mountains and in the palinspastic Ellsworth-Whitmore Mountains crustal block constitute a 200-km-wide belt oblique to the postulated trend of the Transantarctic Basin. In order to construct an orogen-normal cross section, we divided the basin into segments each of which represents the geology of a broad region projected into the line of section (Figs. 3 and 4). Projections are complicated by the probability that sectors of the Transantarctic Mountains have experienced varying amounts of strike-normal offset during the late Mesozoic and Cenozoic rifting of West Antarctica. They are further complicated by the spatial separation of the Ellsworth and Pensacola Mountains from the principal Gondwanan section in the Ross Sea sector. Also, the boundaries of these segments may slightly overlap or have parts missing, but this does not drastically change the postulated configuration of the basin. A further simplifying assumption is that the basin is similar throughout its length. The outermost segments, E and F, include the Thurston Island–Eights Coast crustal blocks and the Antarctic Peninsula. Segment D2 is across the restored Ellsworth-Whitmore Mountains crustal block, and D1 is across the Pensacola Mountains, areas which are correlated by similar styles of folding and similar stratigraphic sequences. Folds die out along the western margin of segment D1 toward the East Antarctic craton. Segment C includes the central Transantarctic Mountains and is linked to segment D1 by similar stratigraphic sequences. The most cratonward segment, A, is across Victoria Land, and is separated from segment C by the postulated Ross basement high (B).

The composite cross section shows the Transantarctic basin (Fig. 4) thinning away from the fold belt margin to wedge out against the East Antarctic craton. A basement high, called the Ross High (segment B on Fig. 3, Fig. 4), is indicated by the lack of stratigraphic continuity from the central Transantarctic Mountains to Victoria Land and by sedimentologic evidence of a basement source in the direction of the Ross Sea in Victoria Land (Collinson, 1990). The Ross High separates the cratonic side of the basin into the Victoria subbasin (South Victoria basin of Elliot, 1975). The Ross High appears to have been episodically active as early as the Devonian and as late as the Early Triassic when it was buried by volcaniclastic sediment from the Panthalassan margin. The composite cross section resulting from projection to a single line from points 3,500 km apart can only be a model for use as a framework for understanding basin evolution.

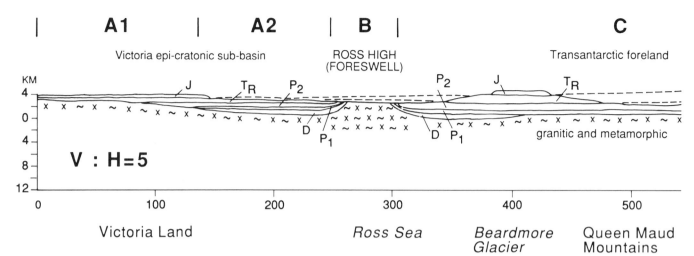

Figure 4. Postulated orogen-normal cross section across West Antarctica and onto East Antarctica just before the Late Jurassic rifting. Segments of the cross section, located in Fig. 3, include the following: northern Victoria Land (A1), southern Victoria Land (A2), the postulated Ross High (B), central Transantarctic Mountains (C), Pensacola Mountains (D1), Ellsworth Mountains (D2), Thurston Island–Eights Coast (E), and Antarctic Peninsula (F). The Ross High is interpreted as a foreswell that divides the Transantarctic basin into a foreland basin on the Panthalassan side and an epicratonic basin on the craton side. Note that this is an idealized cross section, projected onto a single line from points 3,500 km apart. The Ross Sea sector rocks form a coherent set of data points (lines A1 through C), but the other data points are widely separated and are used to indicate possible relations. Ratio of vertical distance to horizontal distance is 5 (V:H = 5).

PRE-GONDWANAN HISTORY

The Transantarctic Mountains reveal a strip of the Late Proterozoic to Late Paleozoic continental margin of Gondwanaland. Consolidation of the paleo-Pacific margin onto the East Antarctic Craton culminated in Ordovician time with the Ross Orogeny. The distribution of pre-Devonian rocks exposed in Antarctica is shown in Figure 2 and the pre-Gondwanan geologic history of the paleo-Pacific margin is summarized in a time-space diagram in Figure 5.

Precambrian

The only place in the Transantarctic Mountains where Late Proterozoic rocks rest on undoubted older Precambrian basement is in the Shackleton Range (Fig. 1). Here metasedimentary rocks (Watts Needle Formation) overlie an unconformity with noticeable relief that is cut into basement gneisses with ages >2,000 Ma (Pankhurst et al., 1983; Marsh, 1983; Kleinschmidt, 1989). With the exception of a single exposure of metamorphic rocks dated by Rb/Sr whole-rock analyses at 1,000 Ma in the Haag Nunataks (Clarkson and Brook, 1977; Millar and Pankhurst, 1987) in part of the Ellsworth-Whitmore Mountains crustal block, no Precambrian rocks are known from West Antarctica. The oldest known rocks in the central Transantarctic Mountains are amphibolite-grade metamorphic rocks in the Miller Range (Fig. 2) that are assigned to the Nimrod Group (Grindley, 1963; Grindley et

al., 1964). Although K-Ar data indicated a mid-Proterozoic age of metamorphism (Grindley and McDougall, 1969), recent studies reflect the uncertainty of this interpretation. Stump et al. (1991) have questioned the age of the Nimrod Group and suggest that it may be the metamorphosed equivalent of part of the Beardmore Group. Some geologic and isotopic evidence, however, indicates Archean-to-Early Proterozoic tectonic events (Borg et al., 1990). Goodge et al. (1991) have documented a thrust-type shear zone indicating major crustal shortening in Middle-to-Late Proterozoic times.

The Beardmore Group in the central Transantarctic Mountains has been interpreted as a turbidite sequence of graywacke and shale deposited along a passive margin (Stump et al., 1986; Smit and Stump, 1986). No satisfactory means has been found to date this sequence (Pankhurst et al., 1988) and the lower contact is unknown. In the Beardmore Glacier region the upper contact is an angular unconformity below the Early Cambrian sedimentary rocks of the Byrd Group. The lithologically similar Patuxent Formation in the Pensacola Mountains is intensely folded and is unconformably overlain by Late Cambrian Limestone (Schmidt and Ford, 1969).

The folding and metamorphism of the Beardmore Group has been called the Beardmore Orogeny (Grindley et al., 1964; Grindley and Laird, 1969). Metamorphism of the Late Proterozoic sequence in the Shackleton Range (except the Watts Needle Formation) may also have occurred at this time (Marsh, 1983). The only constraint on the age of the Beardmore Orogeny is the Early Cambrian age of the overlying sed-

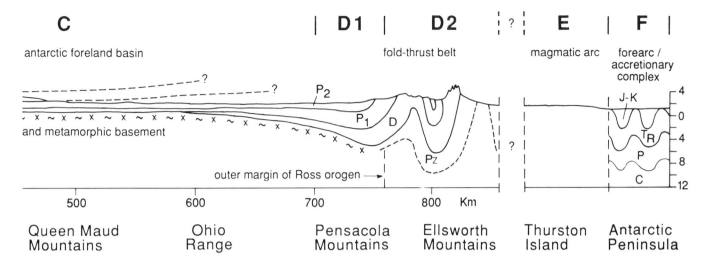

imentary rocks that truncate the folds. Until recently 650 Ma Rb-Sr isochron dates from the Thiel Mountains had been used as a minimum age for the Beardmore Orogeny. Pankhurst et al. (1988) show that these dates are unreliable and point out that the Beardmore Orogeny may be as young as Early Cambrian (Fig. 5).

Cambro-Ordovician

The margin of East Antarctica was characterized by deposition of shallow water detrital sediments in the Ellsworth and Pensacola Mountains and carbonate elsewhere. The margin may have been bordered by a calc-alkaline volcanic belt inboard of a cratonward dipping subduction zone along the paleo-Pacific margin, but as discussed by Rowell and Rees (1989), the outboard volcanic facies may represent allochthonous terranes. Such a possible terrane boundary is shown on Figure 5. Northern Victoria Land and Marie Byrd Land comprise several problematic tectono-stratigraphic terranes.

Ellsworth-Whitmore Mountains crustal block. The only lower Paleozoic sequence that does not seem to have suffered the effects of the Cambro-Ordovician Ross Orogeny is in the Ellsworth Mountains. The Middle to Late Cambrian Heritage Group (7,200 m thick), the Late Cambrian Minaret Formation (0 to 600 m thick), and the Late Cambrian lower Crashsite Quartzite are part of a concordant, but not necessarily complete, 13 km-thick Paleozoic succession that continues into the Permian (Webers and Sporli, 1983). The base of this sequence is not exposed, but the existence of Proterozoic rocks in the nearby crust (Clarkson and Brook, 1977) and magnetic data for that region (Garrett et al., 1987) suggest that the entire region is underlain by continental crust (Clarkson and Brook, 1977). The lower 3,000 m of the Heritage Group consists of volcanic diamictite mostly of terrestrial origin; the rest of the section consists of a variety of fine- to coarse-grained siliciclastic sedimentary rocks and rare limestones, partly of marine origin as indicated by fossil faunas (Webers and Spörli, 1983).

The overlying Minaret Formation, which is almost entirely carbonate, contains a well-preserved shallow water marine fauna of Late Cambrian age (Webers, 1972). The basal part of the 3,200-m-thick Crashsite Quartzite also contains a Late Cambrian shelly fauna (Webers and Spörli, 1983).

Shackleton Range. A predominantly quartzose clastic sequence of low grade (greenschist) metasedimentary rocks is assigned to the Turnpike Bluff Group (Clarkson, 1981). A shale Rb-Sr isochron age of 526 ± 6 Ma, interpreted to represent the age of diagenesis (Pankhurst et al., 1983), and stromatolites (Kleinschmidt, 1989) suggest an Early Cambrian age. Middle Cambrian fossils occur in glacial erratics (Clarkson et al., 1979), but these could have been transported from a distant source.

The Blaiklock Glacier Group, consisting of conglomerate, cross-bedded arkosic sandstone and red siltstone, is interpreted as an early Ordovician molasse deposit related to folding and thrusting in the Ross Orogeny (Buggisch et al., 1990). The succession begins with shallow marine deposits and grades upward into braided fluvial deposits. Three samples of completely unmetamorphosed red shale yielded a poor Rb-Sr isochron age of 482 ± 11 Ma, probably reflecting the age of diagenesis (Pankhurst et al., 1983).

Pensacola Mountains. The Nelson Limestone overlies the Late Proterozoic Patuxent Formation on an angular unconformity. It is 200 to 275 m thick and is characterized by cross-bedded oolitic limestones, small channels, storm lag deposits, bioturbated horizons, and structureless micrite beds (Schmidt et al., 1965). A Middle Cambrian shallow water fauna is dominated by archeocyathids and contains trilobites and inarticulate brachiopods (Palmer and Gatehouse, 1972). Cave deposits represented by flow stone/brecciated horizons and fissures filled by younger units (Elliott, Wiens, and Gambacorta Formations) occur throughout the formation. The Nelson Limestone is in gradational to sharp contact with the overlying Gambacorta Formation, which consists of silicic volcanics and volcaniclas-

Orogen-normal time-space diagram 650-300 Ma : interpretation

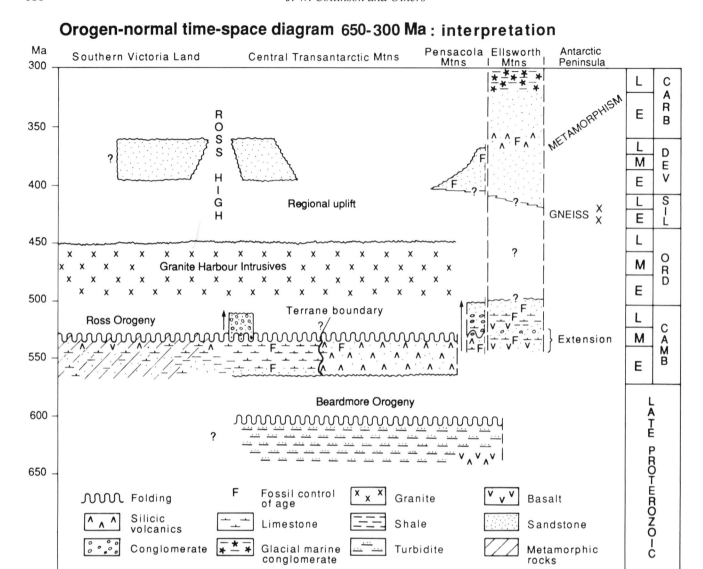

Figure 5. Orogen-normal time-space diagram for the latest Proterozoic to Carboniferous (650 to 300 Ma), along the line of the orogen-normal cross section (Figs. 3 and 4). The vertical arrows indicate synorogenic depositional sequences in the Pensacola Mountains (Macdonald et al., 1991b) and in the Nimrod Glacier region (Rees and Rowell, 1991).

tics suggestive of crustal extension. The Wiens Formation, a fluvial to shallow marine sandstone containing oolitic limestone, overlies the Gambacorta Formation on an angular unconformity. It is truncated by an angular unconformity below the Neptune Group of Late Cambrian or younger age. This sequence, including the basal Neptune Limestone, was folded progressively during deposition (Macdonald et al., 1991b).

Central Transantarctic Mountains. Two major lithofacies occur within Cambrian rocks. Northward from the Beardmore Glacier a predominantly carbonate sequence of Early to Middle Cambrian age was assigned to the Byrd Group (Laird, 1963, 1981). Middle to Late Cambrian silicic volcanic rocks,

widely scattered throughout the Queen Maud Range, are included in the Liv Group (Stump, 1982). Because of the different ages, gross differences in lithologies, and lack of significant amounts of volcanic detritus in the more inboard Byrd Group, Rowell and Rees (1989) hypothesized that these groups represent distinct terranes that were later juxtaposed by strike-slip faulting (Fig. 5).

The major unit in the Byrd Group is the Early Cambrian Shackleton Limestone, which Laird (1981) estimated to be at least 5,400 m thick. Rees et al. (1989) mapped thrust faults and noted that the thickest continuously exposed sections are no more than 200 m thick, implying that the thickness reported

by Laird is due to tectonic stacking. The Shackleton Limestone contains small shelly fossils of Early Cambrian age (Evans and Rowell, 1990) and archeocyathids of late Early to early Middle Cambrian ages (Hill, 1964). The lower contact of the formation is an angular unconformity above the Late Proterozoic Beardmore Group. Laird et al. (1971) reported a variety of basal facies including 500 m of quartzose sandstone south of the Nimrod Glacier and a 1,200-m-thick breccia to the north, but Rees and Rowell (1991) have interpreted all such clastic deposits as fault-bounded remnants of the younger Douglas Conglomerate, which overlies the Shackleton Limestone. Carbonate facies including ecologic reefs of *Epiphyton* and archeocyathids, oolites, and burrow-mottled mudstone all indicate shallow carbonate shelf deposition (Rees et al., 1989). Folded Shackleton Limestone is overlain by synorogenic basin-fill deposits of the Douglas Conglomerate in the upper Byrd Group that have undergone at least three episodes of deformation (Rees and Rowell, 1991; Rowell and Rees, 1989). These deposits accumulated in alluvial fans and braided streams. They locally exceed 1,200 m in thickness, and many of the clasts were derived from the erosion of the underlying Shackleton Limestone.

The Liv Group contains a bimodal suite of volcanics, predominantly rhyolites and dacites, but with subsidiary basalts. The volcanics occur with fine-grained siliciclastic and carbonate sedimentary rocks. In the Queen Maud Mountains silicic volcanic rocks are associated with shallow-water carbonate rocks in which Early Cambrian fossils occur (Yochelson and Stump, 1977).

Southern Victoria Land. Gunn and Warren (1962) assigned metasedimentary rocks in southern Victoria Land to two groups, the Skelton Group in the south and the Koettlitz Group in the north. The less metamorphosed Skelton Group consists of a lower sequence of carbonate and fine-grained siliciclastic sedimentary rocks containing some intermediate to basic volcanic rocks and an upper sequence of intermediate to silicic volcanic and siliciclastic sedimentary rocks (Skinner, 1982). The Koettlitz Group, except for its higher metamorphic grade, would be difficult to distinguish from the Skelton Group (Findlay et al., 1984).

Northern Victoria Land. Three major tectono-stratigraphic terranes have been recognized (Bradshaw et al., 1982; Grindley and Oliver, 1983). A narrow belt of Cambrian sedimentary rocks bounded by northwest-trending faults, the Lanterman fault zone on the west and the Leap Year fault on the east, is called the Bowers Terrane (Fig. 2). Rocks on the cratonic side of this belt are referred to the Wilson Terrane and those toward the Ross Sea, the Robertson Bay Terrane. The relationship of these terranes to each other remains problematic.

Wilson Terrane. Older units consist of highly deformed rocks of varying metamorphic grade, ranging from clastic sedimentary rocks to amphibolites, gneisses, and schists. Protoliths were probably of two interfingering types, a clastic-graywacke suite and a calcareous platform suite (Tessensohn, 1984). Grew

et al. (1984) noted two metamorphic facies: a western belt (Daniels Range metamorphic and igneous complex) that was metamorphosed under low-pressure conditions and one to the east (Lanterman metamorphics) that formed under medium-pressure conditions. These metamorphic belts are separated by the Lanterman Fault on the east side of the Rennick Glacier (Fig. 2). The P/T conditions inferred from the metamorphic assemblages associated with this boundary suggest a crustal thickness that would have been consistent with an orogenic zone involving a continental collision (Grew and Sandiford, 1985). Rocks of the Wilson Terrane have been traditionally regarded as Precambrian (e.g., Gair et al., 1969). Stump et al. (1986) noted that recent isotopic dating has failed to produce any Precambrian ages, suggesting that the Wilson Group may be no older than similar grade metamorphic rocks in southern Victoria Land, which are regarded as latest Precambrian to Cambrian in age.

Bowers Terrane. The Bowers Supergroup is >10 km thick and occurs within a narrow belt 20 to 30 km wide and 350 km long (Laird and Bradshaw, 1983). The supergroup consists (in ascending order) of the Sledgers, Mariner, and Leap Year Groups. Paleocurrent and provenance data from the sedimentary infill in the Bowers basin suggests that only the central remnant of the original basin remains within the Bower Terrane (Laird, 1988).

The Sledgers Group is at least 3.5 km thick and consists of two intertonguing facies, clastic sedimentary and volcanic rocks (Laird and Bradshaw, 1983). Paleocurrents in sedimentary rocks trend toward the southeast. Wodzicki and Robert (1986) interpreted the clastic sedimentary facies as representing a shelf to slope depositional environment; dark mudstones and fine-grained sandstones grade southwestward into turbiditic sandstones and mudstones. This would favor a southwest-facing paleoslope toward East Antarctica. Volcanic rocks, forming the Glasgow Formation, comprise two major suites, submarine tholeiitic basalts in the lower part and mafic to felsic volcanic rocks in the upper part. The latter are interpreted to represent a magmatic arc (Weaver et al., 1984). The base of the group is possibly exposed in the Lanterman Range where a 120-m-thick basal conglomerate overlies undifferentiated metamorphic rocks (Laird and Bradshaw, 1983). A variety of marine fossils suggests that the Sledgers Group is entirely Middle Cambrian in age (Cooper et al., 1983).

The Mariner Group, which may exceed 2.5 km in thickness, appears to conformably overlie the Sledgers Group (Laird and Bradshaw, 1983). The group consists of quartzose sandstone and volcanics, including thin tuffs and a volcanic debris flow near the base, and mudstones, sandstones, and lenticular limestones. Sparse paleocurrent data suggest transport to the southwest and southeast. Channelized limestone breccias with clasts up to 3 m in diameter have been interpreted as debris flow deposits. Marine fossils indicate a range in age from late Middle Cambrian to earliest Ordovician for the group (Cooper et al., 1983; Schmidt-Thome and Wolfart, 1984).

The Leap Year Group overlies an angular unconformity above the Mariner and Sledgers Groups (Laird and Bradshaw, 1983; Wodzicki and Robert, 1986). This sequence, which is at least 4 km thick and consists of a quartzose sandstone and conglomerate, is interpreted as an alluvial deposit (Laird and Bradshaw, 1983). Paleocurrent directions vary from northwesterly to northeasterly. Bradshaw and Laird (1983) have suggested that these coarse clastics were derived during orogenic uplift from an adjacent terrane. The age of the group must be younger than the underlying Early Ordovician Mariner Group.

Robertson Bay Terrane. The Robertson Bay Group is at least 2 km thick and consists of quartzose turbidites and hemipelagic shales (Wright, 1981). Paleocurrent directions average northwest (Field and Findlay, 1983). Sediments have been interpreted as representing a middle fan to basin-plain setting (Findlay, 1986). Allochthonous blocks of shelfal limestone in the uppermost part of the group (Handler Formation) contain trilobite and conodont faunas of latest Cambrian or earliest Ordovician age (Burrett and Findlay, 1984; Wright et al., 1984; Wright and Brodie, 1987).

Marie Byrd Land crustal block. The Swanson Formation, a sequence of well-bedded quartzose sandstone and subsidiary mudstone of unknown thickness that crops out at several localities in Marie Byrd Land has been interpreted as being of turbidite origin (Bradshaw et al., 1983). The occurrence of the trace fossil *Paleodictyon* and a mid-Ordovician date of deformation brackets the formation within the Cambrian to Ordovician (Bradshaw et al., 1983). The formation is lithologically similar to the Robertson Bay Group and has undergone a similar deformational history, but a direct relationship between the two regions, which are now widely separated by the Ross Sea, is impossible to prove (Bradshaw et al., 1983).

Late Cambrian to Ordovician Ross Orogeny

A series of events involving folding, metamorphism, intrusion of Granite Harbour granitoids (Fig. 2), and uplift all along the Transantarctic Mountains have been grouped into the Ross Orogeny. The Ellsworth-Whitmore Mountains crustal block seems to have been unaffected by granitic intrusion, but may have experienced compression (Yoshida, 1982, 1983). An anomalous east-west structural trend in the Shackleton Range, dated at about 500 Ma, has been correlated with the Ross Orogeny (Pankhurst et al., 1983). This large nappe was thrust southwards onto the craton at this time (Buggisch et al., 1990). However, this area was not intruded by granite, possibly because it was cratonward of the granitoid belt (Marsh, 1983). Intrusion also did not occur in the Bowers and Robertson Bay Terranes and the Marie Byrd Land crustal block, which may have been allochthonous and/or on the paleo-Pacific margin side of the granitoid belt.

The deformation of Late Proterozoic–Early Paleozoic sequences is far more complex in Victoria Land than in the Pensacola Mountains and central Transantarctic Mountains. The latter two regions were deformed into broad open folds with fold axes more or less parallel to the older folds of the Beardmore Orogeny and metamorphism was slight (Schmidt and Ford, 1969; Grindley and Laird, 1969). Equivalent rocks in southern Victoria Land experienced three stages of folding (Skinner, 1982; Findlay et al., 1984), and four stages are recognized in the Wilson Group of northern Victoria Land (Kleinschmidt and Skinner, 1981). Folding and metamorphism in the Wilson Group, constrained by Rb-Sr whole rock isochron ages of 530 ± 20 Ma (Middle Cambrian) on schists (Adams, 1986), may represent the effects of continent-continent collision (Grew and Sandiford, 1985).

The rocks of the Bowers and Robertson Bay Terranes appear to have a history different from their present neighbor across the Lanterman Fault Zone. They are relatively unmetamorphosed and are less folded. Findlay (1986) separated these two terranes by the Millen Zone, a narrow belt of schists that parallels the east side of the Leap Year Fault. He suggested that the same major episode of deformation was responsible for similar styles of folding in the Bowers, Millen, and Robertson Bay Terranes; folding was induced by west over east thrusting shortly after deposition of earliest Ordovician sedimentary sequences. Evidence for east over west thrusting, described in the Bowers Terrane by Tessensohn (1984), is considered by Findlay to represent a later event that occurred between the Early Ordovician and the Devonian.

Emplacement and cooling of Granite Harbour intrusions has been narrowed to a range of approximately 500 to 480 Ma in northern Victoria Land (Borg et al., 1986) and 500 to 450 Ma in the central Transantarctic Mountains (Borg, 1983). Wilson Group orthogneisses, which were derived by anatexis of schists, have Rb-Sr isochron ages of 490 ± 33 Ma and are interpreted as representing an early Plutonic phase of the Granite Harbour Intrusives in northern Victoria Land (Adams, 1986). The distribution of S-type and I-type granitoids (cf. White and Chappell, 1977) in the central Transantarctic Mountains and along the eastern margin of the Wilson Terrane in northern Victoria Land indicates westward subduction and most likely a major crustal boundary lying toward the paleo-Pacific margin, possibly the Late Proterozoic–Early Paleozoic continental margin of East Antarctica (Borg, 1983; Borg et al., 1986). Sr and Nd isotopic studies across the central Transantarctic Mountains suggest terrane accretion before the Ross events (Borg et al., 1990), and similar studies in northern Victoria Land suggest accretion of the Bowers and Robertson Bay Terranes in the Paleozoic (Borg and Stump, 1987; Borg et al., 1990).

Synorogenic terrestrial clastics, some of them very coarse, occur throughout the Transantarctic Mountains (Fig. 5), and include the Blaiklock Glacier Group in the Shackleton Range, the Wiens Formation and Neptune Group in the Pensacola Mountains, the Douglas Conglomerate in the central Transantarctic Mountains, and the Leap Year Group in the Bowers Terrane in northern Victoria Land. Progressive fold-

ing occurred along the cratonic margin during Late Cambrian and possibly Early Ordovician time (Rowell and Rees, 1989). The clastics rest with angular unconformity on older units and contain clasts derived from the underlying units.

Ordovician to Devonian

Ellsworth-Whitmore Mountains crustal block. The 3,200-m-thick Crashsite Quartzite contains both Late Cambrian and Devonian marine faunas, with no diagnostic fossils reported from the intervening 3,000 m of section (Webers and Spörli, 1983). The Devonian fossils are summarized by Webers et al. (1992). The location of the Ellsworth Mountains in West Antarctica outboard of the Transantarctic Mountains and the relatively great thickness of Paleozoic rocks are indicative of a continental shelf setting along the paleo-Pacific margin (Fig. 6). As in most continental shelf sequences, disconformities probably divide the succession.

Pensacola Mountains. The Neptune Group and Dover Sandstone comprise a 3-km-thick quartzose sandstone sequence in the Pensacola Mountains (Cathcart and Schmidt, 1977). Stratigraphic units in the Neptune Group range from coarse- to fine-grained siliciclastic sedimentary rocks representing alluvial fan to fluvial to shallow marine settings. Primary phosphate occurs in shallow marine deposits and as reworked clasts in the overlying Dover Sandstone, suggesting a disconformity (Cathcart and Schmidt, 1977). Late Devonian plant fossils have been reported from a similar quartzose sandstone at a remote nunatak in the region (Schopf, 1968).

Central Transantarctic Mountains. Remnants of Devonian to Carboniferous(?) sandstone sequences unconformably overlie basement of the Ross Orogen from the Ohio Range to southern Victoria Land. The upper part of this sequence may have been removed by erosion during Gondwanan glaciation. In the Ohio Range a 45-m-thick sequence of shallow marine sandstone, siltstone and shale contains an Early Devonian shelly fauna (Long, 1964; Doumani et al., 1965; Bradshaw and McCarton, 1983). In the Beardmore Glacier region a quartzose succession, the Alexandra Formation, is up to 300 m thick (Barrett et al., 1986). No shelly fossils have been found, but rare trace fossils and mixed paleocurrent directions suggest a shallow coastal environment.

Southern Victoria Land. A 1,400-m-thick sequence of Devonian quartzose sandstone, siltstone and mudstone has been subdivided into formations of the Taylor Group (McKelvey et al., 1972; McElroy and Rose, 1987). Palynomorphs of Early Devonian age (Kyle, 1977) occur in the lower part of the sequence and Late Devonian palynomorphs (Helby and McElroy, 1969) and fish faunas (Young, 1987) near the top. Abundant trace fossils, particularly in the lower part of the succession, have been used to argue for a shallow marine environment (Bradshaw, 1981; Gevers and Twomey, 1982). Various sedimentary features such as style of bedding, mudcracks, and paleocurrent data, however, have been cited as proof of a terrestrial environment (Barrett and Kohn, 1975; Plume, 1982;

Woolfe, 1990). Trace fossil assemblages suggest a transition from shallow marine to alluvial deposits midway through the sequence (Bradshaw, 1991).

Northern Victoria Land. Scattered local occurrences of plant-bearing, silicic, possibly calc-alkaline, volcanic rocks of Middle to Late Devonian age occur in all three terranes of northern Victoria Land (Grindley and Oliver, 1983; Adams et al., 1986). Grindley and Oliver (1983) have related these to the Admiralty Intrusives of approximately the same age, but Borg and Stump (1987) suggest that no compelling evidence exists to place all the volcanics in the same magmatic province. Whichever the case, terrestrial volcanics were erupting in the various terranes of northern Victoria Land at about the same time while quartzose alluvial sediments were being deposited in adjacent southern Victoria Land.

Granitoids in northern Victoria Land can be differentiated into two probably unrelated groups, the Admiralty Intrusives and the Salamander Granite complex (Borg et al., 1986). The Admiralty Intrusives, which are restricted to the Bowers and Robertson Bay Terranes, are I-type granitoids that were emplaced and cooled in the Devonian to Early Carboniferous (390 to 350 Ma). The Salamander Granite complex, which is Early Carboniferous (350 to 320 Ma), is a hypabyssal, differentiated granite that occurs along a poorly defined part of the tectonic boundary between the Wilson and Bowers Terranes (Borg et al., 1986).

Marie Byrd Land. Possibly calc-alkaline volcanic rocks occur in the Ruppert Coast Metavolcanics (Grindley and Mildenhall, 1983). These rocks overlie, and are probably the same age as, plant-bearing carbonaceous sandstone and siltstone containing a mid-Devonian plant flora (Grindley et al., 1980). The volcanic rocks are similar to those in northern Victoria Land and are the same age.

Early Carboniferous

Fossil evidence of pre-glacial Carboniferous sedimentary rocks is missing from Antarctica. Late Carboniferous to Early Permian glacial deposits disconformably overlie Devonian to Carboniferous(?) sandstone or Early Paleozoic basement throughout most of the Transantarctic Mountains. Sandstones of the upper Crashsite Group in the northern Ellsworth Mountains appear to be conformable with the overlying glacial marine deposits (Ojakangas and Matsch, 1981; Matsch and Ojakangas, 1992). In the Pensacola Mountains and some other parts of the Transantarctic Mountains, grooves in sandstones underlying Gondwanan glacial deposits are interpreted as drag marks in unconsolidated sediment (Frakes, 1981). The contact may represent erosion by grounded ice on a marine shelf in a continuous sedimentary sequence.

Plate Tectonic summary of pre-Gondwanan rocks

Two major tectonic events affected the paleo-Pacific margin in Proterozoic to mid-Paleozoic time, the Late Proterozoic Beardmore Orogeny and the Cambro-Ordovician Ross Orog-

eny. Following these events, the margin, except for northern Victoria Land, remained quiescent, probably representing a passive continental margin, until the Permo-Triassic folding that initiated foreland basin sedimentation. In northern Victoria Land, the margin appears to have been consolidated by the Devonian at the latest, when the Admiralty granites intruded the various terranes.

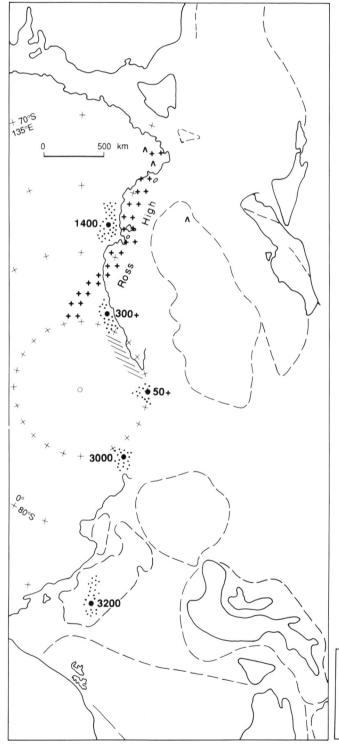

The oldest recognizable sedimentary rocks in the Transantarctic Mountains (Beardmore Group and Patuxent Formation) are turbidite sequences with cratonic provenance, which were probably deposited along a passive margin (Stump et al., 1986). Only recently an intriguing, but unsubstantiated hypothesis was proposed (Moores, 1991; Dalziel, 1991) that would juxtapose the western North American plate margin against the Transantarctic–southeastern Australian plate margin. The geologic histories of these two plate margins are similar. Both were passive margins of great length in the Late Proterozoic-Cambrian and have no specifically identified counterparts.

The Proterozoic thrust-type shear zone in the Miller Range (Fig. 2) and folding and uplift of the Beardmore Group and Patuxent Formation in the Late Proterozoic–earliest Cambrian are part of the evidence for subduction-related processes along the paleo-Pacific margin in virtually all plate tectonic models for Antarctica. Sedimentary evidence for this model is the pervasive occurrence of a calc-alkaline volcaniclastic facies (Liv Group) outboard of thick carbonate and quartzose shelfal facies (Byrd Group), although Rowell and Rees (1989) regard these as two unrelated terranes (Fig. 5). Intraformational conglomerates and other evidence of slumps in a grossly thick carbonate sequence in the central Transantarctic Mountains (Laird, 1981) suggest that a steep carbonate platform developed on the margin of the East Antarctic Craton during Cambrian time. This cycle ended with the Late Cambrian to Early Ordovician Ross Orogeny in which folding, metamorphism, and the intrusion of granitoids occurred along the entire paleo-Pacific margin of East Antarctica. The existence of a petrographic and geochemical boundary within distribution of Granite Harbour Intrusives, I-type granitoids outboard of predominantly S-type granitoids along the margin of the central Transantarctic Mountains and the Wilson Terrane, also suggests proximity of a continental margin (Borg, 1983; Borg et al., 1986).

In northern Victoria Land a much broader segment of the margin of the East Antarctic Craton is exposed than elsewhere in the Transantarctic Mountains. The complexity of the geology of this region suggests that the early Paleozoic history of Antarctica as summarized above may be oversimplified. The relationship of the Bowers and Robertson Bay Terranes to the Wilson Terrane and to each other remains problematic. The

Figure 6. Post-Ross sedimentary rocks (Late Ordovician–Carboniferous). The map area is located in Figure 1.

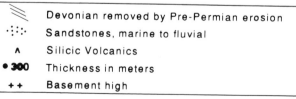

western part of the Wilson Terrane appears to be an extension of the granitic and metamorphic complex that occurs in southern Victoria Land and may represent the Early Cambrian continental margin. More and more evidence is suggesting that the Bowers and Robertson Bay Terranes are allochthonous. They are relatively unmetamorphosed and have generally undergone only one major folding event as compared to the highly metamorphosed and complexly folded Wilson Terrane. Changes in sedimentary facies across terrane boundaries are far too abrupt, indicating extreme telescoping through thrusting or juxtaposition by strike-slip faulting. The deep-water turbiditic sequence of the Robertson Bay Group has a metamorphic provenance even though it is presently separated from the metamorphic rocks of the Wilson Terrane by contemporaneous island-arc deposits within the Bowers Terrane.

Earlier models in which the Bowers Group was deposited in a graben (Laird et al., 1982) have given way to more elaborate models involving subduction and(or) strike-slip faulting and exotic terranes. New models with increasing complexity are required to account for new data appearing from the recent field work underway by several research groups (Bradshaw, J. D., 1987).

Weaver et al. (1984) reviewed various tectonic models and concluded that the best way to account for missing crust between the Wilson and Bowers Terranes and the cratonic provenance of sandstones in the Robertson Bay Group is to bring these terranes into juxtaposition by post-subduction strike-slip faulting. Westward convergence based on structural evidence has been suggested by Gibson and Wright (1985) and Flöttmann and Kleinschmidt (1991). The Bowers Terrane with its backarc basin and volcanic arc was accreted onto the Wilson Terrane. The Robertson Bay Terrane, representing an exotic continental margin prism, was accreted onto the Bowers Terrane by subduction and the closure of an ocean basin. Subsequent strike-slip faulting along the Leap Year Fault (Fig. 2) was suggested as a mechanism to remove direct evidence of subduction between terrane boundaries. Westward dipping subduction zones between each of the three terranes was proposed by Kleinschmidt and Tessensohn (1987). In their model the closure of the Wilson and Bowers Terranes occurred in the Late Cambrian, allowing sediments from the uplifted Wilson Terrane to bypass the Bowers Terrane to the subducting Robertson Bay Terrane, which was not accreted until the mid-Ordovician. Flöttmann and Kleinschmidt (1991) provided further structural evidence in support of westward subduction and accretion.

Other models have considered the Bowers and Robertson Bay Terranes as a single exotic block that was telescoped prior to docking with the Wilson Terrane. Eastward subduction models have been favored by Wodzicki and Robert (1986) and Findlay (1986). Borg and Stump (1987) pointed out that the distribution and geochemistry of Admiralty Intrusives, which they restrict to the Bowers and Robertson Bay Terranes, indicate thicker continental crust toward the present Ross Sea. Their interpretation of the data also casts doubt on the consoli-

dation of terranes before the Devonian, which is indicated in other models.

A unifying feature in all models is that northern Victoria Land was in place prior to the deposition of the Gondwanan Sequence. Depositional patterns in remnants of thick mid-Paleozoic mature quartzose sandstones of cratonic provenance in the Ellsworth Mountains and Transantarctic Mountains, except for northern Victoria Land, are indicative of continental shelf deposition on a passive paleo-Pacific margin, such as the contiguous Cape Basin of South Africa. In northern Victoria Land, however, terrane accretion occurred and the history was markedly different.

GONDWANAN HISTORY

Introduction

Sandstones of the relatively undeformed Devonian to Triassic Beacon Supergroup (Barrett, 1981) were deposited on the eroded roots of the Ross Orogen throughout the Transantarctic Mountains. According to D.I.M. Macdonald (personal communication, 1992; Macdonald and Butterworth, 1990), there was probably continuous accretion and deformation from ?Devonian until Early Tertiary times somewhere on the Antarctic Peninsula. The Late Carboniferous to Jurassic history of the Panthalassan margin is summarized in a time-space diagram that shows our structural interpretation (Fig. 7), and interpreted original distribution of depositional environments is shown in a parallel time-space diagram (Fig. 8). Stratigraphic sections and their correlations are shown in Figures 9 through 13. Preservation of parts of a 3-km-thick clastic wedge of craton-derived sandstone in the Ellsworth and Pensacola Mountains suggests that the Transantarctic basin began as part of a subsiding continental margin during the Devonian (Fig. 6). Remnants of the thinner cratonward part of this wedge occur in various parts of the Transantarctic Mountains.

During the Late Carboniferous, as a continental ice sheet expanded to cover a large area of Gondwanaland, a trough-shaped depression most likely formed on the inner continental shelf adjacent to the continental ice cap as the result of flexural isostatic loading, a phenomenon described by Walcott (1970). Similar troughs, 1.5 to 2 km deep and 50 km wide, lie subparallel to the present coastline of Antarctica (Anderson and Molnia, 1989). During glacial maxima/sea-level minima, Late Cenozoic ice streams were diverted into these lows (marginal channels), scouring and deepening them, commonly to crystalline basement. In the Early Permian a comparable marginal trough would have quickly filled with sediment to above sea level as continental glaciers retreated.

The Transantarctic basin in the Ellsworth Mountains region became a backarc basin by the Early Permian as suggested by the abundance of calc-alkaline volcanic detritus in Early Permian sequences. A paleocurrent reversal and change to volcanic provenance in the central Transantarctic Mountains suggests that a (retroarc) foreland basin began in the Ross Sea

Orogen- normal time-space diagram 300-160 Ma : tectonic interpretation

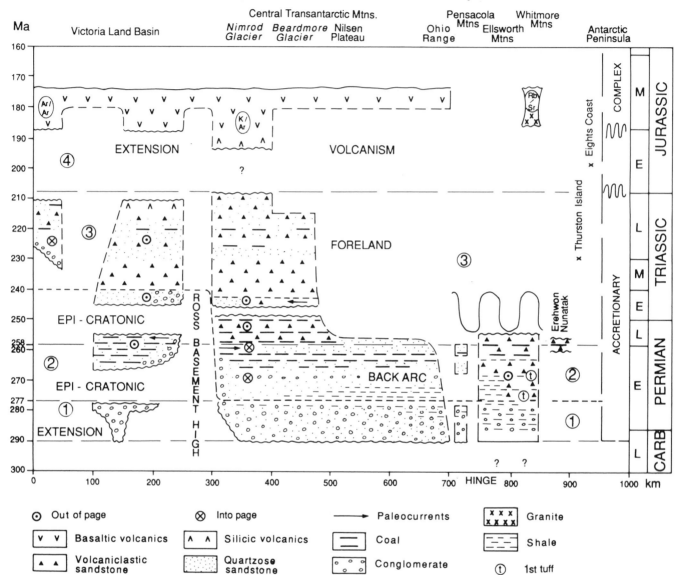

Figure 7. Orogen-normal time-space diagram for the Late Carboniferous–Jurassic (300–160 Ma), showing the tectonic interpretation. The 230 Ma granodiorite at Mount Bramhall on Thurston Island and the 198 Ma granite in the Jones Mountains on the Eights Coast are from Grunow et al. (1991). The four stages of development of the Transantarctic Basin are (1) extension (290–277 Ma); (2) backarc basin (277–258 Ma) behind the volcanic arc, marked by the first occurrence of abundant volcanic detritus, including probable tuff, at 277 Ma, along the Panthalassan margin and an epicratonic basin behind the foreswell of the Ross Basement High; (3) foreland basin (258–240–208 Ma) between the thrust-fold front and the Ross High and a continuing epicratonic basin behind the Ross High until its demise at 240 Ma whereafter the foreland basin stretches cratonward; (4) extension and onset of volcanism (208–170 Ma), at first silicic then mafic.

Orogen-normal time-space diagram 300-160 Ma : depositional environment

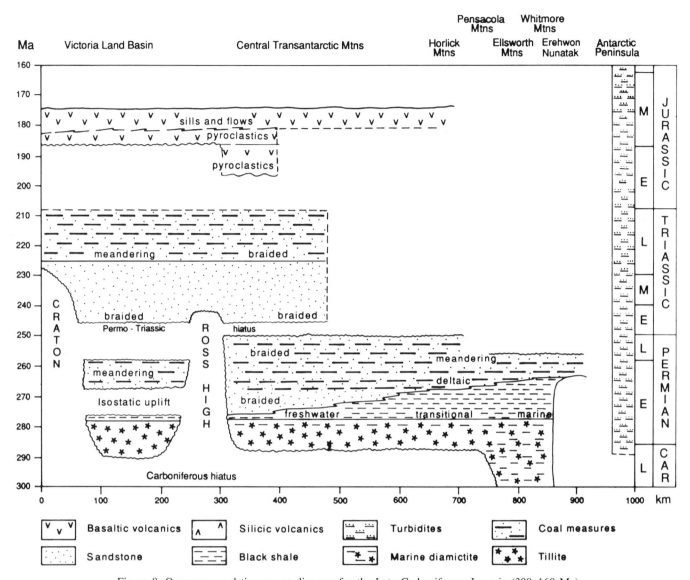

Figure 8. Orogen-normal time-space diagram for the Late Carboniferous-Jurassic (300–160 Ma), showing the interpreted distribution of depositional environments, shown as extending during the Triassic from Victoria Land to the Nimrod Glacier.

sector with the development of a fold belt during the Late Permian (Isbell, 1990a, 1991). Folded Permian sequences are found in the Ellsworth and Pensacola Mountains. Subsidence in the foreland basin continued until at least the latest Triassic. In the Jurassic the basin was affected by extensional tectonism during the intrusion and extrusion of great volumes of tholeiitic basalt related to the breakup of Gondwanaland.

Late Carboniferous-Early Permian

Stratigraphy. Late Paleozoic glacial deposits are widespread throughout the Transantarctic basin (Miller and Waugh, 1991) (Fig. 14). Rare pollen and spores in these sequences are

invariably assigned to the upper Stage 2 microflora in the eastern Australia palynologic zonal scheme (Evans, 1969; Kemp et al., 1977; Kyle and Schopf, 1982), the base of which straddles the Carboniferous/Permian boundary. This microflora, dominated by radiosymmetric monosaccate pollen, probably reflects the cold, inhospitable climate (Playford, 1990).

Victoria Land. Glacial deposits of the Darwin and Metschel Tillites, up to 80 m thick, are scattered throughout southern Victoria Land (Fig. 10) (Barrett and Kyle, 1975). An unnamed diamictite-bearing sequence, up to 350 m thick, is confined to one area just east of the Rennick Glacier in northern Victoria Land (Laird and Bradshaw, 1981). Barrett and

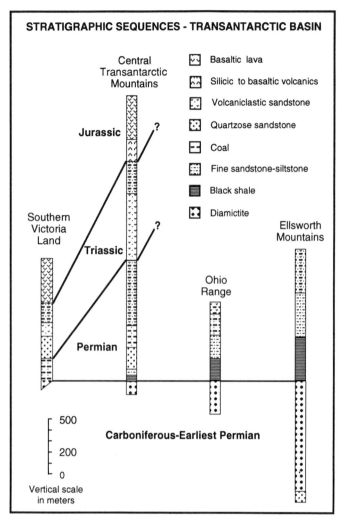

Figure 9. Orogen-normal set of stratigraphic sections. Detailed sections are shown in Figures 10 to 13.

McKelvey (1981) theorized that these terrestrial glacial deposits are discontinuous because they were preferentially preserved as valley fills where they were protected from erosion caused by post-glacial isostatic rebound.

Central Transantarctic Mountains. Glacially deposited sequences of diamictite, sandstone, and mudstone are typical of the Pagoda and Scott Glacier Formations, which form an almost continuous sheet throughout the central Transantarctic Mountains (Fig. 11) (Minshew, 1967; Coates, 1985; Barrett et al., 1986). Thicknesses vary from 0 to 395 m, but are generally 100 to 200 m. The diamictites are associated with striated surfaces. Miller (1989) interpreted cyclical sequences in the Beardmore Glacier region to represent multiple advances and retreats of the ice front in glacial, proglacial-fluvial and glacial-lacustrine environments.

The Buckeye Formation in the Ohio Range (Fig. 12), up to 310 m thick, is interpreted by Aitchison et al. (1988) as a glacial

and glacial-marine deposit. Diamictites at the base were deposited directly by grounded ice, which also formed striated pavements. Above the base, stratified diamictite, sandstone, and mudstone reflect glacial-marine deposition during ice retreat. A marine origin is supported by possible marine acritarchs (Kemp, 1975), the presence of hummocky cross-stratification, and by low iron content of the deposits. Frakes and Crowell (1975) suggested that the low iron content reflects reducing conditions associated with glacial-marine deposition.

Pensacola Mountains. As much as 1,200 m of sandy diamictite, lenticular sandstone, and interstratified mudstone containing dropstones have been assigned to the Gale Mudstone in the Pensacola Mountains (Schmidt and Williams, 1969). Large blocks of Devonian-Carboniferous(?) sandstone occur in the lower 10 to 30 m of the formation. Nelson (1981) described the lower contact as a disconformity. Drag marks on top of the underlying sandstone unit led Frakes et al. (1971) and Frakes (1981) to suggest that the ice moved across unconsolidated sediment on the sea floor. Because striated surfaces occur at the base and within the unit, Nelson (1981) attributed deposition to terrestrial processes. Considering the great thickness, however, and presence of interstratified mudstone containing dropstones, a glacial-marine origin for at least part of the sequence seems likely.

Ellsworth Mountains. The Whiteout Conglomerate, a 1,000 m-thick sequence of diamictite containing abundant dropstones occurs in the northern Ellsworth Mountains (Fig. 13). Although no marine fossils have been identified, the homogeneity and great thickness of this sequence have led to the interpretation that it is a glacial-marine deposit (Frakes and Crowell, 1975; Ojakangas and Matsch, 1981). This sequence appears to be conformable with the underlying Wyatt Earp Formation (Spörli, 1992; Matsch and Ojakangas, 1992), and its base may extend into the Carboniferous, signified by the question marks shown in Fig. 7.

In the southern Ellsworth Mountains a 250 m-thick diamictite with several striated boulder pavements (Ojakangas and Matsch, 1981) is attributed to deposition directly by glaciers and by floating ice near the grounding line (Matsch and Ojakangas, 1992).

Paleocurrents and provenance. Paleo-ice flow trends (Fig. 15) and paleocurrent directions (Fig. 14) are generally parallel to the trend of the Transantarctic Mountains, except in Victoria Land where locally they are perpendicular. This parallelism to the Panthalassan margin of East Antarctica, which began at this time and continued throughout the Permian and Triassic, is part of the evidence for the development of a linear margin-parallel basin.

The provenance of the glacial deposits generally reflects the local bedrock. For example, in southern Victoria Land granite constitutes 70% of the clasts in diamictites (Barrett and McKelvey, 1981). This is probably related to the exposure of granites along the Ross High. In the central Transantarctic Mountains, clast composition is much more mixed, but it is

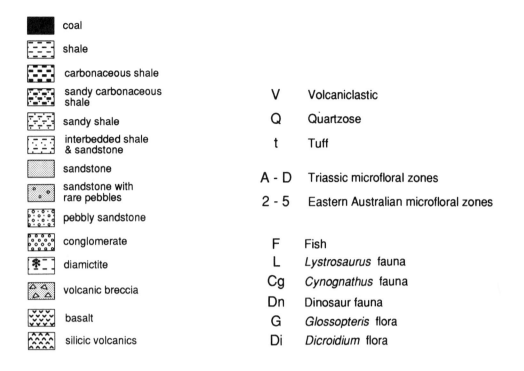

coal
shale
carbonaceous shale
sandy carbonaceous shale
sandy shale
interbedded shale & sandstone
sandstone
sandstone with rare pebbles
pebbly sandstone
conglomerate
diamictite
volcanic breccia
basalt
silicic volcanics

V   Volcaniclastic
Q   Quartzose
t   Tuff

A - D   Triassic microfloral zones
2 - 5   Eastern Australian microfloral zones

F   Fish
L   *Lystrosaurus* fauna
Cg   *Cynognathus* fauna
Dn   Dinosaur fauna
G   *Glossopteris* flora
Di   *Dicroidium* flora

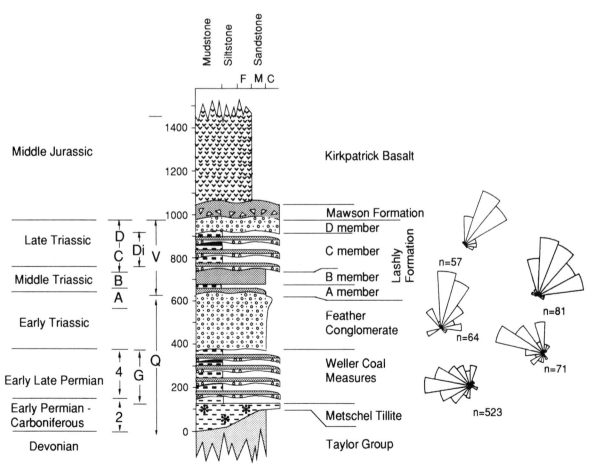

Figure 10. Generalized stratigraphic section for southern Victoria Land. Much of the data, including paleocurrent directions, are from Barrett and Kohn (1975) and Collinson et al. (1983).

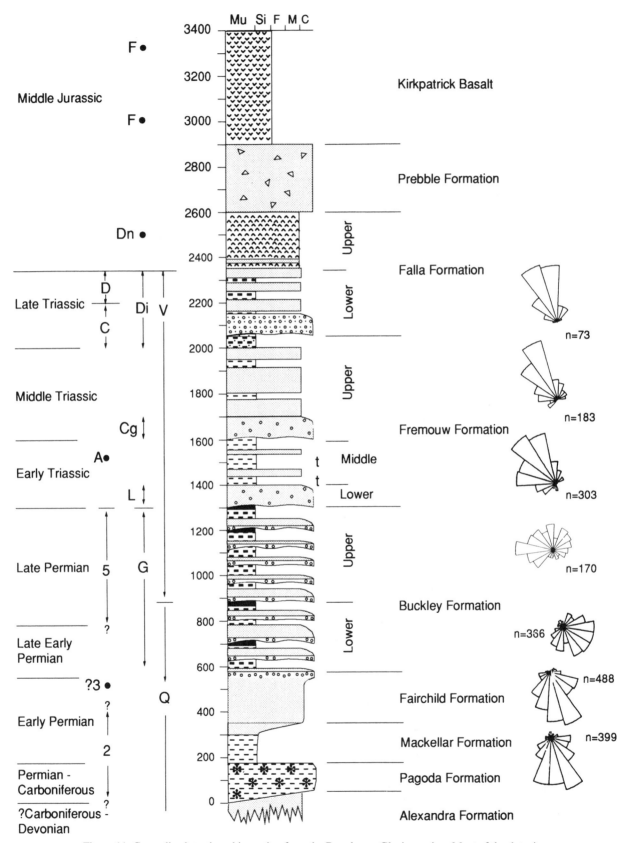

Figure 11. Generalized stratigraphic section from the Beardmore Glacier region. Most of the data, including paleocurrent directions, are from Barrett et al. (1986) and Isbell (1990a). Symbols in Fig. 10.

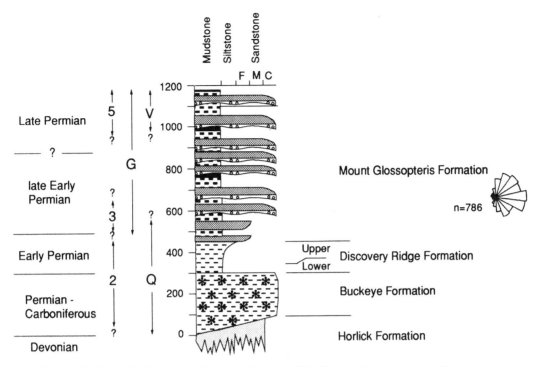

Figure 12. Generalized stratigraphic section from the Ohio Range. All the data, including paleocurrent directions, are from Long (1965). Symbols in Fig 10.

dominated by granite and quartzite from the local basement (Lindsay, 1970; Miller and Waugh, 1991). The lower glacial deposits are very sandy and contain sandstone inclusions, indicating erosion of the underlying Devonian-Carboniferous(?) sequence, probably during glaciation. In the Ellsworth Mountains, clasts derived from the underlying Paleozoic sedimentary rocks dominate the Whiteout Conglomerate, but there is also a significant proportion (17%) of granite clasts (Ojakangas and Matsch, 1981). Pre-Jurassic granite is not known to occur in the Ellsworth-Whitmore Mountains crustal block. These clasts were derived from the Ross Orogenic Belt on the margin of East Antarctica.

Paleogeography. Several factors must be considered in reconstructing the paleogeography of Carboniferous–Early Permian glacial deposits. These include:

1. The continental shelf setting inherited from the mid-Paleozoic continental margin

2. The possible continuity of the mid-Paleozoic section with glacial deposits in the Ellsworth Mountains and Pensacola Mountains

3. The erosional removal of Devonian to Carboniferous(?) sandstones from the Beardmore Glacier region to the Ohio Range, where only a thin remnant of Early Devonian sandstone remains

4. The great thicknesses (>1 km) of glacial deposits in the Ellsworth and Pensacola Mountains

5. The continuity of glacial deposits in the central Transantarctic Mountains

6. The discontinuous nature and valley-fill origin of glacial deposits in Victoria Land (Barrett, 1972) and evidence for an erosional unconformity at the tops of many of these sequences

7. The general parallelism of sub-ice and sedimentary paleocurrent directions to the trend of the East Antarctic continental margin in the central Transantarctic Mountains

8. The location of glacial-marine deposits at the northern end of the Ellsworth Mountains compared to basal tillites and glacial-marine deposits at the southern end of the Ellsworth Mountains and in the Pensacola Mountains and Ohio Range

9. The regional gradation within the Transantarctic basin from glacial-marine deposits in the Ellsworth Mountains to glacial and glacial-lacustrine/marine deposits in the Beardmore Glacier region

10. The vertical gradation from glacial deposits into marine or freshwater inland sea deposits in the central Transantarctic Mountains.

During the Late Carboniferous, a continental ice sheet formed over a large area of Gondwanaland, including the Antarctic sector. A trough-shaped depression most likely formed on the inner continental shelf adjacent to the continental ice cap as the result of flexural isostatic loading. Paleo-iceflow and paleocurrent directions are parallel to this trough-shaped basin. Except for a few thin remnants, the preglacial clastic sequence, which is at least 300 m thick in the Beardmore Glacier region, has been removed from east of the Beardmore Glacier to the Ohio Range, where a thin remnant of Early Devonian sandstone is preserved. These rocks may

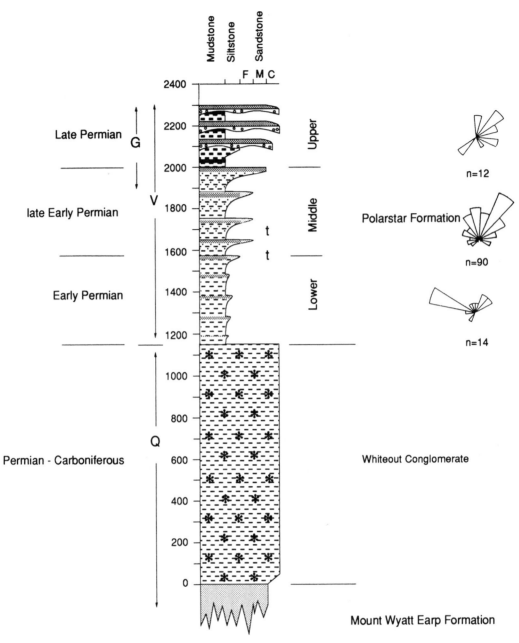

Figure 13. Generalized stratigraphic section from the Ellsworth Mountains. All the data are from Collinson et al. (1992) and Ojakangas and Matsch (1981). Symbols in Fig. 10. Note the tuffs in the Polarstar Formation.

have been removed by basal ice-sheet erosion, although Carboniferous tectonic uplift cannot be discounted.

The thick glacial-marine sequences in the Ellsworth Mountains and Pensacola Mountains, located on the mid-Paleozoic continental shelf, may have accumulated under an extensive ice shelf that bordered the Panthalassan margin of East Antarctica. Deposition may have been in a setting analogous to the 1.2 km of similar Cenozoic glacial-marine sediments in the Ross Sea encountered in DSDP Leg 28 (Hayes and Frakes, 1975). Stratigraphic successions in the southern

Ellsworth Mountains and Pensacola Mountains begin with basal tillites and striated pavements, probably representing the transition from glacial to glacial-marine deposition. Similar sediments are deposited in a broad zone related to the migrating position of the grounding line in modern Antarctic settings (Anderson and Molnia, 1989).

Victoria Land was inland of the erosionally deepened part of the flexural trough and was subject to greater postglacial isostatic uplift. Therefore, glacial deposits were preserved only in glacial valleys. An extensive sheet of glacial deposits was pre-

served within the flexural trough on the inner shelf margin of the continental shelf along the present central Transantarctic Mountains. These deposits most likely represent the final stage of Gondwanan glaciation, whereas the thicker sequences in the Ellsworth Mountains and Pensacola Mountains probably represent a much greater part of Late Carboniferous glaciation.

Early Permian

Stratigraphy. A regressive sequence of post-glacial shale followed by non-coaly sandstone in the central Transantarctic and Ellsworth Mountains was deposited in the Early Permian (Fig. 16). Rocks of this age are apparently missing along a disconformity at many localities in Victoria Land, although Woolfe et al. (1990) have reported possible conformable sequences. At Mount Fleming, diamictite grades upward into *Gangamopteris-* and *Glossopteris*-bearing shale. The post-glacial shale is assigned an Early Permian age on the basis of its gradation with underlying glacial deposits and a microflora that is comparable to upper Stage 2 in the eastern Australia palynologic zonal scheme (Evans, 1969; Kemp et al., 1977; Kyle and Schopf, 1982). In Antarctica this microflora is indistin-

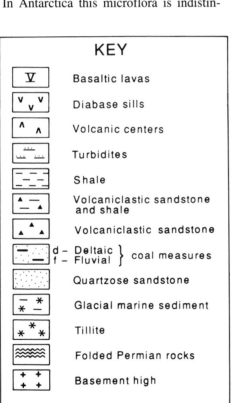

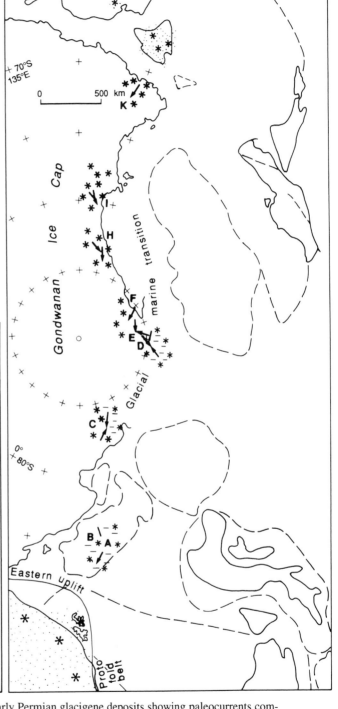

Figure 14. Stage of basin extension. Late Carboniferous–Early Permian glacigene deposits showing paleocurrents compiled by J.M.G. Miller. The map area is located in Figure 1. Key applies to Figs. 14–21. Letters in Figs. 14 and 15 refer to directional data at the following localities: A and B, Ellsworth Mountains (Matsch and Ojakangas, 1992); C, Pensacola Mountains (Frakes et al., 1971); D, Ohio Range (Bradshaw et al., 1984; Frakes et al., 1971); E, Ohio Range (Minshew, 1967); F, Queen Maud Mountains (Minshew, 1967); H, Beardmore Glacier region (Barrett et al., 1986; Lindsay, 1970; Miller and Waugh, 1986); I, southern Victoria Land (Barrett and Kyle, 1975); K, northern Victoria Land (Laird and Bradshaw, 1981). In this and succeeding figures, the data from Tasmania are from Chapter 3, and those from southern Africa and the Falkland Islands from Chapter 5.

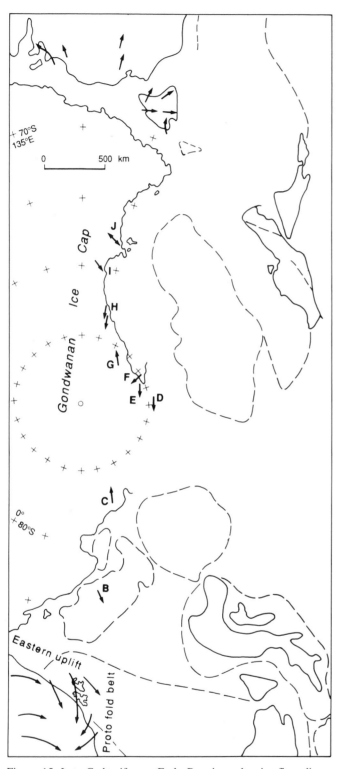
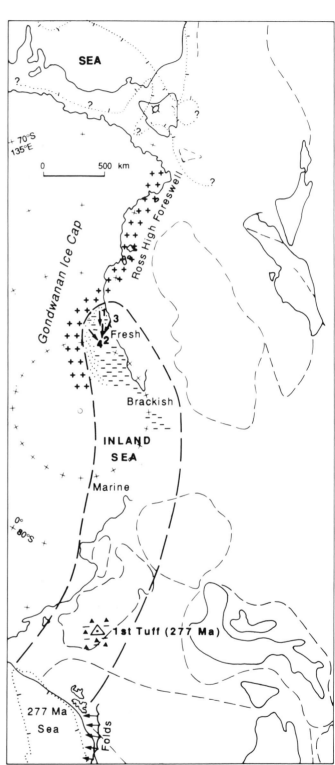

Figure 15. Late Carboniferous–Early Permian paleo–ice-flow directions. The map area is located in Figure 1. Letters refer to ice-flow directions at the localities listed in the description of Figure 14, together with G, Queen Maud Mountains (Coates, 1985); J, southern Victoria Land (Barrett and Kyle, 1975).

Figure 16. Early Permian paleogeography. The map area is located in Figure 1. Numbers refer to paleocurrent directions in the following areas in the Beardmore Glacier region: 1, 160°–166°E, 83°–83°45′S; 2, 166°–168°E, 84°–84°30′S; 3, 166°–168°E, 83°45′–84°S; 4, 162°–166°E, 84°30′–85°S. Data from Barrett et al. (1986), Isbell (1990a), and M. F. Miller. Note the first occurrence of volcanic detritus of probable airfall origin in the Ellsworth Mountains. In Tasmania and southern Africa, lacustrine (?marine) shale and mixtite extend the range of these deposits beyond Antarctica. An early folding event affected southern Africa (Hälbich, 1983). See Figure 14 for symbols.

◄

guishable from the microflora in the underlying glacial deposits. Kyle and Schopf (1982) referred to this low diversity microflora, represented mainly by monosaccate pollen, as a periglacial flora.

Central Transantarctic Mountains. In the Beardmore Glacier region the Mackellar Formation consists of one to three major coarsening upward cycles of interbedded shale and fine-grained sandstone and is typically 60 to 150 m thick (Fig. 11). Upward-fining sandstone beds, consisting of Bouma sequences Tb-e or Tc-e, are interpreted as turbidites. Major coarsening upward sequences are capped by multistory sandstone bodies, each story of which is an upward-fining channelfill sandstone. These coarsening upward sequences are interpreted to represent prograding subaqueous deltas. High carbon/sulfur ratios, a property of sediments deposited in freshwater (Berner and Raiswell, 1984) and a low diversity ichnofauna, including small bilobed endostratal trails (*Isopodichnus*) and small looped endostratal trails, suggest freshwater or slightly brackish conditions in the Beardmore Glacier area (Miller et al., 1991). In contrast, the occurrence of pyrite in the equivalent Weaver Formation in the eastern Queen Maud Range (Minshew, 1967) suggests the low carbon/sulfur ratios characteristic of marine deposition. Abundant burrows, including a possible marine trace fossil, *Paleodictyon,* in the same section also points to a marine environment. Bradshaw et al. (1984) interpreted the equivalent Discovery Ridge Formation in the Ohio Range (Fig. 12) as marine.

A thick sandstone unit without coal, known as the Fairchild Formation in the Beardmore Glacier region (Fig. 11) and as the upper Weaver Formation in the eastern Queen Maud Mountains, gradationally overlies the postglacial shale throughout the central Transantarctic Mountains. Thickness varies from 130 to 230 m. This unit loses its identity in the Ohio Range within the lower Mount Glossopteris Formation (Fig. 12). A microflora comparable to the Stage 3 microflora of Eastern Australia has been reported from the lower part and base of the upper part of the Mount Glossopteris Formation in the Ohio Range (Kyle and Schopf, 1982). The unit is dominated by interconnected channel-form sandstones that occur as sheets, 5 to 20 m thick and tens to hundreds of meters wide (Isbell and Collinson, 1988). In the Beardmore Glacier region

the Fairchild Formation is medium- to coarse-grained and trough cross bedded or fine grained and ripple laminated. In the eastern Queen Maud Mountains the upper Weaver Formation is medium- to coarse-grained sandstone and locally conglomeratic (Minshew, 1967). These sandstones were deposited by sandy braided streams. The paucity of fossil wood suggests that vegetation was sparse.

Ellsworth Mountains. In the Ellsworth Mountains the lower Polarstar Formation is characterized by laminated black shale intercalated with laminated, lenticular, and wavy bedded siltstone and fine-grained volcaniclastic sandstone (Fig. 13) (Collinson and Miller, 1991; Collinson et al., 1992). The lower contact of this unit with glacial deposits is abrupt except for rare dropstones in the lower few meters. An accurate thickness of this unit is not available because of folding and faulting, but a reasonable estimate is 300 to 500 m. The rare occurrence of probable marine trace fossils near the base of the unit suggests a marine environment. The rarity of burrows and the undisturbed nature of laminae suggest deposition under anaerobic conditions in a density stratified basin (Collinson et al., 1992). The base of the middle part of the Polarstar Formation in the Ellsworth Mountains (Fig. 13) contains the first recorded tuff (*see* Appendix 1, Fig. 24, A and B) of the Late Paleozoic volcanic arc; it was deposited in a prodeltaic to deltaic environment (Collinson et al., 1992). The age is estimated to be Stage 3a or 277 Ma (Appendix 1).

Paleocurrents and provenance. Paleocurrent data of significance are reported only for the Beardmore Glacier region. Paleocurrent data from the Mackellar Formation, mostly determined from ripple bedding, vary greatly, but generally agree with data from the overlying Fairchild Formation. Directions are generally southeastward away from or parallel to the margin of the East Antarctic craton (Fig. 16).

Sandstone compositions in the Beardmore Glacier region average 50 to 60% quartz and subequal amounts of plagioclase and K-feldspar (Barrett et al., 1986; Frisch and Miller, 1991). Primary sources were older Paleozoic sandstones and granitic and low-grade metamorphic basement rocks. In contrast, sandstones in the Ellsworth Mountains are dominated by volcanic detritus and contain only a small proportion of quartz. These sandstones were derived from a calc-alkaline source.

Paleogeography. Pinchout of the postglacial shale to the north toward Victoria Land and the gradation from fresh or brackish water conditions to marine conditions toward the Ohio Range and Ellsworth Mountains support the model of a trough-shaped embayment with the closed end in the Beardmore Glacier region. This interior sea quickly filled by the progradation of subaqueous deltaic fans followed by braided fluvial deposits. Clastics in this braided fluvial plain were probably reworked glacial outwash. Paleocurrent data are consistent with those in the underlying glacial deposits, suggesting that they were controlled by the filling of the flexural trough exhumed by glaciers. The paucity of fossil plants and the low diversity of microfloras suggest that the Early Permian climate remained

inhospitable. A much reduced continental ice sheet may have continued to occupy parts of East Antarctica.

The provenance of sandstones in the central Transantarctic Mountains remained the same as during glaciation, offering the possibility that much clastic material was derived from the reworking of glacial material as well as from newly exposed basement rocks. Coarse sandstones and conglomerates in the eastern Queen Maud Range and in the adjoining Ohio Range suggest the presence of a basement high, possibly an extension of the Ross High, just behind the present Transantarctic Mountains. In the Ellsworth Mountains, the volcanic provenance of sandstones contrasts strongly with the sedimentary, granitic and low grade metamorphic source of the underlying glacial sediments. The volcanic source area indicated by the tuff in the middle part of the Polarstar Formation is the earliest evidence of the late Paleozoic volcanic arc along the Panthalassan margin.

Late Early Permian

Stratigraphy. Coal-bearing strata containing a *Gangamopteris-Glossopteris* flora are assigned a late Early Permian (or Artinskian-Kungurian) age. The appearance of widespread coal within the Permian sequences throughout Antarctica (Fig. 17) is interpreted as a chronostratigraphic event, perhaps representing warming at the end of continental glaciation. A fairly diverse Stage 4 microflora has been described from the Weller Coal Measures in southern Victoria Land (Kyle and Schopf, 1982).

Figure 17. Stage of backarc and epicratonic basins. Late Early Permian paleogeography. The map area is located in Figure 1. Numbers refer to paleocurrent directions at the following localities: 1 and 3, northern Victoria Land (Collinson et al., 1986); 2, northern Victoria Land (Walker, 1983); 4, southern Victoria Land (Isbell, 1990a); 5 and 6, southern Victoria Land (Barrett and Kohn, 1975); 7–11, Beardmore Glacier region (Barrett et al., 1986; Isbell, 1990a) (7, 160°–166°E, 83°–83°30′S; 8, 160°–164°E, 83°30′–84°S; 9, 160°–162°E, 84°–84°30′S; 10, 162°–164°E, 84°30′–85°S;11, 162°–166°E, 84°30′–85°S); 12, Ohio Range (Long, 1965); 13, Ellsworth Mountains (Collinson et al., 1992). The first occurrence of tuff (t) or probable airfall volcanic detritus, equivalent to East Australian palynological stage 3a (277 Ma) (Fig. 7), indicates contemporaneous volcanism from an inferred volcanic arc behind the convergent Panthalassan margin. This arc is inferred to continue into the Brook Street magmatic arc (BSMA) of New Zealand, which at this time was an intra-oceanic tholeiitic arc. The Brook Street magmatic arc is located palinspastically by removing an assumed twofold subsequent extension of the back arc region. Data from New Zealand in this and succeeding figures are from Korsch and Wellman (1988). In Tasmania, the Tastubian (277 Ma) marine transgression, with marine invertebrates (illustrated by coils), as in southern Africa, is succeeded by coal measures and finally another set of marine sediments. In southern Africa, the Swartberg Folds represent the first phase of the Cape Orogeny (Hälbich, 1983). See Figure 14 for symbols.

Victoria Land. Sandstones in Victoria Land are characterized by two facies, a generally coarser braided stream facies and a finer meandering stream facies. The former occurs in areas closest to the Ross Sea and the latter more inland. Coal and carbonaceous shale are much more common in the meandering stream facies.

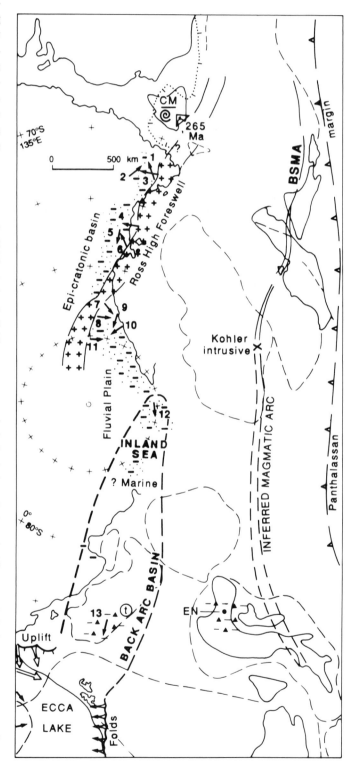

The Takrouna Formation in northern Victoria Land becomes decidedly coarser at its eastward limit of exposures (Collinson et al., 1986). The farthest east sections contain thick basal conglomerates composed of cobbles and small boulders derived from local granitic basement. These sections consist of medium- to coarse-grained, cross-bedded, feldspathic to quartzose sandstone with thin stringers or lenticular beds of carbonaceous or noncarbonaceous silty mudstone and minor coal. Trough cross-bedding predominates, but large-scale planar cross-bedding also occurs. Sandstone bodies are sheet-like, have an erosional base, and contain the multiple internal scours characteristic of braided stream deposits. Westward the Takrouna Formation becomes finer grained and more carbonaceous, although intervals of coarse-grained sandstone are likely to occur in any section. Across the Rennick Glacier the formation is generally fine-grained and contains abundant coal. These sections are dominated by ripple-laminated, fine-grained carbonaceous sandstone with a few trough cross-bedded, medium-grained sandstone units and thin coal beds (Walker, 1983).

Late Early Permian deposition in southern Victoria Land was similar to that in northern Victoria Land. The Weller Coal Measures (Fig. 10) tend to be coarser grained upward in the section and laterally both east an west toward the flanks of the basin (Isbell, 1990a). The lower part of the exposed formation in the Allan Hills is composed of fining upward cycles of trough cross-bedded, medium-grained to fine-grained sandstone, carbonaceous siltstone to mudstone, and coal. These sandstones are characterized by pronounced erosional bases and large-scale lateral accretion beds (Collinson et al., 1983). Surfaces of these gently dipping beds, which have been exposed by erosion, leave no doubt as to their point-bar origin. Paleocurrent directions at each level show relatively little dispersion, but each level is greatly different. Sandstone bodies in the upper part of the Weller Coal Measures in the Allan Hills (*see* Appendix 1, Fig. 24D) become coarser and contain the sand-filled channels and reactivation surfaces overlain by mudstone drapes characteristic of braided stream deposits (Isbell et al., 1990). Paleocurrent readings show far less variability, with directions compatible with a source are from the east.

Central Transantarctic Mountains. Cyclical sequences of sandstone, carbonaceous siltstone/mudstone, and coal can be traced 800 km from the Beardmore Glacier region (Fig. 11, lower Buckley Formation) to the eastern Queen Maud Range (lower Queen Maud Formation) to the Ohio Range (Fig. 12, lower Mount Glossopteris Formation). Throughout their area of exposure, these sequences begin with lenticular beds of well-rounded quartz-pebble conglomerate overlain by coal measures.

In the Beardmore Glacier region these coal measures are interpreted to represent braided streams and flood plains (Isbell, 1990a, 1991). The section is about 70% sandstone, consisting of sheet-like bodies 10 to 20 m thick and 2 to 10+ km wide. Local relief below channels is 1 to 3 m and internally these sand bodies consist of 1- to 2-m-thick scour-

bounded, fining upward cycles. Trough cross-bedded, coarse- to medium-grained sandstones are in some cases overlain by ripple- or horizontal-laminated fine-grained sandstone/siltstone and mud/organic drapes. Abandoned channels are sand-filled and possible lateral accretion surfaces are rare.

Descriptions of the lower Queen Maud Formation in the eastern Queen Maud Range (Minshew, 1967) suggest a similar sequence, probably representing a similar environmental setting. The basal Mount Glossopteris Formation in the Ohio Range, however, was deposited in a deltaic to meandering stream environment. Bradshaw et al. (1984) described abundant burrows and illustrated large-scale lateral accretion bedding. Furthermore, paleocurrent directions reported by Long (1965) show far more dispersion than those in equivalent sandstones in the Beardmore Glacier area.

Theron Mountains and Pensacola Mountains. Coal-bearing rocks of probable late Early Permian age are at least 200 m thick in the Pensacola Mountains (Williams, 1969) and 700 m in the Theron Mountains (Brook, 1972). Scattered outcrops in western Queen Maud Land, including the Heimefrontfjella area, are listed in Elliot (1975). The overall descriptions suggest a deltaic to fluvial-deltaic environment of deposition.

Ellsworth Mountains. In the middle Polarstar Formation, cycles coarsen upward from thin silty laminae in shale to lenticular, wavy and flaser bedded fine-grained sandstone. Convolute bedding and slump structures occur. Cycles increase upward in thickness, ranging from 2 to 20 m, and are capped by channel-form units of medium-grained, trough cross-bedded sandstone containing coaly fragments of plant material. Fine-grained sandstones contain a probable marine trace fossil fauna including *Phycodes, Chondrites,* and *Rhizocorallium* (Collinson et al., 1992). These cycles are interpreted as prograding deltaic lobes. The capping sandstones represent distributary channel deposits.

Marie Byrd Land. Quartz diorite from the Kohler Range has yielded a three-point Rb/Sr isochron date of 274 ± 20 Ma (Halpern, 1972; date corrected for new constants) and somewhat older dates have been reported for granitoids and gneissic rocks from that region and Thurston Island (Lopatin and Polyakov, 1976; Pankhurst, 1990).

Antarctic Peninsula. Glossopterid-bearing sediments have been reported from three tiny exposures near the south end of the Antarctic Peninsula (Laudon et al., 1987; Laudon, 1991). These rocks have been called the Erehwon Formation after Erehwon (*nowhere* spelled backwards) Nunatak. The Erehwon appears to be similar to parts of the upper Polarstar Formation and therefore may be part of a fluvial-deltaic sequence.

Paleocurrents and provenance. Paleocurrent data in Victoria Land show a variety of directions with trends toward the west (inland) from the Ross High and along the trend of the Transantarctic Mountains toward the Weddell Sea (Fig. 17). The dispersion of the data, particularly in southern Victoria Land, is typical of the diversity of transport directions recorded by meandering stream deposits. In the well devel-

oped braided stream deposits of northern Victoria Land, data from any single section show little variation, but mean directions vary considerably from locality to locality (Collinson et al., 1986). Tight control of drainage by structural trends or pre-existing topography may account for the drainage patterns.

In the Beardmore Glacier area trends appear to be opposite those in Victoria Land, mostly toward the south from the craton (Ross High) and the Ross Sea embayment. Directions in the Ohio Range and the rotated Ellsworth-Whitmore Mountains crustal block are consistent with those in the Beardmore Glacier region, trending along the axis of the basin toward the Weddell Sea embayment.

Sandstones throughout Victoria Land and from the Beardmore Glacier region to the Ohio Range contain clasts and grains from a predominantly granitic and metamorphic terrain (Appendix 2). In the Beardmore Glacier region feldspar content of sandstones increases in localities toward the craton (Barrett et al., 1986; Isbell, 1990a). Probable source areas were exposed highs along the early Paleozoic Ross Orogen and the East Antarctic Craton. This is in contrast to the volcaniclastic sandstones and tuff found in the Ellsworth Mountains that were most likely derived from a calc-alkaline volcanic source (Collinson et al., 1992; Appendix 2).

Paleogeography. The entire Panthalassan margin of the East Antarctic Craton appears to have become the site of clastic deposition by the Late Early Permian. Paleocurrent, grain-size, and provenance data suggest that the Ross High was active.

In Victoria Land meandering streams and lakes occupied an extensive valley that paralleled the Ross High. Coarse-grained, compositionally immature sands entered this valley as braided fans from both the Ross High and the western margin of the basin.

On the other side of the Ross High, in the Beardmore Glacier region, the drainage flowed down the axis of a longitudinal valley and entered the inland sea near the Ohio Range. Braided streams in the upper part of the drainage and along the Ross High graded downstream into a meandering river system that fed a major delta. This delta prograded from the Ohio Range to the Ellsworth Mountains. Drainage from the volcanic arc side of the basin and airfall detritus from the arc contributed volcaniclastic debris to the Ellsworth Mountains sequence, but not to sequences overlying the East Antarctic craton. Sandstones in the Ellsworth Mountains are dominated by volcanic detritus of intermediate to silicic composition and contain only a small proportion of quartz. These sandstones were most likely derived from a calc-alkaline source.

Deposition of siliciclastic sediment, mostly of fluvial or fluvial-deltaic origin, throughout the extent of the East Antarctic margin in the late Early Permian suggests subsidence along the entire length of the Transantarctic basin. The original exhumed flexural trough was largely filled by this time, so tectonic origins must be sought for the widespread subsidence. The only evidence for tectonism along the margin at this time, apart from subsidence, is the uplift of the Ross High.

Evidence for the Ross High and uplift in the Late Early Permian includes the following:

1. Paleocurrents that indicate a high to the east in Victoria Land and a high to the west in the Beardmore Glacier region

2. Coarsening upward of the Weller Coal Measures in southern Victoria Land as the paleocurrent directions swing toward the west off the high

3. Influx of rounded quartz pebbles throughout the central Transantarctic Mountains and in Victoria Land

4. The abundance of K-feldspar in sandstones at western localities in the Beardmore Glacier region.

Evidence of a magmatic arc on the Panthalassan margin of Antarctica and on either side along strike includes the following:

1. Airfall tuffs: in the Polarstar Formation of the Ellsworth Mountains (Appendix 2); in the Karoo Sequence of South Africa (Elliot and Watts, 1974; Martini, 1974), together with the sandstone petrology of the Ecca and Beaufort Groups (Johnson, 1991), both treated further in the chapter on Southern Africa; tuffs further afield are treated in the chapters on Eastern Australian (Chapter 3) and South American (Chapter 6)

2. The poorly known granitoid intrusions in Marie Byrd Land

3. The Brook Street Magmatic Arc of New Zealand (Korsch and Wellman, 1988), which includes Early Permian volcanic rocks, mainly basaltic and andesitic but with some silicic tuffs (Blake et al., 1974) forming an island arc erupted onto oceanic crust (Williams, 1979), and scattered plutonic rocks with poorly constrained radiometric dates suggesting possible granitoid emplacement during the Early Permian (Aronson, 1968) and Late Permian/Early Triassic (Aronson, 1965; Devereux et al., 1968). Kimbrough et al. (1992) recently reported the following U/Pb dates from (a) the Brook Street terrane: 265 Ma for hornblende gabbronorite at Bluff Peninsula, associated with a layered mafic-ultramafic intrusion; and ca. 260 Ma for a biotite granite at Oraka Point; (b) Maitai Group: 265 Ma for a granite clast from conglomerate.

The granite clast indicates that the magmatic arc had reached the continental margin of New Zealand by 265 Ma and could have contributed the first known tuff in adjacent Tasmania, dated by fossils as Artinskian, or 265 Ma.

Late Permian

Stratigraphy. Glossopterid-dominated floral assemblages are found in the upper part of the Permian coal measures throughout the central Transantarctic Mountains and Ellsworth Mountains (Fig. 18). Microfloral assemblages comparable to the Eastern Australian Stage 5 assemblages have been reported from the upper Buckley Formation in the Beardmore Glacier region, the upper Queen Maud Formation in the Queen Maud Mountains, the upper Mount Glossopteris Formation in the Ohio Range, and the Amery Group in the Amery Basin (Fig. 1) (Kyle and Schopf, 1982; Farabee et al., 1991). Late Permian sedimentary rocks are apparently missing

from Victoria Land. A ferruginous paleosol typifies the top of the Weller Coal Measures beneath Triassic sandstones in southern Victoria Land. In northern Victoria Land only the lower part of the Permian section has been preserved.

Barrett and Kohn (1975) assigned the lower member of the Feather Conglomerate in southern Victoria Land to the Late Permian, based on a 90° change in paleocurrents, which they correlated with a reversal in paleocurrent directions in the Beardmore Glacier area. More recently, Isbell (1991) has shown that the paleocurrent reversal in the Beardmore Glacier area occurred within the Late Permian sequence, which removes support for a Permian age for the lower Feather. No fossil evidence exists for a Permian age for the Feather; to the contrary, Early Triassic microfloras occur in the upper part of the formation (Kyle and Schopf, 1982).

Central Transantarctic Mountains. The Upper Buckley Formation (Fig. 11) and the equivalent Queen Maud Formation can be traced 800 km along the Transantarctic Mountains from the Beardmore Glacier region to the eastern Queen Maud Mountains. These formations consist of approximately equal proportions of resistant coarse- to medium-grained sandstone and generally less resistant fine-grained sandstone, siltstone, mudstone, and coal. These proportions vary from one locality to another.

The coarser sandstone units are composed of multiple scour-bounded, fining upward packets, 1 to 5 m thick. Mudstone drapes occur on reactivation surfaces. Mudstone within sandstone bodies is common. In some cases mudstone occurs at the top of packets. Scour-bounded packets probably represent migration of bars in individual channels. Individual packets can be traced laterally for tens to hundreds of meters. One to 10 individual packets comprise a sandstone unit.

Paleocurrent dispersion may reflect changing paleogeographic conditions in a foreland basin adjacent to a migrating fold-thrust belt. The progradation of alluvial fans and the shift in the axis of the basin may cause such changes. The finer grained units are composed primarily of carbonaceous siltstone and mudstone. Coal beds, 0.1 to 10 m thick, typically occur in the upper part of fine-grained units, commonly immediately below the next channel-form sandstone. Thin-bedded sheets of fine- to medium-grained sandstone are also more common in the upper part of fine-grained units.

Isbell (1990a, 1991) interpreted these rocks, in spite of

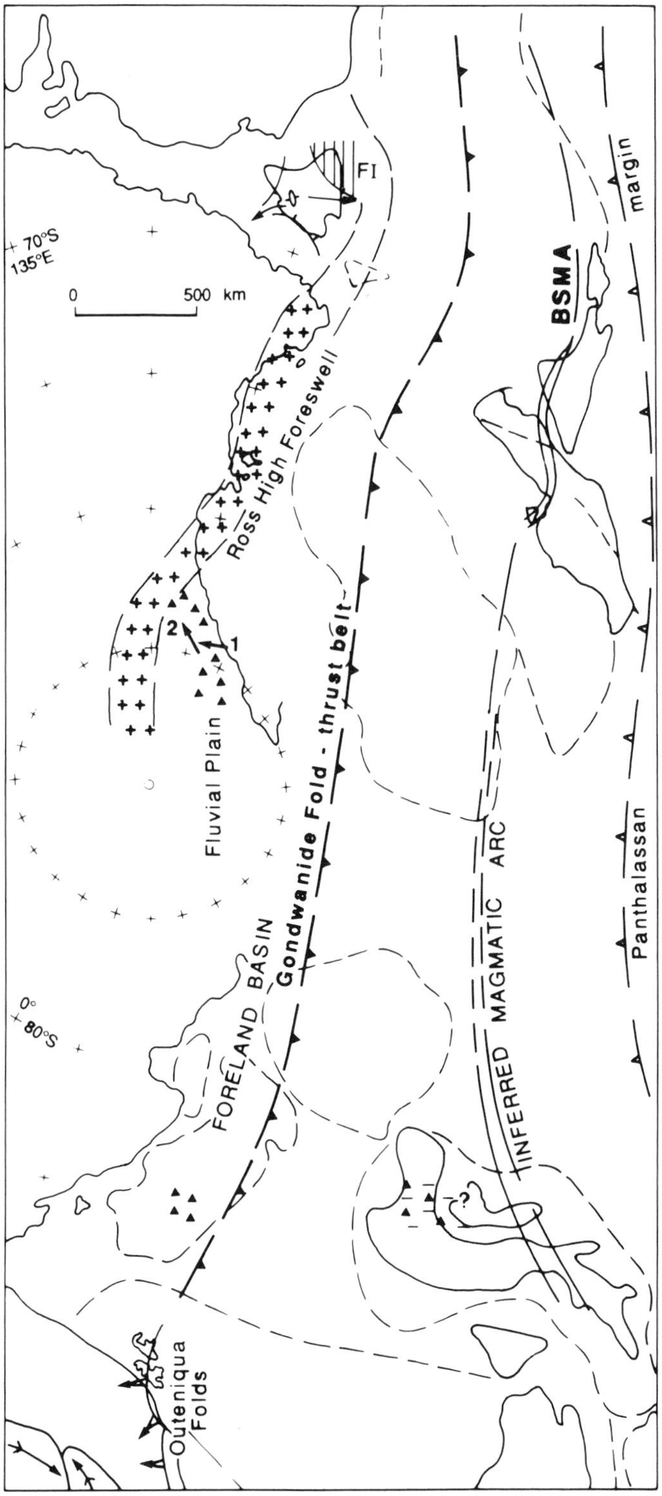

Figure 18. Stage of foreland basin. Late Permian paleogeography. The map area is located in Figure 1. Numbers refer to paleocurrent directions in the following areas of the Beardmore Glacier region (Barrett et al., 1986; Isbell, 1990a): 1, 172°–174°E, 85°–85°30′S; 2, 160°–164°E, 83°30′–84°15′S. The Brook Street magmatic arc with calc-alkaline granitoid emplacement at this time (Korsch and Wellman, 1988) is carried across Antarctica as the magmatic arc that formed the source for the volcaniclastic sandstone in the foreland basin. In Tasmania, quartzose coal measures shed from the western craton were deflected to the east-southeast by a rising foreswell (F I) (*see* Chapter 3), which may have been similar to the Ross High. The front of the fold-thrust belt is inferred to lie between the Brook Street magmatic arc and distant from the foreswell. In the African sector, it is located in the Falkland Islands–Cape Fold Belt according to the position of the Outeniqua folds, a possible source region for the Beaufort Group (*see* Chapter 5). See Figure 14 for symbols.

the high proportion of fine-grained rocks, as deposits of braided streams. Evidence for this interpretation in the coarser grained sandstone units is the large width to thickness ratios of individual sandstone sheets, sandstone fill in abandoned channels, scour-bounded fining upward cycles, abundant reactivation surfaces with mud/organic drapes indicating large discharge fluctuation, current structures indicating flow around emergent bed forms, and paucity of lateral accretion bedding. Fine-grained sediments between sandstone bodies were deposited on flood plains and in flood basins. Thin sheet sandstones resulted from crevasse splays, their upward increase in abundance reflecting emplacement of a new channel in the vicinity. Descriptions of the Queen Maud Formation in the Queen Maud Mountains (Minshew, 1967) and the Ohio Range (Long, 1965) are similar to those of the Buckley Formation, but do not give sufficient detail to make sedimentologic comparisons.

Ellsworth Mountains. The upper part of the Polarstar Formation (Fig. 13) consists of fining-upward cycles of medium- to fine-grained volcaniclastic sandstone (*see* Appendix 1, Fig. 24C), siltstone, and mudstone. Finer grained rocks are typically carbonaceous. Cycles are erosional at the base and typically contain 5 to 10 m of trough cross-bedded medium-grained sandstone. Intraformational conglomerate typically occurs at the base and along scours. Parallel and ripple-laminated fine-grained sandstone and siltstone grade upward into carbonaceous mudstone with abundant plant fossils. The trace fossil *Palaeophycus sulcatus* and simple epichnial grooves occur on some bedding planes (Collinson et al., 1992). The sedimentary sequence from the middle to the upper part of the Polarstar Formation records the transition from prodeltaic to deltaic deposits. The cycles in the uppermost part of this sequence probably represent the deposits of meandering streams on a delta plain.

Paleocurrents and provenance. A reversal in paleocurrent directions occurs within the upper part of the Buckley Formation in the Beardmore Glacier region (Isbell, 1991) (Fig. 18). The shift is from southward toward the present Weddell Sea to northward toward Victoria land. A component appears to come from the direction of the Panthalassan margin rather than East Antarctica. Concomitant with the change in paleocurrent directions is an abrupt change in composition of sandstones from quartzose to volcaniclastic (Fig. 11; *see* Appendix 1, Fig. 24, E and F). Below the change, sandstones average about 85% quartz, above, only 10% (Barrett et al., 1986). Upper Buckley sandstones are dominated by felsic volcanic detritus, plagioclase, and angular monocrystalline quartz with straight extinction, probably of volcanic origin (*see* Appendix 1, Fig. 24F). The provenance of sandstones changed from a completely cratonic source to predominantly calc-alkaline source.

A change from quartzose to volcaniclastic sandstones also occurs in the upper Queen Maud Formation in the eastern Queen Maud Mountains and in the upper Mount Glossopteris Formation in the Ohio Range, but paleocurrent data are insufficient to determine whether a reversal is linked with the change in provenance. Paleocurrent directions reported by Long (1965) from the Ohio Range trend generally toward the Weddell Sea. It may be that the reversal occurred only in the Beardmore Glacier region, but there is no indication of a drainage divide between these regions in terms of facies or thickness changes. Few paleocurrent data are available for the upper Polarstar Formation in the Ellsworth Mountains. Sandstones in this sequence, like those in other parts of the formation, are low in quartz and other grains derived from granitic and metamorphic rocks, and high in volcanic rock fragments of silicic to intermediate composition.

Paleogeography. The paleocurrent reversal and change in provenance from the craton to the volcanic-orogenic Panthalassan margin reflects evidence of uplift and folding along the margin. Folding in the Ellsworth Mountains sequence was later than the deposition of the youngest (Late Permian, 250 Ma) preserved sediment, probably in the Early Triassic, and probably contemporaneous with the third deformational event in the Cape Fold Belt reported by Hälbich (1983), and treated in the Southern African chapter.

Isbell (1991) compared the Upper Buckley Formation to the deposits of modern aggrading braided river systems such as the Kosi (Wells and Dorr, 1987), and Brahmaputra River (Coleman, 1969), the Samalá (Kuenzi et al., 1979), the Rufiji (Anderson, 1961), and the Earp (Knight, 1975). These rivers contain extensive flood plains and large-scale channel migration is by avulsion. Under conditions of rapid subsidence, relative to avulsion periodicity, flood-plain sediments subside below the zone of erosion of migrating channels and are therefore preserved. In active foreland basins, the ratio between subsidence and sedimentation rates determines whether the basin is underfilled or overfilled. Underfilled basins occur when subsidence dominates. Under such conditions the geometry of the basin is controlled by flexural loading of the crust and dispersal patterns are parallel to structural trends. Flow off the mountain belt and the cratonic forebulge is transverse to the axis of the basin, commonly in the form of low-gradient alluvial fans. Regional flow is longitudinal down the axis of the basin. In overfilled basins, sedimentation is greater than subsidence, resulting in dispersal patterns perpendicular to structural trends. The increased proportion of flood-plain components upward in the Buckley Formation is the pattern to be expected in an underfilled foreland basin. Underfilling occurred when subsidence increased due to tectonic loading from the fold-thrust belt (Isbell, 1990a, 1991). The fold-thrust belt is defined by the Late Permian (250 Ma) and older sediments in the Ellsworth and Pensacola Mountains deformed soon after deposition, either in the latest Permian or earliest Triassic (Figs. 7, 19). The fold-thrust belt lay inboard of the magmatic arc, indicated by Late Permian calc-alkaline granitoids in the Brook Street Magmatic Arc (Korsch and Wellman, 1988).

Although stream directions shifted toward Victoria Land, the lack of volcanic detritus in Permian or Early Triassic sand-

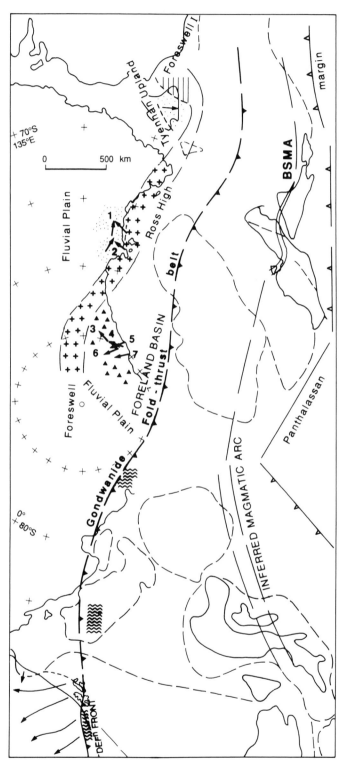

stones in the Victoria subbasin suggests that the Ross High blocked their dispersal. Late Permian drainage probably included the area of the present Ross Sea.

Early Triassic

Stratigraphy. Early Triassic sandstones rest disconformably on Permian coal measures in both southern Victoria Land and the central Transantarctic Mountains (Fig. 19). Local channeling along the basal sandstone and a change from carbonaceous to noncarbonaceous beds are the only physical evidence of a hiatus in the central Transantarctic Mountains. The Early Triassic *Lystrosaurus* Zone fauna, including a variety of reptiles and amphibians, occurs throughout the lower member of the Fremouw Formation (Colbert, 1982; Hammer, 1990). The middle member contains an Early Triassic microflora (Kyle and Fasola, 1978).

In southern Victoria Land noncarbonaceous sandstones of the Feather Conglomerate disconformably overlie a well-developed paleosol at the top of the Early Permian Weller Coal Measures. Most of the Feather is unfossiliferous, but an early Triassic microflora similar to the one in the middle Fremouw Formation occurs near the top of the Formation (Kyle and Schopf, 1982). This microflora extends upward into member A of the Lashly Formation.

Victoria Land. The Feather Conglomerate, a 175- to 250-m-thick braided stream deposit, forms massive cliffs consisting of multiple shallow channel-fills of trough or planar cross-bedded, pebbly sandstone (Fig. 10) (Collinson et al., 1983; Barrett and Fitzgerald, 1985). In the middle part of the formation, 20- to 30-cm-long vertical tubes of *Skolithos* form abundant "pipe-rock." In the lower member, siltstones and mudstone occur only as thin drapes. In the upper (Fleming) member greenish gray siltstone and mudstone occur between pebbly sandstone units (*see* Appendix 2, Fig. 25A). Ferruginous paleosols occur at several levels and at the top of the formation (Barrett and Fitzgerald, 1985). The Feather becomes less conglomeratic and finer grained toward the north (Barrett and Kohn, 1975).

Overlying the Feather is member A of the Lashly

Figure 19. Early Triassic paleogeography. The map area is located in Figure 1. Numbers refer to paleocurrent directions at the following localities: 1, Allan Hills (Collinson et al., 1983); 2, southern Victoria Land (Barrett and Kohn, 1975); 3–5, Beardmore Glacier region (Barrett et al., 1986) (3, 162°–166°E, 84°–85°S; 4, 170°E, 84°30′S; 5, 173°E, 85°30′S); 6 and 7, Queen Maud Mountains (Collinson and Elliot, 1984b).

The paleogeography did not differ from that of the Late Permian but the paleoenvironment did in so much as coal did not form. The Tyennan upland and foreswell I of Tasmania projected southward into the Ross High. In Africa, the volcaniclastic-quartzose sediment of the Beaufort Group continued to be shed off the Outeniqua uplift. See Figure 14 for symbols.

Formation, a 50- to 70-m-thick unit consisting of fining-up-ward cycles of medium- to fine-grained sandstone and greenish gray siltstone and mudstone with abundant root casts (Fig. 10) (Barrett and Kohn, 1975; Collinson et al., 1983). The rippled tops of some sandstone beds are overlain by thin mud-cracked clay drapes. Sandstone units are channel form and contain multiple scours. The unit is interpreted as a braided stream and flood-plain deposit. All these characteristics and little variability in paleocurrent directions suggest that Lashly A represents a low-sinuosity braided stream and flood-plain deposit.

Central Transantarctic Mountains. The lower and middle members of the Fremouw Formation range in thickness from 200 to 280 m in the western Queen Maud Mountains to 330 m in the Beardmore Glacier region (Fig. 11). The basal sandstone, 10 to 30 m thick, channels into the underlying Permian coal measures. Local relief on the Permian-Triassic contact is no greater than at the base of other sandstone units in the underlying Buckley or the lower Fremouw. At localities where carbonaceous beds are absent from the uppermost Buckley, this contact is difficult to recognize.

Coarse- to medium-grained sandstone units in the lower Fremouw typically form extensive channel-based sheets that extend for hundreds of meters laterally. Internally these sandstone bodies contain deep scours preserving evidence of major floods. Channel bases and scours are commonly lined by fine-grained intraformational clasts and scattered vertebrate bones. The tops of some of these sandstone units form resistant platforms, which show many of the features of the ancient stream bottom including dunes and interdune mudstone drapes. Trails are abundant on some of these surfaces. The vertical-tube burrows of *Skolithos* penetrate downward up to a meter into the underlying sandstone. In rare cases mud cracks occur. The various features of these sandstone bodies point to a braided stream origin (Collinson et al., 1981). The tops of sandstone units record avulsion and abandonment of the stream channel.

Fine-grained units representing flood-plain deposits make up approximately half of the lower Fremouw. Thin sheets of fine-grained sandstone, probably representing levee and crevasse-splay deposits, are intercalated within greenish gray siltstone and mudstone flood-plain units. Abundant root traces at many horizons indicate paleosol development (Horner and Krissek, 1991). In the Queen Maud Mountains, upstream from the Beardmore Glacier area, purple and red beds, representing more oxidizing conditions during burial, are mixed with greenish gray units (Collinson and Elliot, 1984b). Almost complete vertebrate skeletons, primarily of mammal-like reptiles, occur in flood-plain deposits, particularly just above major sandstone units.

The middle member contains a much greater proportion of flood-plain to channel deposits. Sandstone bodies are less sheet like and in many cases are only tens of meters across, terminating laterally in "wing-like" appendages. The separation into lower and upper members is arbitrarily defined at the level where fine-grained beds become predominant. Near the

Beardmore Glacier the middle member contains more sandstone than in the Queen Maud Mountains and at some localities is indistinguishable from the lower member. A vertebrate trackway was described from the Beardmore Glacier area (Macdonald et al., 1991a)

Paleocurrents and provenance. Paleocurrent directions changed little from Permian to Triassic time (Fig. 19). In southern Victoria Land (Fig. 19, 1 and 2) an abrupt change of 90° from north to west between the lower member and the upper (Fleming) member of the Feather Conglomerate accompanies an increase in grain size. Paleocurrents in the Beardmore Glacier region (Fig. 19, 3 through 5) are toward the northwest, but in the Queen Maud Range (Fig. 19, 6 and 7) swing toward the southwest.

The Feather Conglomerate is predominantly quartzose, but contains orthoclase and microcline (*see* Appendix 2, Fig. 25A). Conglomerates composed almost entirely of well-rounded quartz pebbles are most abundant in the north. A major change occurs at the base of the Lashly Formation, where volcanic detritus appears in feldspathic sandstones.

Volcaniclastic sandstone and tuffs occur throughout the Early Triassic section in the central Transantarctic Mountains (Collinson and Elliot, 1984b; Barrett et al., 1986). The base of the Fremouw Formation in the Beardmore Glacier region is quartzose (*see* Appendix 2, Fig. 25B), but it becomes increasingly volcaniclastic upward and laterally toward the Queen Maud Mountains (Vavra et al., 1981) (*see* Appendix 2, Fig. 25C).

Paleogeography. Paleocurrent directions in the central Transantarctic Mountains and Victoria Land trend in the same general direction, but the greatly different composition of sediments in these two regions suggests that they were in separate drainage systems until late Early Triassic time. Collinson et al. (1987) suggested an elongate basin into which quartzose sediments fed from the cratonic side and volcaniclastics fed from the Panthalassan margin. According to this model, the axis of the basin migrated toward the craton with the fold-thrust belt with time, and the composition of sandstones depended on which side of the axis of the basin the section was deposited. The complete lack of volcanic detritus in Victoria Land until latest Early Triassic time suggests that the Ross High continued to block the flow of volcanogenic detritus from the nearby Brook Street Magmatic Arc (Korsch and Wellman, 1988).

The 90° shift in paleocurrent directions in southern Victoria Land between the lower and upper (Fleming) members of the Feather Conglomerate and the coarsening of the sequence to the north and upward probably reflect uplift on the Ross High. The sandstone facies, which contains paleocurrent directions toward the north, accumulated in the longitudinal drainage along the axis of the Victoria subbasin. The pebbly sandstone facies, which contains westward paleocurrent directions, may have been deposited by alluvial fans prograding into the basin from the Ross High.

Patterns similar to those in southern Victoria Land occur

in the central Transantarctic Mountains, except that the source area to the east was the volcanic terrain along the Panthalassan margin. Paleocurrent directions in the Beardmore Glacier area are mostly toward the north and the basal sandstones are quartzose, although volcanic detritus is present. Upward in sections and in localities toward the Queen Maud Mountains the proportion of volcanic detritus increases as the paleocurrent directions shift toward the west (Vavra et al., 1981). Also, upstream toward the Queen Maud Mountains many of the greenish flood-plain deposits are replaced by more oxidized purple and redbeds. Channel-form sandstone bodies are smaller and the proportion of fine-grained beds increases.

The major longitudinal drainage passed through the Beardmore Glacier area. Alluvial fans composed of volcanic detritus from the thrust-fold belt prograded into the basin from the Panthalassan margin. The up-section increase in volcaniclastic material represents cratonward migration of the axis of the basin with the fold-thrust belt. Erosion and weathering of volcanic detritus resulted in the delivery of great amounts of fine-grained sediment into the basin. Increasing rates of subsidence in the central Transantarctic Mountains as the axis of the basin migrated cratonward resulted in an upward fining of the sequence as coarser clastics were trapped adjacent to the mountain belt.

Although paleocurrents were generally in the same direction, the source areas during most of the Early Triassic were different in Victoria Land and the central Transantarctic Mountains. The Fremouw drainage must have extended to the sea somewhere in the vicinity of the present Ross Sea. The Victoria Land drainage may have continued into Tasmania, where a similar quartzose sequence was deposited. The abundance of volcanic detritus in member A of the Lashly suggests that the Ross High was buried by latest Early Triassic time and that the Transantarctic foreland basin captured the Victoria subbasin drainage. Middle and Late Triassic stratigraphic units in the two regions are very similar.

The paleontology and sedimentology of Early Triassic deposits in the Transantarctic basin contain several paleoclimate indicators of a seasonal, but equable climate. Particularly significant is the occurrence of a diverse fauna of reptiles and amphibians in a polar region. These types of animals required relatively warm conditions with year-around temperatures above freezing.

The abundance of root traces in the flood-plain deposits and the food requirements of the large herbivores, one of them the size of a cow (Cosgriff et al., 1982), in the *Lystrosaurus* fauna suggest that abundant vegetation was available. Fossil logs occur rarely and only in the Queen Maud Mountains, upstream from the Beardmore Glacier area. The ubiquitous root traces may have represented a herbaceous flora, perhaps the equivalent of modern grasslands before the existence of angiosperms. Gymnosperm forests may have existed along stream channels farther upstream where redbeds indicate better drained, more oxidizing upland conditions.

The sandstones in the Fremouw, particularly in the basal part in the Beardmore Glacier region (*see* Appendix 2, Fig. 25B), are similar to those in the underlying Permian, but most of the organic carbon in Triassic fine-grained beds has been oxidized. Carbonaceous beds are also rare in the lower Feather Conglomerate in southern Victoria Land. The lack of preservation of organic carbon in the Early Triassic may be attributable to greater oxidation under drier climatic conditions.

Deep scours in sandstone bodies show that much of the drainage basin was deeply incised in floods. Because most of these streams appeared to have been ephemeral, *Lystrosaurus* and other aquatic vertebrates would have required pools of standing water at least during dry periods. The upstream change from greenish gray to purple redbeds suggests that water tables were not as high along the flanks of the basin, allowing greater oxidation of sediment.

Middle Triassic

Stratigraphy. The Middle Triassic sequences of southern Victoria Land and the central Transantarctic Mountains are lithologically similar (Fig. 20). They are dominated by thick sandstone units with few fine-grained units. A microflora in member B of the Lashly Formation in southern Victoria Land is comparable to that assigned to the Middle Triassic in eastern Australia (Kyle and Schopf, 1982). A *Cynognathus* fauna, including large reptiles and amphibians, occurs at the base of the upper member of the Fremouw Formation in the Beardmore Glacier area. The *Cynognathus* Zone is usually assigned to the latest Early Triassic, but the Fremouw fauna contains some elements that appear to be more advanced than related vertebrates in South African faunas (Hammer, 1990; Hammer et al., 1990).

Victoria Land. Member B of the Lashly Formation is a 50- to 60-m-thick, medium- to fine-grained, sandstone (Fig. 10) (Collinson et al., 1983). Abundant scours filled by trough cross-bedded sandstone suggest that this unit represents the deposits of a sandy braided stream deposit. Although preservation of carbonaceous material is rare, blocks of silicified peat and wood occur along one horizon in the Allan Hills.

Central Transantarctic Mountains. The Middle Triassic part of the upper member of the Fremouw Formation, a sandy braided-stream deposit, thickens from about 220 m near the Beardmore Glacier to 450 m in the Queen Maud Mountains (Collinson and Elliot, 1984b; Barrett et al., 1986). Similar in lithology to member B of the Lashly, it is composed predominantly of light gray, medium- to fine-grained sandstone that forms extensive resistant slopes (Fig. 11). The basal 30 m is quartzose and forms resistant ledges throughout the Beardmore Glacier region. The top of an 8-m-thick ledge at the base of this unit forms a prominent platform. Fossil vertebrate remains, including large skulls, are concentrated on this surface along with fossil logs and siltstone clasts (Hammer et al., 1990; Krissek and Horner, 1991). Above the vertebrate-bearing horizon, sandstone units up to 100 m thick are separated

by relatively thin siltstone and mudstone units. Sandstone units contain multiple scours that are filled by large-scale, low-angle, in some cases deformed, cross-beds. Greenish gray siltstone and mudstone clasts along with casts of fossil logs occur along scour surfaces. Finer grained units contain paleosols with abundant root casts (Horner and Krissek, 1991).

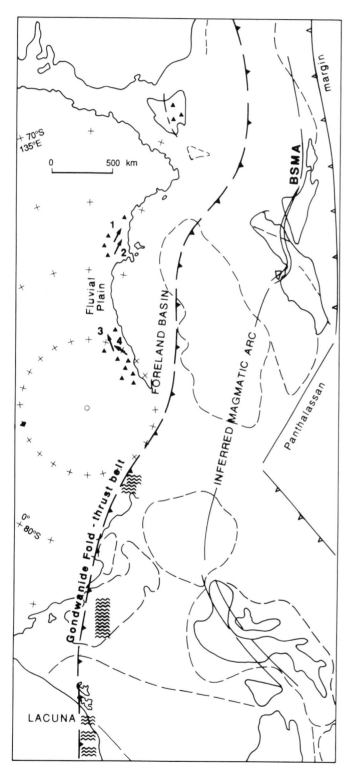

Paleocurrents and provenance. Middle Triassic sandstones are composed of approximately equal amounts of quartz, feldspar, and calc-alkaline volcanic grains, suggesting derivation from basement and arc sources (Barrett et al., 1986; Vavra, 1989). The volcaniclastic content of the sandstone (*see* Appendix 2, Fig. 25, D and E) gives it a greenish cast and accounts for the slope style of weathering. The basal quartzose unit of the upper Fremouw contains microcline and K-feldspar typical of a cratonic basement source. Paleocurrent directions are similar to those in the Early Triassic, generally parallel to the axis of the basin toward the north, but there are fewer from the West Antarctica side of the basin (Fig. 20).

Paleogeography. By the Middle Triassic a wedge of volcaniclastic sediment from the Panthalassan margin, including the Brook Street Magmatic Arc (Korsch and Wellman, 1988) had flooded the entire Transantarctic basin, including the Victoria subbasin. Evidence of any influence on depositional patterns by the Ross High is lacking. The similarity of stratigraphic units suggests a common depositional basin. Braided-stream conditions were produced by the quantity of sediment coming into the basin from the Panthalassan margin as well as by the episodic or seasonal nature of water input into the basin. Conditions were dry enough to oxidize most of the organic carbon in the system during burial. The common occurrence of fossil logs shows that forests were present. The diverse *Cynognathus* fauna, particularly the large amphibians, needed access to pools of water throughout dry periods. The concentration of fossil remains at some localities may be the result of animals crowding around water holes during dry periods.

Late Triassic

Stratigraphy. Conditions suitable for the accumulation of carbonaceous beds returned in the Late Triassic, allowing for the preservation of floras and microfloras (Kyle and Schopf, 1982; Farabee et al., 1989; Taylor et al., 1988). The uppermost Fremouw Formation and the lower Falla Formation in the Beardmore Glacier area, the Lashly Formation in southern Victoria Land, and the Section Peak Formation in northern Victoria Land (Fig. 21) all contain *Dicroidium odontopteroides,* the foliage of a seedfern that is typical of Late Triassic

Figure 20. Middle Triassic paleogeography. The map area is located in Figure 1. Numbers refer to paleocurrent directions at the following localities: 1, Allan Hills (Collinson et al., 1983); 2, southern Victoria Land (Barrett and Kohn, 1975); 3 and 4, Beardmore Glacier region (Barrett et al., 1986) (3, 162°–166°E, 84°–85°S; 4, 168°–174°E, 84°30′–85°30′S).

In Tasmania, deposition of quartzose and lithic sandstone over the former foreswell coincided with the demise of the Ross High in the south so that the foreland basin now migrated westward toward the craton. Southern Africa was uplifted and eroded. See Figure 14 for symbols.

deposits throughout Gondwanaland (Kyle and Schopf, 1982; Tessensohn and Madler, 1987).

Late Triassic pelagic chert has been reported from the South Orkney Islands at the tip of the Antarctic Peninsula (Dalziel et al., 1981). Early Jurassic (Liassic) marine faunas have been reported from a deformed accretionary complex in

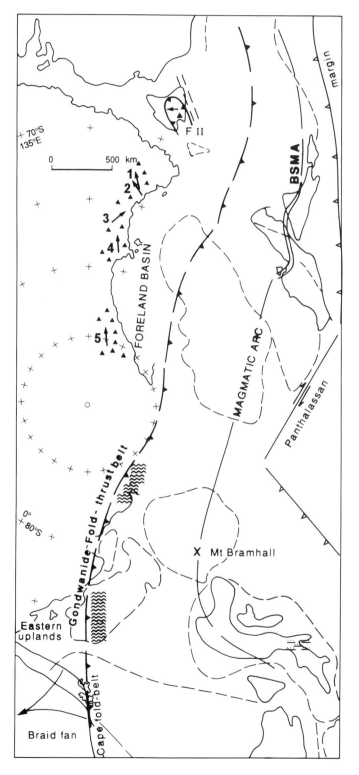

the Antarctic Peninsula (Thomson and Tranter, 1986). Floras from the South Shetland Islands previously regarded as Late Triassic (Schopf, 1973; Lemoigne, 1987) are probably Cretaceous (Rees and Smellie, 1989).

Victoria Land. In southern Victoria Land the upper Lashly Formation (members C and D) ranges from 220 to 300 m thick (Fig. 10) (Barrett and Kohn, 1975; Collinson et al., 1983). Variations in thickness are in part due to a regional unconformity above the Lashly with at least 500 m of local relief (Collinson et al., 1983). Member C consists of fining-upward cycles of medium- to fine-grained volcaniclastic sandstone, carbonaceous siltstone and mudstone, and coal. Large-scale lateral accretion surfaces in sandstones demonstrate the meandering stream origin for these deposits. Member D is similar, but sandstone units are thicker and more resistant. Not enough sedimentologic data are available to interpret the depositional environment of this member, but it may represent a major influx of quartz sand into the basin.

The Section Peak Formation in northern Victoria Land is similar to the uppermost Lashly. This formation is relatively thin, 50 to 100 m, and directly overlies granite basement west of the Rennick Glacier. It consists of cross-bedded, pebbly, coarse- to medium-grained sandstone. Siltstone clasts and coalified fossil logs are concentrated along scour surfaces. The unit has been interpreted as a braided-stream deposit (Collinson et al., 1986).

As described in Chapter 3, the Tasmania epicratonic basin, 500 km north of Section Peak, contains Late Triassic volcaniclastic sandstone with rhyolitic ash-flow tuff and widespread coal deposited in highly sinuous streams on a westerly slope. A recrudescent foreswell (F II in Fig. 21) marked the eastern limit of these deposits and the western limit of a presumed foreland basin that lay behind the Gondwanide Fold-Thrust Belt in New Zealand. According to Bradshaw et al. (1981, p. 218), New Zealand had "lesser [than the Early Cretaceous Rangitata Orogeny] orogenic pulses in the Late Permian and Late Triassic." The calc-alkaline granitoids of the

Figure 21. Late Triassic paleogeography. The map area is located in Figure 1. Reconstruction of Antarctica from Grunow et al. (1991, fig. 5a p. 17,945): "For simplicity, the present continental outlines and sub-ice topography have been used to define the crustal blocks. . . ." Other blocks restored as detailed in the text and in Figure 3. Numbers refer to paleocurrent directions at the following localities: 1 and 2, northern Victoria Land (Collinson et al., 1986); 3, Allan Hills (Collinson et al., 1983); 4, southern Victoria Land (Barrett and Kohn, 1975); 5, Beardmore Glacier region (162°–166°E, 84°–85°S) (Barrett et al., 1986).

In Tasmania, volcaniclastic sediment forming the coal measures was shed westward, presumably from the marginal magmatic arc through or across foreswell II. In Africa, subsidence resumed in the Karoo Basin leading to the deposition of the quartzose-quartzolithic Molteno Coal Measures. See Figure 14 for symbols.

Brook Street Magmatic Arc may extend to the 230 Ma Mount Bramhall granodiorite (Grunow et al., 1991, p. 17,943).

Central Transantarctic Mountains. Included in the Late Triassic is the upper 100 m of the Fremouw Formation, which is composed of coarse-grained, pebbly sandstone with thin beds of carbonaceous mudstone (Fig. 11). Fossil forests with standing tree trunks occur at several localities in the Beardmore Glacier region and the Queen Maud Mountains (Taylor et al., 1991). Spectacular preservation of fossil plants occurs in blocks of permineralized peat (Taylor and Taylor, 1990). These blocks and several large logs along the same horizon were rafted into the middle of a braided-stream channel and buried during a major flood (Taylor et al., 1986).

The lower member of the Falla Formation, which is latest Triassic in age, is composed of cyclical sandstone and shale (Fig. 11). An upper member, which is Jurassic, is dominated by volcanic tuff. The only known complete section of the Falla Formation occurs on Mount Falla. The lower Falla is up to 280 m thick. At the reference section the lower Falla comprises 12 fining upward cycles of coarse- to fine-grained sandstone overlain by carbonaceous shale and thin coal seams, which were deposited by braided streams (Barrett et al., 1986).

Antarctic Peninsula. The Trinity Peninsula Group, an accretionary complex at the northern end of the Antarctic Peninsula, may be as old as Carboniferous (Dalziel, 1982; Pankhurst, 1983). The only convincing fossils reported from this region, however, are part of a Triassic marine fauna in the Legoupil Formation (Thomson, 1975). Edwards (1982) reported Late Triassic and younger marine faunas in the Le May Group on Alexander Island on the west side of the Antarctic Peninsula, and Thomson and Tranter (1986) revised the age to Liassic. Macdonald and Butterworth (1990) indicate that the Le May Group may be as old as Late Triassic. All these deposits constitute turbiditic graywacke-shale sequences.

Paleocurrents and provenance. Paleocurrent directions in the Late Triassic (Fig. 21) did not change significantly from the Middle Triassic. Drainage remained along the axis of the Transantarctic basin toward the north. Paleocurrent data are more variable in Victoria Land, where some of the sequence was deposited by meandering streams.

Volcanic detritus remained the dominant component in sandstones during the early Late Triassic, but quartzose sandstones became important throughout the basin with the initial deposition of the Falla Formation, the upper member of the Lashly Formation, and the Section Peak Formation. Along with quartz and fresh K-feldspar derived from the craton, these units also contain a significant proportion of volcanic detritus. A reworked crystal tuff was identified in the Section Peak Formation.

The sedimentary petrology of the Trinity Peninsula Group shows derivation from coarse-grained plutonic, high-grade metamorphic, and andesitic to rhyolitic volcanic sources, possibly from the dissection of a volcanic arc (Smellie, 1987).

Paleogeography. Lithologic similarities and paleocurrent directions suggest that Late Triassic sandstones were deposited in a single drainage system along the axis of the Transantarctic basin. During the early Late Triassic, streams from the Queen Maud Mountains to the Beardmore Glacier region (upper Fremouw) were braided, but changed downstream in Victoria Land (Lashly member C) to meandering streams. In the latest Triassic with the deposition of the lower Falla, braided-stream deposition may have extended throughout the length of the basin. Fossil forests and preservation of carbonaceous plant material suggest that conditions became more humid than they were in the Early and Middle Triassic.

The western Antarctic Peninsula was probably the site of forearc basin and trench-slope basin deposition (Fig. 21) (Smellie, 1987), in front of a magmatic arc indicated along strike by the 230 Ma granodiorite at Mount Bramhall on Thurston Island (Grunow et al., 1991) and the Late Triassic granitoids of the Brook Street Magmatic Arc (Korsch and Wellman, 1988). Other forearc terranes as old as this are unknown in the rest of West Antarctica, but they are abundant in the New Zealand sector of Gondwanaland.

Paleoclimatic implications of the Gondwanan sequence. The Permian to Jurassic Pangean supercontinent stretched from pole to pole. Paleoclimatic modelers suggest that this distribution of land would have had a profound effect on the world's climate. In all the models, solar radiation is the most important component within large continents. Crowley et al. (1989) predicted seasonal temperature extremes of up to 60° C in Late Permian polar areas of Gondwanaland. Even much warmer ocean temperatures in polar realms would not have greatly altered this effect (Sloan and Barron, 1990). A zonal climate would have been in effect with arid regions in mid-latitudes and polar winds delivering moisture to the Panthalassan margin on Antarctica (Parrish, 1990). The interior of Pangea would have been mostly arid.

Crowley et al. (1989) pointed to the difficulty of reconciling the occurrence of reptilian faunas in South Africa in the Late Permian with the temperature extremes in their climatic model. Even more difficult to explain is the absence of frost rings in Late Permian fossil forests in the Beardmore Glacier area, which was at an even higher latitude (Taylor et al., 1992). South African vertebrate localities in the Late Permian were at approximately the same high latitude (70°) as were central Transantarctic Mountains vertebrate localities in the Early to Middle Triassic (Denham and Scotese, 1988). The reptilian and amphibian faunas in the central Transantarctic Mountains are similar to those in southern Africa and have

some elements in common with those in eastern Australia. Rather than face the possibility that their climatic models might be inaccurate, Crowley et al. (1989) suggested that animals migrated northward to areas of milder climate during winter. A simpler hypothesis would be that the climate in the regions of fossil localities was much milder than climatic models predict.

Climatic modelers have not recognized that all the known reptile localities in polar Gondwanaland are in foreland basins adjacent to the Panthalassan margin. The various climatic models suggest that interior Antarctica, if ice were not present, would be an unlikely place to find fossil reptiles in Permian to Jurassic rocks, because of the increased aridity and seasonality of a continental climate. At least the axial drainage portions of fluvial systems occupying these foreland basins must have been at relatively low elevations and close enough to the sea to have had a maritime climate.

In the mid-Permian before thrusting and folding, the Buckley fluvial system extended from the Beardmore Glacier area to the Ohio Range, a distance of 500 to 1,000 km, and perhaps to the Ellsworth Mountains. The Triassic drainage extended from the Queen Maud Mountains to the Beardmore Glacier area, and beyond into Victoria Land, more than 1,500 km, probably entering the sea somewhere in the eastern Australian sector. The Permian-Triassic Victoria subbasin drainage probably extended into Tasmania, approximately 1,000 km. Triassic vertebrates have been found only in Tasmania at the seaward end of the Victoria-Tasmania basin. In the case of the foreland basin, mild climatic conditions would have had to extend at least 1,500 km up the drainage valley from the sea. Slopes of 0.2 m/km, which are reasonable for river systems of this type (Isbell, 1990a), would suggest elevations of approximately 300 m at vertebrate localities. These elevations would not have been high enough to have greatly influenced temperature. The highlands of the thrust-fold belt, however, would have been substantially cooler and probably supported considerable snow accumulation during the winter. The mountains along the thrust-fold belt were probably an important source of water for the foreland basin especially during times of aridity.

General climatic cycles for the Late Carboniferous to Triassic can be hypothesized from plant floras, vertebrate faunas, the preservation of organic matter and occurrence of coal, and sedimentologic factors. Glacial deposits and the low diversity of microfloras in the Late Carboniferous and much of the Early Permian attest to the cold conditions of an ice age. Increases in diversity in plant floras, the preservation of organic material, and the widespread occurrence of coal in the late Early to Late Permian suggest humid conditions and mild temperatures. Diverse reptilian and amphibian faunas in the Early to Middle Triassic indicate mild temperatures and sufficient plant cover for food, but the lack of preservation of organic matter suggests oxidizing conditions under drier climatic conditions. The return of coal and organic matter in Late Triassic rocks points toward a return to more humid conditions.

In summary, the geologic and paleontologic record does not provide support for the published climatic models for Gondwanaland.

POST-GONDWANAN HISTORY

East Antarctica

Jurassic tholeiites occur as intrusive bodies or extrusive rocks throughout the length of the Transantarctic Mountains (Fig. 22). Those from the Theron Mountains to northern Victoria Land, together with Jurassic tholeiites in Tasmania and South Australia, form a geochemical province characterized by unusually high initial $^{87}Sr/^{86}Sr$ ratios (Kyle, 1980; Hergt et al., 1991). Tholeiitic flood basalts and comagmatic intrusion are commonly associated with Gondwanan rifting. Tholeiites in Antarctica, which today crop out in a narrow belt coincident with the Transantarctic Mountains, were not, however, precursors to continental separation and seafloor spreading; they represent magmatism associated with rifting (Elliot et al., 1985; Elliot, 1992) (Fig. 22), possibly in the failed arm of a triple junction (Ford and Kistler, 1980).

The Kirkpatrick Basalt is the youngest widespread stratigraphic unit in the Ross Sea sector of the Transantarctic Mountains other than Cenozoic glacial deposits. The thickness of the lavas is as much as 780 m in northern Victoria Land (Elliot et al., 1986b), 380 m in southern Victoria Land (Kyle et al., 1981), and 500 m in the Beardmore Glacier region (Barrett et al., 1986). Major sills are commonly 150 to 200 m thick and may be up to 400 m thick (Elliot et al., 1985). The total thickness of sills in a given section may exceed 1,000 m. Dikes are widely distributed, relatively uncommon, and volumetrically insignificant.

Fossils recovered from sedimentary interbeds within the lavas include freshwater fish (Schaeffer, 1972), conchostracans (Tasch, 1987), plants (Townrow, 1967), and a variety of other organisms including ostracods and insect remains. An early Middle Jurassic age for the lavas estimated at 180 ± 5 Ma is based on K-Ar and $^{40}Ar/^{39}Ar$ radiometric dating (Elliot et al., 1985; Elliot and Foland, 1986), recently refined to 176 ± 1 Ma (Heimann et al., 1992). A minimum age of 177 ± 2 Ma has been established for the northern Victoria Land sequence by the application of the $^{40}Ar/^{39}Ar$ dating technique to plagioclase mineral separates (Fleming et al., 1991, 1993).

The only known place where the transition from Gondwanan clastic sedimentation to tholeiitic volcanism is nearly

complete is in the Beardmore Glacier area. Here Late Triassic plant-bearing beds of the lower Falla Formation are overlain by volcaniclastic beds of the upper Falla Formation, which contain a Jurassic dinosaur fauna, including large carnosaur and sauropod remains (Hammer et al., 1991). Upward the upper Falla is increasingly dominated by siliceous tuffs (*see*

Appendix 2, Fig. 25F). A volcanic pebble from near the base of the upper Falla yielded a minimum K-Ar age of 203 ± 3 Ma, and the analysis of five tuffs from the upper Falla produced an array on a Rb-Sr isochron diagram equivalent to an age of 186 ± 9 Ma (Barrett et al., 1986). Both these ages may have been affected by the low-grade metamorphism associated with Ferrar magmatism.

The upper Falla is overlain by the Prebble Formation, a coarse pyroclastic unit. Contacts appear to be abrupt but gradational although at one locality the Prebble is markedly disconformable on Falla beds (Barrett et al., 1986). The lower quartzose and upper volcanic members of the Falla vary locally in thickness by as much as 200 m, suggesting the presence of disconformities. The Prebble also varies considerably in thickness, and exhibits constructional landforms, which are evident from relations with the basaltic lavas.

Much of the pyroclastic material in the Prebble Formation is coarse breccia, which has been interpreted as lahar deposits and debris flows (Barrett et al., 1986). Such deposits were locally derived and transported no more than a few tens of kilometers. Accretionary lapilli tuffs also suggest local volcanic centers, and a contemporaneous volcanic vent has been located in the Beardmore Glacier area (Barrett and Elliot, 1972). The transition to flood basalt volcanism occurs also in Victoria Land. In southern Victoria Land the Lashly Formation (Late Triassic) is disconformably overlain by basaltic volcanic breccia of the Mawson Formation (Borns and Hall, 1969; Bradshaw, M. A., 1987). The Mawson overlies an erosional surface that cuts as deep as 500 m into the Early Permian Weller Coal Measures (Collinson et al., 1983). Similarly, in northern Victoria Land lavas of the Kirkpatrick Basalt are also underlain by a pyroclastic unit, the Exposure Hill Formation, which has both basaltic and silicic components (Elliot et al., 1986a). The lower contact of this unit with the Late Triassic Section Peak Formation is conformable at the only known locality.

The sequence in the central Transantarctic Mountains, transitional from the *Dicroidium*-bearing beds of the lower Falla to the flood basalts of the Kirkpatrick, provides evidence for an Early Jurassic extensional tectonic regime (Elliot and Larsen, 1993). Coarse immature arkosic grits within the upper

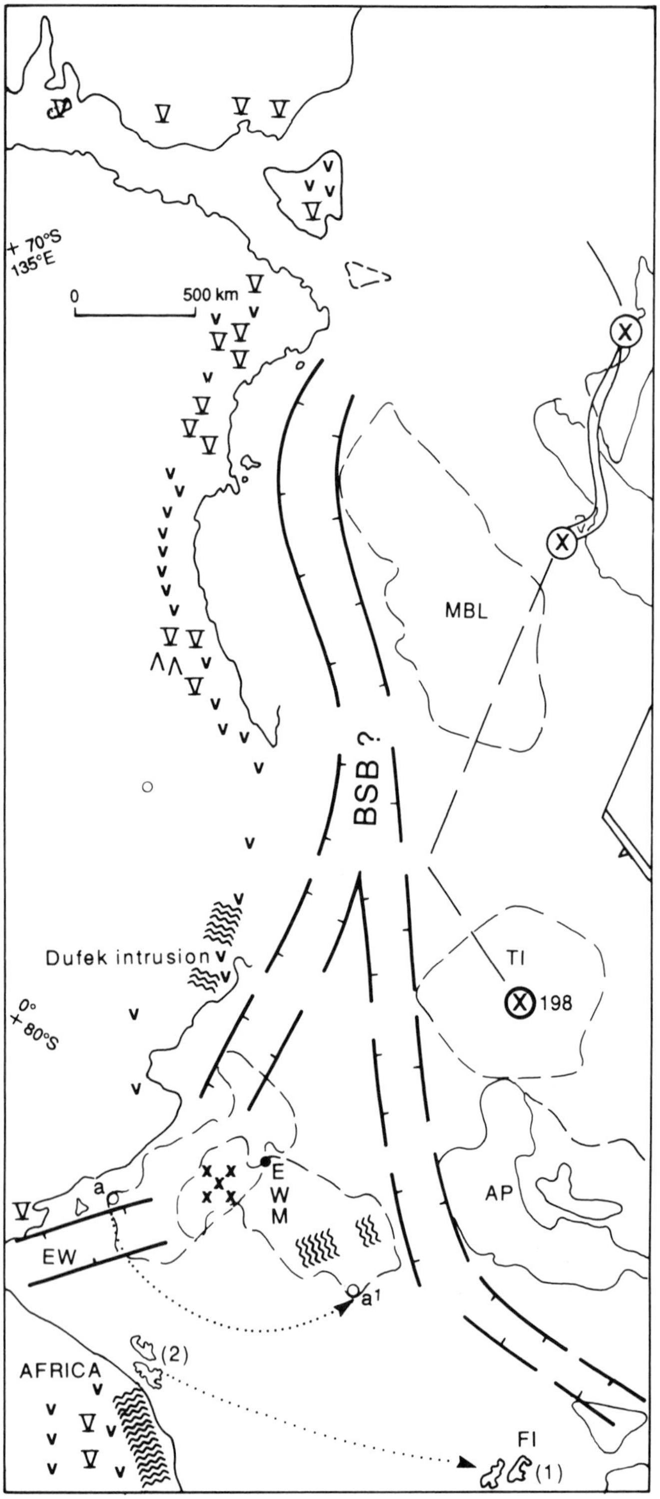

Figure 22. Jurassic paleogeography. The map area is located in Figure 1. Reconstruction for 175 Ma from Grunow et al. (1991, fig. 5b). Abbreviations of crustal blocks (broken lines) and other features are EWM = Ellsworth-Whitmore Mountains; TI = Thurston Island–Eights Coast; EW = Explora Wedge; BSB = Byrd Subglacial Basin; AP = Antarctic Peninsula; FI = Falkland Islands; MBL = Marie Byrd Land; a = 230 Ma position of the Ellsworth-Whitmore Mountains block; a[1] = 175 Ma position of the Ellsworth-Whitmore Mountains block; (1) = present-day position of Falkland Islands block with respect to South America; (2) = ca. 190 Ma position of Falkland Islands block adjacent to Africa. *See* discussion in West Antarctica section of text. See Figure 14 for symbols.

Falla volcanic beds require a proximal source, although paleoflow directions consistent with the underlying Triassic strata suggest the source lay outside the present outcrop area. Volcaniclastic beds in the upper Falla are affected by monoclinal flexuring contemporaneous with Prebble volcanism. Older strata at Coalsack Bluff also exhibit structures with similar orientation and have been interpreted as a faulted monocline into which Ferrar dolerite was intruded (Collinson and Elliot, 1984a). Very thick laharic deposits (>360 m), with a regionally planar upper surface onto which the flood basalts were erupted, suggest depositional control by topography. Ponded lavas up to 170 m thick also require confining topography. These features are attributed to basement block faulting associated with extensional tectonism and represent vertical movements of the crust and development of significant relief. The pre-Mawson erosional relief in southern Victoria Land is consistent with such a regime. The onset of silicic volcanism in the Jurassic marks the termination of foreland basin sedimentation and the establishment of a new tectonic setting for the Transantarctic Mountains.

This sequence further points to a major episode of silicic volcanism that predates, but also overlaps, tholeiitic igneous activity. Granitic plutons in the Ellsworth-Whitmore Mountains crustal block have yielded Rb-Sr isochron ages of 182 ± 5 Ma to 173 ± 3 Ma (Pankhurst et al., 1991).

These plutons, which have some of the characteristics of S-type granites, are probably part of the same magmatic episode as the silicic volcanics in the central Transantarctic Mountains. Crustal anatexis probably played a significant role in the generation of these silicic rocks and was related to the thermal event that culminated with the intrusion and extrusion of tholeiitic magmas along the margin of East Antarctica (Elliot, 1991).

West Antarctica

In Grunow et al.'s (1991, fig. 5b) reconstruction at 175 Ma, shown in Fig. 22, the Ellsworth-Whitmore Mountains block has rotated 90° from its 230 Ma position (dotted line, a) to the 175 Ma position (broken line, a^1) about a pole marked by the filled circle. The cross in the middle of the Thurston Island–Eights Coast block marks the position of the 198 Ma granite in the Jones Mountains. The Explora Wedge represents one margin of a rift basin formed in the early stages of breakup; other possible rift basins include the Byrd Subglacial Basin, and the backarc side of the Antarctic Peninsula block. The Falkland Islands block first underwent a 155° rotation to a position adjacent to Africa [(2) in Fig. 22] after ca. 190 Ma before moving to its present-day position with respect to South America [(1) in Fig. 22] during Middle to Late Jurassic east-west translation in the Falkland Islands Plateau region. The wavy pattern in Africa, the Ellsworth-Whitmore Mountains block and the Pensacola Mountains, shows the dismembered parts of the Gondwanide Fold Belt. Australian data (Tasma-

nian Dolerite and lava, basalt in Victoria and on Kangaroo Island) is from Chapter 3 and Hergt et al. (1991). Off the other side of Antarctica, the Karoo Dolerite and Drakensberg Volcanics covered much of southern Africa north of the Cape Fold Belt (*see* Chapter 5 on the Karoo Basin).

Stratigraphic sequences in the Transantarctic basin provide evidence for the existence of an active Panthalassan plate margin during the Permian and Triassic. Direct evidence of the magmatic arc and forearc is confined to scattered plutons and metamorphic rocks through coastal West Antarctica and the Antarctic Peninsula (Pankhurst, 1990) and in the Antarctic Peninsula to folded forearc and accretionary prism rocks (Dalziel, 1982). The deformed sequence is overlain by relatively undeformed calc-alkaline volcanic rocks and sedimentary sequences of Middle Jurassic to Cretaceous age (Thomson and Pankhurst, 1983). Abundant evidence exists for an active magmatic arc along the Antarctic Peninsula by Late Jurassic time (Thomson et al., 1983; Elliot, 1983). Late Jurassic and younger forearc strata crop out on Alexander Island, west of the Antarctic Peninsula, where they are notably fossiliferous, and on the South Shetland Islands. Backarc strata with Middle and Late Jurassic marine faunas, including ammonites (Quilty, 1970; 1977; Thomson, 1983), occur on the east side of the southern Antarctic Peninsula, with Late Jurassic strata also widespread in the northeast Antarctic Peninsula (Farquharson, 1983).

The apparent break in the record represented by the unconformity in the Antarctic Peninsula coincides with Gondwanan breakup magmatism and an outward shift in the locus of backarc sedimentation which, in contrast to the Beacon, was marine, thus indicating marked subsidence of part of the former arc and backarc region (Elliot, 1991).

ACKNOWLEDGMENTS

Support for this synthesis was provided by Ohio State University and the National Science Foundation (DPP-8716414). We are grateful for the thorough reviews and many helpful suggestions by GSA reviewers Ian Dalziel and David Macdonald. Judith Jenkins kindly offered editorial suggestions. John Veevers made many helpful suggestions throughout the manuscript and drew our attention to connections with along-strike Eastern Australia (Chapter 3) and southern Africa (Chapter 5). Drafting of the figures was done by Judy Davis and David Hilyard (Macquarie University). Collinson made the first draft of this chapter while visiting Macquarie University from 1987 to 1988.

APPENDIX 1. LATE CARBONIFEROUS-JURASSIC CORRELATION CHART FOR ANTARCTICA

J. J. Veevers and J. W. Collinson

The columns of the chart (Fig. 23) are arranged along the Transantarctic Mountains from north Victoria Land to the Ellsworth Mountains and then to the Antarctic Peninsula. The final columns are

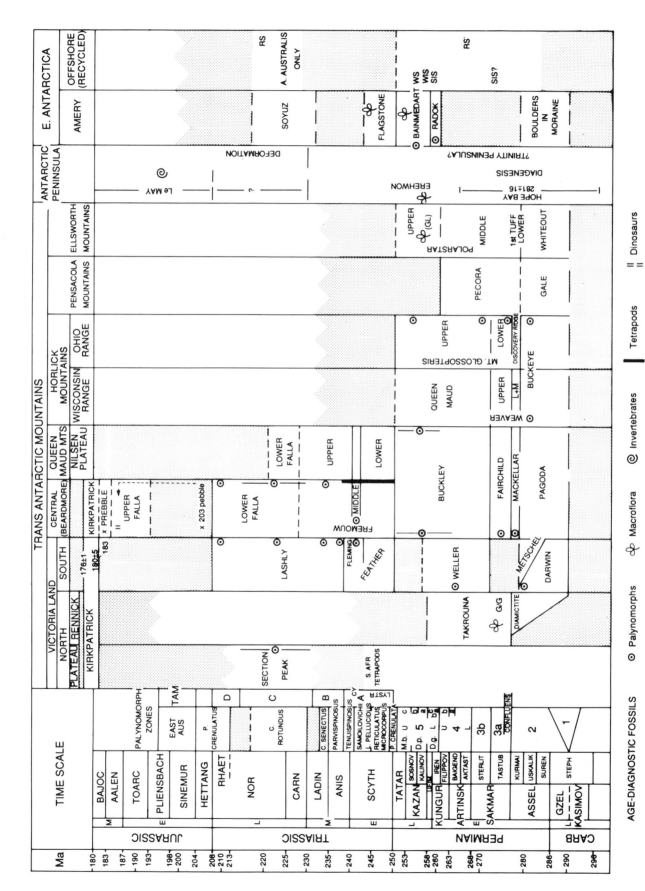

Figure 23. Correlation chart of Antarctica. Abbreviations are G/G—*Glossopteris/Gangamopteris* flora, GL—*Glossopteris*, RS—Ross Sea, SIS—Shackleton Ice Shelf, WIS—West Ice Shelf, WS—Weddell Sea.

for East Antarctica: the Prince Charles Mountains of the Amery Ice Shelf region and the recycled assemblages of the offshore area from the Weddell Sea to the Ross Sea.

The radiometric time scale, from Appendix 1, Chapter 3, has the Permian/Triassic boundary calibrated as 250 Ma.

Biostratigraphical schemes

The formations are correlated to the east Australian palynomorph zones (*see* Fig. 44, Chapter 3) and the subzones of the Victoria Group of the Transantarctic Mountains (Kyle and Schopf, 1982), and to the *Lystrosaurus* and *Cynognathus* tetrapod zones of southern Africa (*see* Fig. 21, Chapter 5), with the range of *Cynognathus* in Antarctica extended into the Middle Triassic (Hammer et al., 1990). The dinosaurs in the upper Falla Formation indicate a Jurassic age (Hammer et al., 1991), consistent with the radiometric dates.

Transantarctic Mountains

Victoria Land. Plateau—Section Peak. The palynomorph assemblage described by Norris (1965) from Timber Peak in northern Victoria Land—the Section Peak Formation of Collinson et al. (1987)—was included in subzones C and D of Kyle and Schopf (1982, p. 656), and, according to Dolby and Balme (1976, p. 136), is part of the Ipswich Microflora or subzone C.

Rennick Graben area. An unnamed barren diamictite at the base of the section is taken to be the same general age (Stage 2) as the other glacial sediments along the Transantarctic Mountains. The overlying Takrouna Formation contains *Glossopteris* associated with *Gangamopteris* and is taken to be younger Early Permian.

South. Palynomorphs in the upper shale beds of the Darwin Tillite indicate upper Stage 2 (Kyle and Schopf, 1982, p. 651). For convenience, the base of the Tillite is taken as the same age as diamictites in adjacent parts of Gondwanaland, namely, the base of Stage 2, or 290 Ma. Its facies equivalent, the Metschel Tillite, is regarded by Kyle and Schopf (1982) as the same age. The middle and upper parts of the Weller Coal Measures contain palynomorphs that indicate Stage 4 (Kyle and Schopf, 1982, p. 653), and the uppermost part, with *Glossopteris,* extends into the Late Permian (T. N. Taylor, personal communication, 1992). The base of the Weller Coal Measures probably extends down to the 3a/3b boundary, level with the top of the Fairchild Formation (Isbell, 1990b). The overlying Feather Conglomerate is barren except for its top, the Fleming Member, which, together with the basal beds of the Lashly Formation, contain palynomorphs of subzone A; subzones B–D occur higher in the Lashly Formation (Kyle and Schopf, 1982, p. 656). We place the noncarbonaceous Feather Conglomerate after a lacuna above the carbonaceous Permian.

Central Transantarctic Mountains: Beardmore Glacier and Queen Maud Mountains–Nilsen Plateau. The Permian and Triassic formations of the central Transantarctic Mountains extend to the Nilsen Plateau of the Queen Maud Mountains. The Mackellar Formation contains upper Stage 2 palynomorphs and the succeeding Fairchild Formation Stage 3a palynomorphs (R. A. Askin, personal communication, 1987). By superposition, the Pagoda Formation is taken as older Stage 2. The Buckley Formation of the Nilsen Plateau (designated the Queen Maud Formation by Kyle and Schopf, 1982) and the Beardmore Glacier region (Farabee et al., 1991) contains Stage 5 palynomorphs (Kyle and Schopf, 1982, p. 654; Taylor et al., 1988); its lower part probably extends, without break, down to the Fairchild Formation. The youngest exposed sediments in the Nilsen Plateau (Nilsen Formation of Kyle and Schopf, 1982) are equated with the Fremouw Formation and lower Falla Formation in the central region. Here, the lower part of the Fremouw Formation contains the tetrapod *Lystrosaurus* (Hammer et al., 1990; Hammer, 1990); the middle part contains subzone A and B palynomorphs (Kyle and Schopf, 1982,

p. 656); the base of the upper part contains the tetrapod *Cynognathus* (Hammer et al., 1990), and the rest of the upper part contains subzone B and C palynomorphs (Kyle and Schopf, 1982, p. 656). The lower Falla Formation contains subzones C and D (Kyle and Schopf, 1982, p. 657). The upper Falla Formation contains dinosaurs of Middle Jurassic aspect (W. R. Hammer, personal communication, 1992). In the upper Falla Formation, a trachyte pebble with a whole-rock K-Ar age of 203 ± 3 Ma and tuffs with a 5-point whole-rock Rb-Sr isochron age of 186 ± 9 Ma have been reported by Barrett et al. (1986); the isochron date may reflect the widespread zeolitization that has affected these rocks. A diabase boulder in the overlying Prebble Formation has given a K-Ar age of 183 ± 10 Ma. The youngest preserved formation, the Kirkpatrick Basalt, has an age best estimated as early Middle Jurassic age from lavas estimated at 180 ± 5 Ma on K-Ar and ^{40}Ar/^{39}Ar radiometric dating (Elliot et al., 1985; Elliot and Foland, 1986), recently refined to 176 ± 1 Ma (Heimann et al., 1992).

Horlick Mountains. Wisconsin Range. At the base, the Buckeye Formation here and in the Ohio Range contains upper Stage 2 palynomorphs (Kyle and Schopf, 1982, p. 651). The Weaver Formation is a facies equivalent of the Discovery Ridge Formation and the lower part of the Mount Glossopteris Formation (Minshew, 1967), and is regarded as the same age. The Queen Maud Formation is the same age as the Buckley Formation for the same reason.

Ohio Range. As mentioned above, the Buckeye Formation is upper Stage 2, as is the basal part of the Mount Glossopteris Formation (Kyle and Schopf, 1982, p. 651), with the result that the intervening Discovery Ridge Formation is upper Stage 2 also. The rest of the Mount Glossopteris Formation contains palynomorphs of Stages 3 and 5 (Kyle and Schopf, 1982, p. 653).

Pensacola Mountains. From its glacial facies, the barren Gale Mudstone is placed in Stage 2. The geographically isolated and presumably stratigraphically higher Pecora Formation is thought to be later Early Permian.

Ellsworth Mountains. The upper part of the glacial Whiteout Conglomerate is likewise placed in Stage 2 and the upper part of the Polarstar Formation, on account of *Glossopteris,* is placed in Stage 5 (Taylor and Taylor, 1992). The first clearly defined tuff (Fig. 24, A and B), 274 m above the base of the Polarstar Formation (Fig. 13), is regarded by facies comparison of the enclosing beds as the same age as the base of the Fairchild, upper Weaver, and lower Mount Glossopteris Formations, equivalent to the base of Stage 3a or 277 Ma. Shale lower in the Polarstar Formation contains fine-grained volcanic detritus, including tiny angular clear quartz and biotite, probably of airfall origin.

Antarctic Peninsula. According to Macdonald and Butterworth (1990), the pre-Mesozoic basement of the Antarctic Peninsula comprises granite orthogneiss of early Paleozoic age (410 ± 15 and 426 ± 12 Ma; Pankhurst, 1990), with a major metamorphic event in the mid-Carboniferous, which may have been associated with plutonism to the east of the Antarctic Peninsula (Milne and Millar, 1989). Accretionary complexes are represented by (1) the Trinity Peninsula Group, probably deposited in Permian-Triassic times (a siltstone in the Hope Bay Formation of the Trinity Peninsula Group has a 9-point Rb/Sr isochron age of 281 ± 16 Ma, interpreted as the age of diagenesis (Pankhurst, 1983) and deformed in Late Triassic or Early Jurassic times (equivalent to Gondwanides); and (2) the LeMay Group, with known ages of Early Jurassic (Thomson and Tranter, 1986) and Early Cretaceous (Burn, 1984), and a possible span of Late Triassic to Paleogene.

East Antarctica

Amery Shelf sector. In the Amery Group of the Prince Charles Mountains of the Amery Shelf sector, the Radok Conglomerate and the Bainmedart Coal Measures contain Stage 5 palynomorph (Trus-

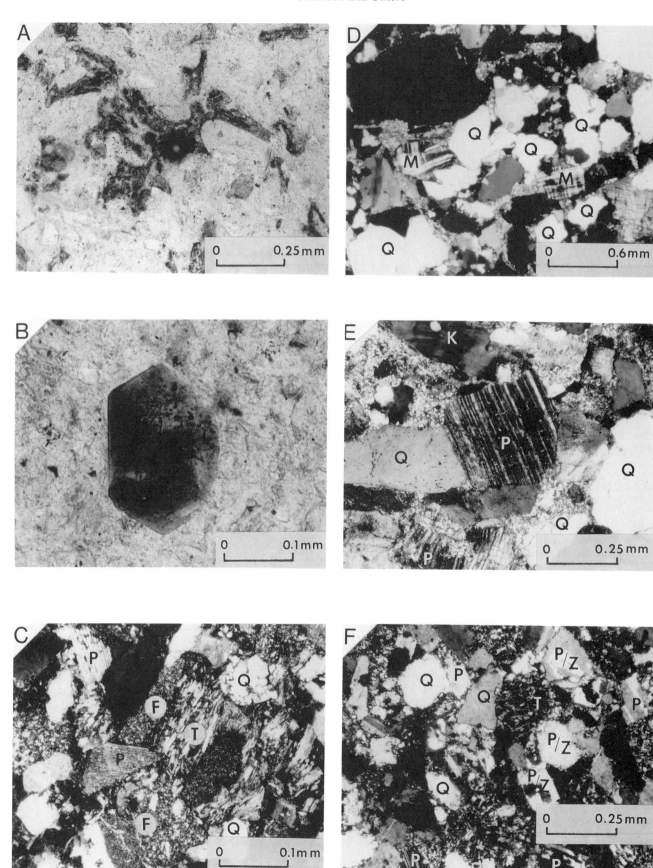

well, 1980, p. 98), and the Bainmedart Coal Measures contain *Glossopteris,* which is replaced in the overlying Flagstone Bench Formation by *Dicroidium,* believed to be Early Triassic (Webb and Fielding 1990, 1993). The Soyuz Formation is probably Late Triassic (R. Helby, personal communication, 1992). Boulders of diamictite in modern moraine indicate the possibility that Late Carboniferous or Early Permian glacigene sediment was deposited in the Lambert Graben, as confirmed by the recycled palynomorphs offshore, west of the Shackleton Ice Shelf.

   Recycled palynomorphs. (a) According to Kemp (1972b) and Truswell (1980), recycled spores in bottom sediments of the Weddell Sea provide the only Antarctic specimens referrable to *Dulhuntispora,* the guide genus to the Australian Stage 5. The source is unknown, "but is probably in Dronning Maud Land, or the Theron or Pensacola Mountains, all regions in which coal-bearing sequences with *Glossopteris* occur" (Truswell, 1980, p. 99) (b) Assemblages from the West Ice Shelf, between 75° and 80°E (Kemp, 1972a), include form species recorded from the Amery Group, or Stage 5. (c) Assemblages west of the Shackleton Ice Shelf, at 95°E, also include taxa found in the Amery Group, or Stage 5, and rare forms suggesting Stages 2 to 4 (Truswell, 1982, p. 345). Truswell (1982) notes the excellent preservation of most of the Permian forms in contrast with the severely carbonized Permian palynomorphs elsewhere in Antarctica, suggesting that the Permian source beds were neither deeply buried nor subjected to thermal metamorphism by igneous intrusions. (d) The oldest recognized palynomorph to the east, off Wilkes Land (the Cape Carr area, 132° to 135°E, and Mertz Glacier area, 145°E) are Cretaceous (Truswell, 1983). (e) According to Truswell (1983) and Truswell and Drewry (1984), Permian spores are present throughout the Ross Sea area and most frequent in the west; they match assemblages described from the Victoria Group in the Transantarctic Mountains by Kyle (1977). Only a single pollen type of Triassic age, *Alisporites australis,* was identified, and this is the only Triassic form recorded from recycled assemblages off Antarctica. This contrasts with the flood of Triassic palynomorphs in the later Mesozoic sequence of the southern margin of Australia. An area of high palynomorph density north of Roosevelt Island is dominated by Late Cretaceous to Early Tertiary palynomorphs.

Figure 24. Photomicrographs of Permian rocks. A: Airfall tuff with bubble-wall shards replaced by zeolite. Sample MW274.5, collected by C. L. Vavra from middle part of Early Permian Polarstar Formation, Mount Weems, northern Ellsworth Mountains. Plane-polarized light. B: Early Permian airfall tuff with hexagonal biotite. Same sample as A. Plane-polarized light. C: Late Permian volcaniclastic sandstone with angular quartz (Q), pilotaxitic volcanic rock fragment (T), felsic volcanic rock fragment (F), and plagioclase (P). Sample MV22.5, collected by C. L. Vavra, upper Polarstar Formation, near Polarstar Peak, northern Ellsworth Mountains. Cross-polarized light. D: Early Permian quartzo-feldspathic sandstone with microcline (M), and quartz (Q). Sample AHW5, collected by J. W. Collinson, Weller Coal Measures, Allan Hills, southern Victoria Land. Cross-polarized light. E: Early Permian quartzo-feldspathic sandstone with quartz (Q), K-feldspar (K), and plagioclase (P). Sample PB009, collected by P. J. Barrett, lower Buckley Formation, Painted Cliffs, Beardmore Glacier area, central Transantarctic Mountains. Cross-polarized light. F: Late Permian volcaniclastic sandstone with pilotaxitic volcanic rock fragment (T), plagioclase (P) in places replaced by zeolite (P/Z), and quartz (Q). Sample BW108, collected by P. J. Barrett, upper Buckley Formation, Wahl Glacier, Beardmore Glacier area, central Transantarctic Mountains. Cross-polarized light.

APPENDIX 2: PHOTOMICROGRAPHS OF PERMIAN AND TRIASSIC-JURASSIC SANDSTONES

J. W. Collinson and D. H. Elliot

   Evidence of active volcanism shows up first as volcaniclastic sandstone and airfall tuff in the Early Permian Polarstar Formation in the Ellsworth Mountains (Fig. 24, A through C). Early Permian rocks in Victoria Land (Fig. 24D) and the central Transantarctic Mountains (Fig. 24E) are quartzofeldspathic. The first volcaniclastic sandstone in the central Transantarctic Mountains is Late Permian (Fig. 24F).

   The Early Triassic in southern Victoria Land is quartzofeldspathic (Fig. 25A). The Early Triassic in the area between the Beardmore Glacier (Fig. 25B) and the Shackleton Glacier area (Fig. 25C) becomes more and more volcaniclastic upsection. Evidence of active volcanism first shows up in southern Victoria Land as volcaniclastic sandstone in the Middle Triassic (Fig. 25D). Tuff and tuffaceous sandstone occur in the Early Triassic (Fig. 25E) and Jurassic (Fig. 25F) of the central Transantarctic Mountains.

REFERENCES CITED

Adams, C. J., 1986, Age and ancestry of metamorphic rocks of the Daniels Range, USARP Mountains, Antarctica, *in* Stump, E., ed., Geological investigations in Northern Victoria Land: Washington, D.C., Antarctic Research Series, American Geophysical Union, v. 46, p. 25–38.

Adams, C. J., Whitla, P. F., Findlay, R. H., and Field, B. F., 1986, Age of the Black Prince volcanics in the central Admiralty Mountains and possibly related hypabyssal rocks in the Millen Range, *in* Stump, E., ed., Geological investigations in Northern Victoria Land: Washington, D.C., Antarctic Research Series, American Geophysical Union, v. 46, p. 203–210.

Aitchison, J. C., Bradshaw, M. A., and Newman, J., 1988, Lithofacies and origin of the Buckeye Formation: Late Paleozoic glacial and glaciomarine sediments, Ohio Range, Transantarctic Mountains, Antarctica: Palaeogeography, Palaeoclimatology, Palaeoecology, v. 64, p. 93–104.

Allen, P. A., Homewood, P., and Williams, G. D., 1986, Foreland basins: An introduction, *in* Allen, P. A., and Homewood, P., eds., Foreland basins: Special publication number 8 of the International Association of Sedimentologists: Oxford, Blackwell Scientific Publications, p. 3–12.

Anderson, B., 1961, The Rufiji Basin, Tanganyika: Preliminary reconnaissance survey Rufiji Basin, soils of the main irrigable areas: Report of the Government of Tanganyika, no. 7, 125 p.

Anderson, J. B., and Molnia, B. F., 1989, Glacial-marine sedimentation: Short course in geology: Washington, D.C., American Geophysical Union, v. 9, 127 p.

Aronson, J. L., 1965, Reconnaissance rubidium-strontium geochronology of New Zealand plutonic and metamorphic rocks: New Zealand Journal of Geology and Geophysics, v. 8, p. 401–423.

Aronson, J. L., 1968, Regional geochronology of New Zealand: Geochimica et Cosmochimica Acta, v. 32, p. 669–697.

Barrett, P. J., 1972, Late Paleozoic glacial valley at Alligator Peak, Southern Victoria Land Antarctica: New Zealand Journal of Geology and Geophysics, v. 15, p. 262–268.

Barrett, P. J., 1981, History of the Ross Sea region during the deposition of the Beacon Supergroup 400–180 million years ago: Journal of the Royal Society of New Zealand, v. 11, p. 447–458.

Barrett, P. J., and Elliot, D. H., 1972, The early Mesozoic volcaniclastic Prebble Formation, Beardmore Glacier area, *in* Adie, R. J., ed., Antarctic geology and geophysics: Oslo, Universitetsforlaget, p. 403–409.

Barrett, P. J., and Fitzgerald, P. G., 1985, Deposition of the lower Feather Conglomerate, a Permian braided river deposit in southern Victoria Land, Antarctica, with notes on regional paleogeography: Sedimentary

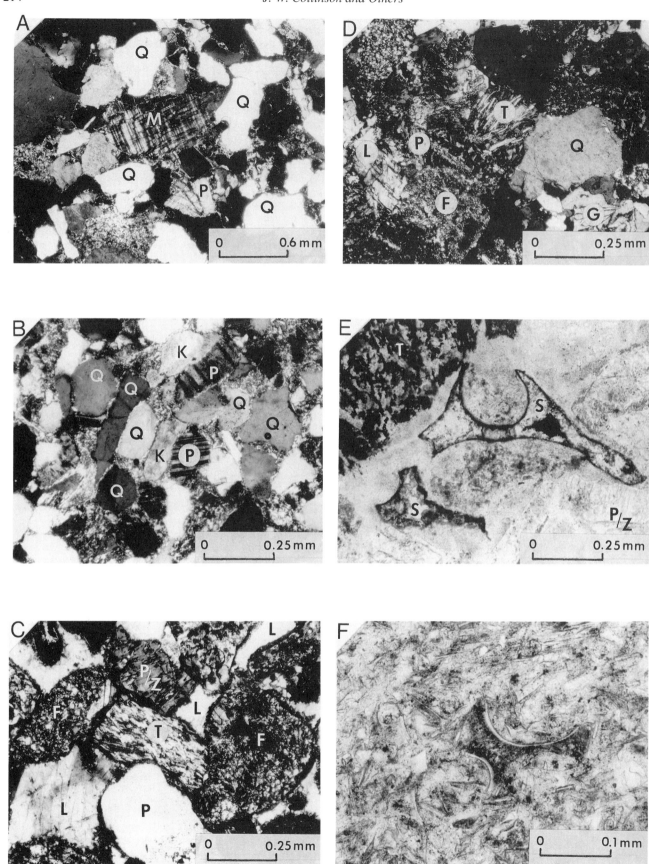

Geology, v. 45, p. 189–208.

Barrett, P. J., and Kohn, B. P., 1975, Changing sediment transport directions from Devonian to Triassic in the Beacon Supergroup of south Victoria Land, Antarctica, *in* Campbell, K.S.W., ed., Gondwanan geology: Canberra, Australian National University Press, p. 329–332.

Barrett, P. J., and Kyle, R. A., 1975, The Early Permian glacial beds of south Victoria Land and the Darwin Mountains, Antarctica, *in* Campbell, K.S.W., ed., Gondwanan geology: Canberra, Australian National University Press, p. 333–346.

Barrett, P. J., and McKelvey, B. C., 1981, Permian tillites of South Victoria Land, Antarctica, *in* Hambrey, J. J., and Harland, W. B., eds., Earth's pre-Pleistocene glacial record: Cambridge, Cambridge University Press, p. 233–236.

Barrett, P. J., Elliot, D. H., and Lindsay, J. F., 1986, The Beacon Supergroup (Devonian-Triassic) and Ferrar Group (Jurassic) in the Beardmore Glacier area, Antarctica, *in* Turner, M. D., and Splettstoesser, J. F., eds., Geology of the central Transantarctic Mountains: Washington, D.C., Antarctic Research Series, American Geophysical Union, v. 36, p. 339–428.

Behrendt, J. C., and Cooper, A. K., 1991, Evidence of rapid Cenozoic uplift of the shoulder escarpment of the Cenozoic West Antarctic rift system and a speculation on possible climate forcing: Geology, v. 19, p. 315–319.

Behrendt, J. C., LeMasurier, W. E., Cooper, A. K., Tessensohn, F., Trehu, A., and Damaske, D., 1991, Geophysical studies of the West Antarctic rift system: Tectonics, v. 10, p. 1257–1273.

Bentley, R. R., 1983, Crustal structure of Antarctica from geophysical evidence—A review, *in* Oliver, R. L., James, P. R., and Jago, J. B., eds., Antarctic Earth science: Canberra, Australian Academy of Science, p. 491–497.

Berner, R. A., and Raiswell, R., 1984, C/S method for distinguishing freshwater from marine sedimentary rocks: Geology, v. 12, p. 365–368.

Blake, M. C., Jones, D. L., and Landis, C. A., 1974, Active continental margins; contrasts between California and New Zealand, *in* Burk, C. A., and Drake, C. L., eds., The geology of continental margins: New York, Springer-Verlag, p. 853–872.

Borg, S. G., 1983, Petrology and geochemistry of the Queen Maud Batholith, Central Transantarctic Mountains, with implications for the Ross Orogeny, *in* Oliver, R. L., James, P. R., and Jago, J. B., eds., Antarctic Earth science: Canberra, Australian Academy of Science, p. 165–169.

Borg, S. G., and Stump, E., 1987, Paleozoic magmatism and associated tectonic problems of northern Victoria Land, Antarctica, *in* McKenzie, G. D., ed., Gondwanan Six: Structure, tectonics, and geophysics: Washington, D.C., Geophysical Monograph Series, American Geophysical Union, v. 40, p. 67–75.

Borg, S. G., Stump, E., and Holloway, J. R., 1986, Granitoids of northern Victoria Land, Antarctica: A reconnaisance study of field relations, petrography, and geochemistry, *in* Stump, E., ed., Geological investigations in northern Victoria Land: Washington, D.C., Antarctic Research Series, American Geophysical Union, v. 46, p. 115–188.

Borg, S. G., DePaolo, D. J., and Smith, B. M., 1990, Isotopic structure and tectonics of the central Transantarctic Mountains: Journal of Geophysical Research, v. 95(5B), p. 6647–6667.

Borns, H. W., and Hall, B. A., 1969, Mawson "Tillite" in Antarctica: Preliminary report of a volcanic deposit of Jurassic age: Science, v. 166(3907), p. 870–872.

Bradshaw, J. D., 1987, Terrane boundaries in North Victoria Land, *in* Proceedings of the meeting, Geosciences in Victoria Land, Antarctica, Siena, 2–3 September 1987: Memorie della Società Geologica Italiana, v. 33, p. 9–15.

Bradshaw, J. D., and Laird, M. G., 1983, The pre-Beacon geology of northern Victoria Land: A review, *in* Oliver, R. L., James, P. R., and Jago, J. B., eds., Antarctic Earth science: Canberra, Australian Academy of Science, p. 98–101.

Bradshaw, J. D., Andrews, P. B., and Adams, C. J., 1981, Carboniferous to Cretaceous on the Pacific margin of Gondwana: The Rangitata phase of New Zealand, *in* Cresswell, M. M., and Vella, P., eds., Gondwana Five: Rotterdam, A. A. Balkema, p. 217–221.

Bradshaw, J. D., Laird, M. G., Wodzicki, A., 1982, Structural style and tectonic history in northern Victoria Land, *in* Craddock, C., ed., Antarctic geoscience: Madison, University of Wisconsin Press, p. 809–816.

Bradshaw, J. D., Andrews, P. B., Field, B. D., 1983, Swanson Formation and related rocks of Marie Byrd Land and a comparison with the Robertson Bay Group of northern Victoria Land, *in* Oliver, R. L., James, P. R., and Jago, J. B., eds., Antarctic Earth science: Canberra, Australian Academy of Science, p. 274–279.

Bradshaw, M. A., 1981, Palaeoenvironmental interpretation and systematics of Devonian trace fossils from the Taylor Group (lower Beacon Supergroup), Antarctica: New Zealand Journal of Geology and Geophysics, v. 24, p. 615–652.

Bradshaw, M. A., 1987, Additional field interpretation of the Jurassic sequence at Carapace Nunatak and Coombs Hills, south Victoria Land, Antarctica: New Zealand Journal of Geology and Geophysics, v. 30, p. 37–49.

Bradshaw, M. A., 1991, The Devonian Pacific margin of Antarctica, *in* Thomson, M.R.A., Crame, J. A., and Thomson, J. W., eds., Geological evolution of Antarctica: Cambridge, Cambridge University Press, p. 193–197.

Bradshaw, M. A., and McCarton, L., 1983, The depositional environment of the Lower Devonian Horlick Formation, Ohio Range, *in* Oliver, R. L., James, P. R., and Jago, J. B., eds., Antarctic Earth science: Canberra, Australian Academy of Science, p. 238–241.

Bradshaw, M. A., Newman, J., and Aitchison, J. C., 1984, Preliminary geological results of the 1983–84 Ohio Range Expedition: New Zealand Antarctic Record, v. 5(3), p. 1–17.

Brook, D., 1972, Stratigraphy of the Theron Mountains: British Antarctic

Figure 25. Photomicrographs of Triassic-Jurassic rocks. A: Early Triassic quartzose sandstone with quartz (Q), microcline (M), and plagioclase (P). Sample AHF138, collected by J. W. Collinson, upper Feather Conglomerate (Fleming Member), Allan Hills, southern Victoria Land. Cross-polarized light. B: Early Triassic quartzose sandstone with quartz (Q), plagioclase (P), and K-feldspar (K). Sample BF001, collected by P. J. Barrett, lower Fremouw Formation, Fremouw Peak, Beardmore Glacier area, central Transantarctic Mountains. Cross-polarized light. C: Early Triassic volcaniclastic sandstone with pilotaxitic volcanic rock fragments (T), fine-grained felsic volcanic rock fragments (F), plagioclase (P) replaced by zeolite (P/Z), and laumontite (L) cement. Sample 86.23, collected by D. H. Elliot, lower Fremouw Formation, Mount Rosenwald, Shackleton Glacier area, central Transantarctic Mountains. Cross-polarized light. D: Middle Triassic volcaniclastic sandstone with pilotaxitic volcanic rock fragments (T), fine-grained felsic volcanic rock fragments (F), granophyric rock fragments (G), quartz (Q) with overgrowths, plagioclase (P), and laumontite (L) after plagioclase. Sample AHL15, collected by J. W. Collinson, Lashly Formation, Member C, Allan Hills, southern Victoria Land. Cross-polarized light. E: Middle Triassic volcaniclastic sandstone with scattered glass shards (S) replaced by zeolite (laumontite), plagioclase strongly altered to laumontite (P/Z), and pilotaxitic volcanic rock fragment (T). Sample VC224.8, collected by C. L. Vavra, middle Fremouw Formation, Shenk Peak, Shackleton Glacier area, central Transantarctic Mountains. Plane-polarized light. F: Jurassic vitric-crystal tuff with numerous tricuspate and bubble-wall shards. Sample 85.19.2, collected by D. H. Elliot, upper Falla Formation, Mount Falla, Beardmore Glacier area, central Transantarctic Mountains. Plane-polarized light.

Survey Bulletin, v. 29, p. 67–89.

Buggisch, W., Kleinschmidt, G., Kreuzer, H., and Krumm, S., 1990, Stratigraphy, metamorphism and nappe-tectonics in the Shackleton Range (Antarctica): Berlin, Geodätische und geophysikalische Veröffentlichungen, Reihe I, v. 15, p. 64–86.

Burn, R. W., 1984, The geology of the Le May Group, Alexander Island: British Antarctic Survey Science Report, v. 109, 65 p.

Burrett, C. F., and Findlay, R. H., 1984, Cambrian and Ordovician conodonts from the Robertson Bay Group, Antarctica and their tectonic significance: Nature, v. 307, p. 723–726.

Cathcart, J. B., and Schmidt, D. L., 1977, Middle Paleozoic sedimentary phosphate in the Pensacola Mountains, Antarctica: Contributions to the Geology of Antarctica, U.S. Geological Survey Professional Paper 456E, p. 1–18.

Clarkson, P. D., 1981, Geology of the Shackleton Range: II, the Turnpike Bluff Group: Bulletin of the British Antarctic Survey, v. 52, p. 109–124.

Clarkson, P. D., and Brook, M., 1977, Age and position of the Ellsworth Mountains crustal fragment, Antarctica: Nature, v. 265, p. 615–616.

Clarkson, P. D., Hughes, C. P., and Thomson, M.R.A., 1979, Geological significance of a Middle Cambrian fauna from Antarctica: Nature, v. 279, p. 791–792.

Coates, D. A., 1985, Late Paleozoic glacial patterns in the central Transantarctic Mountains, Antarctica, *in* Turner, M. D., and Splettstoesser, J. F., eds., Geology of the central Transantarctic Mountains: Washington, D.C., Antarctic Research Series, American Geophysical Union, v. 36, no. 13, p. 325–338.

Colbert, E. H., 1982, Triassic vertebrates in the Transantarctic Mountains, *in* Turner, M. D., and Splettstoesser, J. F., eds., Geology of the central Transantarctic Mountains: Washington, D.C., Antarctic Research Series, American Geophysical Union, v. 36, p. 339–429.

Coleman, J. M., 1969, Brahmaputra River: Channel processes and sedimentation: Sedimentary Geology, v. 3, p. 129–239.

Collinson, J. W., 1990, Depositional setting of Late Carboniferous to Triassic biota in the Transantarctic Basin, *in* Taylor, T. N., and Taylor, E. L., eds., Antarctic paleobiology: Its role in the reconstruction of Gondwana: New York, Springer-Verlag, p. 1–14.

Collinson, J. W., 1991, The palaeo-Pacific margin as seen from East Antarctica, in Thomson, M.R.A., Crame, J. A., Thomson, J. W., eds., Geological evolution of Antarctica: Cambridge, Cambridge University Press, p. 199–204.

Collinson, J. W., and Elliot, D. H., 1984a, Geology of Coalsack Bluff, Antarctica, in Turner, M. D., and Splettstoesser, J. F., eds., Geology of the central Transantarctic Mountains: Washington, D.C., Antarctic Research Series, American Geophysical Union, v. 36, no. 6, p. 97–102.

Collinson, J. W., and Elliot, D. H., 1984b, Triassic stratigraphy of the Shackleton Glacier area, in Turner, M. D., and Splettstoesser, J. F., eds., Geology of the central Transantarctic Mountains: Washington, D.C., Antarctic Research Series, American Geophysical Union, v. 36, no. 6, p. 103–117.

Collinson, J. W., and Miller, M. F., 1991, Comparison of Lower Permian post-glacial black shale sequences in the Ellsworth and Transantarctic Mountains, Antarctica, *in* Ulbrich, H., and Rocha-Campos, A. C., eds., Gondwana Seven Proceedings: Papers presented at the Seventh International Gondwana Symposium: São Paulo, Brazil, Instituto de Geociecias de Universidad de São Paulo, p. 217–231.

Collinson, J. W., Stanley, K. O., and Vavra, C. L., 1981, Triassic fluvial depositional systems in the Fremouw Formation, Cumulus Hills, Antarctica, *in* Cresswell, M. M., and Vella, P., eds., Gondwana Five: Selected papers and abstracts of papers presented at the Fifth International Gondwana Symposium, Wellington, 1980: Rotterdam, A. A. Balkema, p. 141–148.

Collinson, J. W., Pennington, D. C., and Kemp, N. R., 1983, Sedimentary petrology of Permian-Triassic fluvial rocks in Allan Hills, central Victoria Land: Antarctic Journal of the United States, v. 18(5), p. 20–22.

Collinson, J. W., Pennington, D. C., and Kemp, N. R., 1986, Stratigraphy and

petrology of Permian and Triassic fluvial deposits in northern Victoria Land, Antarctica, *in* Stump, E., ed., Geological investigations in northern Victoria Land: Washington, D.C., Antarctic Research Series, American Geophysical Union, v. 46, p. 211–242.

Collinson, J. W., Kemp, N. R., Eggert, J. T., 1987, Comparison of the Triassic Gondwana Sequences in the Transantarctic Mountains and Tasmania, in McKenzie, G. D., ed., Gondwana Six: Stratigraphy, sedimentology, and paleontology: Washington, D.C., Geophysical Monograph Series, American Geophysical Union, v. 41, p. 51–61.

Collinson, J. W., Vavra, C. L., and Zawiskie, J. M., 1992, Sedimentology of the Polarstar Formation, Permian, Ellsworth Mountains, Antarctica, *in* Webers, G. F., Craddock, C., and Splettstoesser, J. F., eds., Geology of the Ellsworth Mountains, Antarctica: Boulder, Colorado, Geological Society of America Memoir 170, p. 63–79.

Cooper, A. K., Davey, F. J., and Behrendt, J. C., 1987, Seismic stratigraphy and structure of the Victoria Land basin, western Ross Sea, Antarctica, *in* Cooper, A. K., and Davey, F. J., eds., The Antarctic continental margin: Geology and geophysics of the western Ross Sea: Houston, Earth Science Series, Circum-Pacific Council for Energy and Mineral Resources, v. 5B, p. 27–76.

Cooper, A. K., Davey, F. J., and Hinz, K., 1991, Crustal extension and origin of sedimentary basins beneath the Ross Sea and Ross Ice Shelf, Antarctica, *in* Thomson, M.R.A., Crame, J. A., and Thomson, J. W., eds., Geological evolution of Antarctica: Cambridge, Cambridge University Press, p. 285–291.

Cooper, R. A., Jago, J. B., Rowell, A. J., and Braddock, P., 1983, Age and correlation of the Cambrian-Ordovician Bowers Supergroup, northern Victoria Land, *in* Oliver, R. L., James, P. R., and Jago, J. B., eds., Antarctic Earth science: Canberra, Australian Academy of Science, p. 128–131.

Cosgriff, J. W., Hammer, W. R., and Ryan, W. J., 1982, The pangaean reptile *Lystrosaurus maccaigi,* in the Lower Triassic of Antarctica: Journal of Paleontology, v. 56, p. 371–385.

Crowley, T. J., Hyde, W. T., and Short, D. A., 1989, Seasonal cycle variations on the supercontinent of Pangaea: Geology, v. 17, p. 457–460.

Dalziel, I.W.D., 1982, The early (pre-Middle Jurassic) history of the Scotia Arc region: A review and progress report, *in* Craddock, C., ed., Antarctic geoscience: Madison, University of Wisconsin Press, p. 111–126.

Dalziel, I.W.D., 1991, Pacific margins of Laurentia and East Antarctica-Australia as a conjugate rift pair: Evidence and implications for an Eocambrian supercontinent: Geology, v. 19(6), p. 598–601.

Dalziel, I.W.D., and Elliot, D. H., 1982, West Antarctica: Problem child of Gondwanaland: Tectonics, v. 1, p. 3–19.

Dalziel, I.W.D., Elliot, D. H., Jones, D. L., Thomson, J. W., Thomson, M.R.A., Wells, N. A., and Zinsmeister, W. J., 1981, The geological significance of some Triassic microfossils from the South Orkney Islands, Scotia Ridge: Geology Magazine, v. 118, p. 15–25.

Denham, C. R., and Scotese, C. R., 1988, Program Terra Mobilis Version 2.0, Austin, Texas.

Devereux, I., McDougall, I., and Watters, W. A., 1968, Potassium-argon mineral dates on intrusive rocks from the Foveaux Strait area: New Zealand Journal of Geology and Geophysics, v. 11, p. 1230–1235.

Dolby, J., and Balme, B. E., 1976, Triassic palynology of the Carnarvon Basin, Western Australia: Review of Palaeobotany and Palynology, v. 22, p. 105–168.

Doumani, G. A., Boardman, G. S., Rowell, A. J., Boucot, A. J., Johnson, J. G., McAlester, A. L., Saul, J., Fisher, D. V., and Miles, R. S., 1965, Lower Devonian fauna of the Horlick Formation, Ohio Range, Antarctica, *in* Hadley, J. B., ed., Geology and paleontology of the Antarctic: Washington, D.C., American Geophysical Union, p. 241–281.

Drewry, D. J., 1972, Subglacial morphology between the Transantarctic Mountains and the South Pole, *in* Adie, R. J., ed., Antarctic geology and geophysics: Oslo, Universitetsforlaget, p. 693–722.

Drewry, D. J., ed., 1983, Antarctica: Glaciological and geophysical folio:

Cambridge, Scott Polar Research Institute, Cambridge University, 9 sheets.

Edwards, G. W., 1982, New paleontologic evidence of Triassic sedimentation in West Antarctica, in Craddock, C., ed., Antarctic geoscience: Madison, University of Wisconsin Press, p. 325–329.

Elliot, D. H., 1975, Gondwana basins in Antarctica, in Campbell, K.S.W., ed., Gondwana geology: Canberra, Australian National University Press, p. 493–536.

Elliot, D. H., 1983, The mid-Mesozoic to mid-Cenozoic active plate margin of the Antarctic Peninsula, in Oliver, R. L., James, P. R., and Jago, J. B., eds., Antarctic Earth science: Canberra, Australian Academy of Science, p. 347–351.

Elliot, D. H., 1991, Triassic-Early Cretaceous evolution of Antarctica, in Thomson, M.R.A., Crame, J. A., and Thomson, J. W., eds., Geological evolution of Antarctica: Cambridge, Cambridge University Press, p. 541–548.

Elliot, D. H., 1992, Jurassic magmatism and tectonism associated with Gondwanaland break-up: An Antarctic perspective, in Storey, B. C., Alabaster, T., and Pankhurst, R. J., eds., Magmatism and the causes of continental breakup: Bath, Geological Society (London) Special Publication 68, p. 165–184.

Elliot, D. H., and Foland, K. A., 1986, Potassium-argon age determinations of the Kirkpatrick Basalt, Mesa Range, in Stump, E., ed., Geological investigations in Northern Victoria Land: Washington, D.C., Antarctic Research Series, American Geophysical Union, v. 46, p. 279–288.

Elliot, D. H., and Larsen, D., 1993, Mesozoic volcanism in the Transantarctic Mountains: Depositional environment and tectonic setting, in Findlay, R. H., Banks, M. R., Veevers, J. J., and Unrug, R., eds., Gondwana Eight: Assembly, evolution, and dispersal: Rotterdam, A. A. Balkema, p. 397–410.

Elliot, D. H., and Watts, D. R., 1974, The nature and origin of volcaniclastic material in some Karroo and Beacon rocks: Transactions of the Geological Society of South Africa, v. 77, p. 109–111.

Elliot, D. H., Fleck, R. J., and Sutter, J. F., 1985, Potassium-argon age determination of Ferrar Group rocks, central Transantarctic Mountains, in Turner, M. D., and Splettstoesser, J. F., eds., Geology of the central Transantarctic Mountains: Washington, D.C., Antarctic Research Series, American Geophysical Union, v. 36, p. 197–223.

Elliot, D. H., Haban, M. A., and Siders, M. A., 1986a, The Exposure Hill Formation, in Stump, E., ed., Geological investigations in Northern Victoria Land: Washington, D.C., Antarctic Research Series, American Geophysical Union, v. 46, p. 267–278.

Elliot, D. H., Siders, M. A., and Haban, M. A., 1986b, Jurassic tholeiites in the region of the upper Rennick Glacier, north Victoria Land, in Stump, E., ed., Geological investigations in Northern Victoria Land: Washington, D.C., Antarctic Research Series, American Geophysical Union, v. 46, p. 249–265.

Evans, K. R., and Rowell, A. J., 1990, Small shelly fossils from Antarctica: An Early Cambrian faunal connection with Australia: Journal of Paleontology, v. 64(5), p. 692–700.

Evans, P. R., 1969, Upper Carboniferous and Permian palynological stages and their distribution in eastern Australia, in Amos, A. J., ed., Gondwanan Stratigraphy: Paris, UNESCO, p. 41–54.

Farabee, M. J., Taylor, E. L., and Taylor, T. N., 1989, Pollen and spore assemblages from the Falla Formation (Upper Triassic), Central Transantarctic Mountains, Antarctica: Review of Palaeobotany and Palynology, v. 61, p. 101–138.

Farabee, M. J., Taylor, E. L., and Taylor, T. N., 1991, Late Permian palynomorphs from the Buckley Formation, central Transantarctic Mountains, Antarctica: Review of Palaeobotany and Palynology, v. 69, p. 353–368.

Farquharson, G. W., 1983, Evolution of late Mesozoic sedimentary basins in the northern Antarctic Peninsula, in Oliver, R. L., James, P. R., and Jago, J. B., eds., Antarctic Earth science: Canberra, Australian Academy of Science, p. 323–327.

Field, B. D., and Findlay, R. H., 1983, The sedimentology of the Robertson Bay Group, northern Victoria Land, in Oliver, R. L., James, P. R., and Jago, J. B., eds., Antarctic Earth science: Canberra, Australian Academy of Science, p. 102–106.

Findlay, R. H., 1986, Structural geology of the Robertson Bay and Millen terranes, northern Victoria Land, Antarctica, in Stump, E., ed., Geological investigations in Northern Victoria Land: Washington, D.C., Antarctic Research Series, American Geophysical Union, v. 46, p. 91–114.

Findlay, R. H., 1989, Silurian and Devonian events in the Tasman orogenic zone, New Zealand and Marie Byrd Land and their comparisons with northern Victoria Land, in Proceedings of the 2nd meeting, Earth Science in Antarctica, Siena, 27–28 September 1988, v. 43, p. 9–32, refs. p. 30–32.

Findlay, R. H., Skinner, D.N.B., and Graw, D., 1984, Lithostratigraphy and structure of the Koettlitz Group, McMurdo Sound, Antarctica: New Zealand Journal of Geology and Geophysics, v. 27, p. 513–536.

Fleming, T. H., Foland, K. A., Elliot, D. H., and Jones, L. M., 1991, A new and improved 40Ar/39Ar age determination of iron-rich tholeiite of the Kirkpatrick basalt from the Mesa Range, north Victoria Land, Antarctica: Eos, Transactions, American Geophysical Union, Spring Meeting 1991, v. 72(17), p. 295.

Fleming, T. H., Elliot, D. H., Foland, K. A., Jones, L. M., and Bowman, J. R., 1993, Disturbance of Rb-Sr and K-Ar isotopic systems in the Kirkpatrick Basalt, northern Victoria Land, Antarctica: Implications for mid-Cretaceous tectonism, in Findlay, R. H., Banks, M. R., Veevers, J. J., and Unrug, R., eds., Gondwana Eight: Assembly, evolution, and dispersal: Rotterdam, A. A. Balkema, p. 411–424.

Flöttmann, T., and Kleinschmidt, G., 1991, Opposite thrust systems in Northern Victoria Land, Antarctica: Imprints of Gondwana's Paleozoic accretion: Geology, v. 19, p. 45–47.

Ford, A. B., and Kistler, R. W., 1980, K-Ar age, composition, and origin of Mesozoic mafic rocks related to Ferrar Group, Pensacola Mountains, Antarctica: New Zealand Journal of Geology and Geophysics, v. 23, p. 371–390.

Frakes, L. A., 1981, Late Paleozoic tillites near the southern Ross Ice Shelf, Antarctica, in Hambrey, J. J., and Harland, W. B., eds., Earth's pre-Pleistocene glacial record: Cambridge University Press, p. 230–232.

Frakes, L. A., and Crowell, J. C., 1975, Characteristics of modern glacial marine sediments: Application to Gondwana glacials, in Campbell, K.S.W., ed., Gondwana geology: Canberra, Australian National University Press, p. 373–380.

Frakes, L. A., Matthews, J. L., and Crowell, J. C., 1971, Late Paleozoic glaciation: Part III, Antarctica: Bulletin of the Geological Society of America, v. 82, p. 1581–1604.

Frisch, R. S., and Miller, M. F., 1991, Provenance and tectonic implications of sandstones within the Permian Mackellar Formation, Beacon Supergroup of East Antarctica, in Thomson, M.R.A., Crame, J. A., and Thomson, J. W., eds., Geological evolution of Antarctica: Cambridge, Cambridge University Press, p. 219–223.

Gair, H. S., Sturm, A., Carryer, S. J., and Grindley, G. W., 1969, The geology of northern Victoria Land, in Bushnell, V. C., and Craddock, C., Antarctic map folio series, folio 12, plate XII: New York, American Geographical Society.

Gevers, T. W., and Twomey, A., 1982, Trace fossils and their environment in Devonian (Silurian?) lower Beacon strata in the Asgard Range, Victoria Land, Antarctica, in Craddock, C., ed., Antarctic geoscience: Madison, University of Wisconsin Press, p. 639–647.

Gibson, G. M., and Wright, T. O., 1985, Importance of thrust faulting in the tectonic development of northern Victoria Land: Nature, v. 315, p. 480–483.

Gleadow, A.J.W., and Fitzgerald, P. G., 1987, Uplift history and structure of the Transantarctic Mountains: New evidence from fission track dating of basement apatites in the Dry Valleys area, southern Victoria Land: Earth and Planetary Science Letters, v. 82, p. 1–14.

Goodge, J. W., Borg, S. C., Smith, B. K., and Bennett, V. C., 1991, Tectonic significance of Proterozoic ductile shortening and translation along the

Antarctic margin of Gondwana: Earth and Planetary Science Letters, v. 102, p. 58–70.

Grew, E. S., and Sandiford, M., 1985, Staurolite in a garnet-hornblende-biotite schist from the Lanterman Range, northern Victoria Land, Antarctica: Neues Jahrbuch für Mineralogie, Monatshefte 1985(9), p. 396–410.

Grew, E. S., Kleinschmidt, G., Schubert, W., 1984, Contrasting metamorphic belts in North Victoria Land, Antarctica: Geologisches Jahrbuch, Reihe B, v. 60, p. 253–263.

Grindley, G. W., 1963, The geology of the Queen Alexandra Range, Beardmore Glacier, Ross Dependency, Antarctica; with notes on the correlation of Gondwana Sequences: New Zealand Journal of Geology and Geophysics, v. 6, p. 307–347.

Grindley, G. W., and Laird, M. G., 1969, Geology of the Shackleton Coast, scale 1:1,000,000, in Bushnell, V. C., and Craddock, C., Antarctic map folio series, folio 12, plate XV: New York, American Geographical Society.

Grindley, G. W., and McDougall, I., 1969, Age and correlation of the Nimrod Group and other Precambrian rock units in the central Transantarctic Mountains, Antarctica: New Zealand Journal of Geology and Geophysics, v. 12, p. 391–411.

Grindley, G. W., and Mildenhall, D. C., 1983, Geological background to a Devonian plant fossil discovery, in Oliver, R. L., James, P. R., and Jago, J. B., eds., Antarctic Earth science: Canberra, Australian Academy of Science, p. 23–30.

Grindley, G. W., and Oliver, P. J., 1983, Post-Ross orogeny cratonization of northern Victoria Land, in Oliver, R. L., James, P. R., and Jago, J. B., eds., Antarctic Earth science: Canberra, Australian Academy of Science, p. 133–139.

Grindley, G. W., McGregor, V. R., and Walcott, R. I., 1964, Outline of the geology of the Nimrod-Beardmore-Axel Heiberg Glaciers region, Ross Dependency, in Adie, R. J., Antarctic geology, in Proceedings of the First International Symposium on Antarctic Geology, Capetown, 16–21 September 1963, p. 206–219.

Grindley, G. W., Mildenhall, D. C., and Schopf, J. M., 1980, A mid-Late Devonian flora from the Ruppert Coast, Marie Byrd Land, West Antarctica: Journal of the Royal Society of New Zealand, v. 10, p. 271–285.

Grunow, A. M., Kent, D. V., and Dalziel, I.W.D., 1991, New paleomagnetic data from Thurston Island: Implications for the tectonics of West Antarctica and Weddell Sea opening: Journal of Geophysical Research, v. 96, p. 17,935–17,954.

Gunn, B. M., and Warren, G., 1962, Geology of Victoria Land between the Mawson and Mulock Glaciers, Antarctica: New Zealand Geological Survey Bulletin, v. 71, 157 p.

Hälbich, I. W., 1983, A tectogenesis of the Cape Fold Belt (CFB), in Söhnge, A.P.G., and Hälbich, I. W., eds., Geodynamics of the Cape Fold Belt: Johannesburg, Special Publications of the Geological Society of South Africa, v. 12, p. 165–175.

Halpern, M., 1972, Rb-Sr total-rock and mineral ages from the Marguerite Bay area, Kohler Range and Fosdick Mountains, in Adie, R. J., ed., Antarctic geology and geophysics: Oslo, Universitetsforlaget, p. 197–204.

Hammer, W. R., 1990, Triassic terrestrial vertebrate faunas of Antarctica, in Taylor, T. N., and Taylor, E. L., eds., Antarctic paleobiology, its role in the reconstruction of Gondwana: New York, Springer-Verlag, p. 15–26.

Hammer, W. R., Collinson, J. W., and Ryan, W. J., Jr., 1990, A new Triassic vertebrate fauna from Antarctica and its depositional setting: Antarctic Science, v. 2(2), p. 163–167.

Hammer, W. R., Hickerson, W. J., Tamplin, J., and Krippner, S., 1991, Therapsids, temnospondyls and dinosaurs from the Fremouw and Falla formations, Beardmore Glacier region, Antarctica: Antarctic Journal of the United States, v. 26(5), p. 19–20.

Hayes, D. E., and Frakes, L. A., 1975, General synthesis: Deep Sea Drilling Project 28, in Hayes, D. E., and Frakes, L. A., eds., Initial reports of the Deep Sea Drilling Project: Washington, D.C., U.S. Government Printing Office, v. 28, p. 919–942.

Heimann, A., Fleming, T. H., Foland, K. A., and Elliot, D. H., 1992, $^{40}Ar/^{39}Ar$ geochronology of the Kirkpatrick Basalt, Transantarctic Mountains, Antarctica: Distribution, time of emplacement, and tectonic implications: Eos, v. 73, p. 279.

Helby, R. J., and McElroy, C. T., 1969, Microfloras from the Devonian and Triassic of the Beacon Group, Antarctica: New Zealand Journal of Geology and Geophysics, v. 12, p. 376–382.

Hergt, J. M., Peate, D. W., and Hawkesworth, C. J., 1991, The petrogenesis of Mesozoic Gondwanan low-Ti flood basalts: Earth and Planetary Science Letters, v. 105, p. 134–148.

Hill, D., 1964, Archaeocyatha from the Shackleton Limestone of the Ross System, Nimrod Glacier area, Antarctica: Transactions of the Royal Society of New Zealand, Geology 2, p. 137–146.

Horner, T. C., and Krissek, L. A., 1991, Permian and Triassic paleosols from the Beardmore Glacier region, Antarctica: Antarctic Journal of the United States, v. 26, p. 7–8.

Isbell, J. L., 1990a, Permian fluvial sedimentology of the Transantarctic basin [Ph.D. thesis]: Columbus, Ohio, Ohio State University, 306 p.

Isbell, J. L., 1990b, Depositional architecture of the Lower Permian Weller Coal Measures, southern Victoria Land: Antarctic Journal of the United States, v. 25(5), p. 28–29.

Isbell, J. L., 1991, Evidence for a low-gradient alluvial fan from the palaeo-Pacific margin in the Upper Permian Buckley Formation, Beardmore Glacier region, Antarctica, in Thomson, M.R.A., Crame, J. A., and Thomson, J. W., eds., Geological evolution of Antarctica: Cambridge, Cambridge University Press, p. 215–217.

Isbell, J. L., and Collinson, J. W., 1988, Fluvial architecture of the Fairchild and Buckley Formations (Permian), Beardmore Glacier area: Antarctic Journal of the United States, v. 23(5), p. 3–5.

Isbell, J. L., Taylor, T. N., Taylor, E. L., Cuneo, N. R., and Meyer-Berthaud, B., 1990, Depositional setting of Permian and Triassic fossil plants in the Allan Hills, southern Victoria Land: Antarctic Journal of the United States, v. 25(5), p. 28–29.

Johnson, M. R., 1991, Sandstone petrography, provenance and plate tectonic setting in Gondwana context of the southeastern Cape-Karoo Basin: South African Journal of Geology, v. 94, p. 137–154.

Kemp, E. M., 1972a, Reworked palynomorphs from the West Ice Shelf area, East Antarctica, and their possible geological and palaeoclimatological significance: Marine Geology, v. 13, p. 145–157.

Kemp, E. M., 1972b, Recycled palynomorphs in continental shelf sediments from Antarctica: U.S. Antarctic Journal, v. 7, p. 190–191.

Kemp, E. M., 1975, The palynology of late Palaeozoic glacial deposits of Gondwanaland, in Campbell, K.S.W., ed., Gondwana geology: Canberra, Australian National University Press, p. 397–413.

Kemp, E. M., Balme, B. E.,Helby, R. J., Kyle, R. A., Playford, G., and Price, P. L., 1977, Carboniferous and Permian palynostratigraphy in Australia and Antarctica: A review: Australian Bureau of Mineral Resources Journal, v. 2, p. 177–208.

Kimbrough, D. L., Mattinson, J. M., Coombs, D. S., Landis, C. A., and Johnston, M. R., 1992, Uranium-lead ages from the Dun Mountain ophiolite belt and Brook Street terrane, South Island, New Zealand: Geological Society of America Bulletin, v. 104, p. 429–443.

Kleinschmidt, G., 1989, Die Shackleton Range im innern der Antarcktus: Die Geowissenschaften, no. 1, p. 10–12.

Kleinschmidt, G., and Skinner, D.N.B., 1981, Deformation styles in the basement rock of North Victoria Land, Antarctica: Geologisches Jahrbuch, B. 41, p. 155–199.

Kleinschmidt, G., and Tessensohn, F., 1987, Early Paleozoic westward directed subduction at the Pacific margin of Antarctica, in McKenzie, G. D., ed., Gondwana Six: Structure, tectonics, and geophysics: Washington, D.C., Geophysical Monograph Series, American Geophysical Union, v. 40, p. 89–105.

Knight, M. J., 1975, Recent crevassing of the Erap River, Papua New Guinea: Australian Geographical Studies, v. 13, p. 77–84.

Korsch, R. J., and Wellman, H. W., 1988, The geological evolution of New Zealand and the New Zealand region, *in* Nairn, A.E.M., Stehli, F. G., and Uyeda, S., eds., The ocean basins and margins: The Pacific Ocean: New York, Plenum, v. 7B, p. 411–482.

Krissek, L. A., and Horner, T. C., 1991, Sedimentology of a vertebrate-bearing bed in the Triassic Fremouw Formation at Gordon Valley, Beardmore Glacier region, Antarctica: Antarctic Journal of the United States, v. 26(5), p. 17–19.

Kuenzi, W. D., Horst, O. H., and McGehee, R. V., 1979, Effect of volcanic activity on fluvial-deltaic sedimentation in a modern arc-trench gap, southwestern Guatemala: Geological Society of America Bulletin, v. 90, p. 827–838.

Kyle, P. R., 1980, Development of heterogeneities in the subcontinental mantle: Evidence from the Ferrar Group, Antarctica: Contributions to Mineralogy and Petrology, v. 73(1), p. 89–104.

Kyle, P. R., Elliot, D. H., and Sutter, J. F., 1981, Jurassic Ferrar Supergroup tholeiites from the Transantarctic Mountains, Antarctica, and their relationship to the initial fragmentation of Gondwana, *in* Cresswell, M. M., and Vella, P., eds., Gondwana Five: Rotterdam, A. A. Balkema, p. 283–287.

Kyle, R. A., 1977, Palynostratigraphy of the Victoria Group of South Victoria Land, Antarctica: New Zealand Journal of Geology and Geophysics, v. 20, p. 1081–1102.

Kyle, R. A., and Fasola, A., 1978, Triassic palynology of the Beardmore Glacier area of Antarctica: Palinologia, v. 1, p. 313–319.

Kyle, R. A., and Schopf, J. M., 1982, Permian and Triassic palynostratigraphy of the Victoria Group, Transantarctic Mountains, *in* Craddock, C., ed., Antarctic geoscience: Madison, University of Wisconsin Press, International Union of Geological Sciences, Series B-4, p. 649–659.

Laird, M. G., 1963, Geomorphology and stratigraphy of the Nimrod Glacier-Beaumont Bay region, southern Victoria Land, Antarctica: New Zealand Journal of Geology and Geophysics, v. 6, p. 465–484.

Laird, M. G., 1981, Lower Palaeozoic rocks of the Ross Sea area and their significance in the Gondwana context: Journal of the Royal Society of New Zealand, v. 11, p. 425–438.

Laird, M. G., 1988, Evolution of the Cambrian-Early Ordovician Bowers basin, northern Victoria Land; and its relationship with the adjacent Wilson and Robertson Bay terranes, *in* Proceedings of the meeting, Geosciences in Victoria Land, Antarctica, Siena, 2–3 September 1987: Memorie della Società Geologica Italiana, v. 33, p. 25–34.

Laird, M. G., and Bradshaw, J. D., 1981, Permian tillites of north Victoria Land, *in* Hambrey, J. J., and Harland, W. B., eds., Earth's pre-Pleistocene glacial record: Cambridge University Press, p. 237–240.

Laird, M. G., and Bradshaw, J. D., 1983, New data on the lower Palaeozoic Bowers Supergroup, northern Victoria Land, *in* Oliver, R. L., James, P. R., and Jago, J. B., eds., Antarctic Earth science: Canberra, Australian Academy of Science, p. 123–126.

Laird, M. G., Mansergh, G. D., and Chappell, J.M.A., 1971, Geology of the central Nimrod Glacier area, Antarctica: New Zealand Journal of Geology and Geophysics, v. 14, p. 427–468.

Laird, M. G., Bradshaw, J. D., and Wodzicki, A., 1982, Stratigraphy of the Upper Precambrian and Lower Paleozoic Bowers Supergroup, northern Victoria Land, Antarctica, *in* Craddock, C., ed., Antarctic geoscience: Madison, University of Wisconsin Press, p. 535–542.

Laudon, T. S., 1991, Petrology of sedimentary rocks from the English Coast, eastern Ellsworth Land, *in* Thomson, M.R.A., Crame, J. A., and Thomson, J. W., eds., Geological evolution of Antarctica: Cambridge, Cambridge University Press, p. 455–460.

Laudon, T. S., Lidke, D. J., Delevoryas, T., and Gee, C. T., 1987, Sedimentary rocks of the English Coast, Eastern Ellsworth Land, Antarctica, *in* McKenzie, G. D., ed., Gondwana Six: Structure, tectonics, and geophysics: Washington, D.C., Geophysical Monograph Series, American Geophysical Union, v. 41, p. 183–192.

Lemoigne, Y., 1987, Confirmation de l'existence d'une flore triasique dans l'Ile Livingston des Shetland du sud (Ouest Antarctique): Paris, Académie des Sciences, Comptes rendus hebdomadaires des séances, Série II, Mars 14, 1987, v. 304(10), p. 543–546.

Lindsay, J. F., 1970, Depositional environment of Paleozoic glacial rocks in the central Transantarctic Mountains: Geological Society of America Bulletin, v. 81, p. 1149–1172.

Long, W. E., 1964, The stratigraphy of the Horlick Mountains, *in* Adie, R. J., International symposium on Antarctic geology: Amsterdam, North-Holland Publishing, p. 352–363.

Long, W. E., 1965, Stratigraphy of the Ohio Range, Antarctica, in Hadley, J. B., ed., Geology and paleontology of the Antarctic: Washington, D.C., Antarctic Research Series, American Geophysical Union, v. 6, p. 71–116.

Lopatin, B. G., and Polyakov, M. M., 1976, Geologiia Zemli Meri Berd i Berega Eitsa (Zapadnaia Antarktida): Moscow, Isdatel'stvo Nauka, 175 p.

Macdonald, D.I.M., and Butterworth, P. J., 1990, The stratigraphy, setting, and hydrocarbon potential of the Mesozoic sedimentary basins of the Antarctic Peninsula, in St. John, B., ed., Antarctica as an exploration frontier: American Association of Petroleum Geologists, Studies in Geology, v. 31, p. 101–125.

Macdonald, D.I.M., Isbell, J. L., and Hammer, W. R., 1991a, Vertebrate trackways from the Triassic Fremouw Formation, Queen Alexandra Range, Antarctica: Antarctic Journal of the United States, v. 26(5), p. 20–22.

Macdonald, D.I.M., Storey, B. C., Dalziel, I.W.D., Grunow, A. M., and Isbell, J. L., 1991b, Early Paleozoic sedimentation and tectonics in the Pensacola Mountains: Abstracts, 6th International Symposium on Antarctic Earth Sciences, National (Japan) Institute of Polar Research, Ranzan-machi, p. 381.

Marsh, P. D., 1983, The Late Precambrian and Early Palaeozoic history of the Shackleton Range, Coats Land, in Oliver, R. L., James, P. R., and Jago, J. B., eds., Antarctic Earth science: Canberra, Australian Academy of Science, p. 190–193.

Martini, J.E.J., 1974, On the presence of ash beds and volcanic fragments in the graywackes of the Karroo System in the southern Cape Province (South Africa): Transactions of the Geological Society of South Africa, v. 77, p. 113–116.

Matsch, C. L., and Ojakangas, R. W., 1992, Stratigraphy and sedimentology of the Whiteout Conglomerate—A late Paleozoic glacigenic sequence in the Ellsworth Mountains, West Antarctica, *in* Webers, G. F., Craddock, C., and Splettstoesser, J. F., eds., Geology of the Ellsworth Mountains, Antarctica: Boulder, Colorado, Geological Society of America Memoir 170, p. 37–62.

McElroy, C. T., and Rose, G., 1987, Geology of the Beacon Heights area, southern Victoria Land, Antarctica: New Zealand Geological Survey Miscellaneous Series Map 15 and Notes, 47 p., scale 1:50,000.

McKelvey, B. C., Webb, P. N., Gorton, M., and Kohn, B. P., 1972, Stratigraphy of the Beacon Supergroup between the Olympus and Boomerang Ranges, Victoria Land, *in* Adie, R. J., ed., Antarctic geology and geophysics: Oslo, Universitetsforlaget, p. 345–358.

Millar, I. L., and Pankhurst, R. J., 1987, Rb-Sr geochronology of the region between the Antarctic Peninsula and the Transantarctic Mountains: Haag Nunataks and Mesozoic granitoids, *in* McKenzie, G. D., ed., Gondwana Six: Structure, tectonics, and geophysics: Washington, D.C., Geophysical Monograph Series, American Geophysical Union, v. 40, p. 151–160.

Miller, J.M.G., 1989, Glacial advance and retreat sequences in a Permo-Carboniferous section, central Transantarctic Mountains: Sedimentology, v. 36, p. 419–430.

Miller, J.M.G., and Waugh, B., 1986, Sedimentology of the Pagoda Formation (Permian), Beardmore Glacier area: Antarctic Journal of the United States, v. 21(5), p. 45–46.

Miller, J.M.G., and Waugh, B., 1991, Permo-Carboniferous glacial sedimentation in the central Transantarctic Mountains and its palaeotectonic implications, *in* Thomson, M.R.A., Crame, J. A., and Thomson, J. W., eds., Geological evolution of Antarctica: Cambridge, Cambridge University Press, p. 205–208.

Miller, M. F., Collinson, J. W., and Frisch, R. S., 1991, Depositional setting and history of a Permian post-glacial shale, Mackellar Formation, Beardmore Glacier area, Antarctica, in Ulbrich, H., and Rocha-Campos, A. C., eds., Gondwana Seven Proceedings: Papers presented at the Seventh International Gondwana Symposium: Sâo Paulo, Brazil, Instituto de Geociências de Universidade de Sâo Paulo, p. 201–215.

Milne, A. J., and Millar, I. L., 1989, The significance of mid-Palaeozoic basement in Graham Land, Antarctic Peninsula: Journal of the Geological Society of London, v. 146, p. 207–210.

Minshew, V. H., 1967, Geology of the Scott Glacier and Wisconsin Range areas, Central Transantarctic Mountains, Antarctica [Ph.D. thesis]: Columbus, Ohio, Ohio State University, 268 p.

Moores, E. M., 1991, Southwest U.S.–East Antarctic (SWEAT) connection: A hypothesis: Geology, v. 19, p. 425–428.

Nelson, W. H., 1981, The Gale Mudstone: A Permian(?) tillite in the Pensacola Mountains and neighboring parts of the Transantarctic Mountains, in Hambrey, J. J., and Harland, W. B., eds., Earth's pre-Pleistocene glacial record: Cambridge, Cambridge University Press, p. 227–232.

Norris, G., 1965, Triassic and Jurassic miospores and acritarchs from the Beacon and Ferrar Groups, Victoria Land, Antarctica: New Zealand Journal of Geology and Geophysics, v. 8, p. 236–277.

Ojakangas, R. W., and Matsch, C. L., 1981, The late Paleozoic Whiteout Conglomerate: A glacial and glaciomarine sequence in the Ellsworth Mountains, West Antarctica, in Hambrey, J. J., and Harland, W. B., eds., Earth's pre-Pleistocene glacial record: Cambridge, Cambridge University Press, p. 241–244.

Palmer, A. R., and Gatehouse, C. G., 1972, Early and Middle Cambrian trilobites from Antarctica: U.S. Geological Survey Professional Paper 456-D, 37 p.

Pankhurst, R. J., 1983, Rb-Sr constraints on the ages of basement rocks of the Antarctic Peninsula, in Oliver, R. L., James, P. R., and Jago, J. B., eds., Antarctic Earth science: Canberra, Australian Academy of Science, p. 367–371.

Pankhurst, R. J., 1990, The Paleozoic and Andean magmatic arcs of West Antarctica and southern South America, in Kay, S. M., and Rapela, C. W., eds., Plutonism from Antarctica to Alaska: Boulder, Colorado, Geological Society of America Special Paper 241, p. 1–7.

Pankhurst, R. J., Marsh, P. D., and Clarkson, P. D., 1983, A geochronological investigation of the Shackleton Range, in Oliver, R. L., James, P. R., and Jago, J. B., eds., Antarctic Earth science: Canberra, Australian Academy of Science, p. 176–182.

Pankhurst, R. J., Storey, B. C., Millar, I. L., Macdonald, D.I.M., 1988, Cambrian-Ordovician magmatism in the Thiel Mountains, Transantarctic Mountains, and implications for the Beardmore Orogeny: Geology, v. 16, p. 246–249.

Pankhurst, R. J., Storey, B. C., and Millar, I. L., 1991, Magmatism related to the break-up of Gondwana, in Thomson, M.R.A., Crame, J. A., and Thomson, J. W., eds., Geological evolution of Antarctica: Cambridge, Cambridge University Press, p. 573–579.

Parrish, J. T., 1990, Gondwanan paleogeography and paleoclimatology, in Taylor, T. N., and Taylor, E. L., eds., Antarctic paleobiology, its role in the reconstruction of Gondwana: New York, Springer-Verlag, p. 15–26.

Playford, G., 1990, Proterozoic and Paleozoic palynology of Antarctica: A review, in Taylor, T. N., and Taylor, E. L., eds., Antarctic paleobiology, its role in the reconstruction of Gondwana: New York, Springer-Verlag, p. 51–70.

Plume, R. W., 1982, Sedimentology and paleocurrent analysis of the basal part of the Beacon Supergroup (Devonian [and older] to Triassic) in South Victoria Land, Antarctica, in Craddock, C., ed., Antarctic geoscience: Madison, University of Wisconsin Press, p. 571–590.

Powell, C. McA., Roots, S. R., and Veevers, J. J., 1988, Pre-breakup continental extension in East Gondwanaland and the early opening of the eastern Indian Ocean: Tectonophysics, v. 155, p. 261–283.

Quilty, P. G., 1970, Jurassic ammonites from Ellsworth Land, Antarctica:

Journal of Paleontology, v. 44, p. 110–116.

Quilty, P. G., 1977, Late Jurassic bivalves from Ellsworth Land, Antarctica: Their systematics and palaeogeographic implications: New Zealand Journal of Geology and Geophysics, v. 44, p. 110–116.

Rees, M. N., and Rowell, A. J., 1991, The pre-Devonian Palaeozoic clastics of the central Transantarctic Mountains: Stratigraphy and depositional setting, in Thomson, M.R.A., Crame, J. A., and Thomson, J. W., eds., Geological evolution of Antarctica: Cambridge, Cambridge University Press, p. 187–192.

Rees, M. N., Pratt, B. R., and Rowell, A. J., 1989, Early Cambrian reefs, reef complexes, and associated lithofacies of the Shackleton Limestone, Transantarctic Mountains: Sedimentology, v. 36(2), p. 341–361.

Rees, P. M., and Smellie, J. L., 1989, Cretaceous angiosperms from an allegedly Triassic flora at Williams Point, Livingston Island, South Shetland Islands: Antarctic Science, v. 1, p. 239–248.

Rowell, A. J., and Rees, M. N., 1989, Early Palaeozoic history of the upper Beardmore Glacier area: Implications for a major Antarctic structural boundary within the Transantarctic Mountains: Antarctic Science, v. 1, p. 249–260.

Schaeffer, B., 1972, Jurassic fish from Antarctica: American Museum Novitates, no. 2495, p. 1–17.

Schmidt, D. L., and Ford, A. B., 1969, Geology of the Pensacola and Thiel Mountains, Folio 12—Geology: American Geographical Society, Antarctic Map Folio Series, plate V.

Schmidt, D. L., and Williams, P. L., 1969, Continental glaciation of late Paleozoic age, Pensacola Mountains, Antarctica, in Amos, A. J., ed., Gondwana stratigraphy: Paris, UNESCO, p. 617–644.

Schmidt, D. L., Williams, P. L., Nelson, W. H., and Ege, J. R., 1965, Upper Precambrian and Paleozoic stratigraphy and structure of the Neptune Range, Antarctica: U.S. Geological Survey Professional Paper 525-D, p. 112–119.

Schmidt-Thome, M., and Wolfart, R., 1984, Tremadocian fauna (trilobites, brachiopods) from Reilly Ridge, north Victoria Land, Antarctica: newsletter on Stratigraphy, v. 13, p. 88–93.

Schopf, J. M., 1968, Studies in Antarctic paleobotany: Antarctic Journal of the United States, v. 3(5), p. 176–177.

Schopf, J. M., 1969, Ellsworth Mountains: Position in West Antarctica due to sea-floor spreading: Science, v. 164, p. 63–66.

Schopf, J. M., 1973, Plant material from the Miers Bluff Formation of the South Shetland Islands: Ohio State University Institute of Polar Studies Report 45, 45 p.

Skinner, D.N.B., 1982, Stratigraphy and structure of low-grade metasedimentary rocks of the Skelton Group, southern Victoria Land—Does Teall graywacke really exist? in Craddock, C., ed., Antarctic geoscience: Madison, University of Wisconsin Press, p. 555–569.

Sloan, L. C., and Barron, E. J., 1990, "Equable" climates during Earth history: Geology, v. 18, p. 489–492.

Smellie, J. L., 1987, Sandstone detrital modes and basinal setting of the Trinity Peninsula Group, northern Graham Land, Antarctic Peninsula: Preliminary survey, in McKenzie, G. D., ed., Gondwana Six: Structure, tectonics, and geophysics: Washington, D.C., Geophysical Monograph Series, American Geophysical Union, v. 40, p. 199–207.

Smit, J. H., and Stump, E., 1986, Sedimentology of the La Gorce Formation, La Gorce Mountains, Antarctica: Journal of Sedimentary Petrology, v. 56, p. 663–668.

Spörli, K. B., 1992, Stratigraphy of the Crashsite Group, Ellsworth Mountains, West Antarctica, in Webers, G. F., Craddock, C., and Splettstoesser, J. F., eds., Geology of the Ellsworth Mountains, West Antarctica: Geological Society of America Memoir 170, p. 21–35.

Stern, T. A., and Ten Brink, U. S., 1989, Flexural uplift of the Transantarctic Mountains: Journal of Geophysical Research, v. 94(B8), p. 10,315–10,330.

Storey, B. C., and Nell, P.A.R., 1988, Role of strike-slip faulting in the tectonic evolution of the Antarctic Peninsula: Journal of the Geological Society of London, v. 145, p. 333–337.

Stump, E., 1982, The Ross Supergroup in the Queen Maud Mountains, *in* Craddock, C., ed., Antarctic geoscience: Madison, University of Wisconsin Press, p. 565–569.

Stump, E., and Fitzgerald, P. G., 1992, Episodic uplift of the Transantarctic Mountains: Geology, v. 20, p. 161–164.

Stump, E., Smit, J. H., and Self, S., 1986, Timing of events during the Late Proterozoic Beardmore Orogeny, Antarctica: Geological evidence from the La Gorce Mountains: Geological Society of America Bulletin, v. 97, p. 953–965.

Stump, E., Korsch, R. J., and Edgerton, D. G., 1991, The myth of the Nimrod and Beardmore orogenies, *in* Thomson, M.R.A., Crame, J. A., and Thomson, J. W., eds., Geological evolution of Antarctica: Cambridge, Cambridge University Press, p. 143–147.

Tasch, P., 1987, Fossil Conchostraca of the Southern Hemisphere and continental drift: Paleontology, biostratigraphy, and dispersal: Geological Society of America Memoir 165, 290 p.

Taylor, E. L., and Taylor, T. N., 1990, Structurally preserved Permian and Triassic floras from Antarctica, *in* Taylor, T. N., and Taylor, E. L., eds., Antarctic paleobiology, its role in the reconstruction of Gondwana: New York, Springer-Verlag, p. 149–163.

Taylor, E. L., Cuneo, R., and Taylor, T. N., 1991, Permian and Triassic fossil forests from the central Transantarctic Mountains: Antarctic Journal of the United States, v. 26(5), p. 23–24.

Taylor, E. L., Taylor, T. N., and Cuneo, R., 1992, The present is not the key to the past: A polar forest from the Permian of Antarctica: Science, v. 257, p. 1675–1677.

Taylor, T. N., and Taylor, E. L., 1992, Permian plants from the Ellsworth Mountains, West Antarctica, *in* Webers, G. F., Craddock, C., and Splettstoesser, J. F., eds., Geology of the Ellsworth Mountains, Antarctica: Boulder, Colorado, Geological Society of America Memoir 170, p. 285–294.

Taylor, T. N., Taylor, E. L., and Collinson, J. W., 1986, Paleoenvironments of Upper Triassic plants from the Fremouw Formation: Antarctic Journal of the United States, v. 21(5), p. 26–27.

Taylor, T. N., Taylor, E. L., and Farabee, M. J., 1988, Palynostratigraphy of the Buckley Formation (Permian), central Transantarctic Mountains: Antarctic Journal of the United States, v. 23(5), p. 21–23.

Tessensohn, F., 1984, Geological and tectonic history of the Bowers structural zone, north Victoria Land, Antarctica: Geologisches Jahrbuch, Reihe B, v. 60, p. 371–396.

Tessensohn, F., and Madler, K., 1987, Triassic plant fossils from north Victoria Land, Antarctica: Geologisches Jahrbuch, Reihe B, v. 66, p. 187–201.

Thomson, M.R.A., 1975, New palaeontological and lithological observations on the Legoupil Formation, north-west Antarctic Peninsula: Bulletin of the British Antarctic Survey, no. 41–42, p. 169–185.

Thomson, M.R.A., 1983, Late Jurassic ammonites from the Orville Coast, Antarctica, *in* Oliver, R. L., James, P. R., and Jago, J. B., eds., Antarctic Earth science: Canberra, Australian Academy of Science, p. 315–319.

Thomson, M.R.A., and Pankhurst, R. J., 1983, Age of post-Gondwanian calc-alkaline volcanism in the Antarctic Peninsula region, *in* Oliver, R. L., James, P. R., and Jago, J. B., eds., Antarctic Earth science: Canberra, Australian Academy of Science, p. 328–333.

Thomson, M.R.A., and Tranter, T. H., 1986, Early Jurassic fossils from central Alexander Island and their geological setting: Bulletin of the British Antarctic Survey, v. 70, p. 23–29.

Thomson, M.R.A., Pankhurst, R. J., and Clarkson, P. D., 1983, The Antarctic Peninsula—A late Mesozoic-Cenozoic arc (review), *in* Oliver, R. L., James, P. R., and Jago, J. B., eds., Antarctic Earth science: Canberra, Australian Academy of Science, p. 289–294.

Townrow, J. A., 1967, Fossil plants from the Allan and Carapace Nunataks and from the upper Mill and Shackleton Glaciers, Antarctica: New Zealand Journal of Geology and Geophysics, v. 10, p. 456–473.

Truswell, E. M., 1980, Permo-Carboniferous palynology of Gondwanaland: Progress and problems in the decade to 1980: Australian Bureau of Mineral Resources Journal, v. 5, p. 95–111.

Truswell, E. M., 1982, Palynology of seafloor samples collected by the 1911–14 Australasian Antarctic Expedition: Implications for the geology of coastal East Antarctica: Journal of the Geological Society of Australia, v. 29, p. 343–356.

Truswell, E. M., 1983, Geological implications of recycled palynomorphs in continental shelf sediments around Antarctica, *in* Oliver, R. L., James, P. R., and Jago, J. B., eds., Antarctic Earth science: Canberra, Australian Academy of Science, p. 394–399.

Truswell, E. M., and Drewry, D. J., 1984, Distribution and provenance of recycled palynomorphs in surficial sediments of the Ross Sea, Antarctica: Marine Geology, v. 59, p. 187–214.

Vavra, C. L., 1989, Mineral reactions and controls on zeolite-facies alteration in sandstone of the central Transantarctic Mountains, Antarctica: Journal of Sedimentary Petrology, v. 59(5), p. 688–703.

Vavra, C. L., Stanley, K. O., and Collinson, J. W.,1981, Provenance and alteration of Triassic Fremouw Formation, Central Transantarctic Mountains, *in* Cresswell, M. M., and Vella, P., eds., Gondwana Five: Selected papers and abstracts of papers presented at the Fifth International Gondwana Symposium, Wellington, 1980: Rotterdam, A. A. Balkema, p. 149–153.

Veevers, J. J., and Eittreim, S. L., 1988, Reconstruction of Antarctica and Australia at breakup (95 ± 5 Ma) and before rifting (160 Ma): Australian Journal of Earth Sciences, v. 35, p. 355–362.

Walcott, R. I., 1970, Isostatic response to loading of the crust in Canada: Canadian Journal of Earth Science, v. 7, p. 716–722.

Walker, B. C., 1983, The Beacon Supergroup of northern Victoria Land, Antarctica, *in* Oliver, R. L., James, P. R., and Jago, J. B., eds., Antarctic Earth science: Canberra, Australian Academy of Science, p. 211–214.

Watts, D. R., and Bramall, A. M., 1981, Palaeomagnetic evidence for a displaced terrain in western Antarctica: Nature, v. 293, p. 638–641.

Weaver, S. D., Bradshaw, J. D., and Laird, M. G., 1984, Geochemistry of Cambrian volcanics of the Bowers Supergroup and implications for the early Paleozoic tectonic evolution of northern Victoria Land, Antarctica: Earth and Planetary Science Letters, v. 68, p. 128–140.

Webb, J. A., and Fielding, C. R., 1990, Once upon a time in the Prince Charles Mountains: ANARE (Australian National Antarctic Research Expeditions) News, v. 61, p. 5–6.

Webb, J. A., and Fielding, C. R., 1993, Late Permian–Early Triassic stratigraphy and basin evolution of the Lambert Graben, northern Prince Charles Mountains, Antarctica, *in* Findlay, R. H., Banks, M. R., Veevers, J. J., and Unrug, R., eds., Gondwana Eight: Assembly, evolution, and dispersal: Rotterdam, A. A. Balkema, p. 357–370.

Webers, G. F., 1972, Unusual Upper Cambrian fauna from West Antarctica, *in* Adie, R. J., ed., Antarctic geology and geophysics: Oslo, Universitetsforlaget, p. 235–237.

Webers, G. F., and Spörli, K. B., 1983, Palaeontological and stratigraphic investigations in the Ellsworth Mountains, West Antarctica, *in* Oliver, R. L., James, P. R., and Jago, J. B., eds., Antarctic Earth science: Canberra, Australian Academy of Science, p. 261–264.

Webers, G. F., Glenister, B., Pojeta, J., and Young, G., 1992, Devonian fossils from the Ellsworth Mountains, *in* Webers, G. F., Craddock, C., and Splettstoesser, J. F., eds., Geology of the Ellsworth Mountains, Antarctica: Geological Society of America Memoir 170, p. 269–284.

Wells, N. A., and Dorr, J. A., 1987, Shifting of the Kosi River, northern India: Geology, v. 15(3), p. 204–207.

White, A.J.R., and Chappell, B. W., 1977, Ultrametamorphism and granitoid genesis: Tectonophysics, v. 43, p. 7–22.

Williams, J. G., 1979, Eglinton Volcanics—Stratigraphy, petrography, and metamorphism: New Zealand Journal of Geology and Geophysics, v. 21, p. 713–732.

Williams, P. L., 1969, Petrology of Upper Precambrian and Paleozoic sandstones in the Pensacola Mountains, Antarctica: Journal of Sedimentary Petrology, v. 39, p. 1455–1468.

Wilson, T. J., 1990, Mesozoic and Cenozoic structural patterns in the Trans-

antarctic Mountains, south Victoria Land: Antarctic Journal of the United States, v. 25(5), p. 31–34.

Wilson, T. J., 1993, Jurassic faulting and magmatism in the Transantarctic Mountains: Implications for Gondwana breakup: *in* Findlay, R. H., Banks, M. R., Veevers, J. J., and Unrug, R., eds., Gondwana Eight: Assembly, evolution, and dispersal: Rotterdam, A. A. Balkema, p. 563–572.

Wodzicki, A., and Robert, R., Jr., 1986, Geology of the Bowers Supergroup, central Bowers Mountains, northern Victoria Land, *in* Stump, E., ed., Geological investigations in northern Victoria Land: Washington, D.C., Antarctic Research Series, American Geophysical Union, v. 46, p. 39–68.

Woolfe, K. J., 1990, Trace fossils as paleoenvironmental indicators in the Taylor Group (Devonian) of Antarctica: Palaeogeography, Palaeoclimatology, Palaeoecology, v. 80, p. 301–310.

Woolfe, K. J., Arnot, M. J., and Zwartz, D. P., 1990, Beacon studies in the Skelton Nevé: New Zealand Antarctic Record, v. 10 (2), p. 15.

Wright, T. O., 1981, Sedimentology of the Robertson Bay Group, northern Victoria Land: Geologisches Jahrbuch, Reihe B, v. 41, p. 127–138.

Wright, T. O., and Brodie, C., 1987, The Handler Formation, a new unit of the Robertson Bay Group, northern Victoria Land, Antarctica, *in* McKenzie, G. D., ed., Gondwana Six: Structure, tectonics, and geophysics: Washington, D.C., Geophysical Monograph Series, American Geophysical Union, v. 40, p. 25–29.

Wright, T. O., Ross, R. J., Jr., and Repetski, J., 1984, Newly discovered youngest Cambrian or oldest Ordovician fossils from the Robertson Bay terrane (formerly Precambrian), northern Victoria Land, Antarctica: Geology, v. 12, p. 301–305.

Yochelson, E. L., and Stump, E., 1977, Discovery of Early Cambrian fossils at Taylor Nunatak, Antarctica: Journal of Paleontology, v. 51, p. 872–875.

Yoshida, M., 1982, Superposed deformation and its implication to the geologic history of the Ellsworth Mountains, West Antarctica: Japan, Memoir of the National Institute of Polar Research, Special Issue 21, p. 120–171.

Yoshida, M., 1983, Structural and metamorphic history of the Ellsworth Mountains, West Antarctica, *in* Oliver, R. L., James, P. R., and Jago, J. B., eds., Antarctic Earth science: Canberra, Australian Academy of Science, p. 266–269.

Young, G. C., 1987, Devonian vertebrates of Gondwana, *in* McKenzie, G. D., ed., Gondwana Six: Stratigraphy, sedimentology and paleontology: Washington, D.C., Geophysical Monograph Series, American Geophysical Union, v. 41, p. 41–50.

Young, G. C., 1991, Fossil fishes from Antarctica, *in* Tingey, R. J., ed., The geology of Antarctica: Oxford, Clarendon Press, p. 538–567.

MANUSCRIPT ACCEPTED BY THE SOCIETY SEPTEMBER 9, 1993

Geological Society of America
Memoir 184
1994

Southern Africa: Karoo Basin and Cape Fold Belt

J. J. Veevers
Australian Plate Research Group, School of Earth Sciences, Macquarie University, North Ryde, N.S.W. 2109, Australia
D. I. Cole
Geological Survey of South Africa P.O. Box 572, Bellville 7535, South Africa
E. J. Cowan*
Australian Plate Research Group, School of Earth Sciences, Macquarie University, North Ryde, N.S.W. 2109, Australia

ABSTRACT

Three basement trends, defined by the 1.0–0.5 Ga foldbelts of weak crust that wrap around the 1 Ga Namaqua-Natal Belt and >2.5 Ga Kaapvaal Province, provide a tub-shaped template that was impressed on succeeding structures up to the Cretaceous breakup of Pangea along the present divergent margins. The pattern is reprinted during the Ordovician-Devonian deposition of the Cape Supergroup in grabens on the northwest and northeast linked by an east-west depositional axis and during the Permian and Triassic development of the Cape Fold Belt along the east-west trend linked with intermittent uplifts to the northwest (Atlantic upland) at a syntaxis around Cape Town and to the northeast (Eastern upland) at a syntaxis in the (restored) Falkland Islands.

The inception of the Karoo (Gondwanan) Sequence in the latest Carboniferous (290 Ma) reflected the Gondwanaland-wide relaxation of the Pangean platform in sags (Karoo terrain) and rifts (Zambezian terrain). The first appearance of tuffs from a convergent arc in the Sakmarian (ca. 277 Ma) marked the onset of a foreland basin. Material derived from the south included a small component of mainly rhyodacitic tuff which persisted to the end of Beaufort deposition, when the presumed southern magmatic arc became extinct. Karoo deposition expanded northward over the interior beyond that of the confined pre-Gondwanan Cape Sequence. The axis of maximum thickness of the Permian-Triassic foredeep remained near the South Crop of the Karoo Basin; the parallel drainage axis migrated northward from an initial distance of 80 km during Dwyka deposition through 400 km during Ecca deposition and 550 km during Beaufort to a final 1,000 km during Stormberg deposition. The increasing separation of foredeep and drainage axis reflects the widening during the growth of the Cape Fold Belt of the southern depositional flank of the Karoo Basin at the expense of the starved northern cratonic side. Only during Stormberg deposition did the northern craton match the Cape Fold Belt as a source of voluminous sediment.

*Present address: Department of Geology, University of Toronto, Toronto, Ontario M5S 3B1, Canada.

Veevers, J. J., Cole, D. I., and Cowan, E. J., 1994, Southern Africa: Karoo Basin and Cape Fold Belt, *in* Veevers, J. J., and Powell, C. McA., eds., Permian-Triassic Pangean Basins and Foldbelts Along the Panthalassan Margin of Gondwanaland: Boulder, Colorado, Geological Society of America Memoir 184.

**Renewed and more extensive deposition in the Late Triassic corresponds to a
singularity in Pangean history: terminal compression of foldbelts (Cape Fold Belt,
Bowen-Sydney Basin, Canning Basin) and widespread subsidence, mainly in rifts
that prefigured the divergent margins of the Atlantic and Indian Ocean regions. The
subsequent Karoo volcanism reflects the increased activity of Pangean hotspots.**

INTRODUCTION

This synopsis of the geologic history of southern Africa,
south of 26°S, focuses on the Permian-Triassic Karoo Sequence of the Karoo Basin in its Gondwanan setting. The Karoo Sequence (Figs. 1–3) is a component of the Pangean
Supersequence of the Gondwanaland cratonic province of
Pangea (Veevers, 1990). In the latest Carboniferous (290 Ma),
the Karoo Sequence started to accumulate in the oval-shaped
Karoo Basin and narrow, fault-affected, Zambezian basins.
From 277 Ma, the Karoo Basin became a foreland basin that
subsided along a zone of inherited (pre-Gondwanan) structure
behind the Panthalassan margin. Information was compiled
from the regional accounts of Kent (1980), Tankard et al.
(1982), Dingle et al. (1983), Söhnge and Hälbich (1983), and
from the primary literature. Original contributions are Cole's
field studies and Cowan's petrographic studies of the Karoo
Basin (Appendix 1). Particular attention is given to an ap-
praisal of the contribution to the Karoo Basin fill by the source
region to the south, comprising the Cape Fold Belt and an inferred magmatic arc generated by the convergence of Panthalassa and Pangea. The data are presented in the figures and
documented in the captions and tables; the text is reserved for
critical evaluation of the pre-Gondwanan inheritance of the
Karoo Basin and Cape Fold Belt and the influence of the inferred magmatic arc.

Reconstruction of the Falkland Islands off southeast Africa

A prerequisite to the analysis is a palinspastic reconstruction of the Falkland Islands off southeast Africa before this
part of Gondwanaland started to break up in the Middle-Late
Jurassic and Early Cretaceous (Ben-Avraham et al., 1993).

On structural and stratigraphical grounds, Adie (1952a)
fitted an inverted Falkland Islands off the Transkei coast, between 33° and 34°S, to complete the truncated Karoo Basin
and Cape Fold Belt. Adie's (1952a) reconstruction was confirmed by Mitchell et al. (1986, p. 131)

with preliminary palaeomagnetic evidence from [Early Jurassic] dolerite dykes on West Falkland which suggests that the Falkland Islands were rotated through ~120° during the early stages of the
break-up and dispersal of the southern part of Gondwanaland. . . .The
first stage of continental drift probably took place during the Jurassic,
as Antarctica separated from southern Africa, the islands moving as a
microplate to a position approximately 500 km south-east of present-
day Cape Town. Subsequently, during the opening of the South
Atlantic, the islands and the Falklands Plateau have drifted to their
present position and undergone a further rotation of ~60°.

The pre-Middle Jurassic reconstruction, shown in Fig. 1,
entails the following connections.

(1) >0.5 Ga: Rex and Tanner (1982) dated gneiss of the
Cape Meredith amphibolite-grade Complex of the Falkland
Islands (Fig. 4) by the K-Ar method on biotite and hornblende
as 963 ± 30 and 987 ± 40 Ma (converted with the new constants). Rex and Tanner (1982) compared these dates with K-
Ar ages of 950–1150 Ma on gneisses from the Natal coast of
the Namaqua-Natal Belt affected by the Kibaran orogenic
event. The Cape Meredith dates straddle the boundary between the Namaqua-Natal Belt (2.1–1.0 Ga) and the Saldanian
Province (1.0–0.5 Ga) so that the possible connections of the
Cape Meredith Complex are (1) with the Natal Belt, either a
southern salient, e.g., the Transkei amphibolite-grade terrane
at 31.3°S on the coast, or an allochthonous block of the Natal
Belt within the Southern Cape Conductive Zone–Saldanian
Belt, or (2) with an extension of the Saldanian Belt. The
strongest connection is through the common amphibolite
grade of the Transkei terrane and the Cape Meredith Complex
(Thomas, 1989; Thomas and Mawson, 1989).

(2) <0.5 Ga: (a) Port Stephens–Fox Bay–Monte Maria
(Port Stanley plus Port Philomel) Formations (Adie, 1952a,
1952b; Frakes and Crowell, 1967; Barrett and Isaacson, 1988).
The Port Stephens Formation, 1,500 m of cross-bedded quartzose sandstone with minor conglomerate and red shale, nonconformably overlies the Cape Meredith Complex and passes
conformably upwards into the Fox Bay Formation, 800 m of interbedded shale and sandstone with Emsian brachiopods and
trilobites of the Malvinokaffric Realm. The conformably overlying 800 m of Monte Maria Formation comprises the Port
Philomel Beds in West Falkland, sandstone and shale with
Lepidodendroid plants, and the Port Stanley Beds in East
Falkland, white quartzite with relatively few intercalated shales.

Adie (1952a, 1952b) correlated the Port Stanley Beds with
the Witteberg Group, the Port Philomel Beds and Fox
Bay Beds with the Bokkeveld Group, and the Port Stephens
Beds with the Table Mountain Group. The only secure correlation is by the Emsian marine fauna in the Fox Bay Formation and in the lower two-thirds of the Bokkeveld Group.
We choose the option that provides a consistent trend of the
isopachs, such that the Fox Bay and Port Stephens Formations
are correlated with the Bokkeveld Group, and the Monte
Maria Formation is correlated with the Witteberg Group (Fig.
5, B and C), while at the same time allowing for the possibility
of Table Mountain Group equivalents (Fig. 5A).

(b) The strongly cleaved Lafonian Diamictite, 350–850 m

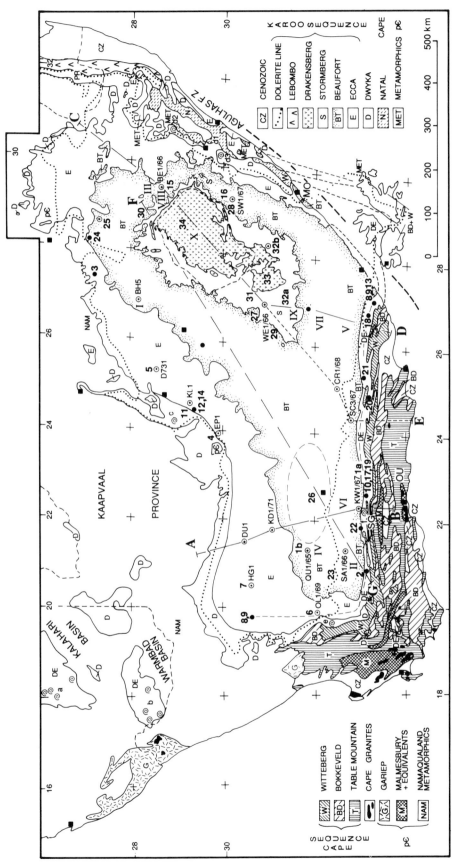

Figure 1. Solid geology of the Karoo Basin (the Karoo Sequence) and the underlying Natal Group in the northeast and the Cape Supergroup in the south (Cape Sequence) and, in turn, the pre-Cape basement, from Geological Map of the Republic of South Africa (1970, 1984) and Kent (1980). Zenithal equal-area meridional projection (Haughton, 1969). Strata dip into the basin except those along the southeastern coast, between 31.5°S and 32.5°S, including an outlier of Molteno Formation (MO), which are involved in the eastern limb of an anticline or in a horst. Also shown are the location of the stratigraphic cross section FG of Figure 10; stratigraphic columns I–X of Figure 8, A–C; AB and CD and boreholes (bullseye) of Figure 2; the time-space diagram CE of Figures 3 and 7; the localities (filled circles and numerals 1–34) with volcanic and volcanogenic material (Tables 2 and 3); and localities of the Dwyka Formation and Prince Albert Formation with marine fossils, shown by coils (McLachlan and Anderson, 1973, 1975) (Table 1): a, western Kalahari Basin; b, southwestern Kalahari (Warmbad); c, north-central Karoo Basin (Douglas); d, northeastern—Pietermaritzburg region; e, western—Tankwa River (actually without invertebrates but correlated with the others by its contained biota); and f, southern—shark and marine microfossils. OU—Outeniqua Mountains, SG—Swartberg Mountains. The Falkland Islands are reconstructed according to the scheme of Adie (1952a) and Mitchell et al. (1986) immediately southeast of the Agulhas Fracture Zone off the Transkei coast. Possible connections through the 1.0 Ga metamorphics, the Bokkeveld and Witteberg equivalents, and the Dwyka and Ecca equivalents are drawn to the mainland. The filled squares represent towns.

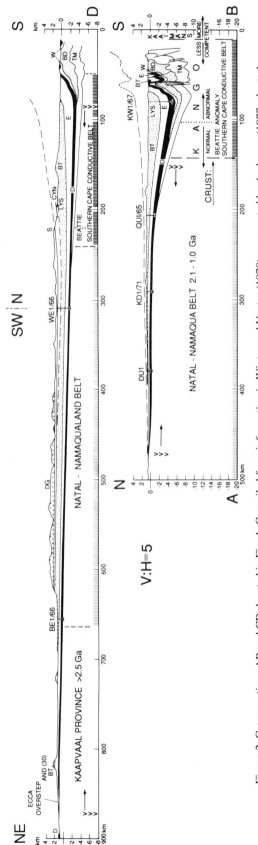

Figure 2. Cross sections AB and CD, located in Fig. 1. Compiled from information in Winter and Venter (1970), augmented by Anderson (1977, charts 1 and 2) and Söhnge and Hälbich (1983). Beaufort Group in south restored (dashes) after Rowsell and de Swardt (1976, p. 84). BD—Bokkeveld, BT—Beaufort, D—Dwyka, DG—Drakensberg, E—Ecca, S—Stormberg, TM—Table Mountain, W—Witteberg. Boundary between *Cynognathus* (CYN) and *Lystrosaurus* (LYS) zones shown by dashed line. V's indicate limit of dolerite intrusions, which have been subtracted from the sections.

of massive gray and brown diamictite, with clasts up to 7 m and stratified intercalations of sandstone and shale, is slightly disconformable to the underlying quartzites, in the same way as the equivalent Dwyka Formation (Adie, 1952a, 1952b; Visser, 1987) (*see* Fig. 11 in a following section) disconformably overlies the Witteberg Group (Du Toit, 1937, p. 119). The conformably overlying sequence (Port Sussex Formation, Terra Mota Sandstone, and Bay of Harbours, Choiseul Sound, Brenton Loch, and West Lafonian Beds) is about 3,000 m thick and comprises gray interbedded shale and sandstone with *Glossopteris*. Adie (1952b) correlated this sequence by the plant fossils with the Ecca and Beaufort Groups. The lack in this sequence of the tetrapods and of the red pigment characteristic of the Beaufort Group suggests to us that the sequence is related to the Ecca Group only. Both have a comparable thickness (*see* Fig. 12 in a following section).

(c) Dolerite dykes cut the Devonian Port Stephens–Monte Maria sequence on West Falkland and the Permian sequence on an island east of Lafonia. As cited by Mitchell et al. (1986), the West Falkland dykes have yielded a K-Ar date of 192 ± 10 Ma, which lies near the old end of the radiometric age range of the Karoo Dolerite (Kent, 1980, p. 566) and the Drakensberg Volcanics. Fig. 1 shows the extension of the Dolerite line to the Falkland Islands.

The Early Jurassic dolerite postdates the Gondwanide deformations. In the Falklands, field evidence indicates the ages of deformation to be Early Permian, shown by the cleaved diamictite, and within the interval Late Permian to Early Jurassic; in southern Africa the deformations are dated as Early Permian, Late Permian, and mid-Triassic (Fig. 3).

In summary, the geologic history of the Falkland Islands followed a course parallel with that of the Cape Fold Belt and the southern crop of the Karoo Basin except that deposition in the Falklands started later, in the Devonian.

Adie's (1952a, b) reconstruction is confirmed by J.E.A. Marshall (1994).

PRE-GONDWANAN HISTORY

Pre-Cape geology

The pre-Cape geology of southern Africa (Fig. 4) comprises (1) the Archean (>2.5 Ga) Kaapvaal Province, (2) the Early-Middle Proterozoic (2.1–1.0 Ga) Namaqua-Natal Belt, and (3) the following Late Proterozoic and epi-Late Proterozoic/Cambrian (1.0–0.5 Ga) belts: (a) Southern Cape Conductive Belt and Beattie Anomaly; (b) the overlapping Saldanian Province, including the Malmesbury Group and the possible equivalents of the Gamtoos Formation and the Kaaimans and Kango Groups, intruded by the Cape Granite; (c) the Gariep Province; (d) Nama and Vanrhynsdorp Basins; and (e) the correlates of the Gariep near Vanrhynsdorp. De Beer et al. (1982) interpreted the Namaqua-Natal Belt as an Andean chain generated behind a trench that terminated 0.8–1.0 Ga and the Southern Cape Conductive Belt (SCCB) as

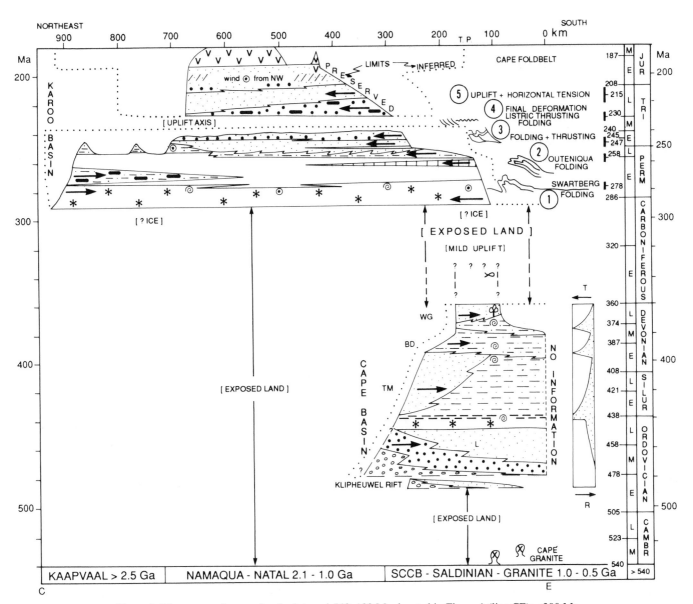

Figure 3. Time-space diagram for the interval 540–180 Ma, located in Figure 1 (line CE); <300 Ma condensed from Figure 7. The youngest reliable age of the Cape Granite Suite is 525 Ma (Kent, 1980, p. 483). The oldest reliable age of the Cape Supergroup is latest Ordovician or Hirnantian (Tankard et al., 1982, p. 335; Theron and Loock, 1989, p. 730) and the base possibly extends to the Early Ordovician. Here we arbitrarily take the base as Middle Ordovician, with the Klipheuwel Formation stretching into the Early Ordovician. The uppermost Table Mountain Group is Pragian/Emsian, the Bokkeveld Group is Emsian and Givetian/Frasnian, and the Witteberg Group is Late Devonian (Boucot et al., 1983; Anderson and Anderson, 1985, p. 20; and other authors cited in Theron and Loock 1989, p. 730). The exposed top of the Cape Supergroup (Witteberg Group) is probably no younger than latest Devonian (Anderson and Anderson, 1985, p. 20), contrary to Gardiner's (1969) view that fish indicate late Early Carboniferous (Visean), as shown by the queries. *See* Figure 7 for symbols, with these additional abbreviations and symbols for the Cape Basin: TM—Table Mountain Group, BD—Bokkeveld Group, WG—Witteberg Group; age-diagnostic megaplants in the Witteberg; environments—alluvial fans (open circles), littoral sediment (L), transgression (T), and regression (R).

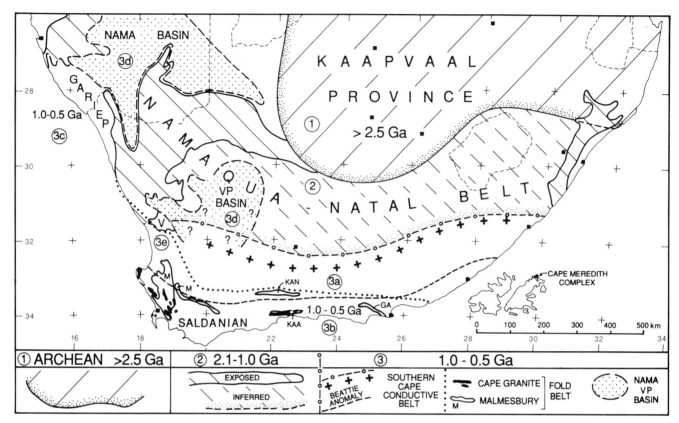

Figure 4. Pre-Cape geology. Precambrian provinces: (1) the Archean (>2.5 Ga) Kaapvaal Province (Hartnady et al., 1985), (2) the Early-Middle Proterozoic (2.1–1.0 Ga) Namaqua-Natal Belt (Tankard et al., 1982), (3) the Late Proterozoic and epi-Late Proterozoic/Cambrian (1.0–0.5 Ga): (a) Southern Cape Conductive Belt and Beattie Anomaly (de Beer et al., 1982); (b) the Saldanian Province (Tankard et al., 1982), including the Malmesbury Group and the possible equivalents of the Gamtoos Formation (GA) and the Kaaimans (KAA) and Kango (KAN) Groups, intruded by the Cape Granite, mapped from Haughton (1969) and Kent (1980); (c) the Gariep Province; (d) the Nama and Vanrhynsdorp (VP) Basins (Hälbich and Hartnady, 1985; Tankard et al., 1982); and (e) the rocks near Vanrhynsdorp (V) correlated with the Gariep (Germs and Gresse, 1991; Tankard et al., 1982). Falkland Islands reconstruction off the Transkei coast from Adie (1952a) and Mitchell et al. (1986), showing the location of the 1.0 Ga Cape Meredith Complex (Rex and Tanner, 1982).

Figure 5. Pre-Gondwanan geology: Cape Supergroup of Ordovician-Devonian (and possibly Early Carboniferous) age. Inset: basement (from Fig. 4) of 1.0–0.5 Ga foldbelts and the reworked Mozambique Province wrapped around the Namaqua-Natal Belt (2.1–1.0 Ga) and Kaapvaal Province (>2.5 Ga). Superimposed on the deep structure are the depositional axes of the Cape Supergroup (Fig. 5D) and of the early Karoo: Dwyka, Ecca, Beaufort (see Fig. 13B).

A. Table Mountain and Natal Groups (Tankard et al., 1982; Hobday and Mathew, 1974; Hobday and Von Brunn, 1979). Isopachs (km) (Visser, 1974), from boreholes (circled dot) and field sections (circled cross); prominent paleocurrents from Visser (1974), Hobday and Mathew (1974), Hobday and Von Brunn (1979), and Turner (1990); facies and depositional environments from Visser (1974), Hobday and Von Brunn (1979), Tankard et al. (1982), Turner (1990), Fuller (1985), Shone (1987), Thamm (1987), Marshall (1988), and Thomas et al. (1992). In the western Cape, the solid black denotes outcrops of the Klipheuwel Formation (Kent, 1980, p. 493), and the broken lines denote the edges of the valley in which the Piekenierskloof Formation was deposited east of a western provenance area that Rust (1973, p. 252) called the Atlantic Highland and southwest

of the Bushman Mountainland. C—Citrusdal.

B. Bokkeveld Group of Early-Middle Devonian age. Isopachs (km) from Rust (1973), extended to the 2.3-km-thick Fox Bay and Port Stephens Formations of the Falkland Islands, paleocurrents and sand-shale ratios (dotted line denotes the 0.5 value) from J. N. Theron (1970), depositional environments from Tankard et al. (1982), and paleogeography from Theron and Loock (1989).

C. Witteberg Group of Middle and Late Devonian (and possibly Early Carboniferous) age. Environments, isopachs (km), and flow pattern from Theron and Loock (1989). The 0.8 km isopach in the Falkland Islands refers to the Monte Maria Formation, and the data point of the 0.5 km isopach in the northeast is the Pondoland (Transkei) quartz arenite (Visser, 1974).

D. Cape Supergroup. Isopachs (km) summed from A, B, and C. Zero isopach of Table Mountain Group indicated by line of open circles, of Bokkeveld Group by filled circles, and of Witteberg Group by triangles. The zero isopach of the Witteberg Group is extended past Port St Johns (PSJ) to accommodate the occurrence of a Late Devonian megaplant in the Pondoland quartz arenite.

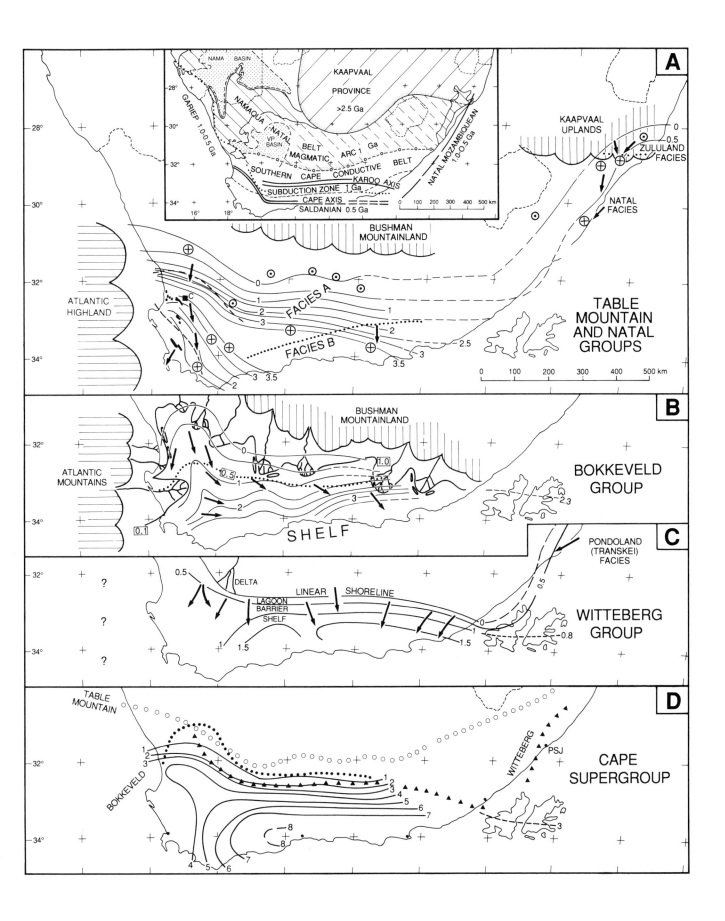

serpentinized basalt obducted against the Namaqua-Natal Belt at the termination of subduction. The overlapping Saldanian Province was deformed further and intruded by granites (0.6–0.5 Ga) during the final closing of a marginal sea. In this manner, the Pan-African fold belts—the Gariep on the west, the Saldanian on the south, and the reworked Natal-Mozambiquean on the east—became wrapped around the cratonic nucleus of the Namaqua-Natal Belt and the Kaapvaal Province. Subsequent extensions and compressions of this anisotropic lithosphere determined depositional structural axes of the succeeding Cape and Karoo Basins. In particular, the weak crust of the SCCB determined the location of the Cape Fold Belt. As described later, this is unlike the situation elsewhere along the Paleo-Pacific margin, where deformation was located along and landward of the weak zone along the magmatic arc. This different location of the foreland basin of the Karoo, distal to the arc, is consistent with the apparently small volume of volcanogenic material found in the Karoo Basin fill. *Note:* A geoscience transect of Southern Africa (Hälbich, 1993) was published when this work was in press.

Cape geology

Table Mountain Group. Preferential subsidence of the 1.0–0.5 Ga foldbelts determined the site of the depositional axis of the Ordovician-Devonian Cape Basin (Fig. 5, inset): a main W–E trend along the Saldanian Province, and branches to the NW (north of Cape Town) and NE (Natal).

According to Tankard et al. (1982, p. 12, 14),

During the Early Paleozoic southern Africa lay at the heart of Gondwana, bounded in the west by South America, in the south by the Falkland Plateau, and to the east by Antarctica. Abortive rifting around the southern and eastern fringe of the Kalahari Province resulted in accumulations of continental and marine clastic successions, known as the Cape Supergroup, in elongate troughs in the southern Cape and Natal. . . . Up to 8 km of sediment accumulated in the Cape basin [Fig. 5D]. The lower 4 km of quartz arenites, mudstones, and conglomerates in the Table Mountain Group [Fig. 5A] record terrestrial and shallow-marine environments and intermittent northward transgression of the Cape sea during the Ordovician and Early Devonian. Prolonged periods of tectonic and eustatic stability are reflected in quartz arenites up to 2,100 m thick, representing one of the greatest known accumulations of quartz sand (Visser, 1974). . . . The Natal embayment developed along a trend parallel to the Pan African Mozambique Province to the north. Proximal coarse alluvial sediments were deposited at the rugged northern end of the embayment, which opened southward into a tide-dominated marine reentrant where considerable thicknesses of marine quartz sand accumulated.

Visser (1974) subdivided the Cape Supergroup of the Cape Province into two facies (Fig. 5A), A, fluvial-coastal plain-beach-neritic facies and latest Ordovician glacigenic facies) in the north and west, and B, beach-neritic facies in the south and east. From the base upward, facies A comprises 1,000 m of alluvial fan to fluvial conglomerate and coarse-

grained sandstone (Piekenierskloof Formation) (thicknesses and formations from Tankard et al., 1982, p. 335); 440 m of paralic interbedded quartz arenite, siltstone, and mudstone, with trace fossils (Graafwater); 1,800 m of open-beach quartz arenite with quartz pebbles and trace fossils (Peninsula); 120 m of glacigene sandstone, conglomerate, and diamictite (Pakhuis), deposited during the Winterhoek glaciation (Tankard et al., 1982, p. 345–348); 140 m of neritic fine-grained sandstone, siltstone, and mudstone, with marine invertebrates, conodonts (Theron et al., 1990), and chitinozoans (Cramer et al., 1974) of latest Ordovician or Hirnantian age (Cedarberg); and 1,100 m of open-beach coarse-grained quartz arenite, with trace fossils (Nardouw). Lacking the basal conglomerate of A, facies B starts with 2,150 m of the open-beach Peninsula Formation and is succeeded by 50 m of the neritic Cedarberg Formation, 640 m of open-beach sandstone, and finally 150 m of the Baviaanskloof Formation: neritic shale, mudstone, and quartz arenite, with Pragian/Emsian marine invertebrates (Theron and Loock, 1989).

The basal formation in the west, the Piekenierskloof Formation, is thickest where it overlies the Klipheuwel Formation (solid black in Fig. 5A), another alluvial fan to fluvial deposit, 2,000 m thick, that wedges out to the north. The Klipheuwel Formation unconformably overlies Malmesbury metasediments and Cape granites.

It becomes progressively less deformed upward, and in its uppermost parts it is probably a facies equivalent of the basal Cape Supergroup. . . . The Klipheuwel succession is envisaged as a series of coalescing and stacked alluvial-fan and interlobe deposits of southward-flowing braided fluvial systems. The coarseness and poor rounding of the conglomerates, their low degree of sorting, and the high proportion of clay matrix suggest limited distances of transportation and local debris flows in a block-faulted terrain such as commonly marks the culminating phase of orogenesis (cf. the Triassic rift basins of the Appalachians) (Tankard et al., 1982, p. 333). [Cross-dip azimuths are widely dispersed about a southeastward mean.] The variety of interrelated facies types and the fan-shaped paleocurrent distribution of the upper Klipheuwel and Piekenierskloof Formations reflect downstream changes in alluvial processes associated with decreasing gradients. High relief was probably maintained along the northern margin of the basin by progressive subsidence relative to adjoining fault-bounded highlands (Tankard et al., 1982, p. 337).

The southeast-trending valley (Rust, 1973) is shown between the broken lines in Fig. 5A. The common southeast trend of the Klipheuwel and Piekenierskloof Formations and their overlap near Citrusdal (32.5°S, 19°E) near the axis of deposition of the entire Table Mountain Group suggests to us that rift-valley grabens (Klipheuwel and Piekenierskloof rifts, *see* Fig. 3) were the site of early crustal extension ("the steer's head") that was followed by down-flexure of the rift shoulders (Atlantic Highland and Bushman Mountainland, Rust, 1973) to accumulate the more widely distributed Graafwater and younger formations ("the horns").

The preponderance of quartz arenite in the Table Mountain Group reflects its source in the interior of Gondwanaland, as do the Ordovician turbidites of southeast Australia (Powell, 1984, p. 293), both indicating deep denudation of the surface of interior Gondwanaland with its maximum apatite fission-track age of 500 Ma (A. Gleadow, personal communication, 1992).

Natal Group. The Natal Group comprises a marked north-south zonation of the facies (Fig. 5A). In Zululand, a boulder conglomerate was deposited in intermontane valleys in the Kaapvaal uplands and passed southward in Natal to red arkosic sandstone deposited in braided channels. Lacking age-diagnostic fossils, the Natal Group was identified as equivalent to the early Paleozoic Table Mountain Group on the basis of similar lithofacies and ichnofacies (Tankard et al., 1982, p. 348), confirmed by Thomas et al.'s (1992) interpretation of Ar-Ar isotopic dates of authigenic muscovite as indicating an age of about 490 Ma, and accordingly we show the Natal Group on the same map as the Table Mountain Group (Fig. 5A).

At about 31°S in Pondoland (Transkei) and southernmost Natal, south of a basement high at 30°30′S (the Dweshula High mentioned by Thomas et al., 1992), clean quartz arenite deposited on a storm- and tide-dominated marine shelf (Visser, 1974; Hobday and Mathew, 1974; Hobday and Von Brunn, 1979; Tankard et al., 1982; Marshall, 1988) was thought to have been part of the Natal Group until Anderson and Anderson (1985, p. 21 and 91) described a megaplant fossil in a quarry 5 km west of Port St Johns (Lock, 1973; Kent, 1980, p. 530) as a new taxon of Late Devonian Lycophyta, but unlike any known species from the Cape sequence. We follow Thomas et al.'s (1992) view that the Pondoland arenite may be equivalent to the lower Witteberg Group (Fig. 5C), and we draw the zero isopach of the Witteberg Group west of Port St Johns (Fig. 5D).

Bokkeveld and Witteberg Groups. In the southern part of the Cape Province, the predominantly arenitic Table Mountain Group is succeeded by the predominantly argillaceous Bokkeveld and Witteberg Groups. The Bokkeveld Group (Fig. 5B) comprises a northern facies of five or six upward-coarsening fluvial-deltaic wedges (sand:shale ratio of 1.0) that pass southward, across the 0.5 sand:shale line (dotted) into a southern shelf facies of homogeneous mudstone to subgraywacke (ratio 0.1). During its time span of 20 m.y., the Bokkeveld Group accumulated at least 3.5 km of sediment at a mean rate of 175 m/m.y. This was about five times the rate of the equally thick but 100-m.y.-long Table Mountain Group and about twice the rate of the 1.5 km thick and equally long Witteberg Group. The southeast-trending depositional axis of what Theron and Loock (1989) called the Clanwilliam Bay, bounded on the west by the Atlantic Mountains, implied by Tankard et al. (1982, figs. 5.10–5.13) from the absence of marine rocks in adjacent South America, and on the north by the Bushman Mountainland (J. N. Theron, 1970, 1972, p. 135), merged eastward into the east-trending depression, both features inherited

from the Table Mountain Group. Theron and Loock (1989, p. 733–735) note that

The marked, almost instantaneous, change from a few thousand metres of supermature sand to between 1500 to 4000 m of predominantly muddy sediments, evidently without marked variation of provenance, climate or environment, indicates that the accelerated downwarp of the Bokkeveld basin was matched by simultaneous increase in the rate of sedimentation because the proximal and medial Bokkeveld sediments certainly do not display the characteristics of deep water deposits. The rapid rate of filling evidently prevented appreciable sorting and halted the process whereby previously most of the fine grained sediments escaped from the basin. Bokkeveld deposition therefore represents a period of cyclic deposition in a marginal cratonic basin and the lateral continuity of the arenitic entities are ascribed to the coalescence of various drainage systems through current and wave action. . . .

The shallowing of the Cape Basin as a whole during the Givetian, as revealed by the upper Bokkeveld units, continued during deposition of the overlying Frasnian sediments of the Witteberg Group [Fig. 5C]. Extensively bioturbated mudstone, siltstone and sandstone with desiccation cracks and rill marks characterize the basal Witteberg units . . . and some thin intercalated shale horizons have yielded marine bivalves, brachiopods and trilobites (Boucot et al., 1983). In the western part of the basin upward coarsening to an extensive sheet-like quartz arenite sequence . . . as well as similar vertical cyclical stacking of litho units as in the Bokkeveld Group, exists. . . . Laterally these . . . units merge with 5 formations in the eastern part of the basin. . . . This medium to coarse grained, cross-bedded, predominantly quartz arenite succession varies from less than 140 m in the west to 850 m eastward. . . . The main basin in broad outline consisted of two major depressions approximately similarly disposed as during the Bokkeveld cycle. . . . The main provenance was still to the north but the limited outcrops to the west and south at present prohibit clarity on the configuration or distribution of these borderlands. Consideration of the paleo ice-flow of the basal Dwyka tillite and its clast composition, however, indicate highland to the southwest undergoing erosion and an eastward directed paleo-slope at the "onset of glaciation." . . . The Witteberg basin at its termination was reduced to only about one-third its original size and can best be described as an open ended marginal intracratonic basin in which shelf deposition took place.

The three major transgressive/regressive events indicated by the appearance of marine fossils (Fig. 3) correspond with the glacio-eustatic transgression at the end-Ordovician (Hirnantian) after the melting of the African ice sheets (Tankard et al., 1982, p. 345–348) and at the Pragian/Emsian start and Givetian/Frasnian peak of the rise in sea level in Euramerica (Johnson et al., 1985, p. 584).

Tectonic setting of the Cape Sequence. The northwest trend of the Clanwilliam Bay in the western part of the Cape Basin denotes an axis of faster sediment accumulation that persisted through the deposition of the Table Mountain Group, the Bokkeveld Group, and possibly the Witteberg Group (Fig. 5A–C). According to Tankard et al. (1982, p. 348), the Natal Group accumulated in an elongate downwarp, the Natal embayment (Fig. 5A), whose axis parallels the present coastline along the trend of the Proterozoic Mozambiquean Province to

the north (Fig. 5, inset). Subsidence and concomitant accumulation of sediment along the northwest and northeast branches of the main west-trending Cape axis are believed to have started probably as long ago as the Ordovician and continued to the end of the Devonian, as indicated by the shape of the zero isopachs and reflected in the isopachs of the total Cape Sequence (Fig. 5D and inset).

We interpret the Cape axis and its northwest and northeast branches as axes of extension concentrated along the deep structural boundaries (Fig. 5, inset) between the east-trending 1.0 Ga Namaqua-Natal Belt and the Southern Cape Conductive Belt and the Pan-African (1.0–0.5 Ga) Saldanian Province on the south and between the Namaqua-Natal Belt and the Gariep Province on the present southwestern margin and the Mozambiquean Province on the southeastern margin (Tankard et al., 1982). The provenance of the Cape Sequence was to the northeast (Bushman Mountainland) and west (Atlantic Mountains).

The 600–500 Ma Cape Granites were emplaced during the final closing of a marginal sea, which was followed by the opening of a rifted ocean or a second marginal sea. The Cape Sequence was deposited across this passive margin (Johnson, 1991) throughout the Ordovician to end-Devonian or Early Carboniferous. Only with the Early Permian development of the overlying structural Karoo Basin is evidence of a southern flank and provenance of deposition seen.

GONDWANAN (KAROO) HISTORY

Introduction

Having sketched the pre-Gondwanan history of southern Africa, we now come to a synopsis of the Gondwanan (Permian-Jurassic) geologic history of the craton and its development by the growth of the Cape Fold Belt as part of the foreland basin behind the Panthalassan margin. This account is illustrated by geological maps (Figs. 1, 6, 9) and cross sections (Figs. 2 and 10), a time-space diagram (Fig. 7), stratigraphic columns (Fig. 8), and paleogeographic and paleotectonic maps (Figs. 11–14).

Setting

Following a lacuna that lasted all but the latest part of the Carboniferous, the succeeding Karoo Sequence, together with the other parts of the Gondwana Facies of the Gondwanaland Province of the newly formed Pangea (Veevers, 1988), started accumulating sediment in the initial sags and grabens of the Pangean cratonic platform during the latest Carboniferous (Gzelian, Eastern Australian palynological stage 2) (Veevers, 1989). The realm of subsidence in southern Africa, hitherto confined to the southern tip, spread northward to the equator (Fig. 6). Subsidence was effected by two tectonic processes (Rust, 1975): (1) the Karoo tectono-sedimentary terrain was

warped into basins (Karoo, Kalahari, Congo, Gabon) and intervening swells, and (2) the Zambezian terrain was faulted into graben-type depositories. In the Early Permian, renewed sinking of the denser material in the northern part of the SCCB between the lighter material north and south of it, together with uplift by folding in the south, initiated the Karoo structural basin. Later in the Permian and Triassic, the SCCB marked the site (south of the Beattie Anomaly, Hälbich et al., 1983) (see Fig. 2) of intense thrusting and folding of the less competent crust (de Beer, 1983, Hälbich, 1983). The site of the deformation of the Karoo Basin was therefore determined by the inherited structure.

Karoo Sequence

The correlation of the latest Carboniferous to Jurassic formations of southern Africa is given in Figure 21 (Appendix 2), and shown in the time-space diagrams (Figs. 3, 7). In brief, deposition started in the latest Carboniferous (Gzelian or 290 Ma) and continued through the Permian into the Early Triassic; after a Middle Triassic lacuna, deposition resumed in the Late Triassic and Early Jurassic, and concluded with the widespread intrusion of dolerite and extrusion of basalt in the Early and Middle Jurassic.

The major units in the Karoo Basin are now discussed in ascending order: Dwyka Formation, Ecca Group, Beaufort Group, Stormberg Group, and Drakensberg Group.

Dwyka Formation. This account of the Dwyka Formation

→

Figure 6. Gondwana (Karoo) basins of Africa south of the Equator, with Malagasy and Falkland Islands in restored positions, showing distribution of Permian and Triassic (and younger) sediment and Jurassic and Cretaceous lavas and intrusions. From de Wit et al. (1988), augmented by information on (1) position of Falkland Islands, as detailed above; (2) distribution of Permian sediment (Frakes and Crowell, 1970; Anderson, 1977, map 2; Kent, 1980; Truswell, 1980; Reimann, 1986), including coal measures (Ryan and Whitfield, 1979; Kreuser and Semkiwa, 1987); (3) distribution of Late Triassic sediment, including coal measures (Dingle et al., 1983; Visser, 1984); (4) the dolerite line in the Karoo Basin and Falkland Islands, from Figure 1; (5) Karoo magmatic events, lavas (circled numbers) and intrusions, according to Fitch and Miller (1984): 1—early events about 204 ± 5 Ma, 2—major about 193 ± 5 Ma, 3—minor about 186 ± 3 Ma, 4—major about 178 ± 5 Ma, 5—events about 165 ± 5 Ma, 6—about 150 ± 5 Ma, 7—about 137 ± 5 Ma, post-Karoo Etendeka lavas about 120 ± 5 Ma; subsurface lavas (circled) in Mozambique from Dingle et al. (1983); and (6) Karoo and Zambezian tectono-sedimentary terrains, from Rust (1975). Heterolithic breccias containing up to 70% of basalt xenoliths occur at Kolonkwaren (KN), Prieska (PA), and Postmarburg (PG) (Eales et al., 1984, p. 3), and xenoliths sampling the entire Karoo column above the Dwyka occur in the Kimberley (KY) pipe (Truswell, 1977). ZI marks the township of Zambesi in the Barotse Basin of western Zambia in the region of a subsurface occurrence of a 2-m-thick arkose and mudstone in which acritarchs of suspected early Paleozoic age were found (Reimann, 1986).

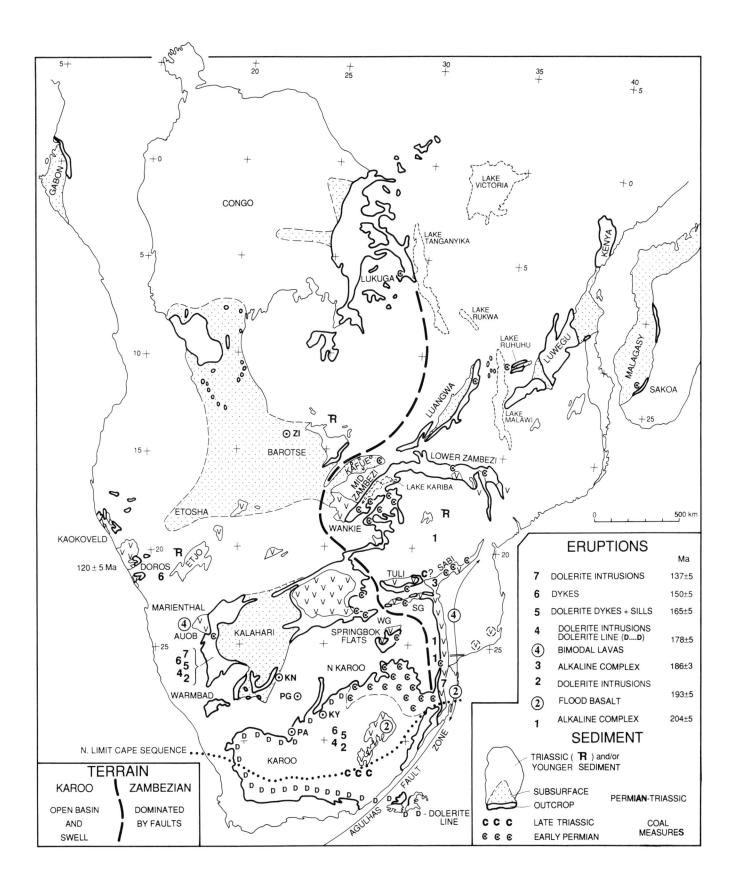

is derived from the monumental work of J.N.J. Visser and colleagues, as summarized in Visser (1979, 1989, 1990).

The Cape Fold Belt comprises an easterly trending southern branch and a northwesterly trending western branch that meet in the southwesterly trending syntaxis (Fig. 13A). Following mild uplift and erosion during the Carboniferous lacuna, the Witteberg Group was disconformably overlain by

the Dwyka Formation along the present southern crop of the Karoo Basin. From east to west in the Southern Branch, the Dwyka Formation steps across the youngest preserved Witteberg Group (Dirkskraal Formation) at Willowmore (23°30′E) down 500 m of stratigraphic relief to the top of the Waaipoort Formation at 23°00′E and continues at this level to Touws River; from here the line of section swings to north and south,

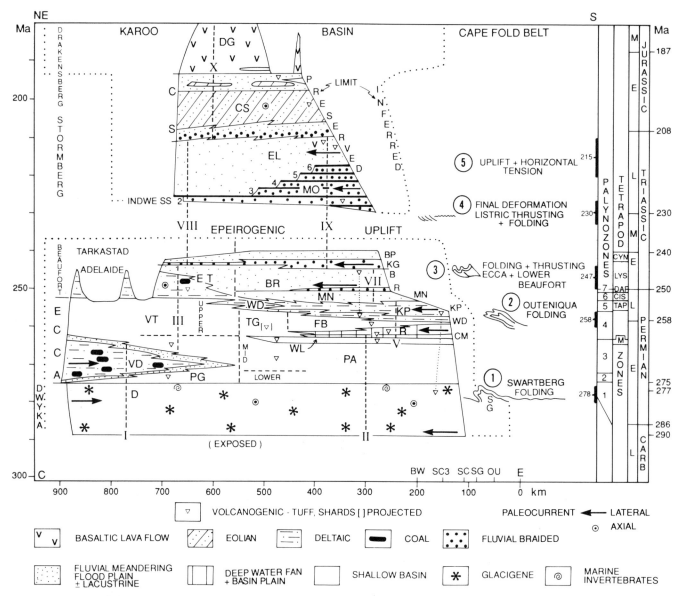

Figure 7. Distribution in time and space of latest Carboniferous, Permian, Triassic, and Jurassic lithostratigraphic units and environments, located on Figure 1 (line CE). Correlation from Fig. 21. Abbreviations: BP—Burgersdorp, BR—Balfour, CM—Collingham, CS—Clarens, D—Dwyka, DG—Drakensberg, EL—Elliot, ET—Estcourt, FB—Fort Brown, KG—Katberg, KP—Koonap, MN—Middleton, MO—Molteno, PA—Prince Albert, PG—Pietermaritzburg, R—Ripon, TG—Tierberg, VD—Vryheid, VT—Volksrust, WD—Waterford, WL—Whitehill. Stratigraphic columns of Figure 8, except IV and VI in the southwest part of the basin, located by fine broken line. Paleocurrents from the side (from data in Figs. 11–14) shown by arrow, southward in the Dwyka and Vryheid of the northern Karoo, northward elsewhere except along the axis (westward) in the upper Dwyka and the Estcourt (Botha and Linström, 1978); northwesterly wind in the Clarens Formation (Visser, 1984) shown by a circled dot. Upward-fining cycles 1–6 of Molteno Formation (Fig. 8, part IX) from Turner (1983, fig. 3). Folding, thrusting, and uplift of the Cape Fold Belt from Hälbich et al. (1983, table 13.3). Localities projected into section are OU—Outeniqua Mountains, SG—Swartberg Mountains, SC—south crop, SC3—drillhole SC3/67, BW—Beaufort West.

and the Dwyka Formation steps down another 500 m of stratigraphic relief to cross the Witpoort Formation (Loock and Visser 1985, p. 168). According to de Beer (1990, p. 585),

Uppermost Witteberg units (Kweekvlei and Floriskraal Formations) occur below Dwyka tillite in two down-faulted grabens situated well into the Western Branch. . . . Erosion of upper Witteberg sediment apparently resulted from uplift of the basin rim to the north and south of the hinge-line of the early Karoo depository, or from a general lowering of sea level. The available field evidence is insufficient to prove uplift along a N-trending zone during post-Witteberg/pre-glaciation times in the west. . . .

No evidence demonstrating independent folding of the Witteberg Group before erosion of its upper part, or before deposition of the Dwyka glacials, could be found in the Tankwa Karoo [west]. Dwyka tillite of the Swartruggens Mountain Range is folded about the same NW–SE axes displayed in the immediately underlying Witpoort Formation. This, together with the points made in the previous paragraph, removes the two cornerstones of the pre-Dwyka folding hypothesis.

Farther south, at 19.5°E, 33.8°S, an outlier of the Dwyka and Ecca rests on the Witteberg (Fig. 1). To the north, the Dwyka Formation steps across the Cape Sequence and the Namaqua-Natal Belt to reach the Kaapvaal Province and beyond (Wopfner and Kreuser, 1986) in an expansion of deposition deep into the interior Karoo terrain to the present Equator (Fig. 6).

The Dwyka Formation was deposited as a sheet of glacial sediment that thickened away from a ragged edge in the contemporary Cargonian Highlands, as indicated by the areas of basement overlapped by the Ecca Group (black on Fig. 11 and double line on Fig. 12), into the Karoo Basin on the south and southwest and into the Kalahari Basin on the north and west. The ice flowed westward into the Kalahari Basin and southwestward (through or across the denuded Atlantic Mountains) into an inferred shallow sea. For the first time, sediment was shed to the northwest from an upland, probably from the rising folds of the Cape Fold Belt. In the Karoo Basin, the sediment merged with the drainage from the north in a westward axis. The Karoo Basin has a platform facies association in the south (70% massive diamictite, 22% bedded diamictite, and 8% mudrock), exemplified by Fig. 8A, part II, and a valley facies association in the north (21% massive diamictite, 37% bedded diamictite, and 42% mudrock, half of which contains ice-rafted debris) (Fig. 8A, part I). A third facies association, glacial debris reworked by water into breccia, conglomerate, and sandstone, represents deposition in upland areas by the removal of fines. A sequence of nine sedimentation units in the southern part of the basin was deposited from ice lobes that flowed from the north, east, and south. Features of the ice lobes—grounded or afloat, level of sediment transport, and basal thermal condition—eventually controlled deposition.

A vital control was sea level. Firm evidence of marine deposition (Table 1) is provided by the marine invertebrates (brachiopods, bivalves, cephalopods) of the *Eurydesma* zone in the basal Prince Albert Formation (formerly called the up-

per Dwyka Shales) in the Kalahari Basin (localities a and b, Figs. 1 and 11). Marine microfossils (arenaceous foraminiferids, radiolarians, and sponge spicules) occur also in the Kalahari Basin, and in the Karoo Basin (topmost Dwyka Formation) at Douglas, locality c, correlated by McLachlan and Anderson (1973, p. 45) with the Kalahari Basin on the basis of Sakmarian fish (*Namaichthys schroederi*) and spores, in the Dwyka Formation of the Pietermaritzburg area (localities d1–3), and with a shark at the boundary between the Dwyka Formation and Prince Albert Formation at Zwartskraal (locality f). The palaeoniscid fishes and coprolites (probably from sharks) associated with the invertebrates at Douglas are found alone in the topmost Dwyka Formation at Tankwa River (locality e), which is inferred to be marine. The familiar association of glendonite concretions in the *Eurydesma* zone of Eastern Australia, as in the Woody Island Siltstone of Tasmania (Clarke and Forsyth, 1989, p. 298), is found also at Douglas (McLachlan and Anderson, 1973), although *Eurydesma* has not been found.

Possible marine indicators are the acritarch *Mycrhystridium* at the base of the Dwyka Formation of the southwest Karoo (Anderson, 1977, p. 51–53). Geochemical research, however, indicates that "the geochemistry of the glacial and related rocks [organic carbon:total sulfur ratio, Fe:Mn ratio] does not give positive evidence for marine conditions in the Dwyka Basin . . . [except] the geochemistry of the mudrocks immediately overlying the glacial beds indicates marine conditions which are confirmed by the palaeontology (Visser, 1989, p. 383).

All this evidence suggests to us the following events:

(1) During the Tastubian a single marine transgression or group of multiple transgressions (comparable to the 20 or so in the Quaternary) crossed the Karoo Basin from west to east during deposition of the topmost Dwyka Formation and the overlying basal Prince Albert and Pietermaritzburg Formations as a local manifestation of the glacioeustatic *Eurydesma* transgression (Dickins, 1984; Veevers and Powell, 1987); in our opinion, Visser's (1990) basal Prince Albert transgression is simply a component of the multiple *Eurydesma* transgression, which we regard as a single stepped rise of sea level in southern Africa during the Tastubian.

(2) Mudrock units traceable over distances of up to 400 km and interpretable as interglacial deposits (Visser 1989, p. 381) were deposited in lakes, as depicted by Visser (1987, fig. 9), and not in the sea, as shown by the dotted lines in figure 6 of Visser (1989).

(3) The rise of sea level corresponded to the disintegration of the ice sheets in the Gondwanaland province (Visser, 1989, p. 387); the return of this melt-water to the ocean (glacio-eustatic) thus caused, at least in part, the sea-level rise. The transgression in the Karoo Basin was possibly augmented by concomitant subsidence of the shelf (Visser, 1991a), generated by the first extensional phase of Pangean history (Veevers, 1990).

Figure 8A–C (on the following three pages). Representative stratigraphic columns through the Karoo sequence showing lithology and depositional environment; symbols and abbreviations key beneath columns I and II. Thickness in meters. Columns located in the map (Fig. 1), cross section FG (*see* Fig. 10), and time-space diagram (Fig. 7).

A: I. Northeast. Dwyka from Visser (1986) and Visser and Kingsley (1982), marine fossils from Table 1; Ecca Group: Pietermaritzburg (PG) and Vryheid, from Borehole 5 in Van Vuuren and Cole (1979).

II. Southwest. Dwyka from Visser (1986, 1988); lower part of Ecca Group: Prince Albert (PA) and Whitehill (WL) from Visser and Loock (1978) and Cole and McLachlan (1991), who correlate the Whitehill Formation with the upper Vryheid Formation. I and II represent deposition (Dwyka Formation) from a retreating ice sheet interrupted by at least one re-advance (Visser, 1982), followed by suspension settling of argillite (Prince Albert/Pietermaritzburg) in the deglacial flooded basin (Visser, 1987), culminating in the deposition of carbonaceous shale (Whitehill) on an anoxic basin floor caused by the growth of cyanobacterial mats (Cole and McLachlan, 1991). In the northeast part of the basin, postglacial rebound of the source areas resulted in the progradation of a clastic wedge (Vryheid) (Van Vuuren and Cole, 1979).

III. Northeast. Upper part of Ecca Group: Volksrust from Cadle and Hobday (1977), Visser and Loock (1978), and Van Vuuren and Cole (1979). Lower part of Beaufort Group: Estcourt from Botha and Linström (1978), Hobday (1978), and Tankard et al. (1982, p. 392). All from Borehole BE1/66. Represents the second progradation (above the Vryheid/Pietermaritzburg couplet) in the northeast.

B: IV. Southwest. Upper part of Ecca Group: Collingham (CM) from Viljoen (1993, unpublished), Vischkuil and Laingsburg from Viljoen (1993, unpublished), Visser and Loock (1978) and Visser et al. (1980); Fort Brown from Visser and Loock (1978) and Visser et al. (1980); Waterford from Visser et al. (1980), Jordaan (1981), and Rubidge (1988), nonmarine bivalves from Cooper and Kensley (1984). This is a thick regressive sequence deposited in an initially deep trough or foredeep (Tankard et al., 1982, p. 379) that shoaled during the deposition of the Fort Brown and Waterford Formations.

V. Southeast. Upper part of Ecca Group and base of Beaufort Group. Collingham and Ripon from Kingsley (1977, 1981) and Viljoen (in preparation); Fort Brown and Koonap from Kingsley (1977, 1981). A thicker facies equivalent of IV.

VI. Southwest. Beaufort Group. Abrahamskraal from Stear (1980, 1985), Jordaan (1981), Turner (1981), Rubidge (1988), and Cole and Wipplinger (1991); Teekloof from Stear (1980, 1985), Turner (1981), and Cole and Wipplinger (in preparation). Initial progradation of fluvial deposits over a delta, followed by decreasing fluvial energy as a result of denudation of the source area (Turner, 1985). The fluvial channels range from low- to high-sinuosity (Stear, 1980, 1985) and the sandstones cluster into packages (Stear, 1980; Cole and Wipplinger, 1991).

C: VII. Southeast. Middle part of Beaufort Group. Middleton from Visser and Dukas (1979) and Kingsley (1981); Balfour from Visser and Dukas (1979), Stavrakis (1980), Turner (1981), and Hiller and Stavrakis (1984). Similar to VI but the migration of the depositional system sourceward due to a decrease in fluvial energy has been interrupted by high-energy influxes at the base and near the top of the Balfour Formation (Visser and Dukas, 1979).

VIII. Northeast. Upper part of Beaufort Group unconformably overlain by the Stormberg Group. Tarkastad from Botha and Linström (1978), Hiller and Stavrakis (1984), and Turner (1986); Molteno cycle 2 or Indwe Sandstone member from Eriksson (1984) and Turner (1983, 1986); Elliot from Eriksson (1985) and Turner (1986). Two northward influxes of fluvial sediments from episodes of tectonic uplift (Hiller and Stavrakis, 1984, p. 2). Each influx was followed by sourceward migration of the distal, fine-grained facies as the source area was progressively denuded.

IX. Southeast. Upper part of Beaufort Group unconformably overlain by the Stormberg Group. Katberg from Visser and Dukas (1979), Stavrakis (1980), and Hiller and Stavrakis (1984); Burgersdorp from Stavrakis (1980) and Hiller and Stavrakis (1984); composite section of Molteno across southern outcrop, showing fining-upward sequences 1–6, from Turner (1975, 1977, 1983, fig. 3); Elliot from Visser and Botha (1980) and Tankard et al. (1982). Similar to VIII except the Molteno Formation contains coal at the top of coarse-grained facies that indicates at least three phases of provenance uplift (Turner, 1983).

X. Northeast. Drakensberg Group and uppermost Stormberg Group. Clarens from Beukes (1970), Eriksson (1979), and Tankard et al. (1982, p. 397); Drakensberg from Lock et al. (1974), Tankard et al. (1982, p. 400), and Visser (1984). Deposition in a desert environment subject to ephemeral floods, followed by lava flows that are interlayered with pyroclastics in the lower third of the Drakensberg.

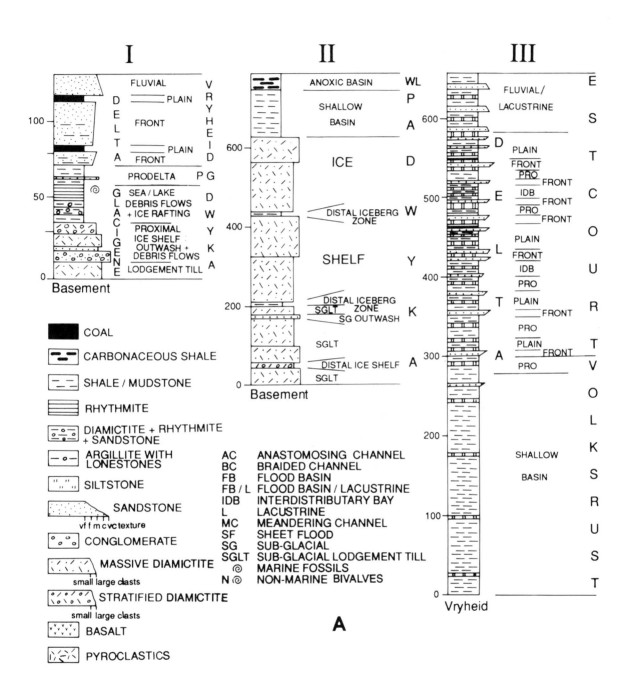

I

		FLUVIAL	V
D		PLAIN	R
E		FRONT	Y
L		PLAIN	H
T		FRONT	E
A		PRODELTA	I D
	G	SEA / LAKE	D G
	L	DEBRIS FLOWS + ICE RAFTING	D
	A	PROXIMAL ICE SHELF OUTWASH + DEBRIS FLOWS	W Y
	C	LODGEMENT TILL	K
	I		A
	G		
	E		
	N		
	E		

100

50

0

Basement

II

	ANOXIC BASIN	WL
	SHALLOW BASIN	P A
	ICE	D
	DISTAL ICEBERG ZONE	W
	SHELF	Y
	DISTAL ICEBERG ZONE	
	SGLT · SG OUTWASH	K
	SGLT	
	DISTAL ICE SHELF	A
	SGLT	

600

400

200

0

Basement

III

	FLUVIAL / LACUSTRINE	E S
D	PLAIN	T
	FRONT	
	PRO · FRONT	C
E	IDB · FRONT	O
	PRO · FRONT	U
L	PLAIN	
	FRONT	R
	IDB	
	PRO	T
T	PLAIN	V
	FRONT	O
	PRO	L
A	PLAIN · FRONT	K
	PRO	S R
	SHALLOW BASIN	U S T

600

500

400

300

200

100

0

Vryheid

Legend:

■ COAL

▭ CARBONACEOUS SHALE

▭ SHALE / MUDSTONE

▤ RHYTHMITE

▤ DIAMICTITE + RHYTHMITE + SANDSTONE

▭ ARGILLITE WITH LONESTONES

▭ SILTSTONE

▭ SANDSTONE
vf f m c vc texture

▭ CONGLOMERATE

▭ MASSIVE DIAMICTITE
small large clasts

▭ STRATIFIED DIAMICTITE
small large clasts

▭ BASALT

▭ PYROCLASTICS

AC ANASTOMOSING CHANNEL
BC BRAIDED CHANNEL
FB FLOOD BASIN
FB / L FLOOD BASIN / LACUSTRINE
IDB INTERDISTRIBUTARY BAY
L LACUSTRINE
MC MEANDERING CHANNEL
SF SHEET FLOOD
SG SUB-GLACIAL
SGLT SUB-GLACIAL LODGEMENT TILL
◎ MARINE FOSSILS
N ◎ NON-MARINE BIVALVES

A

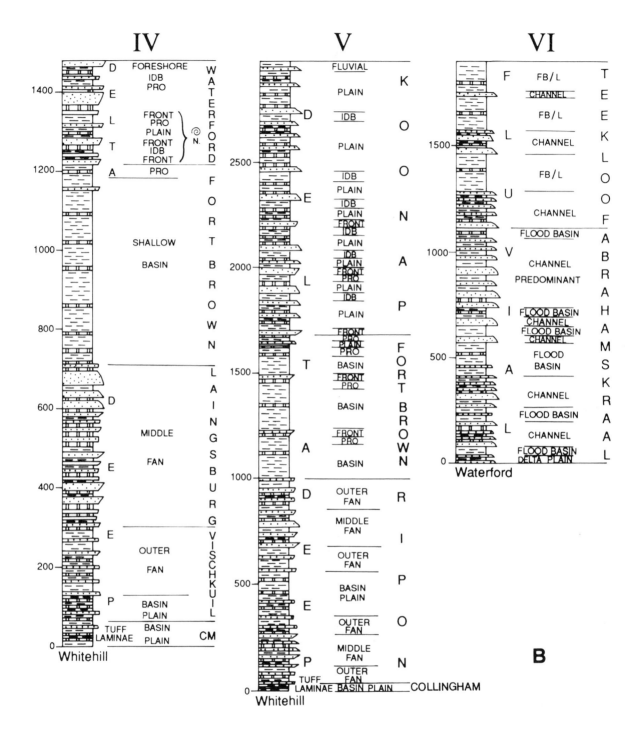

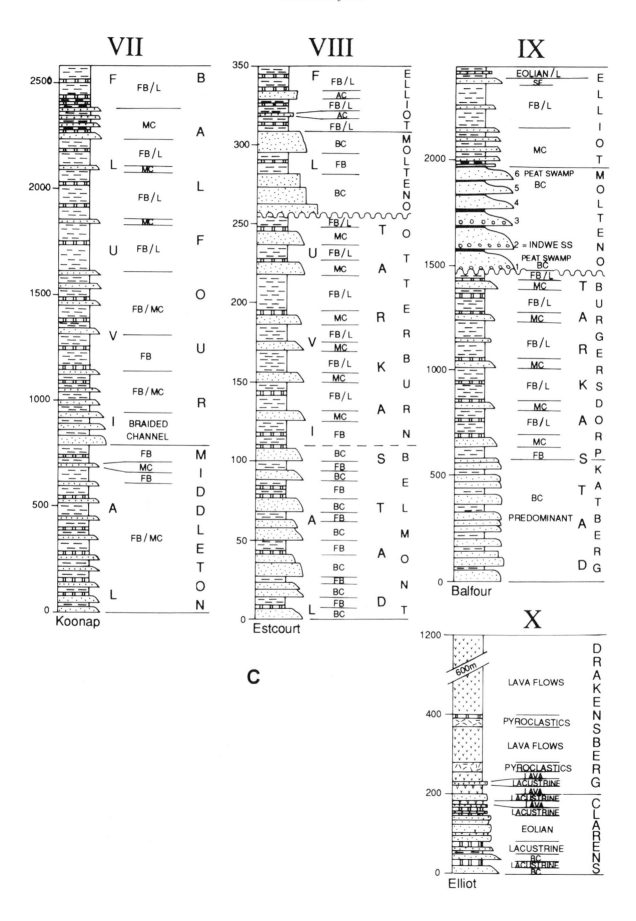

VII

2500 — F   FB/L   B
          MC   A
      L   FB/L   L
          MC   L
2000 —   FB/L   F
      U   MC   O
          FB/L   U
1500 —   FB/MC   R
      V   FB   I
          FB/MC
1000 —   BRAIDED M
          CHANNEL I
          FB   D
          MC   D
          FB   L
500 — A   E
          FB/MC   T
                  O
      L   N
0 —
Koonap

VIII

350 — F   FB/L   E
          AC   L
          FB/L   L
          AC   I
          FB/L   O
300 —   BC   T
      L   FB   M
          BC   O
250 —   FB/L   T   T
          MC   A
      U   FB/L   R   R
          MC   K
200 —   FB/L   E
          MC   R   R
      V   FB/L   B
          MC   U
          FB/L   K   R
150 —   MC   N
          FB/L   A
      I   MC
          FB   S   B
100 —   BC   E   E
          FB   T   L
          BC   M
          FB   O
          BC   N
50 —   FB   A   D
          BC
          FB
          BC   T
          FB
      L   BC   D
0 —
Estcourt

C

IX

          EOLIAN/L   E
          SF   L
          FB/L   L
2000 —   MC   I
          6 PEAT SWAMP O
          5 BC   T
          4   M
          3   O
          2 = INDWE SS L
1500 —   1 PEAT SWAMP T
          BC   E
          FB/L   N
          MC   T   O
          FB/L   B
          MC   A   U
          FB/L   R
          MC   R   G
          FB/L   E
1000 —   MC   K   R
          FB/L   S
          MC   A   D
          FB/L   O
          MC   S   R
          FB   P
500 —   K
          BC   T   A
          PREDOMINANT T
                  A   B
                  B   E
                  D   R
0 —   G
Balfour

X

1200 —
          ╱600m   D
          LAVA FLOWS   R
                       A
400 —   PYROCLASTICS K
          LAVA FLOWS   E
          PYROCLASTICS N
          LAVA   S
          LACUSTRINE   B
200 —   LAVA   E
          LACUSTRINE   R
          LAVA   G
          LACUSTRINE
          EOLIAN   C
          LACUSTRINE   L
          BC   A
          LACUSTRINE   R
          BC   E
0 —   N
Elliot   S

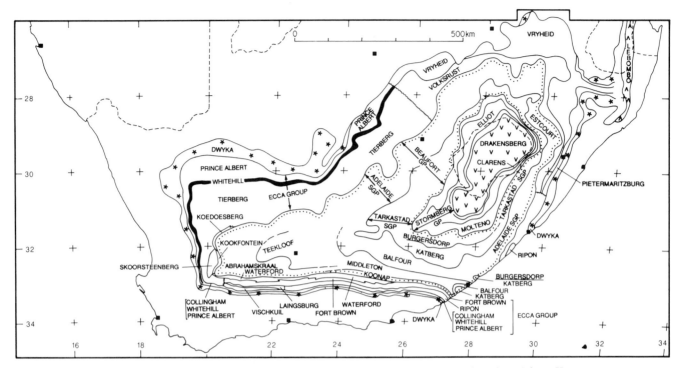

Figure 9. Schematic map distribution of formations in the main Karoo Basin, adapted from Kent (1980, p. 560) as updated by M. R. Johnson (personal communication, 1990). Also shown are subgroups (SGP) and groups (GP).

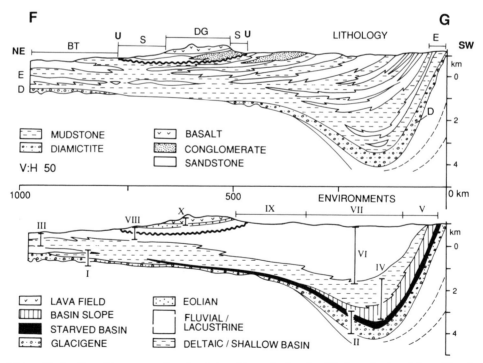

Figure 10. Stratigraphic cross section, located on Figure 1, showing lithology (top) and depositional environments (below) of the Karoo Sequence, modified from fig. 11-2 of Tankard et al. (1982). The lower figure shows also the location (actual or projected) of the stratigraphic columns of Figures 8, A–C. BT—Beaufort, DG—Drakensberg, D—Dwyka, E—Ecca, S—Stormberg, U—unconformity.

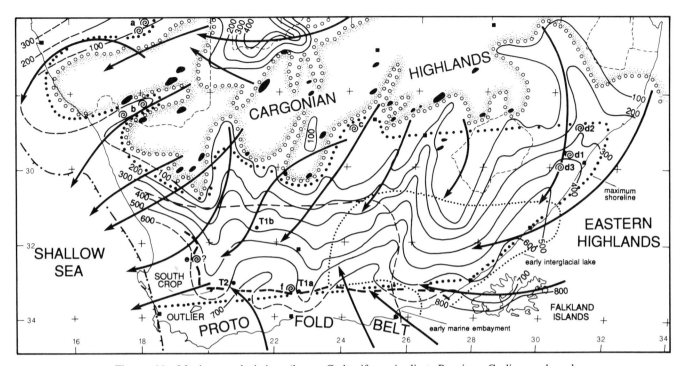

Figure 11. Maximum glaciation (latest Carboniferous/earliest Permian, Gzelian and early Tastubian: 290–277 Ma) in the Kalahari Basin (north of 28°S) and the Karoo Basin (Dwyka Formation), and on the Falkland Islands (Lafonian Diamictite), from Visser (1987, fig. 9), and additionally our interpretation of the Tastubian marine shoreline. Falkland Islands replotted from Mitchell et al. (1986). Shown are isopachs (m), the location (within the Cargonian Highlands) of uplands where younger rocks overlap the glacial deposits (solid), a westward axial ice flow (arrows) during maximum glaciation when the ice sheet was still grounded around a marine embayment (light dashed line) (Visser, 1989, fig. 6), and an early interglacial lake in the east (light dotted line). Coils indicate the location of Tastubian marine invertebrates at localities a and b in the Prince Albert Formation (McLachlan and Anderson, 1973, 1975), at d of marine microfossils in the Dwyka Formation, at c and e of associated fossils, and at f of a shark with radiolarians and arenaceous foraminiferids, as shown also on Figure 1. Heavy dotted line indicates our interpretation of the maximum advance of the Tastubian shoreline, which, limited to the area of known marine fossils, does not include the Falkland Islands. T1a, T1b, and T2 mark the location of tuff in the upper Dwyka Formation (277 Ma) (Table 2). The south crop is indicated by the heavy broken line.

(4) The subsequent fall of sea level in the previously glaciated areas of Gondwanaland (Argentina, southern Africa, India, southern Australia) was due to isostatic rebound.

Our conclusions, broadly in agreement with Visser's (1989, p. 387) history of the Karoo ice sheets, are as follows:

(a) Grounded ice streams from the Cargonian Highlands in the north and from highlands in the south coalesced at the present latitude, 33°S, and flowed westward to a presumed shoreline along the present west coast (heavy broken barbed line in Fig. 11).

(b) During at least two interglacials the basin axis was occupied by a lake (light dotted line in Fig. 11, after Visser, 1987, fig. 9) or a marine embayment (dashed line in Fig. 11).

(c) With the rise of sea level in the Tastubian, the ice sheet flowed through valleys on the southern flank of the Cargonian Highlands at a grounding line (heavy dotted line in Fig. 11) floating as an ice shelf over the basin axis, as suggested also by Von Brunn and Gravenor (1983).

(d) The rapid disintegration of the ice shelf and sheet was followed by an equally rapid isostatic rebound of the areas of formerly grounded ice and regression of the shoreline well past its original position somewhere in the west.

(e) Deposition after the Dwyka Formation and basal Prince Albert and Pietermaritzburg Formations was mainly nonmarine. From time to time the water may have been brackish to form a layer beneath the surface freshwater (cf. the Caspian Sea, Yassini, 1987) as a holdover from the Dwyka/Prince Albert sea, as suggested by phosphatic nodules (Visser, 1991b) and glauconite in the Vryheid coal measures (Van Vuuren and Cole, 1979, p. 109) and the geochemistry of mudrocks in the Ecca and Beaufort Groups (Marchant, 1978; Zawada, 1988; Visser, 1989). But an effectively open connection with the world ocean

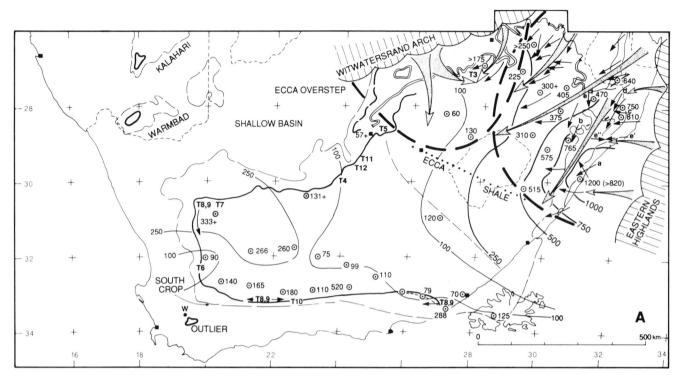

Figure 12 (on this and facing page). Early Permian (late Tastubian–Kalinovian: 275–255 Ma) Ecca Group. The double line denotes the northern feather edge of the Ecca Group that oversteps the Dwyka Formation to rest on basement.

A. Lower Ecca (palynozones 2 and 3). Whitehill Formation (denoted by the solid line, McLachlan and Anderson, 1977; Cole and McLachlan, 1991) and Prince Albert Shale in west and south, and equivalent Vryheid Formation and Pietermaritzburg Shale in north and east. Heavy dotted line denotes the southern limit of sand from the northeast and heavy broken lines the delta complex lobes from the Witwatersrand Arch and the Eastern Highlands (Van Vuuren and Cole, 1979, p. 103, 110). Isopachs (m) from thicknesses of Whitehill Shale + Prince Albert Shale ("Upper Dwyka shales") and Pietermaritzburg ("Lower Ecca shales") + Vryheid Formation ("Middle Ecca") from Anderson (1977, chart 1), Winter and Venter (1970), and unpublished data from D. I. Cole, augmented by thicknesses of Whitehill and Prince Albert on south crop east of 26°E from Kingsley (1979, table 1), and of the presumed equivalent Black Rock member of the Falkland Islands (Frakes and Crowell, 1967, p. 41). Thickness of Whitehill Formation (80 m) at Worcester (W) outlier from Cole and McLachlan (1991). In the north only the first-order isopach trends are shown; details of the valley-fill geometry of the Vryheid Formation are given by Van Vuuren and Cole (1979, figs. 15 and 17). Paleoflow (stippled arrow) from paleocurrents (in north and east only) from Ryan and Whitfield (1979) shown by plain arrow, others by (a) Vryheid Formation (Taverner-Smith, 1982), (b) Vryheid (Middle Ecca) deltaic deposits (Hobday, 1973), (c) Ecca Group (Whateley, 1980), (d) Vryheid Formation (Whateley, 1980), (e) Ecca Group (Hobday et al., 1975), (e′) Vryheid, early flow WNW, later flow (e″) eastward (Mason and Taverner-Smith, 1978).

B. Upper Ecca (palynozone 4), comprising the interval between the Whitehill Formation and the Beaufort Group in the south (Waterford/Fort Brown/Ripon/Laingsburg/Vischkuil/Collingham), the Tierberg, Skorsteenberg, Kookfontein, and Koedoesberg Formations in the northwest, and the Volksrust Shale in the northeast. Isopachs (m) from thicknesses of palynozone 4 in Anderson's (1977) chart 1, except for the south crop east of 26°E, from Kingsley (1979, table 1), and the 3,000-m-thick sequence above the Lafonian Diamictite on the Falkland Islands, from Frakes and Crowell (1967). Paleocurrents from Ryan and Whitfield (1979) except (f) Ripon Formation and (g) Fort Brown Formation (Kingsley, 1979, 1981), and the six localities marked with a circle (Visser et al., 1980), including two from subaqueous fans (heavy arrow), four from the lower delta plain (single-tipped arrows), four from the upper delta plain (double-tipped arrow), and two from the undifferentiated delta plain (triple-tipped arrow). The coil indicates the locality near Prince Albert of the nonmarine (NM) bivalves in the Waterford Formation (Cooper and Kensley, 1984).

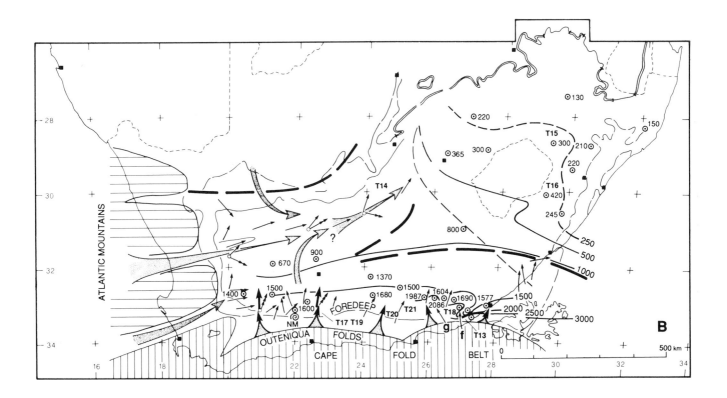

ceased after the Tastubian, and southern Africa has remained above or beyond the sea to the present day, as shown by the lack of indubitably marine sediment.

Three tectonic events took place during deposition of the Dwyka Formation: (1) Clast type and ice-flow direction indicate that there was a change to the high ground ("proto fold belt") of the southern epiclastic provenance (Visser, 1993); this was a reversal of slope from that during Cape deposition. (2) The first juvenile volcanogenic material in the form of siliceous tuff beds in outcropping mudrock appeared within diamictite of the upper Dwyka Formation of the south crop north of Klaarstrom, about 100 km south of Beaufort West (Table 2, 1a). This is the first sign in the Karoo Basin of the activity of the magmatic zone along the Panthalassan margin, presumably well south of the Cape Fold Belt, which itself lacks contemporaneous magmatic rocks. And (3) the first movements in the Cape Fold Belt were manifested in the Swartberg folding dated at 278 ± 2 Ma (Hälbich et al., 1983, table 13.3). These events (Figs. 3 and 7) foreshadow the subsequent development of the orogen that delimits the southern edge of the Karoo structural basin.

Ecca Group. This account of the Ecca Group is derived from data given mainly by Anderson (1977), Van Vuuren and Cole (1979), Anderson and McLachlan (1979), McLachlan (1973, 1977), Visser et al. (1980), Tankard et al. (1982), Cole and McLachlan (1991), and Cole et al. (1990).

Following the collapse and melting of the Dwyka ice sheet

and the concomitant marine transgression (probably due to a combination of shelf subsidence and glacio-eustatic rise in sea level), the formerly glaciated ground rebounded isostatically to form uplands. The Cargonian Highlands in the northeast (Fig. 11) became the Witwatersrand Arch (Ryan and Whitfield, 1979), and the Eastern Highlands persisted in the east (Fig. 12A). The first folding of the Cape Orogeny, the Swartberg event, dated by the 278 ± 2 Ma recrystallization cleavage in the Cedarberg Formation (Hälbich et al., 1983), probably took place during the deposition of the upper part of the Dwyka Formation (Fig. 7). The event was probably deep-seated because no major facies changes indicate uplift and shedding of sediment in the upper Dwyka and lower Ecca in the basin adjacent to the Cape Fold Belt. The basin stretched southward across the site of the Cape Fold Belt at least to the proto fold belt during Dwyka deposition, possibly as far as a magmatic belt along the Panthalassan margin (Visser, 1987, 1991b). It was not until the second, Outeniqua, folding event dated at 258 ± 2 Ma (Hälbich et al., 1983) that the Cape Fold Belt became a major provenance for Karoo sediment, initially supplying the sand and mud of the upper Ecca Group. The lowermost part of the Ecca Group (Prince Albert, Whitehill, Pietermaritzburg, and Vryheid Formations) conformably succeeded the Dwyka Formation except in the northeast where the Vryheid Formation overstepped the valley-fill deposits of the Dwyka to rest directly on basement (double lines in Fig. 12A).

The question of the marine or nonmarine character of the

J. J. Veevers and Others

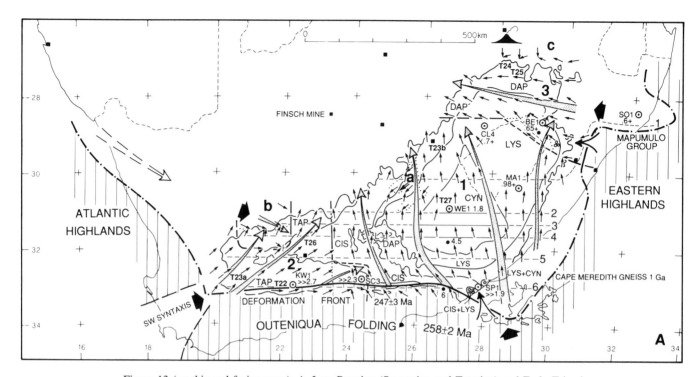

Figure 13 (on this and facing page). A. Late Permian (Sosnovian and Tatarian) and Early Triassic (Scythian) (250–240 Ma) Beaufort Group. Isopachs (km) from drill-hole (circled dot) data (Winter and Venter, 1970, fig. 12, dolerite-free thickness) and inferred original thickness on southern outcrop (filled circle) from Rowsell and de Swardt (1976, p. 84). Tetrapod zones west of 26°E and south of 31°S from Keyser and Smith (1979), rest from Kitching (1970), whose terminology is retained: TAP-*Tapinocephalus,* CIS-*Cistecephalus,* DAP-*Daptocephalus,* LYS-*Lystrosaurus,* CYN-*Cynognathus.* Paleocurrents, from Cole and Wipplinger (1991), are moving averages for each 1° × 1° square progressing in steps of ½°; the vector means were calculated from 13,000 readings taken principally from Theron (1975) and augmented from Botha and Linström (1978), Cole (1980), Hiller and Stavrakis (1980), Kingsley (1970, 1979), Kingsley and Theron (1964), Stavrakis (1980), Turner (1978, 1986), and Visser and Dukas (1979), together with unpublished data from D. I. Cole and various theses and reports. Single arrows show range of paleocurrent directions in each of domains 1–3 (areas of overlap shown by broken and dotted lines) and in enclaves a and b; broad arrows show the general direction of sediment transport. Deformation front of Cape Fold Belt (heavy line) during Beaufort deposition (and deformation) about the southwestern (SW) syntaxis from Söhnge (1983), to the east (with dates) from Hälbich et al. (1983), and on the Falkland Islands from Adie (1952a). Dot-and-dashed line marks foot of highlands. Rudites (circles) in the Katberg Sandstone on the coast near 28°E and in the Belmont Sandstone near the Mooi River from Theron (1975, p. 65). The ash-flow tuff at T24 and T25 probably had a local volcanic source in the northern Karoo Basin. All others had a southern source.

B. Total isopachs (km) of the Dwyka Formation and Ecca and Beaufort Groups, summed from Figs. 11, 12, and 13A. Also shown are the dolerite line (dotted, from Fig. 1), the boundary between the *Cistecephalus* (CIS) and *Tapinocephalus* (TAP) zones (from part A), approximating the boundary between the Teekloof and Abrahamskraal Formations in the west and the Middleton and Koonap Formations in the east, and the Great Escarpment (barred line, from Dingle et al. 1983, fig. 169). BW—Beaufort West (township).

Ecca Group above its marine base is not resolved, and we present the evidence now before continuing the description. From a review of the fossils in the Ecca Group, McLachlan (1973, p. 10) concluded that the water in which the Ecca was deposited "was normally fresh. Certain lines of evidence (the glauconite bands, spinose acritarchs and sponge spicules) suggest that the water was at times saline, but the lack of recog-

nizable marine faunas indicates that a connection with the oceans was unlikely." The only indubitably marine invertebrate recorded from the Ecca Group, an ammonite said to have been from the Vryheid coal measures, was shown to be grossly displaced (McLachlan, 1977). As related above, the marine invertebrates in the upper Dywka Formation and basal Prince Albert Formation constitute the only convincing evi-

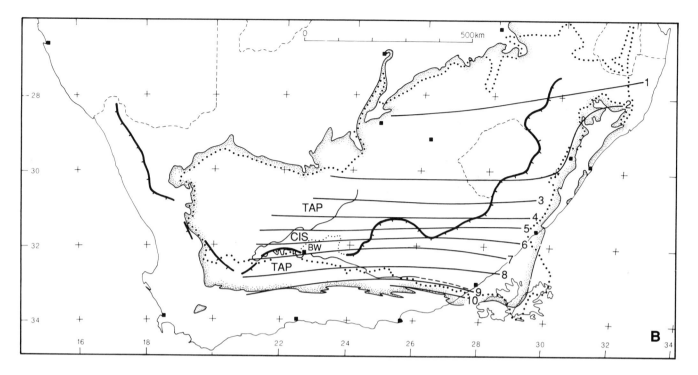

dence of a Permian sea in southern Africa. The water in the later Ecca basin, after the withdrawal of the sea at the end of the Tastubian, may have been brackish to form a dense layer beneath the surface fresh water, and this would reconcile Marchant's (1978) conclusion that the concentration of Ni, Zn, and Cu in organic separations of the Ecca shale favors deposition in saline water. Zawada (1988) used the concentration of Rb, B, V, and Cu, and adsorbable Mg^{2+} and Ca^{2+} to show that the Ecca Group mudrocks were probably deposited in fresh or brackish water. According to Van Vuuren and Cole (1979, p. 109), the glauconite in the Ecca Group is confined to the transgressive top of two of the regressive cycles (B2 and C) of the Vryheid Formation in the 1-m thick poorly sorted sandstone on top of fluvial sandstone. They cite another occurrence of glauconite in South Africa in the soil and calcrete of modern salt pans, in a setting possibly comparable with the occurrence of casts of gypsum in the dolostone of the Whitehill Formation (McLachlan and Anderson, 1977, p. 93). The essential paleogeographic distinction is between the marine and nonmarine realms. Saline water is not uniquely marine, as signified by the term *salina* for a class of lake, but freshwater is uniquely nonmarine.

We believe that the water of the Ecca basin was brackish to fresh except in the northeast, where the fluvial parts of the Vryheid Formation were wholly fresh. Later, as a result of the shrinking of the basin due to uplift along the Cape Fold Belt and an influx of deltaic sediment, the environment became entirely nonmarine. In the southwest, the nonmarine sediments form the uppermost units of the Ecca Group: the Koedoesberg and Waterford Formations (Fig. 9). The Waterford Formation

contains endemic freshwater or nonmarine bivalves (Cooper, 1979; Cooper and Kensley, 1984). Cooper (1979) and Cooper and Kensley (1984) pointed out the connection with the nonmarine fauna of the Estrada Nova Formation of the Parana Basin of South America, which was deposited in an isolated body of brackish water likened by Runnegar and Newell (1971) to the Caspian Sea, a vast lake filled with fresh to brackish water, and isolated from the ocean. In the northern Caspian Sea, fresh surface water from the Volga delta flows over the brackish lake water to form a permanent stratification, and the deeper (>800 m) floor in the central and southern parts is anoxic (Yassini, 1987, fig. 5, table 1).

The Ecca Group is divided into lower and upper sequences. The lower sequence comprises the Prince Albert, Whitehill, Pietermaritzburg, and Vryheid Formations; the upper comprises the Collingham, Vischkuil, Laingsburg, Ripon, Fort Brown, Tierberg, Volksrust, Skoorsteenberg, Kookfontein, Koedesberg, and Waterford Formations (Figs. 7 and 9; Kent, 1980, fig. 7.3.3).

Lower Ecca. In the Sakmarian and Artinskian palynozones 2 and 3 (see Fig. 21 in Appendix 2), the lower Ecca sequence of shale was deposited over the Dwyka Formation in an initially shallow marine and then a shallow brackish-water basin (Visser, 1991b). In the south the shale is called the Prince Albert Formation. It is overlain by the 50-m-thick white-weathering ("white band") pyritic and carbonaceous Whitehill Formation with occasional chert lenses (Fig. 8A, part II; Cole and McLachlan, 1991), with its age in the *Mesosaurus* zone. The Whitehill Formation extends north-northwestward through the Warmbad and Kalahari basins (Fig. 12A) to South America where it is called the Iratí Formation

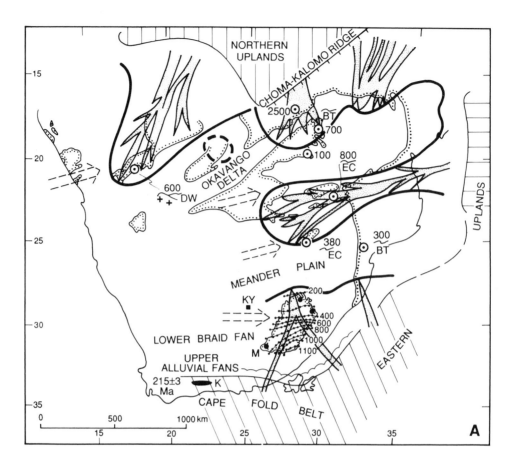

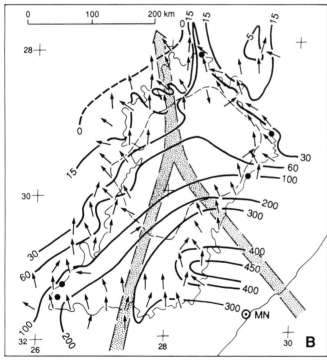

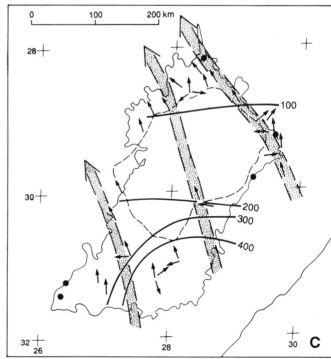

Figure 14. A. Late Triassic–Early Jurassic (Carnian-Pliensbachian: 230–193 Ma) Stormberg Group and equivalents, and overlying volcanics (outlined by parallel full and dotted lines), from Visser (1984, figs. 13 and 20). Sediment thickness (m) at encircled dots is greatest at 2500 m in the Gwenbe district of Zambia in front of the paleoscarp of the Choma-Kalomo granitic ridge (Taverner-Smith, 1962; Rust, 1975). Disconformably underlying rocks are Beaufort equivalent (BT), Ecca (EC), Dwyka (DW), and Precambrian (++). Source areas are ruled. In the recess between the Cape Fold Belt and Eastern Uplands, fluvial facies belts thin downslope from upper alluvial fans (not preserved, from Turner, 1983, fig. 7) through lower braid fans (northern limit of prograding cycle 2 or Indwe Sandstone Member of the Molteno Formation) to meander plains (Elliot Formation), which retrograded 300 km southward during Molteno cycle 6. Isopachs (m) in the Lesotho region from Dingle et al. (1983), and major lines of fluvial sediment transport from Figure 14B. Solid dots in Molteno outcrop indicate clasts >30 cm long, including the 75-cm block of Witteberg quartzite (Table 5) near Molteno township (M). K—Kango inlier with a retrogressive metamorphic date of 215 ± 3 Ma. KY—Kimberley. Other lobate fluvial systems from east and north, from Visser (1984). Broad broken arrows indicate pattern of the winds from which the Clarens Formation (Beukes, 1970) and equivalents (Visser, 1984, fig. 18) were deposited.

B. Late Triassic (Carnian–early Norian) Molteno Formation exposed within the Lesotho mesa. Isopachs (m) and mean maximum length of clasts >30 cm (dot) from Turner (1984) and vector mean cross-bedding azimuths for 89 sample sectors (5,800 measurements at 540 localities) and derived sediment transport pattern, from Turner (1977). MN—Mngazana, an isolated outcrop of Molteno Formation (Dingle et al., 1983, fig. 108).

C. Late Triassic (late Norian) Elliot Formation exposed within the Lesotho mesa. Isopachs (m) from Dingle et al. (1983, fig. 16B). Paleocurrents in northeast (vector means of 63 measurements at 9 localities) from Eriksson (1985) and elsewhere (about 500 measurements at 22 localities) from Botha (1967), indicating a pattern of sediment transport (broad arrows) from the south-southeast.

←——————————————————————————

(Anderson and McLachlan, 1979). In the northeast, the basal shale is called the Pietermaritzburg Formation, and it is overlain by a regressive-transgressive fluvio-deltaic wedge called the Vryheid Formation (Fig. 8A, part I). The productive coal measures of the Karoo coal province (Hobday, 1987) were deposited on the delta plain of this delta complex. In parts of the north, the Vryheid Formation itself steps across both Pietermaritzburg Shale and the Dwyka Formation to rest on basement and drapes the original Dwyka glacial valleys (Van Vuuren and Cole, 1979, figs. 15 and 17). Van Vuuren and Cole (1979) recognize eight individual cycles of delta progradation that shale out by the dotted line in Figure 12A. We interpret the paleocurrents as showing two overlapping delta complexes. The one in the northwest was sourced from the Witwatersrand Arch and is no thicker than 250 m. The complex in the east was sourced from the Eastern Highlands and is at least 1,000 m thick along an axis located on the present coastline in the position of the former Natal trough of the Cape Supergroup. "The abundance of coarse-grained clastics, the increase in the total sand content of the Vryheid Formation and

the increase in the number of recognisable cycles of sedimentation toward the east indicate that the Eastern Highlands were a much more active source than the Witwatersrand Arch" (Van Vuuren and Cole, 1979, p. 110).

Southwest of the line of shale-out (Fig. 12A), the Prince Albert and the Whitehill Formations together thin to <100 m toward the southeast and in a smaller area to the southwest, with the thickest sequence (333+ m) in the west-north-west. Variations in thickness are due partly to the diachronous contact between the Dwyka and Prince Albert Formations. The mud at the base of the Prince Albert Formation was deposited from suspension as the distal part of the glacigenic sediment deposited from tidewater glaciers (Visser, 1982, 1989). The thin sequence in the southwest coincides with the Cape Fold Belt syntaxis, interpreted by de Beer (1990) as a basin swell that rose during the first (278 ± 2 Ma) folding event. A similar situation may have applied in the corresponding syntaxis centered on the restored Falkland Islands (Fig. 12A). The overlying Whitehill Formation thickens to 70 m in the west-north-west and to 80 m in the outlier near Worcester (Cole and McLachlan, 1991), and both trends presumably extended to the proposed South Atlantic embayment that connected the Karoo and Paraná Basins (Oelofsen and Araujo, 1987).

Upper Ecca. During palynozone 4 (Fig. 12B), the situation was reversed. Thick sandy sediment came wholly from the south and west, and the northeastern area of the former Vryheid delta complexes was covered by a shale (Volksrust Formation) that wedged out to the northeast. By the end of Ecca deposition, the brackish lake was almost filled with sediment to become the broad fluvial plain of the Beaufort Group (Smith, 1990).

In the south and west, the Upper Ecca comprises regressive deposits that grade upward through deepwater (<500 m) (Visser and Loock, 1978) subaqueous fans to a delta complex (Cole et al., 1990; Kingsley, 1981). In the southwest (Fig. 8B, part IV), the formations in ascending order are the Collingham, 60 m of interlaminated tuffaceous siltstone and shale; the Vischkuil, 250 m of shale, siltstone, and sandstone of the basin plain and outer fan; the Laingsburg, 400 m of almost equal amounts of shale and sandstone of the middle fan; 500 m of the Fort Brown, deposited in a shallow basin and pro-delta plain; and the Waterford, 250 m of delta-front and pro-delta sandstone and shale that contains nonmarine endemic bivalves. In the southeast (Fig. 8B, part V), the Upper Ecca, almost twice as thick as it is in the west (Fig. 12B), comprises the basin-plain Collingham Formation, the outer-middle fan Ripon Formation, and the outer-deltaic Fort Brown Formation. A third regressive sequence is present in a smaller area in the western part of the basin. It comprises the basin-plain Collingham and Tierberg Formations, which are overlain by 200 m of sandstone and shale corresponding to the subaqueous fan Skoorsteenberg Formation, in turn succeeded by rhythmically bedded, pro-delta mudrock of the Kookfontein Formation, and

TABLE 1. FOSSIL INDICATORS OF MARINE AND NONMARINE ENVIRONMENTS IN THE KAROO AND NEIGHBORING BASINS

Indicator	Reliability	Formation/Group	Locality*		Reference†
MARINE ENVIRONMENTS					
Marine invertebrates	High	Prince Albert	**a**	W. Kalahari	1
		(formerly Dwyka Shales)	**b**	Warmbad	1
Marine microfossils	High	Dwyka	**d1**	Ashburton	2
			d2	Tugela Rand	2
			d3	Mkomazi	2
Fish, coprolites	Low	Topmost Dwyka	**c**	Douglas	2
			e	Tankwa R.	2
Shark, marine microfossils	High	Ecca/Dwyka	**f**	Zwartskraal	3
Acritarchs	Low	Basal Dwyka		S Karoo	1
				Vaal R.	4
Glauconite	Low	Ecca (Vryheid)		northeast	4
NON-MARINE ENVIRONMENTS					
Nonmarine bivalves	High	Waterford		Prince Albert	5

*Letters refer to designations used in Figure 11.
†1 = McLachlan and Anderson, 1973; 2 = Von Brunn and Gravenor, 1983, p. 203; 3 = Oelofsen, 1986; 4 = Anderson, 1977, p. 51; 5 = Cooper and Kensley, 1984.

finally by delta-front and delta-plain sandstone of the Koedoesberg Formation (Fig. 9; Cole et al., 1990, p. 6).

The Upper Ecca Group represents the first pulse of thick sediment shed into the foredeep from the rising proto-Cape Fold Belt, which underwent a second deformation, the Outeniqua folding, at 258 ± 2 Ma (Häblich et al., 1983, table 13.3) toward the end of Ecca deposition. Folding probably began earlier than its climax (de Beer, 1991) during rapid subsidence of the shallow floor of the Whitehill basin. With continued subsidence outstripping deposition, the foredeep reached a maximum depth of about 500 m during the deposition of the subaqueous fans (Visser and Loock, 1978). The isopachs now parallel the strike of the Cape Fold Belt, with the thickest known section near the coast and on the Falkland Islands, interpreted as lying at the immediate foot of the Cape Fold Belt. In the west, sediment was transported down an east-north-eastward paleoslope. A Tankwa sub-basin was separated from a southern Laingsburg sub-basin by a basin-floor swell located along the line of the Cape Fold Belt syntaxis (Cole et al., 1990, p. 6; de Beer, 1990). Sediment in the Tankwa sub-basin was probably derived by consequent drainage down the flank of the resurgent Atlantic Mountains that expanded eastward to the Western Branch of the Cape Fold Belt (de Beer, 1990). Across the syntaxis in the Laingsburg sub-basin and the rest of the foredeep, sediment was derived by consequent drainage from the Southern Branch of the Cape Fold Belt (Fig. 12B). The Atlantic Mountains temporarily interrupted or restricted the previous Tastubian upper Dwyka/basal Prince Albert marine and Artinskian Lower Ecca brackish-lacustrine connection between southern Africa and South America. A connection was restored during the growth of the early Kazanian Waterford-Estrada Nova lake with its endemic bivalves.

Beaufort Group. This account of the Beaufort Group is derived from data given mainly by Anderson (1977), Tankard et al. (1982), Dingle et al. (1983), Cole and Wipplinger (1991), and the works of B. R. Turner that are listed in the References Cited section.

Following the filling of the Ecca basin by deltas that prograded initially southwestward from the northeast and then northward from the south, a much greater volume of sediment from the faster rising Cape Fold Belt prograded diachronously northward in rivers that flowed across the floor of the former brackish to freshwater basin down a 500-km-long piedmont flank into a westward-sloping axis of sediment transport (Fig. 13A). During 15 million years of deposition, the piedmont wedge reached an estimated thickness of 6 km at the south crop. The rapid northward thinning of the Beaufort Group is effected by a combination of younging of the Ecca/Beaufort boundary and deeper erosion of upper Beaufort strata before deposition of the Late Triassic Molteno Formation (Fig. 7). The mud and sand are commonly red, due to the well-drained, hence highly oxidized, fluvial slope on which they were deposited as well as to the onset of a warmer and seasonal global climate. The related poverty of palynomorphs—Anderson

TABLE 2. VOLCANIGENIC MATERIAL IN THE TASTUBIAN TO SCYTHIAN PART OF THE KAROO BASIN INDICATING COEVAL DISTAL VULCANICITY EXCEPT THE PROXIMAL 24 AND 25*

Unit	Locality	Material	Reference
27. Burgersdorp	Herschel	Laumontite	Fuller, 1970
Burgersdorp[†]	**Bf 21**	**44 %**	
Katberg	**Bf 2**	**48 %**	
Balfour	**Bf 16, 17, 20**	**84 %**	
Middleton	**Bf 7, 8**	**72 %**	
Koonap	**Bf 10**	**61 %**	
26. Lower Beaufort (Teekloof, Abrahamskraal)	SW Karoo	i. Shards ii. Sand-sized clasts of alkaline trachyte iii. Drop-like bodies of trachyte iv. Volcanic quartz v. Laumontite	Ho-Tun, 1979 Turner, 1978
25. Lower Beaufort	Blydschap near Frankfort	Ash-flow tuff with pumice lapilli	Keyser and Zawada, 1988
24. Lower Beaufort	Oranje near Heilbron	Ash-flow tuff	Keyser and Zawada, 1988
23. Lower Beaufort Adelaide Subgroup	SW Karoo Edenburg	(a) Shards (b) Shards	Le Roux, 1985 Le Roux, 1985
Waterford	**E3**	**50 %**	
Ripon	**E2, 5**	**56 %**	
22. Base of Beaufort	Prince Albert Road	Shards in silicified tuff	Martini, 1974
21. Fort Brown	Geelhoutboom	Crystal (plagioclase) tuff	Lock and Johnson, 1974
20. Fort Brown	13 km N of Wolwefontein	Shards in felsitic groundmass (SA 10)	Appendix 1; Figure 20D
19. Base Collingham	Remhoogte	Shards plus volcanic quartz Shards (SA 15)	Elliot and Watts, 1974 Appendix 1, Figure 20B
18. Base Collingham	Ecca Pass	Shards (SA 7a)	Appendix 1, Figure 20C
17. Base Collingham Matjiesfontein chert	Remhoogte	Tuff (SA 14c)	Appendix 1; Figure 20E
16. Base Volksrust	SW1/67	Tuff (K-bentonite)	Viljoen, 1990
15. Base Volksrust	BE1/66	Tuff (K-bentonite)	Viljoen, 1990
14. Base Tierberg	Hopetown	Altered tuff (illite-mont. claystone)	McLachlan and Jonker, 1990
13. "Volcanic interval" in Collingham	South Crop	Shards, metabentonite	Lock and Wilson, 1975
12. Whitehill	Hopetown	Altered tuff (illite-mont. claystone)	McLachlan and Jonker, 1990
11. Whitehill	KL1, Hopetown	Shards in crystal tuff	McLachlan and Jonker, 1990
10. Whitehill	Remhoogte	Shards (SA 16a)	Appendix 1, Figure 20A
9. Whitehill	South Crop	Shards in dolostones	McLachlan and Anderson, 1977
8. Base Beaufort to top of Prince Albert	South Crop	Ash beds plus laumontite	Martini, 1974
7. Tierberg to Prince Albert	HG1	Tuff (K-bentonites)	Viljoen, 1990
6. Prince Albert ("Upper Dwyka Shales")	OL1/69	"Volcanoclastic material"	Rowsell and de Swardt, 1976, p. 121
5. Prince Albert	D731	Shards in ash beds	McLachlan and Jonker, 1990
4. Prince Albert	EP1	Shards in claystone	McLachlan and Jonker, 1990
3. Pietermaritzburg	Koppies	Altered ash (mont. claystone)	Schmidt, 1976
2. Laingsburg–Prince Albert	Laingsburg	Tuff beds (K-bentonite)	Viljoen, 1990
1b. Vischkuil–upper Dwyka	QU1/65	Tuff beds (K-bentonite)	Viljoen, 1990
1a. Upper Dwyka	Klaarstrom	Tuff beds (K-bentonite)	Viljoen (1993, unpublished)

*Arranged in stratigraphic order. Located in Figures 1, 7, and 21, and illustrated in Figure 20.
[†]Supplementary data (**bold**), from Johnson, 1991, Table 2, are selected from those southeastern Karoo Basin sandstones with >44% volcanic rock fragments (Fig. 15C). Whether these fragments are derived from juvenile (coeval) pyroclastics or from ancient (pre-Permian) volcanic rocks is unknown, and their inclusion here as pyroclastics is tentative.

(1977) was unable to zone the Early Triassic upper part of the Beaufort Group because it yielded few palynomorphs—is compensated for by the diversity and abundance of tetrapods, which are grouped into five zones (Fig. 13A). The Beaufort Group is exposed over the entire Karoo Basin except in Lesotho where it is covered by the Stormberg Group, and this large outcrop has provided the wealth of paleocurrent vectors shown in Figure 13A. It extended also northwest at least as far as the Finsch mine, 140 km west-northwest of Kimberley, as indicated by xenoliths in a kimberlite pipe (Visser, 1972). The Beaufort Group is subdivided into two parts: the Adelaide Subgroup of greenish gray and grayish-red mudstone and sandstone overlain by the Tarkastad Subgroup with more sandstone and red mudstone (Kent, 1980, p. 538).

Adelaide Subgroup. In the southwest (Fig. 8B, part VI), the Waterford delta is succeeded by the fluvial channel and flood basin deposits of the Abrahamskraal and Teekloof Formations. The channels, which contain uranium (Ryan and Whitfield, 1979), range from low- to high-sinuosity, and the fluvial energy decreases upward due to denudation of the source area (Turner, 1985). In the southeast (Fig. 8A, parts V and VII), the Fort Brown delta is continued upward by the Koonap Formation and then succeeded by the fluvial Middleton Formation and Balfour Formation, which is punctuated by the coarse influx of braided channels at the base. In the northeast (Fig. 8A, part III), the lacustrine Volksrust Shale is succeeded by the Estcourt Formation of deltaic sediment, including thin coal measures (unique in the Beaufort), capped by fluvial-lacustrine sediment (Tankard et al., 1982, p. 393–394).

Tarkastad Subgroup. The Tarkastad Subgroup corresponds to the Early Triassic *Lystrosaurus* and *Cynognathus* tetrapod zones. A strong pulse of fluvial braided channel sandstone is represented by the 500–1,000-m-thick proximal Katberg Formation in the southeast (Fig. 8C, part IX) (Hiller and Stavrakis, 1980, 1984) and 100-m-thick distal Belmont Formation in the northeast (Figs. 7, 8C, part VIII), overlain by the fluvial meandering-channel Burgersdorp Formation and flood basin–lacustrine Otterburn Formation, the tops of which are exposed at the low-angle unconformity at the base of the Late Triassic Molteno Formation. The pulse is concomitant with folding and thrusting along the front of the Cape Fold Belt, involving older Beaufort strata, and dated by Hälbich et al. (1983) as 247 ± 3 Ma (latest Permian) (Fig. 7).

Paleocurrent analysis. The paleocurrent vectors were measured from the exposed parts of the Beaufort Group that range through five tetrapod zones (Fig. 13A). The vectors have been grouped into three domains of uniform trend: (1) a large N to NNW domain covering the entire outcrop except the extreme west and north, centered on the *Cynognathus* zone and sampled from all five zones, with an enclave (a) with SW trend on the northwest; (2) a small NE domain in the west, sampled from the *Tapinocephalus* and *Cistecephalus* zones, with a small area (b) of SE-E trend on its northern edge, restricted to the *Tapinocephalus* zone; and (3) a small WNW-W-

WSW domain in the far north, from the *Daptocephalus* and *Lystrosaurus* zones, with an enclave (c) of southerly trend in the extreme north (Keyser and Zawada, 1988). The paleocurrent patterns apply to specific units or groups of units within the Beaufort Group. The southeastern domain 1, encompassing the entire Beaufort Group from the Koonap Formation to the Burgersdorp Formation (Fig. 7), has a consistent northerly paleocurrent. The paleoslope arrows radiate from a focus near the proposed southeastern syntaxis of the Cape Fold Belt as do those of domain 2 from the southwestern syntaxis (de Beer, 1990). The syntaxes reflect the intersections of the pre-Cape (1.0–0.5 Ga) basement trends of the east-west Saldanian strike with the southwesterly Natal-Mozambiquean strike on the east, and with the southeasterly strike of the Gariep trend on the west (Fig. 5, inset). During deposition of the Beaufort Group, the Saldanian trend was paralleled by the growing Cape Fold Belt of the Outeniqua folds and the deformation front, which incorporated previously deposited Karoo sediment up to the Adelaide Subgroup. The Natal trend is paralleled by the Cape Meredith–Mapumulo trend, which is identified as a resurgent Eastern Highlands (Fig. 12A). The Mapumulo Group of the Natal Metamorphic Province, about 1 Ga old, is specifically identified as a proximal source of the angular to subrounded clasts (<19 cm across) of garnetiferous gneiss, red granite, milky quartz, and large microclines in the Belmont Formation at Mooi River (circles) (Theron, 1975). To the south-south-east, the Falkland Islands are specifically identified as a proximal source of the Katberg Formation near East London (circles), which contains pebble clasts (<12 cm across) of lignite and silicified wood (from the Devonian or Permian or both) and a distal source of the gneiss pebbles (from the 1 Ga Cape Meredith gneiss). The Gariep trend is paralleled by the Western Branch of the Cape Fold Belt (de Beer, 1990), which is identified as the resurgent Atlantic Mountains (Fig. 12B). The source of paleocurrent domain 2 was probably the Baviaanshoek and Hex River anticlinoria of the southwestern syntaxis, and that of enclave b the more distant northwest (de Beer, 1990).

In the northernmost part of the basin, the west-trending arrows of domain 3 are interpreted as showing sediment funneled through a gap or saddle in the Eastern Highlands along an axial paleoslope parallel to the Mapumulo piedmont and at a high angle to the flank slope of the Cape piedmont. This entry point of sediment from the east echoes the situation that pertained during deposition of the Lower Ecca Group (Fig. 12A). Another echo, of the Lower Ecca drainage axial to the Witwatersrand Arch (Fig. 12B), is the southerly drainage in enclave c. The southwesterly drainage in enclave a spans the *Daptocephalus* zone and probably part of the *Lystrosaurus* zone in the central and southern Orange Free State (Kingsley and Theron, 1964). This drainage system was responsible for the deposition of coarse arkose derived from an intrabasinal upland of Precambrian granite and sedimentary inliers. Coarse arkose of the southwesterly drainage interfingered with fine-

grained sandstone of the major N to NNW drainage (Kingsley and Theron, 1964).

Summary of the lower Karoo Sequence. The lower part of the Karoo Sequence—the Dwyka, Ecca, and Beaufort—was deposited continuously during the Permian and Early Triassic. The cross-sectional wedge shape of the lower Karoo (Fig. 13B) has two parts: (1) a steep inclined part, at most 250 km long, from the 10-km isopach at the front of the Cape Fold Belt at about 33°S to the 3-km isopach, and thence (2) a low inclined part, at least 500 km long, from the 3-km to the 1-km isopach. The steep part of the wedge is the foredeep of the foreland basin, seen in section in Fig. 2, south of QU1/65 in AB, and south of WE1/66 in CD.

The upper part of the sequence—the Stormberg Group (Molteno, Elliot, and Clarens Formations) and the Drakensberg Group—is separated from the lower Karoo by a lacuna that occupies the entire Middle Triassic. The disconformity between the lower and upper Karoo is underlain by the Tarkastad Subgroup that ranges from the *Cynognathus* zone over most of the area (Burgersdorp Formation; Welman et al., 1991) to the *Lystrosaurus* zone in the extreme north (Otterburn Formation); it is overlain by successively younger cycles of the Molteno Sandstone so that the Indwe Sandstone Member, the second cycle in the south, oversteps the *Cynognathus* zone to rest on the *Lystrosaurus* zone in the north (Dingle et al., 1983, fig. 20) (Fig. 2, CD; Fig. 7). An epeirogenic uplift brought all southern Africa (including the axis of the Karoo Basin) above base level to produce the Middle Triassic lacuna by erosion down to the *Lystrosaurus* zone. At the end of the Middle Triassic (230 ± 3 Ma), the Cape Fold Belt was deformed for the last time by listric thrusting and folding (Hälbich et al., 1983). At about this time, its piedmont was depressed below base level at about 32°S to resume deposition in the form of the initial formation of the Stormberg Group, the Carnian (230–225 Ma) Molteno Formation.

Stormberg Group. This account of the Stormberg Group is derived mainly from data given by Rust (1975), Dingle et al. (1983), Visser (1984), and Turner (1977, 1983, 1984).

The relaxation of the Middle Triassic plateau uplift by depression of the region north of the terminally deformed Cape Fold Belt along an axis at the Tropic of Capricorn resulted in the deposition of the Stormberg Group in lobes of fluvial (Molteno and Elliot Formations) and eolian-lacustrine (Clarens Formation) sediment (Fig. 7).

Molteno and Elliot Formations. The Molteno Formation, dated as Carnian–early Norian from its megaflora (Anderson and Anderson, 1983), disconformably overlies the Beaufort Group above the Middle Triassic lacuna and is conformably overlain (and in the north overstepped) by the Elliot Formation. The following description is drawn from Turner (1983). Of the six fining-upward cycles of the Molteno Formation (Fig. 8C, part IX), the first prograded some 100 km northward of the southern edge of outcrop; the second, called the Indwe Sandstone Member, prograded at least to the north-

ern edge of outcrop (Fig. 8C, part VIII); the remaining cycles 3–6 receded toward the southern edge (Fig. 7) and are followed by the downslope facies equivalent, the Elliot Formation, which itself finally receded southward beyond the southern edge to replace the entire Molteno Formation. The Molteno and Elliot Formations thin northward individually (Fig. 14B and C) and collectively with the Clarens Formation as the Stormberg Group (Fig. 14A). The marked thinning of the Molteno Formation north of 30°S is due to the wedge-out of the northward extensive Indwe Sandstone Member.

A typical fining-upward sequence of the Molteno Formation comprises, from the base upward:

(1) A mainly matrix-supported conglomerate set in a pebbly very coarse feldspathic sandstone, 5–120 cm thick. Extraformational clasts are sparse (most are found in the Indwe Sandstone member) and are predominantly quartzite, including the 75-cm-long clast of the Witteberg Group at Molteno (Fig. 14A, and *see* Table 5 in a following section); clast size decreases from the southeast to northwest (Fig. 14B) except at the northern edge where the high caliber of clasts is interpreted as due to a funneling of braid channels into a single main channel (Turner, 1984, fig. 5).

(2) Coarse to very coarse multilateral and multistoried sandstone (feldspathic lithic arenite), which forms laterally extensive sheets from 30 to 120 m thick. Trough crossbedding predominates, in cosets up to 1.5 m thick and 6 m long; crossbedding indicates a north-northwest paleoslope except for sequence 1, to the north-northeast.

(3) Finer-grained lenticular sandstone and siltstone, <12 m thick; flat laminated with primary current lineation and minor ripple cross-lamination.

(4) Lenticular and impersistent shale and coal, up to 6 m thick. The shale contains the *Dicroidium* flora (Anderson and Anderson, 1983), specimens of which are most abundant in cycle 2, least in cycles 1 and 6; tree trunks of *Dadoxylon,* some in living position, are fossilized in this and the underlying facies. The coal is lenticular in stringers and seams, thickest in cycle 1, barely developed in cycles 5 and 6; it is up to 4.3 m thick in the Indwe Seam, a composite of equal parts of coal and shale/fireclay (Turner, 1971; Dingle et al., 1983). We interpret tonsteins in the coal measures in the southeast (Heinemann and Buhmann, 1987) as the first sign of proximal Karoo vulcanicity (Table 3).

Turner (1983) interprets the fining-upward sequences as deposits of a braided fluvial system. The basal conglomerate was deposited as a channel lag by a single high-energy short-lived event, probably high seasonal rainfall and the failure of earth- or ice-dammed lakes in the source area. The overlying trough crossbedded sandstone is attributed to dune migration within channels, and the finer-grained sandstone and siltstone deposited from waning bedload and suspension-load currents as a consequence of channel shifting, abandonment, and overbank flooding. The shale was deposited by vertical accretion from overbank floods. "Standing bodies of water on the allu-

Unit	Locality	Material	Reference
34. Lesotho lavas, Karoo dykes	Karoo Basin	Basalt, dolerite	Eales et al., 1984
33. Clarens	Barkly East	Pyroclastics and basalt	Lock et al., 1974
32. Clarens and Elliot	Dordrecht **a** Maclear **b**	Basalt and breccia	Eales et al., 1984, p. 6
31. Clarens and Elliot	Herschel	Laumontite (altered volcanic glass)	Fuller, 1970
30. Elliot	Golden Gate	Glass shards	Bristow and Saggerson, 1983a, p. 1027
29. Elliot	Jamestown	Bentonite	Eales et al., 1984, p. 6
28. Molteno	Maluti	Tonsteins	Heinemann and Buhmann, 1987

vial plain became the locus of plant growth giving rise to thin, *in situ* coals whose development and preservation was encouraged by the abundant though seasonal rainfall, and maintenance of a high water table" (Turner, 1983, p. 82).

Turner (1983) concluded that the Molteno Formation was deposited by perennial high-energy braided streams that drained an extensive alluvial plain on the lower slopes of an alluvial fan complex. In Figure 14A, we show the inferred distribution of the (since eroded) upper alluvial fans and the downslope lower braid fan of the Molteno Formation at its most extensive development in the Indwe Sandstone member. Still further downslope is the meander plain (Elliot Formation) that receded sourceward after the deposition of the Indwe Sandstone member and finally, in the basin axis but transverse to the inferred Eastern Uplands, a fluvial system flowing westward along the present Tropic of Capricorn.

The *Dicroidium* flora and *Dadoxylon* with growth rings suggest a cool temperate climate with marked seasonal rainfall. "The feldspars include both fresh and altered varieties . . . suggesting an uplifted or fault-blocked igneous terrain of high-relief undergoing rapid erosion with mechanical weathering dominant over chemical weathering" (Turner, 1983, p. 83).

The Elliot Formation, formerly the Red Beds Stage, was deposited in the lower tract of the transverse fluvial system. The section that overlies cycle 6 of the Molteno Formation in the south (Fig. 8C, part IX, from Visser and Botha, 1980) is 460 m thick and comprises three parts: (1) a lower interval (0–200 m) of stacked multistoried crossbedded medium-grained sandstone with alternating fine-grained sandstone and mudstone, interpreted as deposited in a meandering channel and flood plain; (2) a middle interval (200–400 m) of alternating very fine-grained sandstone and red mudstone with CO_3 concretions, interpreted as deposits of a flood basin; and (3) an upper interval (400–460 m) of alternating crossbedded fine-grained sandstone and red silty mudstone with CO_3 concretions and conchostracans, deposited in flood fans reworked by the wind in eolian dunes. According to Turner (1972), the lower boundary of the Elliot Formation is denoted by a combi-

nation of abundant reptile remains, persistent red mudstone, no carbonaceous shale, and a low sandstone:shale ratio. In the Elliot Formation, the ubiquitous crossbeds in the underlying Molteno Formation give way to sporadic ones, which indicate a north-northwest paleoslope (Fig. 14C).

In the north, the 40-m-thick Elliot Formation oversteps the outlapping Indwe Sandstone member of the Molteno Formation (Fig. 8C, VIII) to rest on the Otterburn Formation of the Beaufort Group (Botha, 1967; Turner, 1983, fig. 2), and comprises alternating floodbasin-lacustrine and anastomosing channel deposits (Eriksson, 1985).

Bentonites and glass shards (Table 3) indicate a second early phase of proximal Karoo vulcanicity.

Clarens Formation. The Clarens Formation (Fig. 8C, X), formerly the Cave Sandstone, passes upward from the Elliot Formation "at the point where red and purple mudstones and crossbedded sandstones become subordinate to lighter and 'massive' sandstones and grey/green shales" (Dingle et al., 1983, p. 48). The Clarens Formation is a sequence, from the base upward, of three units (Beukes, 1970; Eriksson, 1979; Tankard et al., 1982, p. 397–399):

(I) Planar crossbedded braided-channel medium- to coarse-grained sandstone, up to 50 m thick in the Natal Drakensberg, alternating elsewhere with lacustrine siltstone that contains fish, crustaceans, and dinosaurs. According to Eriksson (1979), the unit was deposited in converging wadi systems.

(II) Fine-grained sandstone in crossbeds up to 10 m high deposited by straight-crested transverse eolian dunes and in crossbeds up to 3 m high deposited by barchan dunes. The unit is up to 60 m thick in the north and was deposited from westerly winds that redistributed sand from the Elliot Formation and older formations.

(III) An alternation of (I) (but without the coarse-grained braided-channel sandstone) and (II), and toward the top lenses of basalt and pyroclastics (Table 3).

Summary of Stormberg Group. According to Turner (1986), the Stormberg Group has a modern analogue in the central Australian Lake Eyre basin, an arid aggrading internal

drainage basin flanked by eolian dunes, in which braided and anastomosing channel sand (Rust, 1981) gives way downslope to the lacustrine mud of Lake Eyre (Veevers and Rundle, 1979).

However, Lake Eyre was warm and wet during deposition of the relict braided sand sheets and tectonism had little influence whereas cool wet tectonically active conditions prevailed during deposition of the equivalent coarse braided channel subfacies in the Karoo sequence. Furthermore, the Lake Eyre channels are essentially braided and ephemeral whereas in the Karoo sequence braided bedload channels passed downslope into low-sinuosity mixed-load channels, both of which were probably perennial. These in turn passed distally into ephemeral low-sinuosity channels and lacustrine flood flats (Turner, 1986, p. 251–252).

Another model is the Okavango Delta of northwest Botswana (Fig. 14A), which "has an important bearing on patterns of fluvial sedimentation in arid regions since it shows many characteristics of temperate, well-vegetated anastomosed fluvial systems despite its location in the Kalahari Desert" (Turner, 1986, p. 231).

The Stormberg Group is now limited to the Lesotho mesa and the nearby Mngazana area at the recess between the Cape Fold Belt and Eastern Uplands but probably extended wider in the Karoo Basin, as suggested by the possible occurrence in the Cretaceous (94–86 Ma) (Dawson,1989, p. 331) Kimberley pipe of all units (Ecca through Drakensberg) above the outcropping Dwyka (Fig. 14A) (Truswell, 1977, p. 172). Within its preserved limits, the entire Stormberg Group thins northwestward and northward from a maximum known thickness of 1,100 m in the southern part of the recess to 200 m in the north (Fig. 14A). The Stormberg Group started to accumulate during the subsidence in the Karoo Basin that accompanied the fourth diastrophic event in the Cape Fold Belt (final deformation and listric thrusting and folding) (Fig. 7). Turner (1983, fig. 6) interpreted the three pulses of upward-fining deposition in the Molteno and Elliot Formations complex as follows: cycle 1 follows initial uplift/denudation and scarp retreat; cycle 2 (Indwe Sandstone) of maximum progradation follows maximum uplift/denudation and scarp retreat; and cycles 3–6 follow lowering of the basin margin by back-faulting so that the equivalent Elliot Formation retrogrades southward over the Molteno Formation. A regressive metamorphic age of 215 ± 3 Ma in metabasic rocks of the basement Kango Group (K in Fig. 14A) is interpreted by Hälbich et al. (1983, table 13.3) as registering uplift and horizontal tension in the Cape Fold Belt, and it correlates with the final phase of braidplain deposition at the base of the Clarens Formation (Fig. 7). Basin subsidence in the north, in front of the Northern Uplands, including the upfaulted Choma-Kalomo Ridge and downfaulted mid-Zambezi valley, produced a similar set of contemporaneous strata on the northern flank of the wide basin, which had a west-sloping axis along the present Tropic. The chief factor in the accumulation of the Stormberg Group and its equivalents in southern Africa was differential subsidence of the Middle

Triassic plateau. Thrusting was important in elevating the source area of the Cape Fold Belt and depressing the coupled foredeep of the Karoo Basin, although the amount of subsidence, measured in hundreds of meters, was much less than the kilometers during the Beaufort deposition. Subsidence over a wider area to the north locally produced a sediment thickness of 2,500 m and was unrelated to this mechanism. In the tectonic synthesis, we suggest that wider subsidence was due to the release of Pangean heat during the Carnian (Veevers, 1989), coincident with the first appearance of proximal Karoo vulcanicity in the Molteno Formation (Table 3).

Drakensberg Group and Lebombo Group. The Lesotho mesa is capped by 1,400 m of Jurassic basaltic flood lavas called the Drakensberg Group that were fed by a complex of dykes and sills called the Karoo dolerite (Eales et al., 1984, p. 9) in the culmination of proximal Karoo vulcanicity. The Karoo dolerite extends over all but the southern edge of the Karoo basin, as delimited by the dolerite line on Fig. 1, implying an original extension of the Drakensberg Volcanics in the Karoo Basin out to the present edge of outcrop and beyond, as shown by abundant basalt xenoliths within younger diatremes at Postmasburg, 150 km WNW of Kimberley (Eales et al., 1984, p. 3) before the plateau surface was worn back during the later Mesozoic and Cenozoic to the present remnant of Lesotho. Sedimentary basins north of the Karoo Basin are capped by lavas or intruded by dolerite of the same Jurassic age; Namibia additionally has lavas of Early Cretaceous age (Figs. 6 and 14A); to the east, the Transantarctic Mountains (*see* Chapter 4) and Tasmania (Hergt et al., 1989, p. 377) contain a similar volcanic sequence of Ferrar-Tasmanian dolerite feeding the surface lavas of the Kirkpatrick Basalt–Ida Bay basalt; to the west, the Amazon Basin contains equally voluminous (340,000 km^3) Late Triassic and Jurassic dolerite (Mosmann et al., 1986); the Paraná (Serra Geral) basalt is Early Cretaceous, all comprising a low-Ti province (Cox, 1988). The flood basalts reflect an enormous drain on the Pangean heat store, between the earlier phases of sagging and rifting and the final phase of seafloor spreading (Veevers, 1989).

Sources of information on the Drakensberg Group drawn on here are Dingle et al. (1983), Bristow and Saggerson (1983a, 1983b), Visser (1984), Eales et al. (1984), and Fitch and Miller (1984).

The Drakensberg Group comprises lava flows interlayered in the basal third with pyroclastics and thin lacustrine sediment (Fig. 8C, part X). At the base, about 200 individual shield volcanoes and explosive vents are known from all but the northeastern part of the Lesotho mesa (Dingle et al., 1983, p. 56). At a well-mapped example in the Moshesh's Ford area near Barkly East, 200 m of basaltic vent fill and extra-vent effusive and ejected material overlie a surface with 100 m relief in the Clarens Formation (Lock et al., 1974). Interlayered with the volcanics are lenses of lacustrine sandstone, most no more than 10 m thick except an exceptional one 130 m thick and 25

km long (Eales et al., 1984, p. 6, 8). Other evidence for aqueous environments comes from the presence of pillow lava. Above the basal section is the Lesotho Formation, a monotonous succession of amygdaloidal basalt lava flows, nearly all of which are of the pahoehoe type. Thicknesses of individual flows range from 0.5 to 50 m (Eales et al., 1984, p. 8), and the aggregate thickness to the exposed top is 1,400 m. The basalt is tholeiitic. Rare silicic lavas in the andesite-dacite range are encountered in the lower part of the Drakensberg Group. "Thus in the north-east Cape, a feature of the overall development of the Karoo igneous events is that the earliest eruptions were characterized by diversity in style and composition of the erupted products. This later evolved into widespread and regular effusion of compositionally monotonous lavas which built the bulk of the volcanic pile" (Eales et al., 1984, p. 8).

The subvolcanic complex of dolerite dykes, sills, "bell-jar intrusions" and irregular bodies that intruded the Karoo sediments represents the feeder channels for the overlying lavas, as shown unequivocally by major and trace element compositional data (Eales et al., 1984, table 1). Within the limits of the dolerite line that skirts basement and the southern edge of the Karoo Basin (Fig. 1), the greatest concentration of dolerite (dolerite:sediment ratio of >0.5) lies in an elliptical zone that trends northeast across northern Lesotho; the dolerite sheets "seem to terminate at a critical distance of a few thousand feet below the basalts" (Winter and Venter, 1970). Dingle et al. (1983, p. 64) explain this observation and the stratigraphic distribution of sills in the Beaufort and Molteno, dolerite plugs in the Molteno, tuff in the Elliot, agglomerate and lava in the Clarens, and pyroclastics and lava at the base of the Drakensberg Group and lava above by the following sequence of events: (1) earliest upward movement of magma in the Late Triassic to produce explosive activity; we believe the Molteno tonsteins (28 in Table 3) are the first sign of proximal Karoo vulcanicity; (2) during Late Triassic–Early Jurassic Elliot and Clarens deposition, an increased scale of upward movement of magma in sills 1 to 2 km beneath the surface, degassing of the magma, and contact with ground water produced explosive action and created diatremes (29–33); (3) main eruption of flood basalt by dyke injection in the Early Jurassic (193 ± 5 Ma), followed by a second major injection in the Middle Jurassic (178 ± 5 Ma), as dated by Fitch and Miller (1984).

In the southern part of the basin (Fig. 13B), the CIS/TAP boundary coincides with the dolerite line, indicating that the dolerite sills intrude only the section above the base of the CIS zone. Given that the sills dilated only at a critically lower pressure related to the thickness of the overburden in the Early Jurassic, we infer that sediment at least above the *Tapinocephalus* zone (upper half of Beaufort, Stormberg, and perhaps some of the Drakensberg Group) had accumulated a thickness in the west equal to that in the east and use this as the basis for extending the isopachs westward from the values estimated from the better preserved section south of Lesotho.

The coincidence of the dolerite line and the Great Escarpment west of Beaufort West (BW) suggests that backwearing of the plateau in this region has been impeded at the southern limit of the durable dolerite.

Karoo magmatic events extended northward to 15°S. Fitch and Miller (1984, p. 263–264) found the following sequence of events in the Jurassic and Early Cretaceous (Fig. 6): (1) 204 ± 5 Ma: intrusion of isolated alkaline complexes west of the Lebombo line and in Zimbabwe; (2) 193 ± 5 Ma: very rapidly emplaced Lesotho eruptive centers covered by flood basalt that extended along the Lebombo to 25°S, dyke swarms and major intrusions of dolerite in Namibia and Karoo Basin; (3) 186 ± 3 Ma: minor alkaline complex intruded east of Tuli; (4) 178 ± 5 Ma: very rapidly emplaced dolerite intrusions in Namibia and Karoo Basin and bimodal lavas in Lebombo north of 25°S, also Marienthal lavas in Namibia; (5) 165 ± 5 Ma: dolerite dykes and sills in Karoo Basin and Namibia; (6) 150 ± 5 Ma: dykes in Karoo Basin and Namibia, and an intrusive complex in Namibia; (7) 137 ± 5 Ma (Early Cretaceous): a dolerite sill in Lesotho and a syenite complex in Lebombo, and a major group of dolerite intrusions in Namibia.

POST-GONDWANAN HISTORY

Classified as post-Karoo are the 120 ± 5 Ma (Early Cretaceous) Etendeka lavas of the Kaokofeld, the 145–115 Ma (Early Cretaceous) micaceous kimberlites, and the 95–80 Ma (Late Cretaceous) "normal" kimberlites (Dawson, 1989; Allsopp et al., 1989). Deposition of sediment after the Stormberg switched to the continental margins that formed by seafloor spreading (the fourth stage of Pangean heat release) that started about 160 Ma (Middle Jurassic) in the western Indian Ocean and about 130 Ma (Early Cretaceous) in the South Atlantic Ocean with related right-lateral transform motion along the Agulhas Fracture Zone. Dingle et al. (1983) summarize these developments thus:

By mid-Jurassic times, large complex depocentres had begun to develop along what were later to become the continental margins of southern Africa, and today, mid-Jurassic-Cretaceous sediments are known only from the coastal fringes and under the continental shelf and slope. This radical shift in depocentres from a previous mid-continental location, was accomplished by horst and graben development (taphrogenic tectonics) around the edges of the old cratonic blocks, and ultimately led to the breakup of this part of Gondwanaland [p. 99]. . . . With the exception of the central Kalahari Basin (non-marine), and the wide Zululand-Mozambique coastal plain, all the extensive deposits of Tertiary strata occur on the continental shelf and slope, where cover is complete apart from nearshore areas in the south and west Cape. . . . The west coast Tertiary forms a vast, coast parallel lens along the whole margin from the Walvis Ridge to the Agulhas Fracture Zone. . . . This reflects the cessation of differential crustal subsidence, and the progressively subordinate role played by terrigenous sedimentation since Late Cretaceous times, with the concomitant increase in the importance of biogenic and authigenic sediment production. . . . Tertiary sediment distribution along the east coast was, by contrast, still dominated by major terrigenous input points: Zambezi, Limpopo, and Tugela Rivers [p. 234].

In a review of the work of F. Dixey and L. C. King on the development of the southern African landscape, Partridge and Maud (1987, p. 179) conclude that

The Great Escarpment [Fig. 13B] was formed following continental rifting, and its survival up to the present as a major topographic feature is, in large measure, a function of the high elevations consequent upon the central position of southern Africa in Gondwanaland prior to this event. . . . The onshore evidence of erosion surfaces is correlated with recent data on offshore sedimentation, and reveals that a single cycle of erosion (interrupted by minor tectonic interludes) prevailed from the time of rifting to the early Miocene. By the end of this period a gentle pediplain (the African surface) extended across most of southern Africa at elevations of 500–600 m. Most erosion and scarp recession occurred during the earlier part of this interval and produced thick late Jurassic and Cretaceous sedimentary sequences, but shelf sedimentation declined during the Tertiary and had virtually ceased by the Oligocene, when interior planation had advanced to a stage where sediment supply to most rivers was minimal. Modest renewed uplift of 150–300 m in the Miocene tilted the continent slightly to the west and initiated a new (Post-African I) landscape cycle. This was accompanied by renewed offshore sedimentation, although at lower rates than during the Cretaceous. The cycle was terminated near the end of the Pliocene, and its relatively short duration resulted in imperfect planation in most areas to levels of no more than 100–300 m below the African surface. A second uplift of major proportions at the end of the Pliocene raised the eastern interior of the sub-continent by as much as 900 m, although much smaller movements characterized the western areas and interior axes of uplift. Major monoclinal warping resulted in the southeastern hinterland. The ensuing Post-African II cycle is manifested chiefly in deep incision of the coastal hinterland and downcutting along major rivers of the interior. Earlier surfaces were severely deformed and dissected, especially in the south-east of the country. Planation was limited to a few areas close to the coast. The resulting sedimentation is evident mainly in the offshore deltas of major rivers owing to the late development of vigorous coastal currents, but a major increase is reflected throughout the deeper ocean basins.

In summary, most of the Karoo Basin was unroofed by scarp retreat in the Late Jurassic and Cretaceous. Renewed uplift in the Miocene (150–300 m) and end-Pliocene (900 m) led to unroofing of the rest of the basin except in the present outlier in Lesotho.

PROVENANCE OF KAROO BASIN SEDIMENTS

Introduction

In the Karoo Basin, paleocurrents inferred from the almost wholly nonmarine or fluvial Karoo sequence indicate the paleoslope of the depositional surface on the flank of a source upland or in the axis between two uplands. The petrology of the coarse fraction of the sediment reflects the petrology of the source area. By these means, authors have distinguished source areas at each of the cardinal points of the compass: (1) the northern cratonic interior of Gondwanaland; (2) an eastern source, now occupied by the continental margin or beyond, possibly a Late Proterozoic fold belt; (3) a southern source, comprising the proximal Cape Fold Belt and an in-

ferred distal convergent Panthalassan magmatic arc; and (4) a western source, now occupied by the continental margin or beyond, possibly a Late Proterozoic foldbelt, as in the east. In studying the foreland basin, our interest centers on the southern distal magmatic source. Estimates of the type of provenance in the south, e.g., magmatic arc, recycled orogen, or continental block (Dickinson and Suczek, 1979), and their age, whether coeval with or older than the sediments, can be made from the petrography of detrital arenites and rudites. Previously published work is supplemented by Cowan's petrographic study of the rock specimens collected by Veevers in 1984 from localities along the southern outcrop of the Karoo Basin (Appendix 1).

To the work mentioned above was added, at the last moment, Johnson's (1991) comprehensive account of sandstone petrography of the southeastern part of the Karoo Basin and preceding Cape Basin, summarized in Figures 15B and 15C. Johnson (1991) reports new finds of abundant volcanic rock fragments in the Ecca Group (Ripon and Waterford Formations) and Beaufort Group (Koonap, Middleton, and Balfour Formations of the Adelaide Subgroup, and the Katberg and Burgersdorp Formations of the Tarkastad Subgroup) (Table 2). Johnson (1991, p. 137) found that "the Ecca Group and Adelaide Subgroup [were derived] from a magmatic arc provenance, and the upper part of the Karoo Sequence [Tarkastad Subgroup, Molteno, Elliot, and Clarens Formations] from a recycled orogenic source (Cape Fold Belt)." He found (p.140) more rock fragments and feldspar than previous workers (Fig. 16 in a following section) for these reasons: (1) "The amount of matrix present in greywackes in particular has often been greatly overestimated in the past, and a great deal of what is called matrix is, in fact, made up of rock fragment grains and minerals larger than 20 microns"; and (2) "By classifying all clear, untwinned grains of the right birefringence as quartz, workers have in the past often ended up grossly underestimating the feldspar content of Karoo sandstones in particular." The age of the magmatic arc source rocks remains uncertain because Johnson (1991) assumed implicitly that the volcanolithic grains are penecontemporaneous with deposition and not derived epiclastically from ancient volcanics. We try to overcome this uncertainty by identifying the unequivocally syndepositional tuffs.

Active volcanic (pyroclastic) provenances

Southern distal Panthalassan margin provenance: Tastubian to Scythian distal vulcanicity (Table 2, Fig. 1). The Karoo Basin, particularly the southern part, contains a minor amount of recognizable juvenile volcanogenic material, mainly in the form of devitrified shards, in the Dwyka Formation (1a and 1b of Table 2), Pietermaritzburg Shale (3), Prince Albert Formation (4–8), Whitehill Formation (9–12), Collingham Formation (13, 17–19), Vischkuil Formation (1), Laingsburg

Formation (2), Fort Brown Formation (20, 21), Tierberg Formation (14), Volksrust Formation (15, 16), Adelaide Subgroup (Lower Beaufort or Abrahamskraal and Teekloof Formations, 23–26), and at the top of the Beaufort Group in the Burgersdorp Formation (27), all located in Fig. 1 and some illustrated in Fig. 20 (Appendix 1). In addition to the almost ubiquitous shards, the material includes sand-sized fragments of tuff crystals (plagioclase and pyroxene) and drop-like bodies, felsite, volcanic quartz, and alkali trachyte, and associated clay minerals regarded as metabentonite, all interpreted as silicic volcanic ash or tuff. In the Ecca and Lower Beaufort of the western part of the southern Karoo Basin, laumontite is the predominant calc-silicate mineral (Martini, 1974; Turner, 1978), and here as well as in the Burgersdorp, Elliot, and Clarens Formations to the east (Fuller, 1970), it is interpreted as derived by load metamorphism of volcanic glass or, as in Antarctica (J. W. Collinson, personal communication, 1988), enhanced by the dolerite acting on volcanogenic precursors. In turn, "the volcanic fragments in the host sandstones and interbedded tuffaceous material [in the uraniferous Lower Beaufort of the southern and central Karoo] suggest a volcanic source for the uranium" (Turner, 1978, p. 844), amended by Cole and Wipplinger (1991) to "some of the uranium." In the Lower Beaufort, the sand-sized fragments make up appreciable to minor amounts of the column and of these "rounded and drop-like bodies (average SiO$_2$ content 52%) of tuffaceous material . . . ejected from exploding volcanoes whilst still in a plastic or semi-liquid state" may constitute up to 80% of the sand-sized volcanogenic fraction in some of the rocks (Ho-Tun, 1979). A more silicic source magma is indicated by the 1–15% of the detrital quartz fraction being volcanic quartz, including the beta form. Also silicic is the possibly dacitic crystal tuff, 0.5 m thick, in the Fort Brown Formation at Geelhoutboom (Lock and Johnson, 1974), with cm-thick bands of plagioclase crystals (>90%) and quartz (<10%), emplaced ultimately by subaqueous flow as suggested by the mixing of the pyroclastics with the underlying mudstone. Another silicic rock is the felsite groundmass with shard "phenocrysts" in the Fort Brown Formation near Wolwefontein. Within the 50 m-thick "volcanic interval" of Lock and Wilson (1975) in and about the Collingham Formation, some 30–40% consists of yellow mudstone which Lock and Wilson (1975) interpret as a metabentonite containing 30–40% quartz. In Appendix 1 Cowan (*see* Fig. 20E in Appendix 1) interprets the scattered angular plagioclase and quartz grains in a microcrystalline quartz groundmass or matrix of the Matjiesfontein Chert at Remhoogte (SA 14C in Table 7) as a tuff. Also at Remhoogte, at the base of the Collingham Formation, Elliot and Watts (1974) found clasts of quartz and feldspar that may be volcanogenic. Elsewhere in the Ecca and lower Beaufort Groups, the detrital plagioclase is interpreted as derived from low-grade metamorphic rocks (Kingsley, 1981, p. 37).

Viljoen (1987) suggested that the widespread occurrence of albite, quartz, illite, and montmorillonite in the tuffs of the Ecca Group could be the result of the following diagenetic and low-grade metamorphic reactions (Iijima, 1978):

$$\text{silicic glass} + H_2O \rightarrow \text{montmorillonite} + \text{opal} \rightarrow$$
$$\text{alkali zeolites} + \text{opal-CT} + \text{montmorillonite} + K^+ \rightarrow$$
$$\text{analcime} + \text{quartz} + \text{illite/montmorillonite} + K^+ \rightarrow$$
$$\text{albite} + \text{illite} + \text{quartz}.$$

The tuffs (K- or meta-bentonites), some of which are multiple fallout units (Viljoen, 1987), occur throughout the Ecca Group and upper Dwyka Formation. When plotted on a Zr/TiO$_2$ versus Nb/Y diagram, the tuffs are seen to have a rhyolitic to dacitic composition (Viljoen, 1987). Furthermore, chert beds from the Collingham and Abrahamskraal Formations also fall within the rhyolite to dacite fields (Fig. 15A). Using the same technique, McLachlan and Jonker (1990) found that Ecca Group tuffs from the Hopetown and Boshof areas had a wider compositional range from rhyolite to andesite. In the Lower Beaufort of the northern Karoo, two isolated deposits interpreted to be ash-fall tuffs plot within the rhyodacite field but because the tuffs have undergone hydrothermal alteration, Keyser and Zawada (1988) suggest that the original magma was trachyandesite based on the chemical analysis of a less altered volcanic bomb embedded in the tuff.

From all this evidence, it is apparent that the bulk of the volcanogenic material is rhyodacitic, and that it was emplaced by air fall from contemporaneous explosive vulcanicity, lightly modified by subaqueous flow. Redistribution by water was probably least in the Ecca Group, where the volcanogenic silt and sand is concentrated in bands within lacustrine mudstone deposited from suspension on an anaerobic lake floor (McLachlan and Anderson, 1977), analogous to the concentration of tuff bands in coal seams and lacustrine sediments in the Late Permian Newcastle Coal Measures of eastern Australia (McDonnell 1983; Jones et al., 1987). All occurrences are inferred to come from a southern source except the coarse-grained tuffs in the lower Beaufort Group of the northern Karoo Basin (24 and 25 in Table 2, Fig. 13A), for which Keyser and Zawada (1988) suggest a minor local source in the north. Redistribution is greatest in the coarse fluvial sediments of the Lower Beaufort Group. Ho-Tun's (1979) observation that volcanic quartz grains constitute up to 15% of the detrital quartz fraction in part of the Beaufort Group suggests to us that, taking into account the expected much greater dilution of originally pyroclastic material in a fluvial section, the source of the pyroclastics compared to that of the Ecca Group was either more proximal or more copious or both. According to Martini (1974, p. 115), "The generally small size of the volcanic clasts [of the ash beds] is in agreement with a source several hundreds of kilometres away." As to the direction, Martini (1974) pointed to a source south of the southern Karoo Basin and beyond the southern Cape Province and Agulhas Bank, narrowed to the Paleo-Pacific (Panthalassan) margin

from Patagonia to West Antarctica, which contains evidence of Permian magmatic activity. Elliot and Watts (1974) independently pointed to the more general Gondwanide Orogen, and Coutinho et al. (1991) suggest a common rhyolitic volcanic source located in central Argentina for the Late Permian ash beds in both the Paraná and Karoo Basins. Johnson (1991, fig. 8) shows the Permian-Triassic magmatic arc some 400 km south of southern Africa.

Viljoen (1987, 1990) and Verwoerd et al. (1990) found that the total thickness of the tuff and the number of tuff beds in the Collingham Formation decreases gradually toward the northeast. This zone of tuffs is still recognizable in stratigraphically equivalent horizons in the northern Karoo Basin at the base of the Volksrust Formation in boreholes BE1/66 (15 in Table 2) and SW1/67 (16) (Viljoen, 1990) and at the base of the Tierberg Formation near Hopetown (14) (McLachlan and Jonker, 1990). The distribution of the tuffs suggests volcanic eruption centers to the south and southwest along the Panthalassan convergent margin (Viljoen, 1990; Verwoerd et al., 1990).

Bristow and Saggerson (1983a, p. 1028) note "the possibility that much of the volcanoclastic material found in the Karoo sediment was derived locally [e.g., from volcanic maars, small vents, and diatremes that weather extremely rapidly], and that the structures from which the material originated were subsequently reworked by erosion." A similar view is expressed by McLachlan and Jonker (1990), who inferred that "volcanoes of Permian age were present within a few hundred kilometres of the tuffs recorded in the northwestern part of the Karoo Basin." A local source is clear for the volcanic material found in the Late Triassic Stormberg Group and in the Beaufort Group of the northern Karoo Basin (24 and 25 in Table 2). But the uniformly fine grain and wide distribution of all the other volcanogenic material, with the possible exception of the crystal tuff in the Whitehill Formation near Hopetown (11), suggest to us that the vents were distant.

The proximal source, particularly that of the lower Beaufort ash-flow tuffs and the Whitehill Formation crystal tuff in the northern Karoo Basin, has a more basic composition than that of the tuffs found in the Dwyka Formation, Ecca Group, and in the lower Beaufort elsewhere. No Permian volcanic vents are known from southern Africa; the only direct evidence of *in situ* contemporaneous igneous activity is a dolerite dated at 260 ± 16 Ma that intrudes the Table Mountain Group in borehole OL1/69 (Fig. 1) (Rowsell and de Swardt, 1976, p. 121).

We conclude that the provenance of the bulk of the Tastubian to Scythian pyroclastic component of the Karoo Basin was located an unknown distance, probably at least several hundred kilometers, south and southwest of southern Africa, presumably on the adjacent sector of the magmatic arc along the convergent margin of Panthalassa/Gondwanaland. The composition of the source was rhyodacite. A minor proximal provenance (of the lapilli trachyandesite tuff of 25) was a volcanic center at or near the northern edge of the Karoo Basin.

Late Triassic to Early and Middle Jurassic proximal Karoo vulcanicity (Table 3, Figs. 1, 6). Bristow and Saggerson (1983a), Visser (1984), and Eales et al. (1984) reviewed the Karoo vulcanicity that extended over a large part of southern Africa. The first occurrence of tuff, in the Molteno Formation (28 of Table 3), was followed by glass shards in the Elliot Formation, and basalt and pyroclastics in the Clarens Formation. Laumontite in both probably originated as volcanic glass (Table 3). The localized coal-tonsteins of the Molteno Formation are thought to represent volcanic ash beds that were rhyolitic in composition (Heinemann and Buhmann, 1987). By about 193 Ma sediment deposition ended with the outpouring of the 1,500-m-thick Drakensberg basaltic lavas and the intrusion of Karoo dolerite that continued intermittently in southern Africa to the Early Cretaceous. For the purpose of this account, it suffices to note that the Mesozoic Karoo vulcanicity, although some of its products such as shards and laumontite were the same as those of the Permian vulcanicity, was proximal and mafic (and in places bimodal: mafic/silicic) whereas the Permian vulcanicity discussed above was mostly distal and silicic with a minor proximal and trachyandesitic contribution.

Provenances indicated by the petrography of the epiclastics

Recycled orogen provenance. A second type of provenance, that of a recycled orogen (Dickinson and Suczek, 1979),

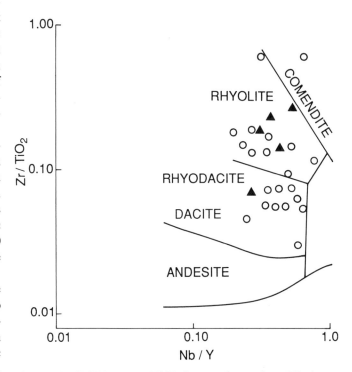

Figure 15A. Zr/TiO$_2$ versus Nb/Y diagram (format from Winchester and Floyd, 1977) for the tuff (circles) and chert (triangles) in the Early Permian strata of the main Karoo Basin (personal communication from J.H.A. Viljoen, in preparation).

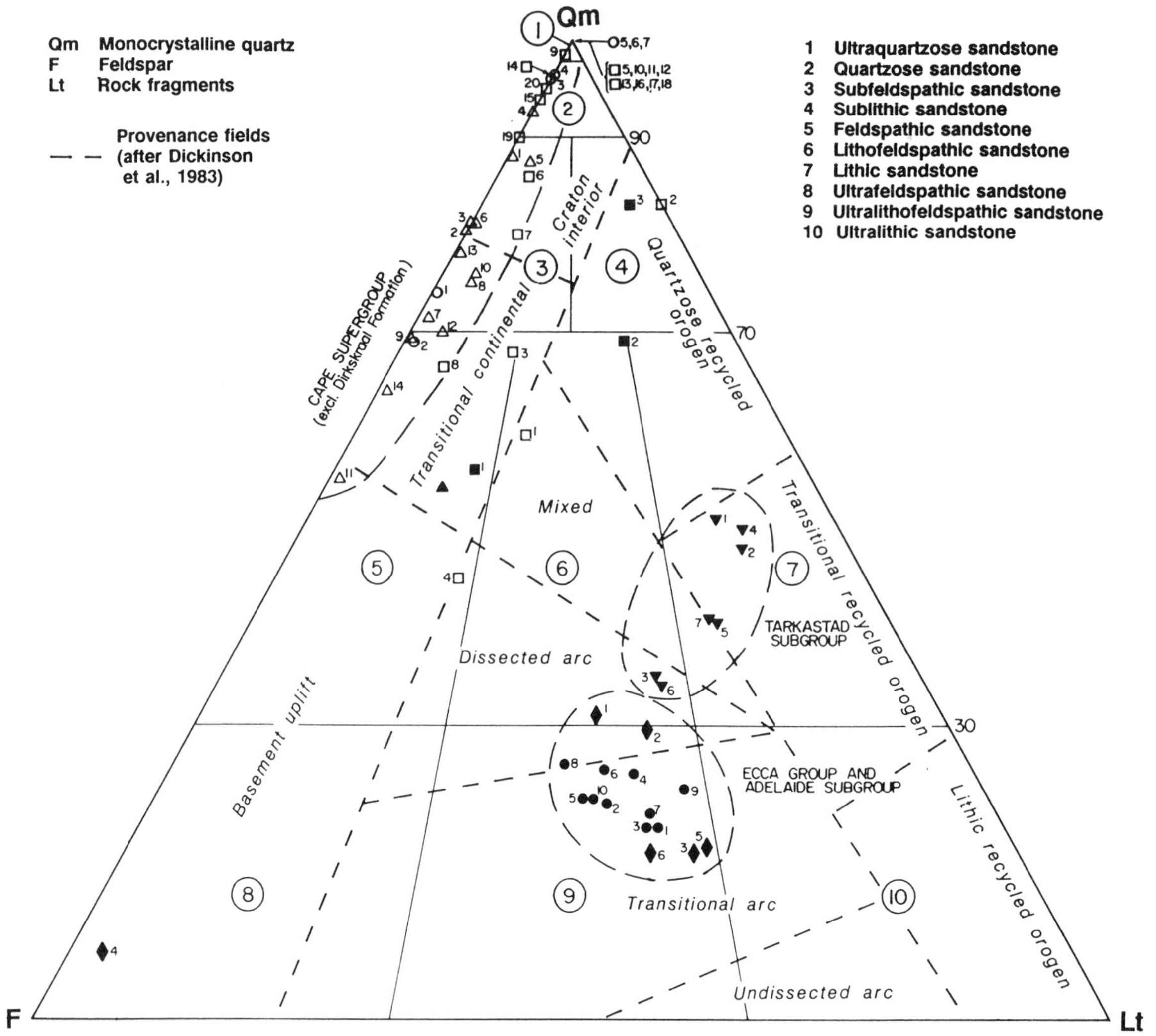
Qm Monocrystalline quartz
F Feldspar
Lt Rock fragments
— — Provenance fields (after Dickinson et al., 1983)
1 Ultraquartzose sandstone
2 Quartzose sandstone
3 Subfeldspathic sandstone
4 Sublithic sandstone
5 Feldspathic sandstone
6 Lithofeldspathic sandstone
7 Lithic sandstone
8 Ultrafeldspathic sandstone
9 Ultralithofeldspathic sandstone
10 Ultralithic sandstone
Qm
F
Lt
90
70
30
Craton interior
Quartzose recycled orogen
Transitional continental
CAPE SUPERGROUP (excl. Dirkskraal Formation)
Mixed
Transitional recycled orogen
TARKASTAD SUBGROUP
Dissected arc
Basement uplift
ECCA GROUP AND ADELAIDE SUBGROUP
Lithic recycled orogen
Transitional arc
Undissected arc

Group	Subgroup	Symbol	Stratigraphic unit[1]	Locality	n
		■ 1	Clarens	C1	3
		■ 2	Elliot	El1	3
		■ 3	Molteno		8
BEAUFORT GROUP	TARKASTAD SUBGROUP	▼ 1	Burgersdorp	Bf 15	5
		▼ 2	Burgersdorp	Bf 21	5
		▼ 3	Katberg (base)	Bf 2	4
		▼ 4	Katberg (top)	Bf 19	3
		▼ 5	Katberg (middle)	Bf 19	8
		▼ 6	Katberg (base)	Bf 19	3
		▼ 7	Katberg (combined)	Bf 19	14
	ADELAIDE SUBGROUP	● 1	Balfour (top)	Bf 3	4
		● 2	Balfour (top)	Bf 9	3
		● 3	Balfour	Bf 12,13,14	9
		● 4	Balfour	Bf 16,17,20	7
		● 5	Middleton (top)	Bf 18	3
		● 6	Middleton	Bf 11	6
		● 7	Middleton	Bf 7,8	4
		● 8	Middleton/Koonap	Bf 4	8
		● 9	Koonap	Bf 5,6	3
		● 10	Koonap	Bf 10	5
	ECCA GROUP	♦ 1	Waterford (top)	E 1,3	3
		♦ 2	Waterford (excl. top)	E 1	4
		♦ 3	Waterford	E 3	4
		♦ 4	Ford Brown (tuff)	E 4	1
		♦ 5	Ripon	E 2	7
		♦ 6	Ripon	E 5	10
		▲	Dwyka	E 5	3

Group	Symbol	Stratigraphic unit[1]	Locality	n
WITTEBERG GROUP	□ 1	Dirkskraal	W 1,5	3
	□ 2	Dirkskraal (?)	W 5(?)	1
	□ 3	Dirkskraal	W 7,10	2
	□ 4	Dirkskraal	W 8	3
	□ 5	Swartwaterspoort	W 3,4,9	4
	□ 6	Miller	W 1,4,5	3
	□ 7	Miller	W 8	3
	□ 8	Waaipoort	W 5	2
	□ 9	Floriskraal	W 5	6
	□ 10	Floriskraal	W 9	3
	□ 11	Skitterykloof[2]	W 5	2
	□ 12	Perdepoort[2]	W 5	5
	□ 13	Rooirand[2] (upper)	W 5	7
	□ 14	Rooirand[2] (lower)	W 5	4
	□ 15	Rooirand[2] (base)	W 5	3
	□ 16	Rooirand[2]	W 11,12	10
	□ 17	Weltevrede	W 5	17
	□ 18	Weltevrede	W 6	4
	□ 19	Driekuilen[3]	W 2	2
	□ 20	Driekuilen[3]	W 6	5

Group	Symbol	Stratigraphic unit[1]	Locality	n
BOKKEVELD GROUP	△ 1	Sandpoort	B 5	1
	△ 2	Adolphspoort	B 4	1
	△ 3	Karies	B 5	1
	△ 4	Boplaas (upper)	B 1,2	6
	△ 5	Boplaas (lower)	B 1	2
	△ 6	Boplaas	B 5	4
	△ 7	Hex River	B 2	2
	△ 8	Hex River	B 4	2
	△ 9	Gamka ("arenite facies")	B 2	5
	△ 10	Gamka ("wacke facies")	B 2	4
	△ 11	Gamka ("arenite facies")	B 3	2
	△ 12	Gamka ("wacke facies")	B 3	4
	△ 13	Gamka ("wacke facies")	B 4	3
	△ 14	Gamka	B 6	6
TABLE MOUNTAIN GROUP	○ 1	Kareedouw[4]	T 2,5	5
	○ 2	Baviaanskloof[5]	T 5	3
	○ 3	Skurweberg	T 5	3
	○ 4	Goudini	T 4	2
	○ 5	Peninsula	T 1	5
	○ 6	Peninsula	T 3	12
	○ 7	Peninsula	T 6	5

1 All units are formations except where otherwise indicated
2 Members of Witpoort Formation
3 Member of Weltevrede Formation
4 Member of Baviaanskloof Formation
5 Excluding Kareedouw Member

Figure 15B. Framework mineralogy, provenance fields (after Dickinson et al., 1983) and sandstone classification (after Johnson, 1976) of the southeastern Cape-Karoo Basin, all from Johnson (1991) and reproduced with the permission of the author and the editor of the *South African Journal of Geology*.

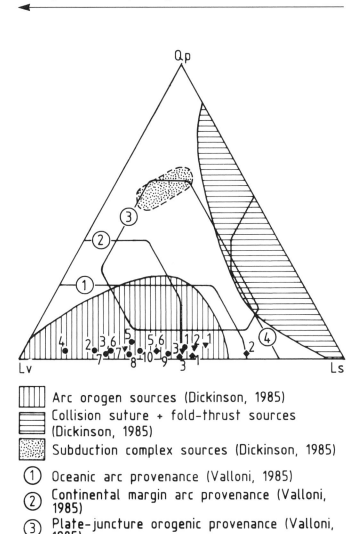

Arc orogen sources (Dickinson, 1985)

Collision suture + fold-thrust sources (Dickinson, 1985)

Subduction complex sources (Dickinson, 1985)

(1) Oceanic arc provenance (Valloni, 1985)

(2) Continental margin arc provenance (Valloni, 1985)

(3) Plate-juncture orogenic provenance (Valloni, 1985)

(4) Fold-thrust belt orogenic provenance (Valloni, 1985)

Figure 15C. QpLvLs plot for Ecca Group and Adelaide Subgroup sandstones. Sample symbols and numbers as in Figure 15B. Qp—polycrystalline quartz, Lv—volcanic rock fragments, and Ls—sedimentary and metasedimentary rock fragments. From Johnson (1991), with acknowledgments as above.

is indicated by the framework modes of detrital or epiclastic sandstones in the Karoo Basin (Table 4, Fig. 16), by dated clasts (Table 5), and by the occurrence of garnet (Table 6).

The framework modes of quartz, feldspar, and lithic fragments show that sandstone samples of the Ecca and lower Beaufort Groups are feldspathic litharenite to lithic feldsaren-

ite, those of the Katberg Formation lithic feldsarenite, and those of the Molteno and Elliot Formations quartzose litharenite (Fig. 16, A–C). As noted above, the different petrographic classification of grain types used by Johnson (1991) shifts his samples toward the feldspar and lithic end members into the magmatic arc field (Fig. 15B), whereas the mean modes of the samples plotted in Figure 16D plot in the recycled orogen provenance with the Ecca Group (1) and lower Beaufort Group (2) on the side near the magmatic arc province.

Most of the lithic fragments in the fine-grained plagioclase litharenite near the base of the Ripon Formation (SA 9, Appendix 1, Tables 7 and 8) are felsite with quartz and ?feldspar phenocrysts, as also found in the Beaufort West sample (SA 30, Appendix 1, Tables 7 and 8) and in the other Beaufort rocks reported by Turner (1978, p. 834, "volcanic fragments of predominantly acid character"). The mineral fraction in the Ripon Formation sample is mainly K-feldspar (plagioclase is minor) and mono- or poly-crystalline quartz. From the Ecca and lower Beaufort, Martini (1974, figs. 2–4) described lithic fragments with a felsitic texture with or without phenocrysts and also fragments with trachytic and pilotaxitic textures; and from the lower Beaufort of the northern Karoo, Keyser and Zawada (1988) described lapilli. The felsite grains are interpreted as having been derived originally as pumice from a silicic volcanic province. Unlike the felsite grains in the Sydney Basin, which can be shown to be epiclastic (Chapter 3, Appendix 2), the Karoo pumice grains are too small to indicate whether they were deposited in a porous or a consolidated state before burial. It remains possible therefore that they were derived from the same provenance that provided the air-fall tuffs, but the framework modes suggest another possible source from an older recycled orogen provenance. A paleocurrent analysis of the Beaufort Group (Cole and Wipplinger, 1991) indicates that this provenance may have been located in the region of the Cape Fold Belt south of the Karoo Basin.

Other lithic fragments seen in the collection (Appendix 1, Table 7), besides the recognizable volcanic ash fragments (shards), are granite, quartzite, and an altered plagioclase-rich mafic rock in the Dwyka Formation (SA 5b), and sedimentary chert (banded chalcedony) (*see* Fig. 20F in Appendix 1) in the Fort Brown Formation. Turner (1975) found that the rock fragments in the sandstones of the Molteno Formation comprise quartzite and polycrystalline quartz, probably derived from metamorphic rocks. The mineral fraction is mainly mono- and poly-crystalline quartz and feldspar, with K-feldspar far more abundant than plagioclase and probably derived from granite (Turner, 1975; *see* also Le Roux, 1985).

As detailed in Table 6, *garnet* is a common heavy mineral (as is zircon) throughout the entire Karoo Basin sequence and is most abundant in the glacigene Dwyka Formation, where it is distinctly chattermarked, as it is in other Early Permian glacigene deposits in Gondwanaland (Gravenor, 1979). In the underlying Cape Supergroup, the only known occurrence of

TABLE 4. RECYCLED OROGEN PROVENANCE, INDICATED BY QFL COMPOSITION OF EPICLASTIC SANDSTONES IN THE SOUTHERN KAROO BASIN (FIG. 16D)

Unit	Locality	Material	Reference
6. Elliot	Eastern Cape and Witsieshoek	Lithic feldspathic graywacke	Le Roux, 1985
5. Molteno	Molteno outcrop	Quartzose feldsarenite	Turner, 1975
4. Katberg	Queenstown, Kidds Beach	Lithic feldsarenite	Johnson, 1966, *in* Dingle et al., 1983, p. 21
3. Beaufort (Teekloof)	Beaufort West	Lithic feldsarenite (SA 30)	Appendix 1, Table 8
2. L. Beaufort, Ecca	Eastern Cape	Sandstones—90% of detrital feldspar is low-temperature sodic plagioclase from low-grade metamorphics	Kingsley, 1981
1. Ripon	Ecca Pass	Feldspathic litharenite (SA 9)	Appendix 1, Table 8
L. Beaufort, Ecca	Southern Karoo	Graywackes, with volcanic fragments: trachytic, pilotaxitic, and felsitic texture	Martini, 1974

garnet is in the glacigene Pakhuis Formation of the Table Mountain Group (de Villiers and Wardhaugh, 1962). In the modern river sands (mouth of the Orange River, the Crocodile River at Thabazimbi, and the Limpopo River at Messina) that drain the craton north of the Karoo Basin, de Villiers and Wardhaugh (1962) found garnet in greater than trace amounts (>2% of the heavy residue) in one sample only. They concluded that "the resultant rocks would not resemble those portions of the Karroo succession which are characterized by an abundance of garnet in the heavy residue" (de Villiers and Wardhaugh, 1962, p. 105). Within the Karoo Basin provenance itself, de Villiers and Wardhaugh (1962, p. 105) found that "during the present cycle of erosion and deposition of the Karroo sediments, most of the garnet is apparently destroyed." Having eliminated the region north of the Karoo Basin as a provenance of the garnet, de Villiers and Wardhaugh (1962) pointed to a garnet provenance in a southerly extension of Africa of composition similar to that of garnetiferous East Antarctica. Subsequent studies, however, have shown that the northern region was a definite provenance of garnet in the Mooi River Formation (*see* the following discussion) and a probable provenance in the case of the garnet-bearing heavy mineral sands in the Vryheid Formation (Behr, 1986).

Gravenor's (1979) analysis of the heavy mineral suites from Early Permian deposits of South Africa, Australia, and Antarctica, dominated by chattermarked garnet, shows that these garnets underwent a much longer distance of transport, probably as a result of continuous recycling, than the Pleistocene glacial sediments of North America. "The dominance of garnets in the heavy mineral suites was caused by a combination of dissolution of chemically unstable minerals by intrastratal solutions, and mechanical abrasion of the softer minerals" (Gravenor, 1979, p. 1149, 1150). These arguments account for the garnet in the glacigene Pakhuis and Dwyka

Formations, but not in the nonglacial part of the Karoo sequence, for which proximity to a voluminous source of garnet is required. The available evidence pinpoints a source for the garnet in one formation only. According to Theron (1975, p. 65; Table 6), cobbles of gneiss in the Mooi River Formation (= Belmont Formation) of Natal contain relatively large amounts of a red garnet, and this kind of rock is common up-paleocurrent 100 km eastward in exposures of the Mapumulo Group in the 1.1 Ga Natal Metamorphic Belt (Fig. 13A). Other formations, in particular the coeval Katberg Formation, shown by paleocurrents to be derived from the south-southeast (Fig. 13A), and the Estcourt Formation, which originated from a source to the east and southeast (Botha and Linström, 1978), could have come from the southern extension of the Natal Belt, perhaps identifiable in the 1.0 Ga Cape Meredith gneiss of the Falkland Islands (Fig. 13A), in the garnetiferous gneiss of the Falkland Plateau (Tarney, 1976), and in the rocks in Dronning Maud Land (2°W to 30°E in Antarctica) that also register the 1.0 Ga tectonic-metamorphic event (Yoshida and Kizaki, 1983).

The age and composition of the proximal sources are best indicated by the rudites. From the list of datable rudite clasts in the southern Karoo Basin in Table 5, we can infer (1) that the southern basal Dwyka Formation came from the south— the nearest known source of the archaeocyathid limestone clasts is Antarctica; (2) that the Katberg Formation came from a province containing granite >390 Ma, and lignite and silicified wood, within the interval Devonian to Permian, both found up-paleocurrent within the Cape Fold Belt, and gneiss of minimum age 535 ± 66 Ma (probably about 1000 Ma) found in the Falkland Plateau (Beckinsdale et al., 1976); and (3) that the Molteno Formation came from a proximal province yielding blocks of Witteberg quartzite, available a short distance south of Molteno in the Cape Fold Belt.

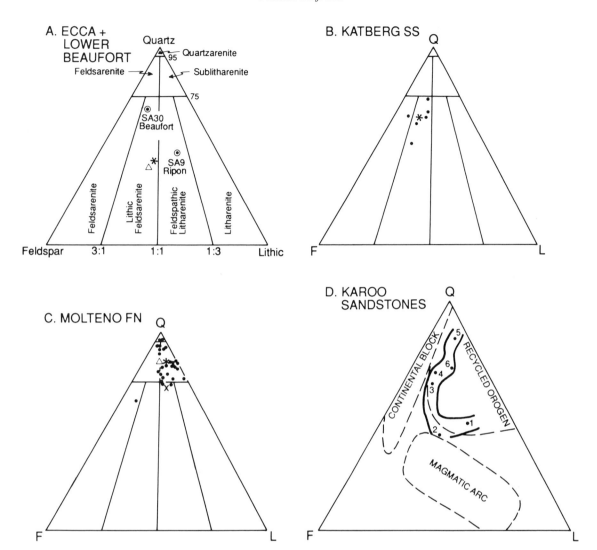

Figure 16. Framework modes of Karoo sandstones plotted on a QFL triangular diagram divided according to Folk et al.'s (1970) scheme, except D, from Dickinson and Suczek (1979). A. Samples SA 9 (Ripon Formation of Ecca Group) and SA 30 (lower Beaufort Group) (Appendix 1) and the mean for the Ecca and lower Beaufort (asterisk) (Kingsley, 1981) and for the lower Beaufort (triangle) (Le Roux, 1985). B. Katberg Formation (upper Beaufort), from Johnson (1966), as given by Dingle et al. (1983 p. 21). The mean is shown by the asterisk. C. Molteno Formation and mean (asterisk) (Turner, 1975), with mean Molteno (triangle) and Elliot Formation (X) from Le Roux (1985). D. Summary. Field of Karoo sandstones represented by 1 (SA9, Ripon Formation), 2 (mean Ecca and Beaufort), 3 (mean Katberg), 4 (SA 30, lower Beaufort), 5 (mean Molteno), 6 (mean Elliot). All lie within the recycled orogen provenance except 2, which lies between this provenance and the magmatic arc provenance.

Other types of provenance, including the northern cratonic provenance. Along the northern feather-edge of the Karoo Basin, paleocurrent directions in the Dwyka Formation, Ecca Group and parts of the Beaufort Group indicate derivation from the northern craton. The large caliber of the abundant glacial stones in the Dwyka (Visser et al., 1986) facilitates the recognition of the proximal cratonic source in the valley fills of the northern Karoo Basin and farther afield, as at Elandsvlei

(Table 6), whereas the sand grade of the northern Ecca Group provides a less distinctive clue to the nature of the provenance. The heavy mineral sands in the Vryheid Formation (Behr, 1986) may have been derived from reworked Dwyka Formation to the northwest or from basement. In the Beaufort Group, except the Mooi River Formation, the Belmont/Katberg and Otterburn/Burgersdorp Formations are from the southern provenance. As mentioned already, the rudite of the Mooi River

TABLE 5. DATED RUDITE AND ARENITE CLASTS IN THE SOUTHERN KAROO BASIN

Unit	Locality	Material	Reference
Molteno	Molteno	75-cm-wide clast of Devonian Witteberg quartzite	Rust, 1962; Turner, 1975
Middle Beaufort (Katberg)	East London	Lignite, silicified wood: Permian or older to Devonian	Stavrakis, 1980
Katberg (245 Ma)	Katberg	Granite: K/Ar 390 to 210 Ma min. age*	Elliot and Watts, 1974
Abrahamskraal†	Beaufort West	Detrital zircons: U/Pb 1050 ± 20 Ma	Burger and Coertze, 1977
Dwyka	Gamkapoort to Willowmore	Archaeocyathid limestone: Early Cambrian	Cooper and Oosthuizen, 1974; Oosthuizen, 1981; Visser et al., 1986.

*Possibly an argon loss event (Martin et al., 1981, p. 300).
†The sandstone bearing these grains was deposited from paleocurrents from the southwest, which does not contain this Namaqua-Natal provenance (see Fig. 4). It is possible that the zircons are from reworked Dwyka whose provenance lay northward in the Namaqua-Natal belt.

Formation (= Belmont Formation in Natal) indicates a unique source in the Natal Metamorphic Belt. The Estcourt/Lower Beaufort is from an east-south-east provenance (Botha and Linström, 1978; Cole and Wipplinger, 1991) and a northern provenance in the Heilbron-Frankfort area (Fig. 13A) (Keyser and Zawada, 1988).

Cole and Wipplinger (1991) show that five major paleo-current directions dominate the Beaufort Group—NE, NNW, ESE, WNW, and SW. The first two suggest derivation of sediment from the Cape Fold Belt region (Recycled Orogen Provenance—*see* previous discussion). The ESE direction indicates transport of sediment from a cratonic provenance centered on Namaqualand, and the WNW direction implies that the Natal Metamorphic Belt was a significant source of Beaufort Group sediment. The SW paleoflow direction was only recently recorded in the Colesberg-Edenburg region (Cole and Wipplinger, 1991). An intrabasinal granitic provenance east of Bloemfontein has been proposed by J. C. Theron (1970).

SYNTHESIS

The analytical data presented above are put together in the form of the series of paleo-tectonic/geographic maps presented in Fig. 17.

Influence of basement structure

Three basement trends (Fig. 17A) existed by the start of pre-Gondwanan (Ordovician-Devonian) deposition. The trends, defined by the 1.0–0.5 Ga foldbelts of weak crust that wrap around the 1 Ga Namaqua-Natal Belt and >2.5 Ga Kaapvaal Province, provide a tub-shaped template that was impressed on succeeding structures up to the Cretaceous breakup of Pangea along the present divergent margins. The pattern is reprinted

during the Ordovician-Devonian deposition of the Cape Supergroup in grabens on the northwest and northeast linked by an east-west depositional axis (Fig. 17, B and C) and during the Permian and Triassic development of the Cape Fold Belt along the east-west trend linked with intermittent uplifts to the northwest (Atlantic upland) at a syntaxis around Cape Town and to the northeast (Eastern upland) at a syntaxis in the (restored) Falkland Islands (Fig. 17, D–H). Finally the pattern was exploited to determine the present shape of southern Africa by the dispersal of the neighboring South America. The southeastern margin of South America was dispersed by sea-floor spreading perpendicular to the southwest margin of Africa, and the Falkland Plateau protuberance of South America was dispersed from the southeastern margin of Africa by transform motion along the Agulhas Fracture Zone, after preliminary rotations of subplates including the Falkland Islands and parts of West Antarctica.

Ordovician-Devonian Cape Sequence

Figure 17B shows the initial rift valleys that formed over the Gariep and Mozambiquean trends. In the southwest, the alluvial-fan to braided-river complex of the Klipheuwel and Piekenierskloof Formations occupied the funnel between the proximal Atlantic and Bushman uplands; in the northeast, the Natal valley was filled with a similar complex of alluvial-fan through braided-river to shallow marine shelf that was deposited longitudinally from and within the proximal Kaapvaal uplands. Subsequent down-flexure of the rift shoulders allowed sediment in a coastal plain-marine shelf complex to thicken southward from at least the northern preserved pinch-out into a trough axis (marked by the 3.5 km isopachs) that crosses the present south coast. The only other known occurrence of possibly early Paleozoic sediment in southern Africa

TABLE 6. OTHER DATA ON PROVENANCE PETROLOGY, INCLUDING THE OCCURRENCE OF GARNET

Unit	Locality	Material	Reference
Elliot	Natal Drakensberg	Sandstones: sedimentary source, from primary igneous/metamorphic source	Eriksson, 1985
Elliot	Clocolan	**Garnet** and other heavy minerals (HM)	Le Roux, 1969, *fide* Theron, 1975
Molteno	Molteno outcrop	**Garnet** (almandine-spessartite) "from granites, pegmatites, and possibly contact altered siliceous rocks"	Turner, 1975, p. 199
Molteno	Molteno outcrop	Two sources: S—Cape Supergroup quartzites; SE—granitic fault-block terrain of high relief	Turner, 1980
Burgersdorp	Natal Drakensberg	**Garnet**	Botha and Linström, 1978
Belmont	Mooi River	<18 cm clasts: gneiss with red **garnet**, vein quartz, microcline, red granite, quartzite—metamorphic terrane	Theron, 1975, p. 65
Katberg	East London*	Pebbles, HM	Hiller and Stavrakis, 1980
Katberg	Eastern Cape	Quartz-feldspar porphyry and non-porphyritic "lava." Also granite, gneiss, quartzite, sandstone, conglomerate, fossil wood	Johnson, 1976; Theron, 1975
Estcourt	Estcourt/Mooi R.†	Sand grains	Botha and Linström, 1978
Beaufort	Karoo Basin	**Garnet**	Theron, 1975
L. Beaufort and Ecca	E. Cape	Modal analysis, feldspar studies, heavy mineral analysis: low-metamorphic to gneissic terrane	Kingsley, 1979
Ecca	Laingsburg, Harrismith	**Garnet**	Nel, 1962
Vryheid	Bothaville	**Garnet**, zircon from granitic terrane in NW	Behr, 1986
Vryheid	Delmas, Carolina, Muden	**Garnet**	Ryan and Whitfield, 1979; Behr, 1986
Dwyka	N. Karoo, Elandsvlei	Proximal valley source: red quartzite plus stromatolitic dolomite from "Ghaap Plateau" 500 km to NE	Visser et al., 1986
Dwyka	S. outcrop	Diorite, gray limestone with archaeocyathids, from S or SE	Cooper and Oosthuizen, 1974; Visser and Loock, 1982; Visser et al., 1986
Dwyka	Pluto's Vale	Heavy mineral suite flooded with colorless and pink **garnet**	Rust, 1962
Dwyka	Karoo Basin	**Garnet**	de Villiers and Wardhaugh, 1962
Dwyka	Karoo Basin	Chattermarked **garnet**: recycled glacigenic sediment	Gravenor, 1979
Weltevrede (Witteberg)	Willowmore	Rutile, zircon, ilmenite, magnetite, monazite, staurolite, sphene. Namaqua-Natal metamorphic igneous source 100 km to north	Cole and Labuschagne, 1983
Pakhuis	W. Cape	Only known occurrence of **garnet** in Cape Supergroup	de Villiers and Wardhaugh, 1962

*"The pebbles...indicate that the sediments were derived from the weathering and erosion of a granitic or granite gneiss terrane which, in places, was associated with a sedimentary cover and intrusive igneous rocks.... The heavy mineral suite, in which pink **garnet**, magnetite, epidote, sphene and zircon are the most common species, would indicate derivation from a high-grade metamorphic or acid igneous source. The feldspars present in the sandstones, mostly microcline and plagioclase in the albite to andesine range, tend to confirm such a source."
†"The plagioclase and microcline in the sandstones were probably derived from gneissic metamorphic rocks and this conclusion is confirmed by the high garnet content.... The quartz in the sandstones has a typical undulatory extinction, which also points to metamorphic source rocks. The bulk of the material was therefore derived from garnet-bearing basement rocks around the south-eastern and eastern periphery of the Karoo Basin."

TABLE 7. SOUTH AFRICAN ROCK SPECIMENS, ARRANGED STRATIGRAPHICALLY*

Formation/ Locality	SA No.	Description
Beaufort Group, 3 km N of Beaufort West	30	Fine-grained plagioclase litharenite (see Table 8 for point count).
Top Fort Brown, Geel-houtboom	11b	Argillite with very fine quartz fragments.
	11c	Very fine sandstone, with a lithic grain of sedimentary chert (banded chalcedony) (Fig. 20F).
	11d	Argillite similar to 11b and 11c.
Fort Brown, 13 km N of Wolwefontein	10	Tuff of shards in felsitic groundmass, part replaced by carbonate (Fig. 20D).
Basal Ripon Formation, Ecca Pass, 2.4 km past Bain Monument	9	Fine-grained plagioclase litharenite (see Table 8 for point count).
Top Collingham, Ecca Pass	8c, d	Argillite with minor angular quartz grains.
Remhoogte	14–16	
Collingham, Matjiesfontein Chert (MC)	14c	Felsitic groundmass with scattered angular plagioclase and quartz, interpreted as a tuff (Fig. 20E).
	14g	Similar to 14c, some parts replaced by carbonate.
	14h	Similar to 14c and 14g. Replacing mineral (high relief) picks up faint traces of ?vitric shard outlines.
Collingham, 8 m above base	15	Argillite, with curvilinear fragments interpreted as shards with triple junctions (Fig. 20B).
Whitehill	16a	Yellow part of slide is argillite. Black part is vitric tuff with shards replaced by clay (Fig. 20A).
Between MC and White-hill, 16 km E. of Matjies-fontein	26	Black argillite.
MC at locality 26	27	Laminated microcrystalline quartz with very fine quartz and plagioclase grains.
Whitehill, 1.3 km N of 5	6	Very fine-grained red siltstone. Wavy lamination with truncation surfaces present.
Six m below Whitehill near Laingsburg R. crossing	29	Argillite.
Basal 5-m Collingham, Ecca Pass	7a	Very fine-grained siltstone with devitrified glass shards. Very few detrital grains present. White patches full of shards (Fig. 20C).
	7b	As above but without shards, and replaced locally by carbonate.
	7c	Argillite mostly replaced by carbonate. Black patches show shard-like grains, but too small for positive identification.
	7d	Argillite replaced by carbonate.
		Poorly sorted medium-grained immature sandstone; quartz, and plagioclase, minor microcline. Clasts of granite, quartzite, altered plagioclase-rich (lathwork) mafic rock.

*Volcanogenic material located on Figure 1 and in Table 2.

TABLE 8. MODAL POINT-COUNT PERCENTAGE COMPOSITION OF SANDSTONES*

Category	SA9 Ripon		SA30 Beaufort	
Counts	700		720	
Grainsize	0.17 mm		0.14 mm	
Whole rock quartz	42.7		56.7	
Common		26.1		55.9
Volcanic		1.1		0.0
Polycrystalline		15.2		0.8
Chert/metachert		0.3		0.0
Interstitial	6.4		13.8	
Feldspar	16.7		17.3	
K-feldspar		11.6		5.1
Plagioclase		5.1		12.2
Mica	2.6		3.1	
Heavy mineral	0.1		0.3	
Lithic grains	31.5		8.8	
Volcanic, felsitic		18.4		2.2
Argillite		10.0		6.3
Schist		3.1		0.3
Total	100.0		100.0	
Quartz (Q)	47.0		68.5	
Feldspar (F)	18.4		20.9	
Lithic grains (L)	34.6		10.6	
Total (Q + F + L)	100.0		100.0	

*Categories from Cowan, 1985, after McDonnell, 1983.

is at Zambezi (14.4°S, 22.5°E) (ZI in Fig. 6) (Reimann, 1986), 1,200 km north of the edge of the map area of Fig. 17B.

During the Early and Middle Devonian (Fig. 17C), accelerated subsidence was accompanied by the southward prograding clastic wedge >3.5 km thick of the Bokkeveld Group, shed presumably from the relic Bushman and Atlantic uplands, in the form of "a major trough with marked eastward axial pitch (Agulhas Trough)" (Theron and Loock, 1989, p. 733). In the Late Devonian (Witteberg Group), a linear clastic shoreline developed east of 20°E, and possibly branched northeastward into the Natal trough.

In view of its setting alongside the Paleo-Pacific Ocean, the trough was on the cratonic side of a marginal or back-arc basin behind a magmatic arc, as suggested for the Devonian of the neighboring Antarctica (Bradshaw and Webers, 1988, p. 792) (Fig. 4 in Chap. 7), and of southern South America and the Antarctic Peninsula (Hiller, 1992, fig. 9). In deriving all its epiclastic sediment from the craton, the Cape Basin has the same kind of fill as a passive margin (Johnson, 1991) but the very different tectonic setting of a back-arc basin within a zone of plate convergence.

Karoo (Gondwanan) Sequence

Figure 17D shows the following superimposed events at the latest Carboniferous–earliest Permian (290–277 Ma) onset of Karoo deposition: (5) end-Tastubian regressive marine shoreline (heavy dotted line) and tuff; (4) Tastubian marine transgressive shoreline (light dotted line); (3) ice flow during final glaciation (arrows); (2) interglacial lake (heavy broken line); (1) uplands isolated by basin subsidence (stippled line: mapped edge of upland; dot-and-dashed line: inferred edge). T1 and T2 are localities of volcanigenic material, mainly tuff, that indicates coeval distal volcanicity.

The commonest stone in the Dwyka diamictite is granite and gneiss but their occurrence in basement on all sides of the Karoo Basin rules out a specific source area. Distinctive stones in the basal diamictite relatable to source are stromatolitic dolomite and reddish quartzite at Elandsvlei (EL in Fig. 17D) from the Proterozoic of the Ghaap Plateau, and diorite, schist, slate, and archeocyathid limestone (Table 5, a), attributed by default (outcrops are not known in southern Africa) to the south and southeast (Visser et al., 1986).

During the Early Permian (275–263 Ma) deposition of the lower Ecca Group (Fig. 17E), lobes of a delta complex with coal measures overlapped subsiding cratonic uplands in the north (Witwatersrand Arch) and in the east (Eastern uplands) and prograde and shale-out to the southwest into a shallow anoxic basin, which continued to the northwest into South America (Whitehill-Iratí shale basin). A distal magmatic arc is indicated by widespread tuff (1–12 of Table 2).

Figure 17F shows the deposition of the Early/Late Permian (263–255 Ma) Upper Ecca Group of southern Africa and the 3-km-thick equivalent above the Lafonian Diamictite–Black Rock Member on the Falkland Islands. The situation is now reversed from that of the Lower Ecca. Coarse and thick sediment came wholly from the south and west, and the northeastern area of the former delta complexes was covered by a shale (Volksrust Formation). By the end of Ecca deposition, the basin was almost filled with sediment to become the broad fluvial plain of the Beaufort Group. The Upper Ecca represents the first pulse of thick sediment shed into a foredeep from the rising Cape Foldbelt, which underwent a second deformation, the Outeniqua folding, towards the end of Ecca deposition. This foredeep developed by rapid subsidence of the shallow floor of the Whitehill lake to a probable maximum depth of 500 m during the deposition of subaqueous fans. The isopachs parallel the strike of the Cape Fold Belt, and the axis pitches eastward into the thickest known section, near the coast and on the Falkland Islands. In the west, sediment was transported down an east-northeastward paleoslope and was probably derived from two sources: (1) by consequent drainage down the flank of the resurgent Atlantic Mountains, and (2) by axial drainage from the Cape Fold Belt. The Atlantic Mountains temporarily interrupted or restricted the previous Tastubian upper Dwyka margin and Artinskian Lower Ecca

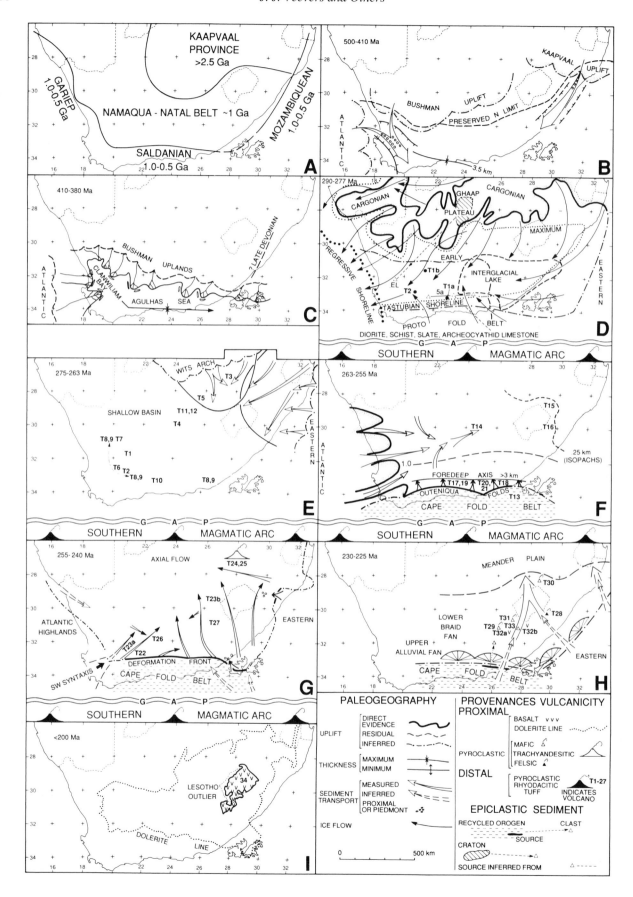

brackish connection between southern Africa and South America that resumed later during the early Kazanian Waterford-Estrada Nova lake with its endemic bivalves.

The framework modes of sandstone in the Ripon Sandstone fall within the recycled orogen field (Fig. 16); the mean of the Ecca and Beaufort Groups (point 2 of Fig. 16) falls between the provenances of the recycled orogen and the magmatic arc, which is represented by widespread tuff (13–21 of Table 2).

Figure 17G shows the Late Permian–Early Triassic (255–240 Ma) deposition of the Beaufort Group. Following the filling of the Ecca basin by northward and eastward prograding deltas, a greater volume of sediment from the faster growing Cape Fold Belt, now including deformed and uplifted early Beaufort strata, prograded diachronously northward in rivers that flowed across the former basin floor down a 500-km-long piedmont flank to an east-trending axis of sediment transport in the northeastern Karoo Basin. During 15 million years of deposition, the piedmont wedge reached an estimated thickness of 6 km at the south crop. The main fan in the southeast radiated from a source region (Theron, 1975) upslope from the syntaxis at the Falkland Islands, as the minor fan in the southwest radiated from a similar syntaxis in the Cape Town–Ceres region. Enclosure of the basin on the northwest is indicated by SE-trending paleocurrents northwest of T26. In the Early Triassic, piedmont conglomerate (circles) was deposited at the foot of uplands in the Falkland Islands and Precambrian Mapumulo Group in the northeast. Proximal sources are an inferred trachyandesite volcanic center near T24, 25, the garnetiferous gneiss of the Mapumulo Group east of the Mooi River (Table 6, g), the Devonian to Permian section of

Figure 17. Paleo-tectonic/geographic maps, showing provenances. Volcanigenic material, mainly tuff that indicates coeval distal vulcanicity during the Permian and Early Triassic, denoted by T1–T23 and T27 (from Table 2); proximal vulcanicity during the Triassic and Jurassic denoted by T24–T25 and T28–T33.

A. Pre-Cape (>2.5–0.5 Ga) basement trends, from Figure 4.

B. Ordovician-Silurian: Table Mountain and Natal Groups, from Figure 5A.

C. Early-Middle Devonian Bokkeveld Group and equivalent in the Falkland Islands, from Figure 5B.

D. Latest Carboniferous–earliest Permian (290–277 Ma) Dwyka Formation in Southern Africa and Lafonian Diamictite on the Falkland Islands, from Figure 11. Magmatic arc in D–G located an unknown distance southward beyond the gap.

E. Early Permian (275–263 Ma) Lower and Middle Ecca Group of southern Africa and Black Rock Member of Falkland Islands, from Figure 12A.

F. Early/Late Permian (263–255 Ma) Upper Ecca Group of southern Africa and the 3-km-thick equivalent above the Lafonian Diamictite/Black Rock Member of the Falkland Islands, from Figure 12B.

G. Late Permian–Early Triassic (255–240 Ma) Beaufort Group, from Figure 13A.

H. Late Triassic (230–225 Ma) Molteno Formation, from Figure 14A.

I. Early Jurassic Karoo vulcanicity.

the Cape Fold Belt upcurrent from the Katberg–East London area (Table 5, d), and an inferred granite >390–210 Ma (Table 5, c). The bulk of the Beaufort Group sediment was recycled from the orogen of the Cape Fold Belt (Fig. 16), but sediment from the distal magmatic arc, including rhyodacitic tuff (22, 23, 26, 27 of Table 2), was sufficient to shift the sandstone mode toward the magmatic arc provenance. In the northeastern Karoo Basin, coarse (lapilli) tuff (24–25 of Table 2) indicates a proximal source in a trachyandesitic volcanic center.

Figure 17H shows the Late Triassic deposition of the Molteno Formation, as detailed in Figure 14A. Downslope from the inferred upper fans at the foot of the Cape Fold Belt and the Eastern uplands is a belt of a lower braid fan and then a meander plain. In the Molteno Formation, tonstein (28 of Table 3), derived from a rhyodacitic tuff in turn derived from a proximal source volcano, was followed in the Elliot and Clarens Formations (29–33) by basalt and mafic pyroclastics during the initial stage of *in situ* Karoo vulcanicity.

Figure 17I shows the Early Jurassic Karoo vulcanicity, including the eruption of flood basalt, now worn back to the Lesotho outlier, and the intrusion of dolerite limited by the dotted line.

Setting within southern Africa

The inception of the Karoo Sequence in the latest Carboniferous–earliest Permian corresponded to the relaxation of the Pangean platform in sags and rifts. The regional map of Figure 18 shows the great northward expansion of Gondwanan deposition over the interior of the Gondwanaland province of Pangea beyond that of the confined pre-Gondwanan Cape Sequence. The latest Carboniferous–earliest Permian (290–277 Ma) Dwyka Formation in southern Africa, Lafonian Diamictite on the Falkland Islands, and equivalents northward to 5°N occupied sags in the Karoo (western) terrain, including the Kalahari Basin between the Windhoek and Cargonian uplands and the Karoo Basin between the Cargonian uplands and the proto Cape Fold Belt. The Zambezian terrain (east of the heavy broken line) was affected by faulting of basins during or after deposition or both. Faulting also affected the eastern part of the Kalahari Basin (Visser, 1987) and, in the western terrain, the entire Warmbad Basin (J.N.J. Visser, personal communication, 1992) at the southern part of the Kalahari Basin (Fig. 12A). The equivalents of the Ecca and Beaufort Groups are known from place to place north of the Karoo Basin, but are not so widespread as those of the Dwyka. Only with the renewal of widespread deposition in the Late Triassic and the onset of voluminous proximal Karoo vulcanism in the Jurassic (Fig. 19) were areas in the north newly covered by sediment and lava. Renewed and more extensive deposition in the Late Triassic corresponds to a singularity in Pangean history: terminal compression of fold belts (Cape Fold Belt, Bowen-Sydney Basin) and widespread subsidence, mainly in rifts that prefigured the

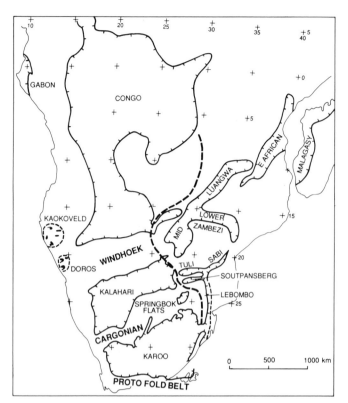

Figure 18. Latest Carboniferous–earliest Permian (290–277 Ma) Dwyka Formation of southern Africa, Lafonian Diamictite of the Falkland Islands, and equivalents northward to 5°N. Map base from Figure 6.

divergent margins of the Atlantic and Indian Ocean regions (Veevers, 1989). The subsequent Karoo vulcanism reflects the increased activity of Pangean hotspots during rifting.

Also shown in Figure 19 are the axes of drainage (marked b) and maximum thickness of the Permian-Triassic foredeep (marked a) for the following intervals:

(I) Tastubian, at end of Dwyka deposition: As shown in section I, the Dwyka Formation was deposited over a basement that comprised the Kaapvaal Province, the Namaqua-Natal Belt (N-N), the Southern Cape Conductive Belt (SCCB), including the Beattie Anomaly (BA), and the Cape Supergroup. The basin subsided between cratonic uplifts and the growing Swartberg fold in the Cape Supergroup and older rocks (near Y in the section). The axis of iceflow drainage (Ib) lay 80 km north of the maximum thickness of 1 km at Ia (wiggly line).

(II) Kazanian, at the end of Ecca deposition: Sediment encroached northward on the remnant Witwatersrand (WITS) Arch. In the south, the foredeep (IIa for the upper Ecca Group) subsided in front of the growing Outeniqua folds of the Cape Fold Belt and filled with 3.5 km of sediment. The drainage axis (IIb) lay 400 km north of the foredeep.

(III) Early Triassic, at end of Beaufort deposition: The

foredeep (IIIa) in front of the further folded and thrusted Cape Fold Belt filled with 6 km of coarse sediment. The drainage axis (IIIb) lay 550 km to the north.

(IV) Late Triassic–Early Jurassic, Stormberg deposition and early part of Karoo vulcanism: The Stormberg Group was deposited during the terminal listric thrusting and epeirogenic uplift of the Cape Fold Belt and Karoo Basin in the south and the onset of rift uplift marked by the Choma–Kalomo Ridge (C-KR) in the north. The drainage basin (IVb) lay 1,000 km from the foot of the foldbelt uplift (IVa), which accumulated at least 1 km of piedmont fans that were subsequently removed during the epeirogenic uplift. The thickest preserved Stormberg equivalent in the north is 2.5 km thick, and was deposited at the foot of the rapidly rising Choma-Kalomo Ridge (C-KR). Karoo vulcanism includes the local volcanic centers in the Stormberg Group and the succeeding flood lava of the Drakensberg Group. To the north, the volcanics have been worn back to smaller outliers or, as in the Lebombo monocline, covered by younger rock. Dolerite intrusions are confined by the dolerite line in the Karoo Basin and (not mapped) within the Kalahari Basin.

As gauged from the vertical datum line in the sections, the foredeep axis shifted 20 km northward from its initial position (I) during Ecca deposition (II) and then a further 40 km northward during Beaufort (III) and Stormberg (IV) deposition by a modest northward encroachment of the deformation front. An initial separation of foredeep (Ia) and drainage (Ib) axes of 80 km increased to 400 km during Ecca deposition, 550 km during Beaufort deposition, and 1,000 km during Stormberg deposition. The increasing separation reflects the widening of the southern depositional flank of the Karoo Basin during the growth of the Cape Fold Belt at the expense of the starved northern cratonic side. Only during Stormberg deposition did the northern craton match the Cape Fold Belt as a source of voluminous sediment. Without this northern source, the drainage axis would have lain even farther than 1,000 km from the Cape Fold Belt.

ACKNOWLEDGMENTS

Veevers acknowledges with thanks the hospitality and guidance given to him by many people during his visit to South Africa in 1984, especially Heidi and John Anderson (Pretoria), Ian McLachlan and Neil Phillips (Johannesburg), Victor Von Brunn (Pietermaritzburg), R. Taverner-Smith and Tom Mason (Durban), Roy Oosthuizen (Klaarstroom), Russell Shone (Port Elizabeth), Richard Dingle and John Rogers (Cape Town), I. W. Hälbich (Stellenbosch), and Johan Loock and Johan Visser (Bloemfontein). Cole acknowledges the help of colleagues Piet Gresse, Hannes Theron, and Jurie Viljoen in commenting on parts of the manuscript. We express our gratitude to the formal reviewers, David Hobday and Johan Visser,

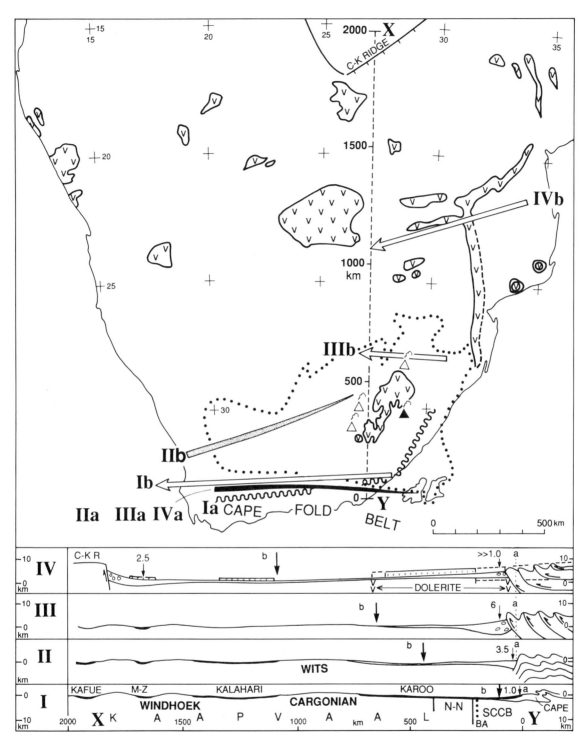

Figure 19. Late Triassic to Middle Jurassic proximal Karoo vulcanicity (V's) (from Figs. 6, 14A, 17H, and 17I), including dolerite (within the dotted line) and, in Mozambique east of the Lebombo monocline, subsurface lavas (doubly encircled V's). Axis of drainage (marked b, open arrows [except stippled II] on map, long arrows on section) and the axis of maximum thickness of the Permian-Triassic foredeep (marked a, solid lines on map, short arrows on the section) for the intervals I–IV, from Figure 17. Restored sections (V:H = 10) along line XY, with sediment <1 km thick shown by solid black. I, end Dwyka; II, end Ecca; III, end Beaufort; IV, Stormberg. BA, Beattie Anomaly; C-K R, Choma-Kalomo Ridge; M-Z, Mid Zambezi, N-N, Namaqua-Natal Belt; SCCB, Southern Cape Conductive Belt; WITS, Witwatersrand Arch.

for their generous and helpful comments on the manuscript. We are grateful to Michael Johnson and the publisher for permission to reproduce one of his figures. The final figures were drafted by Judy Davis.

APPENDIX 1. THIN-SECTION STUDY OF SELECTED SANDSTONES AND SILTSTONES FROM THE KAROO BASIN

E. J. Cowan

Thin sections were cut from rock specimens collected by Veevers in 1984 from localities along the southern crop of the Karoo Basin (Fig. 1, Table 7), including the localities of Elliot and Watts (1974) and Lock and Johnson (1974). Shards were found in the Whitehill Formation (Fig. 20A) and Collingham Formation (Fig. 20B) at Remhoogte, as in Elliot and Watts (1974), and also in the Collingham Formation (Fig. 20C) at Ecca Pass and in the Fort Brown Formation (Fig. 20D) southwest of Waterford. A sample from the Matjiesfontein Chert (Fig. 20E) has a "porphyritic" texture with a groundmass containing feldspar and quartz fragments, all interpreted as a tuff, although in the absence of shards this remains tentative.

Sandstones in the Ripon Formation (SA 9) and Beaufort Group (SA 30) were coarse enough to be point-counted (Table 8) according to the scheme adopted by McDonnell (1983) for Sydney Basin samples.

The sandstones characteristically have fairly abundant feldspar and rather rare nonvolcanic rock fragments, mainly sedimentary chert (Fig. 20F). Nearly all the rock fragments in the point-counted samples are volcanic, but whether they are the same age as or older than deposition cannot be told. As in some sub-labile sandstones of the Sydney Basin, the relatively small amount of volcanic quartz and the abundance instead of common quartz indicate that a volcanic source contributed a small part only of the sand fraction. The volcanogenic lithic fragments indicate that the source contained dacitic/rhyolitic volcanics, again similar to the Sydney Basin sandstones. The lathwork volcanolithic grains found in the Dwyka sample (SA5b) and also in trace amounts in the Late Permian coal measures of the Sydney Basin indicate a minor mafic-intermediate volcanic source.

APPENDIX 2. LATE CARBONIFEROUS–JURASSIC CORRELATION CHART FOR SOUTHERN AFRICA

J. J. Veevers

In Fig. 21 the columns are arranged from south to north through the western or Karoo tectono-sedimentary terrain of open basin and swell structure (Rust, 1975, fig. 38.1) on the left, and from southwest to northeast through the eastern or Zambezian terrain of fault-affected basins (Rust, 1975, fig. 38.1) on the right. The radiometric time scale from Appendix 1, Chapter 3, has the Permian/Triassic boundary calibrated at 250 Ma.

I acknowledge the helpful comments of D. I Cole on a draft of the chart.

Biostratigraphical schemes: Primary anchor points

(1) The Tastubian *Eurydesma* fauna (Dickins, 1961, 1984) of the Kalahari Basin localities a and b (Figs. 1 and 11), correlated by fish and spores (McLachlan and Anderson, 1973, p. 45) with the central Karoo Basin (locality c) (McLachlan and Anderson, 1973, 1975) (Table 1). (2) The early Tatarian (Khachian) and mid-Scythian ammonoid faunas of northern Malagasy (Teichert, 1974, p. 377).

Secondary anchor points

(1) Palynozones 1–4 of the northern Karoo Basin (Anderson, 1977, 1981), tentatively correlated by Truswell (1980, fig. 4) with the eastern Australian palynozones, and firmly correlated by Backhouse's (1991, p. 253–256) palynological study of the Permian Collie Basin, Western Australia, from which this précis is drawn:

"[In] Anderson's (1977) study of the northern Karoo Basin . . . the taxa are comprehensively illustrated and adequate details are provided of their stratigraphic distribution. The study is based on an apparently uninterrupted succession that ranges from Stage 2 to above the *Didecitriletes ericianus* zone [Eastern Australian lower Stage 5b]. . . . Because Anderson's study is more comprehensive in scope than other studies on Gondwanan Permian basins more attention is given here to comparing it with the Collie Basin sequence and in the systematic section of this report the ranges and morphologies of all biostratigraphically significant taxa present in both basins are compared. . . .

The two basins are most similar in the lower parts of the sequences. Up to the lower part of the *M. trisina* zone [Stage 3b] the only major differences in the distribution of index taxa are the absence, or non-recognition, of *Pseudoreticulatispora confluens* and the absence of *Microbaculispora trisina* below Microfloral Zone 3b in the Karoo Basin. The base of the *P. pseudoreticulata* zone [Stage 3a] corresponds here with the base of Microfloral Zone 2a in the Karoo Basin. The *S. fusus* biohorizon [Collie Basin] [upper 3a], which is correlated here with the first appearance of *Striatopodocarpites cancellatus* in the Karoo Basin, and *P. sinuosus* biohorizon [lower Stage 4] are also recognisable on Anderson's distribution chart. Supporting evidence for these correlations is provided by the ranges of *Praecolpatites ovatus* and *Laevigatosporites colliensis* in the Karoo and Collie Basins. The *D. ericianus* biohorizon [lower Stage 5b/c] is the only Collie Basin datum above the *P. sinuosus* biohorizon that can be recognised in the Karoo sequence. *D. ericianus* is first recorded by Anderson from the base of Microfloral Zone 4d."

After R. Helby (personal communication, 1990), I place the *confluens* zone at the base of Stage 3a (Fig. 44 in Chap. 3); Backhouse (1991) places it between Stages 2 and 3a. With this exception, the correlation of the Karoo zones to those of Eastern Australia in Fig. 21 are taken directly from Backhouse (1991, fig. 10). The *Granulatosporites confluens* zone of Foster and Waterhouse (1988) is found at the top of the Dwyka Formation (Anderson's zone 1), so that the rest of the Dwyka Formation, equivalent to the Stockton Formation of the Collie Basin (Fig. 43 in Chap. 3), is Stage 2.

The 12 palynological analyses of C. MacRae, cited by Visser (1990), generally confirm the correlations that Backhouse (1991) made with Anderson (1977).

(2) The middle to ?lower Sakamena Group of southern Malagasy, correlated to the *L. pellucidus* zone of eastern Australia (Foster, 1979, p. 126; Wright and Askin, 1987).

(3) lower Isalo I, correlated to the *S. speciosus* zone (Carnian and younger) of Dolby and Balme (1976) = late Ladinian and Carnian (Helby et al., 1987).

(4) Molteno Formation correlated by mega- and micro-floras to Ipswich Coal Measures (Carnian–early Norian) of eastern Australia (Anderson and Anderson, 1983, p. 8).

(5) Tetrapod zones, in descending order, from Anderson (1981) and Keyser and Smith (1979), given also in Dingle et al. (1983):

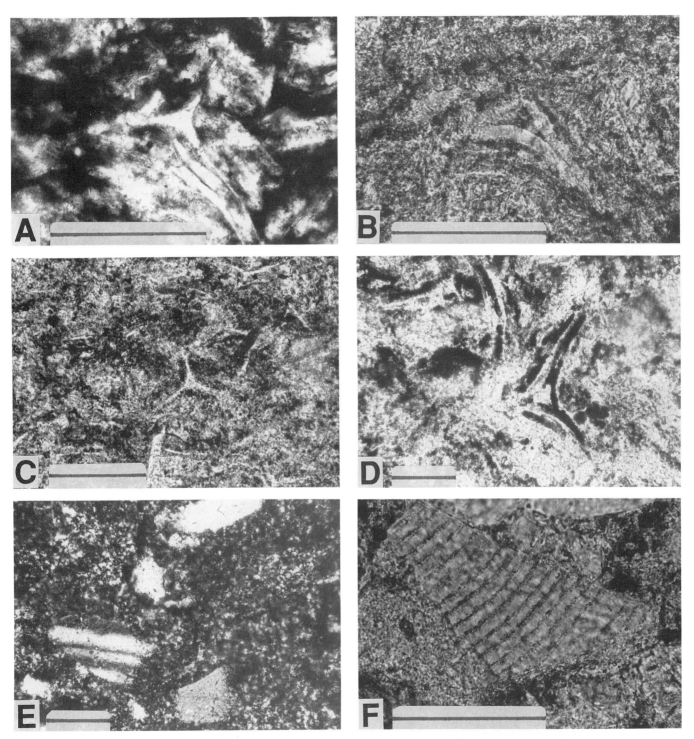

Figure 20. Photomicrographs. The bar scale is 0.1 mm long. A. Vitric tuff with shards replaced by clay, sample SA 16a, Whitehill Formation, Remhoogte. B. Argillite with curvilinear fragments interpreted as shards with triple junctions, SA 15, basal Collingham Formation, Remhoogte. C. Siltstone with devitrified glass shards, SA 7a, basal Collingham Formation, Ecca Pass. D. Tuff of shards in felsitic groundmass partly replaced by carbonate, SA 10, Fort Brown Formation, 13 km north of Wolwefontein. E. Angular feldspar and quartz in a felsitic groundmass, crossed nicols, SA 14c, Matjiesfontein Chert (Collingham Formation), Remhoogte. F. Fine-grained sandstone with a lithic grain of sedimentary chert (banded chalcedony), SA 11c, Fort Brown Formation, Geelhoutboom.

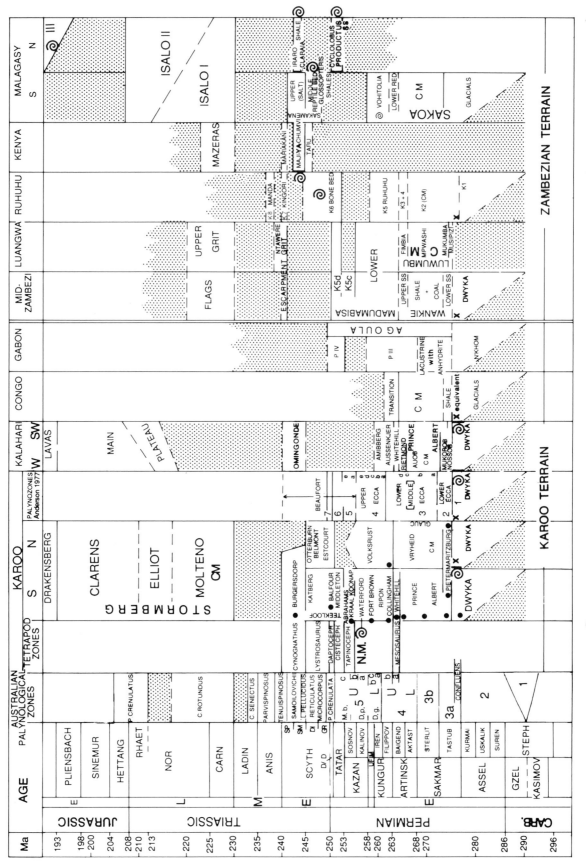

Figure 21. Correlation chart for southern Africa.

Anderson (1981)	**Keyser and Smith (1979)**
Cynognathus	*Kannemeyeria*
Lystrosaurus	*Lystrosaurus*
Daptocephalus	*Dicynodon lacerticeps*
Cistecephalus	*Aulacephalodon baini*
	Tropidostoma microtrema
Tapinocephalus	*Pristerognathus/Diictodon*
	Dinocephalian
Mesosaurus	

According to Dingle et al. (1983), the *Cynognathus* Zone is upper Griesbachian, the *Lystrosaurus* Zone all or partly lower Griesbachian, the *Dicynodon lacerticeps* (= *Daptocephalus*) Zone presumably Dorashamian as well as Djulfian (to which it was restricted by Anderson, 1981) because evidence of a lacuna is lacking. Anderson and Cruickshank (1978) date the *Cynognathus* Zone as late Spathian—note the occurrence of *Paratosaurus* of the *Cynognathus* Zone of South Africa (Dingle et al., 1983) in the Gosford Formation and Hawkesbury Sandstone of eastern Australia (Spathian-Anisian, Fig. 44 in Chap. 3). Since evidence of a lacuna between the *Lystrosaurus* and *Cynognathus* Zones is lacking, I extend each zone to the mid-Scythian, so that the *Lystrosaurus* Zone is taken as Griesbachian and Dienerian, and the *Cynognathus* Zone as Smithian and Spathian.

The ages of Anderson's (1981) Karoo zones 5–7 are calibrated by tetrapods.

Basins

North and northeastern Karoo (Anderson, 1981).

Correlation by palynomorphs encompasses the Dwyka to Molteno, except the Katberg and Burgersdorp, correlated by tetrapods.

Dwyka Formation. According to Anderson (1977, p. 43; 1981), palynozone 1 is restricted to the Dwyka Formation. Truswell (1980) tentatively correlated palynozone 1 with East Australian Stage 2, and this was confirmed by Backhouse (1991). Foster and Waterhouse (1988) correlated the Dwyka Formation (?uppermost part) with the *G. confluens* zone, which I regard as early Tastubian or lower Stage 3a. The fish at the top of the Dwyka at Douglas (locality c in Figs. 1 and 11) also indicate Tastubian. The apparent range of the Dwyka (= South African palynozone 1) in the northern and northeastern Karoo and, according to Anderson (1981), in the southern Karoo (and of the glacials in the Congo and mid-Zambezi Basins), is East Australian Stage 2 and lower 3a (shown by the oblique broken line). The overlying palynozone 2 is placed in the later Tastubian or upper Stage 3a. The biostratigraphical ages underpin Visser's (1990, fig. 6) modeled age of Stephanian to Sakmarian, calculated from sedimentation rates, eustatic events, and facies trends.

Ecca Group. According to McLachlan (1977), the ammonoid recorded from the Ecca Series of Natal is not *in situ*. Glauconite is the only physical indicator of a possible marine depositional environment in the northern Karoo Basin, and it occurs in the Vryheid (or Middle Ecca) Formation, in palynozone 3 (Anderson, 1977, chart 2).

Tetrapods. The *Cynognathus* Zone is restricted to the Burgersdorp Formation, which is overlapped by the Stormberg Group northeast of the line between 27°E, 29°S and 29°E, 32°S (Kitching, 1970; Dingle et al., 1983, p. 14, 18, 22). In northwest Natal, the Estcourt Formation is equivalent to the *Daptocephalus* Zone (Keyser and Smith, 1979, p. 12) and possibly also to the *Lystrosaurus* Zone (Dingle et al., 1983, p. 21). The Otterburn Formation also contains *Lystrosaurus* Zone fossils (Dingle et al.,1983, p. 23) and the Belmont Formation lies within the *Lystrosaurus* Zone (Dingle et al., 1983, p. 23). The Molteno Formation is dated by its macro- and micro-floras (Anderson and Anderson, 1983, p. 8).

Southern and central Karoo (Anderson, 1981).

Formations are correlated by the following:

(a) Those palynomorphs not destroyed by the intense deformation and dolerite intrusion experienced in the southern Karoo provide correlation of the Dwyka, Prince Albert, Fort Brown, and Middleton Formations. The age of the Dwyka Formation in the south is poorly constrained (Anderson, 1977, p. 51, 52) but, as noted already, Anderson (1981) sees no reason to regard it as any older or younger than it is in the north. Visser (1990) developed a facies model in which deposition of the Dwyka Formation becomes younger from shelf-ice deposits in the south to valley deposits in the north, but paleontological evidence of diachronous deposition is lacking.

(b) Tetrapods: (i) in the Whitehill Formation, the vertebrate assemblage of the *Mesosaurus* zone is dated by its position at the boundary between palynozones 3 and 4 (Anderson, 1981), equivalent to the middle of Eastern Australian Stage 4 or Baigendian, as found independently by Anderson and Anderson (1985, p. 29). This age in turn dates the Iratí Formation of the Paraná Basin of Brazil, correlated with the Whitehill through its mesosaurid fauna (Oelofsen, 1987). (ii) Waterford (Anderson and Anderson, 1985, p. 28): the nonmarine bivalve assemblage of the Waterford Formation (Cooper and Kensley, 1984) is dated as mid- to late Kazanian (Kalinovian) from its conformable position beneath the tetrapod-bearing (Sosnovian and younger) Beaufort Group; in turn this dates the Estrada Nova Formation of the Paraná Basin of South America, correlated with the Waterford through its nonmarine bivalve fauna, already described by Runnegar and Newell (1971). (iii) Beaufort (Koonap, Abrahamskraal, Teekloof, Middleton, Balfour, Katberg, Burgersdorp) zoned by tetrapods (Keyser and Smith, 1979) over an interval from the Sosnovian through the Scythian, as detailed above.

(c) mega- and micro-flora : Molteno = Ipswich Coal Measures of Eastern Australia (Anderson and Anderson, 1983, p. 8) (*C. rotundus* zone, Carnian-early Norian). We follow Dingle et al. (1983, p. 28, 29) in discounting Stapleton's (1978) view that the age of the Molteno is indistinguishable from that of the Burgersdorp.

(e) superposition and interpolation: (1) Collingham and Ripon between the Whitehill and Waterford; (2) Elliot and Clarens above the Carnian-early Norian Molteno, and at and below the Toarcian (193 ± 5 Ma, Fitch and Miller, 1984) and younger Drakensberg lavas, consistent with the general age indicated by tetrapods (Dingle et al., 1983, p. 17, 43).

Kalahari.

The upper, marine part of the Dwyka is dated as Tastubian (*Eurydesma* fauna of Dickins, 1961, 1984; McLachlan and Anderson, 1973, 1975). Note the occurrence north of Etosha Pan of Anderson's Zone 2 spores from what are called Dwyka Shales (Truswell, 1980, p. 101), probably the Prince Albert Formation. Overlying formations are named from Kent (1980) and dated by Anderson (1981); Omingonde = Burgersdorp according to Anderson and Anderson (1983, chart 13) and Dingle et al. (1983, p. 72), with additionally Plateau and "Main" Sandstones.

Congo.

From Truswell (1980) and Anderson (1981).

Gabon.

From Truswell (1980).

Mid-Zambezi (Zimbabwe), Luangwa (Zambia), Ruhuhu (Tanzania).

From Truswell (1980), Anderson (1981), Anderson and Anderson (1983), and Foster and Waterhouse (1988). Coil = position of Cox's (1936) marine bivalves in the Kidodi area of Tanzania (Furon, 1963, p. 319) now nonmarine (Yemane and Kelts, 1990).

Kenya.

From Anderson and Anderson (1983), and Anderson (1981). Maji-ya-Chumvi with "une intercalation marine à Poissons et *Esthe-*

ria" (Blant, 1973, p. 197), correlated with the mid-Scythian of northern Malagasy.

Malagasy (Boast and Nairn, 1982).

Southern. Sakoa Group: Coal measures (CM) = Anderson's (1977) palynozone 3 (Anderson, 1981, supported by Truswell, 1980, p. 103). Glacials (barren) by superposition regarded (Anderson, 1981) as equivalent to Zone 1 and part of 2. Vohitolia Limestone Member, with brachiopods = Kungurian (Anderson, 1981).

Sakamena Group, lower part: The *Glossopteris* Shales of Furon (1963, p. 358) contain the same marine bivalves (*Modiolopsis stockleyi* Cox and *Gervillia elianae* Cox) as those in northern Tanzania (Furon, 1963, p. 319). Middle Reptile Bed dated by Anderson (1981) as lower Griesbachian (*see* Battail et al., 1987, for paleobiogeographical correlations with mainland Africa). The middle to ?lower parts are equivalent to the *L. pellucidus* zone of Eastern Australia (Foster, 1979, p. 126; Wright and Askin, 1987). The upper part (Bed 6 of Anderson, 1981) is dated as Dienerian/Smithian.

Isalo I: According to Dolby and Balme (1976, p. 127), there is "clear evidence for equating the *S. speciosus* zone to Goubin's Zone IIINA. It may also represent her Zone IIINB but this correlation is less obvious." Dolby and Balme (1976, fig. 10) show Isalo I ranging from top Carnian to base Rhaetian and overlapping Isalo II. Anderson (1981) and Anderson and Anderson (1983) show all the Isalo with marine invertebrates, contrary to all other sources who regard the Isalo as nonmarine except near the top (Isalo III), which, starting in the Toarcian, is dominantly marine. Dolby and Balme (1976, p. 127) find that Rhaetian-Liassic indices appear first in the upper part of Isalo II, from which I regard the top of Isalo II as end-Triassic.

Northern. The sequence starts with the *Productus* sandstone and *Cyclolobus* shale, equivalent to the Khachian Chhidru Formation of the Salt Range (Teichert, 1974, p. 376). Above a disconformity are the *Claraia* shales, beds with ammonoids, and the Iraro Shale with ammonoids at the top, all mid-Scythian according to Teichert (1974, p. 377). According to Anderson (1981), the Isalo is the same age here as in the south.

Age of distal volcanigenic material

Distal volcanigenic material (shown in Fig. 21 by the filled circles) ranges from the earliest Sakmarian or Tastubian (277 Ma) upper Dwyka Formation (1 in Table 2) to the late Scythian (240 Ma) Burgersdorp Formation (27 in Table 2).

Age of coal measures

Coal measures (CM) are restricted to the Sakmarian and Artinskian (277–263 Ma) of the Early Permian, to the Early Triassic (250–247 Ma)*, and to the Carnian (230–225 Ma) of the Late Triassic.

The Early Permian coal measures occur in the following units, from left to right on Fig. 21: Vryheid (Middle Ecca), Auob, unnamed coal measures in the Congo Basin, Wankie, Luwumbu, K2, and Sakoa. The oldest occurrence is in K2, in the Ruhuhu Basin, from the base of the Sakmarian or Tastubian (277 Ma), and the youngest is in the Vryheid and Sakoa, at the top of the Artinskian (263 Ma). Early Triassic* coal layers are restricted to the Estcourt Formation of the Beaufort Group (Fig. 8A, part III). The Late Triassic coal measures are restricted to the Carnian (230–225 Ma) part of the Molteno Formation.

Pre-Permian sequences in southern Africa

The Pre-Permian sequences beneath the Permo-Triassic basins are entirely Precambrian except the 9-km-thick mid-Paleozoic Cape Supergroup (Rust, 1973) beneath the southern Karoo Basin, the 1-km-thick Natal Group beneath the northeastern Karoo Basin, and the thin (2 m) arkose and mudstone tentatively dated as Ordovician-Silurian (Reimann, 1986) beneath the Barotse lobe of the Congo Basin.

---

*This is in the Estcourt Formation, which, containing *Glossopteris*, is Late Permian (G. J. Retallack, personal communication, 1993).

REFERENCES CITED

Adie, R. J., 1952a, The position of the Falkland Islands in a reconstruction of Gondwanaland: Geological Magazine, v. 89, p. 401–410.

Adie, R. J., 1952b, Representatives of the Gondwana System in the Falkland Islands, *in* Teichert, C., ed., Symposium sur les Séries de Gondwana: Algiers, 19th International Geological Congress, p. 385–392.

Allsopp, H. L., Bristow, J. W., Smith, C. B., Brown, R., Gleadow, A.J.W., Kramers, J. D., and Garvie, O. G., 1989, A summary of radiometric dating methods applicable to kimberlites and related rocks: Geological Society of Australia Special Publication 14, p. 343–357.

Anderson, A. M., and McLachlan, I. R., 1979, The oil-shale potential of the Early Permian White Band Formation in Southern Africa: Geological Society of South Africa Special Publication 6, p. 83–89.

Anderson, J. M., 1977, The biostratigraphy of the Permian and Triassic, Part 3: A review of Gondwana Permian palynology with particular reference to the northern Karoo Basin South Africa: Memoirs of the Botanical Survey of South Africa, v. 41, 67 p., 4 charts, 188 plates.

Anderson, J. M., 1981, World Permo-Triassic correlations: Their biostratigraphic basis, in Cresswell, M. M., and Vella, P., eds., Gondwana Five: Rotterdam, A. A. Balkema, p. 3–10.

Anderson, J. M., and Anderson, H. M., 1983, Palaeoflora of Southern Africa: Molteno Formation Triassic, Volume 1: Rotterdam, A. A. Balkema, 227 p.

Anderson, J. M., and Anderson, H. M., 1985, Palaeoflora of Southern Africa: Prodromus of South African megafloras: Devonian to Lower Cretaceous: Rotterdam, A. A., Balkema, 423 p.

Anderson, J. M., and Cruickshank, A.R.I., 1978, The biostratigraphy of the Permian and Triassic, Part 5: A review of the classification and distribution of Permo-Triassic tetrapods: Palaeontologia Africana, v. 21, p. 15–44.

Backhouse, J., 1991, Permian palynostratigraphy of the Collie Basin, Western Australia: Review of Palaeobotany and Palynology, v. 67, p. 237–314.

Barrett, S. F., and Isaacson, P. E., 1988, Devonian paleogeography of South America, *in* McMillan, N. J., Embry, A. F., and Glass, D. J., eds., Proceedings, International Symposium on the Devonian System, 2nd, Calgary, Volume 1: Regional syntheses: Canadian Society of Petroleum Geologists, p. 655–667.

Battail, R., Beltan, L., and Dutuit, J. M., 1987, Africa and Madagascar during Permo-Triassic time: The evidence of the vertebrate faunas, *in* McKenzie, G. D., ed., Gondwana Six: American Geophysical Union, Geophysical Monograph, v. 41, p. 147–155.

Beckinsale, R. D., Tarney, J., Darbyshire, D.P.F., and Humm, M. J., 1976, Rb-Sr and K-Ar age determinations on samples of the Falkland Plateau basement at Site 330: Initial Reports of the Deep Sea Drilling Project, v. 36, p. 923–927.

Behr, S. H., 1986, Heavy mineral deposits in the Karoo Sequence, *in* Anhaeusser, C. R., and Maske, S., eds., Mineral deposits of southern Africa, Volumes I and II: Johannesburg, Geological Society of South Africa, p. 2105–2118.

Ben-Avraham, Z., Hartnady, C.J.H., and Malan, J. A., 1993, Early tectonic extension between the Agulhas Bank and the Falkland Plateau due to rotation of the Lafonia microplate: Earth and Planetary Science Letters, v. 117, p. 43–58.

Beukes, N. J., 1970, Stratigraphy and sedimentology of the Cave Sandstone Stage, Karroo System: Proceedings and papers, Gondwana Symposium, 2nd: Johannesburg, p. 321–342.

Blant, G., 1973, Structure et paléogeographie du littoral meridional et oriental de l'Afrique, *in* Blant, G., ed., Sedimentary basins of the African coasts: Paris, Association of African Geological Surveys, p. 193–233.

Boast, J., and Nairn, A.E.M., 1982, An outline of the geology of Madagascar, *in* Nairn, A.E.M., and Stehli, F. G., eds., The ocean basins and margins, Volume 6, The Indian Ocean: New York, Plenum, p. 649–696.

Botha, B.J.V., 1967, The provenance and depositional environment of the Red Beds Stage of the Karroo System: Proceedings, Conference on Gond-

wana Stratigraphy: Buenos Aires, p. 763–774.

Botha, B.J.V., and Linström, W., 1978, A note on the stratigraphy of the Beaufort Group in North-western Natal: Geological Society of South Africa Transactions, v. 81, p. 35–40.

Boucot, A. J., Brunton, C.H.C., and Theron, J. N., 1983, Implications for the age of South African Devonian rocks in which Tropidoleptus (Brachiopoda) has been found: Geological Magazine, v. 120, p. 51–58.

Bradshaw, M. A., and Webers, G. F., 1988, The Devonian rocks of Antarctica, *in* McMillan, N. J., Embry, A. F., and Glass, D. J., eds., Proceedings, International Symposium on the Devonian System, 2nd, Calgary, Volume 1: Regional syntheses: Canadian Society of Petroleum Geologists, p. 783–795.

Bristow, J. W., and Saggerson, E. P., 1983a, A general account of Karoo vulcanicity in southern Africa: Geologische Rundschau, v. 72, p. 1015–1059.

Bristow, J. W., and Saggerson, E. P., 1983b, A review of Karoo vulcanicity in southern Africa: Bulletin Volcanologique, v. 46, p. 135–159.

Burger, A. J., and Coertze, F. J., 1977, Summary of age determinations carried out during the period April 1974 to March 1975: Annals of the Geological Society of South Africa, v. 11, p. 317–321.

Cadle, A. B., and Hobday, D. K., 1977, A subsurface investigation of the Middle Ecca and Lower Beaufort in Northern Natal and the South-eastern Transvaal: Geological Society of South Africa Transactions, v. 80, p. 111–115.

Clarke, M. J., and Forsyth, S. M., 1989, Late Carboniferous-Triassic: Geological Society of Australia Special Publication 15, 293–338.

Cole, D. I., 1980, Aspects of the sedimentology of some uranium-bearing sandstones in the Beaufort West area, Cape Province: Geological Society of South Africa Transactions, v. 83, p. 375–390.

Cole, D. I., and Labuschagne, L. S., 1983, Shallow-marine placer deposits in the Witteberg Group near Willowmore, Cape Province: Transactions of the Geological Society of South Africa, v. 86, p. 105–116.

Cole, D. I., and McLachlan, I. R., 1991, Oil potential of the Permian Whitehill Shale Formation in the main Karoo basin, South Africa, *in* Ulbrich, H., and Rocha Campos, A. C., eds., Proceedings, Gondwana Seven: São Paulo, Instituto de Geociências, Universidade de São Paulo, p. 379–390.

Cole, D. I., and Wipplinger, P. E., 1991, Uranium and molybdenum occurrences in the Beaufort Group of the main Karoo basin, South Africa, *in* Ulbrich, H., and Rocha Campos, A. C., eds., Proceedings, Gondwana Seven: São Paulo, Brazil, Instituto de Geociências, Universidade de São Paulo, p. 391–406.

Cole, D. I., Smith, R.M.H., and Wickens, H. de V., 1990, Basin-plain to fluvio-lacustrine deposits in the Permian Ecca and Lower Beaufort Groups of the Karoo Sequence, Guidebook, Geocongress 1990: Cape Town, Geological Society of South Africa, p. 1–83.

Cooper, M. R., 1979, Discussion of J. W. Marchant's "The metal contents of organic separates of some Ecca shales": Geological Society of South Africa Transactions, v. 82, p. 275.

Cooper, M. R., and Kensley, B., 1984, Endemic South American Permian bivalve molluscs from the Ecca of South Africa: Journal of Paleontology, v. 58, p. 1360–1363.

Cooper, M. R., and Oosthuizen, R., 1974, Archaeocyatha-bearing erratics from the Dwyka Subgroup (Permo-Carboniferous) and their importance to continental drift: Nature, v. 247, p. 396–398.

Coutinho, J.M.V., Hachiro, J., Coimbra, A. M., and Santos, P. R., 1991, Ashfall-derived vitroclastic tuffaceous sediments in the Permian of the Parana basin and their provenance, *in* Ulbrich, H., and Rocha Campos, A. C., eds., Proceedings, Gondwana Seven: São Paulo, p. 147–160.

Cowan, E. J., 1985, A basin analysis of the Triassic System, central coastal Sydney Basin [B.Sc. honours thesis]: Sydney, Macquarie University.

Cox, K. G., 1988, The Karoo Province, *in* Macdougall, J. D., ed., Continental flood basalts: Dordrecht, Kluwer, p. 239–271.

Cox, L. R., 1936, Karroo Lamellibranchia from Tanganyika Territory and Madagascar: Quarterly Journal of the Geological Society of London, v. 92, p. 32–57.

Cramer, F. H., Rust, I. C., and de Cramer, M.d.C.R., 1974, Upper Ordovician chitinozoans from the Cedarberg Formation of South Africa: Preliminary note: Geologische Rundschau, v. 63, p. 340–345.

Dawson, J. B., 1989, Geographic and time distribution of kimberlites and lamproites: relationships to tectonic processes: Geological Society of Australia Special Publication 12, 323–342.

de Beer, C. H., 1990, Simultaneous folding in the western and southern branches of the Cape Fold Belt: South African Journal of Geology, v. 93, p. 583–591.

de Beer, C. H., 1991, Author's reply to discussion on "Simultaneous folding in the western and southern branches of the Cape Fold Belt": South African Journal of Geology, v. 94, p. 393–395.

de Beer, J. H., 1983, Geophysical studies in the Southern Cape Province and models of the lithosphere in the Cape Fold Belt: Geological Society of South Africa Special Publication 12, p. 57–64.

de Beer, J. H., van Zijl, J.S.V., and Gough, D. I., 1982, The Southern Cape Conductive Belt (South Africa): Its composition, origin and tectonic significance: Tectonophysics, v. 83, p. 205–225.

de Villiers, J. E., and Wardhaugh, T. G., 1962, A sedimentary petrological study of some sandstones, conglomerates and tillites of the Cape and Karroo Systems: Transactions of the Geological Society of South Africa, v. 65, p. 101–108.

de Wit, M., Jeffery, M., Bergh, H., and Nicolaysen, L., 1988, Geological map of sectors of Gondwana reconstructed to their disposition 150 Ma ago: Tulsa, American Association of Petroleum Geologists, scale 1:10,000,000.

Dickins, J. M., 1961, *Eurydesma* and *Peruvispira* from the Dwyka Beds of South Africa: Palaeontology, v. 4, p. 138–148.

Dickins, J. M., 1984, Late Palaeozoic glaciation: Australian Bureau of Mineral Resources Journal, v. 9, p. 163–169.

Dickinson, W. R., 1985, Interpreting provenance relations from detrital modes of sandstones, *in* Zuffa, G. G., ed., Provenance of arenites: Dordrecht, D. Reidel, p. 333–361.

Dickinson, W. R., and Suczek, C. A., 1979, Plate tectonics and sandstone compositions: American Association of Petroleum Geologists Bulletin 63, p. 2164–2182.

Dickinson, W. R., Beard, L. S., Brakenridge, G. R., Erjavec, J. L., Ferguson, R. C., Inman, K. F., Knepp, R. A., Lindberg, F. A., and Ryberg, P. T., 1983, Provenance of North American Phanerozoic sandstones in relation to tectonic setting: Geological Society of America Bulletin, v. 94, p. 222–235.

Dingle, R. V., Siesser, W. G., and Newton, A. R., 1983, Mesozoic and Tertiary geology of southern Africa: Rotterdam, A. A. Balkema, 375 p.

Dolby, J., and Balme, B. E., 1976, Triassic palynology of the Carnarvon Basin, Western Australia: Reviews of Palaeobotany and Palynology, v. 22, p. 105–168.

Du Toit, A. L., 1937, Our wandering continents: Edinburgh, Oliver and Boyd, 366 p.

Eales, H. V., March, J. S., and Cox, K. G., 1984, The Karoo Igneous Province: An introduction: Geological Society of South Africa Special Publication 13, p. 1–26.

Elliot, D. H., and Watts, D. R., 1974, The nature and origin of volcaniclastic material in some Karoo and Beacon rocks: Geological Society of South Africa Transactions, v. 77, p. 109–111.

Eriksson, P. G., 1979, Mesozoic sheetflow and playa sediments of the Clarens Formation in the Kamberg area of the Natal Drakensberg: Geological Society of South Africa Transactions, v. 82, p. 257–258.

Eriksson, P. G., 1984, A palaeoenvironmental analysis of the Molteno Formation in the Natal Drakensberg: Geological Society of South Africa Transactions, v. 87, p. 237–244.

Eriksson, P. G., 1985, The depositional palaeoenvironment of the Elliot Formation in the Natal Drakensberg and north-eastern Orange Free State: Geological Society of South Africa Transactions, v. 88, p. 19–26.

Fitch, F. J., and Miller, J. A., 1984, Dating Karoo igneous rocks by the conventional K-Ar and ^{40}Ar/^{39}Ar age spectrum method: Geological Society

of South Africa Special Publication 13, p. 247–266.

Folk, R. L., Andrews, P. B., and Laws, D. W., 1970, Detrital sedimentary rock classification and nomenclature for use in New Zealand: New Zealand Journal of Geology and Geophysics, v. 13, p. 937–968.

Foster, C. B., 1979, Permian plant microfossils of the Blair Athol Coal Measures and basal Rewan Formation of Queensland: Queensland Geological Survey, Palaeontological Papers, v. 45.

Foster, C. B., and Waterhouse, J. B., 1988, The Granulatosporites confluens Oppel-zone and Early Permian marine faunas from the Grant Formation on the Barbwire Terrace, Canning Basin, Western Australia: Australian Journal of Earth Sciences, v. 35, p. 135–157.

Frakes, L. A., and Crowell, J. C., 1967, Facies and paleogeography of Late Paleozoic diamictite, Falkland Islands: Geological Society of America Bulletin 78, p. 37–58.

Frakes, L. A., and Crowell, J. C., 1970, Late Paleozoic glaciation: II, Africa exclusive of the Karroo Basin: Geological Society of America Bulletin, v. 81, p. 2261–2286.

Fuller, A. O., 1970, The occurrence of laumontite in strata of the Karoo System, South Africa: Proceedings and Papers, Gondwana Symposium, 2nd: Johannesburg, p. 455–456.

Fuller, A. O., 1985, A contribution to the conceptual modelling of pre-Devonian fluvial systems: Geological Society of South Africa Transactions, v. 88, p. 189–194.

Furon, R., 1963, Geology of Africa: Edinburgh, Oliver and Boyd, 377 p.

Gardiner, B. G., 1969, New palaeoniscoid fish from the Witteberg Series of South Africa: Zoological Journal of the Linnean Society, v. 48, p. 423–452.

Geological Map of the Republic of South Africa, 1970: Pretoria, Department of Mines, scale 1:1,000,000.

Geological Map of the Republic of South Africa, 1984: Pretoria, Department of Mineral and Energy Affairs, scale 1:1,000,000.

Germs, G.J.B., and Gresse, P. G., 1991, The foreland basin of the Damara and Gariep orogens in Namaqualand and southern Namibia: Stratigraphic correlations and basin dynamics: South African Journal of Geology, v. 94, p. 159–169.

Gravenor, C. P., 1979, The nature of the Late Paleozoic glaciation in Gondwana as determined from an analysis of garnets and other heavy minerals: Canadian Journal of Earth Sciences, v. 16, p. 1137–1153.

Hälbich, I. W., 1983, A tectogenesis of the Cape Fold Belt (CFB): Geological Society of South Africa Special Publication 12, p. 165–175.

Hälbich, I. W., compiler, 1993, The Cape Fold Belt–Agulhas Bank transect across Gondwana suture, Southern Africa, Washington, D.C., American Geophysical Union, 18 p.

Hälbich, I. W. and Hartnady, C., 1985, Structural correlation on the Swartland Dome between Riebeek-Kasteel, Malmesbury and Moorreesburg: Geological Society of South Africa Field Guide, 11 p.

Hälbich, I. W., Fitch, F. J., and Miller, J. A., 1983, Dating the Cape orogeny, in Söhnge, A.P.G., and Hälbich, I. W., eds., Geodynamics of the Cape Fold Belt: Geological Society of South Africa Special Publication 12, p. 149–164.

Hartnady, C., Joubert, P., and Stowe, C., 1985, Proterozoic crustal evolution in southwestern Africa: Episodes, v. 8, p. 236–243.

Haughton, S. H., 1969, Geological history of southern Africa: Johannesburg, Geological Society of South Africa, 535 p.

Heinemann, M., and Buhmann, D., 1987, Coal-tonsteins from the Molteno Formation of the Maluti District, Transkei: South African Journal of Geology, v. 90, p. 296–304.

Helby, R., Morgan, R., and Partridge, A. D., 1987, A palynological zonation of the Australian Mesozoic: Association of Australasian Palaeontologists Memoir 4, p. 1–94.

Hergt, J. M., McDougall, I., Banks, M. R., and Green, D. H., 1989, Jurassic dolerite: Geological Society of Australia Special Publication 15, p. 375–381.

Hiller, N., 1992, The Ordovician system in South Africa: A review, in Webby, B. D., and Laurie, J. R., eds., Global perspectives on Ordovician geology: Rotterdam, A. A. Balkema, p. 473–485.

Hiller, N., and Stavrakis, N., 1980, Distal alluvial fan deposits in the Beaufort Group of the eastern Cape Province: Geological Society of South Africa Transactions, v. 83, p. 353–360.

Hiller, N., and Stavrakis, N., 1984, Permo-Triassic fluvial systems in the southeastern Karoo basin, South Africa: Palaeogeography, Palaeoclimatology, Palaeoecology, v. 45, p. 1–21.

Ho-Tun, E., 1979, Volcaniclastic material in lower Beaufort Group, Karoo rocks: Geological Society of South Africa, 18th Congress Abstracts, p. 197–199.

Hobday, D. K., 1973, Middle Ecca deltaic deposits in the Mulden-Tugela Ferry Area of Natal: Transactions of the Geological Society of South Africa, v. 76, p. 309–318.

Hobday, D. K., 1978, Palaeoenvironmental models in the eastern Karoo Basin: Palaeontologia Africana, v. 21, p. 1–13.

Hobday, D. K., 1987, Gondwana coal basins of Australia and South Africa: Tectonic setting, depositional system and resources: Geological Society of London Special Publication 32, p. 219–233.

Hobday, D. K., and Mathew, D., 1974, Depositional environment of the Cape Supergroup in the Transkei: Geological Society of South Africa Transactions, v. 77, p. 223–227.

Hobday, D. K., and Von Brunn, V., 1979, Fluvial sedimentation and paleogeography of an early Paleozoic failed rift, southeastern margin of Africa: Palaeogeography, Palaeoclimatology, Palaeoecology, v. 28, p. 169–184.

Hobday, D. K., Tavener-Smith, R., and Mathew, D., 1975, Markov analysis and the recognition of palaeoenvironments in the Ecca Group near Vryheid, Natal: Geological Society of South Africa Transactions, v. 78, p. 75–82.

Iijima, A., 1978, Geological occurrences of zeolites in marine environments, in Sand, L. B., and Mumpton, F. A., eds., Natural zeolites; occurrences, properties, and use: Oxford, Pergamon Press, p. 175–198.

Johnson, J. G., Klapper, G., and Sandberg, C. A., 1985, Devonian eustatic fluctuations in Euramerica: Geological Society of America Bulletin 96, p. 567–587.

Johnson, M. R., 1966, The stratigraphy of the Cape and Karoo Systems in the eastern Cape Province [M.Sc. thesis]: Grahamstown, South Africa, Rhodes University.

Johnson, M. R., 1976, Stratigraphy and sedimentology of the Cape and Karoo sequences in the eastern Cape Province [Ph.D. thesis]: Grahamstown, South Africa, Rhodes University, 336 p.

Johnson, M. R., 1991, Sandstone petrography, provenance and plate tectonic setting in Gondwana context of the southeastern Cape-Karoo Basin: South African Journal of Geology, v. 94, p. 137–154.

Jones, J. G., Conaghan, P. J., and McDonnell, K. L., 1987, Coal measures of an orogenic recess: Late Permian Sydney Basin, Australia: Palaeogeography, Palaeoclimatology, Palaeoecology, v. 58, p. 203–219.

Jordaan, M. J., 1981, The Ecca-Beaufort transition in the western parts of the Karoo Basin: Transactions of the Geological Society of South Africa, v. 84, p. 19–25.

Kent, L. E., 1980, Stratigraphy of South Africa, Part 1: Lithostratigraphy of the Republic of South Africa, South West Africa/Namibia, and the Republics of Bophutatswana, Transkei and Venda: Geological Survey of South Africa Handbook, v. 8, 690 p.

Keyser, A. W., and Smith, R.M.H., 1979, Vertebrate biozonation of the Beaufort Group with special reference to the western Karoo Basin: Geological Survey of South Africa Annals, v. 12, p. 1–35.

Keyser, N., and Zawada, P. K., 1988, Two occurrences of ash-flow tuff from the Lower Beaufort Group in the Heilbron-Frankfort area, northern Orange Free State: South African Journal of Geology, v. 91, p. 509–521.

Kingsley, C. S., 1970, Palaeocurrent directions in sandstone and arkose of the Beaufort Series of the Hendrik Verwoerd Dam basin: Geological Survey of South Africa Annals, v. 8, p. 87–88.

Kingsley, C. S., 1977, Stratigraphy and sedimentology of the Ecca Group in the eastern Cape Province, South Africa [Ph.D. thesis]: University of Port Elizabeth, 286 p.

Kingsley, C. S, 1979, Evidence for a southern and southeastern provenance during the Permian in South Africa: Proceedings and Papers, Gondwana Symposium, 4th: Calcutta, p. 683–694.

Kingsley, C. S., 1981, A composite submarine fan-delta-fluvial model for the Ecca and Lower Beaufort Groups of Permian age in the Eastern Cape province, South Africa: Geological Society of South Africa Transactions, v. 84, p. 27–40.

Kingsley, C. S., and Theron, J. C., 1964, Palaeocurrent directions in arkose of the Beaufort Series in the Orange Free State: Geological Survey of South Africa Annals, v. 3, p. 71–74.

Kitching, J. W., 1970, A short review of the Beaufort zoning in South Africa: Proceedings and Papers, Gondwana Symposium, 2nd: Johannesburg, p. 309–312.

Kreuser, T., and Semkiwa, P. M., 1987, Geometry and depositional history of a Karoo (Permian) coal basin (Mchuchuma-Ketewaka) in SW-Tanzania: Neues Jahrbuch fur Geologie und Palaontologie Monatshefte, v. 2, p. 69–98.

Le Roux, J. P., 1985, Palaeochannels and uranium mineralization on the main Karoo basin of South Africa [Ph.D. thesis]: University of Port Elizabeth, 250 p.

Lock, B. E., 1973, The Cape Supergroup in Natal and the northern Transkei: Geological Magazine, v. 110, p. 485–486.

Lock, B. E., and Johnson, M. R., 1974, A crystal tuff from the Ecca Group near Lake Mentz, Eastern Cape province: Geological Society of South Africa Transactions, v. 77, p. 373–374.

Lock, B. E., and Wilson, J. D., 1975, The nature and origin of volcaniclastic material in some Karoo and Beacon rocks: Discussion: Geological Society of South Africa Transactions, v. 78, p. 171.

Lock, B. E., Paverd, A. L., and Broderick, T. J., 1974, Stratigraphy of the Karoo volcanic rocks of the Barkly East district: Geological Society of South Africa Transactions, v. 77, p. 117–129.

Loock, J. C., and Visser, J.N.J., 1985, South Africa, in Martinez Diaz, C., ed., The Carboniferous of the world, II: Australia, Indian subcontinent, South Africa, South America, and North Africa: Madrid, Instituto Geológico y Minero de España, p. 167–174.

Marchant, J. W., 1978, The metal contents of organic separates of some Ecca shales: Geological Society of South Africa Transactions, v. 81, p. 173–178.

Marshall, C.G.A., 1988, Some aspects of the Natal Group between Durban and Eshowe: Geological Society of South Africa, 22nd Congress Abstracts, p. 379–382.

Marshall, J.E.A., 1994, The Falkland Islands: A key element in Gondwana paleogeography: Tectonics, v. 13, p. 499–514.

Martin, K. A., Hartnady, C.J.H., and Goodlad, S. W., 1981, A revised fit of South America and south central Africa: Earth and Planetary Science Letters, v. 54, p. 293–305.

Martini, J.E.J., 1974, On the presence of ash beds and volcanic fragments in the Southern Cape Province (South Africa): Geological Society of South Africa Transactions, v. 77, p. 113–116.

Mason, T. R., and Taverner-Smith, R., 1978, A fluvio-deltaic Middle Ecca succession west of Empangeni, Zululand: Geological Society of South Africa Transactions, v. 81, p. 13–22.

McDonnell, K. L., 1983, The Sydney Basin from Late Permian to Middle Triassic, a study focused on the coastal transect [Ph.D. thesis]: Sydney, Macquarie University.

McLachlan, I. R., 1973, Problematic microfossils from the Lower Karoo beds in South Africa: Palaeontologia Africana, v. 15, p. 1–21.

McLachlan, I. R., 1977, An investigation of the Lower Permian Middle Ecca ammonite locality at Alleta, Natal: Palaeontologia Africana, v. 20, p. 53–63.

McLachlan, I. R., and Anderson, A. M., 1973, A review of the evidence for marine conditions in southern Africa during Dwyka times: Palaeontologia Africana, v. 15, p. 37–64.

McLachlan, I. R., and Anderson, A. M., 1975, The age and stratigraphic relationship of the glacial sediments in southern Africa, in Campbell,

K.S.W., ed., Gondwana Geology: Canberra, Australian National University Press, p. 415–422.

McLachlan, I. R., and Anderson, A. M., 1977, Carbonates, "stromatolites" and tuffs in the Lower Permian White Band Formation: South African Journal of Science, v. 73, p. 92–94.

McLachlan, I. R., and Jonker, J. P., 1990, Tuff beds in the northwestern part of the Karoo Basin: South African Journal of Geology, v. 93, p. 329–338.

Mitchell, C., Taylor, G. K., Cox, K. G., and Shaw, J., 1986, Are the Falkland Islands a rotated microplate?: Nature, v. 319, p. 131–134.

Mosmann, R., Falkenhein, F.U.H., Goncalves, A., and Nepomuceno, F., 1986, Oil and gas potential of the Amazon Paleozoic basins, in Halbouty, M. T., ed., Future petroleum provinces of the world: American Association of Petroleum Geologists Memoir 40, p. 207–241.

Nel, H. J., 1962, A petrographic description of the Ecca sediments in the southern Karroo and eastern Orange Free State: Geological Survey of South Africa Annals, v. 1, p. 91–103.

Oelofsen, B. W, 1986, A fossil shark neurocranium from the Permo-Carboniferous (lowermost Ecca Formation) of South Africa, in Uyeno, T., Arai, R., Taniuchi, T., and Matsuura, K., eds., Indo-Pacific fish biology: Tokyo, Ichthyological Society of Japan, p. 107–124.

Oelofsen, B. W., 1987, The biostratigraphy and fossils of the Whitehill and Iratí Shale Formations of the Karoo and Parana Basins, in McKenzie, G. D., ed., Gondwana Six: American Geophysical Union Geophysical Monograph, v. 41, p. 131–138.

Oelofsen, B. W., and Araujo, D. C., 1987, *Mesosaurus tenuidens* and *Stereosternum tumidum* from the Permian Gondwana of both Southern Africa and South America: South African Journal of Science, v. 83, p. 370–372.

Oosthuizen, R.D.F., 1981, An attempt to determine the provenance of the southern Dwyka from palaeontological evidence: Palaeontologia Africana, v. 24, p. 27–29.

Partridge, T. C., and Maud, R. R., 1987, Geomorphic evolution of southern Africa since the Mesozoic: South African Journal of Geology, v. 90, p. 179–208.

Powell, C. McA., 1984, Uluru regime, in Veevers, J. J., ed., Phanerozoic earth history of Australia: Oxford, Clarendon, p. 290–340.

Reimann, K. U., 1986, Prospects for oil and gas in Zimbabwe, Zambia and Botswana: Episodes, v. 9, p. 95–101.

Rex, D. C., and Tanner, P.W.G., 1982, Precambrian age for gneisses at Cape Meredith in the Falkland Islands, in Craddock, C., ed., Antarctic Geoscience: Madison, University of Wisconsin Press, p. 107–108.

Rowsell, D. M., and de Swardt, A.M.J., 1976, Diagenesis in Cape and Karoo sediments, South Africa, and its bearing on their hydrocarbon potential: Geological Society of South Africa Transactions, v. 79, p. 81–145.

Rubidge, B., 1988, The palaeontology and palaeoenvironment of the rocks of the Ecca-Beaufort contact in the southern Karoo: Geological Society of South Africa, 22nd Congress Abstracts, p. 521–524.

Runnegar, B., and Newell, N. D., 1971, Caspian-like relict molluscan fauna in the South American Permian: American Museum of Natural History Bulletin 146, p. 1–66.

Rust, B. R., 1981, Sedimentation in an arid-zone anastomosing fluvial system: Journal of Sedimentary Petrology, v. 51, p. 745–755.

Rust, I. C., 1962, On the sedimentation of the Molteno Sandstones in the vicinity of Molteno, C.P.: Annals of the University of Stellenbosch, v. 37, Series A, p. 165–236.

Rust, I. C., 1973, The evolution of the Palaeozoic Cape Basin, southern margin of Africa, in Nairn, A.E.M., and Stehli, F. G., eds., The ocean basins and margins, Volume 1: The South Atlantic: Plenum, New York, p. 247–276.

Rust, I. C., 1975, Tectonic and sedimentary framework of Gondwana basins in southern Africa, in Campbell, K.S.W., ed., Gondwana geology: Canberra, Australian National University Press, p. 537–564.

Ryan, P. J., and Whitfield, G. G., 1979, Basinal analysis of the Ecca and lowermost Beaufort Beds and associated coal, uranium and heavy mineral beach sand occurrences: Geological Society of South Africa Special

Publication 6, p. 91–101.

Schmidt, E. R., 1976, Clay, *in* Coetzee, C. B., ed., Mineral resources of the Republic of South Africa: Geological Survey of South Africa Handbook, v. 7, p. 275–287.

Shone, R. W., 1987, A resedimented conglomerate and shallow-water turbidite from the Sardinia Bay Formation, Lower Table Mountain Group: South African Journal of Geology, v. 90, p. 86–93.

Smith, R.M.H., 1990, A review of stratigraphy and sedimentary environments of the Karoo Basin of South Africa: Journal of African Earth Sciences, v. 10, p. 117–137.

Söhnge, A.P.G., and Hälbich, I. W., eds., 1983, Geodynamics of the Cape Fold Belt: Geological Society of South Africa Special Publication 12, 184 p.

Stapleton, R. P., 1978, Microflora from a possible Permo-Triassic transition in South Africa: Review of Palaeobotany and Palynology, v. 25, p. 253–258.

Stavrakis, N., 1980, Sedimentation of the Katberg Sandstone and adjacent formations in the south-eastern Karoo Basin: Geological Society of South Africa Transactions, v. 83, p. 361–374.

Stear, W. M., 1980, Channel sandstone and bar morphology of the Beaufort Group uranium district near Beaufort West: Geological Society of South Africa Transactions, v. 83, p. 391–398.

Stear, W. M., 1985, Comparison of the bedform distribution and dynamics of modern and ancient sandy ephemeral flood deposits in the southwestern Karoo region, South Africa: Sedimentary Geology, v. 45, p. 209–230.

Tankard, A. J., Jackson, M.P.A., Eriksson, K. A., Hobday, D. K., and Minter, W.E.L., 1982, Crustal evolution of Southern Africa—3.8 billion years of earth history: New York, Springer-Verlag, 523 p.

Tarney, J., 1976, Petrology, mineralogy, and geochemistry of the Falkland Plateau basement rocks, Site 330, Deep Sea Drilling Project: Initial Reports of the Deep Sea Drilling Project, v. 36, p. 893–927.

Taverner-Smith, R., 1962, Karroo sedimentation in part of the mid-Zambezi valley: Geological Society of South Africa Transactions, v. 65, p. 43–74.

Tavener-Smith, R., 1982, Prograding coastal facies associations in the Vryheid Formation (Permian) at Effingham quarries near Durban, South Africa: Sedimentary Geology, v. 32, p. 111–140.

Teichert, C., 1974, Marine sedimentary environments and their faunas in Gondwana area: American Association of Petroleum Geologists Memoir 23, p. 361–394.

Thamm, A. G., 1987, Palaeoenvironmental significance of clast shape in the Nardouw Formation, Cape Supergroup: South African Journal of Geology, v. 90, p. 94–97.

Theron, J. C., 1970, Some geological aspects of the Beaufort Series in the Orange Free State [Ph.D. thesis]: Bloemfontein, University of the Orange Free State.

Theron, J. C., 1975, Sedimentological evidence for the extension of the African continent southwards during the Late Permian–Early Triassic times, *in* Campbell, K.S.W., ed., Gondwana geology: Canberra, Australian National University Press, p. 61–71.

Theron, J. N., 1970, A stratigraphical study of the Bokkeveld Group: Proceedings and Papers, Gondwana Symposium, 2nd: Johannesburg, p. 197–204.

Theron, J. N., 1972, The stratigraphy and sedimentation of the Bokkeveld Group [D.Sc. thesis]: University of Stellenbosch.

Theron, J. N., and Loock, J. C., 1989, Devonian deltas of the Cape Supergroup, South Africa: Canadian Society of Petroleum Geologists Memoir 14, p. 729–740.

Theron, J. N., Rickards, R. B., and Aldridge, R. J., 1990, Bedding plane assemblages of Promissum pulchrum, a new giant Ashgill conodont from the Table Mountain Group, South Africa: Palaeontology, v. 33, p. 577–594.

Thomas, R. J., 1989, A tale of two tectonic terranes: South African Journal of Geology, v. 92, p. 306–321.

Thomas, R. J., and Mawson, S. A., 1989, Newly discovered outcrops of Proterozoic basement rocks in northeastern Transkei: South African Journal of Geology, v. 92, p. 369–376.

Thomas, R. J., Marshall, C.G.A., Watkeys, M. K., Fitch, F. J., and Miller, J. A., 1992, K/Ar and ^{40}Ar/^{39}Ar dating of the Natal Group, southeast Africa: Post Pan-African molasse?: Journal of African Earth Science, v. 15, p. 453–471.

Truswell, E. M., 1980, Permo-Carboniferous palynology of Gondwanaland: Progress and problems in the decade to 1980: Australian Bureau of Mineral Resources Journal v. 5, p. 95–111.

Truswell, J. F., 1977, The geological evolution of South Africa: Cape Town, Purnell, 220 p.

Turner, B. R., 1971, The geology and coal resources of the north-eastern Cape Province: Geological Survey of South Africa Bulletin 52, 74 p.

Turner, B. R., 1972, Revision of the stratigraphic position of cynodonts from the upper part of the Karroo (Gondwana) System of Lesotho: Geological Magazine, v. 109, p. 349–360.

Turner, B. R., 1975, The stratigraphy and sedimentary history of the Molteno Formation in the main Karoo Basin of South Africa and Lesotho [Ph.D. thesis]: Johannesburg, University of Witwatersrand, 314 p.

Turner, B. R., 1977, Fluviatile cross-bedding patterns in the Upper Triassic Molteno Formation of the Karoo (Gondwana) Supergroup in South Africa and Lesotho: Geological Society of South Africa Transactions, v. 80, p. 241–252.

Turner, B. R., 1978, Sedimentary patterns of uranium mineralisation in the Beaufort Group of the Southern Karoo (Gondwana) Basin South Africa, *in* Miall, A. D., ed., Fluvial sedimentology: Canadian Society of Petroleum Geologists Memoir 5, p. 831–848.

Turner, B. R., 1980, Palaeohydraulics of an Upper Triassic braided river system in the main Karoo Basin, South Africa: Geological Society of South Africa Transactions, v. 83, p. 425–431.

Turner, B. R., 1981, Revised stratigraphy of the Beaufort Group in the Southern Karoo Basin: Palaeontologia Africana, v. 24, p. 87–98.

Turner, B. R., 1983, Braidplain deposition of the Upper Triassic Molteno Formation in the main Karoo (Gondwana) Basin, South Africa: Sedimentology, v. 30, p. 77–89.

Turner, B. R., 1984, Palaeogeographic implications of braid bar deposition in the Triassic Molteno Formation of the eastern Karoo Basin, South Africa: Palaeontologia Africana, v. 25, p. 29–38.

Turner, B. R., 1985, Uranium mineralization in the Karoo basin, South Africa: Economic Geology, v. 80, p. 256–269.

Turner, B. R., 1986, Tectonic and climatic controls on continental depositional facies in the Karoo Basin of northern Natal, South Africa: Sedimentary Geology, v. 46, p. 231–257.

Turner, B. R., 1990, Continental sediments in South Africa: Journal of African Earth Sciences, v. 10, p. 139–149.

Valloni, R., 1985, Reading provenance from modern marine sands, *in* Zuffa, G. G., ed., Provenance of arenites: Dordrecht, D. Reidel, p. 309–332.

Van Vuuren, C. J., and Cole, D. I, 1979, The stratigraphy and depositional environments of the Ecca Group in the northern part of the Karoo basin: Geological Society of South Africa Special Publication 6, p. 103–111.

Veevers, J. J., 1988, Gondwana facies started when Gondwanaland merged in Pangea: Geology, v. 16, p. 732–734.

Veevers, J. J., 1989, Middle/Late Triassic (230 ± 5 Ma) singularity in the stratigraphic and magmatic history of the Pangean heat anomaly: Geology, v. 17, p. 784–787.

Veevers, J. J., 1990, Tectonic-climatic supercycle in the billion-year plate-tectonic eon: Permian Pangean icehouse alternates with Cretaceous dispersed-continents greenhouse: Sedimentary Geology, v. 68, p. 1–16.

Veevers, J. J., and Powell, C. McA., 1987, Late Paleozoic glacial episodes in Gondwanaland reflected in transgressive-regressive depositional sequences in Euramerica: Geological Society of America Bulletin 98, p. 475–487.

Veevers, J. J., and Rundle, A. S., 1979, Channel Country fluvial sands and associated facies of central-eastern Australia: Modern analogues of Mesozoic desert sands of South America: Palaeogeography, Palaeoecology, Palaeoclimatology, v. 26, p. 1–16.

Verwoerd, W. J., Viljoen, J.H.A., and Viljoen, K. S., 1990, Olivine melilites and associated intrusives of the southwestern Cape Province, Guidebook, Geocongress 1990: Cape Town, Geological Society of South Africa, p. 1–60.

Viljoen, J.H.A., 1987, Subaqueous fallout tuffs of the Ecca Group in the Southern Cape Province, *in* Brown, G., and Preston, V. A., compilers, Workshop on pyroclastic volcanism and associated deposits: Pietermaritzburg, Department of Geology and Mineralogy, University of Natal, p. 45–48.

Viljoen, J.H.A., 1990, K-bentonites in the Ecca Group of the south and central Karoo Basin, Abstracts, Geocongress 1990: Cape Town, Geological Society of South Africa, p. 576–579.

Visser, J.N.J., 1972, Sedimentere insluitsels van Karoo-ouderdom in kimberliet van die Finsch-diamantmyn: Suid Africaanse Tydskrif vir Natuurwetenskap, v. 12, p. 32–36.

Visser, J.N.J., 1974, The Table Mountain Group: A study in the deposition of quartz arenites on a stable shelf: Geological Society of South Africa Transactions, v. 77, p. 221–237.

Visser, J.N.J., 1979, Changes in sediment transport direction in the Cape-Karoo basin (Silurian-Triassic) in South Africa: South African Journal of Science, v. 75, p. 72–75.

Visser, J.N.J., 1982, Implications of a diachronous contact between the Dwyka Formation and Ecca Group in the Karoo basin: South African Journal of Science, v. 78, p. 249–251.

Visser, J.N.J., 1984, A review of the Stormberg Group and Drakensberg Volcanics in southern Africa: Palaeontologia Africana, v. 25, p. 5–27.

Visser, J.N.J., 1986, Lateral lithofacies relationships in the glaciogene Dwyka Formation in the eastern and central parts of the Karoo basin: Geological Society of South Africa Transactions, v. 89, p. 373–383.

Visser, J.N.J., 1987, The palaeogeography of part of southwestern Gondwana during the Permo-Carboniferous glaciation: Palaeogeography, Palaeoclimatology, Palaeoecology, v. 61, p. 205–219.

Visser, J.N.J., 1988, Facies interpretation of thick homogeneous diamictite of the Dwyka Formation in the southern Karoo Geological Society of South Africa Congress, 22nd, Abstracts: p. 677–680.

Visser, J.N.J., 1989, The Permo-Carboniferous Dwyka Formation of southern Africa: Deposition by a predominantly subpolar marine ice sheet: Palaeogeography, Palaeoclimatology, Palaeoecology, v. 70, p. 377–391.

Visser, J.N.J., 1990, The age of the late Palaeozoic glacigene deposits in southern Africa: South African Journal of Geology, v. 93, p. 366–375.

Visser, J.N.J., 1991a, Self-destructive collapse of the Permo-Carboniferous marine ice sheet in the Karoo Basin: Evidence from the southern Karoo: South African Journal of Geology, v. 94, p. 255–262.

Visser, J.N.J., 1991b, Geography and climatology of the Late Carboniferous to Jurassic Karoo Basin in south-western Gondwana: Annals of the South African Museum, v. 99, p. 415–431.

Visser, J.N.J., 1993, Sea-level changes in a back-arc–foreland transition: The Late Carboniferous-Permian Karoo Basin of South Africa: Sedimentary Geology, v. 83, p. 115–131.

Visser, J.N.J., and Botha, B.J.V., 1980, Meander channel, point-bar, crevasse splay and aeolian deposits from the Elliot Formation in Barkly Pass, north-eastern Cape: Geological Society of South Africa Transactions, v. 83, p. 55–62.

Visser, J.N.J., and Dukas, B. A., 1979, Upward-fining fluviatile megacycles in the Beaufort Group, north of Graaff-Reinet, Cape Province: Geological Society of South Africa Transactions, v. 82, p. 149–154.

Visser, J.N.J., and Kingsley, C. S., 1982, Upper Carboniferous glacial valley sedimentation in the Karoo Basin, Orange Free State: Transactions of the Geological Society of South Africa, v. 85, p. 71–79.

Visser, J.N.J., and Loock, J. C., 1978, Water depth in the main Karoo Basin, South Africa, during Ecca (Permian) sedimentation: Geological Society of South Africa Transactions, v. 81, p. 185–191.

Visser, J.N.J., and Loock, J. C., 1982, An investigation of the basal Dwyka Tillite in the southern part of the Karoo Basin, South Africa: Geological Society of South Africa Transactions, v. 85, p. 179–187.

Visser, J.N.J., Loock, J. C., and Jordaan, M. J., 1980, Permian deltaic sedimentation in the western half of the Karoo Basin: Geological Society of South Africa Transactions, v. 83, p. 415–424.

Visser, J.N.J., Hall, K. J., and Loock, J. C., 1986, The application of stone counts in the glacigene Permo-Carboniferous Dwyka Formation, South Africa: Sedimentary Geology, v. 46, p. 197–212.

Von Brunn, V., and Gravenor, C. P., 1983, A model for late Dwyka glaciomarine sedimentation in the eastern Karoo basin: Geological Society of South Africa Transactions, v. 86, p. 199–209.

Welman, J., Groenewald, G. H., and Kitching, J. W., 1991, Confirmation of the occurrence of Cynognathus Zone (Kannemeyeria-Diademodon Assemblage-zone) deposits (uppermost Beaufort Group) in the northeastern Orange Free State, South Africa: South African Journal of Geology, v. 94, p. 245–248.

Whateley, M.K.G., 1980, Deltaic and fluvial deposits of the Ecca Group, Nongoma Graben, Northern Zululand: Geological Society of South Africa Transactions, v. 83, p. 345–351.

Winchester, J. A., and Floyd, P. A., 1977, Geochemical discrimination of different magma series and their differentiation products using immobile elements: Chemical Geology, v. 20, p. 325–343.

Winter, H., and Venter, J. J., 1970, Lithostratigraphic correlation of recent deep boreholes in the Karoo-Cape sequence: Proceedings and papers, Gondwana Symposium, 2nd: Johannesburg, p. 395–408.

Wopfner, H., and Kreuser, T., 1986, Evidence for late Palaeozoic glaciation in southern Tanzania: Palaeogeography, Palaeoclimatology, Palaeoecology, v. 56, p. 259–275.

Wright, R. P., and Askin, R. A., 1987, The Permian-Triassic boundary in the southern Morondava Basin of Madagascar as defined by plant microfossils, *in* McKenzie, G. D., ed., Gondwana Six: American Geophysical Union Geophysical Monograph, v. 41, p. 157–166.

Yassini, I., 1987, Ecology, paleoecology and stratigraphy of ostracodes from Late Pliocene and Quaternary deposits of the south Caspian Sea region in north Iran, *in* McKenzie, K. G., ed., Shallow Tethys 2: Rotterdam, A. A. Balkema, p. 475–497.

Yemane, K., and Kelts, K., 1990, A short review of palaeoenvironments for Lower Beaufort (Upper Permian) Karoo sequences from southern to central Africa: A major Gondwana lacustrine episode: Journal of African Earth Sciences, v. 10, p. 169–185.

Yoshida, M., and Kizaki, K., 1983, Tectonic situation of Lutzow-Holm Bay in East Antarctica and its significance in Gondwanaland, *in* Oliver, R. L., James, P. R., and Jago, J. B., eds., Antarctic Earth science: Cambridge, England, Cambridge University Press, p. 36–39.

Zawada, P. K., 1988, Trace elements as possible palaeosalinity indicators for the Ecca and Beaufort Group mudrocks in the southwestern Orange Free State: Geological Society of South Africa Transactions, v. 91, p. 18–26.

MANUSCRIPT ACCEPTED BY THE SOCIETY SEPTEMBER 9, 1993

Geological Society of America
Memoir 184
1994

Southern South America

Oscar R. López-Gamundí* and Irene S. Espejo*
Departamento de Ciencias Geológicas, Facultad de Ciencias Exactas y Naturales, Universidad de Buenos Aires, Ciudad Universitaria, Pabellón 2, Buenos Aires 1428, Argentina
Patrick J. Conaghan and C. McA. Powell*
Australian Plate Research Group, School of Earth Sciences, Macquarie University, N.S.W. 2109, Australia

With a contribution by
J. J. Veevers
Australian Plate Research Group, School of Earth Sciences, Macquarie University, North Ryde, N.S.W. 2109, Australia

ABSTRACT

In central-western Argentina, the basement comprises Cambrian to Devonian sedimentary rocks, deformed and uplifted during the Late Devonian–earliest Carboniferous Chañic orogeny along the Paleo-Pacific margin of South America. Unconformably above basement, the Gondwana cycle comprises two unconformity-bounded sequences. The Visean (350 Ma) to earliest Permian (275 Ma) Lower Sequence started with deposition in the Andean (or western) Calingasta-Uspallata Basin of valley-fill sediments. By the Namurian, alpine glaciation of a basement ridge, the Proto-Precordillera, fed sediment into the marine Calingasta-Uspallata Basin on the west and the nonmarine western Paganzo Basin on the east. The Paganzo Basin received additional sediment shed from basement highs further to the east. Mainly marine (but not glacial) sediment continued to be deposited in the Calingasta-Uspallata Basin into the Early Permian (275 Ma). At the same time, the Paganzo Basin expanded as dominantly nonmarine sediment encroached eastward over the craton. Tuff first appeared in the earliest Permian (~286 Ma) and reflects initial input from the magmatic arc on the Panthalassan margin. At the same time sediment overlapped the Proto-Precordillera. Southward, in the Andean San Rafael Basin, a similar Lower Sequence started with glacial sediment, probably in the Namurian, and likewise concluded at 275 Ma. In east-central Argentina, in the Sauce Grande Basin, deposition did not start until the latest Carboniferous (~290 Ma), with a marine glacial deposit capped by Tastubian transgressive glaciomarine sandstone and shale.

At about 275 Ma, the mild extensional tectonics that had generated the Lower Sequence of quartzofeldspathic petrofacies from the craton, and sedimentary-lithic petrofacies from the Proto-Precordillera, gave way to convergent magmatic-arc tec-

*Present addresses: López-Gamundí, Texaco, Inc., Frontier Exploration Department, 4800 Fournace Place, Bellaire, Texas 77401-2324; Espejo, Amoco Production Company, P.O. Box 3092, Houston, Texas 77253-3092; Powell, Department of Geology, University of Western Australia, Nedlands, W. A. 6009, Australia.

López-Gamundí, O. R., Espejo, I. S., Conaghan, P. J., and Powell, C. McA., 1994, Southern South America, *in* Veevers, J. J., and Powell, C. McA., eds., Permian-Triassic Pangean Basins and Foldbelts Along the Panthalassan Margin of Gondwanaland: Boulder, Colorado, Geological Society of America Memoir 184.

tonics that generated the volcanic and volcaniclastic Upper Sequence, which continued to the end of the Permian. In the magmatic arc, the Choiyoi Group (275 to 250 Ma) of mainly dacitic ignimbrites merged eastward with volcaniclastics in the now-continuous Calingasta-Uspallata/western Paganzo foreland basin. Further east, in the Sauce Grande Basin, the Tunas Formation with tuffaceous interbeds conformably succeeded the glacial sediments. The tuffaceous interbeds reflect the southeastward swing of the arc at 34°S to parallel the axis of the Sauce Grande Basin 250 km away.

The Gondwana cycle concluded with mild compressive deformation and uplift represented by an Early Triassic lacuna. Violent extensional deformation in the Middle and Late Triassic cut a swath of NW-trending grabens across northern Chile to Patagonia. The grabens filled with marine sediment in Chile and nonmarine alluvial-fan and then lacustrine-fluvial sediment in Argentina. The Late Paleozoic and Triassic succession was finally capped by Late Jurassic and Early Cretaceous flood basalt.

INTRODUCTION

This chapter focuses on the paleogeographic and paleotectonic evolution of the sector of the Late Paleozoic foreland basin in southern South America (Fig. 1). The constituent Paganzo (1), Calingasta-Uspallata (2), and San Rafael (3) Basins occupy central-western Argentina between 27°S and 40°S, and the Sauce Grande–Colorado Basin (4) occupies the central-eastern Atlantic region of the Sierras Australes of the Buenos Aires province. The Late Paleozoic basins along the Palo-Pacific margin of southern South America fall into three tectonic segments:

(A) From 24°S in northern Argentina throughout Bolivia and southeastern Peru, the Tarija Basin (7) developed over the axis of an Early Paleozoic basin on the latest Proterozoic suture (Dalmayrac et al., 1980a; Laubacher and Megard, 1985) between the Precambrian Arequipa Massif and the autochthonous Brazilian Shield of Gondwanaland.

(B) Between 27°S and 40°S, the Late Paleozoic basins (1–3) are related to the evolution of a Chilean-type (Uyeda, 1982) continental margin with a well–preserved magmatic arc along the Andes on the Argentine-Chilean border and a back-arc or foreland succession to the east in western Argentina. Westward along coastal Chile pelagic and turbiditic successions of an accretionary prism (Davidson et al., 1987), best preserved in southernmost South America (Hervé, 1988), are tectonically interleaved with mélanges, metamorphic tectonites, and other subduction-related complexes (Dalziel and Forsythe, 1985).

(C) From 43°S southward, Late Paleozoic fore-arc sedimentation along the Paleo-Pacific margin was localized in the Tepuel Basin (8) or fore-arc province of Dalziel and Forsythe (1985). West and southwest of this basin along southern Chile, subduction-related complexes with tectonic slivers of oceanic crust and exotic terranes have been documented by Mpodozis and Forsythe (1983), Dalziel and Forsythe (1985), and Hervé (1988).

In segment B, the Gondwana sequence accumulated in subsiding foreland areas behind the Paleo-Pacific margin during most of the Carboniferous and Permian. Special attention has been given to a coordinated study of paleocurrents and petrofacies of sandstones in an attempt at deciphering changes in regional slope related to magmatic arc activity on the western margin of the basins.

SUB-GONDWANA (PRE-CARBONIFEROUS) BASEMENT

Introduction

The Late Paleozoic foreland basin developed on a variety of basement types. The basins of western Argentina rest on the mobile Paleo-Pacific margin of Gondwanaland, and those on the east rest on the quasi-cratonic basement of the Sierras Pampeanas (Figs. 2 and 3). The final configuration of this part of the basin, considered to continue into the southwestern Paraná Basin (Fig. 1), was controlled by Pampean structures (Bracaccini, 1960; Salfity and Gorustovich, 1983), some of which were reactivated during sedimentation.

Figure 1. Late Paleozoic tectonic setting along the Paleo-Pacific margin with superimposed paleogeography and generalized isopachs of the following basins: 1 Paganzo, 2 Calingasta-Uspallata, 3 San Rafael, 4 Sauce Grande–Colorado, 5 Chaco-Paraná, 6 Paraná, 7 Tarija, 8 Tepuel, 9 Golondrina, 10 Falkland (Malvinas), 11 Arizaro. Isopachs (km) span Early Carboniferous to Early Permian interval for the Calingasta-Uspallata and Paganzo Basins, and Late Carboniferous to Late Permian for the Chaco-Paraná and Paraná Basins. Heavy line indicates inferred mainly Carboniferous basin edges. Three terranes are differentiated along the Paleo-Pacific margin: (A) Precambrian blocks: the Arequipa Massif and parautochthonous southern extensions in an intracontinental, nonsubduction-related context; (B) subduction-related complexes of metamorphic tectonites, and (C) subduction-related complexes of shales and slates with slivers of oceanic crust and accreted exotic terranes. Sources: Russo et al. (1979), Salfity and Gorustovich (1983), Dalziel and Forsythe (1985), and Mpodozis and Ramos (1989). A, Argentina; Bo, Bolivia; Br, Brazil; C, Chile; Pa, Paraguay; Pe, Peru; U, Uruguay. Bipolar oblique conic conformal projection.

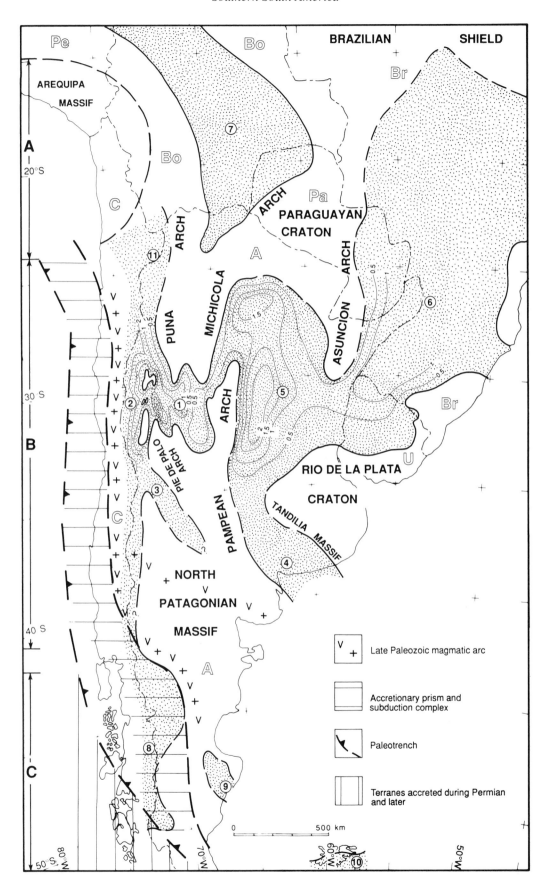

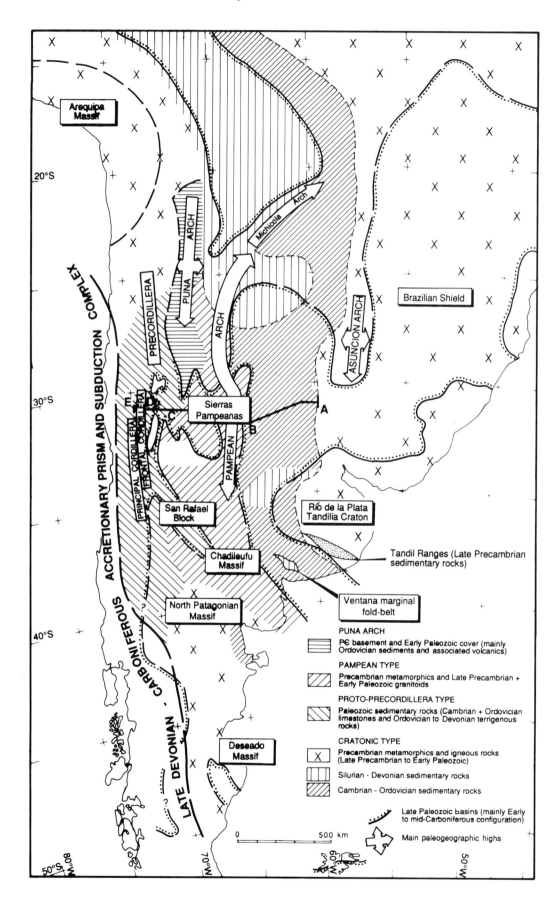

PUNA ARCH
PЄ basement and Early Paleozoic cover (mainly Ordovician sediments and associated volcanics)

PAMPEAN TYPE
Precambrian metamorphics and Late Precambrian + Early Paleozoic granitoids

PROTO-PRECORDILLERA TYPE
Paleozoic sedimentary rocks (Cambrian + Ordovician limestones and Ordovician to Devonian terrigenous rocks)

CRATONIC TYPE
Precambrian metamorphics and igneous rocks (Late Precambrian to Early Paleozoic)

Silurian - Devonian sedimentary rocks

Cambrian - Ordovician sedimentary rocks

Late Paleozoic basins (mainly Early to mid-Carboniferous configuration)

Main paleogeographic highs

0   500 km

Basement of the western or Andean basins

As shown in Fig. 2, the basement is zoned in three north-trending belts that correspond with the present-day morphotectonic units of the Sierras Pampeanas, the Precordillera and San Rafael Block, and the Frontal Cordillera.

Sierras Pampeanas basement. The Sierras Pampeanas basement of Precambrian metamorphic rocks intruded by Paleozoic granitoids resembles the Brazilian Shield and its Argentine extension, the Tandilia Craton. Ramos et al. (1986) described two domains in the Pampean basement: (a) The eastern domain comprises Late Precambrian schists, plagiogneisses, and migmatites intruded by granitic batholiths in the Early Paleozoic. K/Ar radiometric ages suggest phases of recrystallization ranging from 650 to 400 Ma (Gordillo and Lencinas, 1979). Lucero Michaut (1979) found an age of 640 Ma for the basement rocks in the north-central area of the Sierra de Córdoba and suggested that most of the gneisses located in an older (1000 to 1400 Ma) complex of amphibolites and granulites (Cingolani and Varela, 1975) are Late Precambrian. (b) The western domain is characterized by a Precambrian regional metamorphism of low grade coupled in its central part with a high-temperature, middle-pressure metamorphism of Early Paleozoic age (Caminos et al., 1982).

According to Caminos (1985), the three principal phases of plutonic activity in the Pampean terrane during the Paleozoic are Ordovician-Silurian (500–450 Ma), Devonian (400–380 Ma), and Carboniferous (360–330 Ma). Some of the plutonic rocks of the eastern domain, initially regarded as Carboniferous are now assigned to the Devonian (Rapela et al., 1982). The Ordovician-Silurian phase is the most pervasive, and the granitoids and associated migmatites in the western domain have been identified as remnants of a magmatic

arc that probably extended to 38°S in the San Rafael Block (Linares et al., 1980) or as emplaced in a back-arc region east of a magmatic arc located to the west (Rapela et al., 1990). Isotopic studies by Linares et al. (1982) suggest that marbles exposed in the Sierras Pampeanas might be the metamorphic counterpart of the Cambro-Ordovician limestones in the Precordillera region (Fig. 2), and limestones resting on Pampean metamorphic basement have been recorded east of the Precordillera (Bastías et al., 1984).

In summary, the pre-Carboniferous basement of the Sierras Pampeanas comprises >1000 Ma metamorphic nuclei of the Uruçuano Cycle (Almeida, 1971), 900–650 Ma metamorphic rocks of the Brazilian Cycle, and Paleozoic intrusions between 500 and 380 Ma (Caminos et al., 1979).

Precordillera and San Rafael Block basement. The pre-Carboniferous basement of the Precordillera/western San Rafael Block differs from that of the Sierras Pampeanas in containing widespread sedimentary rocks of early and middle Paleozoic age. Cambro-Ordovician limestone is the most conspicuous component of the Precordillera (Figs. 3; 4, B and C) and extends as fossiliferous Ordovician (Llandeilian, Heredia, 1982) limestone to the San Rafael Block at 35°S (Nuñez, 1960). As noted above, limestone is known also in the Sierras Pampeanas region (Bastías et al., 1984), and marble cropping out in the westernmost part of the Sierras Pampeanas has been correlated on the basis of oxygen-isotope studies (Linares et al., 1982) with the Cambro-Ordovician limestone of the eastern Precordillera. The limestone is interpreted as deposited in shallow carbonate platforms (Furque and Cuerda, 1979) and ranges from Early Cambrian (La Laja Formation) to Early Ordovician or Arenigian (San Juan Formation). Stromatolite/thrombolite rhythmic sequences have been identified at the Cambrian-Ordovician transition in the central segment of the Precordillera (Baldis et al., 1981). Early Ordovician deposits in the central and eastern Precordillera are represented by shelfal carbonates that interfinger westward with slope carbonate facies (Cabalieri *in* Baldis et al., 1989) (Figs. 3 and 4C), and both facies contain trilobites and conodonts of Acado-Baltic affinities. Toward the end of the Arenigian (Early Ordovician) the carbonate was replaced by Middle–Late Ordovician siliciclastic marine sediment to form an eastern siliciclastic shelf with subordinate carbonate. Conglomerate wedges are interpreted as reflecting marginal uplifts (Furque and Cuerda, 1979, Baldis et al., 1982). On the eastern part of the Precordillera, evidence of the Late Ordovician (Ashgillian) glaciation is provided by pebbly mudstones with striated clasts and shales and fine-grained sandstones with ice-rafted dropstones (Peralta and Carter, 1990; Büggisch and Astini, 1993). The shallow marine facies grades westward to deep marine turbidites associated with mafic and ultramafic volcanic rocks (Figs. 3 and 4C). Pillow lavas and ophiolites interlayered with fine-grained turbidites have been identified in several localities in the Precordillera, including the Alcaparrosa Formation (Quartino et al., 1971) and the Yerba Loca Formation (Furque,

Figure 2. Pre-Carboniferous basement with superimposed initial configuration of the Late Paleozoic basins and present-day tectonostratigraphic units. The cratonic igneous-metamorphic basement of the Brazilian Shield is fringed by a zone (AB) of Cambrian-Ordovician sediment with outliers of Silurian-Devonian sediment, BC of Pampean-type basement (Precambrian metamorphics and granitoids), CD of Cambrian to Devonian sediments of the Proto-Precordillera, and DE of basement of the Frontal Cordillera. Late Paleozoic basins developed on the Brazilian Shield, Sierras Pampeanas, and Ventana marginal foldbelt. To the west, the basement (Proto-Precordillera type) comprises Early and Middle Paleozoic sedimentary and metasedimentary rocks and the westernmost portions of the Sierras Pampeanas, Precordillera, Frontal Cordillera, and parts of the San Rafael Block, which were covered by Late Paleozoic basins. The Chadileufu Massif with its Pampean-type basement of metamorphic nuclei and associated intrusives is regarded as the southern extension of the Pampean terrane. The North Patagonian Massif of Precambrian–Early Paleozoic granitoids and metamorphics constitutes the main basement for the forearc Tepuel Basin in west central Patagonia (Fig. 1). Bipolar oblique conic conformal projection.

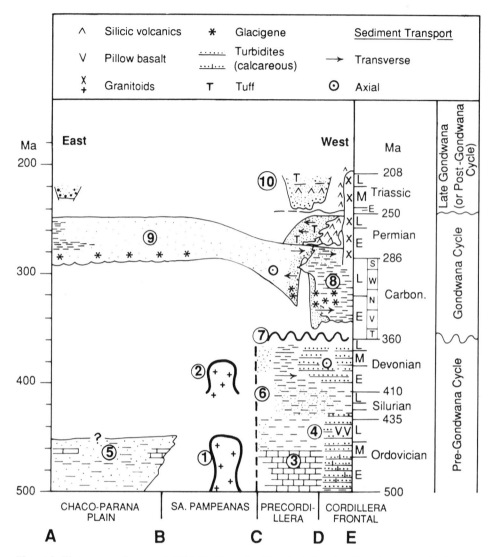

Figure 3. Time-space diagram at 31°S (A–E on Fig. 2), oriented as in other chapters, with the orogen on the right, craton on the left. Time interval from Ordovician (500 Ma) through Triassic (208 Ma), abbreviations are E = Early, M = Middle, L = Late; T = Tournaisian, V = Visean, N = Namurian, W = Westphalian, S = Stephanian. Pre-Gondwana Cycle: two main magmatic pulses can be identified in the Sierras Pampeanas region during the Paleozoic: (1) Ordovician (500–450 Ma) and (2) Devonian (400–380 Ma). Sedimentation in the Precordillera region started with Cambrian and Early Ordovician limestones, which graded westward (western Precordillera) to deeper calcareous turbidites and calcilutites (3). Since early Middle Ordovician (Llanvirnian) a change from calcareous to siliciclastic sedimentation took place in most of the Precordillera with mostly fine-grained, shelf associations toward the east and turbidites, hemipelagites and pillow lavas to the west (4). Marine sedimentation occurred at the same time in cratonic areas of the Chaco-Paraná region (5). The Ordovician strata are succeeded by Silurian shallow marine sandstones (6), succeeded in turn by Early and Middle Devonian turbidites of cratonic (Pampean) provenance. Possible terrane boundaries on either side of the Precordillera are indicated by dashed vertical lines. Sedimentation in the Precordillera and Frontal Cordillera was interrupted by the folding and faulting of the pre-Carboniferous rocks during the Chañic Orogeny (7). Gondwana Cycle: sedimentation commenced in the Precordillera and Frontal Cordillera (8) with glacigenic shallow marine and nonmarine sediments resting unconformably on the Chañic orogen. Fluvial and glacial nonmarine sedimentation prevailed in the Chaco-Paraná Basin (9). The Post-Gondwana or Late Gondwana Cycle comprises the nonmarine fill of grabens (10) separated from the Gondwana Cycle by an unconformity.

1979). Their geochemical setting is that of a juvenile oceanic rift, a transitional ridge segment, or a back-arc basin (Haller and Ramos, 1984; Kay et al., 1984).

Early Silurian (Llandoverian) pelletal iron-bearing sandstones that extend from Peru to Patagonia are interpreted as indicating a warm climate (Baldis et al., 1984). Silurian sedimentation is poorly represented in the Precordillera area by basal shallow marine sandstone, shale, and subordinate conglomerate (Tucunuco Group, Tambolar Formation), deposited in a shelf and with evidence of storm activity (Astini and Piovano, 1992), that grades westward to finer-grained, deeper facies (Calingasta Formation) (Fig. 4D). Graptolites and brachiopods indicate a Late Silurian (Ludlovian-Pridolian) age for the Tucunuco Group (Cuerda, 1969; Amos, 1972; Benedetto et al., 1992). Facies changes suggest a Silurian paleogeography similar to that of the Ordovician, with a shelf in the east onlapping the craton, and a deeper water shelf and slope to the west (Peralta, 1990).

Devonian shelfal and deep marine sediment covered most of the Precordillera, including the northwest (Baldis and Sarudiansky, 1975; Astini, 1991). Shale, shale/sandstone couplets, and pebbly mudstone were deposited by hemipelagic settling and turbidity currents in submarine fans to form the Chinguillos Group and Punilla Formation. Sole marks in the turbidites indicate a flow direction from east-southeast (Fig. 4E). Deep sea fans have been described for the central Precordillera by González-Bonorino (1975). The Punta Negra Formation (Middle to Late Devonian) comprises about 1,000 m of thin-bedded turbidites that thins to the north and south. Paleocurrents radiated to the west and northwest (Figs. 3 and 4E). The sand fraction of the turbidites contains similar proportions of quartz and low grade-metamorphic lithics, and subordinate amounts of mica, plagioclase, titanite, and opaque minerals (González-Bonorino, 1975). Paleocurrents and sandstone composition both indicate a cratonic provenance. The Punta Negra deposits can be correlated with other deep marine facies exposed in the western flank (Ciénaga del Medio Group) and the southern extreme (part of the Villavicencio Group) of the Precordillera and in the San Rafael Block (La Horqueta and Río Seco de los Castaños Formations).

Frontal Cordillera basement. The pre-Carboniferous basement of the Frontal Cordillera has a complex history obscured by several episodes of deformation/recrystallization and overprint of Mesozoic Andean magmatism. Caminos (1979) has recognized four units:

(A) *Metamorphic complex (complejo metamórfico).* This, the older recognizable unit of the Frontal Cordillera, comprises phyllite, mica schist, amphibolite and subordinate marble, and metaquartzite. Originally assigned to the Precambrian, this unit may be a western metamorphic counterpart of the Precordilleran Cambro-Ordovician sediments (Caminos, 1979).

(B) *Ultrabasic belt (faja ultrabásica).* Dunites, harzburguites and serpentinites are exposed in the extreme south of the Frontal Cordillera on south-southwest– to north-north-east–trending belts parallel to the schistosity of the metasedimentary rocks of the metamorphic complex and in the extreme southwest of the Precordillera (Zardini, 1962). Their age is unknown but is tentatively assigned to the Early Paleozoic.

(C) *Tonalitic and granitic intrusions.* Small tonalitic and granitic bodies of Devonian–Early Carboniferous age (400–320 Ma) at several localities in the Frontal Cordillera and southern extremity of the Precordillera (Caminos et al., 1979) are interpreted as part of the Paleozoic plutonic episode in the Sierras Pampeanas region (Caminos, 1985).

(D) *Devonian sedimentary rocks.* The Ciénaga del Medio Group (Amos and Marchese, 1965) crops out on the western flank of the Precordillera and on the eastern flank of the Frontal Cordillera. It comprises graywacke interbedded with siltstone and shale and has been correlated with other Devonian sedimentary units in the Precordillera like the Punta Negra Formation and equivalents (Padula et al., 1967). The Las Lagunitas Formation (Volkheimer, 1978), exposed on the southernmost part of the Frontal Cordillera about 34°S, is similar except for a greater proportion of sandstone and conglomerate; it is at least 2,000 m thick and is correlated with the La Horqueta and Río de Los Castaños Formations in the San Rafael Block (Caminos, 1979).

Basement of the Sauce Grande–Colorado Basin. The pre-Carboniferous basement of the Sauce Grande–Colorado Basin can be subdivided into two parts: (1) igneous-metamorphic basement and (2) Paleozoic sedimentary basement (Figs. 2 and 5).

Igneous-metamorphic basement. This type of basement crops out southwest and northeast of the Sierras Australes of Buenos Aires province. The northeastern outcrop belong to the Tandilia craton, which was the main source of the pre-Carboniferous sedimentary rocks that crop out in the Sierras Australes (Dalla Salda, 1975): (a) Balcarceano Assemblage (2200–2000 Ma)—schist, mylonite, amphibolite, pyroxenite, gneiss, and migmatite; (b) Tandiliano Assemblage (1870–1700 Ma)—granitoid, gneiss, and locally mylonite; (c) Vasconeano Assemblage (1600–1400 Ma)—schist intruded by diabase dykes and subordinate granite; and (d) Migueleano Assemblage (1200–600 Ma)—small granitic bodies.

The igneous-metamorphic basement exposed southwest of the Sierras Australes comprises granite and pegmatite, commonly mylonitized, and rhyolite associated with quartz-mica schist and phyllite (Kilmurray, 1968; Llambías and Prozzi, 1975), the radiometric ages of which have probably been reset by later events. The oldest dates on igneous rocks are 574 ± 10 Ma (whole-rock Rb/Sr) from granite of the Pan de Azúcar and 487 ± 45 Ma from the granite of Aguas Blancas (Cingolani and Varela, 1973). The former age overlaps the younger part of the Migueleano Assemblage of the Tandilia craton. Younger Paleozoic ages (Cingolani and Varela, 1973) are the Cerro Colorado granite with a Rb/Sr whole-rock age of 407 ± 20 Ma, rhyolitic tuff at La Mascota 317 ± 14 and at La Ermita 348 ± 21 Ma, and granite at López Lecube has a K/Ar horn-

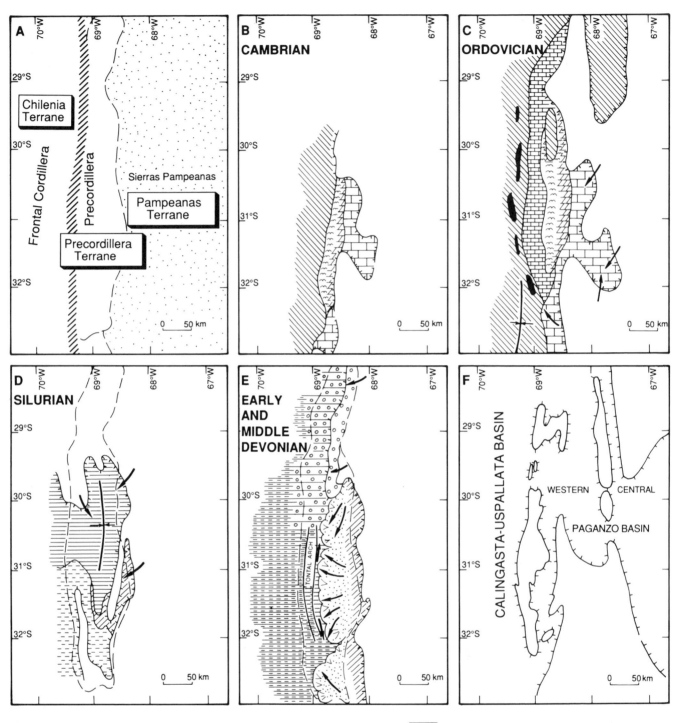

Inner platform carbonates and sandstones
Stromatolitic limestones and dolostones
Outer platform limestones
Black shales
Platform quartzites
Shales and calcareous sandstones
Platform sandstones and shales
Submarine fans
Wackes and shales

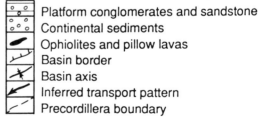

Platform conglomerates and sandstone
Continental sediments
Ophiolites and pillow lavas
Basin border
Basin axis
Inferred transport pattern
Precordillera boundary

blende age of 240 ± 12 Ma and a Rb/Sr whole-rock age of 227 ± 32 Ma (Appendix 1). Andreis et al. (1989) and von Gosen and Büggisch (1989) suggest that the Paleozoic ages reflect shearing, heating and fluid flux that opened the Rb-Sr and K-Ar systems of Precambrian rocks that continue into the adjacent Chadileufu Block (Llambías and Caminos, 1987).

Early Paleozoic sedimentary basement. The pre-Carboniferous sedimentary basement crop out in the Tandilia Ranges on the northeast and in the Ventana marginal fold belt on the southwestern rim of the Sierras Australes (Fig. 2). The Late Precambrian–Early Paleozoic (?Cambrian) metapelite, orthoquartzite, dolomite, micritic limestone, stromatolites, and claystone of La Tinta Formation rest on the igneous-metamorphic complex of the Tandilia craton. Cross-bedding and ripplemark in the orthoquartzite indicate a southward and southwestward flow from positive areas in the north and northeast (Teruggi, 1964). The La Tinta Formation was deposited during alternating deposition of carbonate and siliciclastics on a stable shallow platform.

A thick succession of pre-Carboniferous sediments, the Curamalal and Ventana Groups (Harrington, 1947), crops out in the southwestern rim of the Sierras Australes (Fig. 5). Andreis et al. (1989) regard the Curamalal Group as an undivided transgressive sequence, called Sedimentary Cycle I (Fig. 5), made up of basal clast-supported conglomerate, cross-bedded sandstone, and subordinate bioturbated muddy sandstone, siltstone, and claystone. The basal conglomeratic facies (the La Lola Formation) grades up to sandstone of the overlying

Mascota Formation (Harrington, 1947) but Suero (1972), Andreis and López-Gamundí (1989), Cellini et al. (1986), and Leone (1986) have questioned the validity of this subdivision into formations. Harrington (1947) estimated the thickness of the Curamalal Group as 1,200 m (subsequently measured sections amount to little more than a half) and estimated its age to be Silurian. Pebbles in the conglomerate are 94% quartzite, 4% vein quartz, 2% schist, with a few of mylonitic granite and rhyolite (Andreis, 1965). Most sandstones are orthoquartzites, few feldspathic (Leguizamón and Teruggi, 1985). Paleocurrents from clast imbrication flowed from the southeast, interpreted as longshore littoral currents, while cross bedding suggests currents from the Tandilia Craton (Andreis and López-Gamundí, 1989). The sequence is interpreted as laid down in a high-energy shallow marine environment with sporadic low-energy sedimentation below wave action.

The Ventana Group (Sedimentary Cycle II of Andreis et al., 1989) unconformably overlies the Curamalal Group and has an estimated thickness of 1,400 m (Harrington, 1947). The three lower units (Bravard, Napotá, and Providencia Formations) resemble the Curamalal Group in the predominance of sandstone over finer grained deposits. Conglomerate is confined to the Bravard Formation. Paleocurrents from cross bedding are to the southwest, parallel to that measured by Andreis and López-Gamundí (1989) from cross-bedded sandstone in the lower part of the Curamalal Group. The siltstone and shale of the Lolén Formation in the upper part contains Early Devonian (Emsian) brachiopods (Harrington, 1972). The Ventura Group was deposited on a shallow platform: cross-bedded sandstone during high-energy periods, the siltstone and shale during quiet periods (Andreis et al., 1989).

Summary of pre-Gondwana basement

Ramos et al. (1986) define three major pre-Carboniferous terranes between 29°–33°S (Fig. 4A), from east to west.

(i) the Pampeanas terrane of Precambrian–Early Paleozoic metamorphic rocks and intrusives, cratonized by the onset of the Gondwana Cycle.

(ii) the Precordillera terrane of the Precordillera and part of the San Rafael Block, with thick Cambrian to Devonian marine sediment. The suspect nature of this terrane is supported by its sharp tectonic boundary and stratigraphic contrast with the cratonic Pampeanas terrane to the east.

(iii) the Chilenia terrane corresponds with the present-day morphotectonic units of the Frontal and Principal Cordilleras. Its pre-Carboniferous history is obscured by later magmatic activity. The basement of this terrane comprises amphibolite, pelitic and mylonitic schist, and cataclastic gneiss, and the age of deformation is estimated to be Devonian. The precise extent of the Chilenia terrane, including its suture with the Precordillera, remain obscure. Tightly folded Devonian sediments unconformably overlain by Carboniferous deposits have been recorded from both the Precordillera (Sessarego,

Figure 4. Cambrian-Carboniferous paleogeographic evolution of the Precordillera and surrounding areas (modified from Baldis et al., 1982).

A. Present-day morphotectonic units of western Argentina, with three major pre-Carboniferous terranes as proposed by Ramos et al. (1986) and Ramos (1988): (1) Pampeanas terrane of Precambrian–Early Paleozoic metamorphics and intrusives, cratonized before the onset of Carboniferous sedimentation; (2) Precordillera terrane, corresponding with the present-day Precordillera and part of the San Rafael Block, with an Early and Middle Paleozoic sedimentary and metasedimenary basement; and (3) Chilenia terrane, located along the Frontal and Principal Cordilleras and consisting of a pre-Carboniferous basement of metasedimentary rocks and intrusives intensively deformed on the western flank of the Precordillera terrane by the Late Devonian orogeny.

B. Cambrian platform carbonates.

C. Early and Middle Ordovician platform carbonate (shelfal limestone) in the east, and a Late Ordovician oceanic association of fine-grained turbidites, hemipelagites, ophiolite and pillow lavas in the west.

D. Silurian shallow marine sediments on the east with deeper deposits on the west.

E. Early-Middle Devonian turbidite fans in the southern and central segments of the Precordillera and shallow marine to nonmarine sediment in the northern segments.

F. Carboniferous paleogeographic configuration of the Calingasta-Uspallata and western Paganzo Basins after the Chañic Orogeny.

1988) and the Frontal Cordillera (Caminos et al., 1988). Furthermore, no Precambrian rocks or Silurian/Devonian magmatic rocks have been reported in Chile at the latitude of "Chilenia" (Davidson, 1984; Hervé, 1988), although these were part of the original proposal of Ramos et al. (1986).

In addition to these three major terranes, other morphotectonic units such as the Chadileufu Massif, the Sierras Australes of Buenos Aires Province (the Ventana marginal fold belt of Caminos et al., 1988), and the North Patagonian Massif have different histories and can be considered major tectonostratigraphic units (Fig. 2). The Chadileufu Massif is distinguished from the adjacent blocks by Pampean-type igneous-metamorphic rocks overlain by Late Paleozoic igneous rocks (Linares et al., 1980) and could be the southern extension of Ramos's (1988) Pampeanas terrane. The north-trending basement complex of the north Patagonian Massif–Chadileufu Massif–Pampeanas underwent progressive uplift and cratonization during Late Paleozoic times (Caminos et al., 1988) to provide a first-order drainage divide between the eastern epicratonic basin (Sauce Grande–Colorado Basin) and the western foreland basin (San Rafael Basin) along the Paleo-Pacific continental margin.

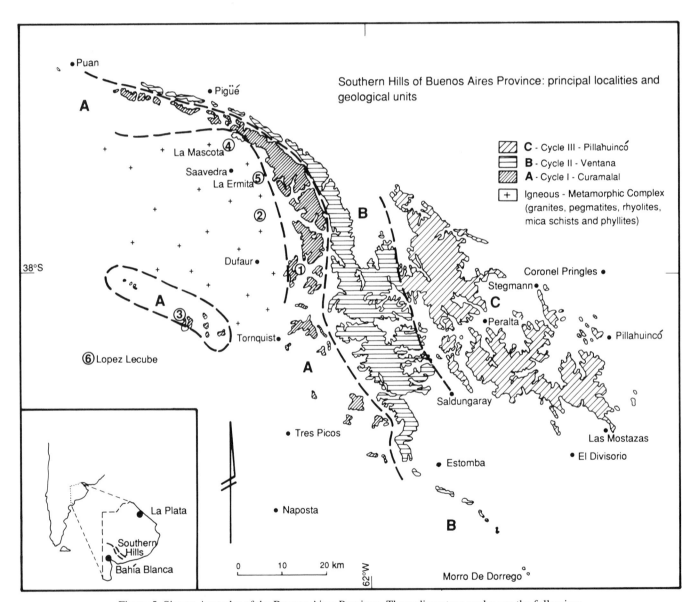

Figure 5. Sierras Australes of the Buenos Aires Province. The sedimentary cycles are the following: A, Curamalal Group of probable Silurian age; B, Ventana Group of Devonian age; and C, Pillahuincó Group of Late Carboniferous-Permian age. An igneous-metamorphic complex lies west of the main belt of outcrops. Localities are 1, Pan de Azúcar; 2, Aguas Blancas; 3, Cerro Colorado; 4, La Mascota; 5, La Ermita; 6, López Lecube. Modified after Andreis et al. (1987), and Andreis and López-Gamundí (1989).

Isopachs in the western Chaco-Paraná Basin parallel the Pampean Arch (Fig. 2) and the axis of the Sauce Grande–Colorado Basin parallels the Tandilia Massif, where a SE-trending Precambrian suture has been proposed by Ramos (1988).

The north-trending structural grain of the pre-Gondwanan sedimentary and metasedimentary basement of the Calingasta-Uspallata and western Paganzo Basins controlled their initial configuration. The basement anisotropy is traceable southward into the western San Rafael Basin and northward parallel to the Arequipa Massif (Dalmayrac et al., 1980b).

GONDWANA CYCLE

Western or Andean basins: Paganzo, Calingasta-Uspallata, and San Rafael

Sedimentary sequences. We divide the Gondwana Cycle into two major sedimentary sequences, equivalent to the unconformity-bounded stratigraphic units or sequences of Chang (1975) and to the allostratigraphic units proposed by the North-American-Commission-on-Stratigraphic Nomenclature

(1983). This approach proved useful in the facies analysis by López-Gamundí et al. (1989) of the deposits of the Paganzo, Calingasta-Uspallata and San Rafael foreland basins and is followed here accordingly. Although by definition not secured by very precise time constraints, the unconformity-bounded sequences studied here together cover most of the Carboniferous and Permian (Fig. 6) and reflect a common paleogeographic evolution.

The lower sequence (megasequence I of López-Gamundí et al., 1989) of marine and nonmarine sediments is bounded below by the Late Devonian–Early Carboniferous unconformity connected with the Chañic movements and above by the Early Permian unconformity of the San Rafael movements. The upper sequence (megasequence II of López-Gamundí et al., 1989) is exclusively volcanics and volcaniclastic sediments emplaced in nonmarine environments separated from the lower sequence by the San Rafael unconformity and succeeded unconformably by Triassic nonmarine sediments.

Late Devonian–Early Carboniferous diastrophism and the initial configuration of the western (Andean) basins. Much of southern South America was deformed during the

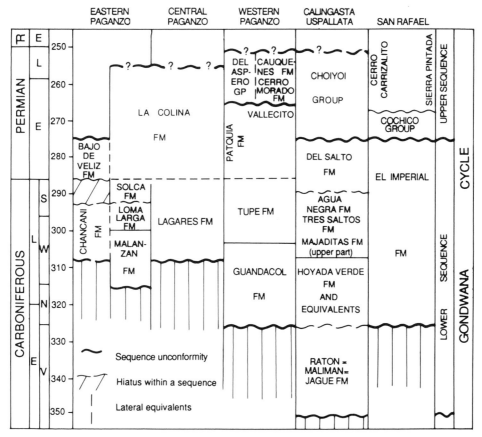

Figure 6. Correlation chart for the Calingasta-Uspallata, San Rafael, and Paganzo Basins of western Argentina, showing the lithostratigraphic (formations and groups) and allostratigraphic (unconformity-bounded sequences) units described in the text. Abbreviations are E = Early, L = Late; V = Visean, N = Namurian, W = Westphalian, S = Stephanian.

Late Devonian–Early Carboniferous. In southeastern Peru and Bolivia, Devonian marine sediments tightly folded during Eohercynian orogenic phase (Dalmayrac et al., 1980b) are overlain by Early Carboniferous coarse deposits (Laubacher and Megard, 1985). In north and central western Argentina, the Chañic Orogeny (Turner and Méndez, 1975) was responsible for the intense north-northwest–trending folding and faulting of the pre-Carboniferous metasedimentary and sedimentary basement in the western Paganzo, Calingasta-Uspallata, and San Rafael regions, along the western Precordillera and San Rafael Block, and determined the initial configuration of these basins within high ground. In the Paganzo Basin, basement lineaments (Pampean Arch, Pie de Palo Arch) (Figs. 2 and 7) controlled the Late Paleozoic sedimentation. The basal contact in these basins is generally evidenced by an angular truncation. The hiatus involved ranges from a minimum of the Late Devonian and Early Carboniferous to a maximum of the Late Devonian through Late Carboniferous (López-Gamundí and Rossello, 1993). The initial stage of these basins, especially in the west, is documented by infrequent Early Carboniferous syntectonic sedimentation. The scarcity of deposits of this age suggests that the Early Carboniferous was dominated by uplift and erosion followed by widespread subsidence during the latest Early and Late Carboniferous.

On the west from 33°S to 29°S (*see* Fig. 9), a north-trending high called the Proto-Precordillera (Amos and Rolleri, 1965)—the Acadian Precordillera of Baldis and Chebli (1969)—emerged during the uplift connected with the Chañic movements on the site of the Devonian submarine Tontal Arch, deduced from paleocurrent patterns (González-Bonorino, 1975; Baldis et al., 1982) (Fig. 4E). The Proto-Precordillera (Figs. 7 and 8) divided the drainage during the Early and mid-Carboniferous as seen from the facies architecture and paleocurrent pattern in the Calingasta-Uspallata and western Paganzo Basins. By the mid-Carboniferous, a major brackish to marine embayment (Guandacol embayment) around its northern tip allowed communication between the mostly nonmarine sedimentation of the Paganzo Basin and the marine sedimentation of the Calingasta-Uspallata Basin (Amos and Rolleri, 1965; Scalabrini Ortiz, 1972). Glacial marine sedimentation took place in this embayment (Milana and Bercowski, 1990; Martínez, 1990). Glacial centers were probably located on the east and south. Available paleocurrent data suggest that north-trending glacial valleys descended northward from the high areas of the Proto-Precordillera. Another seaway is found around the southern tip at about 32°30′S (Fig. 9) (Salfity and Gorustovich (1983). Farther south, a north-trending high in the San Rafael Basin follows the north-trending fold axes and incipient schistosity of the Devonian Horqueta Formation (Dessanti, 1956). The basement ranges from Middle to Late Devonian rocks (Punta Negra Formation and Ciénaga del Medio Group in the Calingasta-Uspallata Basin; La Horqueta and Río Seco de los Castaños in the western segment of the San Rafael Basin) to Ordovician limestone

Figure 7. Time-space diagram for the Calingasta-Uspallata, Paganzo, and western Chaco-Paraná Basins across the Frontal Cordillera, Precordillera, Sierras Pampeanas, and Chaco-Paraná plain, oriented, as in Fig. 3, with the orogen on the right and craton on the left. Time interval is from Early Carboniferous through the Early Triassic, abbreviations are E = Early, L = Late; V = Visean, N = Namurian, W = Westphalian, S = Stephanian. (1) Early Carboniferous (Visean) nonmarine and shallow marine sedimentation (El Ratón, Malimán and Jagüe Formations) commenced on the western flank of th Proto-Precordillera in the Calingasta-Uspallata Basin (Amos and Rolleri, 1965; Rolleri and Baldis, 1969; Scalabrini-Ortiz, 1970, 1972; Sessarego and Cesari, 1986). Glacigene sediments spread over both flanks of the Proto-Precordillera and along north-trending valleys northward into the Guandacol embayment by the Namurian-Westphalian with (2) glaciomarine facies on the west in the Calingasta-Uspallata Basin (González, 1981, 1983; López-Gamundí, 1983, 1987) and (3) glaciolacustrine sediments on the east in the western Paganzo Basin and glaciomarine sediment in a marine embayment north of 31°S (Cuerda and Furque, 1981; Ortiz and Zambrano, 1981; Bossi and Andreis, 1985; Limarino and Cesari, 1988; Milana and Bercowski, 1987; Bercowski and Milana, 1990). At this latitude, the Proto-Precordillera continued to act as a barrier during most of the sedimentation of the Lower Sequence (Early Carboniferous–Early Permian). By Late Carboniferous–early Early Permian times (4) a fluvial-eolian domain prevailed in the Paganzo Basin and linked all the nonmarine depositional areas east of the Proto-Precordillera, which was breached at this time to connect the Panganzo Basin with (4) the fluvial-eolian and shallow marine-deltaic domains of the Calingasta-Uspallata Basin (Sessarego and Limarino, 1987). Tuff beds intercalated in Early Permian sediments in (5) the Paganzo Basin (Teruggi et al., 1969; Coira and Koukharsky, 1970; Di-Paola, 1972; Andreis et al., 1975; Limarino et al., 1986) and (6) the Calingasta-Uspallata Basin reflect the waxing magmatic arc to the west that seems to have been continuous from the Late Carboniferous to the Early Triassic (7) (Nasi and Sepúlveda, 1986). (8) Eolian facies with abundant volcanic detritus came from the west (Limarino, 198; Limarino and Spalletti, 1986). (9) Valley fills in the eastern Paganzo Basin are restricted to small areas during the mid-Late Carboniferous (Andreis et al., 1986) and are independent of (10) the widespread Early Permian sedimentation in the central and western Paganzo regions (Salfity and Gorusovich, 1983; Azcuy et al., 1987) and probably in the Chaco-Paraná Basin further east. The Upper Sequence rests on the Lower Sequence unconformably on the west and disconformably on the east (Furque, 1963; Borrello and Cuerda, 1965; Caminos, 1979) and comprises (11) volcanics and (12) nonmarine volcaniclastic sediments derived from the western arc (López-Gamundí et al., 1986, 1990). The two main glacial episodes are (a) of early–mid-Carboniferous age in the Calingasta-Uspallata (2) and western Paganzo Basin (3), and (b) of Late Carboniferous–Early Permian (290 ma) age (13) in the Chaco-Paraná Basin (Russo et al., 1979) with possibly related glaciolacustrine episodes in (14) the eastern Paganzo Basin (Hünicken and Pensa, 1972; Rodríguez and Fernández-Seveso, 1988).

→

(San Juan Formation) in the western Paganzo Basin. The pre-Carboniferous basement of the central and eastern sectors of the Paganzo Basin comprises plutonic and metamorphic rocks of the Pampean terrain.

Lower Sequence (Carboniferous and earliest Permian). The Lower Sequence (Figs. 6 and 7) is divided into (a) an Early and mid-Carboniferous (*Protocanites* and *Levipustula*

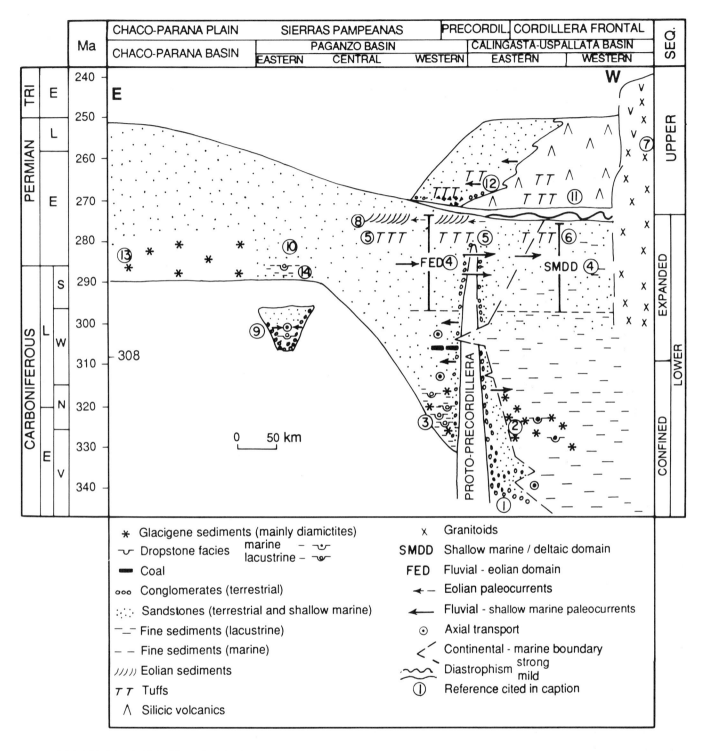

	Ma	CHACO-PARANA PLAIN	SIERRAS PAMPEANAS			PRECORDIL.	CORDILLERA FRONTAL		SEQ.
		CHACO-PARANA BASIN	PAGANZO BASIN			CALINGASTA-USPALLATA BASIN			
			EASTERN	CENTRAL	WESTERN	EASTERN	WESTERN		

Legend:

* Glacigene sediments (mainly diamictites)

~⌣ Dropstone facies marine – ~⌣ lacustrine – ~⊙⌣

▬ Coal

₀₀₀ Conglomerates (terrestrial)

∴∵∴ Sandstones (terrestrial and shallow marine)

⁻‗⁻ Fine sediments (lacustrine)

– – Fine sediments (marine)

/////) Eolian sediments

⊤ ⊤ Tuffs

∧ Silicic volcanics

x Granitoids

SMDD Shallow marine / deltaic domain

FED Fluvial - eolian domain

←‒ Eolian paleocurrents

← Fluvial - shallow marine paleocurrents

⊙ Axial transport

‹ Continental - marine boundary

~~~ Diastrophism strong mild

① Reference cited in caption

zones, 345–308 Ma) initial fill and altitude-controlled glacial sedimentation confined to the west, and (b) later Carboniferous and earliest Permian (308–275 Ma) sedimentation expanded to the east.

Early and mid-Carboniferous (345–308 Ma) initial fill and altitude-controlled glacial sedimentation. Early Carboniferous sediments are known from confined areas in the Calingasta-Uspallata and western Paganzo Basins and sedi-

ments of suspected but not proved Early Carboniferous age from the San Rafael Basin. Early Carboniferous sedimentation was confirmed by the irregular relief shaped by the Chañic diastrophism. The basal fill of the Calingasta-Uspallata Basin and western segment of the Paganzo Basin has numerous lateral facies changes caused by progressive sedimentary fill across the rugged pre-Carboniferous topography. In the Calingasta-Uspallata Basin, the El Ratón Formation is a de-

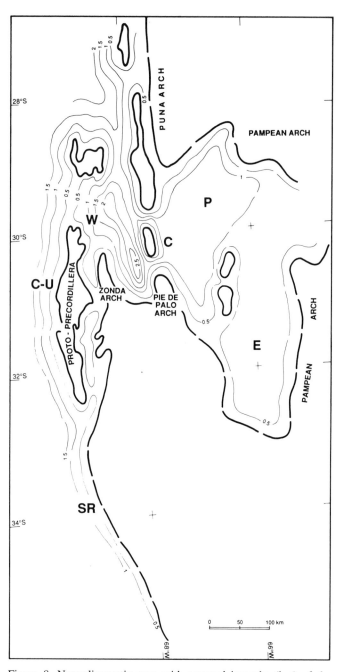

Figure 8. Nonpalinspastic map with restored isopachs (km) of the Lower Sequence (Early Carboniferous–basal Permian) for the Paganzo (P) (W, western; C, central; E, eastern), Calingasta-Uspallata (C-U) and San Rafael (SR) Basins. Paleogeographic features modified from Azcuy and Morelli (1970), Vásquez et al. (198), Salfity and Gorustovich (198), and Alvarez and Fernández Seveso (1987).

posit of alluvial fans succeeded by braided streams that filled small intermontane valleys incised on the flank of the Proto-Precordillera. The flood-plain deposits contain an Early Carboniferous (Visean) flora (Azcuy et al., 1981). The age-equivalent littoral-deltaic Malimán Formation (Fig. 6) on the northwestern flank of the Proto-Precordillera (Fig. 9), with the same flora (*Archeosigillaria-Lepidodendropsis* zone) as the El Ratón Formation (Sessarego and Cesari, 1986), comprises conglomerate, sandstone, and rare shale and coal deposited in a fan-delta (Limarino and Caselli, 1992). A marine intercalation (Scalabrini Ortiz, 1970) of mudstone contains invertebrate fossils of the *Protocanites* zone of Visean (Amos, 1964) or Tournaisian-Visean (Archangelsky et al., 1987) age. The equivalent unit, the basal part of the Jagüé Formation, exposed in the northernmost Calingasta-Uspallata Basin, comprises conglomerate, sandstone, and shale with marine invertebrates and plants (González and Bossi, 1986; Fauqué and Limarino, 1991). In the western Paganzo Basin, turbidite sandstone and pebbly mudstone with slumped intervals present in the basal section of the Guandacol Formation (Fig. 6) have been tentatively assigned to the Early Carboniferous (Bossi and Andreis, 1985) despite the lack of biostratigraphic control. Andreis (1986) regards the slumping as having been induced by the Chañic movements, but we believe that an autocyclic mechanism, as outlined below for the presumably equivalent basal El Imperial Formation, is more likely.

Paleomagnetic studies suggest a fairly high (about 60°S) latitude for the Calingasta-Uspallata, San Rafael, and Paganzo Basins during the Early to mid-Carboniferous *Levipustula* zone (Vilas and Valencio, 1977; Azcuy et al., 1987; López-Gamundí et al., 1987). The high latitude combined with high altitude in the Chañic mountain chain caused the alpine glaciation seen in the Calingasta-Uspallata Basin (Frakes and Crowell, 1969; López-Gamundí, 1987). Widespread glaciomarine deposits on the eastern margin of the Calingasta-Uspallata Basin indicate wet-based glaciers that originated in adjacent high areas (Fig. 9). The glaciomarine facies range from proximal deposits with striated boulder pavements to a wide variety of sediment gravity flows (López-Gamundí, 1987). In the Hoyada Verde Formation, a striated boulder pavement

Figure 9. Early–mid-Carboniferous paleogeography of the Lower Sequence of the Paganzo (P), Calingasta-Uspallata (C-U), and San Rafael (SR) Basins. Marine and glaciomarine sediments are restricted to the Calingasta-Uspallata and San Rafael Basins on the west while nonmarine deposits are dominant in the Paganzo Basin. The nonmarine and marine domains are connected in estuaries or prograding complexes at 29°S and 32°30'S. Paleocurrent information for locality 1—Andreis et al. (1975); 2—Bossi and Andreis (198); 3—Andreis et al. (1986); 4—Scalabrini-Ortiz (1972); 5—González (1981) and López-Gamundí (1984); 6—López-Gamundí and Amos (1985); 7—López-Gamundí (1984); 8—Espejo (1990); 9—Buatois and Mangano (1992).

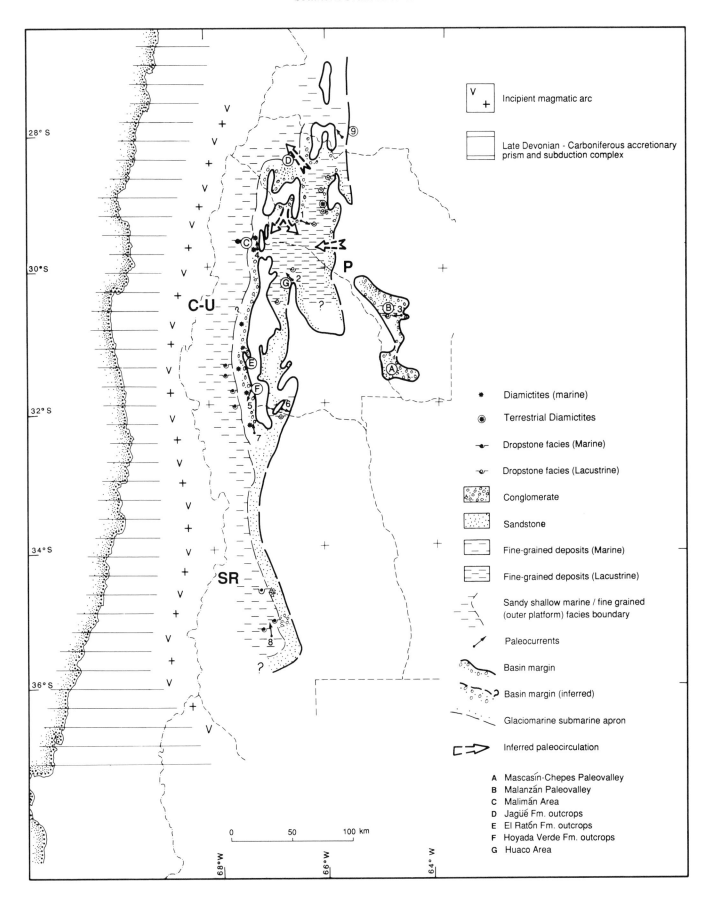

V
+ Incipient magmatic arc

Late Devonian - Carboniferous accretionary
prism and subduction complex

* Diamictites (marine)

⊛ Terrestrial Diamictites

→• Dropstone facies (Marine)

→∘ Dropstone facies (Lacustrine)

Conglomerate

Sandstone

Fine-grained deposits (Marine)

Fine-grained deposits (Lacustrine)

Sandy shallow marine / fine grained
(outer platform) facies boundary

Paleocurrents

Basin margin

Basin margin (inferred)

Glaciomarine submarine apron

Inferred paleocirculation

A Mascasín-Chepes Paleovalley
B Malanzán Paleovalley
C Malimán Area
D Jagüé Fm. outcrops
E El Ratón Fm. outcrops
F Hoyada Verde Fm. outcrops
G Huaco Area

0   50   100 km

(González, 1981) is associated with massive to poorly stratified sandy diamictite, well-stratified muddy diamictite, and pebbly shale (López-Gamundí, 1984, 1987). The pebbly shale, known also as the "dropstone" facies, appears in units of similar age, as is associated with turbidites and other deposits that originated through sediment-gravity flows (López-Gamundí, 1987). This association, along with the presence of dropstones in debris-flow deposits, indicates the intimate relationship between the glaciomarine sedimentation and gravity-induced deposits derived from glacial material. A conservative estimate based on the areal extent of these outcrops would suggest an apron-like configuration of minimal north-south length of 150 km on the eastern border of the Calingasta-Uspallata Basin with glaciated areas located on the Proto-Precordillera highs to the east (Fig. 9; López-Gamundí, 1988). Furthermore, Limarino and Page (1987) reported the presence of glaciomarine sediments in the Cortaderas Formation (Fig. 6) near the Malimán area on the northwestern Precordillera, extending the glaciomarine apron to 30°S (Fig. 9).

Fine shallow glaciomarine deposits in the basal El Imperial Formation represent the initial transgressive episode in the San Rafael Basin in the mid-Carboniferous. In El Imperial Formation (Arias and Azcuy, 1986; Figs. 9 and 10), marine laminated shales with sporadic dropstones are associated with slumped intervals (Volkheimer, 1967; Espejo, 1990; Espejo and López-Gamundí, in preparation). Previously interpreted as tectonically triggered (Volkheimer, 1967), the slumped intervals are interpreted by Arias and Azcuy (1986) as produced by autocyclic rather than allocyclic mechanisms, probably connected with high rates of sedimentation in a prograding facies setting or with gravity remobilization of rain-out tills on a glaciated shelf (Espejo, 1990; Espejo and López-Gamundí, in preparation).

Glacial sedimentation in the northwestern sector of the Paganzo basin at around 30°S (Vásquez et al., 1981; Bossi and Andreis, 1985) took place in restricted marine to lacustrine conditions. Ice rafting is seen in the thin-bedded turbidites and subordinate diamictitic lenses and pebbly shales, the dropstone facies of the Guandacol Formation (Figs. 7, 9, and 11). The dropstone facies association seems to extend to the north and south in the western Paganzo Basin (Fig. 9). In the southwest, several probably lacustrine formations at or near the base of the succession contain dropstones (Cuerda and Furque, 1981; Ortíz and Zambrano, 1981). To the east, glaciolacustrine deposits of the basal part of the Lagares Formation in the north-central Paganzo Basin contain (a) laminated siltstone and claystone with dropstones, (b) interbedded mudstone and sandstone interpreted as turbidite, and (c) diamictite (Limarino and Cesari, 1988). Abundant *Cristatisporites*-like spores indicate that lycophytes dominated the shoreline. The succession is completed by coal-bearing fluvial deposits (Fig. 6).

Probably due to a lower preservation potential than their glaciomarine and glaciolacustrine counterparts, tillites are absent except in the northwestern Paganzo Basin, where massive diamictites with blocks up to 3 m in diameter and striated and faceted clasts in the Agua Colorada are associated with the glaciolacustrine Lagares Formation (Limarino and Cesari, 1988).

The age of this glacial episode is indicated by *Levipustula* zone (Amos and Rolleri, 1965; González, 1985) fossils in the shales capping the glaciomarine strata in the Hoyada Verde Formation and other diamictite-bearing units of the Calingasta-Uspallata Basin. The base of this zone is just above the base of the Namurian, which in eastern Australia is dated by the SHRIMP U-Pb on zircon method as 324 Ma (Roberts et al., 1991; Appendix 1). The *Levipustula* zone is present also in the shale that caps the glaciomarine diamictites of the Tarija Formation of northwestern Argentina and Bolivia (Fig. 1) (Rocha-Campos et al., 1977). Fine-grained marine sediment capping the diamictitic deposits has been interpreted as the product of a sea-level rise, expressed stratigraphically as a transgression (López-Gamundí, 1989). An upper bound on the age of the glacial episode on the west is provided by the Westphalian-Stephanian *Potoniesporites-Lundbladispora* zone (Cesari, 1986) in sediments above the glacigenic deposits in the San Rafael Basin, at the same time as glacigenic deposition started in the east, in the Sauce Grande and Paraná Basins (Appendix 1).

The eastern sector of the Paganzo Basin was a positive area except where sediment was deposited in small intermontane valleys (paleo-valleys in the sense of Andreis et al., 1986) bounded by Pampean basement uplifts (Fig. 9). In the Malanzán Formation (Figs. 6 and 9) basal alluvial-fan conglomerate resting on Pampean-type basement is succeeded by fluvial sandstone and claystone which grade up to lacustrine turbidites with dropstones (Andreis and Bossi, 1981). The Malanzán Formation forms a narrow belt of outcrops which define the original extent of the intermontane valley. Similar deposits have been identified by Fernández-Seveso and Alvarez (1987) in the Mascasín-Chepes paleo-valley (Fig. 9).

Later Carboniferous and earliest Permian: Post-glacial sedimentation and expansion of depositional areas. In the western Paganzo Basin the transition between the lacustrine/brackish Guandacol Formation and the fluvial Tupe Formation is effected by deltaic progradation (Fig. 11) of fine rippled and cross-bedded sandstone (formally included in the uppermost section of the Guandacol Formation) that grade upward to fine cross-bedded sandstone, mudstone, with paleosols and coals of the basal Tupe Formation, interpreted as deposited by river currents in deltaic plains (Bossi and Andreis, 1985). Coal is common in the lower half of the Tupe Formation and in equivalent units in the rest of the Paganzo Basin.

The coals are interbedded with pebbly sandstone and conglomerate deposited in a braided-river system (Limarino, 1985) or with meandering river deposits on the western sector of the Paganzo Basin (Cesari, 1985; Bossi and Andreis, 1985). An important feature of the coal-bearing sediments is plant bioturbation in paleosols indicating an autochthonous or hypo-autochthonous origin. The coal beds are also associated with

kaolinitic claystone described as tonstein (Di-Paola, 1972; Limarino, 1985). The coal seams are some tens of meters above the glacial sediments in the Paganzo Basin (cf. Figs. 7 and 11). A temperate climatic model (López-Gamundí et al., 1993), as proposed by Conaghan (1984) and Martini and Glooschenko (1985) for other parts of Gondwanaland, is preferred to the Late Carboniferous Euroamerican humid-tropical model of coal formation.

A band of coquinas and micritic limestone in the basal part of the Tupe Formation (Fig. 11) is interpreted as a marine ingression or destructive phase during deltaic progradation in the western Paganzo Basin (Fig. 7). The invertebrate fossils belong to the "Zona de Intervalo" (González, 1985), equivalent to the *Nothorhacopteris-Botrychiopsis-Ginkgophyllum* (NBG) zone of Westphalian-Stephanian age (Archangelsky et al., 1987) (Appendix 1). The ingression is found also to the north and west in several localities in the basal portion of the Tupe Formation and equivalents in the northernmost sector of the Calingasta-Uspallata Basin, northwestern Paganzo Basin (Guandacol embayment), and interbedded between continental beds in the Proto-Precordillera (Milana et al., 1987; Lech, 1991). This transgressive peak suggests a temporally brief although areally extensive breach of the Proto-Precordillera at 31° of present latitude (Fig. 7).

The first fluvial facies in the Tupe Formation marks the depositional and paleoclimatic boundary between (a) the dropstone facies of the Guandacol Formation and the coals and deltaic sediments of the lower Tupe Formation and (b) the overlying fluvial deposits of the upper Tupe Formation and Patquía Formation (Bossi and Andreis, 1985; Limarino et al., 1986). The deltaic progradation indicates the maximum expansion of the basin during the later Carboniferous–earliest Permian at 30°S where the western Paganzo deposits grade westward to the marine facies of the Calingasta-Uspallata Basin (Fig. 12). Despite the local and brief drowning of part of its northern and central reaches, the Proto-Precordillera continued to function, especially from 30°S to 32°S, as a barrier that blocked the fluvial/deltaic system of the Paganzo Basin, and only at the southern tip of the Proto-Precordillera can a fluvial-deltaic-marine transition be inferred (Fig. 12).

The rest of the fluvial systems in the eastern flank of the Proto-Precordillera flowed along north-south-trending valleys to become integrated with the northwestward seaway. Paleocurrent data on the cross-bedded fluvial sandstones confirm this axial pattern parallel to the strike of the Proto-Precordillera (Espejo and López-Gamundí, 1984).

The end of the Carboniferous saw a final expansion of deposition, especially in the eastern Paganzo Basin and Chaco-Paraná Basin where sediment laps over crystalline basement (Salfity and Gorustovich, 1983; Alvarez and Fernández-Seveso, 1987) (Fig. 12). Sessarego and Limarino (1987) recognize two depositional domains for the Paganzo and Calingasta-Uspallata Basins during this period: (a) a fluvial-eolian domain and (b) a shallow marine-deltaic domain (Fig. 12). The fluvial-

eolian domain covers the entire Paganzo Basin and starts with coarse alluvium that pass upwards into the facies of high- and low-sinuosity channels of the Tupe Formation and then into the high-sinuosity fluvial deposits of the basal Permian Patquía Formation. In the upper half of the Patquía Formation, the eolian facies (Limarino and Spalletti, 1986) and evaporitic playa-lake deposits (Limarino and Sessarego, 1986; Limarino et al., 1987) indicate aridity (Figs. 7 and 12). Paleocurrent data from the eolian deposits indicate winds from west to east (Fig. 7) (Limarino and Spalletti, 1986; Limarino et al., 1987). Limarino and Spalletti (1986) suggested that the eolian circulation was influenced by a long volcanic arc developed along the western margin of the continent (Fig. 12) which not only deflected winds to the east but also desiccated humid westerly winds in the rain shadow of the mountains. The change in petrofacies at the base of the eolian deposits from quartzose to volcaniclastic (Fig. 11) also indicates the volcanic arc, and the start of the upper sequence at 277 Ma.

The gradual uplift of the arc region is reflected in the shallow marine conditions that prevailed in the Calingasta-Uspallata and San Rafael Basins. The shallow marine–deltaic domain dominated the Calingasta-Uspallata basin (Fig. 12). This domain comprises cross-bedded sandstone and heterolithic successions of thinly interbedded fine sandstone and bioturbated mudstone with marine invertebrates of the Late Carboniferous Zona de Intervalo (González, 1985) and the Early Permian *Cancrinella* zone (Amos and Rolleri, 1965). The shallow marine facies is found in the Agua Negra, Tres Saltos, and Del Salto Formations, and in the upper half of the Agua de Jagüel and Majaditas Formations (Fig. 6). Similar shallow marine conditions have been inferred for the Late Carboniferous intra-arc Las Placetas Formation (29°S–30°S) and the fore-arc Huentelauquen Formation (31°S–32°S) west of the Calingasta-Uspallata Basin in Chile (Nasi and Sepúlveda, 1986). In its upper section the Huentelauquen Formation contains a mixed siliciclastic–calcareous shallow marine facies with a Foramol association (cf. Sellwood, 1978) characteristic of cold to temperate waters (Rivano and Sepúlveda *in* Nasi and Sepúlveda, 1986). The shallow marine–deltaic domain is represented in the San Rafael Basin by littoral wave-dominated cross-bedded sandstone associated with shale in the upper part of the El Imperial Formation (Dessanti, 1956; Giudici, 1971; Espejo, 1987a, 1990) succeeded by high- and low-sinuosity fluvial cross-bedded sandstone in the Early Permian *Gangamopteris* zone (Espejo, 1987b, 1990; Espejo and Cesari, 1987). The vertical facies arrangement suggests a transition from shallow marine to deltaic and fluvial conditions (Fig. 10, Arias and Azcuy, 1986; Espejo, 1990). Sedimentation expanded southeastward toward the Sauce Grande–Colorado Basin by Early Permian times (Fig. 12) as evidenced by the presence of fluvial and shallow-marine deposits with *Gangamopteris* (Melchor, 1990). Preliminary paleocurrent data on fluvial deposits suggest flow to the west and northwest (Fig. 12).

Volcanicity coeval with sedimentation is evidenced by

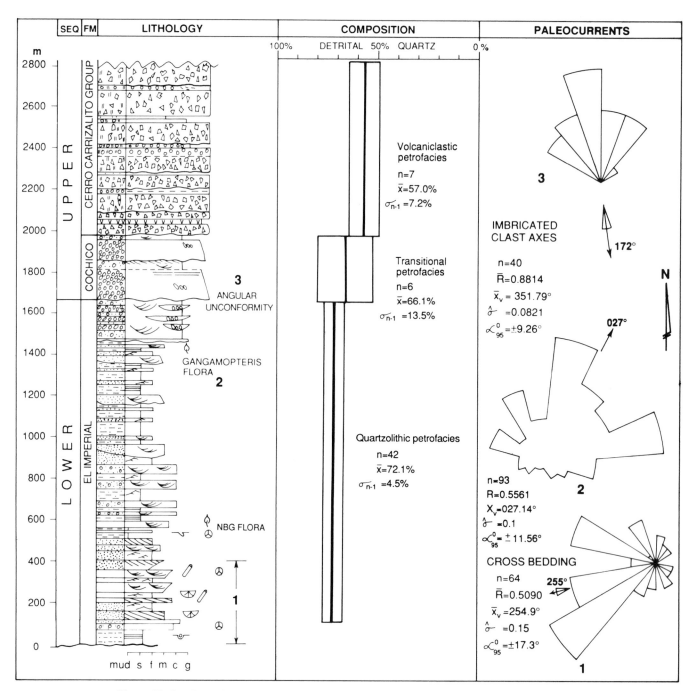

Figure 10. Stratigraphic column for the San Rafael Basin. The Lower Sequence is represented by the El Imperial Formation, with marine deposits at the base (1) (Zona de Intervalo and probably *L. levis* zone), which grade upward to fluvial deposits with NBG and *Gangamopteris* floras (Espejo, 1987a; Espejo and Cesari, 1987). An angular unconformity separates the El Imperial Formation (2) from the volcaniclastic breccias and conglomerates of the Cochicó Group (3), succeeded in turn by tuffaceous sediments and volcanic rocks of the Cerro Carrizalito Group. Note the change in composition from the quartzolithic (mixed) petrofacies of the El Imperial Formation through the transitional Cochicó Group to the volcaniclastic Cerro Carrizalito Group, coupled with a reversal in paleocurrents, suggesting a drastic change in the regional paleoslope (Espejo, 1986b; Stinco, 1986). Explanation of symbols on facing page.

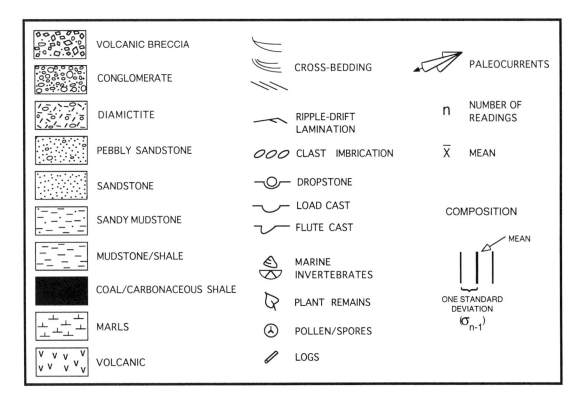

tuffs and subordinate volcanic agglomerates interbedded with fluvial deposits and volcanogenic detritus in the eolian facies (Limarino, 1984; Limarino et al., 1986) in the earliest Permian (286–277 Ma) Patquía Formation and equivalents (Fig. 6) in the Paganzo Basin, in the shallow marine Del Salto Formation in the Calingasta-Uspallata Basin (Fig. 12), and by tuff interbedded in Early Permian shallow marine limestone in the Arizaro Basin (location 11 in Fig. 1) (Donato and Vergani, 1985). In the San Rafael Basin, volcanics were erupted about 280 Ma (Appendix 1) and were involved in the deformation indicated by the lacuna at about 275 Ma. The volcanic episode at the top of the Lower Sequence is a preliminary minor expression of the extensive arc volcanism during deposition of the Upper Sequence (Fig. 7).

Upper Sequence (rest of Early Permian–Late Permian): arc-volcanism invasion of the foreland basin. The Upper Sequence of the Gondwana Cycle (Fig. 6) represents the climax of calc-alkaline arc-volcanism and volcaniclastic sedimentation in nonmarine depositional environments. Radiometric dates suggest the onset of this volcanism in the Calingasta-Uspallata and San Rafael Basins at about 275 Ma in the Early Permian (Fig. 7) although arc magmatism seems to have been continuous on the west from the Late Carboniferous (300 Ma) (Nasi et al., 1985; Mpodozis and Kay, 1992). The discrimination of two subcycles (Early Permian and Late Permian–Early Triassic) has been proposed primarily on the available radiometric information (Ramos and Ramos, 1979; Caminos et al., 1982; Llambías and Caminos, 1987) rather than on field data and stratigraphic relations (Caminos, 1985). The presence,

however, of two plutonic-volcanic bimodal associations of (a) Late Carboniferous (320 ± 18 Ma)–earliest Permian (285 ± 4 Ma) and (b) Early Permian (276 ± 4 Ma)–Late Triassic (207 ± 9 Ma) ages in the magmatic arc terrains located in Chile between 27°S and 31°S (Nasi and Sepúlveda, 1986) west of the Calingasta-Uspallata Basin suggest that two subcycles could have represented the main pulses of magmatic activity. The younger plutonic-volcanic association (Super Unidad Elqui) intruded or unconformably rests on the lower one (Super Unidad Ingaguas), which correlates with the Choiyoi Group (Fig. 6 and 13) in the Argentine flank of the Frontal Cordillera, with Early Permian (275 ± 10 Ma) radiometric ages (Cortés, 1985). The Super Unidad Ingaguas comprises I-type granodiorites, granites, and granitic porphyries with a high SiO_2 content and relatively high $^{87}Sr/^{86}Sr$ initial ratios that suggest partial melting by underplating at the base of the crust or at shallow levels of the mantle (Parada et al., 1981). The Upper Sequence of the Gondwana Cycle (Fig. 6) represents the climax of calc-alkaline arc-volcanism and volcaniclastic sedimentation in nonmarine depositional environments. It rests unconformably (San Rafael orogenic phase) on Lower Sequence deposits along the orogenic flank of the Calingasta-Uspallata Basin (Cortés, 1985) and San Rafael Basin (Espejo, 1990), and it wedges out cratonward where sedimentation is continuous throughout the Early and early Late Permian in the main depocenters of the Paganzo Basin (Limarino and Cesari, 1987; Aceñolaza and Vergel, 1987). Radiometric dates suggest the onset of this vulcanism in the Calingasta-Uspallata and San Rafael Basins at about 275 Ma in the early Permian

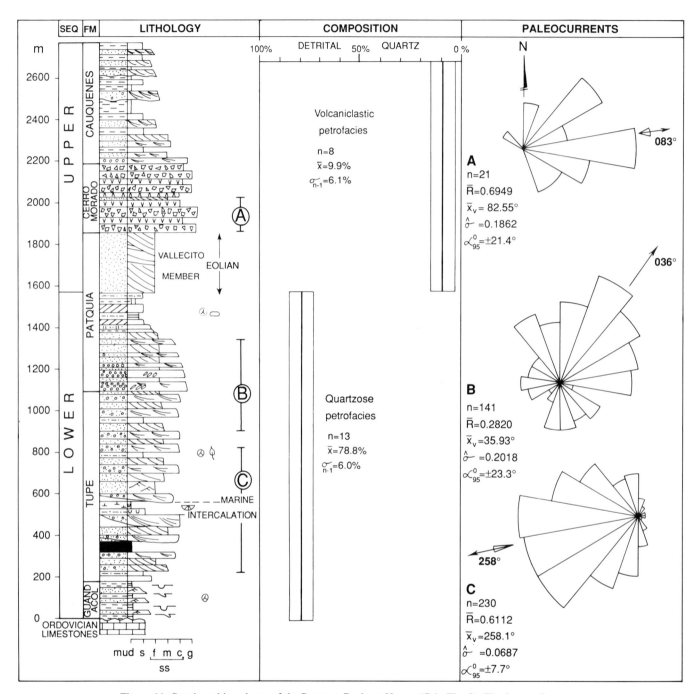

Figure 11. Stratigraphic column of the Paganzo Basin at Huaco (G in Fig. 9). The Lower Sequence comprises the lacustrine Guandacol, fluvial Tupe, and the lower, fluvial part of the Patquía Formation; the Upper Sequence comprises the eolian facies of the Patquía Formation (P. J. Conaghan, unpublished data), the Cerro Morado volcaniclastic breccias and basalt lavas, and the Cauquenes tuffaceous sandstones and mudstones. A cratonic quartzose petrofacies (Guandacol, Tupe and lower half of the Patquía) comprises the Lower Sequence, and a volcaniclastic petrofacies the Upper Sequence (López-Gamundí et al., 1990). As in the San Rafael Basin, the compositional change is coupled with a change in paleoslope (López-Gamundí et al., 1990; P. J. Conaghan and C. O. Limarino, unpublished data). See Figure 10 for symbols.

(Fig. 7) although the arc magmatism seems to have been continuous on the west from the Late Carboniferous (300 Ma) (Nasi and Sepúlveda, 1986). The presence of two plutonic-volcanic bimodal associations of Late Carboniferous–earliest Permian ages in the magmatic arc terrains located in Chile between 27°S and 31°S (Nasi et al., 1985; Nasi and Sepúlveda, 1986) west of the Calingasta-Uspallata Basin suggest that two subcycles could have represented the main pulses of magmatic activity. Furthermore, recent evidence suggests that the older subcycle represents arc magmatism formed along an active margin and contaminated with metasedimentary crustal components probably derived from the accretionary prism. The younger subcycle, characterized by hypersilicic granites and consanguineous volcanics, shows evidence of increasing crustal melting (Llambías and Sato, 1990; Mpodozis and Kay, 1990). This distinction is clear where the magmatic arc was located in the Chilean Andes. There, the oldest Guanta unit of the Ingaguas Superunit comprises S-type calc-alkaline tonalites and granodiorites with relatively high $^{87}Sr/^{86}Sr$ initial ratios (Mpodozis and Kay, 1990). The younger plutonic-volcanic association in the Ingaguas Superunit, I-type peraluminous granitoids and associated rhyolitic volcanics probably formed by melting of a thickened crust (Mpodozis and Kay, 1990), intrudes or unconformably rests on the lower Elqui Superunit. The Elqui Superunit correlates with the Choiyoi Group (Fig. 6 and 13) in the Argentine flank of the Frontal Cordillera and adjacent regions, with Early Permian (275 ± 10 Ma, Cortés, 1985; 267–264 Ma, Llambías and Sato, 1990) radiometric ages. The Ingaguas Superunit granitoids and rhyolites have high SiO_2 and high $^{87}Sr/^{86}Sr$ initial ratios and range in age from Late Permian to Early Triassic. They have been interpreted as indicating extensive crustal melting from partial melting by underplating at the base of the crust or at shallow levels of the mantle (Parada et al., 1981; Nasi et al., 1985; Mpodizis and Kay, 1990). Caminos (1985) and Ramos et al. (1986) have speculated about eastward-younging trends in (a) the Late Carboniferous ages of the granitoids and consanguineous volcanics of the western magmatic arc and in (b) the Late Permian–Early Triassic ages of the volcanics and volcanigenic sediments of the eastern foreland basin along the western Paganzo basin, coupled with a progressive enrichment of crustal components with decreasing age. This cratonward-younging trend is consistent with the diachronous eastward migration of the volcanic front after the San Rafael phase. Whereas Early Permian volcanics and associated volcaniclastics unconformably rest on Late Carboniferous–earliest Permian deposits in the outer portions of the back-arc foreland basins (Frontal Cordillera in Calingasta-Uspallata Basin and in the San Rafael Basin), the same volcanigenic association has a late Permian–earliest Triassic age away from the magmatic arc (western Paganzo Basin) and overlie Early to early Late Permian continental deposits (Fig. 7).

In summary, arc volcanism and its aftermath, identified as the Choiyoi Granite–Rhyolite Province (Kay et al., 1989), cli-

maxed during the late Early and Late Permian, as indicated by the changed configuration of the depositional areas and their flooding by volcanic detritus (Fig. 13).

The radically different paleogeography of the volcanogenic stage was affected by uplift of the arc region west of the Lower Sequence basins and withdrawal of the sea. The arc volcanism covered extensive areas of western Argentina, as shown from the present outcrops in a south-trending belt that bends southeastward at 36°S (Fig. 13), in sympathy with the trend of the preceding Late Carboniferous–Early Permian calc-alkaline granitoids of Coastal Chile (Fig. 12; Beck et al., 1991). The granitoids are intruded into an accretionary prism located outboard of the continent and which underwent contemporaneous metamorphism (Hervé et al., 1988). The southeasternmost exposures of volcanics of similar age and composition are located only 250 km west of the Sierras Australes of the Buenos Aires Province (Fig. 13) (Linares et al., 1980).

The volcanics, seldom associated with sediments, have different local names. The most widely used lithostratigraphic term is the Choiyoi Group (Rolleri and Criado-Roque, 1969). They crop out in the west within the Frontal Cordillera, on the western flank of the Precordillera, and further south in the San Rafael Block, and rest with angular unconformity on pre-Carboniferous to earliest Permian rocks. They represent the consanguineous volcanic cover of the granitoids and rhyolitic volcanics of the Super Unidad Ingaguas and Pastos Blancos Formation (Nasi and Sepúlveda, 1986). They consist of andesitic, rhyolitic, and dacitic lavas, breccias and pyroclastics; and subordinate basalts and basaltic andesites in a trend from mafic rocks at the base to acidic rocks above. In the Frontal Cordillera the volcanics (Portezuelo del Cenizo Formation) rest with angular unconformity on Late Carboniferous marine rocks (Yalguaraz Formation). A K/Ar radiometric date of 275 ± 10 Ma (Vilas and Valencio, 1982) suggests an Early Permian age for the volcanic complex. The equivalent rocks in the San Rafael Basin, the Cochicó Group (Fig. 6), unconformably overlies the Devonian Horqueta and Río de los Castaños Formations and the Late Carboniferous–earliest Permian El Imperial Formation (Fig. 10). The basal Cochicó Group (Dessanti and Caminos, 1967) comprises alluvial-fan, low-sinuosity fluvial deposits and subordinate eolian facies that grade upward to pyroclastic deposits. It represents the basal sedimentary fill and associated volcanism after the uplift and tilt of the Lower Sequence. Radiometric dates of the volcanics suggest an Early Permian age (between 270 and 280 Ma) (Polanski, 1966; Dessanti and Caminos, 1967; Toubes and Spikermann, 1976; Linares et al., 1979), which we interpret as an age younger than the post-*Gangamopteris* zone unconformity (Appendix 1). The succeeding Cerro Carrizalito Group of basalt, andesite, dacite, and more acidic volcanic types with associated volcanic breccias (González-Díaz, 1972) is dated as Late Permian–Early Triassic (between 240 and 268 Ma) (Valencio and Mitchell, 1972; Toubes and Spikermann, 1976;

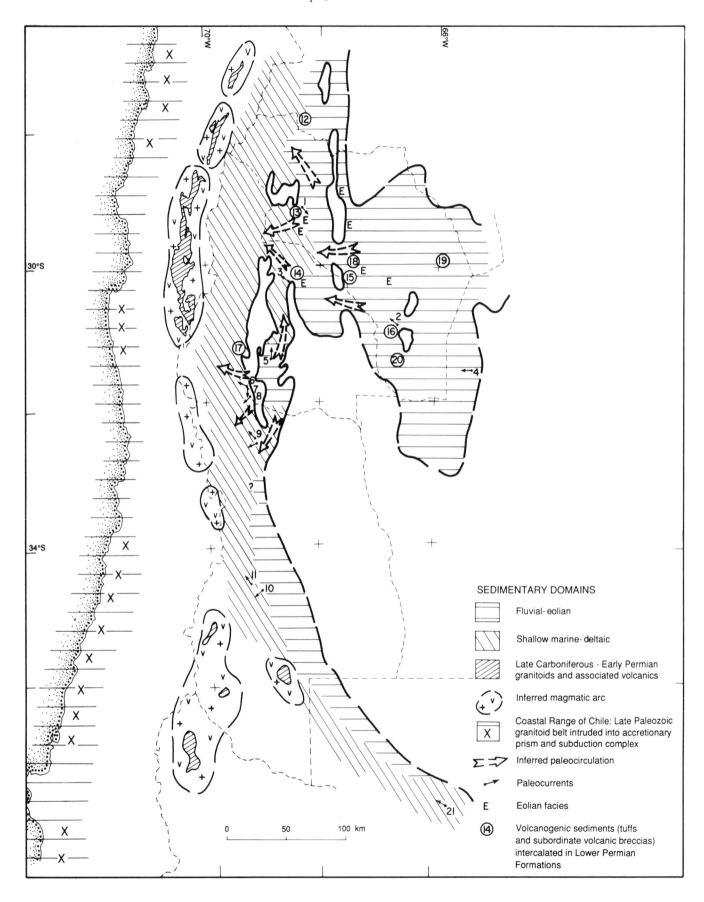

SEDIMENTARY DOMAINS

Fluvial- eolian

Shallow marine- deltaic

Late Carboniferous - Early Permian granitoids and associated volcanics

Inferred magmatic arc

Coastal Range of Chile: Late Paleozoic granitoid belt intruded into accretionary prism and subduction complex

Inferred paleocirculation

Paleocurrents

E   Eolian facies

Volcanogenic sediments (tuffs and subordinate volcanic breccias) intercalated in Lower Permian Formations

Linares et al., 1979). To the east, the Sierra Pintada Group is a plutonic-volcanic complex that rests on the Pampean craton. In the western Paganzo Basin the volcanigenic rocks rest conformably on the Patquía Formation (Fig. 7) and are called the Cerro Morado and Cauquenes Formations (Borrello and Cuerda, 1965) in the Huaco area (Fig. 11) and Del Aspero Group (Furque, 1963) in the Guandacol region. They comprise lava flows and volcanic agglomerates and breccias intercalated with alluvial-fan and low-sinuosity fluvial deposits.

In contrast, Late Carboniferous–Triassic magmatic rocks in northern Chile seem to have formed in a geotectonic transition zone between the Late Paleozoic active margin in the south and the intracontinental rift zone east of the Arequipa Massif in Bolivia and southern Peru (Breitkreuz et al., 1989).

Paleogeographic and paleotectonic evolution: Interplay of arc- and craton-derived detritus in foreland basins. The evolution of the foreland basins of western Argentina is synthesized in two stages which correspond to the Lower and Upper Sequences described above. Studies of sandstone composition indicating the provenance rock type were complemented by paleocurrent studies indicating the provenance location. Three main compositional suites or petrofacies (in the sense of Dickinson and Rich, 1972) are identified for the rocks of the Gondwana cycle. These petrofacies are similar to those defined by López-Gamundí and Espejo (1987) and integrate the modal results of point-count analyses made on the sand-sized fraction of sediments of the Lower and Upper Sequences throughout the Calingasta-Uspallata, San Rafael, and Paganzo Basins. The petrofacies defined are the following:

(a) Quartzofeldspathic: high content of monocrystalline quartz (Qm: 60%–90%) and K-feldspar (K: 5%–25%) and low content of strained polycrystalline quartz (Qp), plagioclase (P), metamorphic (Lm) and sedimentary (Ls) lithic grains. Equivalent to Folk's (1980) subarkose and arkose (K-feldsarenite).

(b) Lithic (sedimentary): high content of sedimentary and metasedimentary lithics (Ls: 20%–60%) and monocrystalline quartz (Qm: 20%–60%), low content of K-feldspar (K: 5%–20%). Folk's (1980) litharenite.

(c) Volcaniclastic: rich in volcanic lithics (Lv: 30%–60%) and plagioclase (P: 30%–50%), low in monocrystalline quartz (Qm: <20%) and K-feldspar (K: <10%). Folk's (1980) feldspathic (plagioclase) litharenite.

Sandstones of the three petrofacies were plotted on the QFL and QmFLt triangular diagrams of Dickinson and Suczek (1979) modified by Dickinson et al. (1983) (Fig. 14, A and A'). The Lower Sequence comprises the quartzofeldspathic and lithic petrofacies (Fig. 15).

The *quartzofeldspathic petrofacies,* dominant in the Paganzo Basin, is related to plutonic and high-grade metamorphic complexes of the craton and fault-bounded uplifted basement of Pampean type that border much of the Paganzo Basin (López-Gamundí and Espejo, 1987). Detailed compositional and paleocurrent studies (Teruggi et al., 1969; Andreis et al., 1975; Espejo and López-Gamundí, 1984; López-Gamundí et al., 1986; Ruzycki and Bercowski, 1987; López-Gamundí and Espejo, 1988; Conaghan and Limarino, 1987, unpublished data) provide further evidence of the derivation of these deposits from the craton. Samples of this petrofacies fall in the continental block field of Dickinson et al. (1983) (Fig. 14, B and B').

The *lithic petrofacies* dominates in the Calingasta-Uspallata Basin and in parts of the westernmost Paganzo Basin. Paleocurrent data suggest that the Proto-Precordillera acted not only as a barrier to the westward dispersal of craton-derived material but also as a sediment source for adjacent areas of deposition (López-Gamundí and Alonso, 1982). The Proto-Precordillera dominated the paleogeography during the Early to mid-Carboniferous. Although bypassed locally by the fluvial/deltaic system in the Late Carboniferous–earliest Permian, the Proto-Precordillera remained an important control for the rest of the Lower Sequence, as evidenced by N-S paleoflow patterns that parallel its structural grain (López-Gamundí and Espejo, 1987). The tectonic significance of the Proto-Precordillera is still in debate. Geotectonic scenarios include a continent-continent collision orogen (Ramos et al., 1986) or a foreland fold-thrust belt in an ocean-continent collision (López-Gamundí and Sessarego, 1988). The mostly sedimentary nature of the Proto-Precordillera (Cambrian-Ordovician limestones, Ordovician turbidites with pillow lavas, Silurian shallow marine mudstones and sandstones, Lower to Middle Devonian turbidites) is reflected in the composition of the Lower Sequence sediments, irrespective of their depositional

Figure 12. Late Carboniferous–Early Permian paleogeography of the Lower Sequence. Two main sedimentary domains have been identified (Sessarego and Limarino, 1987): on the east a fluvial-eolian domain which covered the Paganzo Basin and eastern region of the San Rafael Basin, on the west a shallow marine-deltaic domain covering the Calingasta-Uspallata Basin and western part of the San Rafael Basin. Fluvial drainage (indicated by open arrows) of the Paganzo Basin to the western marine domain was through three areas around 28°S, 30°S, and 32°30'S. Paleocurrents from the southeast in Early Permian fluvial deposits on the southeasternmost edge of the San Rafael Basin suggest an expansion of deposition on the Chadileufu Massif. Carboniferous-Permian granitoids and volcanics from Nasi and Sepúlveda (1986); granitoids in Coastal Range of Chile from Beck et al. (1991) and Hervé (1988). Source of data: 1—Andreis et al. (1975); 2—Andreis et al. (1986); 3—Bossi and Andreis (1986); 4—Hünicken and Pensa (1981); 5—Espejo and López-Gamundí (1984); 6—Damborenea (1974); 7—López-Gamundí (1984); 8—López-Gamundí (1984); 9—López-Gamundí (1989, unpublished data); 10 and 11—Espejo (1990); 12—Turner (1967); 13—Cuerda (1965) and Teruggi et al. (1969); 14—Bracaccini (1946) and Limarino et al. (1986); 15—Azcuy et al. (1979); 16—Alvarez and Fernández-Seveso (1987); 17—Manceñido (1973); 18—Di-Paola (1972); 19—Coira and Koukharsky (1970); 20—Ramos (1982); 21—Melchor (1990).

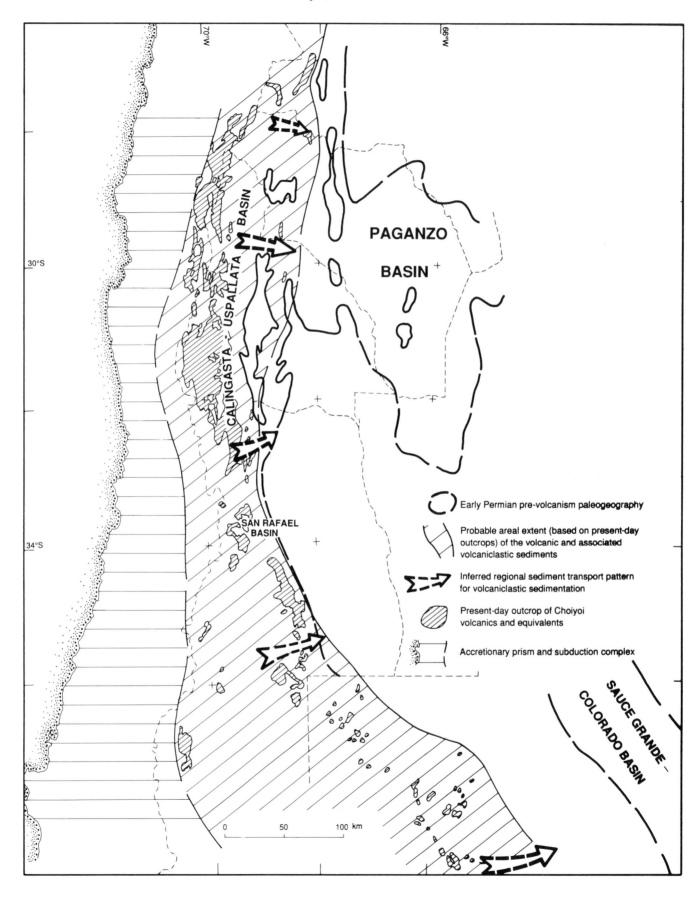

PAGANZO

BASIN

CALINGASTA - USPALLATA   BASIN

SAN RAFAEL
BASIN

SAUCE GRANDE -
COLORADO BASIN

30°S

34°S

70°W

69°W

Early Permian pre-volcanism paleogeography

Probable areal extent (based on present-day outcrops) of the volcanic and associated volcaniclastic sediments

Inferred regional sediment transport pattern for volcaniclastic sedimentation

Present-day outcrop of Choiyoi volcanics and equivalents

Accretionary prism and subduction complex

0   50   100 km

environments. From a provenance study in the Barreal area, López-Gamundí and Alonso (1982) concluded that the Proto-Precordillera terrain was a recycled orogen (Fig. 14, C and C'). The lithic petrofacies is similar to those derived from recycled orogens, specifically from back-arc thrust-belt sources rich in quartz and sedimentary lithics and low in K-feldspar and volcanic lithics (Dickinson, 1985).

On the eastern flank of the Proto-Precordillera (location 6 in Fig. 9), lithic (sedimentary) petrofacies and quartzofeldspathic petrofacies are present in Carboniferous deposits (Bercowski, 1987), indicating the interplay of sedimentary detritus from the Proto-Precordillera on the west and cratonic detritus of a crystalline nature (intrusives and high-grade metamorphics) from the Sierras Pampeanas to the east. Paleocurrent data support the location of the source areas of the Calingasta-Uspallata Lower Sequence to the east.

Farther north, Scalabrini-Ortiz (1972) postulated a similar type of source area for the Carboniferous sediments exposed in the northern Calingasta-Uspallata Basin. He found significant amounts of sedimentary lithic fragments (sandstone, siltstone, and limestone) and subordinate amounts of mafic volcanic lithic fragments in the sand-sized fraction of the Carboniferous rocks of the Malimán area (Fig. 9) and concluded that the material might have been derived from the Proto-Precordillera to the east. Following the ideas of Amos and Rolleri (1965) and Rolleri and Baldis (1969), he also postulated that the Proto-Precordillera acted as a barrier for the east-derived cratonic detritus dominant in the Paganzo Basin.

In the San Rafael Basin, a mixed quartzofeldspathic-lithic petrofacies (Fig. 14, C and C') contains a Pampean-type component (abundant monocrystalline quartz, K-feldspar of plutonic origin, and lithics of microgranites and aplites) admixed with a significant content of sedimentary and metasedimentary lithics (fine micaceous sandstone, slate, and phyllite) derived from the Devonian units (Espejo, 1986a, 1986b, 1990).

The lack of a clear-cut distinction between a Pampean-type cratonic provenance and a Proto-Precordillera-type orogen provenance in the El Imperial Formation sandstones might be related either to poor knowledge of the configuration and related paleogeographic aspects of the basin or to a simpler picture involving the juxtaposition of the Pampean and Proto-Precordillera source types (possibly with the cratonic source

Figure 13. Early Permian–Early Triassic paleogeography of the Upper Sequence. The Early Permian (?Late Carboniferous)–Early Triassic magmatic arc is inferred from the present-day extent of volcanic and associated volcaniclastic deposits of the Choiyoi Group and equivalents. The Late Carboniferous–Early Permian (pre-Choiyoi) configurations of the Calingasta-Uspallata, Paganzo, San Rafael, and Sauce Grande–Colorado Basins are drawn in heavy lines. Note invasion from the west of volcanics and associated deposits overlapping the Late Carboniferous–Early Permian paleogeography.

veneered by sediment) with no evidence of these source types acting independently (Espejo, 1986b, 1990). The latter inference is supported by the fact that the modal compositional analyses of the sandstones in the El Imperial Formation are similar to Recent sands blended from the cratonic plus orogenic belt provenance fields as defined by Dickinson and Valloni (1980).

A *transitional petrofacies* (Fig. 15), with variable amounts of monocrystalline quartz and volcanic lithics and restricted to a short stratigraphic interval around the Lower-Upper Sequence transition, can be identified in several localities of the Paganzo, Calingasta-Uspallata, and San Rafael Basins. The transition is due to local reworking of volcanic material in areas of lithic or quartzofeldspathic petrofacies.

The *volcanic petrofacies* (Fig. 15) succeeds the transition. By the Early Permian, the volcanic activity of the magmatic arc along the Paleo-Pacific margin extended its influence to the foreland basins and marked the transition between the Lower and Upper Sequences of the Gondwana Cycle. Both petrofacies of the Lower Sequence (quartzofeldspathic on the east, lithic on the west) are replaced by the volcaniclastic petrofacies, reflecting the increasing role of arc-volcanism in the foreland basins (Fig. 14, D and D'). The change is accompanied by (1) a reversal of regional paleoslope/paleocurrents, (a) in the Paganzo Basin, from the east in the Lower Sequence, from the west in the Upper Sequence, and (b) in the San Rafael Basin from south to north in the Lower Sequence and from the north in the Upper Sequence (Figs. 7 and 15); (2) a consequent change in sandstone composition: from quartzofeldspathic-lithic to volcaniclastic (Fig. 15); (3) the sea in the west replaced by nonmarine deposition in small basins of internal drainage (*see* Post-Gondwana History section).

Eastern, Sauce Grande–Colorado Basin

Sedimentary sequences. The Early Carboniferous in the North Patagonian Massif and adjacent areas of the Chadileufu Massif and Ventana foldbelt saw the emplacement of granitoids and the subordinate extrusion of consanguineous volcanics followed by uplift and erosion (Cingolani and Varela, 1973; Llambías et al., 1984; Caminos et al., 1988; Rapela and Kay, 1988; Rapela et al., 1992). The magmatism, called the Somuncura phase (Ramos and Ramos, 1979), peaked between 350 and 330 Ma. An extension of the igneous rocks northward into the Sierras Pampeanas is interpreted as a north-trending magmatic tract ("Inner Cordillera" of Rapela and Kay, 1988) that paralleled the western magmatic arc. Only by the latest Carboniferous (290 Ma) did the Ventana region subside to form the Sauce Grande–Colorado Basin.

This is approximately the time that sedimentation started with glacigene deposits north of this basin in the Chaco-Paraná Basin (Russo et al., 1979; Russo and Archangelsky, 1987) and its intracratonic northeastern arm, the Paraná Basin (Daemon and Quadros, 1970; Zalán et al., 1987). In these two

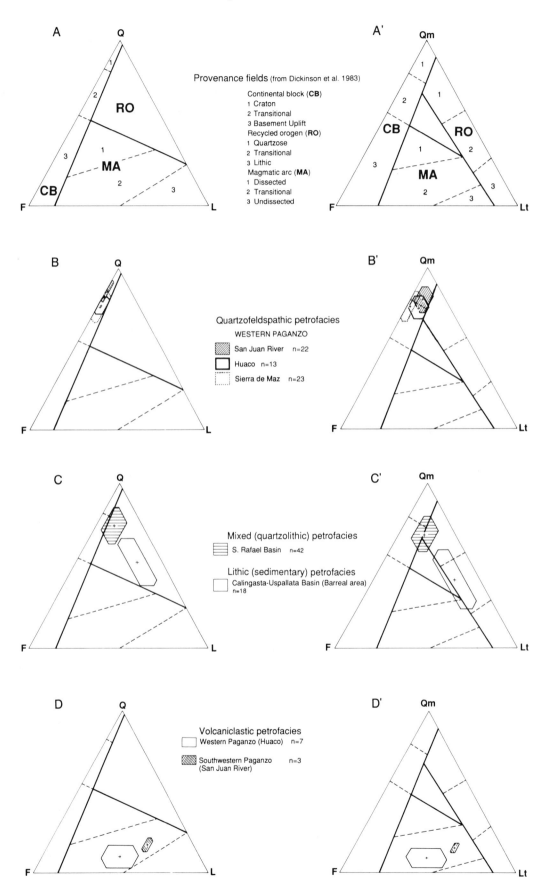

Provenance fields (from Dickinson et al. 1983)

Continental block (**CB**)
1 Craton
2 Transitional
3 Basement Uplift
Recycled orogen (**RO**)
1 Quartzose
2 Transitional
3 Lithic
Magmatic arc (**MA**)
1 Dissected
2 Transitional
3 Undissected

Quartzofeldspathic petrofacies
WESTERN PAGANZO
San Juan River　n=22
Huaco　n=13
Sierra de Maz　n=23

Mixed (quartzolithic) petrofacies
S. Rafael Basin　n=42

Lithic (sedimentary) petrofacies
Calingasta-Uspallata Basin (Barreal area)
n=18

Volcaniclastic petrofacies
Western Paganzo (Huaco)　n=7
Southwestern Paganzo　n=3
(San Juan River)

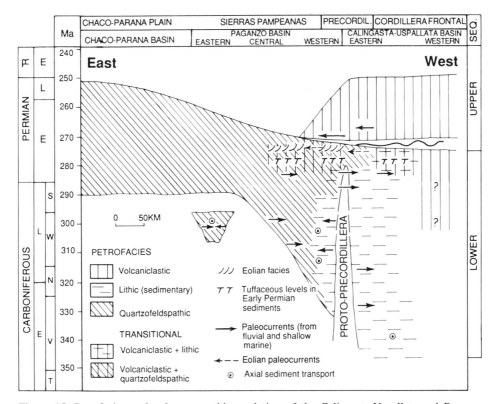

Figure 15. Petrofacies and paleogeographic evolution of the Calingasta-Uspallata and Paganzo Basins shown on the same base as Figure 7. The three petrofacies (López-Gamundí and Espejo, 1987) are (a) quartzofeldspathic petrofacies in the craton-derived sediments of the Paganzo Basin (Teruggi et al., 1969; López-Gamundí et al, 1986; Bercowski, 1987; López-Gamundí and Espejo, 1988); (b) lithic (sedimentary) petrofacies derived from the emergent Proto-Precordillera and identified in the eastern Calingasta-Uspallata Basin and, to a lesser degree, in the westernmost Paganzo Basin (Scalabrini-Ortiz, 1972; López-Gamundí and Alonso, 1982; Bercowski, 1987); and (c) volcaniclastic petrofacies transported from the magmatic-arc terrains by eastward paleocurrents (López-Gamundí et al., 1990; P. J. Conaghan and C. O. Limarino, 1987, unpublished data), dominant on the arc from the Late Carboniferous through Early Triassic. A transitional petrofacies indicates local reworking of tuffs intercalated in Permian sediments (Teruggi et al., 1969; Manceñido, 1973; Limarino et al., 1986) and westerly derived volcanic detritus in eolian beds (Limarino, 1984; Limarino and Spalletti, 1986; López-Gamundí et al., 1986, 1990). The petrofacies evolve through a convergence of a quartzofeldspathic or lithic (sedimentary) suite to a single westerly derived volcaniclastic assemblage (López-Gamundí and Espejo, 1987).

basins, distant from the continental margin, sedimentation, not significantly affected by the tectonic and magmatic activity along the Gondwanan continental margin, continued without major breaks throughout the rest of the Paleozoic. The Chaco-Paraná Basin is characterized by mostly nonmarine basal deposits. Late Carboniferous (Stephanian) to earliest Permian glacigene deposits seem to be widespread, as found in bore-

holes (Russo and Archangelsky, 1987). These deposits are succeeded by late Early and Late Permian continental deposits with rare thin marine intercalations (Archangelsky and Vergel, 1991). In contrast, the Paraná Basin initial fill seems to be mostly glacial marine (França and Potter, 1991), which passed into deltaic and finally continental environments later in the Permian and Triassic (Zalán et al., 1990).

The Gondwana Cycle in the Sauce Grande–Colorado Basin (Figs. 1 and 16) comprises the Pillahuincó Group (Harrington, 1947) of glaciomarine, marine, and probably nonmarine sediment that ranges in age from latest Carboniferous to Early or Late Permian. The Pillahuincó Group lies with regional unconformity on the Early Devonian Lolén Formation and is unconformably overlain by probably Miocene conglomerate in the Sierras Australes of the Buenos Aires Province and by

◄──────────────

Figure 14. Compositional analyses of the sandstones of the Calingasta-Uspallata, San Rafael, and Paganzo Basins. QFL (A) and QmFLt (A′) diagrams are taken from Dickinson et al. (1983). The following petrofacies are defined for the Lower Sequence of the three basins: quartzofeldspathic (B, B′), lithic-sedimentary and mixed (quartzofeldspathic-lithic) (C, C′), and volcaniclastic (D, D′).

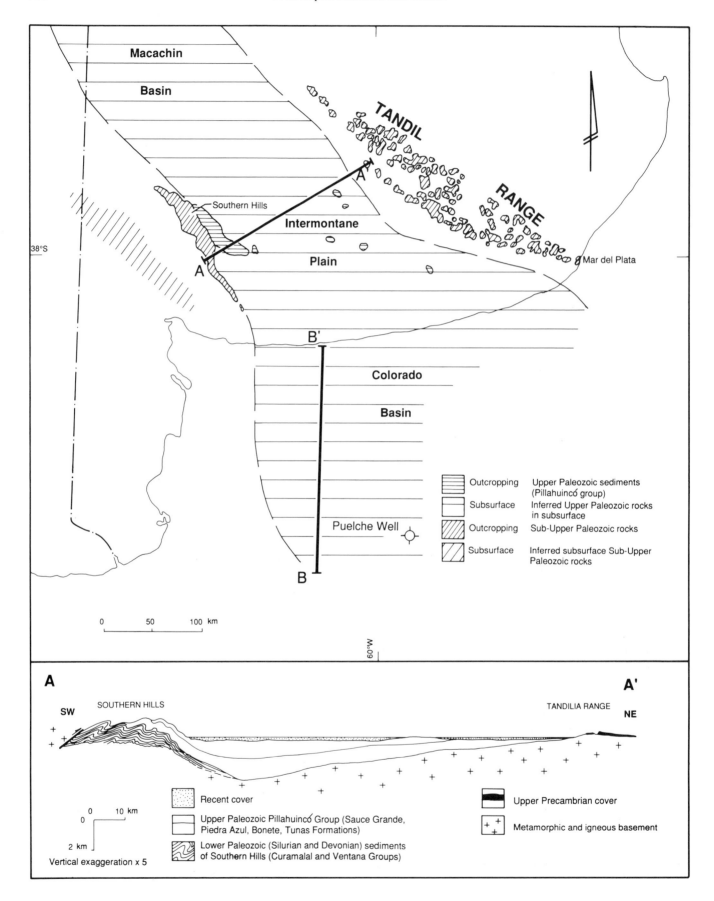

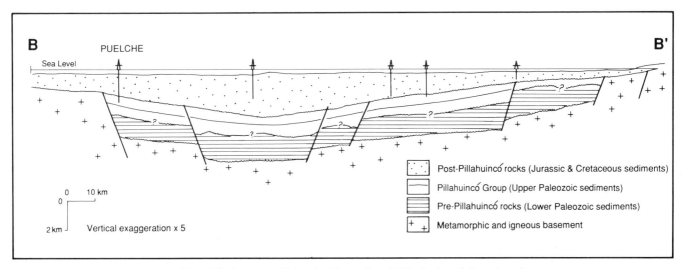

Figure 16 (on this and facing page). Colorado (Sauce Grande) Basin. Late Paleozoic sediments crop out in the Sierras Australes of the Buenos Aires province (Pillahuincó Group, Harrington, 1972) and the intermontane plain (Zambrano, 1974, 1980; Llambías and Prozzi, 1975) and are inferred in the subsurface adjacent to the Sierras Australes onshore in the Colorado and Macachín Basins (Russo et al., 1979; Zambrano, 1974, 1980), and offshore in the Colorado Basin (Lesta et al., 1980). Sections AA' and BB' from Lesta et al. (1980).

Cretaceous sediment in the offshore Colorado Basin (Fig. 16, BB', Puelche well; Lesta et al., 1980). The basal unconformity is variously described as an "imperceptible" regional unconformity (Harrington, 1947) and as a gentle angular discordance (Massabie and Rossello, 1984). The Pillahuincó Group (Sedimentary Cycle III of Andreis et al., 1989) in the Sierras Australes of Buenos Aires Province (Fig. 16) forms a broad northeastward-dipping homocline with superimposed folds which become broader to the northeast (Cobbold et al., 1986). Isolated exposures of the Pillahuincó Group between the Sierras Australes and the Tandilia Range to the northeast (Furque, 1965; Arrondo et al., 1982) and near Mar del Plata (Maack, 1951) define the approximate northeastern margin of the basin. In the Puelche offshore well (Fig. 16 BB'), the basal diamictites of the Gondwana succession (Sauce Grande Formation in the area of the Sierras Australes) were penetrated between 3,598 and 4,063 m below sea level (Lesta et al., 1980; Amos and López-Gamundí, 1981). Subsurface studies suggest Late Paleozoic sediments in the Macachín Basin northwest of the Sierras Australes (Russo et al., 1979). Consequently the total area of Gondwanan sedimentation includes the Macachín Basin to the north, the Colorado Basin to the south, and the southern foothills of the Tandilia Range to the northeast (Fig. 16). The Pillahuincó Group is subdivided into four formations (Harrington, 1947) (Fig. 17), in ascending order, the Sauce Grande, Piedra Azul, Bonete, and Tunas Formations. The age ranges from latest Carboniferous to latest Permian, and the thickness from 2,000 to 4,500 m. The diamictite-bearing Sauce Grande, Piedra Azul, and Bonete Formations (succession 1) (Fig. 17) thin to the north (Harrington, 1947; Suero,

1957; Andreis, 1984; Andreis et al., 1989). The Sauce Grande Formation thins to the north-northeast from 900 m to less than 500 m in a distance of 35 km (Harrington, 1980). In the Piedra Azul Formation, facies changes from sandstone to mudstone (Harrington, 1980) coupled with thickness variations (140–200 m) (Andreis et al., 1989) indicate a proximal-distal trend in the same general direction.

Glaciation and glacioeustatic sea-level rise. Sedimentation in the Sauce Grande–Colorado Basin started with the Sauce Grande Formation (Fig. 17), comprising diamictite (74%), sandstone (24%), and shale (2%) and ranging in thickness from 500 m to 1,100 m (Andreis et al., 1987). The lower half to two-thirds of the Sauce Grande Formation comprises massive diamictite with scattered lenses of sandstone near the base and top. The upper part consists mainly of diamictite with interbeds of rippled sandstone and conglomerate (Amos and López-Gamundí, 1981). Thin shale intervals contain abundant dropstones. At the Puelche offshore well (Fig. 16, BB') diamictites at the base of the well lie beneath 1168 m of shale, siltstone and claystone with intercalated thin sandstone beds assigned to the Piedra Azul and Bonete Formations (Lesta et al., 1980). Spores found in the diamictite-bearing section of the same well suggest a latest Carboniferous–earliest Permian age for the Sauce Grande Formation (Archangelsky et al., 1987). Detailed facies studies (Coates, 1969; Andreis, 1984) indicate abundant soft-sediment deformation associated with the massive diamictites. Faceted and striated clasts have been reported (Keidel, 1916; Du Toit, 1927; Harrington, 1947; Suero, 1957; Andreis, 1965; Coates, 1969). The diamictites are interpreted as glaciomarine from the presence of a solitary

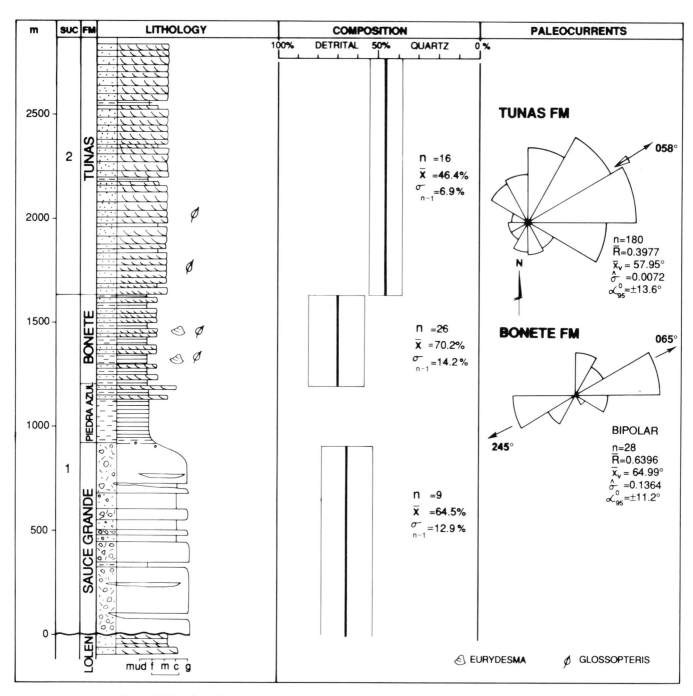

Figure 17. Stratigraphic column for the Sierras Australes of Buenos Aires Province. The Gondwana Cycle comprises four formations: Sauce Grande (glaciomarine and gravity-flow diamictites with rare conglomerates, sandstones, and mudstones), Piedra Azul (marine mudstones), Bonete (shales and cross-bedded sandstones with marine bivalves and plant remains), and Tunas (shallow marine to nonmarine cross-bedded sandstones and subordinate mudstones). The bipolar paleocurrent pattern in the Bonete Formation is reconstructed from Andreis et al. (1979). Paleocurrents in the Tunas Formation are based on unpublished data (1987) from P. J. Conaghan. See Figure 10 for symbols.

marine bivalve (*Astartella? pusilla*) found in the diamictites (Harrington, 1955), and are influenced by sediment gravity flows (Coates, 1969; Harrington, 1980; Andreis, 1984). The upper beds of the Sauce Grande diamictite are transitional to the mudstone of the Piedra Azul Formation (Fig. 17). Harrington (1947) drew the boundary above the highest megaclast in any given section. The gradational transition led Andreis (1965) to propose the integration of the Sauce Grande and Piedra Azul Formations in a single lithostratigraphic unit. The Piedra Azul Formation comprises mudstones and rare cross-bedded fine sandstone and grades upward to the Bonete Formation, bioturbated shale and rippled and cross-bedded sandstone with the *Eurydesma* fauna (González, 1985) of Tastubian age (Appendix 1) and the *Glossopteris* flora (Archangelsky and Cúneo, 1984). Bipolar paleocurrents (Fig. 17) suggest a shallow marine environment with a shoreline running northwest to southeast (Andreis et al., 1979). The fining-upward trend of the Sauce Grande–Piedra Azul–Bonete Formations has been interpreted as the result of the Tastubian glacioeustatic sea-level rise (Dickins, 1984).

Paleocurrent data from the Sauce Grande Formation are complex: the cross-dip azimuth is erratic; ripples have mean crest azimuths of east-northeast to west-southwest and east-southeast to west-northwest (Coates, 1969); and conglomerate channel fills in the diamictite are oriented east to west. The Bonete Formation has a bipolar paleocurrent pattern with northeastward and southwestward azimuths (Andreis et al., 1979; Fig. 17). The sand fraction of the Sauce Grande–Piedra Azul–Bonete succession lies within a compositional field (in the terminology of Folk, 1980) that includes quartz-rich feldsarenites or subarkoses (with individual samples in the quartzose feldsarenite subdivision) and quartz-rich feldsarenites or arkoses (Andreis, 1965; Andreis et al., 1979; Andreis and Cladera, 1992) (Fig. 17). The clasts in the gravel fraction of the Sauce Grande diamictite comprise quartzite (53%), granite (19%), and vein quartz (8%) (Coates, 1969). Despite the widely dispersed paleocurrent azimuths, a northeastern provenance of the Sauce Grande–Piedra Azul–Bonete succession can be inferred from the lithological similarity between its gravel and sand fractions and the cratonic complexes of the Tandilia Massif, even though alternative source areas of similar composition, such as the Pampean terrains on the northwest, cannot be ruled out.

Volcanism/uplift on the southwest and subsidence in the adjacent foreland basin. In contrast to the glaciomarine sedimentation from a NE cratonic provenance and postglacial transgression of the Sauce Grande–Piedra Azul–Bonete succession 1, the Tunas Formation (succession 2) has a less quartzose petrofacies, a northeastward paleocurrent pattern and a deltaic to nonmarine environment of deposition. The paleocurrent and petrofacies data suggest a provenance from the southwest (Fig. 17). The Tunas Formation sandstone has a lower content of quartz and a higher percentage of feldspar (especially plagioclase) and lithics (especially volcanic lithics) than the quartzose

suite of the underlying succession (Fig. 18). Volcanism and uplift present in the adjacent terrain to the southwest explain these changes (Figs. 17 and 18). Furthermore, tuffaceous horizons in the Tunas Formation interpreted as ash falls (Iñiguez et al., 1988) provide evidence of volcanism in adjacent areas during sedimentation. As seen already (Fig. 13), Permian magmatism in western Argentina stretched southeastward to within 250 km of the Sauce Grande–Colorado Basin (Linares et al., 1980). As in the Chilean-Argentine Andes, a clear trend of increasing crustal participation with decreasing age is present in northern Patagonia where two magmatic episodes are identified in the Early–early Late Permian (280–260 Ma) and the Early Triassic (240–230 Ma), both correlatable with the Choiyoi Group. Rocks of both episodes are calc-alkaline, but those of the younger episode become peraluminous, with a tendency to a peralkaline composition (Caminos et al., 1988). Recent studies on the Somuncura batholith of the North Patagonian Massif indicate that several intrusives previously considered of Early and mid-Carboniferous age (Caminos et al., 1988) now have Late Permian ages (Pankhurst et al., 1992) and should be considered part of the younger magmatic episode.

Forsythe (1982) and Uliana and Biddle (1987) relate this volcanism to an Andean-type margin with a wide magmatic-arc/back-arc system southwest of the Sierras Australes, which

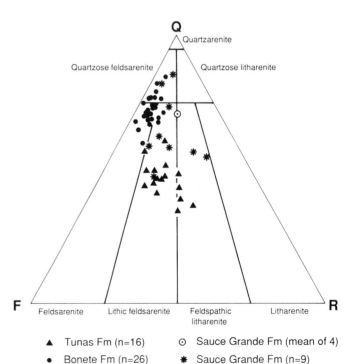

Figure 18. QFR plot of sandstones from Sierra Australes, southern Buenos Aires province. Bonete Formation data (filled circle) from Iñiguez and Andreis (1971) and Sauce Grande Formation (asterisk) from Andreis (1965), with semiquantitative estimate of silt and sand fractions from four thin sections (encircled dot) (Coates, 1969). Tunas Formation compositions (triangle) are from P. J. Conaghan (1987, unpublished data).

we reconstructed (Fig. 13) from the fragmentary outcrops of the Choiyoi Group and equivalents in a belt that bends southeastward at 34°S to 36°S, and with the Calingasta-Uspallata, San Rafael, and Sauce Grande–Colorado Basins behind the arc. The same trend can be seen along central Chile where the Late Carboniferous–Permian belt of calc-alkaline granitoids Hervé et al., 1988) turns around 38°S to trend southeastward along northern Patagonia, probably tracing out the old continental margin (Hervé, 1988). This wide-arc/back-arc system has been subdivided into an outer Cordilleran arc with metaluminous granitoids and an inner arc ("Somuncura inner Cordilleran arc") with peraluminous and weakly metaluminous granitoids and silicic volcanics (Cingolani et al., 1991). This model is consistent with the Andean-type margin proposed by Forsythe (1982) and Uliana and Biddle (1987), among others, and accounts for the extensive magmatic activity as much as 400 km from the paleo-trench. The change in orientation, along with the absence of Late Paleozoic magmatic rocks in Chile south of 38°S (Beck et al., 1991), suggests that the Late Paleozoic magmatic arc and the subsequent Choiyoi magmatism developed along the old Gondwanaland margin, which was broadened after the Late Paleozoic by terrane accretion on its southwestern border. In contrast, Ramos (1984) related the magmatism south of the Sauce Grande–Colorado Basin to a southwestward-dipping subduction zone under a separate Patagonian plate that collided with South America. The suture of this postulated collision has not been found and no reliable paleomagnetic, paleoclimatic, or paleobiogeographic data support the idea of an allochthonous Patagonia and its collision with Gondwanaland by the latest Paleozoic, as originally proposed by Ramos (1984). Furthermore, the available information from radiometric dates and the continuity of the volcanic belt favors the explanation of igneous activity southwest of the Sauce Grande–Colorado Basin as part of arc-magmatism on an Andean-type margin (Fig. 19).

Furque (1973), Harrington (1947, 1980), and Conaghan et al. (in preparation) suggest an upward transition at the base of the Tunas Formation from shallow marine in the underlying Bonete Formation to fluvial and possibly eolian conditions at the top. Andreis and Japas (1991) envisage a deltaic-estuarine environment for the upper part of the Tunas Formation, and indicate a general paleocurrent direction toward the north and northeast, as in our data (Fig. 17). The regression, due to volcanism and associated uplift in the southwest, drove a wedge of lithic (including volcanolithic) sediment northeastward. Volcanism and uplift in the southwest were matched by rapid subsidence in the foreland basin to accommodate the 1,500 m of the Tunas Formation reconstructed by Conaghan from Japas's (1986) cross section, and the 2,500 m estimated by Suero (1957) by removing the subsequent deformation. The syntectonic nature of the Tunas sedimentation is evidenced at two different scales: at a regional scale the Permian sediment thicken from southwest to northeast; and at outcrop scale, Cobbold et al. (1991) describe growth folds with thick de-

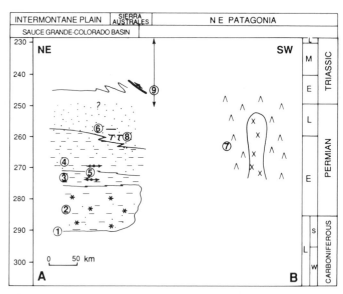

Figure 19. Time-space diagram (AB) for the Sauce Grande–Colorado Basin and NE Patagonia located in Fig. 20 and oriented such that the magmatic arc lies on the right. Sedimentation commenced (1) about the latest Carboniferous (Lesta et al., 1980; Archangelsky and Gamerro, 1981) with glaciomarine sediments of the Sauce Grande Formation (2) (Harrington, 1947; Coates, 1969; Amos and López-Gamundí, 1981; Andreis, 1984), succeeded as a transition by mudstones of the Piedra Azul Formation (3), which in turn grades upward to mudstones and fine sandstones of the Bonete Formation (4). The Piedra Azul and Bonete Formations were deposited in shallow marine environments by bipolar paleocurrents (5) (Andreis et al., 1979). The quartzose petrofacies suggests an easterly provenance from the Tandilia Massif (Andreis et al., 1979). The Bonete Formation contains *Glossopteris* (Harrington, 1933; Archangelsky and Cúneo, 1984) and *Eurydesma* (Harrington, 1955). The Tunas Formation has a high content of volcanic fragments (6), and paleocurrents flowed from the southwest (P. J. Conaghan, unpublished data). In the southwest, magmatic activity in the La Pampa Province (7) (Linares et al., 1980) is reflected in the tuffaceous beds intercalated in the Tunas Formation (8) Iñíguez et al., 1988). Intense folding (9) affected the whole Gondwana succession from the end of deposition in the Early Triassic (Harrington, 1980) or Late Permian (Cobbold et al., 1986; Büggisch, 1987).

pocenters in synclinal axes and condensed sequences over anticlinal axes.

The upper limit of the age of the Tunas Formation is unknown. The *Glossopteris* flora found in the lower half of the formation indicates Early Permian. Harrington (1980) and Azcuy and Caminos (1987) extended the age to the Late Permian by comparison with the age of the comparable succession of the Itararé Group to Rio do Rastro Formation of the Paraná Basin in Brazil (Appendix 1). Rapid subsidence close to the orogenic front in a foreland basin, however, could have accommodated the 1,500 m of the Tunas Formation in the Early Permian–early Late Permian in a similar manner as the last ~1,000 m of the El Imperial Formation were deposited

during a period of fast subsidence probably related to the San Rafael Orogenic phase. The Tunas Formation was deposited during the onset of emplacement of the central part of the Somuncura Batholith in the North Patagonian Massif (Pankhurst et al., 1992) and during the metamorphism observed in illites in the Sierras Australes area, as discussed below.

The age of the deformation in the Sierras Australes is post-Early Permian and pre-Cretaceous. Cobbold et al. (1986, 1991) argue that ductile deformation is Permo-Triassic. From the radiometric data plotted on Fig. 23 (Appendix 1), we group the metamorphic events registered in the Pre-Permian rocks as (1) ~278 Ma (282 ± 3 Ma and 273 ± 8 Ma K/Ar illite), (2) ~262 Ma (265 ± 3 Ma and 260 ± 3 Ma K/Ar illite), (3) 249 ± 8 Ma K-Ar rhyolitic tuff, and (4) ~230 Ma (240 ± 12 Ma K/Ar hornblende, 227 ± 32 Ma WR Rb/Sr granite, 221 ± 6 Ma K/Ar rhyolitic tuff).

The final phase or phases of deformation, postdating the Permian (probably Late Permian) Tunas Formation, are Triassic, and probably correspond to the 249 Ma and 230 Ma metamorphic dates.

POST-GONDWANA HISTORY

Introduction

After the Early Permian to Early Triassic volcanicity and associated volcaniclastic nonmarine sedimentation of the Upper Sequence of the Gondwana cycle, western Argentina underwent uplift/erosion and subsequent extensional relaxation which established a sedimentary regime in fault-bounded depressions or grabens by collapse of portions of the Late Paleozoic fore-arc, arc, and back-arc terrains (Uliana and Biddle, 1988). The Middle-Late Triassic sediments and associated volcanics (Rolleri and Criado Roque, 1968) of Chile and western Argentina were deposited within rapidly subsiding fault-bounded troughs which the sea entered on the west (Fig. 20). In western Argentina, the post-Gondwana Middle to Late Triassic sedimentary succession in the Ischigualasto–Villa Unión, Marayes, Cuyo, and San Luis Basins is entirely non-marine (Figs. 21, 22). In the south the grabens pass between the North Patagonian Massif and the Deseado Massif. Northeast of the North Patagonian Massif adjacent to the Late Paleozoic Sauce Grande–Colorado Basin the Late Permian lacuna between top of the Tunas Formation and the Early Cretaceous base of the Colorado Formation indicates Triassic-Jurassic nondeposition or erosion (Zambrano, 1974, 1980; Lesta et al., 1980).

Depositional setting and evolution of the Triassic basins of western Argentina

The Middle to Late Triassic succession rests on a basement affected by latest Permian–Early Triassic movements

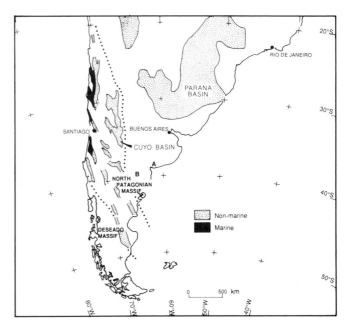

Figure 20. Mid-Late Triassic (230 Ma) paleogeographic reconstruction for southern South America according to Uliana and Biddle (1988) showing the northwest-trending graben depressions with non-marine fill (stipple) in the east, and marine (black) in the west. The northwest-trending troughs formed along a strip outlined by dotted lines between the Late Paleozoic continental margin of arc and back-arc terrains and the cratonic and cratonized terrains to the northeast. Syndepositional igneous activity is represented by silicic and intermediate volcanic and volcaniclastic rocks. The largest Triassic basin is the Cuyo Basin and its northern extension in the San Luis, Marayes and Ischigualasto–Villa Unión regions (*see* Fig. 21). To the south, smaller fault-bounded basins are incised in the North Patagonian and Deseado Massifs. AB is the location of the time-space diagram of Figure 19.

that followed the deposition of the Upper Sequence of the Gondwana Cycle. The common succession is basal conglomerate (Río Mendoza and Las Cabras Formations in the Cuyo Basin and equivalents in other areas) (Fig. 22) on a Precambrian crystalline basement of Pampean affinities, Paleozoic (Cambrian to Early Permian) sedimentary and metasedimentary rocks, and mid-Permian volcanics and volcaniclastics of the Choiyoi Group and equivalents (Rolleri and Fernández Garrasino, 1979; Flores, 1979). The main components of the basement in the Ischigualasto–Villa Unión Basin are Late Paleozoic sediments (Lower Sequence of Gondwana Cycle) and Pampean crystalline basement (Bossi, 1971). The basal coarse sediments are succeeded by widespread organic-rich lacustrine shales capped by fluvial redbeds. Tuff is common throughout the succession (Appendix 1).

The opposite and alternate distribution of the depocenters of the Las Peñas and Cacheuta sub-basins indicates half grabens of opposite geometry separated by interference zones.

The Cuyo is the largest basin (Fig. 21). Outcrops on both

flanks of the Precordillera probably de-limit the northern extent of the Cuyo Basin (Stipanicic, 1979, 1983; Strelkov and Alvarez, 1984; Sessarego, 1986), but the northeasternmost exposures of the Cuyo Basin suggest a potential connection with the Ischigualasto–Villa Unión Basin. Radiometric dates (K/Ar whole rock) on basalts interbedded with the sediments of Las Cabras Formation indicate Early Triassic (Criado Roque and Ibañez, 1979) or Late Middle Triassic (Ramos and Kay, 1991). Biostratigraphic studies based on plant remains (Stip-

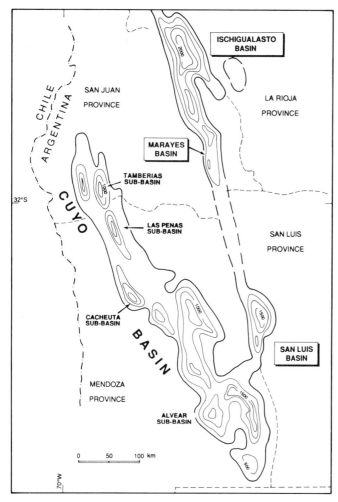

Figure 21. Generalized, nonpalinspastic reconstruction of Triassic basins of western Argentina with isopachs (contour interval: 500 m) (modified from Kokogian, 1991). The San Luis region is an eastward extension of the Cuyo Basin, as shown by Triassic sediments in the subsurface (Flores, 1979). To the north, the distribution of the Triassic exposures on both flanks of the Precordillera in northern Mendoza and southwestern San Juan suggests a connection between the Cuyo Basin and both flanks of the Precordillera region (Stipanicic, 1979; Strelkov and Alvarez, 1984). A possible connection between the Cuyo and Ischigualasto Basins through the San Luis and Marayes Basins is indicated. The opposite and alternate distribution of the depocenters of the Las Peñas and Cacheuta sub-basins indicates half grabens of opposite geometry separated by interference zones.

anicic and Bonetti, 1969), palynology (Yrigoyen and Stover, 1970), and vertebrates (Bonaparte, 1966, 1973), all interpreted by Stipanicic and Bonaparte (1979), indicate Middle to Late Triassic. Subsurface studies indicate a maximum thickness of 3,700 m. The nonmarine sediments have a coarse clastic facies at the base (Río Mendoza and Las Cabras Formations) and grade up to cross-bedded sandstone, shale, bituminous shale, and tuff, all interpreted as deposited in braided-river systems (Potrerillos Formation) (Fig. 22). The fluvial Potrerillos Formation passes upward to the widespread euxinic lacustrine bituminous shale of the Cacheuta Formation, succeeded by the fluvial Río Blanco Formation (Fig. 22), a return to the conditions of the Potrerillos Formation but with more tuff. The Triassic succession of the Cuyo Basin is unconformably overlain by fluvial conglomerate, sandstone, and mudstone of the Early-Middle Jurassic Barrancas Formation (Kokogian and Mancilla, 1989), in turn overlain by the Late Jurassic basalt of the Punta de las Bardas Formation.

Half grabens are common, with the basal coarse fill of the Río Mendoza Formation and some intervals within the Las Cabras Formation reflecting an initial phase of fast tectonic subsidence (synrift phase), and the lacustrine Cacheuta Formation and fluvial Potrerillos and Río Blanco Formations a phase of thermal subsidence (Kokogian and Mancilla, 1989). Most formations described for the Cuyo Basin have been identified in wells drilled in the San Luis Basin (Flores, 1979), suggesting that both areas were part of a single basin (Fig. 21) (Yrigoyen et al., 1989). This correlation probably can be carried further to the Marayes Group of the Marayes Basin (Bossi, 1975). At the base, the Esquina Colorada Formation comprises alluvial-fan clast- and matrix-supported conglomerate which grades up to fluvial cross-bedded sandstone, scarce fine conglomerate, carbonaceous shale, and thin layers of coal of the Carrizal Formation, overlain by the alluvial-fan conglomerate and coarse sandstone of the Quebrada del Barro Formation. The Carrizal Formation correlates in time and facies with the Ischigualasto and Los Rastros Formations of the Ischigualasto–Villa Unión region and the Cacheuta Formation of the Cuyo Basin, suggesting a period of extensive lacustrine sedimentation.

Widely spaced outcrops and subsurface occurrences indicate a wide original extent of the Ischigualasto–Villa Unión Basin (Fig. 21) (Stipanicic and Bonaparte, 1979). At the base, the Early-Middle Triassic Paganzo III of Bodenbender (1911) or the Talampaya and Tarjados Formations (Romer and Jensen, 1966) unconformably overlie Early Permian nonmarine deposits of the Lower Sequence of the Gondwana Cycle and Pampean basement (Andreis, 1969; Mozetic, 1974; Stipanicic and Bonaparte, 1979). They are alluvial-fan conglomerate interbedded with fluvial (braided-river) sandstone and playa-lake fine-grained sediments with tuff and, in places, basaltic flows and volcanic agglomerate (De la Mota, 1946; Andreis, 1969). The overlying Agua de la Peña Group rests at an angu-

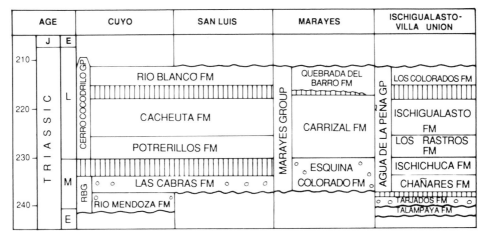

Figure 22. Stratigraphic chart of the Triassic successions of the Cuyo (Rolleri and Fernández-Garrasino, 1979), San Luis (Flores, 1969), Marayes (Bossi, 1975), and Ischigualasto–Villa Unión (Stipanicic and Bonaparte, 1979) Basins. The Río Mendoza, Las Cabras, Esquina Colorada, and Tarjados Formations are coarse basal fills (circles), and the Cachueta, Carrizal, Los Rastros, and Ischigualasto Formations are lacustrine. RBG = Rincón Blanco Group.

lar unconformity on the tilted margins of the Talampaya-Tarjados succession or the Early Permian Lower Sequence (Bossi, 1971). The Agua de la Peña Group is subdivided into five formations: Chañares, Ischichuca, Los Rastros, Ischigualasto, and Los Colorados (Fig. 22). The Chañares/Ischichuca Formation is an alluvial-fan deposit of tuffaceous sheet floods and debris flows (Legarreta and Kokogian, 1986) that grade upward to the conglomerate and sandstone with subordinate shale and coal and then lacustrine black shale associated with channelized cross-bedded sandstone of the Los Rastros Formation (Legarreta and Kokogian, 1986). The Ischigualasto Formation comprises fluvial cross-bedded sandstone and abundant fine sandstone, tuff, and mudstone deposited in a fluvial system of moderate- to high-sinuosity. The succession culminates in Los Colorados Formation of fluvial sandstone and mudstone and eolian sediment (Kokogian et al., 1987).

In the San Rafael Block, the Puesto Viejo Formation (González-Díaz, 1972) probably represents a marginal equivalent of the Cuyo succession. Basal conglomerate grades up to a fluvial cross-bedded sandstone, interbedded in places with tuff, acidic and basic volcanics, and subordinate andesite. Reptiles (Therapsida) found in these deposits belong to the late Early Triassic *Cynognathus* zone, confirmed by the discovery of an Early Triassic spore assemblage (Ottone and García, 1991).

Further south in Patagonia, small Triassic basins cross the western half of the North Patagonian Massif and the Deseado Massif (Spalletti et al., 1988; Franchi et al., 1989). Like the basins to the north, these basins contain abundant tuff in alluvial-fan and fluvial sediment with elements of the Late Triassic *Dicroidium* flora (Stipanicic, 1983), but lack lacustrine deposits.

End of Triassic sedimentation and emplacement of Late Jurassic basalts

Sedimentation ceased by the end of the Triassic and the region was covered by Late Jurassic–(?)Early Cretaceous basalt. Radiometric dates of basalts encountered in boreholes in the Cuyo Basin and in equivalents exposed in the San Luis Basin indicate a range of ages between Middle Jurassic (171–167 Ma) and Early Cretaceous (109–107 Ma). Dates from basalts exposed on the northeastern sector (San Luis Basin) have yielded an average age of 127 Ma (Yrigoyen, 1975), and other dates from the same region range between 161 and 107 Ma (González, 1971; Yrigoyen, 1981). Basalts encountered in boreholes in the Alvear sub-basin range in age between 167 and 107 Ma (Criado-Roque, 1979) and those exposed in the San Luis Basin (González, 1971) are around 160 and 150 Ma. The basalts (Punta de las Bardas Formation), associated with continental sediments, have also been identified in numerous wells drilled in the northern half of the Cuyo Basin (Rolleri and Criado-Roque, 1968). The area of the basalt in the northern half of the Cuyo Basin is approximately 6,500 km^2 with a maximum thickness about 180 m (Rolleri and Fernández-Garrasino, 1979). Tholeiitic basalts in the Ischigualasto–Villa Unión area are possibly also Late Jurassic–Early Cretaceous.

In capping the Permian-Triassic succession, the basalts have the same stratigraphic position as the Serra Geral basalt of Brazil, Drakensberg Volcanics of southern Africa, and Kirkpatrick Basalt of Antarctica. The ages of all these basalts occupy different parts of the Jurassic and Early Cretaceous, as detailed in Chapter 7.

ACKNOWLEDGMENTS

The main sources of regional information are López-Gamundí et al. (1987, 1989), Azcuy et al. (1987), Andreis et al. (1987, 1989), and Espejo et al. (1991), supplemented by individual contributions published in Argentine and international journals, including current research projects by López-Gamundí and Espejo. Original contributions given here are Espejo's field and petrographic studies on the San Rafael Basin succession and Conaghan's work on paleocurrents and composition of sandstones of the Permian Tunas Formation of the Sauce Grande–Colorado Basin, including field work done in December 1986. Powell visited the field during August and September of 1985. Veevers, who compiled the correlation chart (Appendix I), visited López-Gamundí and Espejo in Buenos Aires and in the field during May through July of 1988. López-Gamundí and Espejo visited Macquarie University between November 1987 and March 1988; their visit was supported by the Australian Research Grants Committee and sponsored by the National Research Council of Argentina. Arturo J. Amos and Juvenal J. Zambrano provided insightful reviews. The figures were drafted by John Cleasby.

APPENDIX 1. CARBONIFEROUS-PERMIAN-TRIASSIC CORRELATION CHART

J. J. Veevers

In Figure 23 the columns are arranged from west to east, first at 32°S (Cordillera Frontal–Chaco Paraná and thence into the Paraná Basin) and then southward in a band between 34°S and 40°S. The columns for the Paraná Basin are included here for their links with southern Africa (Chapter 5).

The radiometric time scale, from Appendix 1 in Chapter 3, has the Permian/Triassic boundary calibrated at 250 Ma.

I acknowledge the help of Oscar López-Gamundí in compiling age data for the Argentine formations and igneous bodies.

I. Biostratigraphical schemes

(a) Carboniferous and Permian

Primary anchor points. (1) The Tastubian *Eurydesma* fauna of the Bonete Formation (Dickins, 1984), and probably equivalent marine invertebrates at the top of the Itararé Subgroup and in the San Gregorio Formation, which has a common goniatite genus in Namibia; (2) Archangelsky et al. (1987): (a) the *Protocanites* zone, Visean; (b) the *Levipustula* zone, Namurian (324 Ma) to end-Carboniferous (286 Ma) in eastern Australia (Roberts et al., 1991), Namurian to early Westphalian in Argentina by superposition of (c) the Zona de Intervalo, and (d) the *Cancrinella* zone; (c) and (d) both dated by associated palynomorphs, and the top of (d) overlapping the *Eurydesma* zone.

Secondary anchor points. (1) The endemic South American bivalves in the Estrada Nova and Yaguari Formations equivalent to those in the Waterford Formation of South Africa (Cooper and Kensley, 1984) of Kalinovian age; (2) palynomorph zones (Archangelsky et al., 1987), dated by correlation with Australia (Balme, 1980; Truswell, 1980), such that the *P-L* (*Potonieisporites-Lundbladispora*) zone = Stage 1, the *Cristatisporites* zone = Stage 2 and lower 3a, and the *Striatites* zone to the top of Stage 4. Foster and Waterhouse's (1988) *Granulatisporites confluens* zone, which we date as lower Stage 3a, probably occupies the upper part of the *Cristatisporites* zone. The *NBG* (*Nothorhacopteris ar-*

gentinica–Botrychiopsis weissiana–Ginkgophyllum diazii) megafloral zone = Zona de Intervalo = *Ancistrospora* zone. (3) The South African tetrapod zones of *Mesosaurus* (Oelofsen, 1987), *Cistecephalus, Daptocephalus,* and *Lystrosaurus* found in the Paraná Basin and *Lystrosaurus* and *Cynognathus* in Argentina (Barberena et al., 1985a, 1985b, 1988).

(b) Triassic (Stipanicic and Bonaparte, 1979)

Primary anchor point. Microflora M2 = Ipswich/Molteno (Anderson, 1981) = *C. rotundus* = Carnian and early Norian.

Secondary (local) anchors. Tetrapods and megafloras, as correlated by Anderson (1981) and Anderson and Anderson (1983, p. 10). *See also* Rocha-Campos (1971, p. 407) and Barberena et al. (1985a, 1985b).

II. Basins

Calingasta-Uspallata (Archangelsky et al., 1987, table 14)

Carboniferous and Permian formations correlated by (a) marine invertebrates: Malimán, Hoyada Verde, Agua de Jagüel, Pituil. The marine invertebrates listed in table 2, column 1 of Rocha-Campos and Archangelsky (1985) are attributed to the El Ratón Formation but this attribution must be misplaced because the El Ratón Formation is entirely nonmarine (Archangelsky and Azcuy, 1983, fig. 2). (b) Palynomorphs: *Potonieisporites* zone in the Santa Máxima Formation (Ottone, 1985). (c) Megaflora (Sessarego and Cesari, 1986): El Ratón and Malimán.

Choiyoi Group. (1) a K/Ar minimum age of 251 ± 14 Ma (new constants, from Rocha-Campos et al., 1971); (2) a whole-rock K/Ar age of 278 ± 10 Ma on a rhyolitic porphyry in the Cordillera Frontal (Caminos et al., 1979, specimen 1c). Consanguinous intrusives (granite, granodiorite, syenite) in the Precordillera and Cordillera Frontal de Mendoza range from 291 ± 10 Ma (K/Ar) to 264 ± 8 Ma (Rb/Sr), and have a 4-point Rb/Sr reference isochron of 275 ± 30 Ma (figure in Caminos et al., 1979, p. 17), which suggests that the Rb/Sr ratios of these rocks have not been reset since initial cooling. Another isochron, on 3 points from younger igneous rocks (Caminos et al., 1979, fig. 17), indicates a more precise age of 225 ± 5 (230–220) Ma, in agreement with the age of the Triassic strata nearby, as dated by Stipanicic and Bonaparte (1979), who describe tuff as common in the Ischichuca, Los Rastros, and Ischigualasto Formations.

Paganzo (Archangelsky and Azcuy, 1983; Rocha-Campos and Archangelsky, 1985)

(i) western (a) megaflora: Guandacol, el Tupe, Patquía, with the Patquía extended higher into the Permian (Aceñolaza and Vergel, 1987); (b) microflora: el Tupe; (c) marine invertebrates (intercalated): el Tupe (González, 1985). Eolian facies at top of Patquía (Limarino et al., 1987). Cerro Morado and Cauquenes by facies equivalence with Choiyoi. Triassic from Stipanicic and Bonaparte (1979).

(ii) Central (a) micro- and mega-floras: Lagares (*Ancistrospora, NBG*) and additionally *P-L* (Limarino and Cesari, 1989). La Colina Formation contains in its lower part the *NBG* zone (Archangelsky et al., 1987), and extends to the age of the Bajo de Véliz Formation, both of which contain the *Gangamopteris* zone (Rocha-Campos and Archangelsky, 1985, p. 238). The K/Ar date of 300 ± 6 Ma (new constants, from Thompson and Mitchell, 1972, p. 212) on a flow of tholeiitic basalt 25 m above the base of La Colina Formation (Archangelsky and Azcuy, 1983, p. 271) is anomalously old.

(iii) Eastern. By its microflora, the Bajo de Véliz Formation is placed in the *Cristatisporites* zone (Rocha-Campos and Archangelsky, 1985, p. 243).

Chaco Paraná

(i) Córdoba. Palynomorphs in the Ordoñez Formation indicate the *Potonieisporites-Lundbladispora* (*see also* the Sauce Grande–Colorado Basin), *Cristatisporites,* and *Striatites* zones (Rocha-Campos and Archangelsky, 1985, p. 242, 243) and the *G. confluens* zone (Foster and Waterhouse, 1988). The Victoriano Rodríguez Formation is dated by superposition and by contained palynomorphs (Russo and Archangelsky, 1987).

(ii) Chaco. The Sachayoj Formation is correlated by facies with

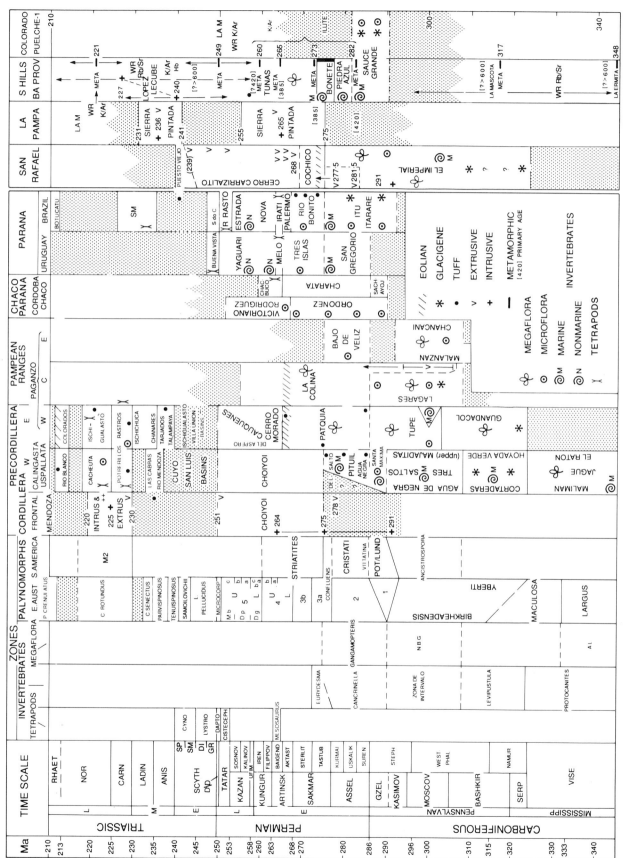

Figure 23. Correlation chart (Appendix 1).

the basal part of the Ordoñez Formation, the Charata Formation with the younger part of the Itararé Subgroup (Rocha-Campos and Archangelsky, 1985, p. 214). The Chacabuco Formation contains *Mesosaurus* and is correlated with the Iratí Formation of Brazil (Padula, 1972, p. 220).

Paraná

(i) Uruguay: Ferrando (1986). From the top down—Buena Vista Formation correlated by amphibians with Sanga do Cabral (S do C) of Brazil (E. Lavina, personal communication, 1988). Yaguari Formation: upper part (presumed to be equivalent to Brazilian Rio do Rasto Formation) contains tetrapods of the *Cistecephalus* zone; bivalves (Ferrando, 1986, p. 49, from Cox, 1934) are presumably at a similar level to the endemic ones of the Estrada Nova Formation, but are not mentioned by Runnegar (1972). The lower part of the Yaguari Formation (called the Paso Aguiar Formation) also contains bivalves, which Mones and Figueiras (1981) correlate with the Estrada Nova fauna. Melo Formation: *Mesosaurus* zone = Iratí Shale. Tres Islas Formation = Rio Bonito by palynomorphs = Stage 3. San Gregorio Formation goniatites not diagnostic of age, but probably Tastubian by comparison with Namibian fauna. The rest probably Stage 2.

(ii) Brazil. From the top, the Serra Geral Formation (not shown) starts within the interval 150 Ma to 130 Ma (Schobbenhaus et al., 1984, p. 349); the Botucatú Formation is latest Triassic and Jurassic by interposition (Schobbenhaus et al., 1984, p. 347). The Caturrita Formation is dated by Carnian tetrapods and the Santa Maria (SM) by Ladinian tetrapods (Barberena et al., 1985a, 1985b). The Sanga do Cabral (S do C) contains Griesbachian amphibians of the *Lystrosaurus* zone (Lavina, 1983; E. L. Lavina, personal communication, 1988). The Rio do Rasto Formation contains *Cistecephalus*- and *Daptocephalus*-zone tetrapods (Barberena et al., 1985a, 1985b). The Estrada Nova, with endemic bivalves, is correlated with the Kalinovian Waterford Formation of South Africa (Cooper and Kensley, 1984). Iratí Shale = *Mesosaurus* zone (Oelofsen, 1987) = Whitehill Formation of southern Africa. Palermo Formation dated by interposition. Rio Bonito = Stage 3 (Truswell, 1980, p. 105). Itararé Subgroup: upper marine fauna of brachiopods, bivalves, and arenaceous foraminifers is probably Tastubian, as shown by Anderson (1981). Palynomorphs in the Itú Shales indicate Stage 2 (Truswell, 1980), elsewhere Stage 1 (Helby and Runnegar, personal communication, *in* Truswell, 1980), confirmed by Rocha-Campos and Archangelsky (1985, p. 215, 240). G–H₁ zones of Itararé = *Potonieisporites-Lundbladispora*, but Rocha-Campos and Archangelsky (1985, p. 240) also seem to assert that the Itararé has the *Ancistrospora* zone, but do not show this on their figure 7 (Rocha Campos and Archangelsky, 1985, p. 218). Their table 11 shows palynomorph assemblages from Itararé and Lagares: only 5 forms are common. I (Veevers) regard the Itararé as no older than the upper part of the *Ancistrospora* zone.

San Rafael

El Imperial Formation contains the Zona de Intervalo, *NBG* (Espejo, 1987a and Espejo and Cesari, 1987) and *Potonieisporites-Lundbladispora* zones (Archangelsky et al., 1987, cuadro 14; Rocha Campos and Archangelsky, 1985) and ranges up into the *Gangamopteris* zone (Espejo, 1987a and Espejo and Cesari, 1987). The overlying Cochicó tuff of the Sierra Pintada Group is dated (K/Ar) as 281.5 ± 13 Ma and 277.5 ± 10 Ma (Toubes and Spikerman, 1976, cuadro 1; Dessanti and Caminos, 1967, p. 154). We draw the boundary at 275 Ma to take account of the unconformity that separates rocks below affected by the San Rafael movement and those above that are unaffected. The oldest K/Ar date is 291 ± 10 Ma on a microtonalite intrusion (sample #54, Toubes and Spikerman, 1976). K/Ar dates from the Cerro Carrizalito Formation range from 268 to 239 Ma (and even apparently younger at 229 Ma and 196 Ma, not plotted) (Toubes and Spikerman, 1976, 1979; Valencio and Mitchell, 1972). The Puesto Viejo Formation is regarded as containing the late Scythian *Cynognathus* zone (Barberena et al., 1985a, 1985b), and sets an upper bound of 245 Ma on the underlying volcanics.

Eruptive rocks of the Province of La Pampa

Linares et al. (1980, p. 122–125, 130, 131) determined two pulses of magmatic activity: (1) 265 ± 10 (275–255) Ma from an Ar⁴⁰/Ar³⁶ versus K⁴⁰/Ar³⁶ isochron on four extrusives and five intrusives and (2) 236 ± 5 (241–231) Ma, a grouped total from (a) an Ar⁴⁰/Ar³⁶ versus K⁴⁰/Ar³⁶ isochron on intrusives of 232 ± 10 Ma (Linares et al., 1980, fig. 15), (b) an 11-point Rb/Sr isochron of 238 ± 5 Ma on extrusives (Linares et al., 1980, fig. 16), and (c) an 18-point Ar⁴⁰/Ar³⁶ versus K⁴⁰/Ar³⁶ isochron of 235 ± 5 Ma.

Southern Hills (Sierras Australes) of Buenos Aires Province and Colorado Basins

The Sauce Grande Formation of the Sierras Australes is barren except for a bivalve, *Astartella ? pusilla,* not diagnostic of precise age, in a diamictite reported by Harrington (1947, 1955), but offshore in the Puelche-1 Borehole of the Colorado Basin contains the *Potonieisporites-Lundbladispora* and *Cristatisporites* zones (Archangelsky et al., 1987, p. 282; Rocha-Campos and Archangelsky, 1985, p. 215, 216, 242). The next dated formation, the Bonete Formation, is Tastubian (*Eurydesma* fauna, Dickins, 1984) so that the Piedra Azul Formation, barren except for a gastropod, is restricted to the late Asselian. The overlying Tunas Formation is correspondingly post-Tastubian, and possibly extends to the Late Permian.

Igneous and metamorphic activity. Tuff is reported from the upper part of the Tunas Formation (Iñiguez et al., 1988). Other ages are interpreted (von Gosen and Büggisch, 1989) as due to metamorphism of Precambrian rocks, such as the barely deformed granite at López Lecube, 50 km west of Tornquist (Fig. 5), with a Rb/Sr whole-rock age of 227 ± 32 Ma and a hornblende K/Ar age of 240 ± 12 Ma, similar in age to the regional metamorphism denoted by K/Ar whole-rock ages of 221 ± 6 Ma and 249 ± 8 Ma of rhyolitic tuff from La Mascota, 10 km southeast of Pigüé, which, together with rhyolite from La Ermita, has a Rb/Sr whole-rock age of 317 to 348 ± 21 Ma (Cobbold et al., 1986). Metamorphic illites from the Ventana Group near Pigüé have K/Ar dates of 273 ± 8 Ma and 265 ± 8 Ma (Varela et al., 1985) and from the Curamalal Group of 260 ± 3 Ma and 282 ± 3 Ma (Büggisch, 1987).

Falkland Islands—Islas Malvinas

Data from the Falkland Islands is included here (but not shown on the chart) because of their present location. The data is used in the chapter on southern Africa, to which the Falklands are presumed to have been connected in the Permian and Triassic.

The only datable fossils are *Glossopteris* spp. in the top 10 m of the Bahia Choiseul Formation, regarded as indicating Early Permian (Jalfin and Bellosi, 1983). Within the Early Permian, the marine transgression peaks near the Port Sussex/Terra Motas boundary, which is inferred to correspond with the main transgression/regression at the Tastubian/Sterlitimakian boundary (Bellosi and Jalfin, 1984). The oldest part of the profile, at the base of the Lafonia Formation, is assumed here to be the same age as the Sauce Grande Formation, as the youngest part, at the top of the Estrecho de San Carlos, and is likewise assumed to be the same age as the Tunas Formation.

APPENDIX 2. SANDSTONES FROM THE CALINGASTA-USPALLATA, PAGANZO, AND SAN RAFAEL BASINS

Irene S. Espejo, Oscar R. López-Gamundí and Patrick J. Conaghan

The photomicrographs of Figures 24 and 25 illustrate the principal grain types of the petrofacies identified in the foreland basins of southern South America. The mineral assemblages correspond to those defined by López-Gamundí and Espejo (1987) as *quartzofeld-*

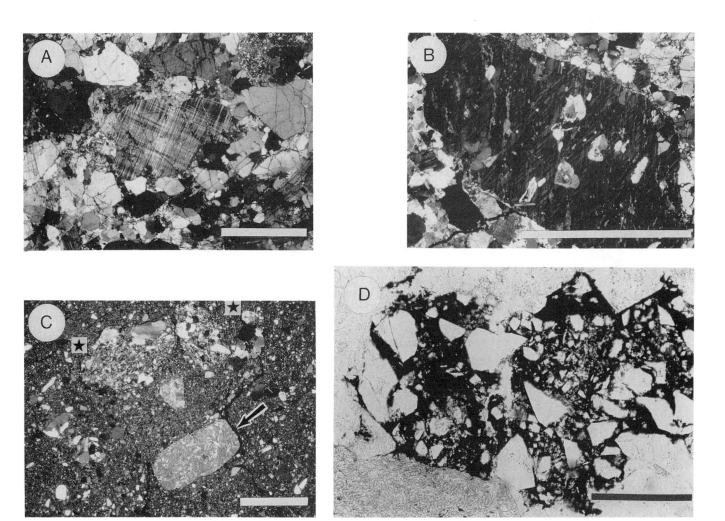

Figure 24 (on this and following page). Quartzose petrofacies: (A) Clast of perthitic microcline, cross-polarized light, scale bar: 1 mm. Tupe Formation, Huaco area, western Paganzo Basin. (B) Clast of quartz and feldspar (granitic) rock, cross-polarized light, scale bar: 0.3 mm. Deposit equivalent to Tupe Formation, San Juan River, southwestern Paganzo Basin.

Lithic (sedimentary) petrofacies: (C) Clasts of micritic limestone (indicated by arrow) and fine sandstone (indicated by stars) in matrix of diamictite, cross-polarized light, scale bar: 1.5 mm. Hoyada Verde Formation, east of Barreal, Calingasta-Uspallata Basin. (D) Clast of coarse quartzose siltstone with ferruginous cement, plane-polarized light, scale bar: 0.3 mm. El Imperial Formation, Agua del Toro Dam, San Rafael Basin. (E) Metasedimentary mica-quartz fragment (slate-schist) with incipient tectonite fabric, cross-polarized light, scale bar: 0.2 mm. El Imperial Formation, Agua del Toro Dam, San Rafael Basin.

Volcaniclastic Petrofacies: (F) Clast of euhedral zoned plagioclase with polysynthetic twinning, cross-polarized light, scale bar: 0.2 mm. Cauquenes Formation, Huaco, western Paganzo Basin. (G) Volcanic clast with pilotaxitic texture and abundant microlites of plagioclase, cross-polarized light, scale bar: 0.3 mm. Cauquenes Formation, Huaco, western Paganzo Basin. (H) Volcanic clast with pilotaxitic (partially fluidal)/hyalopilitic texture and plagioclase microlites, fragment of devitrified shard indicated by arrow, plane-polarized light, scale bar: 0.1 mm. Deposit equivalent to lower half of Patquía Formation, San Juan River, southwestern Paganzo Basin.

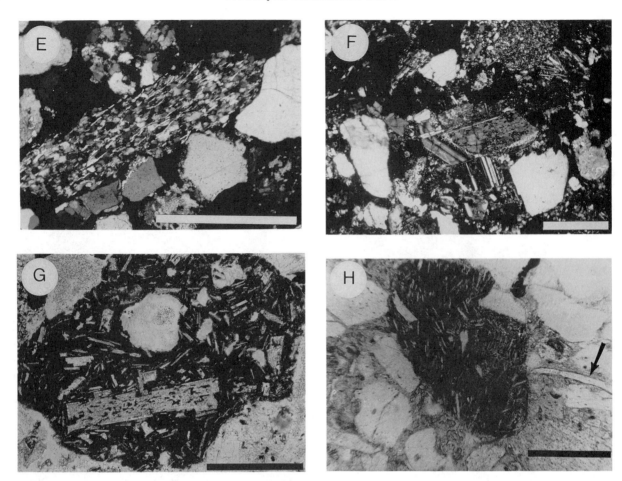

spathic petrofacies, derived from basement highs of Pampean granitoids and associated metamorphics; *lithic (sedimentary) petrofacies,* derived from Early to Middle Paleozoic sedimentary rocks exposed principally in the Proto-Precordillera and analogous terranes as intrabasinal highs; and *volcaniclastic petrofacies,* derived from the volcanic arc terranes along the Paleo-Pacific margin.

REFERENCES CITED

Aceñolaza, F., and Vergel, M., 1987, Hallazgo del Pérmico superior fosilífero en el Sistema de Famatina: 10th Congreso Geológico Argentino, Actas 3, p. 125–129.

Almeida, F.F.M., 1971, Geochronological division of the Precambrian of South America: Revista Brasileira de Geociencias, v. 1, no. 1, p. 13–21.

Alvarez, L. A., and Fernández Seveso, F., 1987, Nueva paleogeografía del ambiente oriental de la Cuenca Paganzo: Yacimientos Petrolíferos Fiscales, Gerencia General, unpublished report.

Amos, A. J., 1964, A review of the marine Carboniferous stratigraphy of Argentina, *in* Proceedings, International Geological Congress, 12th, Calcutta, India: Calcutta International Geological Congress, v. 9, p. 53–72.

Amos, A. J., 1972, Silurian of Argentina, *in* Correlation of the South American silurian rocks: Boulder, Colorado Geological Society of America Special Paper 133, p. 5–16.

Amos, A. J., and López-Gamundí, O. R., 1981, Late Paleozoic tillites and diamictites of the Calingasta-Uspallata and Paganzo basins, San Juan and Mendoza provinces, western Argentina, in Hambrey, M., and Harland, W., eds., Earth's Pre-Pleistocene glacial record: Cambridge, Cambridge University Press, p. 859–868.

Amos, A. J., and Marchese, H. J., 1965, Acerca de una nueva interpretación de la estructura del Carbónico en la Ciénaga del Medio: Revista de la Asociación Geológica Argentina, v. 20, p. 263–270.

Amos, A. J., and Rolleri, E. O., 1965, El Carbónico marino en el valle Calingasta-Uspallata (San Juan–Mendoza): Buenos Aires, Boletín de Informaciones Petroleras, v. 368, p. 50–71.

Anderson, J. M., 1981, World Permo-Triassic correlations: Their biostratigraphic basis, in Cresswell, M. M., and Vella, P., eds., Gondwana five: Rotterdam, A. A. Balkema, p. 3–10.

Anderson, J. M., and Anderson, H. M., 1983, Palaeoflora of Southern Africa: Molteno Formation (Triassic): Rotterdam, A. A. Balkema, v. 1, 227 p.

Andreis, R. R., 1965, Petrografía de las sedimentitas psefíticas paleozoicas de las Sierras Australes bonaerenses: Comisión de Investigaciones Científicas, La Plata, Anales, v. 6, p. 9–63.

Andreis, R. R., 1969, Los basaltos olivínicos del Cerro Guandacol (Sierra de Maz, La Rioja) y su posición estratigráfica: 4th Jornadas Geológicas Argentinas, Actas 1, p. 15–33.

Andreis, R. R., 1984, Análisis litofacial de la Formación Sauce Grande (Carbónico superior?), Sierras Australes, provincia de Buenos Aires: Annual Meeting IUGS–IGCP Project 211 Late Paleozoic of South America, Bariloche, Abstracts, p. 28–29.

Andreis, R. R., 1986, Fases de movilidad en el lineamiento estructural de Valle Fértil (ámbito occidental de la Cuenca Paganzo, San Juan–La Rioja, Argentina): Annual Meeting IUGS–IGCP Project 211 Late Paleozoic of South America, Córdoba, Abstracts, p. 75–76.

Andreis, R. R., and Bossi, G. E., 1981, Algunos ciclos lacustres de la

Figure 25. Volcaniclastic petrofacies: Large-scale cross-bedded eolian sandstone of the Vallecito Member, 50 m below the top of the Patquía Formation, Huaco area, western Pagango Basin, Jachal-Huaco road, west of the camp ground in La Ciénaga del Vallecito. Low-magnification view in (A) plane-polarized and (B) cross-polarized light, showing texture and mineral assemblage; scale bar 0.5 mm. The sand grains are well rounded with thick iron-oxide dust lines and are set in a pervasive zeolite pore-filling cement (clear intergranular areas in A). The rock is a plagarenite with little quartz (q) and essentially no potash-feldspar (QFL = 8.7%, 47.4%, 43.8%). Accessory minerals make up 8.8% of the detrital framework and are dominated by augite (a), with subsidiary hornblende (h). Rock fragments (opaque/semi-opaque iron-stained grains in A) are exclusively volcanic. Higher-magnification views showing details of detrital grain types, (C) in plane-polarized light; scale bar: 0.5 mm, and (D) in cross-polarized light; scale bar: 0.25 mm. Grain codes as for A. All relatively clear grains are fresh plagioclase, except those coded "q"; "p" in C is an altered plagioclase grain. In D the grain at the center is plagioclase with polysynthetic twinning, "zp" is a zoned plagioclase, and the augite grain, "a," at bottom center also shows polysynthetic twinning.

Formación Malanzán (Carbónico superior) en la región de Malanzán, Sierra de los LLanos, provincia de La Rioja: 8th Congreso Geológico Argentino, Actas 4, p. 639–655.

Andreis, R. R., and Cladera, 1992, Las epiclastitas pérmicas de la cuenca Sauce Grande (Sierras Australes, Buenos Aires, Argentina), Parte I: composición y procedencia de los detritos: 4th Reunión Argentina de Sedimentología, Actas 1, p. 127–134.

Andreis, R. R., and Japas, S., 1991, Cuencas de Sauce Grande y Colorado, *in* El Sistema Pérmico en la República Argentina y en la República Oriental del Uruguay: 12th International Congress on Stratigraphy and Geology of the Carboniferous and Permian and Academia Nacional de Ciencias de Córdoba (preprint), p. 45–65.

Andreis, R. R., and López-Gamundí, O. R., 1989, Interpretación paleoambiental de la secuencia paleozoica del cerro Pan de Azúcar, Sierras Australes, Provincia de Buenos Aires: 1st Jornadas Geológicas Bonaerenses, La Plata, Actas, p. 953–965.

Andreis, R. R., Spalletti, L. A., and Mazzoni, M. M., 1975, Estudio geológico del Subgrupo Sierra de Maz, provincia de La Rioja, República Argentina: Revista de la Asociación Geológica Argentina, v. 30, p. 247–273.

Andreis, R. R., Lluch, L. L., and Iñíguez, A. M., 1979, Paleocorrientes y paleoambientes de las Formaciones Bonete y Tunas, Sierras Australes, provincia de Buenos Aires: 6th Congreso Geológico Argentino, Actas 2, p. 207–224.

Andreis, R. R., Leguizamón, R. R., and Archangelsky, S., 1986, El paleovalle

de Malanzán: nuevos criterios para la estratigrafía del Neopaleozoico de la Sierra de los Llanos, La Rioja, República Argentina: Boletín de la Academia Nacional de Ciencias de Córdoba, v. 57, p. 1–119.

Andreis, R. R., Amos, A. J., Archangelsky, S., and González, C. R., 1987, Cuenca Sauce Grande, in Archangelsky, S., ed., El Sistema Carbonífero en la República Argentina: Córdoba, Academia Nacional de Ciencias, p. 213–233.

Andreis, R. R., Iñíguez, A. M., Lluch, L. L., and Rodríguez, S., 1989, Cuenca Paleozoica de Ventania, Sierras Australes, Provincia de Buenos Aires, in Chebli, G. A., and Spalletti, L. A., eds., Las cuencas sedimentarias argentinas: Tucumán, Universidad de Tucumán Serie de Correlación Geológica No. 6, p. 265–289.

Archangelsky, S., and Azcuy, C. L., 1983, Carboniferous palaeobotany and palynology in Argentina, in Martinez Diaz, C., ed., The Carboniferous of the world, II. Australia, Indian subcontinent, South Africa, South America, and North Africa: Madrid, Instituto Geológico y Minero de España, v. 4, p. 267–280.

Archangelsky, S., and Cúneo, R., 1984, Zonación del Pérmico continental argentino sobre la base de sus plantas fósiles: México, 3rd Congreso Latinoamericano de Paleontología, Memorias, p. 143–153.

Archangelsky, S., and Gamerro, J., 1981, Palinomorfos pérmicos del subsuelo de la Cuenca Colorado, en la Plataforma del mar Argentino, Provincia de Buenos Aires: Boletim Instituto de Geociencias, University of Sao Paulo, v. 11, p. 119–124.

Archangelsky, S., and Vergel, M., 1991, Cuenca Chacoparanaense: B. Paleontología, bioestratigrafía y paleoecología, in El Sistema Pérmico en la República Argentina y en la República Oriental del Uruguay: 12th International Congress on Stratigraphy and Geology of the Carboniferous and Permian and Academia Nacional de Ciencias de Córdoba (preprint), p. 40–44.

Archangelsky, S., Azcuy, C. L., González, C. R., and Sabattini, N., 1987, Edad de las biozonas, in Archangelsky, S., ed., El Sistema Carbonífero de la República Argentina: Córdoba, Academia Nacional de Ciencias, p. 293–301.

Arias, W. E., and Azcuy, C. L., 1986, El Paleozoico superior del Cañón del Río Atuel, provincia de Mendoza: Revista de la Asociación Geológica Argentina, v. 61, p. 262–269.

Arrondo, O. G., Iñíguez, A. M., and Dalla Salda, L., 1982, Afloramientos del Paleozoico superior interserrano de la provincia de Buenos Aires y sus relaciones estratigráficas: Annual Meeting IUGS–IGCP Project 42 Late Paleozoic of South America and its boundaries, Abstracts, p. 18.

Astini, R. A., 1991, Sedimentología de la Formación Talacasto: plataforma fangosa del Devónico precordillerano, provincia de San Juan: Revista de la Asociación Geológica Argentina, v. 46, p. 277–294.

Astini, R. A., and Piovano, E. L., 1992, Facies de plataforma terrígena del Silúrico de la Precordillera sanjuanina: Revista de la Asociación Geológica Argentina, v. 47, p. 99–110.

Azcuy, C. L., and Caminos, R., 1987, Diastrofismo, in Archangelsky, S., ed., El Sistema Carbonífero en la República Argentina: Córdoba, Academia Nacional de Ciencias, p. 239–251.

Azcuy, C. L., and Morelli, J. R., 1970, Geología de la comarca Paganzo-Amaná, El Grupo Paganzo, Formaciones que lo componen y sus relaciones: Revista de la Asociación Geológica Argentina, v. 25, p. 405–429.

Azcuy, C. L., Morelli, J. R., Valencio, D., and Vilas, J., 1979, Estratigrafía de la comarca Amaná-Talampaya: 7th Congreso Geológico Argentino, Actas 1, p. 243–246.

Azcuy, C. L., Cesari, S. N., and Longobuco, M., 1981, Las plantas fósiles de la Formación El Ratón (provincia de San Juan): Ameghiniana, v. 22, p. 97–109.

Azcuy, C. L., Andreis, R. R., Cuerda, A., Hünicken, M., Pensa, M., Valencio, D., and Vilas, J., 1987, Cuenca Paganzo, in Archangelsky, S., ed., El Sistema Carbonífero en la República Argentina: Córdoba, Academia Nacional de Ciencias, p. 41–96.

Baldis, B., and Chebli, G., 1969, Estructura profunda de la Precordillera san-

juanina: 4th Jornadas Geológicas Argentina, Actas 1, p. 47–65.

Baldis, B., and Sarudiansky, R., 1975, El Devónico del noroeste de la Precordillera argentina: Revista de la Asociación Geológica Argentina, v. 30, p. 301–330.

Baldis, B., Beresi, M., Bordonaro, O., and Uliarte, E., 1981, Estromatolitos, trombolitos y formas afines en el límite Cámbrico-Ordovícico del oeste argentino: 2nd Congreso Latinoamericano de Paleontología (Brazil), Actas 1, p. 19–30.

Baldis, B., Beresi, M., Bordonaro, O., and Vaca, A., 1982, Síntesis evolutiva de la Precordillera argentina: 5th Congreso Latinoamericano de Geología, Actas 4, p. 399–445.

Baldis, B., Beresi, M., Bordonaro, O., and Vaca, A., 1984, The Argentina Precordillera as a key to the Andean structure: Episodes, v. 7, p. 14–19.

Baldis, B., Bordonaro, O., Armella, C., Beresi, M., Cabalieri, N., Peralta, S., and Bastías H., 1989, La cuenca Paleozoico Inferior de la Precordillera argentina, in Chebli, G. A, and Spalletti, L. A., eds., Cuencas sedimentarias argentinas: Tucumán, Universidad de Tucumán Serie Correlación Geológica No. 6, p. 101–122.

Balme, B. E., 1980, Palynology and the Carboniferous-Permian boundary in Australia and other Gondwana continents: Palynology, v. 4, p. 43–55.

Barberena, M. C., Araujo, D. F., Lavina, E. L., and Azevedo, S. K., 1985a, O estado atual do conhecimento sobre os tetrapodes permianos e triassicos do Brasil meridional: Brasil meridional: Brasilia, Coletania de Trabalhos Paleontologicos, DNPM, serie Geologia 27, Paleontologia e Estratigrafia 2, p. 21–28.

Barberena, M. C., Araujo, D. C., and Lavina, E. L., 1985b, Late Permian and Triassic tetrapods of southern Brazil: National Geographic Review, v. 1, p. 5–20.

Barberena, M. C., Araujo-Barberena, D. C., Lavina, E. L., and Faccini, U. F., 1988, The evidence for close paleofaunistic affinity between South America and Africa, as indicated by Late Permian and Early Triassic tetrapods: Sao Paulo, Brazil, Seventh Gondwana Symposium Abstracts, p. 37.

Bastías, H, Baraldo, J., and Pina, L., 1984, Afloramientos calcáreos en el borde oriental del Valle del Bermejo, provincia de San Juan: Revista de la Asociación Geológica Argentina, v. 39, p. 153–155.

Beck, M. E., García, A., Burmester, R. F., Munizaga, F., Hervé, F., and Drake, R., 1991, Paleomagnetism and geochronology of late Paleozoic granitic rocks from the Lake District of southern Chile: Implications for accretionary tectonics: Geology, v. 19, p. 332–335.

Bellosi, E. S., and Jalfin, G. A., 1984, Litoestratigrafía y evolución paleoambiental Neopaleozoica de las Islas Malvinas, Argentina: Bariloche, 9th Congreso Geológico Argentino, Actas 5, p. 66–86.

Benedetto, J. L., Racheboeuf, P., Herrera, Z., Brussa, E., and Toro, B. A., 1992, Brachiopods siluriens et eodevoniens de la Formation Los Espejos, Precordillera de l'Argentine: Geobios, v. 25, p. 599–637.

Bercowski, F., 1987, Estudio sedimentológico y paleoambiental del Carbonífero de la Quebrada de la Laja, Sierra Chica del Zonda [Ph.D. thesis]: University of Buenos Aires, 156 p.

Bercowski, F., and Milana, J. P., 1990, Sedimentación glacimarina: nueva interpretación para la Formación Guandacol (Carbonífero) en el perfil de Río Francia, Precordillera Central, San Juan: 3rd Reunión Argentina de Sedimentología, Actas, p. 37–42.

Bodenbender, G., 1911, Constitución geológica de la parte meridional de la provincia de La Rioja y regiones limítrofes, Constitución geológica y sus productos minerales: Boletín de la Academia Nacional de Ciencias de Córdoba, v. 39, 220 p.

Bonaparte, J. F., 1966, Cronología de algunas formaciones triásicas argentinas basadas en restos de tetrápodos: Revista Asociación Geológica Argentina, v. 21, p. 20–38.

Bonaparte, J. F., 1973, Edades/reptil para el Triásico de Argentina y Brasil: 5th Congreso Geológico Argentino, Actas 3, p. 93–130.

Borrello, A. V., and Cuerda, A. J., 1965, Grupo Río Huaco, norte de la Precordillera de San Juan, Jáchal-Huaco: Comisión de Investigaciones

Científicas, La Plata, Notas 6, no. 1, p. 3–15.

Bossi, G., 1971, Análisis de la cuenca Ischigualasto-Ischichuca: 1st Congreso Hispano-Lusitano-Americano de Geología Económica, Madrid-Lisboa, Actas 2, sección 1 (Geología), p. 611–626.

Bossi, G., 1975, Geología de la cuenca de Marayes–El Carrizal (provincia de San Juan, República Argentina): 6th Congreso Geológico Argentino, Actas, p. 23–28.

Bossi, G., and Andreis, R. R., 1986, Secuencias deltaicas y lacustres del Carbonífero del centro-oeste argentino: 10th Congrès Internationale de Stratigrafie et Géologie du Carbonifère (Madrid, 1983), Compte Rendu, v. 3, p. 285–309.

Bracaccini, O. I., 1946, Contribución al conocimiento geológico de la Precordillera sanjuanina-mendocina: Buenos Aires, Boletín de Informaciones Petroleras, v. 258/64.

Bracaccini, O. I., 1960, Lineamientos principales de la evolución estructural de la Argentina: Buenos Aires, Petrotecnia, v. 10, p.57–69.

Breitkreuz, C., Bahlburg, H., Delakowitz, B., and Pichowiak, S., 1989, Paleozoic volcanic events in the Central Andes: Journal of South American Earth Sciences, v. 2, p. 171–189.

Buatois, L. A., and Mangano, L. A., 1992, Dinámica de sedimentación en el Carbonífero lacustre del extremo noroccidental de la Sierra de Narváez, Catamarca, Argentina: 4th Reunión Argentina de Sedimentología, Actas 2, p. 223–230.

Büggisch, W., 1987, Stratigraphy and very low grade metamorphism of the Sierras Australes de la Provincia de Buenos Aires (Argentina) and implications in Gondwana correlation: Zentralblatt für Geologie und Paläontologie, v. 1, p. 819–837.

Büggisch, W., and Astini, R. A., 1993, The Late Ordovician Ice Age: New evidence from the Argentine Precordillera, *in* Findlay, R. H., Unrug, H., Banks, M. R., and Veevers, J. J., eds., Gondwana eight: Rotterdam, A. A. Balkema, p. 439–448.

Caminos, R., 1979, Cordillera Frontal: Córdoba, Geología Regional Argentina, Academia Nacional de Ciencias, v. 1, p. 398–453.

Caminos, R., 1985, El magmatismo neopaleozoico en la Argentina. Síntesis y principales problemas: Annual Meeting IUGS–IGCP Project 211 Late Paleozoic of South America, Proceedings, p. 1–15.

Caminos, R., Cordani, U. G., and Linares, E., 1979, Geología y geochronología de las rocas metamórficas y eruptivas de la Precordillera y Cordillera Frontal de Mendoza, República Argentina: 2nd Congreso Geológico Chileno, Actas 1, p. F43–F61.

Caminos, R., Cingolani, C., Hervé, F., and Linares, E., 1982, Geochronology of the Pre-Andean metamorphism and magmatism in the Andean Cordillera between latitudes 30° and 36°S: Earth Science Review, v. 18, p. 333–352.

Caminos, R., Llambías, E., Rapela, C., and Parica, C., 1988, Late Paleozoic–Early Triassic magmatic activity of Argentina and the significance of new Rb-Sr ages from northern Patagonia: Journal of South American Earth Sciences, v. 1, p. 137–145.

Cellini, N., Rodríguez, S., González, G., Balod, M., Guerin, D., Silva, D., and Vega, V., 1986, Interpretación de las relaciones de facies de las secuencias epiclásticas paleozoicas del Cerro Curamalal Grande, Sierras Australes bonaerenses: 1st Reunión Argentina de Sedimentología, Actas, p. 197–200.

Cesari, S. N., 1985, Bioestratigrafía y aspectos paleoambientales de la Formación Tupe en el faldeo oriental de la Sierra de Maz, provincia de La Rioja, República Argentina [Ph.D. thesis]: Buenos Aires, University of Buenos Aires, 259 p.

Cesari, S. N., 1986, Zonación palinológica del Carbonífero tardío de Argentina: 4th Congreso Argentino de Paleontología, Actas 1, p. 227–230.

Chang, K. L., 1975, Unconformity-bounded stratigraphic units: Geological Society of America Bulletin, v. 86, p. 1544–1552.

Cingolani, C., and Varela, R., 1973, Exámen geocronológico por el método Rubidio-Estroncio de las rocas ígneas de las Sierras Australes Bonaerenses: 5th Congreso Geológico Argentino, Actas 1, p. 349–371.

Cingolani, C., and Varela, R., 1975, Geocronología rubidio-estroncio de rocas ígneas y metamórficas de las Sierras Chica y Grande de Córdoba, República Argentina: 2nd Congreso Iberoamericano de Geología Económica, Anales, v. 1, p. 9–35.

Cingolani, C., Dalla Salda, L., Hervé, F., Munizaga, F., Pankhurst, R. J., Parada, M. A., and Rapela, C. W., 1991, The magmatic evolution of northern Patagonia; New impressions of pre-Andean and Andean tectonics, *in* Harmon, R. S., and Rapela, C. W., eds., Andean magmatism and its tectonic setting: Boulder, Colorado, Geological Society of America Special Paper 265, p. 29–43.

Coates, D. A., 1969, Stratigraphy and sedimentation of the Sauce Grande Formation, Sierra de la Ventana, southern Buenos Aires Province, Argentina: 1st Symposium on Geology and Paleontology of Gondwana, v. 2, p. 799–816.

Cobbold, P., Massabie, A. C., and Rossello, E. A., 1986, Hercynian wrenching and thrusting in the Sierras Australes Foldbelt, Argentina: Hercynica, v. 2, p. 135–148.

Cobbold, P., Gapais, D., and Rossello, E., 1991, Partitioning of transpressive motions within a sigmoidal foldbelt: The Variscan Sierras Australes, Argentina: Journal of Structural Geology, v. 13, p. 743–758.

Coira, B., and Koukharsky, M., 1970, Geología y petrología de la Sierra Brava, provincia de La Rioja, República Argentina: Revista de la Asociación Geológica Argentina, v. 25, p. 444–466.

Conaghan, P. J., 1984, Aapamire (string-bog) origin for stone-roll swarms and associated "fluvio-deltaic" coals in the Late permian Illawara Coal Measures of the southern Sydney Basin: Climatic, geomorphic and tectonic implications: Geological Society of Australia, Abstracts, v. 12, p. 106–109.

Cooper, M. R., and Kensley, B., 1984, Endemic South American Permian bivalve molluscs from the Ecca of South Africa: Journal of Paleontology, v. 58, p. 1360–1363.

Cortés, J. M., 1985, Vulcanitas y sedimentitas lacustres en la base del Grupo Choiyoi al sur de la Estancia Tambillos, Mendoza, Argentina: 4th Congreso Geologico Chileno, Actas 1, p. 89–108.

Cox, L. R., 1934, Lamelibranquios de los estratos gondwánicos del Uruguay: Montevideo, Boletín del Instituto de Geología y Perforaciones, v. 21, p. 3–13.

Criado Roque, P., 1979, Subcuenca de Alvear (Provincia de Mendoza), *in* Leanza, A. F., ed., Geología Regional Argentina: Córdoba, Academia Nacional de Ciencias, v. 1, p. 811–836.

Criado Roque, P., and Ibañez, G., 1979, Provincia geológica Sanrafaelino-Pampeana: Córdoba, Geología Regional Argentina, Academia Nacional de Ciencias, v. 1, p. 837–869.

Cuerda, A., 1965, Estratigrafía de los depósitos neopaleozoicos de la Sierra de Maz (provincia de La Rioja): 2nd Jornadas Geológicas Argentinas, Actas 3, p. 79–94.

Cuerda, A., 1969, Sobre las graptofaunas del Silúrico de San Juan: Revista de la Asociación Geológica Argentina, v. 6, p. 223–235.

Cuerda, A., and Furque, G., 1981, Depósitos carbónicos de la Precordillera de San Juan, Parte I, Comaca del Cerro La Chilca (Río Francia): Revista de la Asociación Geológica Argentina, v. 36, p. 187–196.

Daemon, R. F., and Quadros, L. P., 1970, Bioestratigrafia do Neopaleozoico da Bação do Paraná: 24th Congresso Brasileiro de Geologia, Sociedade Brasileira de Geologia, Proceedings, p. 355–412.

Dalla Salda, L., 1975, Geología y petrología del basamento cristalino en el área del Cerro El Cristo e Isla Martín García: Provincia de Buenos Aires, República Argentina [Ph.D. thesis]: Universidad Nacional La Plata, 276 p.

Dalmayrac, B., Laubacher, G., and Marocco, R., 1980a, Géologie des Andes Péruviennes [Travaille de doctorat]: Paris, ORSTOM, v. 122, 501 p.

Dalmayrac, B., Laubacher, G., Marocco, R., Martinez, C., and Tomasi, P., 1980b, La chaîne hecynienne d'Amérique du Sud: Structure et évolution d'un orogène intracratonique: Geologische Rundschau, v. 69, p. 1–21.

Dalziel, I.W. D., and Forsythe, R. D., 1985, Andean Evolution and the

Terrane Concept, *in* Howell, D. G., ed., Tectonostratigraphic terranes of the circum-Pacific region: Houston, Circum-Pacific Council for Energy and Mineral Resources, Earth Sciences Series, No. 1, p. 565–581.

Damborenea, S., 1974, Geología del Cerro Colorado del cementerio Barreal, provincia de San Juan (República Argentina): Revista de la Asociación Geológica Argentina, v. 29, p. 249–263.

Davidson, J., 1984, Introducción a la geología de Chile: Seminario de Actualización de la Geología de Chile: Apuntes, Servicio Nacional de Geología y Minería, miscelánea No. 4, p. B1–B24.

Davidson, J., Mpodozis, C., Godoy, E., Hervé, F., Pankhurst, R., and Brook, M., 1987, Late Paleozoic accretionary complexes on the Gondwana margin of southern Chile; Evidence from the Chonos Archipelago, *in* McKenzie, G. D., ed., Gondwana six: Structure, tectonics and geophysics: Washington, D.C., America Geophysical Union Monograph Series 40, p. 221–227.

De la Mota, H. F., 1946, Estudio geólogico en el Cerro Bola, al sur de Villa Unión, depto, General Lavalle, provincia de La Rioja [Ph.D. thesis]: La Plata, Universidad Nacional de La Plata, 124 p.

Dessanti, N. R., 1956, Descripción geológica de la Hoja 27c "Cerro Diamante" (Provincia de Mendoza): Buenos Aires, Boletín Dirección Nacional Geología y Minería, no. 85, 79 p.

Dessanti, N. R., and Caminos, R., 1967, Edades Potasio-Argón y posición estratigráfica de algunas rocas ígneas y metamórficas de la Precordillera, Cordillera Frontal y Sierras de San Rafael, provincia de Mendoza: Revista de la Asociación Geológica Argentina, v. 22, p. 135–162.

Dickins, J. M., 1984, Late Paleozoic glaciation: Australian Bureau of Mineral Resources Journal, v. 9, p. 162–169.

Dickinson, W. R., 1985, Interpreting provenance relations from detrital modes sandstones, *in* Zuffa, G. G., ed., Provenance of arenites: Dordrecht, Reidel Publication Company, p. 333–361.

Dickinson, W. R., and Rich, E., 1972, Petrologic intervals and petrofacies in the Great Valley sequence, Sacramento Valley, California: Geological Society of America Bulletin, v. 83, p. 3007–3024.

Dickinson, W. R., and Suczek, C. A., 1979, Plate tectonics and sandstone compositions: American Association of Petroleum Geologists Bulletin 63, p. 2164–2182.

Dickinson, W. R., and Valloni, R., 1980, Plate settings and provenance of sands in modern oceans: Geology, v. 8, p. 82–86.

Dickinson, W. R., Beard, L. S., Brakenridge, G. R., Erjavec, J. L., Ferguson, R. C., Innman, K. F., Knepp, R. A., Linberg, F. A., and Ryberg, P. T., 1983, Provenance of North American Phanerozoic sandstones in relation to tectonic setting: Geological Society of America Bulletin, v. 94, p. 222–235.

Di Paola, E. C., 1972, Litología de la sección media del Grupo Paganzo en las comarcas Paganzo-Amaná y Olta-Malanzán, provincia de La Rioja, República Argentina: Revista de la Asociación Geológica Argentina, v. 27, p. 179–187.

Donato, E. O., and Vergani, G., 1985, Geología del Devónico y Neopaleozoico de la Zona del Cerro Rincón, Provincia de Salta, Argentina: 4th Congreso Geológico Chileno, Actas, p. 262–283.

Du Toit, A. L., 1927, A geological comparison of South America with South Africa: With a palaeontological contribution by F. R. Cowper Reed: Carnegie Institute Publication 381, p. 1–157.

Espejo, I. S., 1986a, Análisis litofacial y composicional preliminar de la Formación Agua del Toro (Carbónico inferior?), Bloque de San Rafael, Mendoza: Annual Meeting IUGS–IGCP Project 211 Late Paleozoic of South America, Abstracts, p. 70–71.

Espejo, I. S., 1986b, Estudio composicional y paleoambiental de las formaciones Agua del Toro e Imperial (Neopaleozoico), Río Diamante, Mendoza: Buenos Aires, Informe Preliminar Beca de Iniciación, CONICET, unpublished.

Espejo, I. S., 1987a, Presencia de la zona NBG en la Cuenca San Rafael, República Argentina: 7th Simposio Argentino Paleobotánica y Palinología Actas, p. 55–58.

Espejo, I. S., 1987b, Estudio composicional y paleoambiental de las Formaciones Agua del Toro e Imperial (Neopaleozoico), río Diamante, Mendoza: Buenos Aires, Informe Final Beca de Iniciación, CONICET, unpublished.

Espejo, I. S., 1990, Análisis estratigráfico, paleoambiental y de proveniencia de la Formación El Imperial en los alrededores de los ríos Diamante y Actuel (Provincia de Mendoza) [Ph.D. thesis]: Buenos Aires, University of Buenos Aires, 338 p.

Espejo, I. S., and Cesari, S. N., 1987, Primer hallazgo de flora pérmica en la Cuenca San Rafael: Revista Asociación de la Geológica Argentina, v. 62, p. 472–474.

Espejo, I. S., and López-Gamundí, O. R., 1984, Depósitos continentales del Paleozoico superior en el sector central de la Precordillera sanjuanina: 9th Congreso Geológico Argentino, Actas, 5, p. 258–273.

Espejo, I. S., Andreis, R. R., and Mazzoni, M., 1991, Cuenca San Rafael, in El Sistema Pérmico en la República Argentina y en la República Argentina y en la República Oriental del Uruguay: 12th International Congress on Stratigraphy and Geology of the Carboniferous and Permian and Academia Nacional de Ciencias de Córdoba (preprint), p. 177–185.

Fauqué, L. E., and Limarino, C. O., 1991, El Carbonifero de Agua de Carlos (Precordillera de La Rioja), su importancia tectónica y paleoambiental: Revista Asociación de la Geológica Argentina, v. 46, p. 103–114.

Fernández-Seveso, F., and Alvarez, L., 1987, Interpretación de las secuencias deposicionales del ámbito oriental de la Cuenca Paganzo. Sector Paleovalle de Mascasín-Chepes, Chacaní y Bajo de Véliz: Buenos Aires, Contribution to Simposio Cuencas Sedimentarias, Yacimientos Petrolíferos Fiscales, Gerencia General, unpublished.

Ferrando, L. A., 1986, Estado actual del Paleozoico en Uruguay, *in* Amos, A. J., and Archangelsky, S., eds., Late Paleozoic of South America: Córdoba, Abstracts 3rd Meeting of Working Group, p. 44–54.

Flores, M. A., 1969, El bolsón de Las Salinas en la provincia de San Luis: 4th Jornadas Geológicas Argentinas, Actas 1, p. 311–327.

Flores, M. A., 1979, Cuenca de San Luis: Córdoba, Geología Regional Argentina, Academia Nacional de Ciencias, v. 1, p. 745–769.

Folk, R. L., 1980, Petrology of sedimentary rocks (second edition): Austin, Texas, Hemphill Publishing, 182 p.

Forsythe, R., 1982, The Late Paleozoic to Early Mesozoic evolution of southern South America: A plate tectonic interpretation: Geological Society London Journal, v. 139, p. 671–682.

Foster, C. B., and Waterhouse, J. B., 1988, The *Granulatosporites confluens* Oppel-zone and Early Permian marine faunas from the Grant Formation on the Barbwire Terrace, Canning Basin, Western Australia: Australian Journal of Earth Sciences, v. 35, p. 135–157.

Frakes, L. A., and Crowell, J. C., 1969, Late Paleozoic Glaciation: I. South America: Geological Society of America Bulletin, v. 80, p. 1007–1042.

França, A. B., and Potter, E. D., 1991, Stratigraphy and reservoir potential of glacial deposits of the Itararé Group (Carboniferous-Permian), Paraná basin, Brazil: American Association of Petroleum Geologists Bulletin 75, p. 62–85.

Franchi, M. R., Panza, J. L., and De Barrio, R. E., 1989, Depósitos triásicos y jurásicos de la Patagonia extraandina, *in* Chebli, G. A., and Spalletti, L. A., eds., Las cuencas sedimentarias argentinas: Tucumán, Universidad de Tucumán Serie de Correlación Geológica No. 6, p. 347–378.

Furque, G., 1963, Descripción geológica de la Hoja 17b-Guandacol (Provincia de La Rioja–San Juan): Buenos Aires, Boletín Dirección Nacional de Geología y Minería, No. 92.

Furque, G., 1965, Nuevos afloramientos del Paleozoico en la provincia de Buenos Aires: Revista Museo de La Plata, v. 5 (Geología 35), p. 239–243.

Furque, G., 1973, Descripción geológica de la Hoja 34m, Sierra de Pillahuincó, provincia de Buenos Aires: Buenos Aires, Boletín Dirección Nacional de Geología y Minería, No. 141, 71 p.

Furque, G., 1979, Descripción geológica de la Hoja 19c, Jáchal, provincia de San Juan: Buenos Aires, Boletín Dirección Nacional de Geología y Minería, No. 164, 79 p.

Furque, G., and Cuerda, A., 1979, Precordillera de La Rioja, San Juan and Mendoza: Córdoba, Geología Regional Argentina, Academia Nacional de Ciencias, v. 1, p. 455–522.

Giudici, A. R., 1971, Geología de las adyacencias del Río Diamante al este del cerro homónimo, provincia de Mendoza, República Argentina: Revista Asociación de la Geológica Argentina, v. 26, p. 439–458.

González, C. R., 1981, Pavimento glaciario en el Carbónico de la Precordillera: Revista de la Asociación Geológica Argentina, v. 36, p. 262–266.

González, C. R., 1983, Evidence of the neopaleozoic glaciation in Argentina, *in* Evenson, E., Schuchter, C., and Rabassa, J., eds., Tills and related deposits: Rotterdam, A. A. Balkema, p. 271–272.

González, C. R., 1985, Esquema bioestratigráfico del Paleozoico superior marino de la Cuenca Uspallata-Iglesia, República Argentina: Actas Geológica Lilloana, v. 16, p. 231–244.

González, C. R., and Bossi, G., 1986, Los depósitos del oeste de Jagüé, La Rioja: 4th Congreso Argentino de Paleontología y Bioestratigrafía, Actas 1, p. 231–236.

González, R. R., 1971, Edades radimétricas de algunos cuerpos eruptivos de Argentina: Revista de la Asociación Geólogica Argentina, v. 26, p. 411–412.

González Bonorino, G., 1975, Sedimentología de la Formación Punta Negra y algunas consideraciones sobre la geología regional de la Precordillera de San Juan y Mendoza: Revista Asociación Geológica Argentina, v. 30, p. 223–246.

González Díaz, E. F., 1972, Descripción geológica de la Hoja 27d, San Rafael, provincia de Mendoza: Buenos Aires, Boletín Dirección Nacional de Geología Nacional de Geología y Minería, No. 132, 127 p.

Gordillo, C., and Lencinas, A., 1979, Sierras Pampeanas de Córdoba y San Luis: Córdoba, Geología Regional Argentina, Academia Nacional de Ciencias, v. 1, p. 577–650.

Haller, M., and Ramos, V., 1984, Las ofiolitas famatinianas (eopaleozoico) de las provincias de San Juan y Mendoza: 9th Congreso Geológico Argentino, Actas 2, p. 66–83.

Harrington, H. J., 1933, Sobre la presencia de restos de flora de *Glossopteris* en las Sierras Australes de Buenos Aires: Revista Museo de La Plata, v. 34, p. 303–338.

Harrington, H. J., 1947, Explicación de las Hojas 33m y 34m, Sierras de Curamalal y de la Ventana, provincia de Buenos Aires: Buenos Aires, Boletín Dirección Nacional de Geología y Minería, No. 61.

Harrington, H. J., 1955, The Permian *Eurydesma* fauna of eastern Argentina: Journal of Paleontology, v. 24, p. 112–128.

Harrington, H. J., 1972, Sierras Australes de Buenos Aires, in Leanza, A. F., ed., Geología Regional Argentina: Córdoba, Academia Nacional de Ciencias, p. 395–405.

Harrington, H. J., 1980, Sierras Australes de la Provincia de Buenos Aires: Córdoba, Geología Regional Argentina, Academia Nacional de Ciencias, v. 2, p. 967–983.

Heredia, S. E., 1982, *Pygodus anserynus* Lamont y Lindstrom (Conodonte) en el Llandeiliano de la Formación Ponón Trehué, provincia de Mendoza, Argentina: Ameghiniana, v. 19, p. 229–233.

Hervé, F., 1988, Late Paleozoic Subduction and Accretion in Southern Chile: Episodes, v. 11, p. 183–188.

Hervé, F., Munizaga, F., Parada, M. A., Brook, M., Pankhurst, R. J., Snelling, N. J., and Drake, R., 1988, Granitoids of the Coast Range of central Chile: Geochronology and geologic setting: Journal of South American Earth Sciences, v. 1, p. 185–194.

Hünicken, M., and Pensa, M. V., 1972, Algunas novedades estratigráficas y tectónicas sobre los depósitos gondwánicos del Bajo de Véliz (San Luis): Boletín Asociación Geológica Córdoba, v. 1, p. 138.

Hünicken, M., and Pensa, M. V., 1981, Sedimentitas Paleozoicas: 8th Congreso Geológico Argentino, Relatorío, p. 55–77.

Iñíguez, A. M., and Andreis, R. R., 1971, Carácteres sedimentológicos de la Formación Bonete, Sierras Australes de la Provincia de Buenos Aires, *in* Simposio Sierras Australes: Buenos Aires, Comisión de Investigaciones

Científicas y Técnicas Provincia de Buenos Aires, p. 103–120.

Iñíguez, A. M., Andreis, R. R., and Zalba, P., 1988, Eventos piroclásticos en la Formación Tunas (Pérmico), Sierras Australes, provincia de Buenos Aires, República Argentina: 2nd Jornadas Geológicas Bonaerenses, Actas, p. 383–395.

Jalfin, G. A., and Bellosi, E. S., 1983, Analisis estratigráfico de la Formación Bahía Choiseul, Permico de la Isla Soledad, Islas Malvinas—República Argentina: Revista de la Asociación Geológico Argentina, v. 38, p. 248–262.

Japas, S., 1986, Caracterización geométrica-estructural del Grupo Pillahuincó: I. Perfil del Arroyo Atravesado, Sierra de Las Tunas, Sierras Australes de la Provincia de Buenos Aires: Anales Academia de Ciencias Exactas, Físicas y Naturales, v. 38, p. 145–156.

Kay, S. M., Ramos, V. A., and Kay, R., 1984, Elementos mayoritarios y trazas de las vulcanitas ordovícicas en la Precordillera occidental: Basaltos de rift oceánicos tempranos(?) próximos al márgen continental: 9th Congreso Geológico Argentino, Actas 2, p. 48–65.

Kay, S. M., Ramos, V. A. Mpodozis, C., and Sruoga, P., 1989, Late Paleozoic to Jurassic silicic magmatism at the Gondwana margin: Analogy to the Middle Proterozoic in North America?: Geology, v. 17, p. 324–328.

Keidel, J., 1916, La geología de las sierras de la provincia de Buenos Aires y sus relaciones con las montañas de Sudáfrica y Los Andes: Buenos Aires, Ministerio de Agricultura, Sección Geología, Anales 11, No. 3.

Kilmurray, J. O., 1968, Petrología de las rocas cataclásticas y el skarn del anticlinal del Cerro Pan de Azúcar (partido de Saavedra, Buenos Aires): 3rd Jornadas Geológicas Argentinas, Actas 3, p. 217–238.

Kokogian, D. A., 1991, Estratigrafía secuencial en cuencas continentales, cuenca cuyana, Argentina: 4th Simposio Bolivariano Exploración Petrolera en las Cuencas Subandinas, Bogotá, Actas, trabajo 1.

Kokogian, D. A., and Mancilla, O., 1989, Análisis estratigráfico secuencial de la cuenca cuyana, *in* Chebli, G. A., and Spalletti, L. A., eds., Las cuencas sedimentarias argentinas: Tucumán, Universidad Nacional de Tucumán Serie de Correlación Geológica, No. 6, p. 169–202.

Kokogian, D. A., Fernández-Seveso, F., and Legarreta, L., 1987, Cuenca Ischigualasto–Villa Unión: Análisis estratigráfico y caracterización paleoambiental: Contribution to Simposio de Cuencas Sedimentarias: Yacimientos Petrolíferos Fiscales, Gerencia General, unpublished.

Laubacher, G., and Megard, F., 1985, The Hercynian basement: A review, *in* Pitcher, W. S., Atherton, M. P., Cobbing, E. J., and Beckinsale, R. D., eds., Magmatism at a plate edge—The Peruvian Andes: New York, John Wiley and Sons, p. 29–35.

Lavina, E. L., 1983, *Procolophon pricei* sp. n., um novo reptil procolofonideo do Triassico do Rio Grande do Sul.: Porto Alegre, Iheringia, Serie Geología, v. 9, p. 51–78.

Lech, R. R., 1991, The marine transgression at the Carboniferous-Permian boundary from western central Argentina: 12th International Congress on Carboniferous and Permian Geology and Stratigraphy, Buenos Aires, Abstracts, p. 56.

Legarreta, L., and Kokogian, D. A., 1986, Carácter regional e interregional de las discontinuidades triásicas: Cuencas de Ischigualasto–Villa Unión, Las Salinas y Cuyana: Secuencias deposicionales: Yacimintos Petrolíferos Fiscales, Gerencia General, unpublished.

Leguizamón, M. A., and Teruggi, M., 1985, Contribución al conocimiento petrológico-estructural de las rocas de la Formación Hinojo, Sierras Australes de la provincia de Buenos Aires: 1st Jornadas Geológicas Bonaerenses, Resúmenes, p. 229.

Leone, E. M., 1986, Geología de los cerros Tornsquist y Recreo, Sierra de la Ventana, provincia de Buenos Aires: Revista Asociación Geológica Argentina, v. 41, p. 117–123.

Lesta, P., Mainardi, E., and Stubelj, R., 1980, Plataforma continental argentina: Córdoba, Geología Regional Argentina, Academia Nacional de Ciencias, v. 2, p. 1577–1602.

Limarino, C. O., 1984, Areniscas eólicas en la Formación La Colina (Paleozoico superior), Cuenca Paganzo, República Argentina: Revista Asocia-

ción Geológica Argentina, v. 39, p. 58–67.

Limarino, C. O., 1985, Estratigrafía y paleoambientes de sedimentación del Grupo Paganzo en el Sistema de Famatina [Ph.D. thesis]: Buenos Aires, University of Buenos Aires, 356 p.

Limarino, C. O., and Caselli, A., 1992, Análisis paleoambiental de los conglomerados del Miembro inferior de la Formación Cortaderas (Carbonífero Inferior de la cuenco Río Blanco): Implicancias tectónicas y estratigráficas: 4th Reunión Argentina de Sedimentología, Actas 1, p. 143–150.

Limarino, C. O., and Cesari, S. N., 1987, Consideraciones sobre la edad de la sección superior del Grupo Paganzo (Paleozoico Superior), República Argentina: Bolivia, 4th Congreso Latinoamericano de Paleontología, Actas 1, p. 315–330.

Limarino, C. O., and Cesari, S. N., 1988, Paleoclimatic significance of the lacustrine Carboniferous deposits in Northwest Argentina: Palaeogeography, Palaeoclimatology, Palaeoecology, v. 65, p. 115–131.

Limarino, C. O., and Page, R. F., 1987, Nuevos depósitos de diamictitas en unidades carboníferas del noroeste de Argentina: Annual Meeting of the Argentine Working Group, IUGS–IGCP Project 211 Late Paleozoic of South America, Abstracts, p. 20–22.

Limarino, C. O., and Sessarego, H., 1986, Depósitos lacustres de las Formaciones Ojo de Agua y De la Cuesta (Pérmico), Provincias de San Juan y La Rioja: 1st Reunión Sedimentología Argentina, Actas, p. 145–148.

Limarino, C. O., and Spalletti, L., 1986, Eolian Permian deposits in west and northwest Argentina: Sedimentary Geology, v. 49, p. 109–127.

Limarino, C. O., Sessarego, H., Cesari, S. N., and López-Gamundí, O. R., 1986, El perfil de la Cuesta de Huaco, estratotipo de referencia (hipoestratotipo) del Grupo Paganzo en la Precordillera Central: Buenos Aires, Anales, Academia Nacional de Ciencias Exactas, Físicas y Naturales, v. 38, p. 81–109.

Limarino, C. O., Sessarego, H., López-Gamundí, O. R., Gutiérrez, P., and Cesari, S. N., 1987, Las formaciones Ojo de Agua ;y Vallecito en el área de la Ciénaga Oeste de Huaco, Provincia de San Juan: Estratigrafía y paleoambientes sedimentarios: Revista de la Asociación Geológica Argentina, v. 62, p. 153–167.

Linares, E., Manavella, M. A., and Piñeiro, A., 1979, Geocronología de las rocas efusivas de las zonas de los yacimientos "Dr. Baulíes" y "Los Reyunos," Sierra Pintada de San Rafael, Mendoza, República Argentina: 7th Congreso Geológico Argentino, Actas, 2, p. 13–21.

Linares, E., Llambías, E. J., and Latorre, C. O., 1980, Geología de la provincia de La Pampa, República Argentina y geocronología de sus rocas metamórficas y eruptivas: Revista de la Asociación Geológica Argentina, v. 35, p. 87–146.

Linares, E., Panarello, H. O., Valencio, S. A., and Garcia, C. M., 1982, Isótopos del carbono y oxígeno y el origen de las calizas de las Sierras Chica de Zonda y Pie de Palo, provincia de San Juan: Revista de la Asociación Geológica Argentina, v. 3;7, p. 80–90.

Llambías, E. J., and Caminos, R., 1987, El Magmatismo neopaleozoico en la Argentina, in Archangelsky, S., ed., El Sistema Carbonífero en la República Argentina: Córdoba, Academia Nacional de Ciencias, p. 253–279.

Llambías, E. J., and Prozzi, C. E., 1975, Ventania: Relatorio Geología de la provincia de Buenos Aires: 6th Congreso Geológico Argentino, p. 70–101.

Llambías, E. J., and Sato, A. M., 1990, El batolito de Colangüil (29°–31°S): Estructura y maco tectónico: Revista Geológica de Chile, v. 17, p. 89–108.

Llambías, E. J., Caminos, R., and Rapela, C. W., 1984, Las plutonitas y vulcanitas del ciclo eruptivo gondwánico: 9th Congreso Geológico Argentino, Relatorio, v. 1, p. 85–117.

López-Gamundí, O. R., 1984, Origen y sedimentología de las diamictitas del Paleozoico superior de la República Argentina (con especial referencia a la Cuenca Calingasta-Uspallata) [Ph.D. thesis]: Buenos Aires, Universidad de Buenos Aires, 262 p.

López-Gamundí, O. R., 1987, Depositional models for the glaciomarine sequences of Andean Late Paleozoic basins of Argentina: Sedimentary; Geology, v. 52, p. 109–126.

López-Gamundí, O. R., 1988, Glaciomarine apron in the Calingasta-Uspallata basin: Annual Meeting IUGS–IGCP Project 260 Earth's Glacial Record, Abstracts, p. 14.

López-Gamundí, O. R., 1989, Postglacial transgressions in Late Paleozoic basins of western Argentina: A record of glacio-eustatic sea level rise: Palaegeography, Palaeoclimatology, Palaeoecology, v. 71, p. 257–270.

López-Gamundí, O. R., and Alonso, M. S., 1982, Areas de proveniencia de las sedimentitas carbónicas (F. Majaditas, F. Leoncito y F. Hoyada Verde), Barreal, provincia de San Juan, Argentina: Buenos Aires, 5th Congreso Latinoamericano de Geología, Actas 2, p. 481–490.

López-Gamundí, O. R., and Amos, A. J., 1985, Consideraciones paleoambientales de las secuencias carbónicas del sector precordillerano de la cuenca Calingasta-Uspallata, San Juan y Mendoza: 1st Jornadas sobre Geología de la Precordillera, Serie "A," Monografías y Reuniones, No. 2, Asociación Geológica Argentina, Actas, p. 289–294.

López-Gamundí, O. R., and Espejo, I. S., 1987, Petrofacial analysis of Late Paleozoic sandstones of western Argentina: Its paleotectonic significance: AAPG Annual Convention, Abstracts, American Association Petroleum Geologists Bulletin, v. 71, p. 585.

López-Gamundí, O. R., and Espejo, I. S., 1988, Análisis petrofacial de las epiclastitas neopaleozoicas en los alrededores del río Sassito (San Juan): áreas de proveniencia e implicancia paleogeográfica: Revista de la Asociación Argentina, v. 43, p. 91–105.

López-Gamundí, O. R., and Rossello, E. A., 1993, The Devonian-Carboniferous unconformity in Argentina and its relation to the Eo-Hercynian orogeny in southern South America: Geologische Rundschau, v. 82, p. 136–147.

López-Gamundí, O. R., and Sessarego, H., 1988, Devonian to Permian paleogeographic history of the central segment of the Precordillera (San Juan province, Argentina): Implications for the geotectonic evolution of the paleo-Pacific margin of Gondwanaland: Sao Paulo, 7th Gondwana Symposium Abstracts, p. 113.

López-Gamundí, O. R., Page, S., Ramos, A., Remesal, M., and Espejo, I. S., 1985, Redefinición litoestratigráfica del Grupo Río Huaco: Características genéticas de las Formaciones Cerro Morado y Cauquenes (Triásico) en la Ciénaga del Vallecito, San José de Jáchal, San Juan: 1st Jornadas sobre Geología de la Precordillera, Serie "A," Monografias y Reuniones, No. 2, Asociación Geológica Argentina, Actas 1, p. 65–70.

López-Gamundí, O. R., Espejo, I. S., and Alonso, M. S., 1986, Evolución de las modas detríticas de las secuencias neopaleozoicas y triásicas de la zona de Huaco, San Juan, República Argentina: 1st Reunión Argentina Sedimentología, Actas, p. 256–258.

López-Gamundí, O. R., Azcuy, C. L., Cuerda, A., Valencio, D., and Vilas, J., 1987, Cuencas Río Blanco y Uspallata, in Archangelsky, S., ed., El Sistema Carbonífero en la República Argentina: Córdoba, Academia Nacional de Ciencias, p. 101–132.

López-Gamundí, O. R., and 9 others, 1989, Cuencas intermontanas, in Chebli, G. A., and Spalletti, L. A., eds., Las cuencas sedimentarias argentinas: Tucumán, Universidad de Tucumán Serie Correlación Geológica No. 6, p. 123–168.

López-Gamundí, O. R., Espejo, I. S., and Alonso, M. S., 1990, Sandstone composition changes and paleocurrent reversal in the Upper Paleozoic and Triassic deposits of the Huaco area, western Paganzo Basin, west-central Argentina: Sedimentary Geology, v. 66, p. 99–111.

López-Gamundí, O. R., Cesari, S. N., and Limarino, C. O., 1993, Paleoclimatic significance and age constraints of the Carboniferous coals of Paganza basin, western Argentina, in Findlay, R. H., Unrug, H., Banks, M. R., and Veevers, J. J., eds., Gondwana eight: Rotterdam, A. A. Balkema, p. 291–298.

Lucero Michaut, H. N., 1979, Sierra Pampeanas del norte de Córdoba, sur de Santiago del Estero, borde oriental de Catamarca y ángulo sudeste de Tucumán: Córdoba, Geología Regional Argentina, Academia Nacional de Ciencias, v. 1, p. 293–347.

Maack, R., 1951, Comentarios sobre "Geological map of South America"

1950: Arquivas de Biologia e Tecnologia, V-VI, v. 15, p. 173–196.

Manceñido, M., 1973, La fauna de la Formación del Salto (Paleozoico superior de la provincia de San Juan), Parte I: Introducción y estratigrafía: Ameghiniana, v. 10, p. 235–253.

Martínez, M., 1990, Marine fauna occurrence in the Guandacol Formation (Carboniferous) at Agua Hedionda, San Juan, northeastern Precordillera: Buenos Aires, 12th International Congress on Carboniferous and Permian Geology and Stratigraphy Abstracts, p. 3.

Martini, P., and Glooschenko, W., 1985, Cold climate peat formation in Canada, and its relevance to Lower Permian coal measures of Australia: Earth Science Reviews, v. 22, p. 1–107.

Massabie, A. C., and Rossello, E. A., 1984, La discordancia pre–Formación Sauce Grande y su entorno estratigráfico, Sierras Australes de la provincia de Buenos Aires: 9th Congreso Geológico Argentino, Actas I, p. 337–352.

Melchor, R., 1990, Sedimentitas plantíferas Eopérmicas de la Formación Carapacha en las cercanías de Puelches, provincia de La Pampa: Análisis Paleoambiental e Importancia: 3rd Reunión Argentina de Sedimentología, Actas, p. 366–371.

Milana, J. P., and Bercowski, F., 1987, Rasgos erosivos y deposicionales glaciales en el Paleozoico de Precordillera Central, San Juan, Argentina: Annual Meeting of the Working Group, IUGS–IGCP Project 211 Late Paleozoic of South America, Abstracts, p. 56–59.

Milana, J. P., and Bercowski, F., 1990, Facies y geometría de depósitos glaciales en un paleovalle carbonífero de Precordillera Central, San Juan, Argentina: 3rd Reunión Argentina de Sedimentología, Actas, p. 199–204.

Milana, J. P., Bercowski, F., and Lech, R. R., 1987, Análisis de una secuencia marino-continental neopaleozoica en la región del río San Juan, Precordillera Central, Argentina: 10th Congreso Geológico Argentino, Actas III, p. 113–116.

Mones, A., and Figueiras, A., 1981, A geo-paleontological synthesis of the Gondwana formations of Uruguay, *in* Cresswell, M. M., and Vella, P., eds., Gondwana five: Rotterdam, A. A. Balkema, p. 47–52.

Mozetic, A., 1974, El Triásico de los aledaños al valle del río Bermejo: Provincia de La Rioja y San Juan [Ph.D. thesis]: Universidad de Buenos Aires, 129 p.

Mpodozis, C., and Forsythe, R. D., 1983, Stratigraphy and geochemistry of accreted fragments of the ancestral Pacific Ocean floor in southern South America: Palaeogeography, Palaeoclimatology, Palaeoecology, v. 41, p. 103–124.

Mpodozis, C., and Kay, S. M., 1990, Provincias magmáticas ácidas y evolución tectónica de Gondwana: Andes Chilenos (28°–31°S): Revista Geológica de Chile, v. 17, p. 153–180.

Mpodozis, C., and Kay, S. M., 1992, Late Paleozoic to Triassic evolution of the Gondwana margin: Evidence from Chilean Frontal Cordilleran batholiths (28° to 31°S): Geological Society of America Bulletin 104, p. 999–1014.

Mpodozis, C., and Ramos, V., 1989, The Andes of Chile and Argentina, in Ericksen, G. E., Cañas Pinochet, M. T., and Reinemund, J. A., eds., Geology of the Andes and its relation to hydrocarbon and mineral resources: Houston, Circum-Pacific Council for Energy and Mineral Resources, Earth Science Series, no. 11, p. 59–88.

Nasi, C., and Sepúlveda, P., 1986, Avances en el conocimiento del Carbonífero en el norte de Chile: Annual Meeting of the Working Group, IUGS–IGCP Project 211 Late Paleozoic of South America, Abstracts, p. 27–43.

Nasi, C., Mpodozis, C., Cornejo, P., Moscoso, R., and Maksaev, V., 1985, El Batolito Elqui-Limari (Paleozoico superior-Triásico): Características petrográficas, geoquímicas y significado tectónico: Revista Geológica de Chile, v. 24/25, p. 77–111.

North American Commission on Stratigraphic Nomenclature, 1983, North American Code: American Association of Petroleum Geologists Bulletin 67, p. 841–875.

Nuñez, E., 1960, Sobre la presencia del Paleozoico inferior fosilífero en el Bloque de San Rafael: 1st Jornadas Geológicas Argentinas, Anales 2, p. 185–189.

Oelofsen, B. W., 1987, The biostratigraphy and fossils of the Whitehill and Iratí Shale Formations of the Karoo and Paraná Basins: American Geophysical Union, Geophysical Monograph 41, p. 131–138.

Ortíz, A., and Zambrano, J., 1981, La provincia geológica de Precordillera Oriental: 8th Congreso Geológico Argentino, Actas 3, p. 59–74.

Ottone, E. G., 1985, Estudo actual del conomiento paleofloristico de la Formación Santa Máxima, Paleozoico superior, provincia de Mendoza: Tucumán, 6th Simposio Argentino Paleobotanica y Palinología, p. 8.

Ottone, E. G., and Garcia, B., 1991, A Lower Triassic microspore assemblage from the Puesto Viejo Formation, Argentina: Palebotanical and Palynological Review, v. 68, p. 217–232.

Padula, E. L., 1972, Subsuelo de la Mesopotamia y regiones adyacentes, *in* Leanza, A. F., ed., Geología Regional Argentina: Cordoba, Academia Nacional de Ciencias, p. 213–235.

Padula, E. L., Rolleri, E. O., Mingramm, A. R., Criado Roque, P., Flores, M. A., and Baldis, B. A., 1967, Devonian of Argentina: Calgary, International Symposium on the Devonian System, Proceedings, v. 2, p. 165–193.

Pankhurst, R. J., Rapela, C. W., Caminos, R., Llambías, E., and Parica, C., 1992, A revised age for the granites of the central Somuncura batholith, North Patagonian Massif: Journal of South American Earth Sciences, v. 5, p. 321–326.

Parada, M. A., Munizaga, F., and Kawashita, K., 1981, Antecedentes cronológicos de los ríos Elqui-Limari: Revista Geológica de Chile, v. 13/14, p. 87–93.

Peralta, S. H., 1990, Silúrico de la Precordillera del oeste argentino, in Chebli, G. A., and Spalletti, L. A., eds., Las cuencas sedimentarias argentinas: Tucumán, Universidad de Tucumán Serie Correlación Geológica No. 6, p. 113–117.

Peralta, S. H., and Carter, C., 1990, La glaciación gondwánica del Ordovícico tardío: Evidencias en fangolitas guijosas de la Precordillera de San Juan, Argentina: 10th Congreso Geológico Argentino, Actas 2, p. 181–185.

Polanski, J., 1966, Edades eruptivas suprapaleozoicas asociadas en el diastrofismo varíscico: Revista de la Asociación Geológica Argentina, v. 21, p. 5–19.

Quartino, B., Zardini, R., and Amos, A., 1971, Estudio y exploración de la región Barreal-Calingasta, provincia de San Juan, República Argentina: Asociación Geológica Argentina Monografía 1, 184 p.

Ramos, E. D., and Ramos, V. A., 1979, Los ciclos magmáticos de la República Argentina: 8th Congreso Geológico Argentino, Actas 1, p. 771–786.

Ramos, V. A., 1982, Descripción geológica de la Hoja 20f "Chepes," provincia de La Rioja: Servicio Geológico Nacional, Boletín 171.

Ramos, V. A., 1984, Patagonia: Un continente a la deriva?: 9th Congreso Geológico Argentino, Actas 2, p. 311–328.

Ramos, V. A., 1988, Late Proterozoic–Early Paleozoic of South America—A collisional history: Episodes, v. 11, p. 168–174.

Ramos, V. A., and Kay, S. M., 1991, Triassic rifting and associated basalts in the Cuyo basin, central Argentina, *in* Harmon, R. S., and Rapela, C. W., eds., Andean magmatism and its tectonic setting: Boulder, Colorado, Geological Society of America Special Paper 265, p. 79–91.

Ramos, V. A., Jordan, T., Allmendinger, R., Mpodozis, C., Kay, S. M., Cortés, J., and Palma, M., 1986, Paleozoic terranes of the central Argentine-Chilean Andes: Tectonics, v. 5, p. 855–880.

Rapela, C. W., and Kay, S. M., 1988, Late Paleozoic to Recent magmatic evolution of northern Patagonia: Episodes, v. 11, p. 175–182.

Rapela, C. W., Heaman, L. M., and McNutt, R. H., 1982, Rb-Sr geochronology of granitoid rocks from the Pampean Ranges, Argentina: Journal of Geology, v. 90, p. 574–582.

Rapela, C. W., Toselli, A., Heaman, L., and Saavedra, J., 1990, Granite plutonism of the Sierras Pampeanas: An inner cordilleran Paleozoic arc in the southern Andes, *in* Kay, S. M., and Rapela, C. W., eds., Plutonism from Antarctica to Alaska: Boulder, Colorado, Geological Society of America Special Paper 241, p. 77–90.

Rapela, C. W., Pankhurst, R. J., and Harrison, S. M., 1992, Triassic "Gondwana"

granites of the Gastre district, North Patagonian Massif: Royal Society of Edinburgh Transactions, Earth Sciences, v. 83, p. 291–304.

Roberts, J., Claoué-Long, J. C., and Jones, P. J., 1991, Calibration of the Carboniferous and Early Permian of the southern New England Orogen by SHRIMP ion microprobe zircon analyses: Newcastle Symposium on Advances in the Study of the Sydney Basin, Department of Geology, University of Newcastle, v. 25, p. 38–43.

Rocha-Campos, A. C., 1971, Upper Paleozoic and Lower Mesozoic paleogeography, and paleoclimatological and tectonic events in South America, *in* Logan, A., and Hills, L. V., eds., The Permian and Triassic Systems and their mutual boundary: Canadian Society of Petroleum Geologists Memoir 2, p. 398–424.

Rocha-Campos, A. C., and Archangelsky, S., 1985, South America, *in* Martinez Diaz, C., ed., The Carboniferous of the world, II. Australia, Indian subcontinent, South Africa, South America, and North Africa: Madrid, Instituto Geológico y Minero de España, v. 2, p. 175–270.

Rocha-Campos, A. C., Amaral, G., and Aparicio, E. P., 1971, Algunas edades K-Ar de la "Serie porfirítica" en la Precordillera y Cordillera Frontal de Mendoza, Republica Argentina: Revista de la Asociación Geológica Argentina, v. 26, p. 311–316.

Rocha-Campos, A. C., De Carvalho, R., and Amos, A. J., 1977, A Carboniferous (Gondwana) fauna from Subandean Bolivia: Sao Paulo, Boletim Instituto de Geologia, v. 6, p. 181–191.

Rodríguez, E., and Fernández Seveso, F., 1988, Origen y estructura de las ritmitas del lago eopérmico Bajo de Veliz, San Luis: 2nd Reunión Argentina de Sedimentología, Actas, p. 227–231.

Rolleri, E. O., and Baldis, B. A., 1969, Paleogeography and distribution of Carboniferous deposits in Argentine Precordillera: Paris, Gondwana Stratigraphy, IUGS–UNESCO, v. 2, p. 1005–1024.

Rolleri, E. O., and Criado Roque, P., 1968, La cuenca triásica del norte de Mendoza: 3rd Jornadas Geológicas Argentinas, Actas 1, p. 1–76.

Rolleri, E. O., and Criado Roque, P., 1969, Geología de la provincia de Mendoza: 4th Jornadas Geológicas Argentinas, Actas 2, p. 1–60.

Rolleri, E. O., and Fernández Garrasino, C. A., 1979, Comarca septentrional de Mendoza: Geología Regional Argentina: Córdoba, Academia Nacional de Ciencias, v. 1, p. 771–809.

Romer, A. S., and Jensen, J. A., 1966, The Chañares (Argentina): Triassic reptiles and fauna II: Sketch of the geology of the Río Chañares–Río Gualo region: Breviora, Cambridge, 407 p.

Runnegar, B., 1972, Late Palaeozoic bivalvia from South America: Provincial affinities and age: Anais da Academia Braziliera de Ciencias, v. 44 (suplemento), p. 295–312.

Russo, A., and Archangelsky, S., 1987, Cuenca Chacoparanaense, *in* Archangelsky, S., ed., El Sistema Carbónifero en la República Argentina: Córdoba, Academia Nacional de Ciencias, v. 1, p. 139–184.

Russo, A., Ferello, E., and Chebli, G., 1979, Llanura Chaco-Pampeana: Geología Regional Argentina: Córdoba, Academia Nacional de Ciencias, v. 1, p. 139–183.

Russo, A., Archangelsky, S., and Gamerro, J. C., 1980, Los depósitos suprapaleozoicos en el subsuelo de la llanura Chaco Paranaense, Argentina: Buenos Aires, 2nd Congreso Argentino de Paleontología y Bioestratigrafía and 1st Congreso Latinoamericano de Paleontología, Actas 4, p. 157–173.

Ruzycki, L., and Bercowski, F., 1987, Variaciones composicionales en psamitas neopaleozoicas de la quebrada de Rio de Agua, Precordillera Sanjuanina, Argentina: Annual Meeting IUGS–IGCP Project 211 Late Paleozoic of South America, Abstracts, p. 60–63.

Salfity, J., and Gorustovich, S., 1983, Paleogeografía de la cuenca del Grupo Paganzo (Paleozoico superior): Revista de la Asociación Geológica Argentina, v. 38, p. 437–453.

Scalabrini Ortiz, J., 1970, Litología, facies y biofacies del Carbónico marino en el norte de la Precordillera sanjuanina (zona del Ró Blanco) [Ph.D. thesis]: Buenos Aires, University of Buenos Aires, 234 p.

Scalabrini Ortiz, J., 1972, El Carbónico en el sector septentrional de la Precordillera sanjuanina: Revista de la Asociación Geológica Argentina, v. 27, p. 351–377.

Schobbenhaus, C., Campos, D. deA., Derze, G. R., and Asmus, H. E., eds., 1984, Geologia do Brasil: Brazilia, Departamento Nacional da Produçao Mineral (DNPM), 501 p.

Sellwood, B., 1978, Shallow-water carbonate environments, *in* H., Reading, ed., Sedimentary environments and facies: Oxford, Blackwell Scientific Publishers, p. 259–303.

Sessarego, H., 1986, Nuevos depósitos triásicos en la márgen norte del río San Juan, quebrada del Tigre, provincia de San Juan: estratigrafía y paleoambientes sedimentarios: Revista Asociación Argentina de Mineralogía, Petrología y Sedimentología, v. 17, p. 67–79.

Sessarego, H., 1988, Estratigrafí de las secuencias epiclásticas devónicas a triásicas aflorantes al norte del ró San Juan y al oeste de la sierra del Tigre, provincia de San Juan [Ph.D. thesis]: Buenos Aires, University of Buenos Aires, 324 p.

Sessarego, H., and Cesari, S. N., 1986, La zona (de conjunto) *Archaeosigillaria-Lepidodendropsis* del Carbónifero temprano de Argentina: Annual Meeting IUGS–IGCP Project 211, Late Paleozoic of South America Abstracts, p. 69–70.

Sessarego, H., and Limarino, C. O., 1987, Evolución del relleno sedimentario de las cuencas ubicadas detrás del arco magmático durante el Carbónico tardío-Pérmico: Annual Meeting IUGS–IGCP Project 211 Late Paleozoic of South America, Abstracts, p. 42–45.

Spalletti, L., Arrondo, O., MOrel, E., and Ganuza, D., 1988, Los depósitos fluviales de la cuenca triásica superior en el noroeste del Macizo Norpatagónico: Revista de la Asociación Geológica Argentina, v. 43, p. 544–557.

Stinco, L. P., 1986, Análisis litoestratigráfico del Miembro inferior de la Formación Cochicó [Trabajo Final de Licenciatura]: Universidad de Buenos Aires.

Stipanicic, P. N., 1979, El Triásico del Valle del Río de los Patos (provincia de San Juan): Córdoba, Geología Regional Argentina, Academia Nacional de Ciencias, v. 1, p. 695–774.

Stipanicic, P. N., 1983, The Triassic of Argentina and Chile, *in* Moullade, M., and Nairn, A. E., eds., The Phanerozoic geology of the world, II: The Mesozoic, B: Amsterdam, The Netherlands, Elsevier, p. 181–199.

Stipanicic, P. N., and Bonaparte, J. F., 1979, Cuenca triásica de Ischigualasto–Villa Unión (provincias de La Rioja y San Juan): Córdoba, Geología Regional Argentina, Academia Nacional de Ciencias, v. 1, p. 523–575.

Stipanicic, P. N., and Bonetti, M., 1969, Consideraciones sobre la cronología de los terrenos triásicos argentinos: Paris, Gondwana Stratigraphy, IUGS–UNESCO, v. 2, p. 1081–1119.

Strelkov, E. E., and Alvarez, L. A., 1984, Análisis estratigráfico y evolutivo de la Cuenca triásica Mendocina-Sanjuanina: 9th Congreso Geológico Argentino, Actas 3, p. 115–130.

Suero, T., 1957, Geología de la Sierra de Pillahuincó (Sierras Australes de la provincia de Buenos Aires): Partidos Coronel Pringles y Coronel Suárez: La Plata, Publicación LEMIT, ser. 2, v. 74, 31 p.

Suero, T., 1972, Compilación geológica de las Sierras Australes de la provincia de Buenos Aires: La Plata, LEMIT, Anales 3, p. 135–147.

Teruggi, M., 1964, Paleocorrientes y paleogeografía de las ortocuarcitas de la Serie La Tinta (provincia de Buenos Aires): Comisión Investigaciones Científicas de la Provincia de Buenos Aires, Anales 5, p. 1–27.

Teruggi, M., Andreis, R. R., Iñíguez, M., Abait, J., Mazzoni, M., and Spalletti, L., 1969, Sedimentology of the Paganzo beds at cerro Guandacol, province of La Rioja: Paris, Stratigraphy, IUGS-UNESCO, v. 2, p. 857–880.

Thompson, R., and Mitchell, J. G., 1972, Palaeomagnetic and radiometric evidence for the age of the lower boundary of the Kiaman magnetic interval in South America: Geophysical Journal of the Royal Astronomical Society, v. 27, p. 207–214.

Toubes, R. O., and Spikermann, J. P., 1976, Algunas edades K-Ar para la Sierra Pintada, provincia de Mendoza: Revista de la Asociación Geo-

lógica Argentina, v. 31, p. 118–126.

Toubes, R. O., and Spikermann, J. P., 1979, Nuevas edades K/Ar para la Sierra Pintada, provincia de Mendoza: Revista de la Asociación Geológica Argentina, v. 34, p. 73–79.

Truswell, E. M., 1980, Permo-Carboniferous palynology of Gondwanaland: Progress and problems in the decade to 1980: Australian Bureau of Mineral Resources Journal, v. 5, p. 95–111.

Turner, J. C., 1967, Descripción geológica de la Hoja 13b "Chaschuil" (provincias de Catamarca y La Rioja): Boletín Dirección Nacional de Geología y Minería, No. 106, 78 p.

Turner, J. C., and Méndez, V., 1975, Geología del sector oriental de los departamentos de Santa Victoria e Iruya, provincia de Salta, Republica Argentina: Córdoba, Boletín Academia Nacional de Ciencias, v. 51, p. 11–24.

Uliana, M., and Biddle, K., 1987, Permian to Late Cenozoic evolution of Northern Patagonia: Main tectonics events, magmatic activity and depositional trends, *in* McKenzie, G. D., ed., Gondwana Six: Structure, Tectonics and Geophysics: Washington, D.C., American Geophysical Union Monograph Series 40, p. 271–286.

Uliana, M., and Biddle, K., 1988, Mesozoic-Cenozoic paleogeographic and geodynamic evolution of southern South America: Revista Brasileira de Geociencias, v. 18, p. 172–190.

Uyeda, S., 1982, Subduction zones: An introduction to comparative subductology: Tectonophysics, v. 81, p. 1183–1185.

Valencio, D. A., and Mitchell, J., 1972, Edad potasio-argón y paleomagnetismo de rocas ígneas de las Formaciones Quebrada del Pimiento y Las Cabras, provincia de Mendoza, República Argentina: Revista de la Asociación Geológica Argentina, v. 27, p. 170–178.

Varela, R., Dalla Salda, L., and Cingolani, C., 1985, Estructura y composición geológica de la Sierras Colorado, Chasico y Cortapie, Sierras Australes de Buenos Aires: Revista de la Asociación Argentina, v. 60, p. 254–261.

Vásquez, J. R., Gorroño, R. A., and Ivorra, J., 1981, El Paleozoico superior de las provincias de San Juan y La Rioja: Revista de la Asociación Argentina, v. 36, p. 89–98.

Vilas, J. F., and Valencio, D. A., 1977, Paleomagnetism of South American rocks and the Gondwana continent: Calcutta, 4th International Gondwana Symposium Papers, p. 923–930.

Vilas, J. F., and Valencio, D. A., 1982, Implicancias geodinámicas de los resultados paleomagnéticos de formaciones asignadas al Paleozoico Tardío–Mesozoico del centro-oeste argentino: Buenos Aires, 5th Congreso Latinoamericano de Geología, Actas 3, p. 743–758.

Volkheimer, W., 1967, Herpolitas en el Carbónico de la Sierra Pintada (con un perfil del río Atuel), provincia de Mendoza: Revista de la Asociación Geológica Argentina, v. 22, p. 75–78.

Volkheimer, W., 1978, Descripción geológica de la Hoja 27b "Cerro Sosneado" (provincia de Mendoza): Boletón Dirección Nacional de Geología y Minería, No. 151, 80 p.

von Gosen, W., and Büggisch, W., 1989, Tectonic evolution of the Sierras Australes Fold and Thrust Belt (Buenos Aires Province, Argentina)—An outline: Zentralblatt für Geologie und Palaontologie, v. 5/6, p. 947–958.

Yrigoyen, M., 1975, La edad cretácica del Grupo Gigante (San Luis) y su relación con cuencas circumvecinas: 1st Primer Congreso Argentino de Paleontología y Bioestratigrafía, Actas 2, p. 29–56.

Yrigoyen, M., 1981, Síntesis: Geología y Recursos Naturales de la Provincia de San Luis: 8th Congreso Geológico Argentino, Relatorio, p. 7–32.

Yrigoyen, M., and Stover, L. N., 1970, La palinología como elemento de correlación del Triásico en la Cuenca Cuyana: 4th Jornadas Geológicas Argentinas, Actas 2, p. 427–447.

Yrigoyen, M., Ortíz, A., and Manoni, R., 1989, Cuencas sedimentarias de San Luis, *in* Chebli, G. A., and Spalletti, L. A., eds., Las cuencas sedimentarias argentinas: Universidad de Tucumán Serie Correlación Geológica No. 6, p. 203–220.

Zalán, P. V., Wolff, S., Conceiçao, J. C., Astolfi, I., Vieira, S., Appi, V., and Zanotto, O., 1987, Tectónica e sedimentaçao da Baçía do Paraná: 3rd Simposio Sul-Brasileiro de Geologia, Curitiba, Atas 1, p. 441–477.

Zalán, P. V., Wolff, S., Astolfi, M. A., Vieira, I. S., Conceiçao, J. C., Appi, V. T., Neto, E. V., Cerqueira, J. R., and Marques, A., 1990, The Paraná Basin, Brazil, *in* Leighton, M. W., Kolata, D. R., Oltz, D. F., and Eidel, J. J., eds., Interior cratonic basins: American Association Petroleum Geologists Memoir 51, p. 681–708.

Zambrano, J. J., 1974, Cuencas sedimentarias en el subsuelo de la provincia de Buenos Aires y zonas adyacentes: Revista de la Asociación Geológica Argentina, v. 29, p. 443–469.

Zambrano, J. J., 1980, Comarca de la cuenca cretácica de Colorado: Córdoba, Geología Regional Argentina, Academia Nacional de Ciencias, v. 2, p. 1033–1070.

Zardini, R. A., 1962, Significado geológico de las serpentinitas de Mendoza: 1st Jornadas Geológicas Argentinas, Anales 2, p. 437–442.

MANUSCRIPT ACCEPTED BY THE SOCIETY SEPTEMBER 9, 1993

Geological Society of America
Memoir 184
1994

Synthesis

J. J. Veevers and C. McA. Powell*
Australian Plate Research Group, School of Earth Sciences, Macquarie University, North Ryde,
N.S.W. 2109, Australia
J. W. Collinson
Department of Geological Sciences and Byrd Polar Research Center, Ohio State University,
Columbus, Ohio 43210
O. R. López-Gamundí*
Departamento de Ciencias Geológicas, Facultad de Ciencias Exactas y Naturales, Universidad de
Buenos Aires, Ciudad Universitaria, Pabellón 2, Buenos Aires 1428, Argentina

ABSTRACT

The Permian-Triassic (Gondwanan) basins and foldbelts along the Panthalas-san margin of the Gondwanaland province of Pangea developed on a basement of Proterozoic and Paleozoic rocks in Antarctica and southern Africa and on a basement of foldbelts terminally deformed at the end of the Devonian (360 Ma) in southern South America and in the mid-Carboniferous (320 Ma) in eastern Australia. With the latest Carboniferous (290 Ma) onset of Pangean extension I, deposition resumed after a lacuna in Gondwanaland with glacigenic sediment. Together with post-Hercynian Europe on the other side of Pangea, postorogenic eastern Australia was subjected to continuing dextral transtension that produced an orocline, related pull-apart basins, and widespread volcanism. At the other end of the Panthalassan margin of Gondwanaland, a new magmatic arc and yoked foreland basin arose in southern South America at about 290 Ma, and by 275 Ma had propagated 4,000 km by migration of a junction of subduction parallel and normal to the margin to reach a point opposite Africa and the Ellsworth Mountains of Antarctica. This (Sakmarian) time saw an ephemeral postglacial marine transgression that flooded much of eastern and southern Australia, the south Atlantic margins of southern Africa and South America, and possibly the Transantarctic basin. The following regression was marked by widespread deposition of coal in all parts of the margin except southern South America. By 265 Ma, the magmatic arc and foreland basin had reached the Bowen Basin in northeastern Australia, and from 258 Ma to the 250 Ma end of the Permian, the foreland basin in Antarctica and Australia subsided rapidly beneath the load of the overthrusting magmatic orogen to accumulate a piedmont of thick tuffaceous coal measures.

Both coal and tuff disappeared in Antarctica and Australia at the Permian-Triassic boundary (250 Ma) and were succeeded by barren measures with redbeds,

*Present addresses: Powell, Department of Geology and Geophysics, University of Western Australia, Nedlands, W. A. 6009, Australia; López-Gamundí, Texaco, Inc., Frontier Exploration Department, 4800 Fournace Place, Bellaire, Texas 77401-2324.

Veevers, J. J., Powell, C. McA., Collinson, J. W., and López-Gamundí, O. R., 1994, Synthesis, *in* Veevers, J. J., and Powell, C. McA., eds., Permian-Triassic Pangean Basins and Foldbelts Along the Panthalassan Margin of Gondwanaland: Boulder, Colorado, Geological Society of America Memoir 184.

all probably as a result of the global greenhouse warming generated by the eruption of the Siberian Traps. The magmatic arc continued its northward migration, and plutonic activity in eastern Australia continued unabated. The intermittent thrusting of the foldbelt and adjacent foreland basin during the Permian (Gondwanides I) was followed in the mid-Triassic (235–230 Ma) by terminal thrusting along the entire margin (Gondwanides II). Pangean extension II in the Carnian (230 Ma) generated basins in the foldbelt upland, notably in southern South America and eastern Australia, as well as in the sump between the craton and the orogenic upland. Deposition of coal (oil shale in southern South America) resumed after an Early and Middle Triassic gap of 20 million years.

The Permian-Triassic (Gondwanan) sedimentary and foldbelt successions were capped in the Jurassic by a flood of silicic volcanics in southern South America and by an even bigger flood of tholeiitic basalt in southern Africa, Antarctica, and Tasmania, and scattered volcanics in southeastern Australia. East Antarctica was rifted from West Antarctica on one side, from Australia on another, and on yet another drifted from Africa by seafloor spreading.

INTRODUCTION

After summarizing information about New Zealand, we compile from the preceding regional chapters (1) time-space diagrams along the Panthalassan margin of the chief Permian-Triassic tectonic events and environments, (2) a set of paleo-tectonic/geographic stage maps, including a sketch of the paleogeography of the rest of Gondwanaland. To avoid ambiguity, we denote ages derived from stratigraphical evidence by Ma[*] (from Palmer, 1983, with the Permian-Triassic boundary at 250 Ma), and directions and latitudes in modern coordinates.

New Zealand

Information about New Zealand is compiled here and in Figure 1, and incorporated in Figures 2 and 3, which appear in following sections, and on the paleogeographic maps.

New Zealand is divided by the Median Tectonic Line (MTL) into an old Western Province and a young Eastern Province (Fig. 1, Korsch and Wellman, 1988). In the Western Province, ~265 Ma[*] shallow marine detrital sediment of the Parapara Group rests unconformably on Ordovician rocks (Korsch and Wellman, 1988, p. 449). The only other pre-Jurassic rocks are the infaulted Topfer Coal Measures (TCM), which Raine (1980) dated from miospores as Middle or Late Triassic. In the Eastern Province, the Brook Street terrane contains ~280 Ma ophiolite, generated in a small-ocean basin, succeeded by sediment with mid-Sakmarian (~275 Ma[*]) brachiopods, and ~260 Ma biotite granite (southern Longwoods SL); the Maitai Group near Waipahi (W) contains a clast of ~265 Ma granite (Kimbrough et al., 1992, p. 442). At the base of the Late Triassic–Jurassic Murihiku Supergroup on North Island, the Carnian to early Norian Moeatoa Conglomerate (Fig. 1) contains 226 ± 6 Ma granite clasts derived from a magmatic arc (Graham and Korsch, 1990). On South Island, a coastal facies in the Torlesse Supergroup, probably trench-

slope deposits ponded above older accretionary-wedge material (R. J. Korsch, personal communication, 1993), contains Ladinian (235–230 Ma[*]) coal measures (Retallack, 1987). The fluvial conglomerate in the coal measures reflects a rapidly rising upland (Retallack, 1987), which links with Bradshaw et al.'s (1981) model of the Rangitata I orogeny. According to Bradshaw et al. (1981, p. 220), Caples terrane (mainly Permian volcaniclastic sediments and metavolcanics, probably of trench-slope basin origin) and the Rahaia submarine fan of late Carboniferous or Permian to late Triassic mainly quartzo-feldspathic sandstone and mudstone of marginal marine to submarine fan origin "were folded and metamorphosed to form the Haast Schist by the end of the Triassic," with cooling ages up to 200 Ma, which we place on Fig. 2 as Ladinian (235–230 Ma[*]), at the time of Gondwanides II. In the Jurassic, "the Pahau sediments accumulate on Pacific facing slope to the SE of a new trench. The fore-arc basin continues to fill in the SW." Bradshaw et al. (1981) noted another orogenic pulse in the Late Permian, which we place at ~250 Ma[*], at the end of Gondwanides I.

TECTONICS, CHIEFLY SHORTENING DEFORMATION, EXTENSION, AND CONVERGENT-ARC MAGMATISM

Notable isochronous events or groups of events (Fig. 2)

(post–post-g). Flood basalt of the Paraná Basin (130 Ma Serra Geral) and Namibia (120 Ma Etendeka), the continental products of the Tristan da Cunha hotspot (Stothers, 1993). Contemporaneous are the Rajmahal Traps of India and the Bunbury Basalt of southwestern Australia (130–135 Ma[*]), continental products of the Amsterdam–St Paul hotspot.

(post-g). Grouped within the range of 200–175 Ma, from South America to Australia, are the following:

(1) Late Triassic (220 Ma) to Jurassic/Cretaceous (140

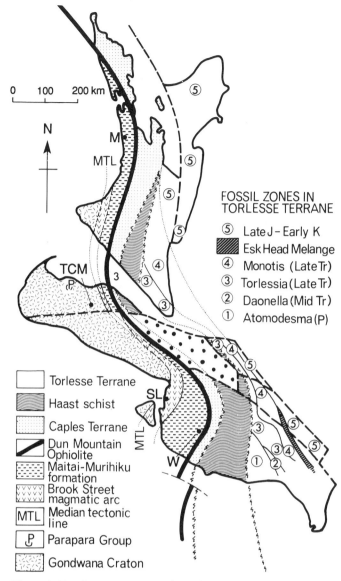

Figure 1. Pre-Cretaceous restoration of New Zealand, from Korsch and Wellman (1988) by permission of the authors and publisher. MTL = Median Tectonic Line; M = Moeatoa Conglomerate; P = Parapara Group; SL = southern Longwoods; TCM = Topfer Coal Measures; W = Waipahi.

Ma) diabase sills in the Amazon Basin, with a large volume of 0.34×10^6 km^3 (Mosmann et al., 1986, fig. 3);

(2) Karoo 178 Ma dolerite sills and 193 Ma flow basalt (Drakensberg Volcanics);

(3) Ferrar (and Tasmanian) dolerite sills/Kirkpatrick basalt flows 180–175 Ma, preceded by 190 Ma acid volcanics, all representing magmatism associated with rifting (Elliot et al., 1985) or with the Bouvet hotspot (Stothers, 1993), or both (Storey and Alabaster, 1991);

(4) 191 Ma volcanic (Mt Dromedary) complex and 190 Ma basalt in SE Australia (Hergt et al., 1991);

(5) 208–193 Ma* maar volcanism, Sydney Basin;

(6) 220–160 Ma Garrawilla Volcanics (flows, sills), Gunnedah Basin; by superposition, the upper limit of the Garrawilla Volcanics is probably not much younger than about 200–190 Ma*;

(7) 210–205 Ma granitoids, Lorne Basin;

(8) 200 Ma* inception of late Innamincka magmatic arc (Veevers, 1984, fig. 230).

(g) 230 Ma, Carnian: Onset of Pangean extension II. Represented by the start of Pangean rifting, including the rift basins about the North Atlantic and eastern Indian Oceans (Veevers, 1989, 1990a), and the start of Stage G in Australia, represented by the Ipswich Coal Measures with bimodal volcanics and the main volcaniclastic coal measures in Tasmania. Other coal measures of this age were deposited in the rest of Gondwanaland (*see* Fig. 3 in a later section), including the Molteno Coal Measures with ?tuffs (tonsteins). Follows immediately after Gondwanides II.

(f) 230+ Ma Gondwanides II. Terminal deformation of the Samfrau Geosyncline and yoked foredeep (Du Toit, 1937). Age constrained in South Africa by 230 ± 3 Ma metamorphic minerals in the Cape Fold Belt, with "final deformation of all pre-Beaufort rocks by kink bands and lower Beaufort rocks by listric thrusts" (Hälbich et al., 1983, p. 158), and the lacuna between the 240 Ma* youngest preserved folded Beaufort Group and the overlying flat 230 Ma* Molteno Coal Measures; in eastern Australia by (i) 233–230 Ma age of the transpressional Demon Fault; (ii) the lacuna between the 234 Ma* Moolayember Formation, the youngest deformed, and the 230 Ma* Ipswich Coal Measures, deposited during extension of the foldbelt. Presumed age of folding in the Sierra de la Ventana, Ellsworth Mountains, and New Zealand. Outside the foldbelt, transpressive folding in the Canning Basin of northwestern Australia is also dated as 230 Ma* (Veevers, 1990b). Less narrowly dated is the Middle to Late Triassic final coalescence of Pangea that involved the accretion by collision of South China and Cimmeria to the Paleo-Tethyan margin (Veevers, 1990a).

(e) 241–235 Ma extension. The definitive 230 Ma extension and coupled Gondwanides II were anticipated in places by a preliminary extensional event, such as the development of the volcanic Esk Trough in southeastern Queensland and the initial deposits (Talampaya), with tuffs, in western Argentina, the tuffaceous Burgersdorp of South Africa, the volcaniclastic Middle Fremouw and Lashly of Antarctica, and equivalents in Tasmania and in the Gunnedah Basin.

(d) 250–241 Ma tuff gap. The gap coincides with the lower half of the 250–230 Ma* coal gap (*see* Fig. 3 in a later section) and ends at 243 Ma* in Tasmania and the Gunnedah Basin. The evidence in South America is negative because this is a general lacuna, but no evidence of tuff from this interval is found in continuous sections of the Beaufort Group, Fremouw Formation (Fig. 11 in Chapter 4), and in the Tasmania, Sydney, and Bowen Basins. Our interpretation of the tuff gap is given in the discussion later of the environment.

(d/c) ~250 Ma Permian/Triassic boundary: End of Gond-

wanides I. Closely dated by the 247 ± 3 Ma second Outeniqua folding in South Africa, bracketed by 250 Ma (BJ—Black Jack Formation) and 245 Ma[*] sediments in the Gunnedah Basin, and by 250 Ma[*] thrusting in the Bowen Basin, and bracketed by ?250 Ma Choiyoi Formation below and 241 Ma[*] Cuyo Basin above in the Callingasta-Uspallata Basin. An orogenic pulse in New Zealand is loosely dated at this time. Deformation at this time or at *f* (Gondwanides II) or both affected the Ellsworth and Pensacola Mountains.

(b) 265 Ma. First appearance of convergent-arc magmatism in New Zealand and eastern Australia, from Tasmania north to the Bowen Basin. Dated in New Zealand by ~260 Ma biotite granite (southern Longwoods), and an indication of a ~265 Ma granite from a clast in the Maitai Group (MG). The first known tuff in Tasmania is 265 Ma[*], in the Sydney Basin 260 Ma[*], in the Gunnedah Basin 259 Ma[*], and in the Bowen Basin 256 Ma[*]. Granitoid dates in the adjacent New England orogen are older: examples are (i) the Barrington Tops Granodiorite and (ii) the Taromeo and Ridgelands plutons.

(i) Barrington Tops Granodiorite (BT) dates are K/Ar hornblende: 265 ± 2 Ma, 269 ± 2 Ma (Roberts and Engel, 1987; Roberts et al., 1991); K/Ar biotite: 262 ± 5 Ma (Cooper et al., 1963); Rb/Sr biotite: 262 Ma (Hensel et al., 1985); and U/Pb dilution zircon: 265 ± 8 Ma (Collins et al., 1993), 281 ± 10 (Kimbrough et al., 1993).

The youngest rock intruded by the Barrington Tops Granodiorite is the Manning Group (Roberts et al., 1991), which in the Barrington Tops area ranges from Briggs's (1993) zones A to H (281–264 Ma[*], Fig. 45 in Chapter 3). The Granodiorite intruded and cooled therefore at or after 264 Ma[*]. An independent check on the cooling age of the Barrington Tops Granodiorite is provided by the Greta Coal Measures, dated as eastern Australian Palynological Zone Upper 4a = 265–263 Ma[*]. The coarse regressive Greta Coal Measures prograde southward into the Sydney Basin; clasts decrease in size from an uplift to the north in the Barrington Tops area (Fig. 16B in Chapter 3), consistent with the view that the Granodiorite was intruded and stripped at this time. The two pieces of biostratigraphical evidence confirm the 265 Ma age adopted by Roberts et al. (1991) and used here (Fig. 2). We reconcile the slightly older date of 281 ± 10 Ma (i.e., 291–271 Ma) by suggesting that it may contain a small amount of inherited zircon.

A second anomalous feature of the Barrington Tops Granodiorite is its "very primitive isotopic signature [that] suggests a mantle derivation and very little contamination by crustal material in the source region of the intrusion . . . and suggests that the rocks . . . formed above oceanic crust" (Kimbrough et al., 1993). This confirms Leitch's (1988) interpretation of the Manning Group (*see* Chapter 3) as occupying a pull-apart (transtensional) structure within the Barnard Basin, probably with ocean floor at its base.

Uplift and stripping of the Barrington Tops Granodiorite was accompanied by the initial growth of the Muswellbrook

and Lochinvar Anticlines (Fig. 9 in Chapter 3), further deformation at 258 Ma, the Nambucca (NA) deformation at 255 Ma, and the final Hunter deformation at 253 Ma[*];

(ii) The Taromeo (TA) pluton in SE Queensland is dated at 264 Ma, and the Ridgelands (RI) pluton alongside the Bowen Basin is followed by Bowen deformations at 262 Ma[*] and 250 Ma[*].

In eastern Australia, we regard the oldest convergent-related granitoids and accompanying or following deformation as defining the inception of the magmatic arc/orogen. As described later, these dates define a flat segment at 265 Ma of the diachronous trend (*see* notable diachronous events below) of dates, mainly on tuffs, that ranges from 286 Ma[*] in South America to 244 Ma in New Guinea.

(a) 290 Ma[*] Pangean Extension I. This is expressed by Stage A in eastern Australia, part of the basal Gondwana facies of the Gondwanaland Province (Veevers, 1988), in turn part of the basal Pangean Super-sequence or Pan-sequence of Pangea deposited during the stage of sagging that followed the lacuna on the initial Pangean platform (Veevers, 1989, 1990a). The age of 290 Ma[*] is the numerical equivalent of the eastern Australian palynological stage 2, which dates the basal (glaciogenic) sediment of the Gondwanan basins (*see* Fig. 3 in a later section). Mainland eastern Australia was a recently formed (345–310 Ma) foldbelt-upland, so that the Sydney-Gunnedah-Bowen basin that developed by right-lateral extension of part of the upland, as in Europe the Rotliegend basins developed over the Variscan foldbelt-upland, accumulated thick extensional volcanics and voluminous plutons, including the Bundarra Granite (BU), in an intra-montane setting (Veevers et al., 1994b). Elsewhere sediments alone were deposited.

(pre-a) 320–290 Ma. Lacuna during thermal uplift of Pangean platform, Stage 1 of the release of Pangean heat (Veevers 1990a). On the Panthalassan margin, South America was affected by the ~360 Ma Chanic orogeny; within the resulting mountain belt the Calingasta-Uspallata Basin started to subside in the Visean ~344 Ma[*] and the western Paganzo Basin ~305 Ma[*], some 15 m.y. before the 290 Ma[*] definitive Pangean Extension I. Eastern Australia followed a similar path. The Lachlan/Thomson Foldbelt was terminally deformed between 355Ma[*] (Famennian, the age of the youngest sediment involved) and 325 Ma (Bathurst postorogenic granite) to form the Kanimblides and terminally intruded 325–310 Ma by the Bathurst I–type granite in New South Wales and by the 316 ± 15 Ma Connors–Auburn I–type granite in Queensland, ending the Late Devonian–Carboniferous magmatic arc (Arc D-C *in* Veevers, 1984, fig. 230). The D-C fore-arc basin was succeeded by the Late Carboniferous–Permian transtensional basins; the D-C arc itself was followed at 300 Ma by the Hillgrove S-type granitoid and the extensional alkaline-acid ignimbrite of the Bulgonunna Volcanics and biotite granite in New Guinea. The Hillgrove and Bulgonunna magmas were drawn into the crust by the onset of the right-lateral transtensional forces that formed the initial orocline at ~300 Ma and

persisted into the Permian to reinforce the effect of the 290 Ma[*] Pangean Extension I. In Antarctica, the 306 ± 9 Ma (Rb/Sr errorchron) orthogneiss in eastern Thurston Island and the older part of the 300 to 250 Ma range of Rb/Sr and K/Ar mineral dates from the Kohler Range and Bear Peninsula in eastern Marie Byrd Land (Pankhurst, 1990) fall in this range.

Notable diachronous event (Fig. 2)

286–244 Ma inception of Panthalassan magmatic arc and Gondwanides I. Indicated by the first appearance of juvenile arc-generated tuff or volcaniclastic sediment (triangle) or granitoid (square), and coupled shortening-deformation (sawtooth), all terminating *d/c* about 250 Ma[*], at the Permian/Triassic boundary. From South America to New Guinea, the dates of the event are as follows:

(a) *South America:* (i) earliest Permian (286 Ma[*]) tuff in the Del Salto Formation and 291 ± 10 Ma (K/Ar granitoid in the Cordillera, (ii) the 275 Ma[*] San Rafael deformation, (iii) the ~250 Ma top of the Choiyoi Formation and granitoids, and (iv) presumed ?immediately following post-Choiyoi deformation.

The La Pampa magmatic arc was active from 275–255 Ma and 243–233 Ma, and the North Patagonian magmatic arc at 259 and 239 Ma (Pankhurst et al., 1992).

In the Sierra de la Ventana, metamorphic illites from the Ventana Group near Pigüé have K/Ar dates of 273 ± 8 Ma and 265 ± 8 Ma (Varela et al., 1985) and from the Curamalal Group of 260 ± 3 Ma and 282 ± 3 Ma (Büggisch, 1987).

Tuff is reported from the upper part of the Tunas Formation (Iñíguez et al., 1988). Other ages are interpreted (von Gosen and Büggisch, 1989) as due to metamorphism of Precambrian rocks, such as the barely deformed granite at Lopez Lecube, 50 km west of Tornquist (Fig. 5 in Chapter 6), with a Rb/Sr whole-rock age of 227 ± 32 Ma and a hornblende K/Ar age of 240 ± 12 Ma, similar in age to the regional metamorphism denoted by K/Ar whole-rock ages of 221 ± 6 Ma and 249 ± 8 Ma of rhyolitic tuff from La Mascota, 10 km southeast of Pigüé, which, together with rhyolite from La Ermita, has a Rb/Sr whole-rock age of 317 to 348 ± 21 Ma (Cobbold et al., 1986). Alternatively, the granite at Lopez Lecube could be regarded as postorogenic, with a primary age indicated by the 227 and 240 Ma dates and La Mascota tuff by the 221 and 249 Ma dates, both indicating a heating event at ~235 Ma. If correct, these dates together with the 236 Ma date of the Sierra Pintada igneous rocks incline us to the possibility of a terminal heating event at ~247 Ma, which we regard as Gondwanides II. Tuff was deposited in the Tunas Formation from 260 Ma[*], and the entire column, up to the exposed top of the Tunas Formation (?Late Permian) was terminally deformed in the Triassic, probably by the ca. 235 Ma event but possibly also by an event in the earliest Triassic. Both events are shown in Figure 2.

In the Paraná Basin, tuff ranges from 280 Ma[*] in the Itararé Group to 255 Ma[*] in the Estrada Nova Formation (Coutinho et al., 1991, p. 158). The "Tardihercynian orogeny" at 275 Ma[*] marks the boundary between the Itararé Group and the Rio Bonito Formation (Zalán et al., 1991; Eyles et al., 1993).

(b) *South Africa:* Tuff ranges from 277 Ma[*] (equivalent to eastern Australian palynological zone 3a, Tastubian) near the top of the Dwyka Formation "just deposited but consolidating during this [278 Ma Swartberg folding] event" (Hälbich et al., 1983, p. 158) through the 258 ± 2 Ma first Outeniqua folding to the 254 Ma[*] tuff in the Balfour Formation and is followed by the 247 ± 3 Ma second Outeniqua folding, including "mega-folding of southernmost lower Beaufort and Ecca groups" (Hälbich et al., 1983, p. 158).

(c) *Ellsworth Mountains:* In the Polarstar Formation, the first tuff, 277 Ma[*] (equivalent to eastern Australian palynological zone 3a, Tastubian), is followed by volcaniclastic sandstone to 255 Ma[*] and shortly afterward (~250 Ma or later, ~230 Ma, or both) is followed by intense folding.

(d) *Ohio Range:* The only possible tuff is 255–250 Ma[*] in volcaniclastic sandstone in the uppermost Mt Glossopteris Formation.

(e) *Central Transantarctic Mountains:* The only possible tuff is 258–250 Ma[*] in volcaniclastic sandstone in the upper Buckley Formation, succeeded by the lower Fremouw Formation, which in the Queen Maud Mountains includes sandstone containing glass shards and pumice lapilli, so filling the "tuff gap."

(f) *South Victoria Land:* The first influx of volcanic detritus in South Victoria Land is recorded at 238 Ma[*] in the Triassic Lashly Formation (Korsch, 1974), immediately after the tuff gap.

(g) *New Zealand (NZ) to Bowen Basin:* Synchronous event b (*see* previous discussion) at ca. 265 Ma is defined mainly by granitoids. In New Zealand, granite continues into the Late Triassic (226 ± 6 Ma plutonic boulders, Graham and Korsch, 1990) and later (Korsch and Wellman, 1988). In Tasmania tuff continues from 265 Ma[*] to 258 Ma[*], and in mainland eastern Australia tuff extends to the tuff gap at 250 Ma[*] from starts at 260 Ma[*] in the Sydney Basin, at 259 Ma[*] in the Gunnedah Basin, and 256 Ma[*] in the Bowen Basin. As noted, granitoids start earlier, and persist almost to the end of the Middle Triassic (230 Ma).

(h) *North Queensland:* Latest Permian (250 Ma[*]) tuffaceous coal measures occur in North Queensland.

(i) *New Guinea:* The 244 Ma Kubor Granite is in New Guinea.

DEPOSITIONAL ENVIRONMENTS, CHIEFLY THE CHARACTERISTIC GONDWANAN GLACIAL, COAL-FORMING, AND REDBED FACIES, AND SUPERIMPOSED MARINE FACIES

Notable isochronous events (Fig. 3)

(g') 220 Ma[*], Norian, replacement of coal measures or bituminous shale in South America and South Africa by redbeds and eolian sandstone.

(g) 230 Ma[*], Carnian (Pangean extension II): Coal II, renewed deposition of coal and, in South America only, bituminous shale and redbeds. Veevers et al. (1994a) record the same event in the Indian Gondwana basins, China, and North America. Coal extends to the end of the Triassic except in South America and South Africa.

(f) 230 Ma[*], end of coal gap except the very thin seams in the Ladinian (235–230 Ma[*]) Moolayember Formation and Wianamatta Group of eastern Australia and the coal measures

in the Torlesse Supergroup of the Eastern Province of New Zealand and possibly the Middle or Late Triassic Topfer coal measures of the Western Province.

(e') 235 Ma[*], Anisian/Ladinian, end of redbeds in eastern Australia.

(e) 241 Ma[*], late Scythian, end of tuff gap (see Fig. 3).

(d) 250–241 Ma[*] tuff gap, except in the Early Triassic lower Fremouw Formation in the Queen Maud Mountains and the 243 Ma[*] tuff in Tasmania and the Gunnedah Basin. The

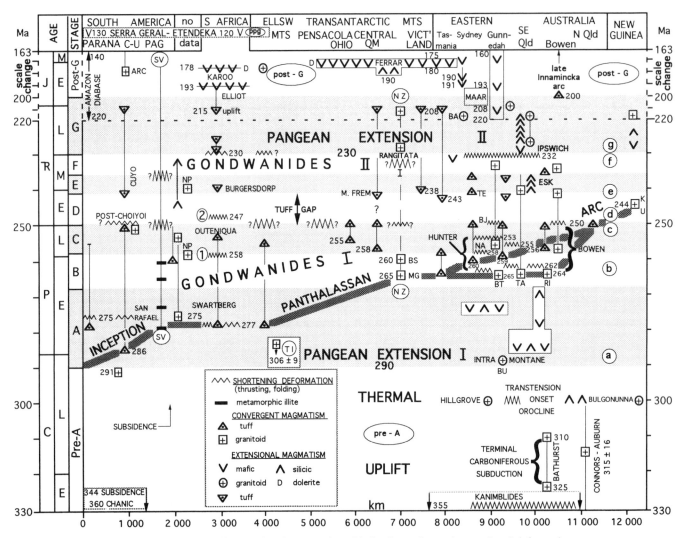

Figure 2. Time-space diagram showing tectonics, chiefly shortening and extensional deformation, and extensional and convergent-arc magmatism. Note change of scale at 220 Ma. Distances along bottom line marked on Figure 6. Information from Figures 42 and 44 in Chapter 3, Figures 7 and 23 in Chapter 4, Figures 7 and 21 in Chapter 5, and Figures 7 and 23 in Chapter 6, and for New Zealand in accompanying text. Ages: J = Jurassic; R = Triassic; P = Permian; C = Carboniferous; L = Late; M = Middle; E = Early. Granitoids: ARC = coastal arc of South America (Pankhurst, 1990); BA = Benambra Complex; BS = Brook Street; BT = Barrington Tops; BU = Bundarra; KU = Kubor; MG = granite clast in Maitai Group; RI = Ridgelands; TA = Taromeo. Deformations: BJ = Black Jack; NA = Nambucca. Places, formations: C-U = Calingasta-Uspallata Basin; Pag = Paganzo Basin; Ellsw Mts = Ellsworth Mountains; M. Frem = middle Fremouw; NP = North Patagonian massif; NZ = New Zealand; QM = Queen Maud Land; SV = Sierra de la Ventana; TE = Terrigal Formation, denotes a 2-cm thick kaolin band, a suspected volcanic ash (Brynes, 1983); Vict' Land = Victoria Land; SE Qld = southeastern Queensland; N Qld = North Queensland.

tuff gap coincides with the older half of the 250–230 Ma coal gap. South America has a general lacuna from 250–241 Ma. Elsewhere tuff is apparently lacking within the continuous interval in the Beaufort Group and in the Tasmania, Sydney, and Bowen Basins.

Interpretation of the tuff gap. Air-fall tuff accumulates only under these conditions: (a) a suitably quiet depositional environment, e.g., one with coal or shale, not rudite; (b) downwind of the vent; (c) minimal weathering in the depositional environment. The possible causes of the tuff gap are (a) no production of pyroclastics in the magmatic arc, or (b) production of pyroclastics but (i) no suitable depositional environment (not applicable to Early Triassic of eastern Australia, which has quiet shale deposition); (ii) weathered away in the depositional environment; and/or (iii) the wrong wind. We favor (ii), weathered away, because the sudden onset of the

tuff gap coincides with the onset of the coal gap, caused by a rapid change (to warming) in the surface environment.

(d/c) 250 Ma[*] Permian/Triassic (P/Tr) boundary, end of Coal I and equivalent start of coal gap in Antarctica and Australia by barren strata with redbeds replacing coal measures. In Australia, the P/Tr boundary is correlated with the 250 Ma Chinese P/Tr stratotype by the drop in δ^{13}carbon in kerogen (Morante et al., 1994). The coal in the Estcourt Formation of the Beaufort Group of South Africa is the only known occurrence of coal within the coal gap,[1] which extends outside the Panthalassan margin of Gondwanaland to India,

---

[1]G. J. Retallack (personal communication, 1993) points out that the Estcourt Formation contains *Glossopteris* and is thus Permian and *outside* the gap, as shown in Figure 3.

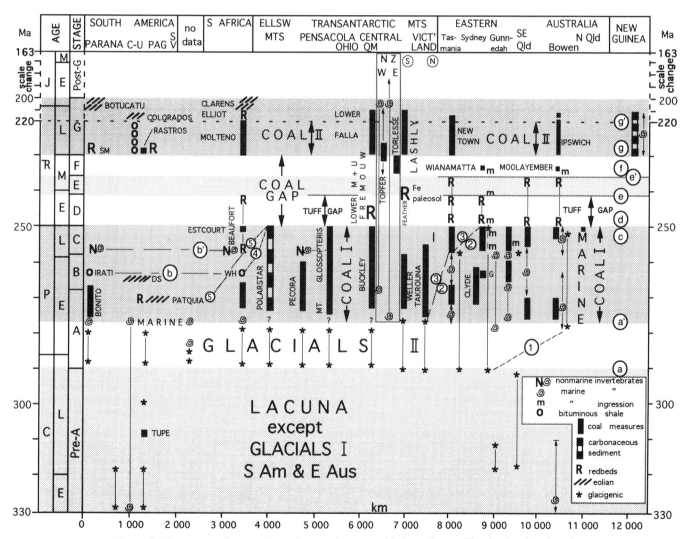

Figure 3. Time-space diagram (same base and source of information as Fig. 2) showing depositional environments, chiefly the characteristic Gondwanan glacial, coal-forming, and redbed facies, and superimposed marine facies. Note change of scale at 220 Ma. DS = Del Salto; SM = Santa Maria; WH = Whitehill; *see* Figure 2 for other abbreviations.

North America, and China (Conaghan et al., 1994). The tuff gap also starts at 250 Ma.

(b') 256 Ma[*] lake with endemic nonmarine bivalves covers the Paraná and Karoo Basins.

(b) 263 Ma[*] end of Coal I (oil shale of Iratí and Whitehill) in South America and South Africa.

(a') 277 Ma[*] (Tastubian) end of Glacials II with post-glacial marine transgression followed by regression (except in eastern Australia where the sea and glacigenic sediment remained until 252 Ma[*], glacial episode IIID of Veevers and Powell, 1987) and start of Coal I.

(a) 290 Ma[*] Pangean Extension I, inception of basin subsidence with start of Glacials II (glacial episode IIIB of Veevers and Powell, 1987).

(pre-a) 320–290 Ma[*] Lacuna on platform during Pangean thermal uplift (Veevers, 1990a) and continental glaciation. The lacuna is filled by Glacials I (glacial episode IIIA of Veevers and Powell, 1987) and marine deposition in the marginal Calingasta-Uspallata Basin and western Paganzo Basin (with Tupe coal) in southern South America and in the Tamworth-Yarrol Trough in eastern Australia.

Notable diachronous events (Fig. 3)

(5) Start of redbeds from 270 Ma[*] in southern South America through 258 Ma[*] in South Africa, to 250 Ma[*] in Antarctica and Australia.

(4) End of Coal I with 263 Ma[*] Iratí and Whitehill bituminous shale in South America and South Africa to 250 Ma[*] coal in Antarctica and Australia.

(3) Youngest marine sediment 277 Ma[*] in South America, South Africa, and ?Antarctica to 258 Ma[*] Tasmania to 251 Ma[*] Sydney and Bowen Basins (and ultimately to 235 Ma[*] ingressions in the Wianamatta Group and Moolayember Formation).

(2) Youngest Glacials II 277 Ma[*] in South America, South Africa, and Antarctica to 258 Ma[*] glacial dropstones in Tasmania to 252 Ma[*] in the Sydney Basin and 254 Ma[*] in the Bowen Basin.

(1) Oldest Glacials II in eastern Australia from 290 Ma[*] at Mudgee, west of the Sydney Basin, to 280 Ma in the Bowen Basin.

PRE-STAGE A (>290 Ma)

End-Devonian (~360 Ma)

The paleogeography at the end of the Devonian (Fig. 4) shows an Andean-type magmatic arc along the eastern continental edge of Australia, with the inboard Lambian and Drummond foreland basins, and a fore-arc basin and subduction complex lying farther east in the New England Fold Belt (Powell, 1984a, Korsch et al., 1990). Provenance and paleo-

currents in the eastern Lambie Basin indicate a mixture of mature quartzose sediments derived from the Gondwanaland craton to the south and west with immature volcanogenic sediment derived from the magmatic arc to the east (Powell, 1984a). In the Drummond Basin a trunk stream parallel to the bordering volcanic cordillera drained northward into the Broken River Province (Olgers, 1972; Wyatt and Jell, 1980; Lang, 1988). In far North Queensland, there was major uplift and folding at the Devonian-Carboniferous boundary (Henderson, 1980). In the Lambie foreland basin, streams drained northward from the subsiding Tabberabberan highlands in the southern Lachlan Fold Belt into a trunk stream system flowing east (Powell, 1984a, figs. 212; 221) before being deflected southeastward by the input of volcanogenic material close to the marginal cordillera (Powell, 1984a, fig. 222). A sinistral transtensional basin in eastern Victoria created local fault-controlled basins (Powell, 1984a, fig. 222; Marsden, 1988).

In central Australia, movement along faults in the Precambrian Arunta complex on the northern margin of the Amadeus Basin led to uplift and deposition of the upward-coarsening Brewer Conglomerate and correlatives (Jones, 1972, 1991; Lindsay and Korsch, 1991; Shaw et al., 1991). Carbonate deposition in and adjacent to the Devonian "Great Barrier Reef" of the Canning Basin (Playford, 1980) persisted into the earliest Carboniferous, but was overwhelmed by terrigenous clastics in the later Tournaisian and Visean (Middleton, 1990).

The continuation of the volcanic cordillera farther south along the Transantarctic Mountains can be inferred from the presence of extensive Devonian-Carboniferous granitoids in northeastern Tasmania (Williams et al., 1989) and northern Victoria Land (Grindley and Oliver, 1983; Borg and Stump, 1987). In the Central Transantarctic Mountains, the Devonian-Carboniferous marks the beginning of a lacuna that includes all but the last part of the Carboniferous, presumably representing an interval of mild uplift. In southern Africa, the youngest preserved part of the Witteberg Group was deposited at the top of the southward-prograding Cape Basin, before a lacuna during all but the latest Carboniferous. In Chile and the western parts of southern Argentina, the Devonian-Carboniferous was a time of major deformation, the Chanic Orogeny, which formed the framework of the Early Carboniferous basins of the Precordillera (PC) (López-Gamundí and Rossello, 1993). Inboard, an apparent hiatus in the Sierra de la Ventana (SV) marks the end of a paralic stable continental margin-style of sedimentation prior to the onset of the Early Permian foreland basin.

Sedimentary facies along the Paleo-Pacific margin reflect a wide range of paleolatitudes. The South Pole was situated in southern Brazil, and the Equator ran through northern Australia (Fig. 4). Carbonates and redbeds, including eolianites and deposits indicative of seasonal desiccation, were deposited in Australia, India, and Pakistan. In contrast, in South America and adjacent Africa, glaciogenic sediment was deposited in the Congo, Parnaíba, Amazonas, and Solimões Basins (Caputo

and Crowell, 1985). In southern South America, mountain glaciers and small ice centers developed along the orogen bordering the Paleo-Pacific margin, and there could have been small ice sheets in Brazil and central South Africa (López-Gamundí et al., 1992; Visser, 1993). An extensive continental ice sheet did not become established, however, until the mid-Carboniferous (Powell and Veevers, 1987).

Mid-Carboniferous (ca. 340 to 320 Ma)

The mid-Carboniferous was a time of mountain building along much of the Paleo-Pacific margin, as part of the global Hercynian orogeny related to the final closure of the ocean between Gondwanaland and Laurussia (Li et al., 1989; Chen et al., 1993; Villeneuve et al., 1993). In eastern Australia, ter-

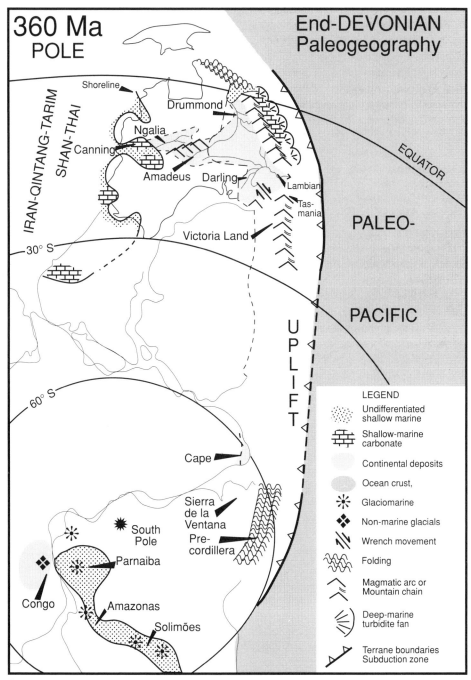

Figure 4. Paleo-tectonic map of the end-Devonian, with the pole interpolated at ca. 360 Ma (Table 3 in Chapter 2). Most of the information comes from the Pre-Gondwanan history given for each sector, supplemented by information in the text.

minal deformation of the Lachlan and Thomson Fold Belts folded sediments in the Lambian and Drummond Basins between 350 and 325 Ma (Olgers, 1972; Powell et al., 1977; Powell, 1984a; Shaw et al., 1982). The deformation extended through central Australia into the Amadeus, Ngalia, and Canning Basins (Wells and Moss, 1983, Shaw et al., 1991). Southward overthrusting along the northern margin of the Ngalia Basin postdates the early Visean (~340 Ma[*]; Chen et al., 1993) and could be as young as Namurian/Westphalian (~330 to 310 Ma), consistent with the Rb/Sr whole-rock/mica ages determined from rocks in the Arunta Inlier involved in deformation (Mortimer et al., 1987; older data summarized in Powell and Veevers, 1987). East-west shortening in the Thomson and Lachlan Fold Belts extended as far west as the edge of the Precambrian shield in western New South Wales, where deformation was restricted mainly to reactivation along old fault lines and local folding of younger cover successions. The intensity of deformation was greatest adjacent to the site of the Late Devonian–Early Carboniferous Andean arc along the Paleo-Pacific margin where tight folds overturned to the east formed (Powell, 1984a, fig. 222B).

Compression in central Australia was in a north-south direction, and involved up to 100 km of intracontinental shortening on deep-seated shear zones cutting the entire continental crust (Goleby et al., 1989). Gentle deformation in the Canning Basin was confined to tilting and uplift. In the Lachlan-Thomson Fold Belts, megakinks deforming the regional meridional fold trends (Powell, 1983; Powell et al., 1985) were interpreted by Powell (1984b) to be caused by mid-Carboniferous compression (~320 ± 10 Ma)—the same north-south compression which produced the crustal shortening in central Australia. An alternative explanation by Stubley (1989) is that the megakinks were produced by a continuation of the same east-west Kanimblan compression which formed the main meridional fold trends in the Lachlan-Thomson Fold Belts. In both interpretations, the megakinks predate the post-tectonic Late Carboniferous granites in the northeastern Lachlan Fold Belt, and the onset of sedimentation in the Sydney Basin.

There is little evidence of deformation in the Tamworth Belt of the New England Fold Belt (NEFB) (*see* Zone 7, Fig. 2 in Chapter 3) at this time. Sedimentation in the Tamworth fore-arc basin continued throughout the Carboniferous. Although there is a change in the composition of the detritus shed from the west, with Lachlan Fold Belt detritus coming into the Tamworth Basin in the Late Carboniferous (Leitch, 1974), there is no folding or angular unconformity to reflect the major, widespread deformation occurring in the Lachlan and Thomson Fold Belts to the west. Farther east in the subduction complex and abyssal plain regions of the NEFB (*see* Fig. 2 in Chapter 3, Zones 8 and 9), metamorphism and deformation in the Wongwabinda Complex and Coffs Harbour Block about 320 Ma could have been associated with the accretion of exotic or displaced terranes into the subduction complex (Fig. 5 in Chapter 3). Deformation was not extensive

throughout the main part of the NEFB, however, leading some authors (e.g., Harrington and Korsch, 1985) to suggest that the New England and Lachlan Fold Belts did not amalgamate finally until the latest Carboniferous.

The widespread deformation, some of which involved the entire continental crust, uplifted much of eastern and central Australia, thereby posing the conundrum of where the substantial volume of sediment lay that must have been eroded from the highlands. The Late Carboniferous in continental Australia is remarkable for its lack of sediment. The solution appears to have been the existence of a continental ice sheet over much of Gondwanaland which prevented any sediments being laid down on the continent until after the final melt-out (Powell and Veevers, 1987; Veevers and Powell, 1987). Glacially derived sediment could be expected in the marine fringes, but there would have been no permanent deposits on the continent while the ice sheet existed. Glacial dropstones are known to exist in Namurian marine strata in the Werrie Basin of eastern Australia, and also in the offshore Bonaparte Basin in northwest Australia, supporting other evidence that the ice sheet was probably formed at the Mississippian-Pennsylvanian boundary in North America (Veevers and Powell, 1987).

The absence of mid-Carboniferous sediment in most other parts of the Paleo-Pacific margin of Gondwanaland can be explained in the same way. Tasmania and Victoria Land had been deformed in the Middle Devonian and intruded by granites from the Middle Devonian to the Early Carboniferous. In Tasmania, there was substantial stripping, with the removal of about 5 km of rock before the latest Carboniferous sediments were laid down. Similarly, in the Cape region of South Africa and the adjacent Sierra de la Ventana there is a lacuna, but no folding, during most of the Carboniferous before the latest Carboniferous glacigenic deposits were laid down. In the Sierra de la Ventana, the underlying Late Devonian sediments are shallow marine and the overlying glacigenic sediments are also marine, arguably deposited below wave influence.

Major folding, metamorphism and intrusion of mid-Carboniferous age has been recognized in the Andean regions of Peru and Bolivia to the north of Argentina (Carlier et al., 1982; Mégard, 1978), and this can be extended into the Mauritanides of northwestern Africa (Villeneuve et al., 1993). In the Precordillera of Argentina, the deformation appears to have been somewhat older around the Devonian-Carboniferous boundary (López-Gamundí and Rossello, 1993).

After the widespread mid-Carboniferous folding and uplift, the Andean magmatic arc was re-established, presumably indicating subduction of the Panthalassan ocean beneath Gondwanaland. Along the Argentinian sector, the newly established magmatic arc continued into the Permian, where a yoked foreland basin was established in the earliest Permian. Along the Australian margin, Andean-type magmatism and subduction of Panthalassa continued until ~310 Ma, when the margin changed to dextral transtension. Subduction along the Australian Panthalassan margin was not re-established until the late Early

Permian. In between, in the Cape and Transantarctic regions, subduction of Panthalassa resumed diachronously, in the Early Permian in the Antarctic Peninsula and Cape regions and younger toward Victoria Land and Tasmania (Fig. 2).

STAGE A (290–268 Ma)

Fig. 5 encompasses Stage A (290–268 Ma), latest Carboniferous (Gzelian) and earliest Permian (Asselian and Sakmarian), eastern Australian palynological Stages 2 and 3, from the 290 Ma[*] inception of the Gondwana basins with basal glacials through the first phase of coal deposition (Aramac coal measures) in the Galilee Basin (2) and Ecca coal measures in the Karoo Basin (30). Most of the information (Table 1) pertains to the Tastubian (277 Ma[*]). This time plane is defined by the postglacial marine transgression with the marine *Eurydesma* fauna (Dickins, 1984) (equivalent to east Australian palynological Stage 3a) that covered parts of the Paleo-Tethyan and Panthalassan margins and extended deep into the plate interior in south-central Australia and over the site of the future South Atlantic Ocean. The Antarctic margin is devoid of marine invertebrates but is regarded by Collinson et al. (Chapter 4) as an inland (or epeiric) sea because trace fossils and C/S ratios suggest fresh or brackish water. We show this inland sea as a broad gulf restricted by a postulated uplift (XVIII) south of the more open sea along the eastern Australian margin and landward of the postulated transform-faulted spreading segments from New Zealand to the leading point of the magmatic arc of the Panthalassan margin. The other feature specific to 277 Ma[*] is the leading point of the magmatic arc (filled circle), taken to lie oceanward of the tuff of this age in the Ellsworth Mountains and in South Africa. Older tuffs are the earliest Permian (286 Ma[*]) tuff in the Del Salto Formation of the Calingasta-Uspallata Basin and the 280 Ma[*] tuff in the Itararé Group of the Paraná Basin. The magmatic arc in South America is defined by the latest Carboniferous–Early Permian to Triassic granites and rhyolites of northern and central Chile (e.g., 291 Ma in the Frontal Cordillera) and Argentina (275 Ma intrusives and associated younger volcanics in La Pampa) that constitute the Choiyoi province of Kay et al. (1989) and Mpodozis and Kay (1992). From 35°S near the coast, the northern edge of the province trends southeastward across the North Patagonian massif and southwest of the Sierra de la Ventana. From 38°S near the coast, at Lago Raco, which has 295 Ma (309–282 Ma) granitoids (Beck et al., 1991), the southern edge trends southeastward for 500 km into northern Patagonia. As related in Chapter 6, the magmatic province is probably a wide back-arc complex behind the arc itself, which traces out the old continental margin, since broadened by an accretionary wedge of deep-sea sediment and accreted terranes. The arc is sketched south of Africa to its postulated leading point opposite the Ellsworth Mountains tuff. Already completed during this stage are a 275 Ma[*] San Rafael (SR) and 278 Ma Sierra de la Ventana (SV) deformation in Argentina, the 275 Ma

Swartberg deformation in the Cape Fold Belt, and the 275 Ma[*] uplift of the Asunción Arch (IX) that shed detritus into the Río Bonito coal measures of the Paraná Basin.

The 650 km of right-lateral movement involved in the eastern Australian orocline (single arrows) was transformed by the same amount of oblique spreading at a half rate of 2.5 cm/year between 290 Ma and 277 Ma in the small ocean basins (we draw four of them within the very loose constraints) that produced the 280 Ma Dun Mountain ophiolite (Kimbrough et al., 1992).

The position of the triple junction is determined at the limit of calc-alkaline magmatism in the arc or arc-derived tuff in the foreland basin. As the onset of calc-alkaline magmatism moves toward Australia with time, so the triple junction is shown to migrate toward Australia. On the Australia side of the triple junction there is transcurrent motion between Panthalassa and Gondwanaland, and the relative plate-motion vector is thus shown parallel to the transform margin. On the South America side of the triple junction, Panthalassa is subducted beneath the Gondwanaland margin and the relative plate-motion vector is thus shown as orthogonal to the subduction zone. Because there is a progressive increase in the length of the subducting margin, there must have been intraoceanic subduction between the segments of Panthalassa on either side of the triple junction, which thus had a T-T-F configuration. The surface trace of the outer edge of the subduction complex on the South American side of the triple junction is offset in map view from the triple junction, which is positioned where the magmatic arc meets the intraoceanic subduction zone. This reflects the strain partitioning in accretionary complexes (cf. the present-day interaction between the Australia-India plate and Southeast Asia along the Sunda Arc).

Information from the rest of Gondwanaland, in particular the onset of basin subsidence in Pangean stage 1 (Veevers, 1990a), is sketched for Australia (Veevers, 1984, p. 240), India and Antarctica (Tewari and Veevers, 1993), Africa (Wopfner, 1993; Fig. 18 in Chapter 5), and South America, in particular, 280 Ma[*] tuff near the top of the Itararé Group (Coutinho et al., 1991), the extent of the post-glacial marine transgression in the Paraná Basin (Eyles et al., 1993), and the uplift of the Asunción Arch (IX) during the Tardihercynian orogeny and the regressive deposition of the 257 Ma[*] coal-bearing Bonito Formation (Zalán et al., 1991). According to Tewari and Veevers (1993), Gondwanaland was dominated by an elongate upland across East Antarctica, centered on the present Gamburtsev Subglacial Mountains and with a 14,000 km periphery, which was drained first by ice streams and then by rivers from which the Gondwana facies was deposited, as shown by the paleoflow arrows.

STAGE B (268–258 Ma)

The leading point of the magmatic arc advanced 6,500 km (at 500 km/m.y.) from a 277 Ma position opposite the Ellsworth Mountains to a 264 Ma position off the northern

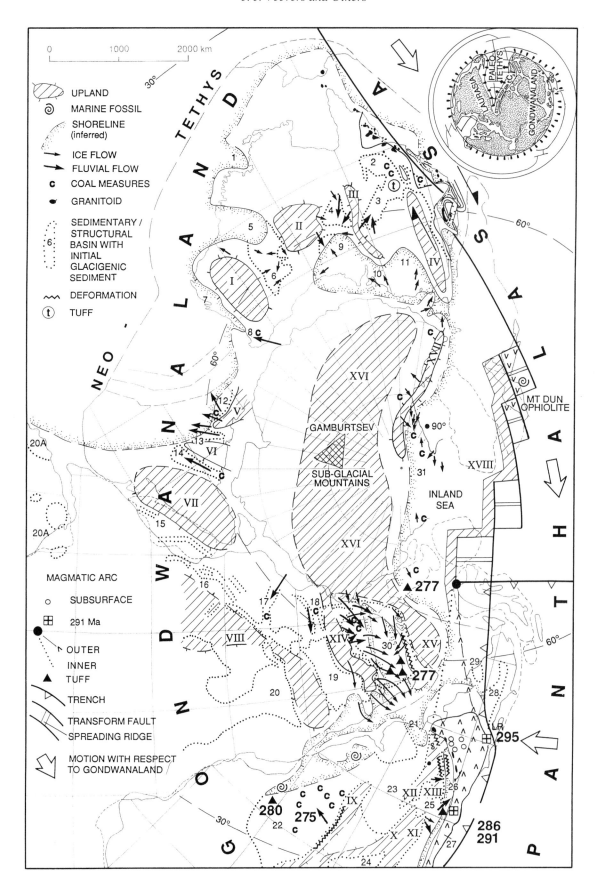

Figure 5. Paleo-tectonic/geographic map encompasses Stage A (290–268 Ma) but most of the information pertains to the Tastubian (277 Ma[*]), including the 280 Ma interpolated pole and paleolatitudes (Chapter 2). Please note that all references to the South Pole and latitudes correspond to present coordinates. The global setting (top right corner, from de Wit et al., 1988) is such that Gondwanaland, the southern province of Pangea, lies between Panthalassa on one side and the equatorial Paleo-Tethys and Laurasia on the other. Shortly after the 320 Ma amalgamation of Pangea, the Cimmerian continent (C) was transferred from the Gondwanaland margin to Laurasia by the generation of Neo-Tethys and the consumption of Paleo-Tethys (Veevers, 1988). The base map of the Permian-Triassic reconstruction of Gondwanaland and paleolatitudes come from Chapter 2. Geological detail of the Panthalassan margin from Chapters 3–6 and other sources, as specified in Table 1.

◄ ─────────────────────────────

Bowen Basin (Fig. 6). In Australia, the following tracts, from craton to margin, are identified: (i) an epicontinental basin with lake sediment (Murteree and Roseneath Shales) in the Cooper Basin at the other end of Gondwanaland from the Iratí-Whitehill Shales, and coal measures in the western Gunnedah Basin and Oaklands (sub-Murray) Basin; (ii) <262 Ma[*] Foreswell I, that shed quartzose sediment into (iii) the initial foreland basin; (iv) an orogen/magmatic arc from 265 Ma with 262 Ma[*] intense deformation in the Gogango Overfolded Zone (GOZ) and in northern New South Wales on the east at the same time as mild deformation in the Denison Trough on the western side of the Bowen Basin and the rise of the foreswell. The 265 Ma Barrington Tops Granodiorite was unroofed to supply gravel to the regressive Greta Coal Measures that constituted the initial (preliminary) deposit of the foreland basin.

TABLE 1. SOURCE OF INFORMATION SHOWN IN FIGURE 5

Chapter 3, Figure 41A (eastern Australia Stage A)

Chapter 4, Figures 14 and 15 (ice) and 16 (post-ice)

Chapter 5, Figures 17D (Dwyka, ice) and 17E (lower Ecca, coal)

Chapter 6, Figure 1 (magmatic arc defined in Fig. 7, which shows the Proto-Precordillera overlapped at this time), Figure 12 (shoreline and cratonic arrows), Figure 16 (Sauce Grande Basin). 280 Ma* tuff in Itararé of Paraná Basin (Coutinho et al., 1991, p. 158).

In the area outside the Panthalassan margin are shown *new (italic)* or renewed basins (e.g., 1) and uplands (I) in Australia (Veevers, 1984, p. 239, 240; BMR Palaeogeographic Group, 1990):

I	Ancestral Great Western Plateau
II	*Central Australia*
III	*Ancestral South Australian Highlands*
IV	*Ancestral eastern Australian foreswell*
1	Bonaparte Gulf
2	*Galilee*
3	*Cooper*
4	*Pedirka*
5	Canning
6	*Officer*
7	Carnarvon-*Perth*
8	*Collie*
9	*Arckaringa*
10	*Troubridge*
11	*Oaklands, Renmark*

India (Tewari and Veevers, 1993)

V	Chhota-Nagpur
VI	Chhatisgarh
12	*Damodar*
13	*Mahanadi*
14	*Godavari*

East Africa (Wopfner, 1993)

15	*Morondava*
16	*Tanzania Karoo*
VII	Malagasy–?southwest India (Rust, 1975, p. 541, 543, 544; Tewari and Veevers, 1993)

Central Africa

VIII	Congo-Kaokoveld core of ice sheet on ?upland (Rust, 1975, p. 541; Crowell, 1983; Frakes, 1979, p. 142)
17	*Lower Zambezi* (Rust, 1975, p. 541, 543, 544)
18	*Waterberg* (Rust, 1975, p. 541, 543, 544)
19	*Kalahari-Botswana* (Rust, 1975, p. 541, 543, 544)
20	*Congo* (Rust, 1975, p. 541, 543, 544)
20A	*Arabia*, Al Khlata (Levell et al., 1988)

South America (Chapter 6, Fig. 1)

21	*Sauce Grande*
22	Paraná
23	*Chaco-Paraná*
24	*Tarija*
25	*Paganzo*
26	*San Rafael*
27	Calingasta-Uspallata
28	Tepuel
29	Golondrina
IX	Asuncion
X	Michicola
XI	Puna
XII	Pampean
XIII	Pie de Palo

Southern Africa (Chapter 5)

30	*Karoo*
XIV	Cargonian
XV	*Proto-Fold Belt*

Antarctica

31	Transantarctic
XVI	*East Antarctic* (Tewari and Veevers, 1993), centered on the Gamburtsev Subglacial Mountains (Drewry, 1983)
XVII	Ross High (Chapter 4)
XVIII	Postulated Antarctic margin

The first identified tuff 260 Ma[*] from the magmatic arc is apparently some 5 m.y. after the cooling age of the first magma. Possibly behind a low foreswell, Tasmania was a marine gulf with tuff from the distant line of the <265 Ma presumably effusive magmatic arc that crops out in the Brook Street Magmatic Arc (BSMA) in New Zealand. The Western Province of New Zealand subsided, probably as an intra-arc basin, and was flooded by the sea.

In Antarctica, the Ross High, possibly a continuation of the Australian Foreswell I, is inferred to have extended almost to the present South Pole. The Ross High separated coal measures on the craton from the foreland basin in the same configuration as in Australia except for the present wide (>1,000 km) separation of arc and basin, which in the Permian, however, may have been closer before the subsequent extension (Sahagian and Collinson, 1993).

Southward axial drainage was fed by east-flowing rivers from the Ross High and SSW-flowing rivers from an eastern orogen at the head of a basin that formed an "inland sea" across the rest of Antarctica and possibly extended into the Whitehill (Karoo-Kalahari)–Iratí (Paraná) basin described by Anderson and McLachlan (1979). If the sea entered part of this vast basin, it must have been ephemeral because it left no clear trace. The South American part of this basin, bounded oceanward by nonmarine sediment shed from the magmatic arc, is unlikely to have provided a connection with the sea, nor, as suggested by Oelofsen (1987), the ground between South America and Africa, which in our interpretation was an upland. Between the Panthalassan shoreline in South America and that in New Zealand is the gulf (in geography and information) of West Antarctica, which by default is the place for the sea (if any) to have entered. This is confirmed, at least for Stage C, by Cooper and Kensley (1984), who concluded that the tenuous oceanic connection of the Ecca trough lay not to the northeast (Visser and Loock, 1978) but to the east, that is, in Antarctica. The indelibly marine mark made by the earlier (277 Ma[*], Tastubian) transgression in South America and southern Africa (Fig. 5) is lacking. The Tastubian sea retreated definitively with the depositional regression represented by the Ecca and Bonito coal measures. During the late Artinskian (265 Ma[*]), the Whitehill and Iratí Formations were deposited in interconnected parts of a basin with a nonmarine biota (Anderson and McLachlan, 1979; Cole and McLachlan, 1991). And during the Kazanian (256 Ma), the Waterford (uppermost Ecca) and Estrada Nova Formations were linked by

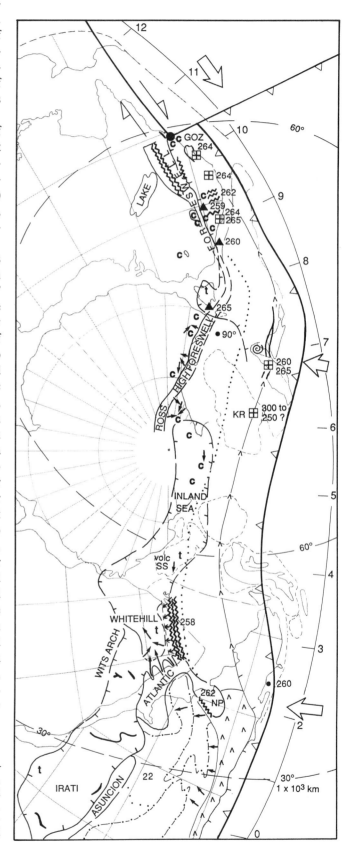

Figure 6. Paleo-tectonic/geographic map: Stage B (268–258 Ma). Data from Figures 41B in Chapter 3, 16 and 17 in Chapter 4, 17E and 17F in Chapter 5, and 13 in Chapter 6. Pole interpolated at 260 Ma. GOZ = Gogango Overfolded Zone; KR = Kohler Range; SS = sandstone; NP = North Patagonian massif. *See* Figure 5 for symbols and Figure 2 for other abbreviations.

an endemic bivalve fauna, which is unknown as yet in Antarctica, although other nonmarine bivalves occur there (Bradshaw, 1984) (Fig. 7).

Cooper and Kensley (1984) found a fourfold biogeographic significance in the bivalves: (1) the fauna evolved in isolation and is thus endemic to the Ecca/Paraná basin, while retaining affinities with the marine Permian of Australia; (2) the Ecca/Estrada Nova sedimentation occurred within a huge inland sea of brackish water with complex salinity gradients, as well as temperature and oxygen stratification; (3) the tenuous oceanic connection of the Ecca trough lay not to the northeast (Visser and Loock, 1978) but to the east and was insufficient to allow the development of normal marine conditions; and (4) the bivalves indicate a mid- to early late Kazanian age (~256 Ma).

The Whitehill basin was bordered on the north by the Wits Arch (XIV) and on the south by the growing Cape Foldbelt, which climaxed at 258 Ma in the first phase of Outeniqua folding. The foldbelt flexed the Karoo Basin basement into a foredeep, which filled rapidly with 3 km of lake and piedmont sediment (upper Ecca). We postulate that a foreswell branched off the Cape Foldbelt as the Atlantic upland to continue into the North Patagonian (NP) massif in the direction of the Asunción Arch (IX). The Atlantic upland shed subaqueous fans into the western part of the Karoo Basin. The magmatic arc/orogen in the North Patagonian massif shed volcanogenic sediment into the Sauce Grande Basin, as the arc farther to the north shed sediment into the San Rafael and Calingasta-Uspallata Basins. The position of the arc/back arc between data points in South America and New Zealand is unconstrained because the 300–250 Ma mineral dates for mostly dioritic granitoids from the Kohler Range (KR) and Bear Peninsula in eastern Marie Byrd Land are probably reset ages of older rocks, such as the 306 ± 9 Ma calc-alkaline orthogneiss in Thurston Island (Pankhurst, 1990). A doubtful data point is provided by the speculation that the ages were reset from intrusions that mark the position of the arc, as shown also in Figure 7. We draw the arc/back arc as a 500 km wide band alongside the early Gondwanides of South America and Africa past New Zealand and into the Australian Gondwanides. In this scheme, the 260 Ma metamorphic rock in the accretionary wedge of the Chonos Peninsula at 45°S (David-

Figure 7. Paleo-tectonic/geographic map: Stage C (258–250 Ma). Endemic nonmarine (N) bivalves in the Estrada Nova Formation from Runnegar and Newell (1971) and in the Waterford Formation from Cooper and Kensley (1984). Other nonmarine bivalves in the Ohio Range (Antarctica) from Bradshaw (1984). All other data from the regional chapters, including 250 Ma mean pole. *See* Figure 5 for symbols.

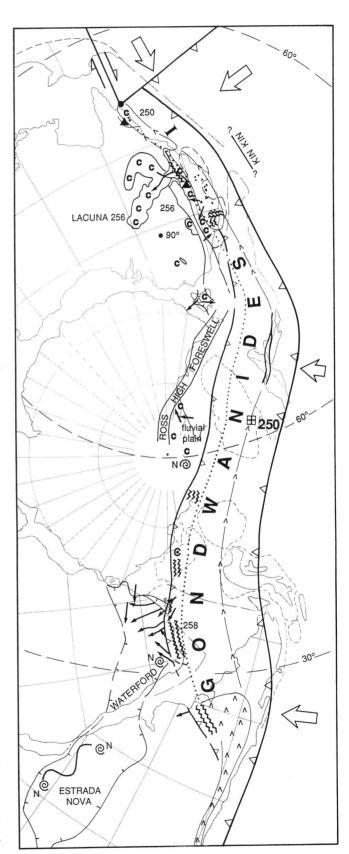

son et al., 1987) and the glossopterid-bearing sediment at Erehwon Nunatak (E) lie between trench and arc.

The Gondwanides I were further developed by metamorphism in the Sierra de la Ventana area, the first phase (258 Ma) of the Outeniqua folding, and the 264 Ma[*] Greta uplift and 262 Ma[*] folding in the southern New England orogen (early part of Hunter orogeny) and in the Galilee and Bowen Basins (Aldebaran-Rainbow Creek phases of the Bowen orogeny).

STAGE C (258–250 Ma)

The magmatic arc advanced to a position opposite the oldest dated Permian tuff in north Queensland, drawing behind it the Late Permian coal measures, which climaxed in the foreland basin and epicontinental basin that stretched 6,000 km to the South Pole and another 2,000 km to the carbonaceous sediment in the Ellsworth Mountains (Fig. 7). In South Africa, the Waterford–Estrada Nova lake with endemic nonmarine bivalves (Runnegar and Newell, 1971; Cooper and Kensley, 1984) was filled in by the coarse Beaufort orogenic sediment flushed from the first phase of Outeniqua folding in the Cape Fold Belt and from uplift in the east and west. Orogenic sediment (Tunas Formation) likewise was shed from the foldbelt in South America and was deformed by the Permian/Triassic terminal deformation of Gondwanides I at 249 Ma, as was the Cape Fold Belt at 247 Ma in the second phase of Outeniqua folding, and possibly at this time the Ellsworth and Pensacola Mountains successions. A foreswell extended from the present South Pole almost to 20°S in Australia, and marked the eastern limit of epicontinental coal measures in Tasmania, at Coorabin and in the western Gunnedah Basin in New South Wales, and in the Galilee Basin and, after a lacuna from 258–256 Ma[*], in the Cooper Basin in Queensland. The southern New England orogen and adjacent Gunnedah and Sydney Basins were deformed in the late phase of Gondwanides I at 258 Ma[*], 255 Ma (Nambucca Slate Belt, NA), and 253 Ma[*], and the Queensland part of the orogen at 250 Ma[*] to form an upland, which shed copious orogenic sediment into the foreland basin and additionally in the north across the foreswell into the distal Galilee Basin in a definitive regression. Base-surge deposits from volcanoes near Newcastle entered the Sydney Basin and added to the load of redistributed juvenile magmatic rock in the foreland basin.

The magmatic arc/back arc is interpolated between eastern Australia and New Zealand and thence to South America, except the possible data point in the Kohler Range (KR).

STAGE D (250–241 Ma)

In Stage D, the arc advanced to its final known position in New Guinea (244 Ma) (Fig. 8). The same tectonics applied as in Stage C, but the surface environment was different. The copious coal measures and tuff of Stage C are replaced by measures barren of both coal and tuff to constitute coal and tuff

gaps, and redbeds first appear in Antarctica and Australia (Fig. 3). This change, at the Permian/Triassic boundary, calibrated at 250 Ma[*], coincides with the global drop in $\delta^{13}C_{org}$ in marine and coastal sediment (Morante, 1993; Morante et al., 1994), which we (Conaghan et al., 1994) attribute to the deleterious effect of the eruption of the Siberian Traps (Campbell et al., 1992), and the elimination of tuff by weathering in a warmer climate. The tectonic setting was unchanged, with continuation of the Beaufort deposition in South Africa after the final phase of Gondwanides I in the Sierra de la Ventana-Cape Fold Belt and Ellsworth-Pensacola Mountains, and deposition on either side of the foreswell of Antarctica and Australia. In the eastern Australian foreland basin, cratonic and orogenic drainage joined in axial flow from a saddle just north of 30°S. The emplacement of plutons climaxed at the same time as tuff disappeared (we believe from weathering) from the sedimentary succession. Another possibility is that explosive volcanism ceased in the NEFB, as in parts of the present-day Andes, which is segmented into volcanic and nonvolcanic parts.

The Kin Kin terrane probably amalgamated with southeastern Queensland at this time.

STAGE E (241–235 Ma)

Most of this stage was a lacuna in South America and South Africa, probably more from nondeposition than from later stripping during the impending Gondwanides II uplift (Fig. 9). Volcanolithic sandstone was deposited in Antarctica in the foreland basin and in Tasmania in the epicontinental basin or distal part of the foreland basin, and in mainland eastern Australia. A drainage divide (foreswell II) arose between the Bowen foreland basin and the Galilee epicontinental basin as indicated by the divergent sediment flow in the Glenidal Formation. Sediment flow in the Bowen Basin was now wholly southerly and was joined later in the stage by the cratonic Expedition Sandstone flowing easterly from the Cooper-Galilee Basin across the former foreswell. In the Gunnedah Basin, a debris-flow fan was deposited at the foot of foreswell II, and in the Sydney Basin the Hawkesbury Sandstone swept across the basin to the northeast from a voluminous source in the foreswell, which also fed another quartz sandstone in Tasmania.

STAGE F (235–230 Ma)

The complex of events called Gondwanides II took place during this stage (Fig. 10). In South America, the earliest rifting of the Cuyo Basin preceded the (poorly dated) deformation in the Sierra de la Ventana along strike from the 230 Ma terminal deformation of the Cape Fold Belt and (poorly dated) Ellsworth and Pensacola Mountains folding. In New Zealand, as outlined at the beginning of this chapter, Rangitata I of Bradshaw et al. (1981) is attributable to this time. In eastern

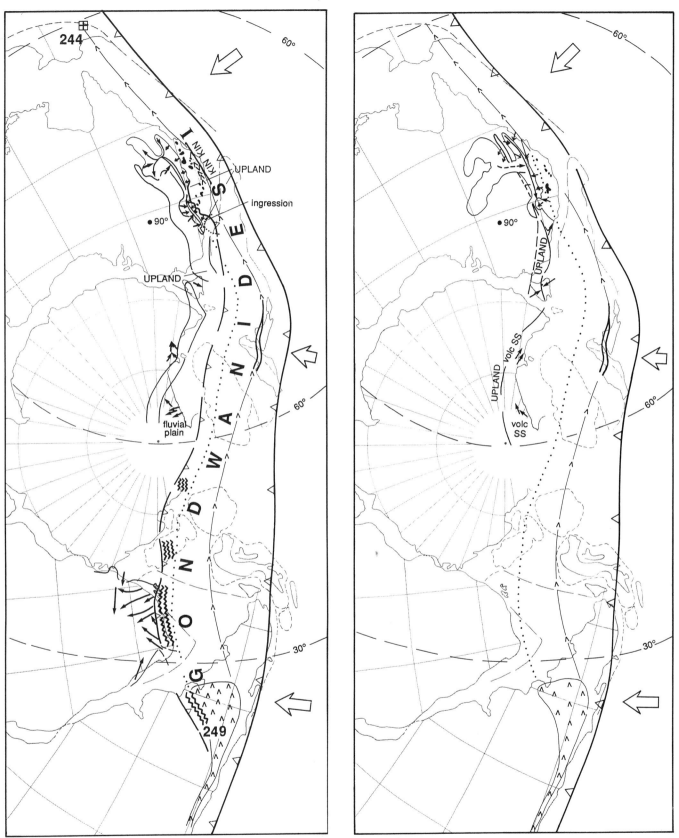

Figure 8. Paleo-tectonic/geographic map: Stage D (250–241 Ma), with 250 Ma mean pole. *See* Figure 5 for symbols.

Figure 9. Paleo-tectonic/geographic map: Stage E (241–235 Ma), with 250 Ma mean pole. *See* Figure 5 for symbols.

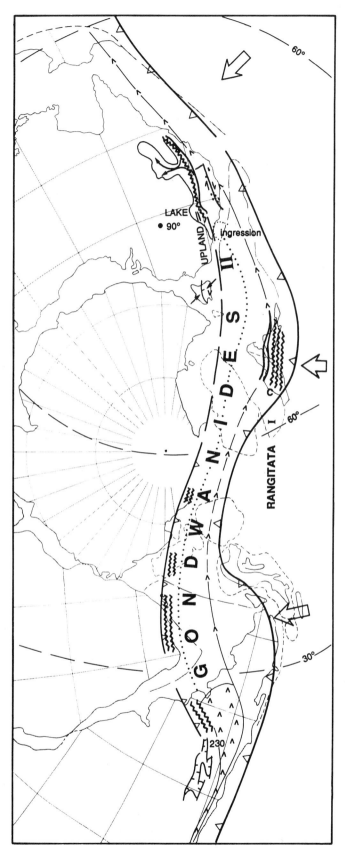

Figure 10. Paleo-tectonic/geographic map: Stage F (235–230 Ma), with 250 Ma mean pole. *See* Figure 5 for symbols.

Australia, the final deposits of the foreland basin, the Wiana-matta Group and the Moolayember Formation, were followed by intense thrusting and in the southern Sydney Basin by left-lateral faulting. In the foldbelt, plutons were intruded along the Demon Fault, which then underwent right-lateral transpression. Epicontinental Tasmania remained unaffected by Gondwanides II except perhaps for the extrusion of basalt at St Marys.

STAGE G (230–200 Ma)

The compressional deformation of Gondwanides II gave way to extension coincident with Pangean Extension II or "the initial rifting stage" of Storey and Alabaster (1991)(Fig. 11). In South America, diabase sills were erupted in the Amazon Basin; intra-montane rift basins south of 25°S were succeeded by the first part of the Patagonian Chon Aike province of rhy-olites and granites, with a date of 207 Ma, and probably ex-tended to the 230 Ma Mt Bramhall intrusion in the Thurston Island block. Storey and Alabaster (1991, p. 1274) interpret the change from Gondwanide compression to lithospheric ex-tension in this region as due to "a change from shallow to steeply dipping subduction and to a slowing of subduction rates caused possibly by a decreasing age of the subducting plate." We show this postulated steep dip of the subducting slab by the double barbs on the trench in Figure 11. The recon-struction shown in Figure 11 is from Grunow et al. (1991, fig. 5a), including the position of the trench and its transform off-set opposite Marie Byrd Land. We change the position of the trench in front of New Zealand by following Bradshaw et al.'s (1981) model of the oceanward migration of the trench after Rangitata I. In southern Africa, the uplifted Cape Fold Belt and an uplift to the northeast funneled sand through piedmont fans across a braidplain and meander belt. Tuffs were de-posited from nearby volcanoes. In mainland Australia, south-east Queensland and adjacent New South Wales were crossed by granitoids and felsic volcanics. Maar diatremes (208–193 Ma[*]) were erupted in the Sydney Basin, and the 212 Ma Benambra syenite complex was emplaced in Victoria. Coal and other carbonaceous sediment resumed deposition in Coal II after the Early and Middle Triassic coal gap (Fig. 3). New intra-montane fault-bounded basins include the Ipswich and Tarong coal basins. New craton basins include the Leigh Creek coal basin and the basal formation of the Eromanga Basin, the carbonaceous Peera Peera Formation. Coal meas-ures in Tasmania (New Town), Victoria Land, the Central Transantarctic Mountains, and South Africa (Molteno), and carbonaceous sediment (oil shale) in South America suc-ceeded the coal gap in the rest of Gondwanaland.

STAGE POST-G (200–160 Ma)

Our detailed analysis of the regions, except Antarctica, stopped with Stage G. Much of the information shown on Figure 12 is therefore drawn from other sources. The 130–120

Ma Etendeka-Serra Geral flood basalt (broken lines) of post–post-G (*ppg* in Fig. 2) and post-G continental flood basalt and dolerite are from Hergt et al. (1991, fig. 1), augmented by the dolerite line (heavy dots) around the Karoo Basin (Chapter 5) and the indicated 220–140 Ma diabase sills in the Amazon Basin (Mosmann et al., 1986). The Late Triassic granite to Middle Jurassic rhyolite of Patagonia and vicinity (Chon Aike group) are from Kay et al. (1989, fig. 2a) and Gust et al. (1985, fig. 1). Jurassic granitoids in South America, Antarctic Peninsula, Jones Mountains of Thurston Island (197 ± 4 Ma), and Ellsworth-Whitmore Mountains (180 Ma, within-plate and S-type) are from Pankhurst (1990). The granitoid in New Zealand is from Korsch and Wellman (1988). On mainland Australia, basalt at Kangaroo Island (174 Ma) and western Victoria (195 Ma), basic breccia pipes near Delegate and the Myalla Road Syenite and basaltic dykes near Cooma (172–167 Ma), and intrusions of syenite, dolerite, basanite, and trachyte in the Mittagong area (199–181 Ma) are from McDougall and Wellman (1976). Immediately east of Mitta-gong, at Kiama, the 191 Ma basanite dyke is from Wass and Shaw (1984). Other information is from Chapter 3, including diatremes in the Sydney Basin, the Garrawilla Volcanics, and the 200 Ma overstep of the foldbelt by the cratonic Precipice Sandstone. The 160 Ma arc volcanics (Vs) and derived volcanogenic sediment (dot-and-dashed line) are from Veevers (1984, p. 265).

The reconstruction of the West Antarctic continental blocks is for 175 Ma (Grunow et al., 1991, fig. 5b, as given in Fig. 22 in Chapter 4), and involves the rotation of the Ellsworth-Whitmore block and the Falkland Islands. They postulate lithospheric extension along the Explora Wedge (EW) and Byrd Subglacial Basin (BSB) and its continuation alongside the Transantarctic Mountains and between South America and the Antarctic Peninsula. The 160 Ma inception of continental rifting between Antarctica and Australia and between Tasmania and the Australian mainland is from Powell et al. (1988), and the slightly later (150 Ma) spreading between Africa and Antarctica is from Grunow et al. (1991, fig. 5c).

By the Early Jurassic (200 Ma*) in eastern Australia, the Eromanga-Surat Basin expanded across the foldbelt to a limit shown by the full line and arrows in Fig. 12. The magmatic arc resurged from 190 Ma*:

The start of the late Innamincka stage is marked by resumption of magmatic arc activity in the Early Jurassic, 190 Ma ago, in a position

Figure 11. Paleo-tectonic/geographic map: Stage G (230–200 Ma), with 210 Ma interpolated pole. Reconstruction from Grunow et al. (1991, fig. 5a), including the position of the trench and its transform offset opposite Marie Byrd Land. Position of trench in front of New Zealand slightly modified from Bradshaw et al.'s (1981) model of the oceanward migration of the trench after Rangitata I. *See* Figure 5 for symbols.

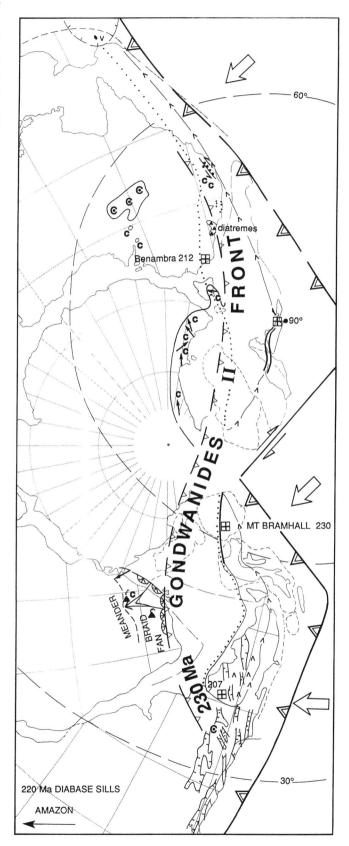

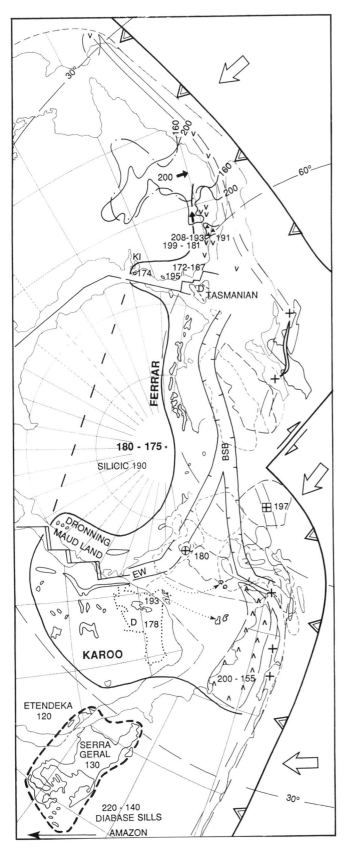

inferred from the occurrence of volcanogenic sediment in the Maryborough Basin, to lie beneath the outer Barrier Reef in offshore Queensland, and northward to New Guinea where fragments of it are exposed. By about 160 Ma ago, this arc shed volcanogenic sediment (Birkhead Formation) back across the craton as far west as the Birdsville Track Ridge (Veevers, 1984, p. 265).

Figure 12 shows the 160 Ma (Jurassic) arc volcanics (Vs) and derived volcanogenic sediment (dot-and-dashed line). The magmatic arc is traced southward into New Zealand, the Thurston Island block, the Antarctic Peninsula, and South America. The intra-plate (back-arc) volcanics have a cratonic limit marked by the heavy line; the broken line is interpolated between outcrops at Kangaroo Island (KI) and Dronning Maud Land. Intra-plate magmatism extends in Australia from the Garrawilla Volcanics to the north and the Kangaroo Island basalt to the west through the Tasmanian dolerite (D) into the Ferrar Group around the East Antarctic margin including the 180 Ma S-type granitoid in the Ellsworth Mountains block through Dronning Maud Land into the Karoo province of southern Africa and finally into the Chon Aike province of Patagonia. The indicated diabase sills in the Amazon Basin probably include rocks of this age. Also shown are the 120–130 Ma Etendeka-Serra Geral basalts (broken line) erupted from the Tristan da Cunha hotspot (Stothers, 1993).

Storey and Alabaster (1991, p. 1287)

link the change from Gondwanide compression to extension as well as the inferred change from shallow to steeply dipping subduction in the Jurassic to a slowing of subduction rates caused possibly by the decreasing age of the subducting plate, culminating in ridge-trench interaction. At least along the proto-Pacific margin this may have resulted in thermally weakened crust, lithospheric melting and increased magmatic production rates, a broad linear extensional zone, and, ultimately, lithospheric rupture. The extent to which magmatism and extension in the interior of the supercontinent were controlled by plate margin processes is uncertain. Clearly, other factors like crustal thickening, absolute plate motions, and the presence of mantle plumes may also be important.

ACKNOWLEDGMENTS

We acknowledge the generous time and effort that the GSA reviewers, R. J. Korsch and E. Scheibner, put into constructive criticism. Figure 1 is reproduced by permission of Korsch and Academic Press. Figures 2 and 3 were drawn by Adam Bryant and Ian Percival, and Figure 4 by Maree Corcoran and Marnie Pascoe, and the set of maps were produced by Judy Davis.

Figure 12. Paleo-tectonic/geographic map: Stage post-G (200–160 Ma), 175 Ma mean pole. The reconstruction of the West Antarctic continental blocks is for 175 Ma (Grunow et al., 1991, fig. 5b, as given in Fig. 22 in Chapter 4). Also shown is the younger (130–120 Ma) Etendeka-Serra Geral flood basalt (broken lines) of post–post-G (ppg in Fig. 2). BSB = Byrd Subglacial Basin; D = Dolerite line; EW = Explora Wedge; KI = Kangaroo Island. *See* Figure 5 for symbols.

REFERENCES CITED

Anderson, A. M., and McLachlan, I. R., 1979, The oil-shale potential of the Early Permian White Band Formation in Southern Africa: Geological Society of South Africa Special Publication, v. 6, p. 83–89.

Beck Jr., M. E., Garcia, R. A., and Burmester, R. F., 1991, Paleomagnetism and geochronology of late Paleozoic granitic rocks from the Lake District of southern Chile: Implications for accretionary tectonics: Geology, v. 19, p. 332–335.

BMR Palaeogeographic Group, 1990, Evolution of a continent: Australia: Canberra, Australian Bureau of Mineral Resources, 97 p.

Borg, S. G., and Stump, E., 1987, Paleozoic magmatism and associated tectonic problems of northern Victoria Land, Antarctica, *in* McKenzie, G. D., ed., Gondwana six: Structure, tectonics, and geophysics: Washington, D.C., American Geophysical Union Geophysical Monograph, v. 40, p. 67–75.

Bradshaw, J. D., Andrews, P. B., and Adams, C. J., 1981, Carboniferous to Cretaceous on the Pacific margin of Gondwana: The Rangitata Phase of New Zealand, *in* Cresswell, M. M., and Vella, P., eds., Gondwana five: Rotterdam, A. A. Balkema, p. 217–221.

Bradshaw, M. A., 1984, Permian nonmarine bivalves from the Ohio Range, Antarctica: Alcheringa, v. 8, p. 305–309.

Briggs, D.J.C., 1993, Time control in the Permian of the Sydney-Bowen Basin and the New England orogen, *in* Findlay, R. H., Banks, M. R., Veevers, J. J., and Unrug, R., eds., Gondwana eight: Assembly, evolution and dispersal: Rotterdam, A. A. Balkema, p. 371–383.

Büggisch, W., 1987, Stratigraphy and very low grade metamorphism of the Sierras Australes de la Provincia de Buenos Aires (Argentina) and implications in Gondwana correlation: Zentralblat für Geologie und Paläontologie, v. 1, p. 819–837.

Byrnes, J. G., 1983, Thin kaolin band from the Terrigal Formation in North-West Oil and Minerals Longley DDH 1 near Somersby: NSW Geological Survey Report, v. GS 1983/037, 2 p. (unpublished).

Campbell, I. H., Czamanske, G. K., Fedorenko, V. A., Hill, R. I., and Stepanov, V., 1992, Synchronism of the Siberian Traps and the Permian-Triassic boundary: Science, v. 258, p. 1760–1763.

Caputo, M. V., and Crowell, J. C., 1985, Migration of glacial centers across Gondwana during Paleozoic Era: Geological Society of America Bulletin, v. 96, p. 1020–1036.

Carlier, G., Grandin, G., Laubacher, G., Marocco, R., and Mégard, F., 1982, Present knowledge of the magmatic evolution of the Eastern Cordillera of Peru: Earth-Science Reviews, v. 18, p. 253–283.

Chen, Z., Li, Z. X., Powell, C. McA., and Balme, B., 1993, An Early Carboniferous paleomagnetic pole for Gondwanaland: New results from diamond-drill core materials of the Mount Eclipse Sandstone in the Ngalia Basin, central Australia: Journal of Geophysical Research (in press).

Cobbold, P., Massabie, A. C., and Rossello, E. A., 1986, Hercynian wrenching and thrusting in the Sierras Australes Foldbelt, Argentina: Hercynica, v. 2, p. 135–148.

Cole, D. I., and McLachlan, I. R., 1991, Oil potential of the Permian Whitehill Shale Formation in the main Karoo basin, South Africa, *in* Ulbrich, H., and Rocha-Campos, A. C., eds., Gondwana Seven Proceedings: São Paulo, Brazil, Instituto de Geociências, Universidade de São Paulo, p. 379–390.

Collins, W. J., Offler, R., Farrell, T. R., and Landenberger, B., 1993, A revised Late Palaeozoic–Early Mesozoic tectonic history for the southern New England Fold Belt: NEO 93 Conference Proceedings: Armidale, Australia, Department of Geology, University of New England, p. 69–84.

Conaghan, P. J., Shaw, S. E., and Veevers, J. J., 1994, Sedimentary evidence of the Permian/Triassic global environmental crisis induced by the Siberian hotspot, *in* Beauchamp, B., and Embry, A., eds., Carboniferous to Jurassic Pangea: Canadian Society of Petroleum Geologists Memoir (in press).

Cooper, J. A., Richards, J. R., and Webb, A. W., 1963, Some potassium-argon ages in New England, New South Wales: Journal of the Geological Society of Australia, v. 10, p. 313–316.

Cooper, M. R., and Kensley, B., 1984, Endemic South American Permian bivalve molluscs from the Ecca of South Africa: Journal of Paleontology, v. 58, p. 1360–1363.

Coutinho, J.M.V., Hachiro, J., Coimbra, A. M., and Santos, P. R., 1991, Ash-fall-derived vitroclastic tuffaceous sediments in the Permian of the Paraná Basin and their provenance, *in* Ulbrich, H., and Rocha-Campos, A. C., eds., Gondwana Seven Proceedings: São Paulo, Brazil, Instituto de Geosciencias, Universidade de São Paulo, p. 147–160.

Crowell, J. C., 1983, Ice ages recorded on Gondwanan continents: Geological Society of South Africa Transactions, v. 86, p. 237–262.

Davidson, J., Mpodozis, C., Godoy, E., Hervé, F., Pankhurst, R., and Brook, M., 1987, Late Paleozoic accretionary complexes on the Gondwana margin of Southern Chile: Evidence from the Chonos Archipelago: American Geophysical Union Geophysical Monograph, v. 40, p. 221–227.

de Wit, M., Jeffery, M., Bergh, H., and Nicolaysen, L., 1988, Geological map of sectors of Gondwana reconstructed to their disposition 150 Ma ago: Lambert equal area projection: Tulsa, Oklahoma, American Association of Petroleum Geologists, scale 1:10,000,000.

Dickins, J. M., 1984, Late Palaeozoic glaciation: Australian Bureau of Mineral Resources Journal, v. 9, p. 163–169.

Drewry, D. J., 1983, ed., Antarctica: Glaciological and geophysical folio: England, Scott Polar Research Institute, University of Cambridge, 9 sheets.

Du Toit, A. L., 1937: Our wandering continents: Edinburgh, Oliver and Boyd, 366 p.

Elliot, D. H., Fleck, R. J., and Sutter, J. F., 1985, Potassium-argon age determination of Ferrar Group rocks, central Transantarctic Mountains, *in* Turner, M. D., and Splettstoesser, J. F., eds., Geology of the central Transantarctic Mountains: Washington, D.C., American Geophysical Union Antarctic Research Series, v. 36, p. 197–223.

Eyles, C. H., Eyles, N., and Franca, A. B., 1993, Glaciation and tectonics in an active intracratonic basin: The Late Paleozoic Itararé Group, Paraná Basin, Brazil: Sedimentology, v. 40, p. 1–25.

Frakes, L. A., 1979, Climates throughout geologic time: Amsterdam, Elsevier, 310 p.

Goleby, B. R., Shaw, R. D., Wright, C., Kennett, B.L.N., and Lambeck, K., 1989, Geophysical evidence for thick-skinned crustal deformation in central Australia: Nature, v. 337, p. 325–330.

Graham, I. J., and Korsch, R. J., 1985, Rb-Sr geochronology of coarse-grained greywackes and argillites from the Coffs Harbour Block, eastern Australia: Chemical Geology (Isotope Geoscience Section), v. 58, p. 45–54.

Graham, I. J., and Korsch, R. J., 1990, Age and provenance of granitoid clasts in Moeatoa Conglomerate, Kawhia Syncline, New Zealand: Journal of the Royal Society of New Zealand, v. 20, p. 25–39.

Grindley, G. W., and Oliver, P. J., 1983, Post-Ross orogeny cratonization of northern Victoria Land, *in* Oliver, R. L., James, P. R., and Jago, J. B., eds., Antarctic Earth science: Canberra, Australian Academy of Science, p. 133–139.

Grunow, A. M., Kent, D. V., and Dalziel, I.W.D., 1991, New paleomagnetic data from Thurston Island: Implications for the tectonics of West Antarctica and Weddell Sea opening: Journal of Geophysical Research, v. 96, p. 17935–17954.

Gust, D. A., Biddle, K. T., Phelps, D. W., and Uliana, M. A., 1985, Associated Middle to Late Jurassic volcanism and extension in southern South America: Tectonophysics, v. 116, p. 223–253.

Hälbich, I. W., Fitch, F. J., and Miller, J. A., 1983, Dating the Cape orogeny, *in* Söhnge, A.P.G., and Hälbich, I. W., eds., Geodynamics of the Cape Fold Belt: Geological Society of South Africa Special Publication, v. 12, p. 149–164.

Harrington, H. J., and Korsch, R. J., 1985, Tectonic model for the Devonian to middle Permian of the New England Orogen: Australian Journal of Earth Sciences, v. 32, p. 163–179.

Henderson, R. A., 1980, Structural outline and summary geological history for northeastern Australia, *in* Henderson, R. A., and Stephenson, P. J., eds., The geology and geophysics of Northeastern Australia: Brisbane,, Geological Society of Australia, Queensland Division, p. 1–26.

Hensel, H. D., McCulloch, M. T., and Chappell, B. W., 1985, The New England Batholith: Constraints on derivation from Nd and Sr isotopic studies of granitoids and country rocks: Geochimica et Cosmochimica Acta, v. 49, p. 369–384.

Hergt, J. M., Peate, D. W., and Hawkesworth, C. J., 1991, The petrogenesis of Mesozoic Gondwana low-Ti flood basalts: Earth and Planetary Science Letters, v. 105, p. 134–148.

Iñiguez, A. M., Andreis, R. R., and Zalba, P., 1988, Eventos piroclásticos en la Formación Tunas (Pérmico), Sierras Australes, provincia de Buenos Aires, República Argentina: 2nd Jornadas Geológicas Bonaerenses, Actas, p. 383–395.

Jones, B. G., 1972, Upper Devonian to Lower Carboniferous stratigraphy of Pertnjara Group, Amadeus Basin, Central Australia: Journal of the Geological Society of Australia, v. 19, p. 229–249.

Jones, B. G., 1991, Fluvial and lacustrine facies of the Middle-Late Devonian Pertnjara Group, *in* Korsch, R. J., and Kennard, J. M., eds., Geological and geophysical studies in the Amadeus Basin: Australian Bureau of Mineral Resources Bulletin, v. 236, p. 338–348.

Kay, S. M., Ramos, V. A., Mpodozis, C., and Sruoga, P., 1989, Late Paleozoic to Jurassic silicic magmatism at the Gondwana margin: Analogy to the middle Proterozoic in North America: Geology, v. 17, p. 324–328.

Kimbrough, D. L., Mattinson, J. M., Coombs, D. S., Landis, C. A., and Johnston, M. R., 1992, Uranium-lead ages from the Dun Mountain ophiolite belt and Brook Street terrane, South Island, New Zealand: Geological Society of America Bulletin, v. 104, p. 429–443.

Kimbrough, D. L., Cross, K. C., and Korsch, R. J., 1993, U-Pb isotopic ages for zircons from the Pola Fogal and Nundal granite suites, southern New England Orogen: NEO 93 Conference Proceedings: Armidale, Australia, Department of Geology, University of New England, p. 403–412, and errata sheet 412a.

Korsch, R. J., 1974, Petrographic comparison of the Taylor and Victoria groups (Devonian to Triassic) in South Victoria Land, Antarctica: New Zealand Journal of Geology and Geophysics, v. 17, p. 523–541.

Korsch, R. J., and Wellman, H. W., 1988, The geological evolution of New Zealand and the New Zealand region, *in* Nairn, A.E.M., Stehli, F. G., and Uyeda, S., eds., The ocean basins and margins, Volume 7B: The Pacific Ocean: New York, Plenum, p. 411–482.

Korsch, R. J., Harrington, H. J., Murray, C. G., Fergusson, C. L., and Flood, P. G., 1990, Tectonics of the New England Orogen: Australian Bureau of Mineral Resources Bulletin, v. 232, p. 35–52.

Lang, S. C., 1988, Devonian to Early Carboniferous history of the Broken River and Bundock Creek Groups, *in* Withnall, I. W., and 11 others, eds., Stratigraphy, sedimentology, biostratigraphy and tectonics of the Ordovician to Carboniferous, Broken River Province, North Queensland: Sydney, Geological Society of Australia, Australian Sedimentologists Group, Field Guide Series, v. 5, p. 97–104.

Leitch, E. C., 1974, The geological development of the southern part of the New England Fold Belt: Journal of the Geological Society of Australia, v. 21, p. 133–156.

Leitch, E. C., 1988, The Barnard Basin and the Early Permian development of the southern part of the New England Fold Belt, *in* Kleeman, J. D., ed., New England Orogen tectonics and metallogenesis: Armidale, Australia, Department of Geology and Geophysics, University of New England, p. 61–67.

Levell, B. K., Braakman, J. H., and Rutten, K. W., 1988, Oil-bearing sediments of Gondwana glaciation in Oman: American Association of Petroleum Geologists Bulletin, v. 72, p. 775–796.

Li, Z. X., Powell, C. McA., and Schmidt, P. W., 1989, Syndeformational remanent magnetization of the Mount Eclipse Sandstone, central Australia: Geophysical Journal International, v. 99, p. 205–222.

Lindsay, J. F., and Korsch, R. J., 1991, The evolution of the Amadeus Basin, central Australia, *in* Korsch, R. J., and Kennard, J. M., eds., Geological and geophysical studies in the Amadeus Basin: Australian Bureau of Mineral Resources Bulletin, v. 236, p. 7–32.

López-Gamundí, O. R., and Rossello, E. A., 1993, The Devonian-Carboniferous unconformity in Argentina and its relation to the Eo-Hercynian orogeny in southern South America: Geologische Rundschau, v. 82, p. 136–147.

López-Gamundí, O. R., Limarino, C. O., and Cesari, S. N., 1992, Late Paleozoic paleoclimatology of central-west Argentina: Palaeogeography, Palaeoclimatology, Palaeoecology, v. 91, p. 305–329.

Marsden, M.A.H., 1988, Upper Devonian-Carboniferous, *in* Douglas, J. G., and Ferguson, J. A., eds., Geology of Victoria: Melbourne, Geological Society of Australia, Victorian Division, p. 147–194.

McDougall, I., and Wellman, P., 1976, Potassium-argon ages for some Australian Mesozoic igneous rocks: Journal of the Geological Society of Australia, v. 23, p. 1–9.

Mégard, F., 1978, Étude géologique des Andes du Pérou Central: contribution à l'étude des Andes, 1: Memoir ORSTOM, v. 86, 310 p.

Middleton, M. F., 1990, Canning basin: Western Australia Geological Survey Memoir 3, p. 425–457.

Morante, R., 1993, Determining the Permian/Triassic boundary in Australia through C-isotope chemostratigraphy, *in* Flood, P. G., and Aitchison, J. C., eds., New England Orogen, eastern Australia: Armidale, Department of Geology and Geophysics, University of New England, p. 293–298.

Morante, R., Veevers, J. J., Andrew, A. S., and Hamilton, P. J., 1994, Determination of the Permian-Triassic boundary in Australia from carbon isotope stratigraphy: Australian Petroleum Exploration Association Journal, v. 34, p. 330–336.

Mortimer, G. E., Cooper, J. A., and James, P. R., 1987, U-Pb and Rb-Sr geochronology and geological evolution of the Harts Range ruby mine area of the Arunta Inlier, central Australia: Lithos, v. 20, p. 445–467.

Mosmann, R., Falkenhein, F.U.H., Goncalves, A., and Nepomuceno, F., 1986, Oil and gas potential of the Amazon Paleozoic basins, *in* Halbouty, M. T., ed., Future petroleum provinces of the world: American Association of Petroleum Geologists Memoir 40, p. 207–241.

Mpodozis, C., and Kay, S. M., 1992, Late Paleozoic to Triassic evolution of the Gondwana margin: Evidence from Chilean Frontal Cordilleran batholiths (28° to 31°S): Geological Society of America Bulletin, v. 104, p. 999–1014.

Oelofsen, B. W., 1987, The biostratigraphy and fossils of the Whitehill and Iratí Shale Formations of the Karoo and Paraná Basins, *in* McKenzie, G. D., ed., Gondwana six: American Geophysical Union Geophysical Monograph, v. 41, p. 131–138.

Olgers, F., 1972, Geology of the Drummond Basin, Queensland: Australian Bureau of Mineral Resources Bulletin, v. 132.

Palmer, A. R., 1983, The Decade of North American Geology 1983 geologic time scale: Geology, v. 11, p. 503–504.

Pankhurst, R. J., 1990, The Paleozoic and Andean magmatic arcs of West Antarctica and southern South America: Boulder, Colorado, Geological Society of America Special Paper 241, p. 1–7.

Pankhurst, R. J., Rapela, C. W., Caminos, R., Llambias, E., and Parica, C., 1992, A revised age for the granites of the central Somuncura Batholith, North Patagonian Massif: Journal of South American Earth Sciences, v. 5, p. 321–325.

Playford, P. E., 1980, Devonian "Great Barrier Reef" of Canning Basin, Western Australia: American Association of Petroleum Geologists Bulletin, v. 64, p. 814–840.

Powell, C. McA., 1983, Tectonic relationship between the Late Ordovician and Late Silurian palaeogeographies of southeastern Australia: Journal of the Geological Society of Australia, v. 30, p. 353–373.

Powell, C. McA., 1984a, Ordovician to Carboniferous, *in* Veevers, J. J., ed., Phanerozoic Earth history of Australia: Oxford, Clarendon Press, p. 290–340.

Powell, C. McA., 1984b, Terminal fold-belt deformation: Relationship of mid-Carboniferous megakinks in the Tasman Fold Belt to coeval thrusts

in cratonic Australia: Geology, v. 12, p. 546–549.

Powell, C. McA., and Veevers, J. J., 1987, Namurian uplift in Australia and South America triggered the main Gondwanan glaciation: Nature, v. 236, p. 177–179.

Powell, C. McA., Edgecombe, D. R., Henry, N. M., and Jones, J. G., 1977, Timing of regional deformation of the Hill End trough: A reassessment: Journal of the Geological Society of Australia, v. 23, p. 407–421.

Powell, C. McA., Cole, J. P., and Cudahy, T. J., 1985, Megakinking in the Lachlan Fold Belt, Australia: Journal of Structural Geology, v. 7, p. 281–300.

Powell, C. McA., Roots, S. R., and Veevers, J. J., 1988, Pre-breakup continental extension in East Gondwanaland and the early opening of the eastern Indian Ocean: Tectonophysics, v. 155, p. 261–283.

Raine, J. I., 1980, Palynology of the Triassic Topfer Formation, Reefton, South Island, New Zealand: Wellington, New Zealand, Fifth Gondwana Symposium, Abstracts, 1 unnumbered page.

Retallack, G. J., 1987, Triassic vegetation and geography of the New Zealand portion on the Gondwana supercontinent: American Geophysical Union Geophysical Monograph, v. 41, p. 9–39.

Roberts, J., and Engel, B. A., 1987, Depositional and tectonic history of the southern New England Orogen: Australian Journal of Earth Sciences, v. 34, p. 1–20.

Roberts, J., Engel, B., and Chapman, J., 1991, Geology of the Camberwell, Dungog, and Bulahdelah 1:100,000 Sheets 9133, 9233, 9333: Sydney, New South Wales Geological Survey, 382 p.

Runnegar, B., and Newell, N. D., 1971, Caspian-like relict molluscan fauna in the South American Permian: Bulletin of the American Museum of Natural History, v. 146, p. 1–66.

Rust, I. C., 1975, Tectonic and sedimentary framework of Gondwana basins in southern Africa, *in* Campbell, K.S.W., ed., Gondwana geology: Canberra, Australian National University Press, p. 537–564.

Sahagian, D. L., and Collinson, J. W., 1993, Gondwanan foreland basin along the Panthalassan margin of Antarctica, *in* Findlay, R. H., Banks, M. R., Veevers, J. J., and Unrug, R., eds., Gondwana eight: Assembly, evolution and dispersal: Rotterdam, A. A. Balkema, p. 497–506.

Shaw, R. D., Etheridge, M. A., and Lambeck, K., 1991, Development of the Late Proterozoic to mid-Paleozoic intracratonic Amadeus Basin: A key to understanding tectonic forces in plate interiors: Tectonics, v. 10, p. 688–721.

Shaw, S. E., Flood, R. H., and Riley, G. H., 1982, The Wologorong Batholith, New South Wales, and the extension of the I-S line of the Siluro-Devonian granitoids: Journal of the Geological Society of Australia, v. 29, p. 41–48.

Storey, B. C., and Alabaster, T., 1991, Tectonomagmatic controls on Gondwana break-up models: Evidence from the Proto-Pacific margin of Antarctica: Tectonics, v. 10, p. 1274–1288.

Stothers, R. B., 1993, Hotspots and sunspots: Surface tracers of deep mantle convection in the Earth and Sun: Earth and Planetary Science Letters, v. 116, p. 1–8.

Stubley, M. P., 1989, Fault and kink band relationships at Mystery Bay, Australia: Tectonophysics, v. 158, p. 75–92.

Tewari, R. C., and Veevers, J. J., 1993, Gondwana basins of India occupy the middle of a 7500 km sector of radial valleys and lobes in central-eastern Gondwanaland, *in* Findlay, R. H., Banks, M. R., Veevers, J. J., and Unrug, R., eds., Gondwana eight: Assembly, evolution and dispersal: Rotterdam, A. A. Balkema, p. 507–512.

Varela, R., Dalla Salda, L., and Cingolani, C., 1985, Estructura y composición geológica de las Sierras Colorado, Chasico y Cortapie, Sierras Australes de Buenos Aires: Revista Asociacion Geológica Argentina, v. 60, p. 254–261.

Veevers, J. J., ed., 1984, Phanerozoic Earth history of Australia: Oxford, Clarendon Press, 418 p.

Veevers, J. J., 1988, Gondwana facies started when Gondwanaland merged in Pangea: Geology, v. 16, p. 732–734.

Veevers, J. J., 1989, Middle/Late Triassic (230 ± 5 Ma) singularity in the stratigraphic and magmatic history of the Pangean heat anomaly: Geology, v. 17, p. 784–787.

Veevers, J. J., 1990a, Tectonic-climatic supercycle in the billion-year plate-tectonic eon: Permian Pangean icehouse alternates with Cretaceous dispersed-continents greenhouse: Sedimentary Geology, v. 68, p. 1–16.

Veevers, J. J., 1990b, Development of Australia's post-Carboniferous sedimentary basins: Petroleum Exploration Society of Australia Journal, v. 16, p. 25–32.

Veevers, J. J., and Powell, C. McA., 1987, Late Paleozoic glacial episodes in Gondwanaland reflected in transgressive-regressive depositional sequences in Euramerica: Geological Society of America Bulletin, v. 98, p. 475–487.

Veevers, J. J., Conaghan, P. J., and Shaw, S. E., 1994a, Turning point in Pangean environmental history at the Permian/Triassic (P/Tr) boundary: Boulder, Colorado, Geological Society of America Special Paper 288, p. 187–196.

Veevers, J. J., Clare, A., and Wopfner, H., 1994b, Neocratonic magmatic-sedimentary basins of post-Variscan Europe and post-Kanimblan eastern Australia generated by right-lateral transtension of Permo-Carboniferous Pangea: Basin Research (in press).

Villeneuve, M., Cornée, J. J., and Muller, J., 1993, Orogenic belts, sutures and block faulting on the northwestern Gondwana margin, *in* Findlay R. H., Banks, M. R., Veevers, J. J., and Unrug, R., eds., Gondwana eight: Assembly, evolution and dispersal: Rotterdam, A. A. Balkema, p. 43–53.

Visser, J.N.J., 1993, A reconstruction of the late Palaeozoic ice sheet on southwestern Gondwana, *in* Findlay R. H., Banks, M. R., Veevers, J. J., and Unrug, R., eds., Gondwana eight: Assembly, evolution and dispersal: Rotterdam, A. A. Balkema, p. 449–458.

Visser, J.N.J., and Loock, J. C., 1978, Water depth in the main Karoo Basin, South Africa, during Ecca (Permian) sedimentation: Geological Society of South Africa Transactions, v. 81, p. 185–191.

von Gosen, W. and Büggisch, W., 1989, Tectonic Evolution of the Sierras Australes Fold and Thrust belt (Buenos Aires Province, Argentina)—An Outline: Zentralblatt für Geologie und Paläontologie, v. 5/6, p. 947–958.

Wass, S. Y., and Shaw, S. E., 1984, Rb/Sr evidence for the nature of the mantle, thermal events and volcanic activity of the southeastern Australian continental margin: Journal of Volcanology and Geothermal Research, v. 21, p. 107–117.

Wells, A. T., and Moss, F. J., 1983, The Ngalia Basin, Northern Territory: Stratigraphy and structure: Australian Bureau of Mineral Resources Bulletin, v. 212, 88 p.

Williams, E., McClenaghan, M. P., and Collins, P.L.F., 1989, Mid-Palaeozoic deformation, granitoids and ore deposits, *in* Burrett, C. F., and Martin, E. L., eds., Geology and mineral resources of Tasmania: Geological Society of Australia Special Publication, v. 15, p. 238–292.

Wopfner, H., 1993, Structural development of Tanzanian Karoo basins and the break-up of Gondwana, *in* Findlay, R. H., Banks, M. R., Veevers, J. J., and Unrug, R., eds., Gondwana eight: Assembly, evolution and dispersal: Rotterdam, A. A. Balkema, p. 531–539.

Wyatt, B. W., and Jell, J. S. 1980, Devonian and Carboniferous stratigraphy of the northern Tasman Orogenic Zone in the Townsville hinterland, north Queensland, *in* Henderson, R. A., and Stephenson, P. J., eds., The geology and geophysics of northeastern Australia: Brisbane, Geological Society of Australia, Queensland Division, p. 201–228.

Zalán, P. V., Wolff, S., Conceicao, J.C.J., Astolfi, A. M., Vieira, I. S., Appi, V. T., Zanotto, A., and Marques, A., 1991, Tectonics and sedimentation of the Paraná Basin, *in* Ulbrich, H., and Rocha-Campos, A. C., eds., Gondwana Seven Proceedings: Universidade de São Paulo, p. 83–117.

MANUSCRIPT ACCEPTED BY THE SOCIETY SEPTEMBER 9, 1993

Index

[Italic page numbers indicate major references]